T0142150

Lecture Notes in Computer Science 12566

Founding Editors

Gerhard Goos
Karlsruhe Institute of Technology, Karlsruhe, Germany
Juris Hartmanis
Cornell University, Ithaca, NY, USA

Editorial Board Members

Elisa Bertino
Purdue University, West Lafayette, IN, USA
Wen Gao
Peking University, Beijing, China
Bernhard Steffen
TU Dortmund University, Dortmund, Germany
Gerhard Woeginger
RWTH Aachen, Aachen, Germany
Moti Yung
Columbia University, New York, NY, USA

More information about this subseries at http://www.springer.com/series/7409

Giuseppe Nicosia · Varun Ojha ·
Emanuele La Malfa · Giorgio Jansen ·
Vincenzo Sciacca · Panos Pardalos ·
Giovanni Giuffrida · Renato Umeton (Eds.)

Machine Learning, Optimization, and Data Science

6th International Conference, LOD 2020
Siena, Italy, July 19–23, 2020
Revised Selected Papers, Part II

 Springer

Editors
Giuseppe Nicosia ⓘ
University of Catania
Catania, Italy

Varun Ojha ⓘ
University of Reading
Reading, UK

Emanuele La Malfa ⓘ
University of Oxford
Oxford, UK

Giorgio Jansen
University of Cambridge
Cambridge, UK

Vincenzo Sciacca
ALMAWAVE
Rome, Italy

Panos Pardalos ⓘ
University of Florida
Gainesville, FL, USA

Giovanni Giuffrida ⓘ
University of Catania
Catania, Italy

Renato Umeton ⓘ
Harvard University
Cambridge, MA, USA

ISSN 0302-9743   ISSN 1611-3349   (electronic)
Lecture Notes in Computer Science
ISBN 978-3-030-64579-3   ISBN 978-3-030-64580-9   (eBook)
https://doi.org/10.1007/978-3-030-64580-9

LNCS Sublibrary: SL3 – Information Systems and Applications, incl. Internet/Web, and HCI

© Springer Nature Switzerland AG 2020
This work is subject to copyright. All rights are reserved by the Publisher, whether the whole or part of the material is concerned, specifically the rights of translation, reprinting, reuse of illustrations, recitation, broadcasting, reproduction on microfilms or in any other physical way, and transmission or information storage and retrieval, electronic adaptation, computer software, or by similar or dissimilar methodology now known or hereafter developed.
The use of general descriptive names, registered names, trademarks, service marks, etc. in this publication does not imply, even in the absence of a specific statement, that such names are exempt from the relevant protective laws and regulations and therefore free for general use.
The publisher, the authors and the editors are safe to assume that the advice and information in this book are believed to be true and accurate at the date of publication. Neither the publisher nor the authors or the editors give a warranty, expressed or implied, with respect to the material contained herein or for any errors or omissions that may have been made. The publisher remains neutral with regard to jurisdictional claims in published maps and institutional affiliations.

This Springer imprint is published by the registered company Springer Nature Switzerland AG
The registered company address is: Gewerbestrasse 11, 6330 Cham, Switzerland

Preface

The 6th edition of the International Conference on Machine Learning, Optimization, and Data Science (LOD 2020), was organized during July 19–23, 2020, in Certosa di Pontignano (Siena) Italy, a stunning medieval town dominating the picturesque countryside of Tuscany. LOD 2020 was held successfully online and onsite to meet challenges posed by the worldwide outbreak of COVID-19. Since 2015, the LOD conference brings academics, researchers, and industrial researchers together in a unique multidisciplinary community to discuss the state of the art and the latest advances in the integration of machine learning, optimization, and data science to provide and support the scientific and technological foundations for interpretable, explainable, and trustworthy AI. Since 2017, LOD adopted the Asilomar AI Principles.

LOD is an annual international conference on machine learning, computational optimization, and big data that includes invited talks, tutorial talks, special sessions, industrial tracks, demonstrations, and oral and poster presentations of refereed papers.

LOD has established itself as a premier multidisciplinary conference in machine learning, computational optimization, and data science. It provides an international forum for presentation of original multidisciplinary research results, as well as exchange and dissemination of innovative and practical development experiences.

The LOD conference manifesto is the following:

"The problem of understanding intelligence is said to be the greatest problem in science today and "the" problem for this century – as deciphering the genetic code was for the second half of the last one. Arguably, the problem of learning represents a gateway to understanding intelligence in brains and machines, to discovering how the human brain works, and to making intelligent machines that learn from experience and improve their competences as children do. In engineering, learning techniques would make it possible to develop software that can be quickly customized to deal with the increasing amount of information and the flood of data around us."

The Mathematics of Learning: Dealing with Data
Tomaso Poggio (MOD 2015 & LOD 2020 Keynote Speaker) and Steve Smale

"Artificial Intelligence has already provided beneficial tools that are used every day by people around the world. Its continued development, guided by the Asilomar principles of AI, will offer amazing opportunities to help and empower people in the decades and centuries ahead."

The AI Asilomar Principles

The AI Asilomar principles have been adopted by the LOD conference since their initial formulation, January 3–5, 2017. Since then they have been an integral part of the manifesto of LOD community (LOD 2017).

LOD 2020 attracted leading experts from industry and the academic world with the aim of strengthening the connection between these institutions. The 2020 edition of LOD represented a great opportunity for professors, scientists, industry experts, and research students to learn about recent developments in their own research areas and to learn about research in contiguous research areas with the aim of creating an environment to share ideas and trigger new collaborations.

As chairs, it was an honor to organize a premiere conference in these areas and to have received a large variety of innovative and original scientific contributions.

During LOD 2020, 16 plenary talks were presented by the following leading experts from the academic world:

Yoshua Bengio, Université de Montréal, Canada (A.M. Turing Award 2018)
Tomaso Poggio, MIT, USA
Pierre Baldi, University of California, Irvine, USA
Bettina Berendt, Technische Universität Berlin, Germany
Artur d'Avila Garcez, City, University of London, UK
Luc De Raedt, KU Leuven, Belgium
Marco Gori, University of Siena, Italy
Marta Kwiatkowska, University of Oxford, UK
Michele Lombardi, University of Bologna, Italy
Angelo Lucia, University of Rhode Island, USA
Andrea Passerini, University of Trento, Italy
Jan Peters, Technische Universität Darmstadt, Max Planck Institute for Intelligent Systems, Germany
Raniero Romagnoli, Almawave, Italy
Cristina Savin, Center for Neural Science, New York University, USA
Maria Schuld, Xanadu, University of KwaZulu-Natal, South Africa
Naftali Tishby, The Hebrew University of Jerusalem, Israel
Ruth Urner, York University, Canada
Isabel Valera, Saarland University, Max Planck Institute for Intelligent Systems, Germany

LOD 2020 received 209 submissions from 63 countries in 5 continents, and each manuscript was independently reviewed by a committee formed by at least 5 members. These proceedings contain 116 research articles written by leading scientists in the fields of machine learning, artificial intelligence, reinforcement learning, computational optimization, and data science presenting a substantial array of ideas, technologies, algorithms, methods, and applications.

At LOD 2020, Springer LNCS generously sponsored the LOD Best Paper Award. This year, the paper by Cole Smith, Andrii Dobroshynskyi, and Suzanne McIntosh titled "Quantifying Local Energy Demand through Pollution Analysis" received the LOD 2020 Best Paper Award.

This conference could not have been organized without the contributions of exceptional researchers and visionary industry experts, so we thank them all for participating. A sincere thank you also goes to the 47 subreviewers and the Program Committee, formed by more than 570 scientists from academia and industry, for their valuable and essential work of selecting the scientific contributions.

Finally, we would like to express our appreciation to the keynote speakers who accepted our invitation, and to all the authors who submitted their research papers to LOD 2020.

September 2020

Giuseppe Nicosia
Varun Ojha
Emanuele La Malfa
Giorgio Jansen
Vincenzo Sciacca
Panos Pardalos
Giovanni Giuffrida
Renato Umeton

Organization

General Chairs

Giorgio Jansen   University of Cambridge, UK
Emanuele La Malfa   University of Oxford, UK
Vincenzo Sciacca   Almawave, Italy
Renato Umeton   Dana-Farber Cancer Institute, MIT, USA

Conference and Technical Program Committee Co-chairs

Giovanni Giuffrida   University of Catania, NeoData Group, Italy
Varun Ojha   University of Reading, UK
Panos Pardalos   University of Florida, USA

Tutorial Chair

Vincenzo Sciacca   Almawave, Italy

Publicity Chair

Stefano Mauceri   University College Dublin, Ireland

Industrial Session Chairs

Giovanni Giuffrida   University of Catania, NeoData Group, Italy
Vincenzo Sciacca   Almawave, Italy

Organizing Committee

Alberto Castellini   University of Verona, Italy
Piero Conca   Fujitsu, Ireland
Jole Costanza   Italian Institute of Technology, Italy
Giuditta Franco   University of Verona, Italy
Marco Gori   University of Siena, Italy
Giorgio Jansen   University of Cambridge, UK
Emanuele La Malfa   University of Oxford, UK
Gabriele La Malfa   University of Cambridge, UK
Kaisa Miettinen   University of Jyväskylä, Finland
Giuseppe Narzisi   New York University, USA
Varun Ojha   University of Reading, UK

Steering Committee

Giuseppe Nicosia   University of Cambridge, UK, and University
                                  of Catania, Italy
Panos Pardalos   University of Florida, USA

Technical Program Committee

Jason Adair   University of Stirling, UK
Agostinho Agra   Universidade de Aveiro, Portugal
Hernan Aguirre   Shinshu University, Japan
Kerem Akartunali   University of Strathclyde, UK
Richard Allmendinger   The University of Manchester, UK
Paula Amaral   NOVA University Lisbon, Portugal
Hoai An Le Thi   Université de Lorraine, France
Aris Anagnostopoulos   Sapienza Università di Roma, Italy
Fabrizio Angaroni   University of Milano-Bicocca, Italy
Davide Anguita   University of Genova, Italy
Alejandro Arbelaez   University College Cork, Ireland
Danilo Ardagna   Politecnico di Milano, Italy
Roberto Aringhieri   University of Turin, Italy
Takaya Arita   Nagoya University, Japan
Ashwin Arulselvan   University of Strathclyde, UK
Martin Atzmueller   Tilburg University, The Netherlands
Martha L. Avendano   Universidad Veracruzana, Mexico
  Garrido
Bar-Hen Avner   Cnam, France
Chloe-Agathe Azencott   Institut Curie Research Centre, France
Kamyar Azizzadenesheli   University of California Irvine, USA
Ozalp Babaoglu   Università di Bologna, Italy
Jaume Bacardit   Newcastle University, UK
Rodolfo Baggio   Bocconi University, Italy
James Bailey   The University of Melbourne, Australia
Marco Baioletti   Università degli Studi di Perugia, Italy
Elena Baralis   Politecnico di Torino, Italy
Xabier E. Barandiaran   University of the Basque Country, Spain
Cristobal Barba-Gonzalez   University of Malaga, Spain
Helio J. C. Barbosa   Laboratório Nacional de Computação Científica, Brazil
Thomas Bartz-Beielstein   IDEA, TH Köln, Germany
Mikhail Batsyn   Higher School of Economics, Russia
Peter Baumann   Jacobs University Bremen, Germany
Lucia Beccai   Istituto Italiano di Tecnologia, Italy
Marta Belchior Lopes   NOVA University Lisbon, Portugal
Aurelien Bellet   Inria, France
Gerardo Beni   University of California, Riverside, USA
Katie Bentley   Harvard Medical School, USA

Erik Berglund	AFRY, Sweden
Heder Bernardino	Universidade Federal de Juiz de Fora, Brazil
Daniel Berrar	Tokyo Institute of Technology, Japan
Adam Berry	CSIRO, Australia
Luc Berthouze	University of Sussex, UK
Martin Berzins	SCI Institute, The University of Utah, USA
Manuel A. Betancourt Odio	Universidad Pontificia Comillas, Spain
Hans-Georg Beyer	FH Vorarlberg University of Applied Sciences, Austria
Rajdeep Bhowmik	Binghamton University, SUNY, USA
Mauro Birattari	IRIDIA, Université Libre de Bruxelles, Belgium
Arnim Bleier	GESIS - Leibniz Institute for the Social Sciences, Germany
Leonidas Bleris	The University of Texas at Dallas, USA
Maria J. Blesa	Universitat Politècnica de Catalunya, Catalonia
Christian Blum	Spanish National Research Council, Spain
Martin Boldt	Blekinge Institute of Technology, Sweden
Fabio Bonassi	Politecnico di Milano, Italy
Flavia Bonomo	Universidad de Buenos Aires, Argentina
Gianluca Bontempi	Université Libre de Bruxelles, Belgium
Ilaria Bordino	UniCredit R&D, Italy
Anton Borg	Blekinge Institute of Technology, Sweden
Anna Bosman	University of Pretoria, South Africa
Paul Bourgine	Ecole Polytechnique Paris, France
Darko Bozhinoski	Delft University of Technology, The Netherlands
Michele Braccini	Università di Bologna, Italy
Juergen Branke	University of Warwick, UK
Ulf Brefeld	Leuphana University of Lüneburg, Germany
Will Browne	Victoria University of Wellington, New Zealand
Alexander Brownlee	University of Stirling, UK
Marcos Bueno	Radboud University, The Netherlands
Larry Bull	University of the West of England, UK
Tadeusz Burczynski	Polish Academy of Sciences, Poland
Robert Busa-Fekete	Yahoo! Research, USA
Adam A. Butchy	University of Pittsburgh, USA
Sergiy I. Butenko	Texas A&M University, USA
Sonia Cafieri	ENAC, France
Luca Cagliero	Politecnico di Torino, Italy
Stefano Cagnoni	University of Parma, Italy
Yizhi Cai	The University of Edinburgh, UK
Guido Caldarelli	IMT Lucca, Italy
Alexandre Campo	Université Libre de Bruxelles, Belgium
Mustafa Canim	IBM Thomas J. Watson Research Center, USA
Salvador Eugenio Caoili	University of the Philippines Manila, Philippines
Timoteo Carletti	University of Namur, Belgium
Jonathan Carlson	Microsoft Research, USA
Luigia Carlucci Aiello	Sapienza Università di Roma, Italy

Celso Carneiro Ribeiro	Universidade Federal Fluminense, Brazil
Alexandra M. Carvalho	Universidade de Lisboa, Portugal
Alberto Castellini	University of Verona, Italy
Michelangelo Ceci	University of Bari, Italy
Adelaide Cerveira	Universidade de Tras-os-Montes e Alto Douro, Portugal
Uday Chakraborty	University of Missouri-St. Louis, USA
Lijun Chang	The University of Sydney, Australia
Antonio Chella	University of Palermo Italy
Rachid Chelouah	Université Paris Seine, France
Haifeng Chen	NEC Labs, USA
Keke Chen	Wright State University, USA
Mulin Chen	Xidian University, China
Steven Chen	University of Pennsylvania, USA
Ying-Ping Chen	National Chiao Tung University, Taiwan, China
John W. Chinneck	Carleton University Ottawa, Canada
Gregory Chirikjian	Johns Hopkins University, USA
Miroslav Chlebik	University of Sussex, UK
Sung-Bae Cho	Yonsei University, South Korea
Stephane Chretien	ASSP/ERIC, Université Lyon 2, France
Anders Lyhne Christensen	University of Southern Denmark, Denmark
Andre Augusto Cire	University of Toronto Scarborough, Canada
Philippe Codognet	Sorbonne University, France, and The University of Tokyo, Japan
Carlos Coello Coello	IPN, Mexico
George Coghill	University of Aberdeen, UK
Pierre Comon	Université Grenoble Alpes, France
Sergio Consoli	EC Joint Research Centre, Belgium
David Cornforth	Newcastle University, UK
Luís Correia	Universidade de Lisboa, Portugal
Paulo J. da Costa Branco	Universidade de Lisboa, Portugal
Sebastian Daberdaku	Sorint.Tek, Italy
Chiara Damiani	University of Milano-Bicocca, Italy
Thomas Dandekar	University of Würzburg, Germany
Ivan Luciano Danesi	Unicredit Bank, Italy
Christian Darabos	Dartmouth College, USA
Elena Daraio	Politecnico di Torino, Italy
Raj Das	RMIT University, Australia
Vachik S. Dave	WalmartLabs, USA
Renato De Leone	Università di Camerino, Italy
Kalyanmoy Deb	Michigan State University, USA
Nicoletta Del Buono	University of Bari, Italy
Jordi Delgado	Universitat Politecnica de Catalunya, Spain
Rosario Delgado	Universitat Autonoma de Barcelona, Spain
Mauro Dell'Amico	University of Modena and Reggio Emilia, Italy
Brian Denton	University of Michigan, USA

Ralf Der	MPG, Germany
Clarisse Dhaenens	University of Lille, France
Barbara Di Camillo	University of Padova, Italy
Gianni Di Caro	IDSIA, Switzerland
Luigi Di Caro	University of Turin, Italy
Giuseppe Di Fatta	University of Reading, UK
Luca Di Gaspero	University of Udine, Italy
Mario Di Raimondo	University of Catania, Italy
Tom Diethe	Amazon Research Cambridge, UK
Matteo Diez	CNR, Institute of Marine Engineering, Italy
Ciprian Dobre	University POLITEHNICA of Bucharest, Romania
Stephan Doerfel	Micromata GmbH, Germany
Carola Doerr	CNRS, Sorbonne University, France
Rafal Drezewski	University of Science and Technology, Poland
Devdatt Dubhashi	Chalmers University, Sweden
George S. Dulikravich	Florida International University, USA
Juan J. Durillo	Leibniz Supercomputing Centre, Germany
Omer Dushek	University of Oxford, UK
Nelson F. F. Ebecken	University of Rio de Janeiro, Brazil
Marc Ebner	Ernst Moritz Arndt - Universität Greifswald, Germany
Tome Eftimov	Jožef Stefan Institute, Slovenia
Pascale Ehrenfreund	George Washington University, USA
Gusz Eiben	VU Amsterdam, The Netherlands
Aniko Ekart	Aston University, UK
Talbi El-Ghazali	University of Lille, France
Michael Elberfeld	RWTH Aachen University, Germany
Michael T. M. Emmerich	Leiden University, The Netherlands
Andries Engelbrecht	University of Pretoria, South Africa
Anton Eremeev	Sobolev Institute of Mathematics, Russia
Roberto Esposito	University of Turin, Italy
Giovanni Fasano	Università Ca' Foscari, Italy
Harold Fellermann	Newcastle University, UK
Chrisantha Fernando	DeepMind, UK
Cesar Ferri	Universidad Politecnica de Valencia, Spain
Paola Festa	University of Napoli Federico II, Italy
Jose Rui Figueira	Instituto Superior Tecnico, Portugal
Lionel Fillatre	Université Côte d'Azur, France
Steffen Finck	FH Vorarlberg University of Applied Sciences, Austria
Christoph Flamm	University of Vienna, Austria
Salvador A. Flores	Center for Mathematical Modelling, Chile
Enrico Formenti	Université Côte d'Azur, France
Giorgia Franchini	University of Modena and Reggio Emilia, Italy
Giuditta Franco	University of Verona, Italy
Emanuele Frandi	Cogent Labs, Japan
Piero Fraternali	Politecnico di Milano, Italy
Alex Freitas	University of Kent, UK

Valerio Freschi	University of Urbino, Italy
Enrique Frias Martinez	Telefonica Research, Spain
Walter Frisch	University of Vienna, Austria
Nikolaus Frohner	TU Wien, Austria
Rudolf M. Fuchslin	Zurich University of Applied Sciences, Switzerland
Antonio Fuduli	University of Calabria, Italy
Ashraf Gaffar	Arizona State University, USA
Carola Gajek	University of Augsburg, Germany
Marcus Gallagher	The University of Queensland, Australia
Claudio Gallicchio	University of Pisa, Italy
Patrick Gallinari	University of Paris 6, France
Luca Gambardella	IDSIA, Switzerland
Jean-Gabriel Ganascia	Pierre and Marie Curie University, France
Xavier Gandibleux	Universite de Nantes, France
Alfredo Garcia Hernandez-Diaz	Pablo de Olavide University, Spain
Jose Manuel Garcia Nieto	University of Malaga, Spain
Jonathan M. Garibaldi	University of Nottingham, UK
Paolo Garza	Politecnico di Torino, Italy
Romaric Gaudel	ENSAI, France
Nicholas Geard	The University of Melbourne, Australia
Martin Josef Geiger	Helmut Schmidt University, Germany
Marius Geitle	Østfold University College, Norway
Michel Gendreau	Polytechnique Montréal, Canada
Philip Gerlee	Chalmers University, Sweden
Mario Giacobini	University of Turin, Italy
Kyriakos Giannakoglou	National Technical University of Athens, Greece
Onofrio Gigliotta	University of Naples Federico II, Italy
Gail Gilboa Freedman	IDC Herzliya, Israel
David Ginsbourger	Idiap Research Institute, University of Bern, Switzerland
Giovanni Giuffrida	University of Catania, Neodata Group, Italy
Aris Gkoulalas-Divanis	IBM Watson Health, USA
Giorgio Stefano Gnecco	IMT School for Advanced Studies, Italy
Christian Gogu	Universite Toulouse III, France
Faustino Gomez	IDSIA, Switzerland
Teresa Gonçalves	University of Evora, Portugal
Eduardo Grampin Castro	Universidad de la Republica, Uruguay
Michael Granitzer	University of Passau, Germany
Alex Graudenzi	University of Milano-Bicocca, Italy
Vladimir Grishagin	Lobachevsky State University of Nizhni Novgorod, Russia
Roderich Gross	The University of Sheffield, UK
Mario Guarracino	ICAR-CNR, Italy
Francesco Gullo	UniCredit R&D, Italy
Vijay K. Gurbani	Illinois Institute of Technology, USA

Steven Gustafson	Noonum Inc., USA
Abbas Haider	Ulster University, UK
Jin-Kao Hao	University of Angers, France
Simon Harding	Machine Intelligence Ltd, Canada
Kyle Robert Harrison	University of New South Wales Canberra, Australia
William Hart	Sandia National Laboratories, USA
Inman Harvey	University of Sussex, UK
Mohammad Hasan	Indiana University Bloomington and Purdue University, USA
Geir Hasle	SINTEF Digital, Norway
Glenn Hawe	Ulster University, UK
Verena Heidrich-Meisner	Kiel University, Germany
Eligius M. T. Hendrix	Universidad de Malaga, Spain
Carlos Henggeler Antunes	University of Coimbra, Portugal
J. Michael Herrmann	The University of Edinburgh, UK
Jaakko Hollmen	Stockholm University, Sweden
Arjen Hommersom	Radboud University, The Netherlands
Vasant Honavar	Penn State University, USA
Hongxuan Huang	Tsinghua University, China
Fabrice Huet	University of Nice Sophia Antipolis, France
Hiroyuki Iizuka	Hokkaido University, Japan
Hisao Ishibuchi	Osaka Prefecture University, Japan
Peter Jacko	Lancaster University Management School, UK
Christian Jacob	University of Calgary, Canada
David Jaidan	Scalian-Eurogiciel, France
Hasan Jamil	University of Idaho, USA
Giorgio Jansen	University of Cambridge, UK
Yaochu Jin	University of Surrey, UK
Colin Johnson	University of Kent, UK
Gareth Jones	Dublin City University, Ireland
Laetitia Jourdan	University of Lille, CNRS, France
Janusz Kacprzyk	Polish Academy of Sciences, Poland
Theodore Kalamboukis	Athens University of Economics and Business, Greece
Valery Kalyagin	Higher School of Economics, Russia
George Kampis	Eötvös Loránd University, Hungary
Jaap Kamps	University of Amsterdam, The Netherlands
Dervis Karaboga	Erciyes University, Turkey
George Karakostas	McMaster University, Canada
Istvan Karsai	ETSU, USA
Branko Kavsek	University of Primorska, Slovenia
Zekarias T. Kefato	KTH Royal Institute of Technology, Sweden
Jozef Kelemen	Silesian University, Czech Republic
Marie-Eleonore Kessaci	University of Lille, France
Didier Keymeulen	NASA - Jet Propulsion Laboratory, USA
Michael Khachay	Ural Federal University, Russia
Daeeun Kim	Yonsei University, South Korea

Timoleon Kipouros	University of Cambridge, UK
Lefteris Kirousis	National and Kapodistrian University of Athens, Greece
Zeynep Kiziltan	University of Bologna, Italy
Mieczysaw Kopotek	Polish Academy of Sciences, Poland
Yury Kochetov	Novosibirsk State University, Russia
Elena Kochkina	University of Warwick, UK
Min Kong	Hefei University of Technology, China
Hennie Kruger	North-West University, South Africa
Erhun Kundakcioglu	Özyegin University, Turkey
Jacek Kustra	ASML, The Netherlands
Dmitri Kvasov	University of Calabria, Italy
Halina Kwasnicka	Wrocław University of Science and Technology, Poland
C. K. Kwong	Hong Kong Polytechnic University, Hong Kong, China
Emanuele La Malfa	University of Oxford, UK
Gabriele La Malfa	Quantitative Finance, Germany
Renaud Lambiotte	University of Namur, Belgium
Doron Lancet	Weizmann Institute of Science, Israel
Pier Luca Lanzi	Politecnico di Milano, Italy
Niklas Lavesson	Jönköping University, Sweden
Alessandro Lazaric	Facebook Artificial Intelligence Research (FAIR), France
Doheon Lee	KAIST, South Korea
Eva K. Lee	Georgia Tech, USA
Jay Lee	Center for Intelligent Maintenance Systems UC, USA
Tom Lenaerts	Université Libre de Bruxelles, Belgium
Rafael Leon	Universidad Politecnica de Madrid, Spain
Carson Leung	University of Manitoba, Canada
Peter R. Lewis	Aston University, UK
Rory Lewis	University of Colorado, USA
Kang Li	Google, USA
Lei Li	Florida International University, USA
Shuai Li	University of Cambridge, UK
Xiaodong Li	RMIT University, Australia
Weifeng Liu	China University of Petroleum, China
Joseph Lizier	The University of Sydney, Australia
Giosue' Lo Bosco	University of Palermo, Italy
Daniel Lobo	University of Maryland, USA
Fernando Lobo	University of Algarve, Portugal
Daniele Loiacono	Politecnico di Milano, Italy
Gianfranco Lombardo	University of Parma, Italy
Yang Lou	City University of Hong Kong, Hong Kong, China
Jose A. Lozano	University of the Basque Country, Spain
Paul Lu	University of Alberta, Canada
Angelo Lucia	University of Rhode Island, USA

Gabriel Luque	University of Malaga, Spain
Pasi Luukka	Lappeenranta-Lahti University of Technology, Finland
Dario Maggiorini	Università degli Studi di Milano, Italy
Gilvan Maia	Universidade Federal do Ceará, Brazil
Donato Malerba	University of Bari, Italy
Anthony Man-Cho So	The Chinese University of Hong Kong, Hong Kong, China
Jacek Madziuk	Warsaw University of Technology, Poland
Vittorio Maniezzo	University of Bologna, Italy
Luca Manzoni	University of Trieste, Italy
Marco Maratea	University of Genova, Italy
Elena Marchiori	Radboud University, The Netherlands
Tiziana Margaria	University of Limerick, Lero, Ireland
Magdalene Marinaki	Technical University of Crete, Greece
Yannis Marinakis	Technical University of Crete, Greece
Omer Markovitch	University of Groningen, The Netherlands
Carlos Martin-Vide	Rovira i Virgili University, Spain
Dominique Martinez	CNRS, France
Aldo Marzullo	University of Calabria, Italy
Joana Matos Dias	Universidade de Coimbra, Portugal
Nikolaos Matsatsinis	Technical University of Crete, Greece
Matteo Matteucci	Politecnico di Milano, Italy
Stefano Mauceri	University College Dublin, Ireland
Giancarlo Mauri	University of Milano-Bicocca, Italy
Antonio Mauttone	Universidad de la Republica, Uruguay
Mirjana Mazuran	Politecnico di Milano, Italy
James McDermott	National University of Ireland, Ireland
Suzanne McIntosh	NYU Courant, NYU Center for Data Science, USA
Gabor Melli	Sony Interactive Entertainment Inc., USA
Jose Fernando Mendes	University of Aveiro, Portugal
Lu Meng	University at Buffalo, USA
Rakesh R. Menon	University of Massachusetts Amherst, USA
David Merodio-Codinachs	ESA, France
Silja Meyer-Nieberg	Universität der Bundeswehr München, Germany
Efren Mezura-Montes	University of Veracruz, Mexico
George Michailidis	University of Florida, USA
Martin Middendorf	Leipzig University, Germany
Kaisa Miettinen	University of Jyväskylä, Finland
Orazio Miglino	University of Naples Federico II, Italy
Julian Miller	University of York, UK
Marco Mirolli	ISTC-CNR, Italy
Mustafa Misir	Istinye University, Turkey
Natasa Miskov-Zivanov	University of Pittsburgh, USA
Carmen Molina-Paris	University of Leeds, UK
Shokoufeh Monjezi Kouchak	Arizona State University, USA

Sara Montagna	Università di Bologna, Italy
Marco Montes de Oca	Clypd Inc., USA
Rafael M. Moraes	Viasat Inc., USA
Monica Mordonini	University of Parma, Italy
Nima Nabizadeh	Ruhr-Universität Bochum, Germany
Mohamed Nadif	University of Paris, France
Hidemoto Nakada	NAIST, Japan
Mirco Nanni	CNR-ISTI, Italy
Valentina Narvaez-Teran	Cinvestav Tamaulipas, Mexico
Sriraam Natarajan	Indiana University Bloomington, USA
Chrystopher L. Nehaniv	University of Hertfordshire, UK
Michael Newell	Athens Consulting, LLC, USA
Binh P. Nguyen	Victoria University of Wellington, New Zealand
Giuseppe Nicosia	University of Catania, Italy
Sotiris Nikoletseas	University of Patras, CTI, Greece
Xia Ning	IUPUI, USA
Jonas Nordhaug Myhre	UiT The Arctic University of Norway, Norway
Wieslaw Nowak	Nicolaus Copernicus University, Poland
Eirini Ntoutsi	Leibniz Universität Hannover, Germany
David Nunez	University of the Basque Country, Spain
Varun Ojha	University of Reading, UK
Michal Or-Guil	Humboldt University of Berlin, Germany
Marcin Orchel	AGH University of Science and Technology, Poland
Mathias Pacher	Goethe University Frankfurt, Germany
Ping-Feng Pai	National Chi Nan University, Taiwan, China
Pramudita Satria Palar	Bandung Institute of Technology, Indonesia
Wei Pang	University of Aberdeen, UK
George Papastefanatos	Athena RC/IMIS, Greece
Luís Paquete	University of Coimbra, Portugal
Panos Pardalos	University of Florida, USA
Rohit Parimi	Bloomberg LP, USA
Konstantinos Parsopoulos	University of Ioannina, Greece
Andrea Patane'	University of Oxford, UK
Remigijus Paulaviius	Vilnius University, Lithuania
Joshua Payne	University of Zurich, Switzerland
Clint George Pazhayidam	Indian Institute of Technology Goa, India
Jun Pei	Hefei University of Technology, China
Nikos Pelekis	University of Piraeus, Greece
David A. Pelta	Universidad de Granada, Spain
Dimitri Perrin	Queensland University of Technology, Australia
Milena Petkovi	Zuse Institute Berlin, Germany
Koumoutsakos Petros	ETH, Switzerland
Juan Peypouquet	Universidad Técnica Federico Santa María, Chile
Andrew Philippides	University of Sussex, UK
Stefan Pickl	Universität der Bundeswehr München, Germany
Fabio Pinelli	Vodafone Italia, Italy

Joao Pinto	Technical University of Lisbon, Portugal
Vincenzo Piuri	Università degli Studi di Milano, Italy
Alessio Pleb	University Messina, Italy
Nikolaos Ploskas	University of Western Macedonia, Greece
Agoritsa Polyzou	University of Minnesota, USA
George Potamias	Institute of Computer Science - FORTH, Greece
Philippe Preux	Inria, France
Mikhail Prokopenko	The University of Sydney, Australia
Paolo Provero	University of Turin, Italy
Buyue Qian	IBM T. J. Watson, USA
Chao Qian	Nanjing University, China
Michela Quadrini	University of Padova, Italy
Tomasz Radzik	King's College London, UK
Gunther Raidl	TU Wien, Austria
Helena Ramalhinho Lourenço	Pompeu Fabra University, Spain
Palaniappan Ramaswamy	University of Kent, UK
Jan Ramon	Inria, France
Vitorino Ramos	Technical University of Lisbon, Portugal
Shoba Ranganathan	Macquarie University, Australia
Zbigniew Ras	The University of North Carolina, USA
Jan Rauch	University of Economics, Czech Republic
Steffen Rebennack	Karlsruhe Institute of Technology (KIT), Germany
Wolfgang Reif	University of Augsburg, Germany
Patrik Reizinger	Budapest University of Technology and Economics, Hungary
Guillermo Rela	Universidad de la Republica, Uruguay
Cristina Requejo	Universidade de Aveiro, Portugal
Paul Reverdy	University of Arizona, USA
John Rieffel	Union College, USA
Francesco Rinaldi	University of Padova, Italy
Laura Anna Ripamonti	Università degli Studi di Milano, Italy
Franco Robledo	Universidad de la Republica, Uruguay
Humberto Rocha	University of Coimbra, Portugal
Marcelo Lisboa Rocha	Universidade Federal do Tocantins, Brazil
Eduardo Rodriguez-Tello	Cinvestav-Tamaulipas, Mexico
Andrea Roli	Università di Bologna, Italy
Massimo Roma	Sapienza Università di Roma, Italy
Vittorio Romano	University of Catania, Italy
Pablo Romero	Universidad de la Republica, Uruguay
Andre Rosendo	University of Cambridge, UK
Samuel Rota Bulo	Mapillary Research, Austria
Arnab Roy	Fujitsu Laboratories of America, USA
Alessandro Rozza	Parthenope University of Naples, Italy
Valeria Ruggiero	University of Ferrara, Italy
Kepa Ruiz-Mirazo	University of the Basque Country, Spain

Florin Rusu	University of California, Merced, USA
Conor Ryan	University of Limerick, Ireland
Jakub Rydzewski	Nicolaus Copernicus University, Poland
Nick Sahinidis	Carnegie Mellon University, USA
Lorenza Saitta	University of Piemonte Orientale, Italy
Isak Samsten	Stockholm University, Sweden
Andrea Santoro	Queen Mary University of London, UK
Giorgio Sartor	SINTEF, Norway
Claudio Sartori	University of Bologna, Italy
Frederic Saubion	Université Angers, France
Khaled Sayed	University of Pittsburgh, USA
Robert Schaefer	AGH University of Science and Technology, Poland
Andrea Schaerf	University of Udine, Italy
Alexander Schiendorfer	University of Augsburg, Germany
Rossano Schifanella	University of Turin, Italy
Christoph Schommer	University of Luxembourg, Luxembourg
Oliver Schuetze	IPN, Mexico
Martin Schulz	Technical University of Munich, Germany
Bryan Scotney	Ulster University, UK
Luís Seabra Lopes	University of Aveiro, Portugal
Natalia Selini Hadjidimitriou	University of Modena and Reggio Emilia, Italy
Giovanni Semeraro	University of Bari, Italy
Alexander Senov	Saint Petersburg State University, Russia
Andrea Serani	CNR, Institute of Marine Engineering, Italy
Roberto Serra	University of Modena and Reggio Emilia, Italy
Marc Sevaux	Université de Bretagne-Sud, France
Kaushik Das Sharma	University of Calcutta, India
Nasrullah Sheikh	IBM Research, USA
Vladimir Shenmaier	Sobolev Institute of Mathematics, Russia
Leonid Sheremetov	Mexican Petroleum Institute, Mexico
Ruey-Lin Sheu	National Cheng Kung University, Taiwan, China
Hsu-Shih Shih	Tamkang University, Taiwan, China
Kilho Shin	Gakushuin University, Japan
Zeren Shui	University of Minnesota, USA
Patrick Siarry	Université Paris-Est Creteil, France
Sergei Sidorov	Saratov State University, Russia
Alkis Simitsis	HP Labs, USA
Yun Sing Koh	The University of Auckland, New Zealand
Alina Sirbu	University of Pisa, Italy
Konstantina Skouri	University of Ioannina, Greece
Elaheh Sobhani	Université Grenoble Alpes, France
Johannes Sollner	Sodatana e.U., Austria
Giandomenico Spezzano	CNR-ICAR, Italy
Antoine Spicher	Université Paris-Est Creteil, France
Claudio Stamile	Université Claude Bernard Lyon 1, France

Pasquale Stano	University of Salento, Italy
Thomas Stibor	GSI Helmholtz Centre for Heavy Ion Research, Germany
Catalin Stoean	University of Craiova, Romania
Johan Suykens	KU Leuven, Belgium
Reiji Suzuki	Nagoya University, Japan
Domenico Talia	University of Calabria, Italy
Kay Chen Tan	National University of Singapore, Singapore
Letizia Tanca	Politecnico di Milano, Italy
Sean Tao	Carnegie Mellon University, USA
Katarzyna Tarnowska	San Jose State University, USA
Erica Tavazzi	University of Padova, Italy
Charles Taylor	UCLA, USA
Tatiana Tchemisova Cordeiro	University of Aveiro, Portugal
Maguelonne Teisseire	INRAE, UMR, TETIS, France
Fabien Teytaud	Université du Littoral Côte d'Opale, France
Tzouramanis Theodoros	University of the Aegean, Greece
Jon Timmis	University of York, UK
Gianna Toffolo	University of Padova, UK
Gabriele Tolomei	Sapienza Università di Roma, Italy
Michele Tomaiuolo	University of Parma, Italy
Joo Chuan Tong	Institute of High Performance Computing, Singapore
Gerardo Toraldo	Università degli Studi di Napoli Federico II, Italy
Jaden Travnik	University of Alberta, Canada
Nickolay Trendafilov	Open University, UK
Sophia Tsoka	King's College London, UK
Shigeyoshi Tsutsui	Hannan University, Japan
Elio Tuci	University of Namur, Belgium
Ali Emre Turgut	IRIDIA-ULB, France
Karl Tuyls	The University of Liverpool, UK
Gregor Ulm	Fraunhofer-Chalmers Research Centre for Industrial Mathematics, Sweden
Jon Umerez	University of the Basque Country, Spain
Renato Umeton	Dana-Farber Cancer Institute, MIT, USA
Ashish Umre	University of Sussex, UK
Olgierd Unold	Politechnika Wrocławska, Poland
Rishabh Upadhyay	Innopolis University, Russia
Alexandru Uta	VU Amsterdam, The Netherlands
Giorgio Valentini	Università degli Studi di Milano, Italy
Sergi Valverde	Pompeu Fabra University, Spain
Werner Van Geit	Blue Brain Project, EPFL, Switzerland
Pascal Van Hentenryck	University of Michigan, USA
Ana Lucia Varbanescu	University of Amsterdam, The Netherlands
Carlos Varela	Rensselaer Polytechnic Institute, USA
Iraklis Varlamis	Harokopio University, Greece

Eleni Vasilaki	The University of Sheffield, UK
Apostol Vassilev	NIST, USA
Richard Vaughan	Simon Fraser University, Canada
Kalyan Veeramachaneni	MIT, USA
Vassilios Verykios	Hellenic Open University, Greece
Herna L. Viktor	University of Ottawa, Canada
Mario Villalobos-Arias	Univesidad de Costa Rica, Costa Rica
Marco Villani	University of Modena and Reggio Emilia, Italy
Susana Vinga	INESC-ID, Instituto Superior Técnico, Portugal
Mirko Viroli	Università di Bologna, Italy
Katya Vladislavleva	Evolved Analytics LLC, Belgium
Robin Vogel	Telecom, France
Stefan Voss	University of Hamburg, Germany
Dean Vuini	Vrije Universiteit Brussel, Belgium
Markus Wagner	The University of Adelaide, Australia
Toby Walsh	UNSW Sydney, Australia
Harry Wang	University of Michigan, USA
Jianwu Wang	University of Maryland, USA
Lipo Wang	Nanyang Technological University, Singapore
Longshaokan Wang	North Carolina State University, USA
Rainer Wansch	Fraunhofer IIS, Germany
Syed Waziruddin	Kansas State University, USA
Janet Wiles	The University of Queensland, Australia
Man Leung Wong	Lingnan University, Hong Kong, China
Andrew Wuensche	University of Sussex, UK
Petros Xanthopoulos	University of Central Florida, USA
Ning Xiong	Mälardalen University, Sweden
Chang Xu	The University of Sydney, Australia
Dachuan Xu	Beijing University of Technology, China
Xin Xu	George Washington University, USA
Gur Yaari	Yale University, USA
Larry Yaeger	Indiana University Bloomington, USA
Shengxiang Yang	De Montfort University, USA
Xin-She Yang	Middlesex University London, UK
Li-Chia Yeh	National Tsing Hua University, Taiwan, China
Sule Yildirim-Yayilgan	Norwegian University of Science and Technology, Norway
Shiu Yin Yuen	The City University of Hong Kong, Hong Kong, China
Qi Yu	Rochester Institute of Technology, USA
Zelda Zabinsky	University of Washington, USA
Luca Zanni	University of Modena and Reggio Emilia, Italy
Ras Zbyszek	The University of North Carolina, USA
Hector Zenil	University of Oxford, UK
Guang Lan Zhang	Boston University, USA
Qingfu Zhang	City University of Hong Kong, Hong Kong, China
Rui Zhang	IBM Research, USA

Zhi-Hua Zhou	Nanjing University, China
Tom Ziemke	Linköping University, Sweden
Antanas Zilinskas	Vilnius University, Lithuania
Julius Ilinskas	Vilnius University, Lithuania

Subreviewers

Agostinho Agra	Hussain Kazmi
Alessandro Suglia	Kamer Kaya
Amlan Chakrabarti	Kaushik Das Sharma
Andrea Tangherloni	Luís Gouveia
Andreas Artemiou	Marco Polignano
Anirban Dey	Paolo Di Lorenzo
Arnaud Liefooghe	Rami Nourddine
Artem Baklanov	S. D. Riccio
Artyom Kondakov	Sai Ji
Athar Khodabakhsh	Sean Tao
Bertrand Gauthier	Shyam Chandramouli
Brian Tsan	Simone G. Riva
Christian Hubbs	Tonguc Yavuz
Constantinos Siettos	Vincenzo Bonnici
David Nizar Jaidan	Xianli Zhang
Dmitry Ivanov	Xiaoyu Li
Edhem Sakarya	Xin Sun
Farzad Avishan	Yang Li
Gaoxiang Zhou	Yasmine Ahmed
Guiying Li	Yujing Ma
Hao Yu	Zuhal Ozcan
Hongqiao Wang	

Best Paper Awards

LOD 2020 Best Paper Award

"Quantifying Local Energy Demand through Pollution Analysis"
Cole Smith[1], Andrii Dobroshynskyi[1], and Suzanne McIntosh[1,2]
[1] Courant Institute of Mathematical Sciences, New York University, USA
[2] Center for Data Science, New York University, USA
Springer sponsored the LOD 2020 Best Paper Award with a cash prize of EUR 1,000.

Special Mention

"Sparsity Meets Robustness: Channel Pruning for the Feynman-Kac Formalism Principled Robust Deep Neural Nets"
Thu Dinh, Bao Wang, Andrea Bertozzi, Stanley Osher and Jack Xin
University of California Irvine, University of California, Los Angeles (UCLA).

"State Representation Learning from Demonstration"
Astrid Merckling, Alexandre Coninx, Loic Cressot, Stephane Doncieux and Nicolas Perrin
Sorbonne University, Paris, France

"Sparse Perturbations for Improved Convergence in SZO Optimization"
Mayumi Ohta, Nathaniel Berger, Artem Sokolov and Stefan Riezler
Heidelberg University, Germany

LOD 2020 Best Talks

"A fast and efficient smoothing approach to LASSO regression and an application in statistical genetics: polygenic risk scores for Chronic obstructive pulmonary disease (COPD)"
Georg Hahn, Sharon Marie Lutz, Nilanjana Laha, Christoph Lange
Department of Biostatistics, T.H. Chan School of Public Health, Harvard University, USA

"Gravitational Forecast Reconciliation"
Carla Freitas Silveira, Mohsen Bahrami, Vinicius Brei, Burcin Bozkaya, Selim Balcsoy, Alex "Sandy" Pentland
University of Bologna, Italy - MIT Media Laboratory, USA - Federal University of Rio Grande do Sul Brazil and MIT Media Laboratory, USA - New College of Florida, USA and Sabanci University, Turkey - Sabanci University, Turkey - Massachusetts Institute of Technology - MIT Media Laboratory, USA

"From Business Curated Products to Algorithmically Generated"
Vera Kalinichenko and Garima Garg
University of California, Los Angeles - UCLA, USA and FabFitFun, USA

LOD 2019 Best Paper Award

"Deep Neural Network Ensembles"
Sean Tao
Carnegie Mellon University, USA

LOD 2018 Best Paper Award

"Calibrating the Classifier: Siamese Neural Network Architecture for End-to-End Arousal Recognition from ECG"
Andrea Patané* and Marta Kwiatkowska*
University of Oxford, UK

MOD 2017 Best Paper Award

"Recipes for Translating Big Data Machine Reading to Executable Cellular Signaling Models"
Khaled Sayed*, Cheryl Telmer**, Adam Butchy* & Natasa Miskov-Zivanov*
*University of Pittsburgh, USA
**Carnegie Mellon University, USA

MOD 2016 Best Paper Award

"Machine Learning: Multi-site Evidence-based Best Practice Discovery"
Eva Lee, Yuanbo Wang and Matthew Hagen
Eva K. Lee, Professor Director, Center for Operations Research in Medicine and HealthCare H. Milton Stewart School of Industrial and Systems Engineering, Georgia Institute of Technology, USA

MOD 2015 Best Paper Award

"Learning with discrete least squares on multivariate polynomial spaces using evaluations at random or low-discrepancy point sets"
Giovanni Migliorati
Ecole Polytechnique Federale de Lausanne – EPFL, Switzerland

Contents – Part II

Contents – Part I

Multi-kernel Covariance Terms in Multi-output Support Vector Machines

Elisa Marcelli[✉][iD] and Renato De Leone[iD]

Department of Mathematics, School of Sciences and Technology,
University of Camerino, via Madonna Delle Carceri 9, Camerino, MC, Italy
{elisa.marcelli,renato.deleone}@unicam.it

Abstract. This paper proposes a novel way to learn multi-task kernel machines by combining the structure of classical Support Vector Machine (SVM) optimization problem with multi-task covariance functions developed in Gaussian process (GP) literature. Specifically, we propose a multi-task Support Vector Machine that can be trained on data with multiple target variables simultaneously, while taking into account the correlation structure between different outputs. In the proposed framework, the correlation structure between multiple tasks is captured by covariance functions constructed using a Fourier transform, which allows to represent both auto and cross-correlation structure between the outputs. We present a mathematical model and validate it experimentally on a rescaled version of the Jura dataset, a collection of samples representing the amount of seven chemical elements into several locations. The results demonstrate the utility of our modeling framework.

Keywords: Support vector machine · Multi task learning · Gaussian processes · Kernel learning

1 Introduction

The problem of simultaneously solving more than one task, i.e., outputs, is named multi-task learning and has gained a significant amount of notoriety over the last years. Unlike classical supervised machine learning techniques where a function is sought to map input values to a specific appropriate output, i.e., one continuous value for regression problems and one discrete label for classification problems, multi-task learning aims to, given a specific input, learn multiple outputs at once. First of all, multi-output and multi-task problems are different even if they may be mistakenly considered to be the same or swapped with each other. Specifically, the specific goal of multi-task learning (MTL) is to simultaneously learn multiple tasks while considering all the possible existing relationships between them. Therefore, for every given input a multidimensional output is assigned, corresponding to each task of the problem. The main hypothesis of MTL is that each task, or at the very least a subset of them, has inter-correlation factors, making necessary an algorithm that explicitly considers this correlation.

© Springer Nature Switzerland AG 2020
G. Nicosia et al. (Eds.): LOD 2020, LNCS 12566, pp. 1–11, 2020.
https://doi.org/10.1007/978-3-030-64580-9_1

On the contrary, in multi-output learning (MOL), each input is associated to an output which value can be assumed from a vector space and, thus, several possible choices of outputs are possible but only one is finally paired to each input. Moreover, even if an inter-output correlation is not excluded, it does not constitute a key factor of MOL. Anyway, for a proper reading of our paper it is important to specify that we have decided to name our model multi-output Support Vector Machine.

In literature, a fair number of review works exists, showing how to approach multi-task learning in different research areas: multi-output regression with real value outputs [3,7]; classifications problems [2,10]; overall frameworks [12]. The simplest approach to solve a generic multi-task problem would be to consider every task as independent from the others and, therefore, separately solving every problem and combine the solutions found to obtain a general result. In this way though, since independent problems are solved, it is very likely to ignore significant multi-output correlation factors, thus resulting with improper results not taking care of relationships between different tasks [1]. As a direct consequence, doing so all the benefits that make multi-task learning particularly suitable for describing real problems, i.e., problem where the inner task correlation factors are not only present but significantly important, are canceled. Of course, if the given multi task problem is defined by tasks that are not related with each other, the methods described above fit the purpose correctly. An attractive approach dealing with multi-task learning makes use of reproducing kernel Hilbert spaces (RKHS) of vector–valued functions [9]. In this case we talk about matrix–valued kernels, a rightful extension of the idea of kernels, where a kernel matrix is defined with respect to every input value and based on kernel Hilbert space.

With regard to Support Vector Machine (SVM), over the last years several approaches have been proposed for solving a multi-task problem based on the typical SVM framework. An interesting approach is defined by a multi-view multi-task Support Vector Machine approach [14] combining multi task learning with data from different backgrounds; regularization based approaches such as [6] using minimization of regularization functionals can be found in literature; a sequential minimal optimization approach [4], decomposing the problem into a set of minimal subproblems, exists. In [13], starting from the basic formulation, a generalized multi-task SVM framework is proposed and solved with the use of Krylow space methods.

In Gaussian processes (GPs) the multi-task learning problem is solved with the use of multi-task covariance functions. Such functions are able to solve the critical issue of multi-task correlations looking at all the possible connections between different tasks. Specifically, this approach requires the covariance function to operate on a cross-correlation factor between tasks in addition to the classical correlation factor of each task. Several approaches were introduced over the years, such as in the geostatical environment [5,11] known under the name of cokriging. Several cokriging methods exist: simple, ordinary, collocated, and indicator. They mainly differ from each other because of the exploited mean value, e.g., simple cokriging employs a clearly defined beforehand mean, making it globally constant throughout the whole process, while similarly ordinary cokriging opts for locally constant means based on specific points. Usually such methods use the same covariance function for different tasks. An innovative approach is proposed in [8], that will be explored in more detail in Sect. 2.

In this paper we propose a novel approach combining a multi-task SVM framework with valid covariance terms: we develop a MT Support Vector Machine with multi-output correlation factors and we use the covariance approach proposed in [8] to solve it.

The rest of the paper is organized as follows. Initially, in Sect. 2, we summarize the basic idea behind the covariance structure in use throughout the paper. In Sect. 3 we describe our multi-output SVM model, explaining the basic construction details. Then, in Sect. 4 we present some preliminary results on the new proposed model. Finally, in Sect. 5 we draw conclusions.

2 Related Work

As briefly introduced in Sect. 1, our model relies on the multi-output covariance function template presented in [8]. Each task is associated with a different kernel function and multi-task factors are modelled with the use of cross covariance terms. Specifically, the paper relies on the idea of using a convolution structure for each single task: given their basis function g_i, i.e., smoothing kernel, a single task auto covariance function k_{ii} can be written as follows

$$k_{ii}(x^i, x^j) = \int_{-\infty}^{+\infty} g_i(x^i - u)g_i(x^j - u)\,\mathrm{d}u \tag{1}$$

Generalizing (1), we get the following form for the covariance matrix K

$$K(x^i, x^j) = \int_{-\infty}^{+\infty} g_i(x^i - u)g_j(x^j - u)\,\mathrm{d}u \tag{2}$$

where x^i is in the i-th task and x^j is in the j-th task. Note that, the covariance function defined in (2) has the characteristic of being a positive semidefinite (PSD) matrix. Therefore, the cross-covariance terms may be found using (2) with the main challenge of computing the basis functions g_i.
A general form to compute basis functions is given by

$$g(\tau) = \frac{1}{(2\pi)^{1/4}} \mathcal{F}_{s\to\tau}^{-1}[\sqrt{\mathcal{F}_{\tau\to s}[k(\tau)]}] \tag{3}$$

where $\tau = x^i - x^j$ and $\mathcal{F}_{\tau\to s}[k(\tau)]$ is the Fourier transformation of an arbitrary stationary covariance function $k(\tau)$.

Moreover, using the scheme just described, in [8] three examples of cross covariance functions are provided, starting with the definitions of the squared exponential, the Matérn and the sparse covariance functions as a basis. The Matérn covariance function is computed with parameter $\nu = 3/2$ and both auto covariance, i.e., when the same covariance function is used, and cross covariance functions, i.e., when two different basis covariance functions are considered, are calculated. Note that, as it will be clarified in the next section, since we have operated using Matérn 3/2 and squared exponential, we are going to present only the covariance terms including such functions. Specifically, given the following covariance functions

$$k_{SE}(r, l_{SE}) = \exp\left[-\frac{1}{2}\left(\frac{r}{l_{SE}}\right)^2\right] \qquad (4)$$

$$k_M(r, l_M) = \left(1 + \frac{\sqrt{3}r}{l_M}\right)\exp\left(-\frac{\sqrt{3}r}{l_M}\right) \qquad (5)$$

with $r = |x^i - x^j|$ and l_M and l_{SE} respectively length scale for Matérn 3/2 and squared exponential function, from Eqs. (4) and (5), auto covariance and cross covariance terms are built.

For the squared exponential in (4), we have

$$k_{SE_1 \times SE_2}(r, l_{SE_1}, l_{SE_2}) = \sqrt{\frac{2l_{SE_1}l_{SE_2}}{l_{SE_1}^2 + l_{SE_2}^2}}\exp\left(-\frac{r^2}{l_{SE_1}^2 + l_{SE_2}^2}\right) \qquad (6)$$

where r is defined as above and l_{SE_1} and l_{SE_2} are the respective length scales. Analogously, using (5) the Matérn auto covariance term is given by

$$k_{M_1 \times M_2}(r, l_{M_1}, l_{M_2}) = \sigma_{M_1 M_2}(l_{M_1}e^{-\sqrt{3}\frac{r}{l_{M_1}}} - l_{M_2}e^{-\sqrt{3}\frac{r}{l_{M_2}}}) \qquad (7)$$

with $\sigma_{M_1 M_2} = 2\sqrt{l_{M_1}l_{M_2}}/(l_{M_1}^2 l_{M_2}^2)$ and l_{M_1}, l_{M_2} lenght scales.

Finally, the squared exponential-Matérn cross covariance term is

$$k_{SE \times M}(r, l_{SE}, l_M) = \sqrt{\lambda}(\pi/2)^{1/4}e^{\lambda^2}\left[2\cosh\left(\frac{\sqrt{3}r}{l_M}\right)\right.$$
$$\left. e^{\frac{\sqrt{3}r}{l_M}}\operatorname{erf}\left(\lambda + \frac{r}{l_{SE}}\right) - e^{-\frac{\sqrt{3}r}{l_M}}\operatorname{erf}\left(\lambda - \frac{r}{l_{SE}}\right)\right] \qquad (8)$$

where $\lambda = \frac{\sqrt{3}l_{SE}}{2l_M}$ and "erf" is the Gauss error function.

3   The Proposed Model

To be able to use a kernel matrix like the one defined by Eqs. (6)–(8) in a SVM framework, we need to construct a multi-output SVM taking care of correlation factors of every output. Given a generic dataset $\{x^i, y_i\}_{i=1}^N$, with $x^i \in \mathbb{R}^n$ and $y^i \in \{-1, +1\}^m$, starting from the classical SVM structure, our goal is to define a problem with a multi-output cross factor. To do so, we propose some modifications to the classical SVM problem. As far as the use of kernels, we decided to consider more than one feature map: specifically, for every output we consider a feature map $\phi_p(\cdot) \in \mathbb{R}^l$, $\forall p = 1, \ldots, m$ of fixed dimension l. Concerning the normal vector to the hyperplane w, we substitute it with a more complex formulation: from every couple of outputs $\{i, j\}$, $i, j = 1, \ldots, m$ we consider a vector w_{ij} of dimension l, i.e., the dimension of each feature map, leading to a three dimensional tensor W of dimension $m \times m \times l$ with elements w_{kpq}. As an additional condition on W, we require $w_{ij} = w_{ji}$, $\forall i, j, = 1, \ldots, m$. For a sparse representation of SVM, w is substituted by its expansion in terms of the dual variable. In this case the objective function contains a term imposing sparsity of the solution. In the same spirit, similarly to classical SVM problem, we propose to minimize the sum of the square of every element present in the three dimensional tensor W.

The optimization problem is the following

$$
\begin{aligned}
\min \quad & \frac{1}{2} \sum_{k,p=1}^{m} \sum_{q=1}^{l} w_{kpq}^2 + C \sum_{k=1}^{m} \sum_{i1}^{N} \xi_k^i \\
\text{s. t.} \quad & y_k^i \left(\sum_{p=1}^{m} \sum_{q=1}^{l} w_{kpq} \phi_{pq}(x^i) + b_k \right) \geq 1 - \xi_k^i \\
& w_{kpq} = w_{pkq}, \quad k, p = 1, \ldots, m, \quad q = 1, \ldots, l \\
& \xi_k^i \geq 0 \ i = 1, \ldots, N, \ k = 1, \ldots, m
\end{aligned}
\tag{9}
$$

Note that, $\sum_{p=1}^{m} \sum_{q=1}^{l} w_{kpq} \phi_{pq}(x^i)$ is equivalent to $\sum_{p=1}^{m} w_{kp}^{\mathrm{T}} \phi_p(x^i)$, where the notation ϕ_{pq} is used to describe the q-th component of vector ϕ_p.

The Lagrangian function associated to (9) is

$$
\begin{aligned}
\mathcal{L} = {} & \frac{1}{2} \sum_{k,p=1}^{m} \sum_{q=1}^{l} w_{kpq}^2 + C \sum_{k=1}^{m} \sum_{i=1}^{N} \xi_k^i \\
& - \sum_{i=1}^{N} \sum_{k=1}^{m} \lambda_k^i \left[y_k^i \left(\sum_{p=1}^{m} \sum_{q=1}^{l} w_{kpq} \phi_{pq}(x^i) + b_k \right) - 1 + \xi_k^i \right] \\
& + \sum_{k=1}^{m} \sum_{i=1}^{N} \mu_k^i \xi_k^i + \sum_{k,p=1}^{m} \sum_{q=1}^{l} \theta_{kpq}(w_{kpq} - w_{pkq}).
\end{aligned}
\tag{10}
$$

The Karush–Kuhn–Tucker conditions are now

$$
\frac{\partial \mathcal{L}}{\partial w_{kpq}} = 0 \implies w_{kpq} = \sum_{i=1}^{N} \lambda_k^i y_k^i \phi_{pq}(x^i) - \theta_{kpq} + \theta_{pkq},
\tag{11a}
$$

$$
\frac{\partial \mathcal{L}}{\partial b_k} = 0 \implies \sum_{i=1}^{N} \lambda_k^i y_k^i = 0,
\tag{11b}
$$

$$
\frac{\partial \mathcal{L}}{\partial \xi_k^i} = 0 \implies C = \lambda_k^i - \mu_k^i.
\tag{11c}
$$

Substituting (11a)-(11c) into (10) we get

$$
\begin{aligned}
\mathcal{L} = & -\frac{1}{2} \sum_{k,p=1}^{m} \sum_{q=1}^{l} w_{kpq}^2 + \sum_{k,p=1}^{m} \sum_{q=1}^{l} w_{kpq} \left(w_{kpq} - \sum_{i=1}^{N} \lambda_k^i y_k^i \phi_{pq}(x^i) - \theta_{kpq} + \theta_{pkq} \right) \\
& + \sum_{k=1}^{m} \sum_{i=1}^{N} \xi_k^i \left(C - \lambda_k^i + \mu_k^i \right) - \sum_{k=1}^{m} \left(\sum_{i=1}^{N} \lambda_k^i y_k^i \right) + \sum_{i=1}^{N} \sum_{k=1}^{m} \lambda_k^i = \\
& -\frac{1}{2} \sum_{k,p=1}^{m} \sum_{q=1}^{l} w_{kpq}^2 + \sum_{i=1}^{N} \sum_{k=1}^{m} \lambda_k^i.
\end{aligned}
\tag{12}
$$

Now, recalling from (11a) that

$$
w_{kpq} = \sum_{i=1}^{N} \lambda_k^i y_k^i \phi_{pq}(x^i) - \theta_{kpq} + \theta_{pkq} \tag{13a}
$$

$$
w_{pkq} = \sum_{i=1}^{N} \lambda_p^i y_p^i \phi_{kq}(x^i) - \theta_{pkq} + \theta_{kpq} \tag{13b}
$$

and since $w_{kpq} - w_{pkq} = 0$, we have that

$$
w_{kpq} = \frac{1}{2} \left(\sum_{i=1}^{N} \lambda_k^i y_k^i \phi_{pq}(x^i) + \sum_{i=1}^{N} \lambda_p^i y_p^i \phi_{kq}(x^i) \right) \tag{14}
$$

Substituting (14) into (10) we finally obtain

$$
\begin{aligned}
\mathcal{L} = & \sum_{i=1}^{N} \sum_{k=1}^{m} \lambda_k^i - \frac{1}{2} \sum_{k,p=1}^{m} \left[\frac{1}{2} \left(\sum_{i=1}^{N} \lambda_k^i y_k^i \phi_p(x^i) + \sum_{i=1}^{N} \lambda_p^i y_p^i \phi_k(x^i) \right) \right]^2 = \\
= & \sum_{i=1}^{N} \sum_{k=1}^{m} \lambda_k^i - \frac{1}{8} \sum_{k,p=1}^{m} \left[\sum_{i=1}^{N} \sum_{i'=1}^{N} \lambda_k^i \lambda_k^{i'} y_k^i y_k^{i'} \left\langle \phi_p(x^i), \phi_p(x^{i'}) \right\rangle + \right. \\
& \left. \sum_{i=1}^{N} \sum_{i'=1}^{N} \lambda_p^i \lambda_p^{i'} y_p^i y_p^{i'} \left\langle \phi_k(x^i), \phi_k(x^{i'}) \right\rangle + 2 \sum_{i=1}^{N} \sum_{i'=1}^{N} \lambda_p^i \lambda_k^{i'} y_p^i y_k^{i'} \left\langle \phi_k(x^i), \phi_p(x^{i'}) \right\rangle \right] = \tag{15} \\
= & \sum_{i=1}^{N} \sum_{k=1}^{m} \lambda_k^i - \frac{1}{4} \sum_{k,p=1}^{m} \left[\sum_{i=1}^{N} \sum_{i'=1}^{N} \lambda_k^i \lambda_k^{i'} y_k^i y_k^{i'} \left\langle \phi_p(x^i), \phi_p(x^{i'}) \right\rangle + \right. \\
& \left. \sum_{i=1}^{N} \sum_{i'=1}^{N} \lambda_p^i \lambda_k^{i'} y_p^i y_k^{i'} \left\langle \phi_k(x^i), \phi_p(x^{i'}) \right\rangle \right]
\end{aligned}
$$

for the final form of the Lagrangian function of Problem (9). Note that, since both indexes k and p depend on dimension m, which was chosen not to change throughout the process, we have that

$$\sum_{i=1}^{N}\sum_{i'=1}^{N}\lambda_p^i\lambda_p^{i'}y_p^iy_p^{i'}\left\langle\phi_k(x^i),\phi_k(x^{i'})\right\rangle = \sum_{i=1}^{N}\sum_{i'=1}^{N}\lambda_k^i\lambda_k^{i'}y_k^iy_k^{i'}\left\langle\phi_p(x^i),\phi_p(x^{i'})\right\rangle$$

leading to the final formulation (15).

Therefore, the Wolfe-dual problem of (9) is

$$\min_{\lambda} \quad -\mathcal{L}$$
$$\text{s. t.} \quad 0 < \lambda_k^i < C$$
$$\sum_{i=1}^{N}\lambda_k^iy_k^i = 0, \quad k = 1,\ldots,m \tag{16}$$

with $\mathcal{L}$ specified as in (15). Note that, both cross covariance and auto covariance terms are present in (15). The term $\left\langle\phi_k(x^i),\phi_k(x^{i'})\right\rangle$ is relative to the same output relations, while $\left\langle\phi_k(x^i),\phi_p(x^{i'})\right\rangle$ takes into account the inter output correlation factors.

Now, using the notation $\left\langle\phi_p(x^i),\phi_k(x^{i'})\right\rangle = \mathcal{K}_{pk}(x^i,x^{i'})$, once Problem (16) is solved and a finite solution λ^* is found, the classifier function for a given input $\hat{x}$ is computed as follows

$$f_k(\hat{x}) = \sum_{p=1}^{m}\left[\frac{1}{2}\sum_{i=1}^{N}\lambda_k^{*i}y_k^i\phi_p(x^i) + \lambda_k^{*i}y_k^i\phi_k(x^i)\right]^T\phi_p(\hat{x}) + b_k$$
$$= \frac{1}{2}\sum_{p=1}^{m}\sum_{i=1}^{N}\left[\lambda_k^{*i}y_k^i\mathcal{K}_{pp}(x^i,\hat{x}) + \lambda_k^{*i}y_k^i\mathcal{K}_{kp}(x^i,\hat{x})\right] + b_k. \tag{17}$$

4 Results and Discussion

4.1 Jura Dataset

The dataset used for this first phase is the Jura dataset, an open-source collection of data freely available online[1], utilized for experimental results in [8] as well. It is defined by two sets of data: a training set containing 259 inputs and a test set made of 100 items with the aim of outlying the specific amount of three elements, i.e., nickel, lead and zinc, into a distinctive new piece of land. Each sample is described by a set of eight features: local coordinates of the item, i.e., x and y, rock type and land use of the area and the chemical elements present in it. Primarily, we needed to modify the dataset in order to be able to use it in our model and to get valid information. For one thing, the Jura dataset is commonly exploited in regression frameworks, since the outputs may assume a continuous range of values. In order to use it in our classification problem we had to modify the output values. For each one of the three outputs we computed the output

---

[1] https://sites.google.com/site/goovaertspierre/.

mean value. Then, we decided to assign a new value for every item as follows. If the original output of a sample was grater than the mean value, then the new discrete output assigned to it is +1, otherwise, i.e., if the original value is less than the mean value, we assign the value of −1. We apply such process to any output leading to a new multi-output classification dataset. Moreover, since the feature ranges were widely dissimilar, we decided to normalize the data. Specifically, as briefly described above, Jura dataset is defined by: two spatial coordinates determining the precise position of the sample; two categorical features specifying the land use where the sample lies, which can assume a value between 1 and 4, and the rock type, in the range of 1 to 5; four features describing the amount of four specific chemical elements in the ground. Despite the issue of having categorical features, these two features have a small range of possible values making them not worth normalizing. It is a different matter when considering the spatial features and the four features evaluating the specific amount of the four chemical elements of the sample. Such features may assume very different values. Doing so, when computing the Euclidian distance in the kernel matrix as in Formulas (6), (7) and (8), the whole results would be compromised, letting the solution be strongly influenced by such four features. To solve such issue, we decided to perform data normalization only on the four features requiring it. Specifically, we applied a min-max normalization, rescaling the features in the range [0,1] using the following general formula

$$\text{new value} = \frac{\text{old value} - \text{min value}}{\text{max value} - \text{min value}}.$$

4.2   Preliminary Results

Since our dataset consists of three outputs, i.e., nickel, lead and zinc, we designed a series of experiments to evaluate the efficiency of our model. Considering each possible combination of the given three outputs, i.e., output 1 and output 2, output 1 and output 3, output 2 and output 3, we analyzed the results achieved for each scenario. In this way, we derived three datasets, each one made of two output values, from the original one. To apply our model, we used a kernel matrix using Eqs. (6), (7) and (8) given by

$$\begin{bmatrix} \mathcal{K}_{M \times M} & \mathcal{K}_{SE \times M} \\ \mathcal{K}_{M \times SE} & \mathcal{K}_{SE \times SE} \end{bmatrix}$$

where $\mathcal{K}_{SE \times M} = \mathcal{K}_{M \times SE}$ by definition. Therefore, when auto-covariance functions are required, $\mathcal{K}_{M \times M}$ and $\mathcal{K}_{SE \times SE}$ are used, corresponding to elements in the main diagonal; otherwise, for cross-covariance terms elements in the secondary diagonal are used.

After splitting the dataset as described above, we executed our algorithm both on the training set, containing 259 samples, and on the test set, with 100 data. The model accuracy results are summarized in Table 1.

The program is coded in MatLab version R2017b and executed on a laptop. The dual constrained optimization problem (16) is solved using the MatLab

Table 1. Performance of the method for the scaled Jura dataset.

	Output 1	Output 2	Output 1	Output 3	Output 2	Output 3
Training set	0.74	0.52	0.72	0.70	0.52	0.56
Test set	0.60	0.72	0.48	0.92	0.73	0.93

fmincon function. The computational cost incurred for the proposed method may be considered comparable with respect to other well known kernel methods. After the optimal solution λ^* has been calculated, (17) is used for the final classification. As a next step, we hope to extend our model to the next level and obtain a fully evaluation of it. Specifically, we aim to make use of other more complex datasets and to compare the outcomes achieved with those obtained using both different multi-task models and task independent strategies combined.

4.3   Discussion

We presented a novel supervised technique linking Support Vector Machines and Gaussian processes. Our approach aims to use a multi-output covariance function for Gaussian processes based on [8] in a SVM framework. The optimization problem utilizes a combination of different kernels. In this way, when dealing with multi-output datasets, it is possible to deal with multi output-correlation factors, shaping multi output connections thought the use of cross covariance terms. We built our model starting from the classical primal and dual single output SVM structure. Beginning with the well known classical framework, we described a multi-output SVM environment, from its primal formulation to the Wolfe dual, by the use of the Lagrangian function. In this manner, we obtained a multi-output SVM structure with cross terms, which made the use of multi-output covariance functions possible. Applying our method to the scaled Jura dataset, we think we achieved some interesting promising results. To the best of our knowlwdge, this is the first time that multi-output Gaussian processes developed covariance functions are utilized in Support Vector Machines structures. Our computational results cast a new light on a little exploited direction in kernel learning.

5   Conclusions

In this paper, we proposed a new method that combines a classical supervised machine learning structure, i.e., SVM, with an essential factor of multi output Gaussian process predictors: multiple covariance functions. This problem consists in defining a novel approach dealing with multi-output datasets. In contrast to classical single output problems where algorithms are defined to associate each input sample with a unique output value, in multi-output problems, multiple outputs are associated to input data, increasing the challenges of the classification problems. The proposed model is a new approach opposed to the currently deployed multi-target support vector models.

We first described the basic idea behind the concept of multi-target (multi-output) learning. The proposed model is based on a novel multi-output Gaussian process approach taking care of multi output correlation factors [8] with the use of an innovative multi-kernel covariance structure.

Starting from the classical Support Vector Machine structure, the proposed model defines a multi-output support vector approach with the characteristic of having inter output correlation factors in it. Specifically, our model presents cross covariance terms that allowed us to use the multi-output covariance approach in it.

The model was tested using a well known dataset applying however some modifications to it in order to obtain meaningful results. We obtain some preliminary still promising results, letting us believe in the core idea of our model.

In future research, we will work on improving our results with the Jura dataset and expanding our model with the use of further datasets, hoping to obtain better results. Specifically, we will focus our attention of the two categorical features: we believe that some of our poor results may be caused by the way Support Vector Machines deal with categorical data. This difficulty can be overcame using encoding techniques such as one-hot encoding or binary encoding.

Acknowledgement. The authors would like to thank Maruan Al-Shedivat for his precious advises and suggestions and Professor Eric Xing for his guidance while one of the authors was visiting the Machine Learning Department at Carnegie Mellon University.

References

1. Ben-David, S., Schuller, R.: Exploiting task relatedness for multiple task learning. In: Schölkopf, B., Warmuth, M.K. (eds.) COLT-Kernel 2003. LNCS (LNAI), vol. 2777, pp. 567–580. Springer, Heidelberg (2003). https://doi.org/10.1007/978-3-540-45167-9_41
2. Bielza, C., Li, G., Larranaga, P.: Multi-dimensional classification with Bayesian networks. Int. J. Approximate Reasoning **52**(6), 705–727 (2011)
3. Borchani, H., Varando, G., Bielza, C., Larrañaga, P.: A survey on multi-output regression. Wiley Interdisci. Rev. Data Min. Knowl. Disc. **5**(5), 216–233 (2015)
4. Cai, F., Cherkassky, V.: Generalized SMO algorithm for SVM-based multitask learning. IEEE Trans. Neural Netw. Learn. Syst. **23**(6), 997–1003 (2012)
5. Cressie, N.: Statistics for spatial data. Terra Nova **4**(5), 613–617 (1992)
6. Evgeniou, T., Pontil, M.: Regularized multi-task learning. In: Proceedings of the Tenth ACM SIGKDD International Conference on Knowledge Discovery and Data Mining, pp. 109–117 (2004)
7. Melki, G., Cano, A., Kecman, V., Ventura, S.: Multi-target support vector regression via correlation regressor chains. Inf. Sci. **415**, 53–69 (2017)
8. Melkumyan, A., Ramos, F.: Multi-kernel gaussian processes. In: Twenty-Second International Joint Conference on Artificial Intelligence (2011)
9. Micchelli, C.A., Pontil, M.: Kernels for multi-task learning. In: Advances in neural Information Processing Systems, pp. 921–928 (2005)

10. Vembu, S., Gärtner, T.: Label ranking algorithms: a survey. In: Fürnkranz, J., Hüllermeier, E. (eds.) Preference Learning, pp. 45–64. Springer, Heidelberg (2010). https://doi.org/10.1007/978-3-642-14125-6_3
11. Wackernagel, H.: Multivariate Geostatistics: An Introduction with Applications. Springer, Heidelberg (2013). https://doi.org/10.1007/978-3-662-05294-5
12. Xu, D., Shi, Y., Tsang, I.W., Ong, Y.S., Gong, C., Shen, X.: Survey on multi-output learning. IEEE Trans. Neural Networks Learn. Syst. 31(7), 2409–2429 (2019)
13. Xu, S., An, X., Qiao, X., Zhu, L.: Multi-task least-squares support vector machines. Multimedia Tools Appl. 71(2), 699–715 (2013). https://doi.org/10.1007/s11042-013-1526-5
14. Zhang, J., He, Y., Tang, J.: Multi-view multi-task support vector machine. In: Shi, Y., et al. (eds.) ICCS 2018. LNCS, vol. 10861, pp. 419–428. Springer, Cham (2018). https://doi.org/10.1007/978-3-319-93701-4_32

Generative Fourier-Based Auto-encoders: Preliminary Results

Alessandro Zonta$^{(\boxtimes)}$, Ali El Hassouni, David W. Romero,
and Jakub M. Tomczak

Vrije Universiteit Amsterdam, Amsterdam, The Netherlands
{a.zonta,d.w.romeroguzman,a.el.hassouni,j.m.tomczak}@vu.nl,
alessandro.zonta@tno.nl

Abstract. This paper presents a new general framework for turning any auto-encoder into a generative model. Here, we focus on a specific instantiation of the auto-encoder that consists of the Short Time Fourier Transform as an encoder, and a composition of the Griffin-Lim Algorithm and the pseudo inverse of the Short Time Fourier Transform as a decoder. In order to allow sampling from this model, we propose to use the probabilistic Principal Component Analysis. We show preliminary results on the UrbanSound8K Dataset.

Keywords: Auto-encoders · Generative modelling · Fourier Transform

1 Introduction

Obtaining a good data representation is claimed to be a crucial task of any AI system [1]. Similarly, for audio analysis and synthesis training representations is of high importance for many learning problems, e.g., classification, and specifically for audio generation. In the context of audio generation, there are powerful models like WaveNet [6] and WaveRNN [4] that represent data in an autoregressive manner. However, the sequential or autoregressive nature of these models comes at a cost: the high sampling cost results in computational inefficiency which is undesirable for edge or real-time computing.

In this paper, we propose a new framework that combines a deterministic auto-encoder and a generative model on the latent space. Our main claim is that we can easily turn classic methods for audio analysis into generative models by combining them with latent-variable models.

2 Problem Formulation

Audio generation problem requires to fit a distribution $p_\theta(\mathbf{x})$ to observed data $\mathcal{D} = \{\mathbf{x}_n\}_{n=1}^N$, where $\mathbf{x}_n \in \mathbb{R}^T$ is a single audio of length T. We denote the

AZ is co-financed by the Netherlands Organisation for Applied Scientific Research (TNO). AEH is co-financed by Mobiquity Inc. DWR is co-financed by the Dutch Research Council (NWO) and Semiotic Labs.

© Springer Nature Switzerland AG 2020
G. Nicosia et al. (Eds.): LOD 2020, LNCS 12566, pp. 12–15, 2020.
https://doi.org/10.1007/978-3-030-64580-9_2

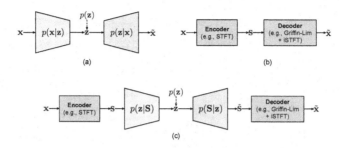

Fig. 1. (a) A latent variable model. (b) An auto-encoder. (c) Our approach: A combination of an auto-encoder and a latent-variable model.

data (empirical) distribution by $p_{data}(\mathbf{x}) = \frac{1}{N}\sum_{n=1}^{N}\delta(\mathbf{x} - \mathbf{x}_n)$, where $\delta(\cdot)$ is the Dirac's delta. We assume that datapoints are *iid*.

Here, we focus on the problem of generating data, and calculating the likelihood of a single audio is not of our interest. Therefore, we can alleviate the problem and consider a broader set of models than prescribed models. As a result, we focus on a specific class of models, i.e., Auto-Encoders (AE), instead of Generative Adversarial Networks [3] or Variational Auto-Encoders [5].

We focus on a deterministic AE with an encoder $e : \mathbb{R}^T \to \mathbb{S}$, where $\mathbf{S} \in \mathbb{S}$ is the code space (e.g., a space of spectrograms), and a decoder $d : \mathbb{S} \to \mathbb{R}^T$. The objective of the AE is to minimize a reconstruction error: $\mathcal{L}_{AE}(\theta) = \frac{1}{N}\sum_n \|\mathbf{x}_n - d(e(\mathbf{x}_n))\|^2$, where $\|\cdot\|$ is a norm, e.g., the ℓ_2-norm. In order to be able to sample new audio, we fit a distribution in the $\mathbb{S}$-space, namely, $p_\vartheta(\mathbf{S})$. We assume a *real* distribution $p(\mathbf{S}) = \iint p(\mathbf{S}|e(\mathbf{x}))p(\mathbf{x})\mathrm{d}\mathbf{S}\mathrm{d}\mathbf{x}$. Since we do not have access to $p(\mathbf{x})$, we approximate it with the data distribution that together with the deterministic AE yield $p_{data}(\mathbf{S}) = \frac{1}{N}\sum_{n=1}^{N}\delta(\mathbf{S} - e(\mathbf{x}))$. Eventually, we can learn $p_\theta(\mathbf{z})$ by minimizing the cross-entropy: $\mathcal{L}_s(\vartheta) = \mathbb{CE}[p_{data}(\mathbf{S})\|p_\vartheta(\mathbf{S})]$. As a result, our objective function is to find such parameters $\{\theta, \vartheta\}$ that minimize the composition of the reconstruction error and the cross-entropy, that is ($\beta > 0$): $\mathcal{L}(\theta, \vartheta) = \mathcal{L}_{AE}(\theta) + \beta\mathcal{L}_s(\vartheta)$. This idea is similar in spirit to [2], however, our motivation and our approach differ significantly from it. More specifically, we employ a non-trainable auto-encoder whereas they have a trainable auto-encoder.

3   Our Framework

We present our framework[1] by utilizing well-known and fast methods as a part of the AE, namely, **Short-Time Fourier Transform** (STFT) to encode data (amplitude only), and a composition of the **Griffin-Lim algorithm** (GLA) to recover phase and **the pseudo inverse of STFT** (iSTFT) to eventually return an original data format. STFT provides a spectrogram $\mathbf{S} \in \mathbb{R}^{T \times \Xi}$ that is easier to analyze and process than raw audio. GLA, on the other hand, allows

---

[1] https://github.com/AleZonta/genfae.

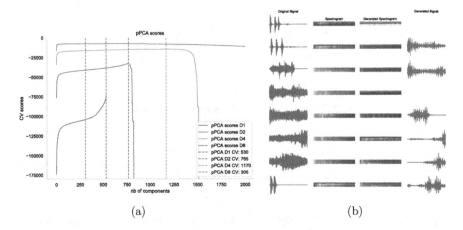

(a)   (b)

Fig. 2. a: Comparison of the models likelihood on three different data-split, b: Examples of real data and spectrograms (left two columns) and generated spectrograms and audio (right two columns). First four graphs generated from low resolution pPCA with 4 components, last four graphs generated from pPCA with 305 components

to estimate the phase of the spectrogram. After applying GLA, iSTFT could be used to turn **S** and the recovered phase into the audio format. In order to verify our generative approach, we propose to take the **linear latent variable model**, namely, probabilistic PCA (pPCA). We apply pPCA to the spectrogram to further compress it, and also to be able to sample from the model. We present our framework in Fig. 1(c). As a result, we can consider the proposed framework as a two-level AE where the first latent level **S** is given by a standard encoding method, e.g., STFT, and the second latent level is **z** is given by a generative model, e.g., the pPCA. We want to highlight that the manner these two levels are composed and our objective make our proposition different than standard AEs. Furthermore, we notice that since STFT and GLA do not contain any trainable parameters, we can focus on the objective in minimizing the cross-entropy alone.

4   Experiments

Setup. The UrbanSound8K Dataset[2] contains 8732 labeled sound excerpts ($\leq$ 4 s) of urban sounds from 10 different classes [7]. The sampling rate, bit depth, and the number of channels are the same as those of the file uploaded to the original source and hence may vary from file to file. To test our approach we selected the *dog_barking* class, limiting the dataset to 1000 elements, with a fixed sampling-rate (8000 Hz) and a constant length of the signals of 4 s (end zero-padding). The dataset is split into training and test-set following the ratio 80%–20%.

---

[2] https://urbansounddataset.weebly.com/urbansound8k.html.

Results and Discussion. Figure 2a shows the comparison of the various models' likelihood on four different data-split, i.e., original signals and signals divided into small chunks to speed up computation. The optimal number of components that maximise the likelihood is highlighted by vertical dash lines. Smaller signals, i.e., pPCA D4 in the graph, and, therefore, a bigger dataset, produce higher likelihood compared to larger signals, i.e., pPCA D1 in the graph. The two just mentioned splits also do not result in a drop in likelihood when the number of components reaches the number of examples in the training set, compared to pPCA D2 and pPCA D4 that have a significant drop in likelihood.

Figure 2b shows the pPCA D8 and a low resolution pPCA generating new data together with some example of real data. The low resolution pPCA is obtained with the number of components explaining the 45% of the variance. In the small sample shown in the graph, the first four lines are form the low resolution pPCA, and the last four from the high resolution. Where in both cases there are still problems with the signal generated, the main difference is visible in the generated spectrograms. Low resolution pPCA is not able to generate invertible spectrograms, with artifacts in the signal, as visible in the first and fourth row of the picture. High resolution pPCA, on the other hand, is able to generate correct spectrograms, that are invertible to signals closer to the real ones. Comparing the original spectrograms with the generated ones, horizontal features are present on both, meanwhile, the gradient of colour from left to right visible in the four centre original examples is missing from the generated spectrograms. A reason for the horizontal features to be present in the generated spectrograms but the colour gradient not, can be found in how the data is given as input to the pPCA. The matrix representing the spectrogram is flatten into a vector before feeding it to the model, therefore horizontal features are easier to grasp from the representation used compared to colour gradients.

References

1. Bengio, Y., et al.: Learning deep architectures for AI. Found. Trends ® Mach. Learn. **2**(1), 1–127 (2009)
2. Ghosh, P., Sajjadi, M.S., Vergari, A., Black, M., Schölkopf, B.: From variational to deterministic autoencoders. arXiv preprint arXiv:1903.12436 (2019)
3. Goodfellow, I., et al.: Generative adversarial nets. In: Advances in Neural Information Processing Systems, pp. 2672–2680 (2014)
4. Kalchbrenner, N., et al.: Efficient neural audio synthesis. In: International Conference on Machine Learning, pp. 2410–2419 (2018)
5. Kingma, D.P., Welling, M.: Auto-encoding variational Bayes. arXiv preprint arXiv:1312.6114 (2013)
6. Oord, A.V.D., et al.: Wavenet: a generative model for raw audio. arXiv preprint arXiv:1609.03499 (2016)
7. Salamon, J., Jacoby, C., Bello, J.P.: A dataset and taxonomy for urban sound research. In: Proceedings of the 22nd ACM International Conference on Multimedia. MM 2014, New York, NY, USA, pp. 1041–1044. Association for Computing Machinery (2014)

Parameterized Structured Pruning
for Deep Neural Networks

Günther Schindler[1(✉)], Wolfgang Roth[2], Franz Pernkopf[2], and Holger Fröning[1]

[1] Institute of Computer Engineering, Ruprecht Karls University,
Heidelberg, Germany
guenther.schindler@ziti.uni-heidelberg.de
[2] Signal Processing and Speech Communication Laboratory,
Graz University of Technology, Graz, Austria

Abstract. As a result of the growing size of Deep Neural Networks (DNNs), the gap to hardware capabilities in terms of memory and compute increases. To effectively compress DNNs, quantization and pruning are usually considered. However, unconstrained pruning usually leads to unstructured parallelism, which maps poorly to massively parallel processors, and substantially reduces the efficiency of general-purpose processors. Similar applies to quantization, which often requires dedicated hardware.

We propose Parameterized Structured Pruning (PSP), a novel technique to dynamically learn the shape of DNNs through structured sparsity. PSP parameterizes structures (e.g. channel- or layer-wise) in a weight tensor and leverages weight decay to learn a clear distinction between important and unimportant structures. As a result, PSP maintains prediction performance, creates a substantial amount of sparsity that is structured and, thus, easy and efficient to map to a variety of massively parallel processors, which are mandatory for utmost compute power and energy efficiency.

1 Introduction

Deep Neural Networks (DNNs) are widely used for many applications including object recognition, speech recognition and robotics. The ability of modern DNNs to excellently fit training data is suspected to be due to heavy over-parameterization, i.e., using more parameters than the total number of training samples, since there always exists parameter choices that achieve a training error of zero. While over-parameterization is essential for the learning ability of neural networks, it results in extreme memory and compute requirements for training (development) as well as inference (deployment). Recent research showed that training can be scaled to up to 1024 accelerators operating in parallel, resulting in a development phase not exceeding a couple of minutes, even for large-scale image classification. However, the deployment has usually much harder constraints than the development, as energy, space and monetary resources are scarce in mobile devices.

© Springer Nature Switzerland AG 2020
G. Nicosia et al. (Eds.): LOD 2020, LNCS 12566, pp. 16–27, 2020.
https://doi.org/10.1007/978-3-030-64580-9_3

Model compression techniques are targeting this issue by training an over-parameterized model and compressing it for deployment. Popular compression techniques are pruning, quantization, knowledge distillation, and low-rank factorization, with the first two being most popular due to their extreme efficiency. Pruning connections [3] achieves impressive theoretical compression rates through fine-grained sparsity (Fig. 1a) without sacrificing prediction performance, but has several practical drawbacks such as indexing overhead, load imbalance and random memory accesses: (i) Compression rates are typically reported without considering the space requirement of additional data structures to represent non-zero weights. For instance, using indices, a model with 8-bit weights, 8-bit indices and 75% sparsity saves only 50% of the space, while a model with 50% sparsity does not save memory at all. (ii) It is a well-known problem that massively parallel processors show notoriously poor performance when the load is not well balanced. Unfortunately, since the end of Dennard CMOS scaling, massive parallelization is mandatory for a continued performance scaling. (iii) Sparse models increase the amount of randomness in memory access patterns, preventing caching techniques, which rely on predictable strides, from being effective. As a result, the amount of cache misses increases the average memory access latency and the energy consumption, as off-chip accesses are 10–100 time higher in terms of latency, respectively 100–1000 times higher in terms of energy consumption. Quantization has recently received plenty of attention and reduces the computational workload, as the complexity of additions and multiplications scales approximately linearly and quadratically with the number of bits, respectively. However, in comparison, pruning avoids a computation completely.

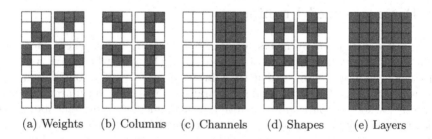

(a) Weights   (b) Columns   (c) Channels   (d) Shapes   (e) Layers

Fig. 1. Illustration of fine-grained (a) and several structured forms of sparsity (b–d) for a 4-dimensional convolution tensor. The large squares represent the kernels, and the corresponding horizontal and vertical dimensions represent the number of input feature and output feature maps, respectively. The computation of all structured forms of sparsity can be lowered to matrix multiplications (independent of stride and padding).

Structured pruning methods can prevent these drawbacks by inducing sparsity in a hardware-friendly way: Fig. 1b–e illustrate exemplary a 4-dimensional convolution tensor (see [2] for details on convolution lowering), where hardware-friendly sparsity structures are shown as channels, layers, etc. However, pruning whole structures in a neural network is not as trivial as pruning individual connections and usually causes high accuracy degradation under mediocre compression constraints.

Structured pruning methods can be roughly clustered into two categories: re-training-based and regularization-based methods (see Sect. 4 for details). Re-training-based methods aim to remove structures by minimizing the pruning error in terms of changes in weight, activation, or loss, respectively, between the pruned and the pre-trained model. Regularization-based methods train a randomly initialized model and apply regularization, usually an ℓ_1 penalty, in order to force structures to zero.

This work introduces a new regularization-based method leveraging learned parameters for structured sparsity without substantial increase in training time. Our approach differs from previous methods, as we explicitly parameterize certain structures of weight tensors and regularize them with weight decay, enabling a clear distinction between important and unimportant structures. Combined with threshold-based magnitude pruning and a straight-through gradient estimator (STE) [1], we can remove a substantial amount of structure while maintaining the classification accuracy. We evaluate the proposed method based on state-of-the-art Convolutional Neural Networks (CNNs) like ResNet [4] and DenseNet [8], and popular datasets like CIFAR-10/100 and ILSVRC2012.

The remainder of this work is structured as follows: In Sect. 2 we introduce the parameterization and regularization approach together with the pruning method. We present experimental results in Sect. 3. Related work is summarized in Sect. 4, before we conclude in Sect. 5.

2 Parameterized Pruning

DNNs are constructed by layers of stacked processing units, where each unit computes an activation function of the form $z = g(\mathbf{W} \oplus \mathbf{x})$, where $\mathbf{W}$ is a weight tensor, $\mathbf{x}$ is an input tensor, $\oplus$ denotes a linear operation, e.g., a convolution, and $g(\cdot)$ is a non-linear function. Modern neural networks have very large numbers of these stacked compute units, resulting in huge memory requirements for the weight tensors $\mathbf{W}$, and compute requirements for the linear operations $\mathbf{W} \oplus \mathbf{x}$. In this work, we aim to learn a structured sparse substitute $\mathbf{Q}$ for the weight tensor $\mathbf{W}$, so that there is only minimal overhead for representing the sparsity pattern in $\mathbf{Q}$ while retaining computational efficiency using dense tensor operations. For instance, by setting all weights at certain indices of the tensor to zero, it suffices to store the indices of non-zero elements only once for the entire tensor $\mathbf{Q}$ and not for each individual dimension separately. By setting all weights connected to an input feature map to zero, the corresponding feature map can effectively be removed without the need to store any indices at all.

2.1   Parameterization

Identifying the importance of certain structures in neural networks is vital for the prediction performance of structured-pruning methods. Our approach is to train the importance of structures by parameterizing and optimizing them together with the weights using backpropagation. Therefore, we divide the tensor $\mathbf{W}$ into subtensors $\{\mathbf{w}_i\}$ so that each $\mathbf{w}_i = (w_{i,j})_{j=1}^m$ constitutes the m weights of structure i. During forward propagation, we substitute $\mathbf{w}_i$ by the structured sparse tensor $\mathbf{q}_i$ as

$$\mathbf{q}_i = \mathbf{w}_i \alpha_i \tag{1}$$

where α_i is the structure parameter associated with structure i. Following the chain rule, the gradient of the structure parameter α_i is:

$$\frac{\partial E}{\partial \alpha_i} = \sum_{j=1}^m \frac{\partial E}{\partial w_{i,j}} \, , \tag{2}$$

where E represents the objective function. Thus, the dense structure parameters α_i descend towards the predominant direction of the structure weights. As a result, the structure parameters are optimized together with the weights of structure i but can be regularized and pruned independent to the weights of structure i. Training the structures introduces additional parameters, however, during inference they are folded into the weight tensors, resulting in no extra memory or compute costs.

2.2   Regularization

Reducing the complexity of a neural network by limiting the growth of parameters is highly advantageous for their generalization abilities. It can be realized by adding a term to the cost function that penalizes parameters. Most commonly the ℓ_1 or ℓ_2 norm are used as penalty term, extending the cost function to $E_{\ell_1}(\alpha_i) = E(\alpha_i) + \lambda|\alpha_i|$ or $E_{\ell_2}(\alpha_i) = E(\alpha_i) + \frac{\lambda}{2}\alpha_i^2$, respectively. Applying ℓ_1 regularization to the structure parameters α_i changes the update rule as: $\Delta\alpha_i(t+1) = -\eta\frac{\partial E}{\partial \alpha_i} - \lambda\eta sign(\alpha_i)$, where η is the learning rate and λ is the regularization strength. For gradient-based optimizations, ℓ_1 regularization only considers the sign of the parameters while the gradient in zero is undefined. Hence, it acts as a feature selector since certain weights are reduced to zero, resulting in sparse structure parameters.

For ℓ_2 regularization (weight decay), the update rule of the structure parameters is: $\Delta\alpha_i(t+1) = -\eta\frac{\partial E}{\partial \alpha} - \lambda\eta\alpha_i$. While ℓ_1 regularization only considers the direction of the parameters, weight decay also takes the magnitude of the parameters into account. This makes weight decay the standard regularization method since it significantly improves the learning capabilities of SGD based neural networks, resulting in faster convergence and better generalization. The benefits of weight decay can be best visualized using the distributions of the structure parameters α_i (corresponding to different layers) in Fig. 2.

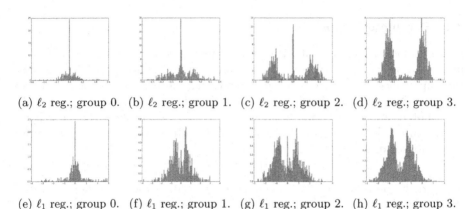

(a) ℓ_2 reg.; group 0. (b) ℓ_2 reg.; group 1. (c) ℓ_2 reg.; group 2. (d) ℓ_2 reg.; group 3.

(e) ℓ_1 reg.; group 0. (f) ℓ_1 reg.; group 1. (g) ℓ_1 reg.; group 2. (h) ℓ_1 reg.; group 3.

Fig. 2. Different distributions of column-wise structure parameters with weight decay (ℓ_2) and ℓ_1 regularization of a fully trained ResNet with 18 layers on ImageNet. The distributions correspond to the first convolution in the first block in the respective group (g0–g3). Note that peaks visually close to zero are not exactly zero.

Parameterizing structures and regularization ultimately shrink the complexity (variance of the layers) of a neural network in a structured way. We observe that weight decay produces unimodal, bimodal and trimodal distributions (Fig. 2a–2d), indicating different complexities, with a clear distinction between important and unimportant structure parameters. In contrast, ℓ_1 regularization (Fig. 2e–Fig. 2h) lacks the ability to form this clear distinction. Second, ℓ_1 regularized structure parameters are roughly one order of magnitude larger than parameters trained with weight decay, making them more sensitive to small noise in the input data and reducing the effective learning rate. Third, even though ℓ_1 implicitly produces sparse models, weight decay reduces more parameters close to zero and therefore achieves better pruning potential. Consequently, we use weight decay for regularizing the structure parameters and perform pruning explicitly.

2.3 Pruning

The explicit pruning can be performed by a simple threshold-based magnitude pruning method. Let ν_i be the regularized *dense structure parameter* associated with structure i, then the *sparse structure parameter* α_i are obtained as:

$$\alpha_i(\nu_i) = \begin{cases} 0 & |\nu_i| < \epsilon \\ \nu_i & |\nu_i| \geq \epsilon \end{cases}, \tag{3}$$

where ϵ is a tuneable pruning threshold. As the threshold function is not differentiable at $\pm\epsilon$ and the gradient is zero in $[-\epsilon, \epsilon]$, we approximate the gradient of ν_i by defining an STE as $\frac{\partial E}{\partial \nu_i} = \frac{\partial E}{\partial \alpha_i}$. We use the sparse parameters α_i for forward and backward propagation and update the respective dense parameters ν_i

based on the gradients of α_i. Updating the dense structure parameters ν_i instead of the sparse parameters α_i is beneficial because improperly pruned structures can reappear if ν_i moves out of the pruning interval $[-\epsilon, \epsilon]$, resulting in faster convergence to a better performance.

2.4   Hardware-Friendly Structures in CNNs

We consider CNNs with $R \times S$ filter kernels, C input and K output feature maps. Different granularities of structured sparsity yield different flexibilities when mapped to hardware. In this work, we consider only coarse-grained structures such as layer, channel and column pruning, that can be implemented using off-the-shelf libraries on general-purpose hardware or shape pruning for direct convolutions on re-configurable hardware.

Layer pruning simply removes unimportant layers and ultimately shrinks the depth of a network (Fig. 1e), making the hardware mapping extremely efficient on every processor type. Note that layer pruning is only applicable to multi-branch architectures (e.g. DenseNet). **Channel pruning** refers to removing input or output channels in a convolutional layer and the respective input or output feature maps (Fig. 1c). Hence, it shrinks the width of a network and, similar to layer pruning, is applicable to every processor type. **Shape pruning** targets to prune filter kernels per layer equally (Fig. 1d), which can be mapped onto re-configurable hardware when direct convolution operations are in use. Convolutions are usually lowered onto matrix multiplications in order to explore data locality and the massive amounts of parallelism in general-purpose GPUs, CPUs or specialized processors like TPUs. This lowering is performed using the *im2col* approach, where discrete input blocks (depending on filter size and stride) are duplicated and reshaped into columns of a two dimensional matrix. The reader may refer to the work of [2] for a detailed explanation. Although layer, channel and shape pruning can be easily mapped to matrix multiplication, the potential sparsity is higher when a finer granularity is used. Thus, **column pruning** sparsifies weight tensors in a way that a whole column of the flattened weight tensor and the respective row of the input data can be removed (Fig. 1b), achieving finer granularity while allowing the efficient mapping to matrix multiplication.

Table 1. Representation of the dense structure parameters and the gradient calculation.

Pruning method	Structure parameter	Gradient
Layer pruning	$\alpha \in \mathbb{R}$	$\partial E/\partial \alpha = \sum_{k=1}^{K} \sum_{c=1}^{C} \sum_{r=1}^{R} \sum_{s=1}^{S} \partial E/\partial W_{k,c,r,s}$
Channel pruning	$\alpha \in \mathbb{R}^{C}$	$\partial E/\partial \alpha_c = \sum_{k=1}^{K} \sum_{r=1}^{R} \sum_{s=1}^{S} \partial E/\partial W_{k,c,r,s}$
Shape pruning	$\alpha \in \mathbb{R}^{R \times S}$	$\partial E/\partial \alpha_{r,s} = \sum_{k=1}^{K} \sum_{c=1}^{C} \partial E/\partial W_{k,c,r,s}$
Column pruning	$\alpha \in \mathbb{R}^{R \times S \times C}$	$\partial E/\partial \alpha_{r,s,c} = \sum_{k=1}^{K} \partial E/\partial W_{k,c,r,s}$

The structure parameters for individual structures and their corresponding gradients are shown in Table 1. PSP is not restricted to these forms of granularities; arbitrary structures and combinations of different structures are possible as well as other layer types such as recurrent or dense layers. For instance, blocks can be defined and pruned in order to enable efficient processor vectorization or tiling techniques for cache and register blocking. Or layer and channel pruning can be combined when a simple hardware mapping is targeted.

3   Experiments

We use the CIFAR10/100 and the ILSVRC 2012 (ImageNet) datasets on ResNet [4] and DenseNet [8] architectures. Both networks can apply 1×1 convolutions as bottleneck layers before the 3×3 convolutions to improve compute and memory efficiency. DenseNet further improves model compactness by reducing the number of feature maps at transition layers. If bottleneck and transition compression is used, the models are labeled as *ResNet-B* and *DenseNet-BC*, respectively. Removing the bottleneck layers in combination with our compression approach has the advantage of reducing both, memory/compute requirements and the depth of the networks. We apply PSP to all convolutional layers except the sensitive input, output, transition and shortcut layers, which have negligible impact on overall memory and compute costs.

We use a weight decay of 10^{-4} and a momentum of 0.9 for weights and structure parameters throughout this work. We use the initialization introduced by [5] for the weights and initialize the structure parameters randomly using a zero-mean Gaussian with standard deviation 0.1.

3.1   Pruning Different Structures

We compare the performance of the different structure granularities using DenseNet on CIFAR10 (Table 2, with 40 layers, a growth rate of $k = 12$ and a pruning threshold of $\epsilon = 0.1$). We report the required layers, parameters and Multiply-Accumulate (MAC) operations.

While all structure granularities show a good prediction performance, with slight deviations compared to the baseline error, column- and channel-pruning achieve the highest compression ratios. Shape pruning results in the best accuracy but only at a small compression rate, indicating that a higher pruning threshold is more appropriate. It is worth noticing that PSP is able to automatically remove structures, which can be seen best when comparing layer pruning and a combination of layer and channel pruning: layer pruning removes 12 layers from the network but still requires 0.55M parameters and 0.14G MACs, while the combination of layer and channel pruning removes only 7 layers but requires only 0.48M parameters and 0.12G MACs.

Table 2. Layer-, channel, shape- and column-pruning using PSP, validated on DenseNet40 ($k = 12$) on the CIFAR10 dataset. M and G represents 10^6 and 10^9, respectively.

Model	Layers	Params	MACs	Error
Baseline	40	1.02M	0.27G	5.80%
Layer pruning	28	0.55M	0.14G	6.46%
Channel pruning	40	0.35M	0.09G	5.61%
Layer+channel	33	0.48M	0.12G	6.39%
Shape pruning	40	0.92M	0.23G	5.40%
Column pruning	40	0.22M	0.05G	5.76%

3.2   CIFAR10/100 and ImageNet

To validate the effectiveness of PSP, we now discuss results from ResNet and DenseNet on CIFAR10/100 and ImageNet. We use column pruning throughout this section, as it offers the highest compression rates while preserving classification performance.

Table 3 reports results for CIFAR10/100. As can be seen, PSP maintains classification performance for a variety of networks and datasets. This is due to the ability of self-adapting the pruned structures during training, which can be best seen when changing the network topology or dataset: for instance, when we use the same models on CIFAR10 and the more complex CIFAR100 task,

Table 3. ResNet and DenseNet on CIFAR10/100 using column pruning for PSP.

	Layers	Param. [10^6]	MAC [10^9]	Error [%]
CIFAR10				
ResNet	56	0.85 (1.0×)	0.13 (1.0×)	**6.35 (+0.00)**
ResNet-PSP	56	**0.21 (4.0×)**	**0.03 (4.3×)**	6.55 (+0.20)
DenseNet	40	1.02 (1.0×)	0.27 (1.0×)	5.80 (+0.00)
DenseNet-PSP	40	**0.22 (4.6×)**	**0.05 (5.3×)**	5.76 (−0.03)
DenseNet	100	6.98 (1.0×)	1.77 (1.0×)	**4.67 (+0.00)**
DenseNet-PSP	100	**0.99 (7.1×)**	**0.22 (8.0×)**	4.87 (+0.20)
CIFAR100				
ResNet	56	0.86 (1.0×)	0.13 (1.0×)	27.79 (+0.00)
ResNet-PSP	56	**0.45 (1.9×)**	**0.07 (1.9×)**	**27.15 (−0.64)**
DenseNet	40	1.06 (1.0×)	0.27 (1.0×)	26.43 (+0.00)
DenseNet-PSP	40	**0.37 (2.9×)**	**0.08 (3.4×)**	26.30 (−0.13)
DenseNet	100	7.09 (1.0×)	1.77 (1.0×)	**22.83 (+0.00)**
DenseNet-PSP	100	**1.17 (6.1×)**	**0.24 (7.4×)**	23.42 (+0.41)

we can see that PSP is able to automatically adapt as it removes less structure from the network trained on CIFAR100. Furthermore, if we increase the number of layers by 2.5× from 40 to 100, we also increase the over-parameterization of the network and PSP automatically removes 2.4× more structure.

The same tendencies can be observed on the large-scale ImageNet task as shown in Table 4; when applying PSP, classification accuracy can be maintained (with some negligible degradation) and a considerable amount of structure can be removed from the networks (e.g. 2.6× from ResNet18 or 1.8× from DenseNet121). Furthermore, PSP obliterates the need for 1×1 bottleneck layers, effectively reducing network depth and MACs. For instance, removing the bottleneck layers from the DenseNet121 network in combination with PSP removes 2.6× parameters, 4.9× MACs and 1.9× layers, while only sacrificing 2.28% top-5 accuracy.

Table 4. ResNet and DenseNet on ImageNet using column pruning.

Model	Layers	Param. [10^6]	MAC [10^9]	Top-1 [%]	Top-5 [%]
ResNet-B	18	11.85 (1.0×)	1.82 (1.0×)	**29.60 (+0.00)**	**10.52 (+0.00)**
ResNet-B-PSP	18	**5.65 (2.1×)**	**0.82 (2.2×)**	30.37 (+0.67)	11.10 (+0.58)
ResNet-B	50	25.61 (1.0×)	4.09 (1.0×)	**23.68 (+0.00)**	6.85 (+0.00)
ResNet-B-PSP	50	**15.08 (1.7×)**	**2.26 (1.8×**	24.07 (+0.39)	**6.69 (−0.16)**
DenseNet-BC	121	7.91 (1.0×)	2.84 (1.0×)	**25.65 (+0.00)**	8.34 (+0.00)
DenseNet-BC-PSP	121	**4.38 (1.8×)**	**1.38 (2.1×)**	25.95 (+0.30)	**8.29 (−0.05)**
DenseNet-C	63	10.80 (1.0×)	3.05 (1.0×)	**28.87 (+0.00)**	**10.02 (+0.00)**
DenseNet-C-PSP	63	**3.03 (3.6×)**	**0.58 (5.3×)**	29.66 (+0.79)	10.62 (+0.60)
DenseNet-C	87	23.66 (1.0×)	5.23 (1.0×)	**26.31 (+0.00)**	**8.55 (+0.00)**
DenseNet-C-PSP	87	**4.87 (4.9×)**	**0.82 (6.4×)**	27.46 (+1.15)	9.15 (+0.40)

3.3   Ablation Experiments

We end the experiments with an ablation experiment to validate methods and statements made in this work. This experiment is evaluated on the ResNet architecture, using column pruning, with 56 layers using the CIFAR10 dataset (Fig. 3). We report the validation error for varying sparsity constraints, and with the baseline error set to the original unpruned network, with some latitude to filter out fluctuations: 6.35% ± 0.25. The dashed vertical lines indicate the maximum amount of sparsity while maintaining the baseline error.

A common way [13] to estimate the importance of structures is the ℓ_1 norm of the targeted structure in a weight tensor $A_{norm} = ||\mathrm{W}||_1$, which is followed by pruning the structures with the smallest norm. We use this rather simple approach as a baseline, denoted as ℓ_1 *norm*, to show the differences to the proposed parameterized structure pruning. The parameterization in its most basic form is denoted as *PSP (fixed sparsity)*, where we do not apply regularization ($\lambda = 0$)

and simply prune the parameters with the lowest magnitude. As can be seen, the parameterization achieves about 10% more sparsity compared to the baseline (ℓ_1 norm) approach, or 1.8% better accuracy under a sparsity constraint of 80%.

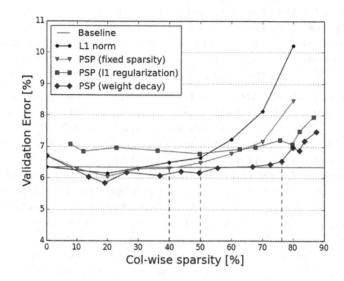

Fig. 3. ResNet network with 56 layers on CIFAR10 and column pruning.

Furthermore, we observe that regularized dense structure parameters are able to learn a clear distinction between important and unimportant structures (Sect. 2.2). Thus, it seems appropriate to use a simple threshold heuristic (Eq. 3) rather than pruning all layers equally (as compared to *PSP (fixed sparsity)*).

We also show the impact of the threshold heuristic in combination with ℓ_1 regularization and weight decay in Fig. 3. These methods are denoted as *PSP (ℓ_1 regularization)* and *PSP (weight decay)*, respectively. We vary the regularization strength for ℓ_1 regularization, since it induces sparsity implicitly, while we vary the threshold parameter for weight decay: for *PSP (ℓ_1 regularization)*, we set the threshold $\epsilon = 10^{-3}$ and the initial regularization strength $\lambda = 10^{-10}$, which is changed by an order of magnitude ($\times 10$) to show various sparsity levels. For *PSP*, we set the regularization strength $\lambda = 10^{-4}$ and the initial threshold $\epsilon = 0.0$ and increase ϵ by $2 \cdot 10^{-2}$ for each sparsity level. Both methods show higher accuracy for high sparsity constraints (sparsity $\geq$ 80%), but only weight decay achieves baseline accuracy.

4 Related Work

Re-training-Based mMethods: [10] evaluate the importance of filters by calculating its absolute weight sum. [13] prune structures with the lowest ℓ_1 norm. Channel Pruning (CP) [7] uses an iterative two-step algorithm to prune each layer by a LASSO regression based channel selection and least square reconstruction. Structured Probabilistic Pruning (SPP) [14] introduces a pruning probability for each weight where pruning is guided by sampling from the pruning probabilities. Soft Filter Pruning (SFP) [6] enables pruned filters to be updated when training the model after pruning, which results in larger model capacity and less dependency on the pre-trained model. ThiNet [12] shows that pruning filters based on statistical information calculated from the following layer is more accurate than using statistics of the current layer. Discrimination-aware Channel Pruning (DCP) [17] selects channels based on their discriminative power.

Regularization-Based Methods: Group Lasso [16] allows predefined groups in a model to be selected together. Adding an ℓ_1 penalty to each group is a heavily used approach for inducing structured sparsity in CNNs [9,15]. Network Slimming [11] applies ℓ_1 regularization on coefficients of batch-normalization layers in order to create sparsity in a structured way.

In contrast to Group Lasso, the explicit parameterization allows to optimize and regularize certain structures independently to the weights of a tensor. Different to the regularization on coefficients of batch-normalization layers, the parameterization allows arbitrary structure selection within a tensor and, therefore, more efficient hardware mappings and structure granularities. Previous works leverage ℓ_1 regularization in order to enforce sparsity on structures although weight decay has superior convergence and generalization abilities for neural networks. The parameterization in combination with threshold-based magnitude pruning and straight-through estimation allows PSP to leverage the superior weight decay as regularizer.

5 Conclusion

We have presented PSP, a novel approach for compressing DNNs through structured pruning, which reduces memory and compute requirements while creating a form of sparsity that is inline with massively parallel processors.

Our approach exhibits parameterization of arbitrary structures (e.g. channels or layers) in a weight tensor and uses weight decay to force certain structures towards zero, while clearly discriminating between important and unimportant structures. Combined with threshold-based magnitude pruning and backward approximation, we can remove a large amount of structure while maintaining prediction performance.

Experiments using state-of-the-art DNN architectures on real-world tasks show the effectiveness of our approach. As a result, the gap between DNN-based application demand and capabilities of resource-constrained devices is reduced, while this method is applicable to a wide range of processors.

References

1. Bengio, Y., Léonard, N., Courville, A.C.: Estimating or propagating gradients through stochastic neurons for conditional computation. Technical Report, Universite de Montreal (2013). http://arxiv.org/abs/1308.3432

2. Chetlur, S., et al.: CUDNN: efficient primitives for deep learning (2014). http:// arxiv.org/abs/1410.0759
3. Han, S., Pool, J., Tran, J., Dally, W.: Learning both weights and connections for efficient neural network. In: Advances in Neural Information Processing Systems (NIPS) (2015). http://papers.nips.cc/paper/5784-learning-both-weights-and-connections-for-efficient-neural-network.pdf
4. He, K., Zhang, X., Ren, S., Sun, J.: Deep residual learning for image recognition. In: Conference on Computer Vision and Pattern Recognition (CVPR) (2016). http:// arxiv.org/abs/1512.03385
5. He, K., Zhang, X., Ren, S., Sun, J.: Delving deep into rectifiers: surpassinghuman-level performance on imagenet classification. In: International Conference on Computer Vision (ICCV) (2016)
6. He, Y., Kang, G., Dong, X., Fu, Y., Yang, Y.: Soft filter pruning for accelerating deep convolutional neural networks. In: International Joint Conference on Artificial Intelligence (IJCAI) (2018). http://arxiv.org/abs/1808.06866
7. He, Y., Zhang, X., Sun, J.: Channel pruning for accelerating very deep neural networks. In: International Conference on Computer Vision (ICCV) (2017). http:// arxiv.org/abs/1707.06168
8. Huang, G., Liu, Z., Weinberger, K.Q.: Densely connected convolutional networks. In: Conference on Computer Vision and Pattern Recognition (CVPR) (2017). http://arxiv.org/abs/1608.06993
9. Lebedev, V., Lempitsky, V.S.: Fast convnets using group-wise brain damage. In: Conference on Computer Vision and Pattern Recognition (CVPR) (2016). http:// arxiv.org/abs/1506.02515
10. Li, H., Kadav, A., Durdanovic, I., Samet, H., Graf, H.P.: Pruning filters for efficient convnets. In: International Conference on Learning Representations (ICLR) (2017), http://arxiv.org/abs/1608.08710
11. Liu, Z., Li, J., Shen, Z., Huang, G., Yan, S., Zhang, C.: Learning efficient convolutional networks through network slimming. In: International Conference on Computer Vision (ICCV) (2017). http://arxiv.org/abs/1708.06519
12. Luo, J., Wu, J., Lin, W.: Thinet: A filter level pruning method for deep neural network compression. In: International Conference on Computer Vision (ICCV) (2017). http://arxiv.org/abs/1707.06342
13. Mao, H., et al.: Exploring the regularity of sparse structure in convolutional neural networks. In: Advances in Neural Information Processing Systems (NIPS) (2017). http://arxiv.org/abs/1705.08922
14. Wang, H., Zhang, Q., Wang, Y., Hu, R.: Structured probabilistic pruning for deep convolutional neural network acceleration. In: British Machine Vision Conference (BMVC) (2018). http://arxiv.org/abs/1709.06994
15. Wen, W., Wu, C., Wang, Y., Chen, Y., Li, H.: Learning structured sparsity in deep neural networks. In: Advances in Neural Information Processing Systems (NIPS) (2016). http://arxiv.org/abs/1608.03665
16. Yuan, M., Lin, Y.: Model selection and estimation in regression with groupedvariables. J. Roy. Stat. Soc. (2006)
17. Zhuang, Z., et al.: Discrimination-aware channel pruning for deep neural networks. In: Advances in Neural Information Processing Systems (NIPS) (2018). http:// arxiv.org/abs/1810.11809

FoodViz: Visualization of Food Entities Linked Across Different Standards

Riste Stojanov[1], Gorjan Popovski[2,3], Nasi Jofce[1], Dimitar Trajanov[1],
Barbara Koroušić Seljak[2], and Tome Eftimov[2(✉)]

[1] Faculty of Computer Science and Engineering, Ss. Cyril and Methodius, University, Skopje, North Macedonia
{riste.stojanov,nasi.jofce,dimitar.trajanov}@finki.ukim.mk
[2] Computer Systems Department, Jožef Stefan Institute, Ljubljana, Slovenia
{gorjan.popovski,barbara.korousic,tome.eftimov}@ijs.si
[3] Jožef Stefan International Postgraduate School, Ljubljana, Slovenia

Abstract. Many research questions from different domains involve combining different data sets in order to explore a research hypothesis. One of the main problems that arises here is that different data sets are structured with respect to different domain standards and ensuring their interoperability is a time-consuming task. In the biomedical domain, the Unified Medical Language System supports interoperability between biomedical data sets by providing semantic resources and Natural Language Processing tools for automatic annotation. This allows users also to understand the links between different biomedical standards. While there are extensive resources available for the biomedical domain, the food and nutrition domain is relatively low-resourced. To make the links between different food standards understandable by food subject matter experts we propose the FoodViz. It is a web-based framework used to present food annotation results from existing Natural Language Processing and Machine Learning pipelines in combination with different food semantic data models. Using this framework, users would become more familiar with the links between different food semantic data models.

Keywords: Visualization · Food named-entity recognition · Big data on food and nutrition

1 Introduction

Recently published studies have shown that food is one of the most important environmental factors that is interrelated with human health [1,2,19]. Due to this, the United Nations have determined "End hunger, achieve food security and improved nutrition and promote sustainable agriculture" as one of its sustainable development goals to be achieved by the target date of 2030 [22]. In support of this goal, many research studies are conducted across the globe, producing big amounts of heterogeneous data whose analysis may find answers to complex questions related to agri-food, nutrition/health, and the environment. However,

© Springer Nature Switzerland AG 2020
G. Nicosia et al. (Eds.): LOD 2020, LNCS 12566, pp. 28–38, 2020.
https://doi.org/10.1007/978-3-030-64580-9_4

a synergy between already conducted studies first needs to be achieved. This requires a fusion of their data sets that is a challenge due of their diversity and the diversity of their meta data. Although food is a common every-day concept, it is described by different concepts and entities in the domains of agri-food, nutrition/health, and environment.

One of the most challenging issues that should be solved before starting the modelling process is to make food- and nutrition-related data sets from the domains of agri-food, nutrition/health, and environment interoperable. This implicates that each of them uses different, already established, food standards to describe the food data. However the links between these different standards are not explored in detail [29].

Conversely, in the past two decades, a huge amount of work and effort have been done in the biomedical domain. Many standards have already been inter-linked by developing the Unified Medical Language System (UMLS) [9]. The UMLS is a collection of several health and biomedical vocabularies and stan-dards, supported with lexical and semantic similarity tools for data normaliza-tion of biomedical entities [7], as well as with different visualization tools to make subject matter experts familiar with the information that is presented in it. All of this contributes to the rapid developments in predictive healthcare, which utilize these resources in combination with state-of-the-art artificial intelligence (AI)-based methods.

Compared to the biomedical domain, the domains of agri-food, nutri-tion/health, and environment have several food standards that can be used to describe the food data depending on the research question that is being addressed. Some of them are either still being developing or under exploration in order to discover their utility. However, it is a clear goal that links between them should be established in order to combine different data sets (e.g., on food intake and lifestyle, food composition and bioactivity, food safety, food authenticity and traceability, dietary guidelines available as text, electronic health records, agri-cultural data, environmental data etc.) and to answer more research questions (e.g. from foodomics). Additionally, the links between different food standards should be well-understood by food subject matter experts in order to know how to code their collected research data, to make it further interoperable with other research data sets.

To make subject matter experts familiar with the links that exist between different food standards, we present FoodViz, a user-friendly tool for visual-ization of automatically annotated text on foods. It helps the process of food data interoperability and additionally offers a possibility of correcting the results given by automatic extraction and data normalization.

The rest of the paper is organized as follows. Section 2 provides an overview of food named-entity recognition methods, followed by Sect. 3 in which food semantic resources along with food data normalization methods are presented. Section 4 presents the FoodViz tool. Finally, the conclusions of the paper are presented in Sect. 5.

2   Food Named-Entity Recognition

To extract information related to a specific domain from raw textual data, information extraction (IE) should be applied. IE is a task from natural language processing (NLP) [3], where the main goal is to automatically extract domain entities from text, which are specified by subject matter experts. This further helps the process of extracting relations between different entities and additionally eases the process of extracting events associated with the extracted entities.

Named-entity recognition (NER) is a task from IE, which deals with automatically detecting and identifying phrases from the text that represent domain entities. With regard to the pipeline according to which they are developed, there exist several types:

- Terminological- or dictionary-based [39] - where the extraction is related with a domain specific dictionary. This means that only the entities that are mentioned in the dictionary can be extracted. Additionally, different search heuristics are also applied in order to solve problems that appear with synonyms and different lexical forms of the same entity.
- Rule-based [18,26] - where apart from the dictionary, additional rules in form of regular expressions are developed. These rules define the characteristics of the entitites, which are also domain specific (e.g., in the chemical domain many of the entities names have known prefixes and suffixes).
- Corpus- or Machine Learning (ML)-based [5,23] - where the extraction is done using a trained machine learning model. It requires an annotated corpus, whose creation can be a time-consuming task where subject matter experts need to manually annotated the entities of interest in the textual data. Once the corpus is prepared, supervised ML methods are used to train a model.
- Active Learning (AL)-based [34] - where semi-supervised learning is applied. It starts with a small annotated corpus to train a NER model. Then, an unannotated corpus is used as input and the NER model is retrained. In these cases, the training process iteratively interacts with subject matter experts, who annotate sentences queried from the corpus.
- Deep Learning (DL)-based [24] - where deep neural networks are used to train a model. They typically require large amounts of annotated data.

Despite the existence of several types of NER methods, their application in different domains depends on the available resources (e.g., dictionaries, semantic data models, annotated corpus) in that specific domain. If we look at the biomedical domain, there is a huge amount of work done in this direction [8,16,36]. This would not be possible without the existence of diverse biomedical vocabularies and standards [9], which play a crucial role in understanding biomedical information, coupled with the collection of a large amount of biomedical data (e.g., drug, diseases and other treatments) from numerous sources and shared NLP workshops [6,21,32,37,38].

However, if we look at food and nutrition from the perspectives of agriculture, health and environment, it seems to be low-resourced. There are only a few food semantic data models (i.e. ontologies) that are developed only as a solution to a specific application problem [10]. Additionally, the links between them are not well explored, which causes interoperability difficulties in combining different food-related data sets that are described using them. As a consequence of the resource limitation, there are only a few NERs that can be used for food information extraction. A rule-based NER, known as drNER [14], is developed to extract information from evidence-based dietary recommendations. Within the entities of interest, drNER additionally extracts phrases that consist of food entities. It was further extended and improved by developing FoodIE [27], where the focus is only on food entities. FoodIE is also a rule-based NER, where the rules are combination of computational linguists properties with food semantic information from Hansard corpus [4]. Another way to extract food entities is to use the NCBO Annotator, which is a web service that annotates text by using relevant ontology concepts [20]. It is a part of the BioPortal software services [25], which means that in the food domain it can be combined with food-related ontologies such as FoodOn [17], OntoFood, and SNOMED-CT [11]. More details about a comprehensive comparison of four food NERs (i.e. FoodIE, NCBO (SNOMED CT), NCBO (OntoFood), and NCBO (FoodON)) are available in [30].

3   Food Data Normalization

Data normalization is a crucial task that allows interoperability between data sets that are described using different standards. By applying data normalization, we are mapping entities between different vocabularies, standards, or semantic resources.

In the food domain, there are several resources that can be used for food data normalization. Some of them are:

- FoodEx2 - a description and classification system, proposed by the European Food Safety Agency (EFSA) [12].
- FoodOn - provides semantics for food safety, food security, agricultural and animal husbandry practices linked to food production, culinary, nutritional and chemical ingredients and processes [17].
- OntoFood - a nutrition ontology for diabetes.
- SNOMED CT - a standardized, multilingual vocabulary of clinical terminology that is used by physicians and other health care providers for the electronic health records [11]. It also consists of a *Food* concept.
- The Hansard corpus - a collection of text and concepts created as a part of the SAMUELS project [4,33]. It consists of 37 higher level semantic groups; one of them is *Food and Drink*.

There are also different food data normalization methods that are developed in a combination with some of the above-mentioned resources.

StandFood [13] is a semi-automatic system for classifying and describing foods according to FoodEx2. It is combination of machine learning and Natural Language Processing methods in order to address the lexical similarity in the food domain. Further, it was also extend in the context of semantic similarity, by applying graph-based embedding methods in order to find a unique representation for each food entity available in the FoodEx2 hierarchy [15].

The information extraction done with FoodIE is also related to the Hansard corpus, since FoodIE uses rules that include the Hasnard food semantic information. Using this information, the FoodBase corpus [31] was recently created, which is one of the first annotated corpora with food entities. It consists of 1000 recipes (curated version), and for each one all food entities that are mentioned are extracted and annotated using the Hansard food semantic tags. The recipe categories that are included are: Appetizers and snacks, Breakfast and Lunch, Dessert, Dinner, and Drinks. This version was manually checked by subject matter experts, so the false positive food entities were removed, while the false negative entities were manually added in the corpus. It additionally provides an uncurated version that consists of annotations for around 22,000 recipes, which were not manually checked by subject matter experts.

Additionally, the results presented in [30] show that the NCBO annotator in combination with food ontology can also be used for food data normalization process. For example, using it with FoodOn, the extracted food entities can also be annotated with the FoodOn semantic tags.

Beside different method that are developed and can be used for food data normalization to a specific food standard, the interoperability still remains an open question. To support an initiative similar to the UMLS in the food domain, a recently published study proposed FoodOntoMap [28]. The FoodOntoMap data set consists of food entites extracted from recipes and normalized to different food standards (i.e. Hansard corpus, FoodOn, OntoFood, and SNOMED CT). It also provides a link between the food ontologies, where to each food entity the semantic information from each resource is available. The FoodOntoMap is available in machine readable format, which supports and enables interoperability between computer systems.

4   FoodViz

To make the links between different food standards understandable by food subject matter experts and to make them familiar with the interoperability process using different standards, we develop FoodViz, which is a web-based framework used to present food annotation results from existing Natural Language Processing and machine learning pipelines in conjunction with different food semantic data resources. Currently, a lot of work can already be done in an automatic way, but it is very important that the results are presented to experts in a concise way so that they can check and approve (or disapprove) the results. To show the utility of FoodViz, we visualize the results that are already published in the FoodOntoMap resource. The results consist of recipes that are coming from the curated and uncurated version of FoodBase, which was constructed by using the food NER method FoodIE.

The FoodViz[1] is a single page application developed with React[2], served by a back-end application programming interface (API) developed in Flask[3]. The back-end API serves pre-processed recipes annotated in our previous work [31] and the annotation mappings from [28].

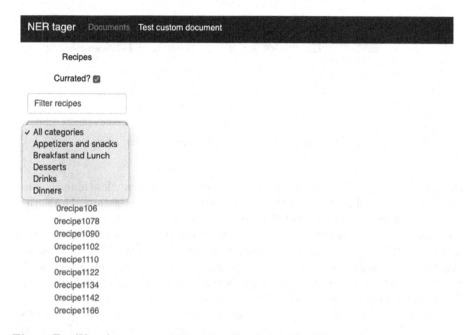

Fig. 1. FoodViz. A new visualization tool for presenting the results published in the FoodOntoMap resource to food subject matter experts.

The home page of FoodViz is presented in Fig. 1. There exist three different parts that can be explored: "NER tagger", "Documents", and "Test custom document".

The "Documents" part displays the curated and uncurated recipes of the FoodBase corpus. There are 1000 curated recipes, 200 per each recipe category, and more than 22.000 uncurated recipes available in FoodViz. The curated version is a ground truth data set, because in the process of developing it, the missing food entities were manually included, while the false positive entities were manually excluded from the corpus [31].

FoodViz allows users to filter the recipes by name, by the recipe category and between the curated and uncurated recipes. Next, the user can select a recipe, for which the semantic annotations are shown. Figure 2 presents an example for a selected curated recipe. The recipe belongs to Appetizers and snacks. Up in

---

[1] http://foodviz.ds4food.ijs.si/fbw/#/recipes.
[2] https://reactjs.org/.
[3] https://flask.palletsprojects.com/en/1.1.x/.

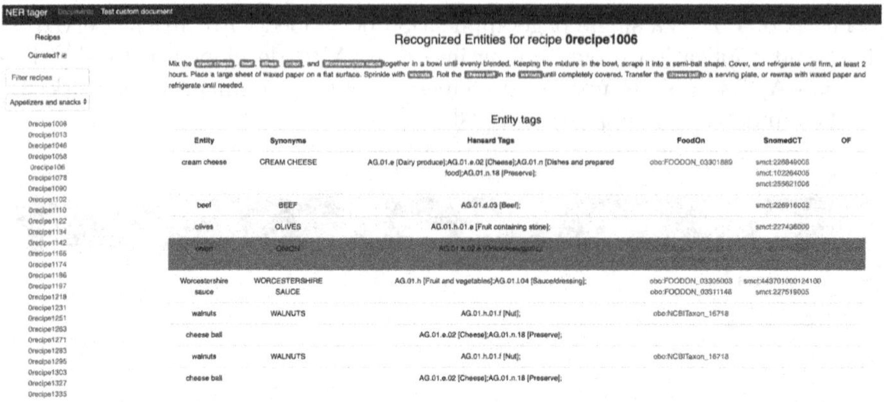

Fig. 2. FoodViz annotation for a recipe from curated corpus.

the top, the recipe description is presented, where all food entities (nine entities) that are mentioned in it are highlighted. FoodViz allows a selection of an entity, which is displayed in the table below. In our case, we selected "onion". Further, for each extracted food entity the synonyms are presented, which are the food names available in different food semantic resources, followed by the semantic tags from Hansard corpus, FoodOn, SNOMED CT, and OntoFood. Additionally, users can further explore the semantic tags from the FoodOn, SNOMED CT and OntoFood, which are linked to their original semantic definitions.

Figure 3 presents an example for a selected uncurated recipe. The uncurated version of FoodBase does not include the false negatives entities and does not exclude the false positive food entities, since it is created from a collection of around 22,000 recipes. With this, subject matter experts can help the process of annotations, by removing the false positives, and including the false negatives, or the FoodViz tool can be also used as annotation tool. By applying this, we will be able to create a much bigger annotated corpus that will allow training on more robust NER based on deep neural networks. Therefore, FoodViz allows manual removal of the false positives, and adding of the false negatives. In this process, as shown in Fig. 3 the removed entities are highlighted in red in the text, and are removed from the table. The only difference in the interface for the curated recipes is that this interaction is not available, since they are already validated.

Using FoodViz, subject matter experts can understand the links between different food standards. It can easily be seen which semantic tag from one food semantic resource is equivalent with a semantic tag from another resource. With this, we also perform food ontology alignment. Additionally, let us assume that information about dietary intake is collected by some dietary assessment tool. This information is normalized using some food semantic data model. Further, if we want this data to be explored and exploited in combination with some health data, its transformation to SNOMED CT semantic tags will be required,

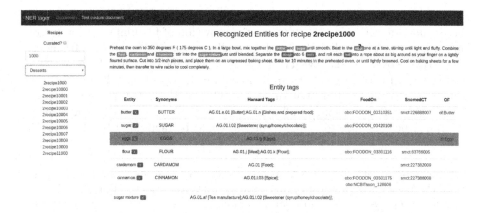

Fig. 3. FoodViz annotation for a recipe from uncurated corpus.

since SNOMED CT is one of the most commonly used semantic resources in the biomedical domain.

For future work, the existence of FoodViz opens different directions for future work in order to make the ML results more closer to subject matter experts. In this direction, we are planning to allow users to select which food NER method they want to use for their data in order to extract the food information. Additionally, the part "Test custom document" will allow users to provide their own text and based on the selection of the NER, the extracted results will be shown.

All in all, we are able to aid and understand the process of food data interoperability, which is a crucial task that should be done as a pre-processing step before involving the data in more advanced data analyses.

5   Conclusion

Information coming from raw textual data is very important albeit difficult to be understand by both humans and machines. There have been debates who beats whom and it seems that machines are becoming better and better in understanding the written word [35]. Several steps need to be performed to support such a complex task, and one of them is a presentation of entities automatically identified in selected texts by ML and NLP. In this paper, we presented the user-friendly tool, named FoodViz, whose goal is visualization of automatically annotated text in the domain of food. Additionally, it can be used as annotation tool where subject matter experts can check the results from automatic entity extraction and data normalization methods.

To illustrate this with an example, let us present the health-related problem that is creating dietary menus in hospitals. For instance, in a menu suitable for patients with an egg allergy, eggs and egg products need to excluded, which is not so difficult to follow because of the regulation relating to the provision of information on substances or products causing allergies or intolerances (e.g. the Regulation (EU) No 1169/2011 on the provision of food information to consumers). However, people suffering the egg allergy, especially children, frequently need to avoid other foods as well (e.g. honey, stock cube etc.), which are not specified in the list of allergens but can be found on the list of ingredients which are usually be written in an unstructured way. Composing a dietary menu requires knowledge of all ingredients of all food items that are to be included in a menu, which can be challenge for a dietitian and could be facilitated by the FoodViz tool.

Acknowledgements. This work was supported by projects from the Slovenian Research Agency [research core grant number P2-0098], and the European Union's Horizon 2020 research and innovation programme (FNS-Cloud, Food Nutrition Security) [grant agreement 863059].

The information and the views set out in this publication are those of the authors and do not necessarily reflect the official opinion of the European Union. Neither the European Union institutions and bodies nor any person acting on their behalf may be held responsible for the use that may be made of the information contained herein.

References

1. Food in the anthropocene: the eat-lancet commission on healthy diets from sustainable food systems. Lancet **393**(10170), 447–492 (2019)
2. Climate change and land: an IPCC special report on climate change, desertification, land degradation, sustainable land management, food security, and greenhouse gas fluxes in terrestrial ecosystems (2020). https://www.ipcc.ch/site/assets/uploads/2019/08/4.-SPM_Approved_Microsite_FINAL.pdf
3. Aggarwal, C.C., Zhai, C.: Mining Text Data. Springer, Heidelberg (2012). https://doi.org/10.1007/978-1-4614-3223-4
4. Alexander, M., Anderson, J.: The hansard corpus, 1803-2003 (2012)
5. Alnazzawi, N., Thompson, P., Batista-Navarro, R., Ananiadou, S.: Using text mining techniques to extract phenotypic information from the phenochf corpus. BMC Med. Inform. Decis. Mak. **15**(2), 1 (2015)
6. Arighi, C.N., et al.: Overview of the biocreative iii workshop. BMC Bioinformat. **12**(8), 1 (2011)
7. Aronson, A.R.: MetaMap: mapping text to the UMLs metathesaurus. Bethesda, MD: NLM, NIH, DHHS **1**, 26 (2006)
8. Boag, W., Wacome, K., Naumann, T., Rumshisky, A.: Cliner: a lightweight tool for clinical named entity recognition. AMIA Joint Summits on Clinical Research Informatics (poster) (2015)
9. Bodenreider, O.: The unified medical language system (UMLs): integrating biomedical terminology. Nucleic Acids Res. **32**(Suppl-1), D267–D270 (2004)
10. Boulos, M.N.K., Yassine, A., Shirmohammadi, S., Namahoot, C.S., Brückner, M.: Towards an "internet of food": Food ontologies for the internet of things. Future Internet **7**(4), 372–392 (2015)
11. Donnelly, K.: SNOMED-CT: the advanced terminology and coding system for eHealth. Stud. Health Technol. Informat. **121**, 279 (2006)
12. (EFSA), E.F.S.A.: The food classification and description system foodex 2 (revision 2). EFSA Support. Publ. **12**(5), 804E (2015)

13. Eftimov, T., Korošec, P., Koroušić Seljak, B.: Standfood: standardization of foods using a semi-automatic system for classifying and describing foods according to FoodEx2. Nutrients **9**(6), 542 (2017)
14. Eftimov, T., Koroušić Seljak, B., Korošec, P.: A rule-based named-entity recognition method for knowledge extraction of evidence-based dietary recommendations. PLoS ONE **12**(6), e0179488 (2017)
15. Eftimov, T., Popovski, G., Valenčič, E., Seljak, B.K.: Foodex2vec: New foods' representation for advanced food data analysis. Food Chem. Toxicol. **138**, 111169 (2020)
16. Gligic, L., Kormilitzin, A., Goldberg, P., Nevado-Holgado, A.: Named entity recognition in electronic health records using transfer learning bootstrapped neural networks. arXiv preprint arXiv:1901.01592 (2019)
17. Griffiths, E.J., Dooley, D.M., Buttigieg, P.L., Hoehndorf, R., Brinkman, F.S., Hsiao, W.W.: Foodon: a global farm-to-fork food ontology. In: ICBO/BioCreative (2016)
18. Hanisch, D., Fundel, K., Mevissen, H.T., Zimmer, R., Fluck, J.: Prominer: rule-based protein and gene entity recognition. BMC Bioinformat. **6**(1), S14 (2005)
19. Johan, F., Owen, G.: Exponential roadmap (2020). https://exponentialroadmap. org/wp-content/uploads/2019/09/Exponential-Roadmap-1.5-September-19-2019. pdf
20. Jonquet, C., Shah, N., Youn, C., Callendar, C., Storey, M.A., Musen, M.: NCBO annotator: semantic annotation of biomedical data. In: International Semantic Web Conference, Poster and Demo Session, vol. 110 (2009)
21. Kim, S., et al.: Biocreative v bioc track overview: collaborative biocurator assistant task for biogrid. Database 2016, baw121 (2016)
22. Lartey, A.: End hunger, achieve food security and improved nutrition and promote sustainable agriculture. UN Chronicle **51**(4), 6–8 (2015)
23. Leaman, R., Wei, C.H., Zou, C., Lu, Z.: Mining patents with tmChem, GNormPlus and an ensemble of open systems. In: Proceedings The Fifth BioCreative Challenge Evaluation Workshop, pp. 140–146 (2015)
24. Lopez, M.M., Kalita, J.: Deep learning applied to nlp. arXiv preprint arXiv: 1703.03091 (2017)
25. Noy, N.F., et al.: Bioportal: ontologies and integrated data resources at the click of a mouse. Nucleic Acids Res. **37**(suppl_2), W170–W173 (2009)
26. Petasis, G., Vichot, F., Wolinski, F., Paliouras, G., Karkaletsis, V., Spyropoulos, C.D.: Using machine learning to maintain rule-based named-entity recognition and classification systems. In: Proceedings of the 39th Annual Meeting on Association for Computational Linguistics, pp. 426–433. Association for Computational Linguistics (2001)
27. Popovski, G., Kochev, S., Koroušić Seljak, B., Eftimov, T.: Foodie: a rule-based named-entity recognition method for food information extraction. In: Proceedings of the 8th International Conference on Pattern Recognition Applications and Methods, (ICPRAM 2019), pp. 915–922 (2019)
28. Popovski, G., Koroušić Seljak, B., Eftimov, T.: Foodontomap: linking food concepts across different food ontologies. In: Proceedings of the 11th International Joint Conference on Knowledge Discovery, Knowledge Engineering and Knowledge Management, KEOD, vol. 2, pp. 195–202. INSTICC, SciTePress (2019). https:// doi.org/10.5220/0008353201950202
29. Popovski, G., Paudel, B., Eftimov, T., Seljak, B.K.: Exploring a standardized language for describing foods using embedding techniques. In: 2019 IEEE International Conference on Big Data (Big Data), pp. 5172–5176. IEEE (2019)

30. Popovski, G., Seljak, B.K., Eftimov, T.: A survey of named-entity recognition methods for food information extraction. IEEE Access **8**, 31586–31594 (2020)
31. Popovski, G., Seljak, B.K., Eftimov, T.: FoodBase corpus: a new resource of annotated food entities. Database 2019, November 2019. https://doi.org/10.1093/database/baz121
32. Rastegar-Mojarad, M., et al.: Biocreative/OHNLP challenge 2018. In: Proceedings of the 2018 ACM International Conference on Bioinformatics, Computational Biology, and Health Informatics, pp. 575–575. ACM (2018)
33. Rayson, P., Archer, D., Piao, S., McEnery, A.: The UCREL semantic analysis system, (2004)
34. Settles, B.: Active learning literature survey. University of Wisconsin, Madison **52**(55–66), 11 (2010)
35. Wang, A., Singh, A., Michael, J., Hill, F., Levy, O., Bowman, S.R.: Glue: a multi-task benchmark and analysis platform for natural language understanding (2020). https://gluebenchmark.com
36. Wang, Q., Zhou, Y., Ruan, T., Gao, D., Xia, Y., He, P.: Incorporating dictionaries into deep neural networks for the Chinese clinical named entity recognition. J. Biomed. Informat. 103133 (2019)
37. Wang, Q., et al.: Overview of the interactive task in biocreative v. Database 2016, baw119 (2016)
38. Wei, C.H., et al.: Assessing the state of the art in biomedical relation extraction: overview of the biocreative v chemical-disease relation (CDR) task. Database 2016, baw032 (2016)
39. Zhou, X., Zhang, X., Hu, X.: MaxMatcher: biological concept extraction using approximate dictionary lookup. In: Yang, Q., Webb, G. (eds.) PRICAI 2006. LNCS (LNAI), vol. 4099, pp. 1145–1149. Springer, Heidelberg (2006). https://doi.org/10.1007/978-3-540-36668-3_150

Sparse Perturbations for Improved Convergence in Stochastic Zeroth-Order Optimization

Mayumi Ohta[1]([⊠]), Nathaniel Berger[1], Artem Sokolov[1], and Stefan Riezler[1,2]

[1] Department of Computational Linguistics, Heidelberg University,
Im Neuenheimer Feld 325, 69120 Heidelberg, Germany
{ohta,berger,sokolov,riezler}@cl.uni-heidelberg.de
[2] Interdisciplinary Center for Scientific Computing (IWR), Heidelberg University,
Im Neuenheimer Feld 325, 69120 Heidelberg, Germany
https://www.cl.uni-heidelberg.de/statnlpgroup/

Abstract. Interest in stochastic zeroth-order (SZO) methods has recently been revived in black-box optimization scenarios such as adversarial black-box attacks to deep neural networks. SZO methods only require the ability to evaluate the objective function at random input points, however, their weakness is the dependency of their convergence speed on the dimensionality of the function to be evaluated. We present a sparse SZO optimization method that reduces this factor to the expected dimensionality of the random perturbation during learning. We give a proof that justifies this reduction for sparse SZO optimization for non-convex functions. Furthermore, we present experimental results for neural networks on MNIST and CIFAR that show empirical sparsity of true gradients, and faster convergence in training loss and test accuracy and a smaller distance of the gradient approximation to the true gradient in sparse SZO compared to dense SZO.

Keywords: Nonconvex optimization · Gradient-free optimization · Zeroth-order optimization

1 Introduction

Zeroth-order optimization methods have gained renewed interest for solving machine learning problems where only the zeroth-order oracle, i.e., the value of the objective function but no explicit gradient, is available. Recent examples include black-box attacks on deep neural networks where adversarial images that lead to misclassification are found by approximating the gradient through a comparison of function values at random perturbations of input images [3]. The advantage of simple random search for scalable and reproducible gradient-free optimization has also been recognized in reinforcement learning [21,24] and hyperparameter tuning for deep neural networks [6], and it is a mainstay in optimization of black-box systems and in simulation optimization [10,22]. While

© Springer Nature Switzerland AG 2020
G. Nicosia et al. (Eds.): LOD 2020, LNCS 12566, pp. 39–64, 2020.
https://doi.org/10.1007/978-3-030-64580-9_5

zeroth-order optimization applies in principle even to non-differentiable functions, in practice Lipschitz-smoothness of the black-box function being evaluated can be assumed. This allows to prove convergence for various zeroth-order gradient approximations [1,5,7,11,22,26,31]. However, even in the optimal case, zeroth-order optimization of n-dimensional functions suffers a factor of $\sqrt{n}$ in convergence rate compared to first-order gradient-based optimization. The goal of our paper is to show theoretically and practically that *sparse perturbations* in stochastic zeroth-order (SZO) optimization can improve convergence speed considerably by replacing the dependency on the dimensionality of the objective function by a dependency on the expected dimensionality of the random perturbation vector. This approach can be motivated by an observation of empirical sparsity of gradients in our experiments on neural networks, or by natural sparsity of gradients in applications to linear models with sparse input features [27]. We give a general convergence proof for non-convex stochastic zeroth-order optimization for Lipschitz-smooth functions that is independent of the dimensionality reduction schedule applied, and shows possible linear improvements in iteration complexity. Our proof is based on [22] and fills in the necessary gaps to verify the dependency of convergence speed on the dimensionality of the perturbation instead of on the full functional dimensionality. We present experiments that perform dimensionality reduction on the random perturbation vector by iteratively selecting parameters with high magnitude for further SZO tuning, and freezing other parameters at their current values. Another dimensionality reduction technique selects random masks by the heldout performance of selected sub-networks, and once a sub-network architecture is identified, it is further fine-tuned by SZO optimization. In our experiments, we purposely choose an application that allows optimization with standard first-order gradient-based techniques, in order to show improved gradient approximation by our sparse SZO technique compared to standard SZO with full perturbations. Our experimental results confirm a smaller distance to the true sparse gradient for the gradient obtained by sparse SZO, and improved convergence in training loss and test accuracy. Furthermore, we compare our technique to a zeroth-order version of iterative magnitude pruning [8,13]. This technique zeros-out unimportant parts of the weight vector while our proposed technique zeros-out only perturbations and freezes unimportant parts of the weight vector at their current values. We find similar convergence speed, but a strong overtraining effect on the test set if weights are pruned to zero values. An estimation of the local Lipschitz constant for the trained models shows that it is growing in the number of iterations for the pruning approach, indicating that zeroing-out low magnitude weights may lead the optimization procedure in a less smooth region of the search space. We conjecture that a search path with a small local Lipschitz constant might have a similar desirable effect as flat minima [15] in terms of generalization on unseen test data. Our code is publicly available[1].

---

[1] https://github.com/StatNLP/sparse_szo.

2   Related Work

SZO techniques in optimization date back to the finite-difference method for gradient estimation of [16] where the value of each component of a weight vector is perturbed separately while holding the other components at nominal value. This technique has since been replaced by more efficient methods based on simultaneous perturbation of all weight vector components [18,28,29]. The central idea of these approaches can be described as approximating non-differentiable functions by smoothing techniques, and applying first-order optimization to the smoothed function [7]. Several works have investigated different update rules and shown the advantages of two-point or multi-point feedback for improved convergence speed [1,5,11,20,22,26,31]. Recent works have investigated sparsity methods for improved convergence in high dimensions [2,30]. These works have to make strong assumptions of function sparsity or gradient sparsity. A precursor to our work that applies sparse SZO to linear models has been presented by [27].

Connections of SZO methods to evolutionary algorithms and reinforcement learning have first been described in [29]. Recent work has applied SZO techniques successfully to reinforcement learning and hyperparameter search for deep neural networks [6,21,23–25]. Similar to these works, in our experiments we apply SZO techniques to Lipschitz-smooth deep neural networks.

Recent practical applications of SZO techniques have been presented in the context of adversarial black-box attacks on deep neural networks. Here a classification function is evaluated at random perturbations of input images with the goal of efficiently finding images that lead to misclassification [3,4]. Because of the high dimensionality of the adversarial attack space in image classification, heuristic methods to dimensionality reduction of perturbations in SZO optimization have already put to practice in [3]. Our approach presents a theoretical foundation for these heuristics.

3   Sparse Perturbations in SZO Optimization for Nonconvex Objectives

We study a stochastic optimization problem of the form

$$\min_{\mathbf{w}} f(\mathbf{w}), \text{ where } f(\mathbf{w}) := \mathbb{E}_{\mathbf{x}}[F(\mathbf{w}, \mathbf{x})], \tag{1}$$

and where $\mathbb{E}_{\mathbf{x}}$ denotes the expectation over inputs $\mathbf{x} \in \mathcal{X}$, and $\mathbf{w} \in \mathbb{R}^n$ parameterizes the objective function F. We address the case of non-convex functions F for which we assume Lipschitz-smoothness[2], i.e., $F(\mathbf{w}, \mathbf{x}) \in C^{1,1}$ is Lipschitz-smooth iff $\forall \mathbf{w}, \mathbf{w}', \mathbf{x} : \|\nabla F(\mathbf{w}, \mathbf{x}) - \nabla F(\mathbf{w}', \mathbf{x})\| \le L(F)\|\mathbf{w} - \mathbf{w}'\|$, where $\|\cdot\|$ is the L2-norm and $L(F)$ denotes the Lipschitz constant of F. This condition is equivalent to

$$|F(\mathbf{w}') - F(\mathbf{w}) - \langle \nabla F(\mathbf{w}), \mathbf{w}' - \mathbf{w} \rangle| \le \frac{L(F)}{2}\|\mathbf{w} - \mathbf{w}'\|^2. \tag{2}$$

---

[2] $C^{p,k}$ denotes the class of p times differentiable functions whose k-th derivative is Lipschitz continuous.

It directly follows that $f \in C^{1,1}$ if $F \in C^{1,1}$.

[22] show how to achieve a smooth version of an arbitrary function $f(\mathbf{w})$ by Gaussian blurring that assures continuous derivatives everywhere in its domain. In their work, random perturbation of parameters is based on sampling a n-dimensional Gaussian random vector $\mathbf{u}$ from a zero-mean isotropic multivariate Gaussian with unit $n \times n$ covariance matrix $\Sigma = \boldsymbol{I}$. The probability density function $\mathrm{pdf}(\mathbf{u})$ is defined by

$$\mathbf{u} \sim \mathcal{N}(\mathbf{0}, \Sigma), \tag{3}$$

$$\mathrm{pdf}(\mathbf{u}) := \frac{1}{\sqrt{(2\pi)^n \cdot \det \Sigma}} e^{-\frac{1}{2}\mathbf{u}^\top \Sigma^{-1} \mathbf{u}}. \tag{4}$$

A Gaussian approximation of a function f is then obtained by the expectation over perturbations $\mathbb{E}_{\mathbf{u}}[f(\mathbf{w} + \mu\mathbf{u})]$, where $\mu > 0$ is a smoothing parameter. Furthermore, a Lipschitz-continuous gradient can be derived even for a non-differentiable original function f by applying standard differentiation rules to the Gaussian approximation, yielding $\mathbb{E}_{\mathbf{u}}[\frac{f(\mathbf{w}+\mu\mathbf{u})-f(\mathbf{w})}{\mu}\mathbf{u}]$.

The central contribution of our work is to show how a sparsification of the random perturbation vector $\mathbf{u}$ directly enters into improved bounds on iteration complexity. This is motivated by an observation of empirical sparsity of gradients in our experiments, as illustrated in the supplementary material (Figs. 7 and 8). Concrete techniques for sparsification will be discussed as parts of our algorithm (Sect. 5), however, our theoretical analysis applies to any method for sparse perturbation. Let $\mathbf{m}^{(t)} \in \{0,1\}^n$ be a mask specified at iteration t, and let $\odot$ denote the componentwise multiplication operator. A sparsified version $\bar{\mathbf{v}}^{(t)}$ of a vector $\mathbf{v}^{(t)} \in \mathbb{R}^n$ is gotten by

$$\bar{\mathbf{v}}^{(t)} := \mathbf{m}^{(t)} \odot \mathbf{v}^{(t)}. \tag{5}$$

The number of nonzero parameters of $\bar{\mathbf{v}}^{(t)}$ is defined for a given mask $\mathbf{m}^{(t)}$ as $\bar{n}^{(t)} := \|\mathbf{m}^{(t)}\|$. Given the definition of a sparsification mask stated above, we can define a sparsified Gaussian random vector drawn from $\mathcal{N}(\mathbf{0}, \boldsymbol{I})$ by $\bar{\mathbf{u}} := \mathbf{m} \odot \mathbf{u}$. Using this notion of sparse perturbation, we redefine the smooth approximation of f, with smoothing parameter $\mu > 0$, as

$$f_\mu(\mathbf{w}) := \mathbb{E}_{\bar{\mathbf{u}}}\left[f(\mathbf{w} + \mu\bar{\mathbf{u}})\right]. \tag{6}$$

Note that f_μ is Lipschitz-smooth with $L(f_\mu) < L(f)$.

4   Convergence Analysis for Sparse SZO Optimization

[22] show that for Lipschitz-smooth functions F, the distance of the gradient approximation to the true gradient can be bounded by the Lipschitz constant and by the norm of the random perturbation. The term $\mathbb{E}_{\mathbf{u}}[\|\mathbf{u}\|^p]$ can itself be bounded by a function of the exponent p and the dimensionality n of the function space. This is how the dependency on n enters iteration complexity bounds and

where an adaptation to sparse perturbations enter the picture. The simple case of the squared norm of the random perturbation given below illustrates the idea. For $p = 2$, we have

$$
\begin{aligned}
\mathbb{E}_{\bar{\mathbf{u}}}\left[\|\bar{\mathbf{u}}\|^2\right] &= \mathbb{E}_{\bar{\mathbf{u}}}\left[\bar{u}_1^2 + \bar{u}_2^2 + \cdots + \bar{u}_n^2\right] \\
&= \mathbb{E}_{\bar{\mathbf{u}}}\left[\bar{u}_1^2\right] + \mathbb{E}_{\bar{\mathbf{u}}}\left[\bar{u}_2^2\right] + \cdots + \mathbb{E}_{\bar{\mathbf{u}}}\left[\bar{u}_n^2\right] \\
&= \mathbb{V}_{\bar{\mathbf{u}}}\left[\bar{u}_1\right] + \mathbb{V}_{\bar{\mathbf{u}}}\left[\bar{u}_2\right] + \cdots + \mathbb{V}_{\bar{\mathbf{u}}}\left[\bar{u}_n\right] \\
&= \bar{n}.
\end{aligned}
\tag{7}
$$

Intuitively this means that if a coordinate i in the parameter space is not perturbed, no variance $\mathbb{V}_{\bar{\mathbf{u}}}[\bar{u}_i]$ is incurred. The smaller the variance, the smaller the factor $\bar{n}$ that directly influences iteration complexity bounds. The central Lemma 1 gives a bound on the expected norm of the random perturbations for the general case of $p \geq 2$: Intuitively, if several coordinates are masked and thus not perturbed, the determinant of the covariance matrix reduces to a product of variances of the unmasked parameters. This allows us to bound the perturbation factor for each input by $\bar{n} \ll n$ where the sparsity pattern is given by the masking strategy determined in the algorithm (see Sect. 5).

Lemma 1. *Let $\bar{\mathbf{u}} = \mathbf{m} \odot \mathbf{u}$ with $\mathbf{u} \sim \mathcal{N}(\mathbf{0}, \mathbf{I})$, then*

$$
\mathbb{E}_{\bar{\mathbf{u}}}\left[\|\bar{\mathbf{u}}\|^p\right] \leq (\bar{n} + p)^{p/2} \text{ for } p \geq 2.
\tag{8}
$$

Given the reduction of the perturbation factor to $\bar{n} \ll n$, we can use the reduced factor in further Lemmata and in the main Theorem that shows improved iteration complexity of sparse SZO optimization. In the following we will state the Lemmata and Theorems that extend [22] to the case of sparse perturbations. For completeness, full proofs of all Lemmata and Theorems can be found in the supplementary material.

Lemma 2 applies standard differentiation rules to the function $f_\mu(\mathbf{w})$, yielding a Lipschitz-continuous gradient.

Lemma 2. *Let $F \in C^{1,1}$ and $\bar{\mathbf{u}} = \mathbf{m} \odot \mathbf{u}$ with $\mathbf{u} \sim \mathcal{N}(\mathbf{0}, \mathbf{I})$. Then we have*

$$
\nabla f_\mu(\mathbf{w}) = \mathbb{E}_{\bar{\mathbf{u}}}\left[\frac{f(\mathbf{w} + \mu\bar{\mathbf{u}})}{\mu}\bar{\mathbf{u}}\right]
\tag{9}
$$

$$
= \mathbb{E}_{\bar{\mathbf{u}}}\left[\frac{f(\mathbf{w} + \mu\bar{\mathbf{u}}) - f(\mathbf{w})}{\mu}\bar{\mathbf{u}}\right]
\tag{10}
$$

$$
= \mathbb{E}_{\bar{\mathbf{u}}}\left[\frac{f(\mathbf{w} + \mu\bar{\mathbf{u}}) - f(\mathbf{w} - \mu\bar{\mathbf{u}})}{2\mu}\bar{\mathbf{u}}\right].
\tag{11}
$$

In the rest of this paper we define our update rule as

$$
g_\mu(\mathbf{w}) := \frac{f(\mathbf{w} + \mu\bar{\mathbf{u}}) - f(\mathbf{w})}{\mu}\bar{\mathbf{u}}
\tag{12}
$$

from (10). Lemma 2 implies that $g_\mu(\mathbf{w})$ is an unbiased estimator of $\nabla f_\mu(\mathbf{w})$.

The next Lemma shows how to bound the distance of the gradient approximation to the true gradient in terms of μ, $L(f)$, and the number of perturbations $\bar{n}$.

Table 1. Data statistics and system settings.

Dataset	MNIST	CIFAR-10
Trainset size	50,000	40,000
Devset size	10,000	10,000
Testset size	10,000	10,000
Architecture	Feed-forward NN	CNN
Model size	266,610	4,301,642
Batch size	64	64
Learning rate h	0.2	0.1
Smoothing μ	0.05	0.05
Variance	1.0	1.0
Total epochs	100	100
Sparsification interval	5 epochs	5 epochs

Lemma 3. *Let $f \in C^{1,1}$ Lipschitz-smooth, then*

$$\|\nabla f(\mathbf{w})\|^2 \leq 2\|\nabla f_\mu(\mathbf{w})\|^2 + \frac{\mu^2 L^2(f)}{2}(\bar{n} + 4)^3. \tag{13}$$

The central Theorem shows that the iteration complexity of an SZO algorithm based on update rule (12) is bounded by the Lipschitz constant L and the expected dimensionality $\hat{n}$ of sparse perturbations with respect to iterations:

Theorem 1. *Assume a sequence $\{\mathbf{w}^{(t)}\}_{t \geq 0}$ be generated by Algorithm 1. Let $\mathcal{X} = \{\mathbf{x}^{(t)}\}_{t \geq 0}$ and $\bar{\mathcal{U}} = \{\bar{\mathbf{u}}^{(t)}\}_{t \geq 0}$. Furthermore, let $f_\mu^\star$ denote a stationary point such that $f_\mu^\star \leq f_\mu(w^{(t)}) \ \forall \ t > 0$ and $\hat{n} \geq \mathbb{E}_t[\bar{n}^{(t)}] := \frac{1}{T+1} \sum_{t=0}^{T+1} \bar{n}^{(t)}$ the upper bound of the expected number of nonzero entries of $\bar{\mathcal{U}}$. Then, choosing learning rate $\hat{h} = \frac{1}{4(\hat{n}+4)L}$ and $\mu = \Omega\left(\frac{\epsilon}{\hat{n}^{3/2}L}\right)$ where $L(f) \leq L$ for all $f(\mathbf{w}^{(t)})$, we have*

$$\mathbb{E}_{\bar{\mathcal{U}},\mathcal{X}}\left[\left\|\nabla f(\mathbf{w}^{(T)})\right\|^2\right] \leq \epsilon^2 \tag{14}$$

for any $T = \mathcal{O}\left(\frac{\hat{n}L}{\epsilon^2}\right)$.

Note that $\hat{n} \geq \mathbb{E}_t\left[\bar{n}^{(t)}\right]$ is much smaller than $n = \mathbb{E}_t\left[n^{(t)}\right] = \frac{1}{T+1} \sum_{t=0}^{T+1} n$, if $\bar{n}^{(t)}$ is chosen such that $\bar{n}^{(t)} \ll n$ at each iteration $t > 0$ of Algorithm 1. This Theorem implies that both strong sparsity patterns and the smoothness of the objective function can boost the convergence speed up to linear factor.

5   Algorithms for Sparse SZO Optimization

5.1   Masking Strategies

Algorithm 1 starts with standard SZO optimization with full perturbations in the first iteration. This is done by initializing the mask to a n-dimensional vector

---

Algorithm 1. Sparse SZO Optimization

1: INPUT: dataset $\mathcal{X} = \{\mathbf{x}_0, \cdots, \mathbf{x}_T\}$, sequence of learning rates h, sparsification interval s, number of samples k, smoothing parameter $\mu > 0$
2: Initialize mask: $\mathbf{m}^{(0)} = \mathbf{1}^n$ $(\bar{n}^{(0)} = n)$
3: Initialize weights: $\mathbf{w}^{(0)}$
4: **for** $t = 0, \cdots, T$ **do**
5:     **if** $\mod(t, s) = 0$ **then**
6:         Update mask $\mathbf{m}^{(t')} = \text{get\_mask}(\mathbf{w}^{(t)})$
7:         **if** pruning **then**
8:             $\mathbf{w}^{(t)} = \mathbf{m}^{(t')} \odot \mathbf{w}^{(t)}$
9:         **end if**
10:     **end if**
11:     Observe $\mathbf{x}^{(t)} \in \mathcal{X}$
12:     **for** $j = 1, \cdots, k$ **do**
13:         Sample unit vector $\mathbf{u}^{(t)} \sim \mathcal{N}(\mathbf{0}, \mathbf{I})$
14:         Apply mask $\bar{\mathbf{u}}^{(t)} = \mathbf{m}^{(t')} \odot \mathbf{u}^{(t)}$ $(\bar{n}^{(t)} \ll n)$
15:         Compute $g_\mu^{(j)}(\mathbf{w}^{(t)})$
16:     **end for**
17:     Average $g_\mu(\mathbf{w}^{(t)}) = \underset{j}{\text{avg}}(g_\mu^{(j)}(\mathbf{w}^{(t)}))$
18:     Update $\mathbf{w}^{(t+1)} = \mathbf{w}^{(t)} - h^{(t)} g_\mu(\mathbf{w}^{(t)})$
19: **end for**
20: OUTPUT: sequence of $\{\mathbf{w}^{(t)}\}_{t \geq 0}$

---

of coordinates with value 1 (line 2). Guided by a schedule that applies sparsification at every s-th iteration, the function get_mask($\cdot$) is applied (line 6). We implemented a first strategy called *magnitude masking*, and another one called *random masking*. In *magnitude masking*, we sort the indices according to their L1-norm magnitude, set a cut-off point, and mask the weights below the threshold. In *random masking*, we sample 50 random mask patterns and select the one that performs best according to accuracy on a heldout set. Following [9], the sparsification interval corresponds to reducing 20% of the remaining unmasked parameters at every 5 epochs.

5.2   Sparse Perturbations with Pruning or Freezing

Algorithm 1 is defined in a general form that allows *pruning* and *freezing* of masked parameters. In the *pruning* variant, the same sparsification mask that is applied to the Gaussian perturbations (line 14) is also applied to the weight vector itself (line 8). That is, we keep the weight values from the previous iteration at the index whose mask value is one, and reset the weight values to zero at the index whose mask value is zero. This can be seen as a straightforward sparse-SZO extension of the iterative magnitude pruning method with rewinding to the value of the previous iteration as proposed by [9]. In the *freezing* variant, we apply the sparsification mask only to the Gaussian perturbations (line 14), and inherit all the weight values from the previous iteration. That is, we keep

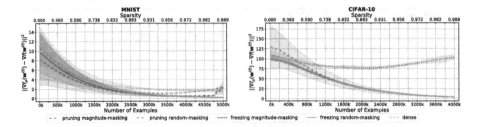

Fig. 1. Distance of gradient approximation to true gradient on training set, interpolation factor 0.99.

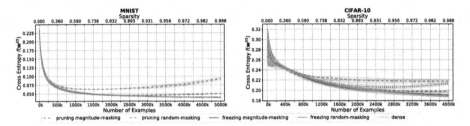

Fig. 2. Cumulative values for cross-entropy loss of sparse and dense SZO algorithms on the training set.

the weight value unchanged at the index whose mask value is zero, and allow the value to be updated at the index whose mask value is one.

6   Experiments

We purposely chose experimental tasks where the true gradient can be calculated exactly in order to experimentally verify the improved gradient approximation by sparse perturbations in SZO. While the convergence rate in SZO is always is suboptimal compared to first-order optimization due to the used gradient approximation (see [5] or [26]), our proofs show that sparse perturbations improve the distance of the approximate gradient to the true gradient (Lemma 3) and thus the convergence rate (Theorem 1). The experimental results given in the following confirm our theoretical analysis.

6.1   Task and Datasets

We apply our sparse SZO algorithm to standard image classification benchmarks on the MNIST [19] and CIFAR-10 [17] datasets. Since the SZO optimizers are less common in an off-the-shelf API package bundled in a framework, we implemented our custom SZO optimizers utilizing `pytorch`, and the data was downloaded via torchvision package v0.4.2. We followed the pre-defined train-test split, and partitioned 20% of the train set for validation. Model performance

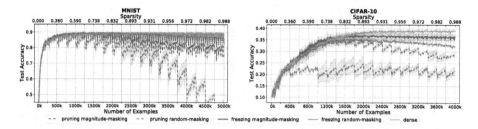

Fig. 3. Multi-class classification accuracy of sparse and dense SZO algorithms on test sets.

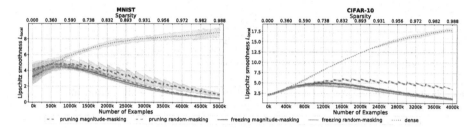

Fig. 4. Estimated local Lipschitz smoothness L_{local} of sparse and dense SZO algorithms along search path, interpolation factor 0.99.

is measured by successive evaluation of the gradient norm on the training set, multi-class classification accuracy on the test set, cumulative cross-entropy loss on the training set, and the local function smoothness, estimated by the local Lipschitz constant computed as

$$L_{\text{local}} := \frac{\left\| \nabla f(\mathbf{w}^{(t-1)}) - \nabla f(\mathbf{w}^{(t)}) \right\|}{\left\| \mathbf{w}^{(t-1)} - \mathbf{w}^{(t)} \right\|}. \tag{15}$$

6.2 Architectures and Experimental Settings

For the MNIST experiments, we constructed a fully-connected feed-forward neural network models with 3 layers including batch normalization [14] after each layer. For the CIFAR-10 experiments, we built a convolutional neural network with 2 convolutional layers with kernel size 3 and stride 1. After taking max pooling of these two convolutional layers, we use 2 extra fully-connected linear layers on top of it. We employed the xavier-normal weights initialization method [12] for all setups. Each model is trained by minimizing a standard cross-entropy loss for 100 epochs with batch size 64. We used a constant learning rate of 0.2 for MNIST and 0.1 for CIFAR, a constant smoothing parameter of 0.05, and two-sided perturbations in the SZO optimization for both datasets. We sampled 10 perturbations per update from a zero-centered Gaussian distribution with variance 1.0, computed the gradients for each, then the average was taken

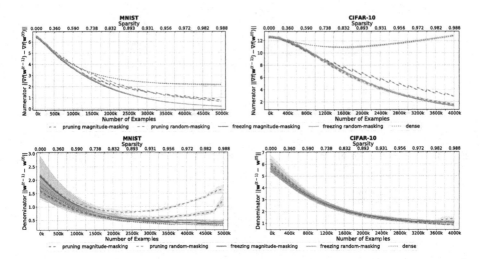

Fig. 5. Denominator and numerator of estimated local Lipschitz smoothness L_{local} along search path, interpolation factor 0.99.

before update. Applying the 20% reduction schedule 20 times throughout the training, the active model parameters were reduced from 266,610 to 3,844 in MNIST experiments, from 4,301,642 to 61,996 in CIFAR experiments. We computed sparsity as $1 - \frac{\bar{n}^{(t)}}{n^{(t)}}$ every time we update the mask pattern. The sparsity value along iteration is given in the upper x-axis of each plot. The lower x-axis shows the number of examples visited during training in thousands. If indicated, plotted curves are smoothed by linear interpolation with the previous values. All plots show mean and standard deviation of results for 3 training runs under different random seeds. A summary of data statistics and system settings can be found in Table 1.

6.3   Experimental Results

Our experimental task was purposely chosen as an application that allows optimization with standard first-order (SFO) gradient-based techniques for comparison with SZO methods. Figure 1 shows that the distance of the approximate gradient to the true gradient converges to zero at the fastest rate for freezing methods, proving empirical support for (34) in Lemma 3. All four sparse SZO variants approach the true gradients considerably faster in most training steps on both training datasets than dense SZO (i.e., the full dimension of parameters is perturbed in each training step). We see that dense SZO is diverging from the true gradient on the CIFAR dataset where more than 4 million parameters are trained.

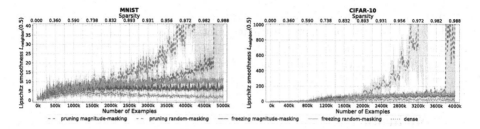

Fig. 6. Estimated local Lipschitz smoothness L_{neighbor} on test set in neighborhood of $w^{(t)}$, interpolation factor 0.8.

Figure 2 gives cumulative values of the cross-entropy loss function on the respective training sets. We see that convergence in cross-entropy loss is similar for sparse and dense SZO variants on MNIST and advantageous for sparse SZO on CIFAR, with both weight pruning methods showing a slight divergence in cross-entropy loss.

Figure 3 shows the accuracy of multi-class classification on MNIST and CIFAR-10 for the sparse and dense SZO variants. We see the fastest convergence in test accuracy for freezing methods, compared to pruning and dense SZO. Interestingly, we see that the test accuracy decreases after a while for all pruning variants, while it continues to increase for the freezing variants and the dense SZO algorithm.

One possibility to analyze this effect is to inspect the estimate of the local Lipschitz smoothness L_{local} defined in (15) along the current search path on the training set. Figure 4 shows that L_{local} decreases considerably along the search trajectory for sparse SZO optimization, with a faster decrease for freezing than for pruning methods. However, it increases considerably for the dense variant, leading the optimization procedure into less smooth areas of the search space. In order to verify that this is not a trivial effect due to pruned weight vectors, we consider numerator and denominator of L_{local} separately in Fig. 5. We see that the denominator and numerator are decreasing at different speeds for the sparse variants, with relatively larger differences between gradients causing an increase in the Lipschitz quotient. The numerator $\left\| \nabla f(w^{(t-1)}) - \nabla f(w^{(t)}) \right\|$ is successively reduced in the sparse methods, with a faster reduction for freezing than for pruning. The reduction indicates that sparse SZO methods find a path converging to the optimal point, while the dense variant seems not to be able to find such a path, but instead stays at the same distance or even increases the distance from the previous step.

Finally, we evaluate an estimate of the local Lipschitz smoothness in a neighborhood $\mathbf{v} \sim \text{Uniform}(-0.5, 0.5)$ around the current parameter values $\mathbf{w}^{(t)}$ on the test set. L_{neighbor} is estimated by taking the maximum of 10 samples per batch as

$$L_{\text{neighbor}} := \max_{\mathbf{v}} \frac{\left\| \nabla f(\mathbf{w}^{(t)}) - \nabla f(\mathbf{w}^{(t)} + \mathbf{v}) \right\|}{\|\mathbf{v}\|}. \tag{16}$$

Figure 6 shows that $L_{neighbor}$ is staying roughly constant on the test set in the neighborhood of the parameter estimates obtained during training for freezing and dense SZO. However, it is increasing considerably for both variants of pruning, indicating that pruning methods converge to less smooth regions in parameter space.

7   Conclusion

SZO methods are flexible and simple tools for provably convergent optimization of black-box functions only from function evaluations at random points. A theoretical analysis of these methods can be given under mild assumptions of Lipschitz-smoothness of the function to be optimized. However, even in the best case, such gradient-free methods suffer a factor of $\sqrt{n}$ in iteration complexity, depending on the dimensionality n of the evaluated function, compared to methods that employ gradient information. This makes SZO techniques always second-best if gradient information is available, especially in high dimensional optimization. However, SZO techniques may be the only option in black-box optimization scenarios where it is desirable as well to improve convergence speed. Our paper showed that the dimensionality factor can be reduced to the expected dimensionality of random perturbations during training, independent of the dimensionality reduction schedule employed. This allows considerable speedups in convergence of SZO optimization in high dimensions. We presented experiments with masking schedules based on L1-magnitude and random masking, both confirming our theoretical result of an improved approximation of the true gradient by a sparse SZO gradient. This result is accompanied by an experimental finding of improved convergence in training loss and test accuracy. A further experiment compared sparse SZO where masked parameters are frozen at their current values to a variant of sparse SZO where masked parameters are pruned to zero values. This technique can be seen as a sparse SZO variant of the well-known technique of iterative magnitude pruning [8,13]. Our results show that pruning techniques perform worse in the world of SZO optimization than fine-tuning unmasked parameters while freezing masked parameters. By further inspection of the local Lipschitz smoothness along the search path of different algorithms we find that pruning in the SZO world can lead the optimization procedure in a locally less smooth search area, while freezing seems to lead to a smoother search path even compared to dense SZO optimization.

Acknowledgements. We would like to thank Michael Hagmann for proofreading the theoretical analysis, and the anonymous reviewers for their useful comments. The research reported in this paper was supported in part by the German research foundation (DFG) under grant RI-2221/4-1.

Appendix

A   Notation

In order to adapt the theoretical analysis of [22] to the case of sparse perturbations, the probability density function $\mathrm{pdf}_{\bar{\mathbf{u}}}(\bar{\mathbf{u}})$ and expectations $\mathbb{E}_{\bar{\mathbf{u}}}[f(\bar{\mathbf{u}})]$ of random variables with respect to it have to be carefully defined. Given a sparsification mask $\mathbf{m} \in \{0,1\}^n$, let $I_{\mathrm{masked}} := \{i | m_i = 0\}$ be a set of indices whose mask value is zero. Likewise, let $I_{\mathrm{unmasked}} := \{i | m_i = 1\}$ be a set of indices whose mask value is one. The probability density function of random perturbations $\mathbf{u} \sim \mathcal{N}(\mathbf{0}, \boldsymbol{I})$ can be expressed as a product of univariate probability density functions $\mathrm{pdf}(u_i)$, where $u_i \sim \mathcal{N}(0,1)$. Then the probability density function $\mathrm{pdf}_{\bar{\mathbf{u}}}(\bar{\mathbf{u}})$ can be defined as a conditional probability density function $\mathrm{pdf}_{\mathbf{u}|u_i=0, i \in I_{\mathrm{masked}}}(\mathbf{u})$, conditioned on zero-valued entries $u_i = 0$ corresponding to masked indices $i \in I_{\mathrm{masked}}$. Since the number of nonzero parameters is determined by the sparsification mask as $|I_{\mathrm{unmasked}}| = \bar{n}$, we can reduce the product over the full dimensionality $i \in \{1, \cdots n\}$ to the product over unmasked entries $i \in I_{\mathrm{unmasked}}$. The conditional probability density function $\mathrm{pdf}_{\mathbf{u}|u_i=0, i \in I_{\mathrm{masked}}}(\mathbf{u})$ is then defined as

$$\mathrm{pdf}_{\mathbf{u}|u_i=0, i \in I_{\mathrm{masked}}}(\mathbf{u}) = \prod_{i \in I_{\mathrm{unmasked}}} \mathrm{pdf}(u_i) \tag{17}$$

$$= \left(\frac{1}{\sqrt{2\pi}}\right)^{\bar{n}} \prod_{i \in I_{\mathrm{unmasked}}} e^{-\frac{1}{2}u_i^2}. \tag{18}$$

The expectation with respect to $\bar{\mathbf{u}}$ is defined as conditional expectation using aforementioned conditional probability density function. Let $f : \mathbb{R}^n \to \mathbb{R}$ be an arbitrary function, let $\int_{\mathbf{u}\upharpoonright} \cdot d\mathbf{u}$ denote the integral $\int_{\mathbf{u}|u_i=0, i \in I_{\mathrm{masked}}} \cdot d\mathbf{u}$, and let $\mathrm{pdf}_{\mathbf{u}\upharpoonright}$ denote the conditional probability density function $\mathrm{pdf}_{\mathbf{u}|u_i=0, i \in I_{\mathrm{masked}}}$, and let $\kappa := \sqrt{(2\pi)^{\bar{n}}}$. Then we have

$$\mathbb{E}_{\bar{\mathbf{u}}}[f(\bar{\mathbf{u}})] := \mathbb{E}_{\mathbf{u}|u_i=0, i \in I_{\mathrm{masked}}}[f(\mathbf{u})] \tag{19}$$

$$= \int_{\mathbf{u}\upharpoonright} f(\mathbf{u}) \cdot \mathrm{pdf}_{\mathbf{u}\upharpoonright}(\mathbf{u}) d\mathbf{u} \tag{20}$$

$$= \frac{1}{\kappa} \int_{\mathbf{u}\upharpoonright} f(\mathbf{u}) \cdot e^{-\frac{1}{2}\|\mathbf{u}\|^2} d\mathbf{u}. \tag{21}$$

In the following, we use the notation $\int_{\mathbf{u}\upharpoonright} \cdot d\mathbf{u}$ for $\int_{\mathbf{u}|u_i=0, i \in I_{\mathrm{masked}}} \cdot d\mathbf{u}$, and $\mathrm{pdf}_{\mathbf{u}\upharpoonright}$ for $\mathrm{pdf}_{\mathbf{u}|u_i=0, i \in I_{\mathrm{masked}}}$.

B   Proof for Lemma 1

Proof. Define by $\tilde{\mathbf{u}} \in \mathbb{R}^{\bar{n}}$ a reduced vector that removes all zero entries from $\bar{\mathbf{u}} \in \mathbb{R}^n$. Taking an expectation over $\bar{\mathbf{u}} \in \mathbb{R}^n$ results in the same value as taking an expectation over $\tilde{\mathbf{u}} \in \mathbb{R}^n$. This can be seen as follows:

$$
\begin{aligned}
\mathbb{E}_{\bar{\mathbf{u}}}\left[\|\bar{\mathbf{u}}\|^p\right] &= \frac{1}{\kappa} \int_{\mathbf{u}|u_i=0, i \in I_{\text{masked}}} \|\mathbf{u}\|^p e^{-\frac{1}{2}\mathbf{u}^\top \Sigma \mathbf{u}} d\mathbf{u} \\
&= \frac{1}{\kappa} \int_{\mathbf{u}|u_i=0, i \in I_{\text{masked}}} \left(\sum_{j=1}^{n} |u_j|^p\right) e^{-\frac{1}{2}\mathbf{u}^\top \Sigma \mathbf{u}} d\mathbf{u} \\
&= \frac{1}{\kappa} \sum_{i \in I_{\text{unmasked}}} \left(\int_{u_i} |u_i|^p e^{-\frac{1}{2}u_i^2} du_i\right) \\
&= \frac{1}{\kappa} \int_{\tilde{\mathbf{u}}} \left(\sum_{i=1}^{\bar{n}} |u_i|^p\right) e^{-\frac{1}{2}\tilde{\mathbf{u}}^\top \tilde{\Sigma} \tilde{\mathbf{u}}} d\tilde{\mathbf{u}} \\
&= \frac{1}{\kappa} \int_{\tilde{\mathbf{u}}} \|\tilde{\mathbf{u}}\|^p e^{-\frac{1}{2}\tilde{\mathbf{u}}^\top \tilde{\Sigma} \tilde{\mathbf{u}}} d\tilde{\mathbf{u}} \\
&= \mathbb{E}_{\tilde{\mathbf{u}}}\left[\|\tilde{\mathbf{u}}\|^p\right]
\end{aligned}
$$

where $\tilde{\Sigma}$ is a $\bar{n} \times \bar{n}$ unit covariance matrix with $\tilde{\Sigma} = \mathbf{I}$.

Let $p \geq 2$ and $\tau \in (0,1)$. Recalling that $\bar{\mathbf{u}} \in \mathbb{R}^n$, $\tilde{\mathbf{u}} \in \mathbb{R}^{\bar{n}}$, and $\tilde{\Sigma} \in \mathbb{R}^{\bar{n} \times \bar{n}}$, we have

$$
\begin{aligned}
\mathbb{E}_{\bar{\mathbf{u}}}\left[\|\bar{\mathbf{u}}\|^p\right] &= \mathbb{E}_{\tilde{\mathbf{u}}}\left[\|\tilde{\mathbf{u}}\|^p\right] \\
&= \frac{1}{\sqrt{(2\pi)^{\bar{n}}}} \int_{\tilde{\mathbf{u}}} \|\tilde{\mathbf{u}}\|^p e^{-\frac{\tau+(1-\tau)}{2}\tilde{\mathbf{u}}^\top \tilde{\mathbf{u}}} d\tilde{\mathbf{u}} \\
&= \frac{1}{\sqrt{(2\pi)^{\bar{n}}}} \int_{\tilde{\mathbf{u}}} \|\tilde{\mathbf{u}}\|^p e^{-\frac{\tau}{2}\tilde{\mathbf{u}}^\top \tilde{\mathbf{u}}} e^{-\frac{1-\tau}{2}\tilde{\mathbf{u}}^\top \tilde{\mathbf{u}}} d\tilde{\mathbf{u}} \\
&\leq \frac{1}{\sqrt{(2\pi)^{\bar{n}}}} \int_{\tilde{\mathbf{u}}} \left(\frac{p}{\tau e}\right)^{\frac{p}{2}} e^{-\frac{1-\tau}{2}\tilde{\mathbf{u}}^\top \tilde{\mathbf{u}}} d\tilde{\mathbf{u}} && (22) \\
&= \left(\frac{p}{\tau e}\right)^{\frac{p}{2}} \frac{1}{\sqrt{(2\pi)^{\bar{n}}}} \int_{\tilde{\mathbf{u}}} e^{-\frac{1}{2}\tilde{\mathbf{u}}^\top \left(\frac{\tilde{\Sigma}}{1-\tau}\right)^{-1} \tilde{\mathbf{u}}} d\tilde{\mathbf{u}} \\
&= \left(\frac{p}{\tau e}\right)^{\frac{p}{2}} \frac{1}{\sqrt{(2\pi)^{\bar{n}}}} \sqrt{(2\pi)^{\bar{n}} \cdot \det\left(\frac{\tilde{\Sigma}}{1-\tau}\right)} && (23) \\
&= \left(\frac{p}{\tau e}\right)^{\frac{p}{2}} (1-\tau)^{-\frac{\bar{n}}{2}} && (24) \\
&\leq (\bar{n}+p)^{\frac{p}{2}}. && (25)
\end{aligned}
$$

The first inequality (22) follows from $t^p e^{-\frac{\tau}{2}t^2} \leq \left(\frac{p}{\tau e}\right)^{\frac{p}{2}}$ for $t > 0$. The second inequality (25) follows by minimizing $\left(\frac{p}{\tau e}\right)^{\frac{p}{2}} (1-\tau)^{-\frac{p}{2}}$ in $\tau \in (0,1)$. Full proofs of these inequalities are given in Subsect. F.

C   Proof for Lemma 2

Proof.

$$\nabla_{\mathbf{w}} f_\mu(\mathbf{w}) = \nabla_{\mathbf{w}} \mathbb{E}_{\bar{\mathbf{u}}} \left[f(\mathbf{w} + \mu \bar{\mathbf{u}}) \right]$$

$$= \nabla_{\mathbf{w}} \int_{\mathbf{u}\upharpoonright} \text{pdf}_{\mathbf{u}\upharpoonright}(\mathbf{u}) \, f(\mathbf{w} + \mu \mathbf{u}) \, d\mathbf{u}$$

$$= \int_{\mathbf{u}\upharpoonright} \nabla_{\mathbf{w}} \, \text{pdf}_{\mathbf{u}\upharpoonright}(\mathbf{u}) \, f(\mathbf{w} + \mu \mathbf{u}) \, d\mathbf{u} \tag{26}$$

$$= \frac{1}{\kappa} \int_{\mathbf{u}\upharpoonright} \nabla_{\mathbf{w}} \, e^{-\frac{1}{2}\|\mathbf{u}\|^2} f(\mathbf{w} + \mu \mathbf{u}) \, d\mathbf{u}$$

$$= \frac{1}{\kappa} \int_{\mathbf{y}\upharpoonright} \nabla_{\mathbf{w}} \, e^{-\frac{1}{2}\|\frac{\mathbf{y}-\mathbf{w}}{\mu}\|^2} f(\mathbf{y}) \frac{1}{\mu^n} \, d\mathbf{y} \tag{27}$$

$$= \frac{1}{\kappa} \int_{\mathbf{y}\upharpoonright} \frac{\mathbf{y}-\mathbf{w}}{\mu^2} e^{-\frac{1}{2\mu^2}\|\mathbf{y}-\mathbf{w}\|^2} f(\mathbf{y}) \frac{1}{\mu^n} \, d\mathbf{y}$$

$$= \frac{1}{\kappa} \int_{\mathbf{u}\upharpoonright} \frac{\mathbf{u}}{\mu} e^{-\frac{1}{2}\|\mathbf{u}\|^2} f(\mathbf{w} + \mu \mathbf{u}) \, d\mathbf{u} \tag{28}$$

$$= \mathbb{E}_{\bar{\mathbf{u}}} \left[\frac{f(\mathbf{w} + \mu \bar{\mathbf{u}})}{\mu} \bar{\mathbf{u}} \right]$$

Since $f \in C^{1,1}$, we can interchange ∇ and the integral in (26). Furthermore, we substituted $\mathbf{y} = \mathbf{w} + \mu \mathbf{u}$ in (27) and put it back in (28).

Now, we show (10) $\Leftrightarrow$ (9).

$$\mathbb{E}_{\bar{\mathbf{u}}} \left[\frac{f(\mathbf{w} + \mu \bar{\mathbf{u}}) - f(\mathbf{w})}{\mu} \bar{\mathbf{u}} \right] \tag{29}$$

$$= \frac{1}{\kappa} \int_{\mathbf{u}\upharpoonright} \frac{f(\mathbf{w} + \mu \mathbf{u}) - f(\mathbf{w})}{\mu} \mathbf{u} \, e^{-\frac{1}{2}\|\mathbf{u}\|^2} d\mathbf{u}$$

$$= \frac{1}{\kappa} \int_{\mathbf{u}\upharpoonright} \frac{f(\mathbf{w} + \mu \mathbf{u})}{\mu} \mathbf{u} \, e^{-\frac{1}{2}\|\mathbf{u}\|^2} d\mathbf{u} - \frac{f(\mathbf{w})}{\mu\kappa} \underbrace{\int_{\mathbf{u}\upharpoonright} \mathbf{u} \, e^{-\frac{1}{2}\|\mathbf{u}\|^2} d\mathbf{u}}_{=0}$$

$$= \mathbb{E}_{\bar{\mathbf{u}}} \left[\frac{f(\mathbf{w} + \mu \bar{\mathbf{u}})}{\mu} \bar{\mathbf{u}} \right] \tag{30}$$

Lastly, we show (11) $\Leftrightarrow$ (9). The expectation doesn't change even if we shift $\mathbf{w}$ by $\mu \bar{\mathbf{u}}$, hence

$$\mathbb{E}_{\bar{\mathbf{u}}} \left[\frac{f(\mathbf{w}) - f(\mathbf{w} - \mu \bar{\mathbf{u}})}{\mu} \bar{\mathbf{u}} \right] = \mathbb{E}_{\bar{\mathbf{u}}} \left[\frac{f(\mathbf{w} + \mu \bar{\mathbf{u}}) - f(\mathbf{w})}{\mu} \bar{\mathbf{u}} \right] \tag{31}$$

$$\overset{(30)}{=} \mathbb{E}_{\bar{\mathbf{u}}} \left[\frac{f(\mathbf{w} + \mu \bar{\mathbf{u}})}{\mu} \bar{\mathbf{u}} \right]. \tag{32}$$

$$\mathbb{E}_{\bar{\mathbf{u}}}\left[\frac{f(\mathbf{w}+\mu\bar{\mathbf{u}})-f(\mathbf{w}-\mu\bar{\mathbf{u}})}{2\mu}\bar{\mathbf{u}}\right]$$

$$=\frac{1}{2}\left(\mathbb{E}_{\bar{\mathbf{u}}}\left[\underbrace{\frac{f(\mathbf{w}+\mu\bar{\mathbf{u}})}{\mu}\bar{\mathbf{u}}-\frac{f(\mathbf{w})}{\mu}\bar{\mathbf{u}}+\frac{f(\mathbf{w})}{\mu}\bar{\mathbf{u}}-\frac{f(\mathbf{w}-\mu\bar{\mathbf{u}})}{\mu}\bar{\mathbf{u}}}_{=0}\right]\right)$$

$$=\frac{1}{2}\left(\mathbb{E}_{\bar{\mathbf{u}}}\left[\frac{f(\mathbf{w}+\mu\bar{\mathbf{u}})-f(\mathbf{w})}{\mu}\bar{\mathbf{u}}\right]+\mathbb{E}_{\bar{\mathbf{u}}}\left[\frac{f(\mathbf{w})-f(\mathbf{w}-\mu\bar{\mathbf{u}})}{\mu}\bar{\mathbf{u}}\right]\right)$$

$$\overset{(32)}{=}\frac{1}{2}\left(\mathbb{E}_{\bar{\mathbf{u}}}\left[\frac{f(\mathbf{w}+\mu\bar{\mathbf{u}})}{\mu}\bar{\mathbf{u}}\right]+\mathbb{E}_{\bar{\mathbf{u}}}\left[\frac{f(\mathbf{w}+\mu\bar{\mathbf{u}})}{\mu}\bar{\mathbf{u}}\right]\right)$$

$$=\mathbb{E}_{\bar{\mathbf{u}}}\left[\frac{f(\mathbf{w}+\mu\bar{\mathbf{u}})}{\mu}\bar{\mathbf{u}}\right]$$

D Proof for Lemma 3

Proof.

$$\left\|\nabla f_{\mu}(\mathbf{w})-\nabla f(\mathbf{w})\right\| \tag{33}$$

$$=\left\|\frac{1}{\kappa}\int_{\mathbf{u}\uparrow}\left(\frac{f(\mathbf{w}+\mu\mathbf{u})-f(\mathbf{w})}{\mu}-\langle\nabla f(\mathbf{w}),\mathbf{u}\rangle\right)\mathbf{u}\,e^{-\frac{1}{2}\|\mathbf{u}\|^{2}}d\mathbf{u}\right\| \tag{34}$$

$$\leq\frac{1}{\kappa\mu}\int_{\mathbf{u}\uparrow}|f(\mathbf{w}+\mu\mathbf{u})-f(\mathbf{w})-\mu\langle\nabla f(\mathbf{w}),\mathbf{u}\rangle|\|\mathbf{u}\|\,e^{-\frac{1}{2}\|\mathbf{u}\|^{2}}d\mathbf{u} \tag{35}$$

$$\overset{(2)}{\leq}\frac{\mu L(f)}{2\kappa}\int_{\mathbf{u}\uparrow}\|\mathbf{u}\|^{3}\,e^{-\frac{1}{2}\|\mathbf{u}\|^{2}}d\mathbf{u}$$

$$=\frac{\mu L(f)}{2}\mathbb{E}_{\bar{\mathbf{u}}}\left[\|\bar{\mathbf{u}}\|^{3}\right]$$

$$\overset{(8)}{\leq}\frac{\mu L(f)}{2}(\bar{n}+3)^{3/2} \tag{36}$$

In the first equality (34), we used $\nabla f(\mathbf{w})\overset{(57)}{=}\mathbb{E}_{\bar{\mathbf{u}}}\left[\langle\nabla f(\mathbf{w}),\bar{\mathbf{u}}\rangle\bar{\mathbf{u}}\right]$ $=\frac{1}{\kappa}\int_{\mathbf{u}\uparrow}\langle\nabla f(\mathbf{w}),\mathbf{u}\rangle\mathbf{u}\,e^{-\frac{1}{2}\|\mathbf{u}\|^{2}}d\mathbf{u}$ (see Lemma 7). The first inequality (35) follows because $\left\|\int f(x)dx\right\|\leq\int\|f(x)\|\,dx$ (the triangle inequality for integrals).

Setting $\mathbf{a}\leftarrow\nabla f(\mathbf{w})$ and $\mathbf{b}\leftarrow\nabla f_{\mu}(\mathbf{w})$ in Lemma 4, we get

$$\left\|\nabla f(\mathbf{w})\right\|^{2}\overset{(53)}{\leq}2\left\|\nabla f(\mathbf{w})-\nabla f_{\mu}(\mathbf{w})\right\|^{2}+2\left\|\nabla f_{\mu}(\mathbf{w})\right\|^{2}$$

$$\overset{(36)}{\leq}\frac{\mu^{2}L^{2}(f)}{2}(\bar{n}+3)^{3}+2\left\|\nabla f_{\mu}(\mathbf{w})\right\|^{2}$$

$$\leq\frac{\mu^{2}L^{2}(f)}{2}(\bar{n}+4)^{3}+2\left\|\nabla f_{\mu}(\mathbf{w})\right\|^{2}$$

E   Proof for Theorem 1

Proof.

$$
\begin{aligned}
f_\mu(\mathbf{w}) - f(\mathbf{w}) &= \mathbb{E}_{\bar{\mathbf{u}}}\left[f(\mathbf{w} + \mu\bar{\mathbf{u}})\right] - f(\mathbf{w}) \\
&= \mathbb{E}_{\bar{\mathbf{u}}}\left[f(\mathbf{w} + \mu\bar{\mathbf{u}}) - f(\mathbf{w}) - \mu\langle\nabla f(\mathbf{w}), \bar{\mathbf{u}}\rangle\right] \qquad (37) \\
&= \frac{1}{\kappa}\int_{\mathbf{u}\upharpoonright}\left[f(\mathbf{w} + \mu\mathbf{u}) - f(\mathbf{w}) - \mu\langle\nabla f(\mathbf{w}), \mathbf{u}\rangle\right]e^{-\frac{1}{2}\|\mathbf{u}\|^2}d\mathbf{u} \\
&\overset{(2)}{\leq} \frac{1}{\kappa}\int_{\mathbf{u}\upharpoonright}\frac{\mu^2 L(f)}{2}\|\mathbf{u}\|^2 e^{-\frac{1}{2}\|\mathbf{u}\|^2}d\mathbf{u} \\
&= \frac{\mu^2 L(f)}{2}\mathbb{E}_{\bar{\mathbf{u}}}\left[\|\bar{\mathbf{u}}\|^2\right] \\
&\overset{(7)}{\leq} \frac{\mu^2 L(f)}{2}\bar{n} \qquad (38)
\end{aligned}
$$

The second equality (37) follows because $\mathbb{E}_{\bar{\mathbf{u}}}[\bar{\mathbf{u}}] = \mathbf{0}$. Moreover, (38) doesn't change even if we shift $\mathbf{w}$ by $\mu\bar{\mathbf{u}}$, therefore we have

$$
\begin{aligned}
&\left[(f_\mu(\mathbf{w}) - f(\mathbf{w})) - (f_\mu(\mathbf{w} + \mu\mathbf{u}) - f(\mathbf{w} + \mu\mathbf{u}))\right]^2 \qquad (39) \\
&\overset{(54)}{\leq} 2\left[f_\mu(\mathbf{w}) - f(\mathbf{w})\right]^2 + 2\left[f_\mu(\mathbf{w} + \mu\mathbf{u}) - f(\mathbf{w} + \mu\mathbf{u})\right]^2 \\
&\overset{(38)}{\leq} \frac{\mu^4 L^2(f)}{2}\bar{n}^2 + \frac{\mu^4 L^2(f)}{2}\bar{n}^2 \\
&= \mu^4 L^2(f)\bar{n}^2 \qquad (40)
\end{aligned}
$$

Setting $\mathbf{a} \leftarrow f_\mu(\mathbf{w} + \mu\mathbf{u}) - f_\mu(\mathbf{w})$ and $\mathbf{b} \leftarrow \mu\langle\nabla f_\mu(\mathbf{w}), \mathbf{u}\rangle$ in Lemma 4, we get

$$
\begin{aligned}
&\left[f_\mu(\mathbf{w} + \mu\mathbf{u}) - f_\mu(\mathbf{w})\right]^2 \qquad (41) \\
&\overset{(53)}{\leq} 2\left[f_\mu(\mathbf{w} + \mu\mathbf{u}) - f_\mu(\mathbf{w}) - \mu\langle\nabla f_\mu(\mathbf{w}), \mathbf{u}\rangle\right]^2 + 2\left[\mu\langle\nabla f_\mu(\mathbf{w}), \mathbf{u}\rangle\right]^2 \\
&\overset{(2)}{\leq} \frac{\mu^4 L^2(f_\mu)}{2}\|\mathbf{u}\|^4 + 2\mu^2\langle\nabla f_\mu(\mathbf{w}), \mathbf{u}\rangle^2 \\
&\leq \frac{\mu^4 L^2(f)}{2}\|\mathbf{u}\|^4 + 2\mu^2\|\nabla f_\mu(\mathbf{w})\|^2\|\mathbf{u}\|^2 \qquad (42)
\end{aligned}
$$

The last inequality follows because $L(f_\mu) < L(f)$ and the Cauchy-Schwarz inequality $\langle\mathbf{a}, \mathbf{b}\rangle^2 \leq \|\mathbf{a}\|^2\|\mathbf{b}\|^2$.

Again, setting $\mathbf{a} \leftarrow f(\mathbf{w} + \mu\mathbf{w}) - f(\mathbf{w})$ and $\mathbf{b} \leftarrow f_\mu(\mathbf{w} + \mu\mathbf{w}) - f_\mu(\mathbf{w})$ in Lemma 4, we obtain

$$
\begin{aligned}
&\left[f(\mathbf{w} + \mu\mathbf{u}) - f(\mathbf{w})\right]^2 \qquad (43) \\
&\overset{(53)}{\leq} 2\left[(f_\mu(\mathbf{w}) - f(\mathbf{w})) - (f_\mu(\mathbf{w} + \mu\mathbf{u}) - f(\mathbf{w} + \mu\mathbf{u}))\right]^2 + 22\left[f_\mu(\mathbf{w} + \mu\mathbf{u}) - f_\mu(\mathbf{w})\right]^2 \\
&\overset{(40),(41)}{\leq} 2\mu^4 L^2(f)\bar{n}^2 + \mu^4 L^2(f)\|\mathbf{u}\|^4 + 4\mu^2\|\nabla f_\mu(\mathbf{w})\|^2\|\mathbf{u}\|^2 \qquad (44)
\end{aligned}
$$

Now, we evaluate the expectation of $\left\|g_\mu(\mathbf{w})\right\|^2$ wrt. $\mathbf{x}$ and $\bar{\mathbf{u}}$.

$$\mathbb{E}_{\bar{\mathbf{u}},\mathbf{x}}\left[\left\|g_\mu(\mathbf{w})\right\|^2\right] \overset{(12)}{=} \mathbb{E}_{\bar{\mathbf{u}}}\left[\left\|\frac{f(\mathbf{w}+\mu\bar{\mathbf{u}})-f(\mathbf{w})}{\mu}\bar{\mathbf{u}}\right\|^2\right]$$

$$= \mathbb{E}_{\bar{\mathbf{u}}}\left[\frac{1}{\mu^2}\left[f(\mathbf{w}+\mu\bar{\mathbf{u}})-f(\mathbf{w})\right]^2\cdot\left\|\bar{\mathbf{u}}\right\|^2\right] \tag{45}$$

$$\overset{(44)}{\leq} \mathbb{E}_{\bar{\mathbf{u}}}\left[2\mu^2 L^2(f)\bar{n}^2\left\|\bar{\mathbf{u}}\right\|^2 + \mu^2 L^2(f)\left\|\bar{\mathbf{u}}\right\|^6 + 4\left\|\nabla f_\mu(\mathbf{w})\right\|^2\left\|\bar{\mathbf{u}}\right\|^4\right]$$

$$\overset{(8)}{\leq} 2\mu^2 L^2(f)\bar{n}^3 + \mu^2 L^2(f)(\bar{n}+6)^3 + 4(\bar{n}+4)^2\left\|\nabla f_\mu(\mathbf{w})\right\|^2$$

$$\leq 3\mu^2 L^2(f)(\bar{n}+4)^3 + 4(\bar{n}+4)^2\left\|\nabla f_\mu(\mathbf{w})\right\|^2 \tag{46}$$

The last inequality follows because $2\bar{n}^3 + (\bar{n}+6)^3 \leq 3(\bar{n}+4)^3$.

From the Lipschitz-smoothness assumption (2) of f_μ, we have

$$f_\mu(\mathbf{w}^{(t+1)}) \overset{(2)}{\leq} f_\mu(\mathbf{w}^{(t)}) + \langle\nabla f_\mu(\mathbf{w}^{(t)}), \mathbf{w}^{(t+1)} - \mathbf{w}^{(t)}\rangle + \frac{L(f_\mu)}{2}\left\|\mathbf{w}^{(t)} - \mathbf{w}^{(t+1)}\right\|^2$$

$$= f_\mu(\mathbf{w}^{(t)}) - h^{(t)}\langle\nabla f_\mu(\mathbf{w}^{(t)}), g_\mu(\mathbf{w}^{(t)})\rangle + \frac{(h^{(t)})^2 L(f_\mu)}{2}\left\|g_\mu(\mathbf{w}^{(t)})\right\|^2 \tag{47}$$

In (47), we used the update rule $\mathbf{w}^{(t+1)} = \mathbf{w}^{(t)} - h^{(t)}g_\mu(\mathbf{w}^{(t)})$. Taking the expectation wrt. $\mathbf{x}^{(t)}$ and $\bar{\mathbf{u}}^{(t)}$, we have

$$\mathbb{E}_{\bar{\mathbf{u}},\mathbf{x}}\left[f_\mu(\mathbf{w}^{(t+1)})\right]$$

$$\leq \mathbb{E}_{\bar{\mathbf{u}},\mathbf{x}}\left[f_\mu(\mathbf{w}^{(t)})\right] - h^{(t)}\mathbb{E}_{\bar{\mathbf{u}},\mathbf{x}}\left[\left\|\nabla f_\mu(\mathbf{w}^{(t)})\right\|^2\right] + \frac{(h^{(t)})^2 L(f_\mu)}{2}\mathbb{E}_{\bar{\mathbf{u}},\mathbf{x}}\left[\left\|g_\mu(\mathbf{w}^{(t)})\right\|^2\right]$$

$$\overset{(46)}{\leq} \mathbb{E}_{\bar{\mathbf{u}},\mathbf{x}}\left[f_\mu(\mathbf{w}^{(t)})\right] - h^{(t)}\mathbb{E}_{\bar{\mathbf{u}},\mathbf{x}}\left[\left\|\nabla f_\mu(\mathbf{w}^{(t)})\right\|^2\right]$$

$$+ \frac{(h^{(t)})^2 L(f)}{2}\left(4(\bar{n}^{(t)}+4)\mathbb{E}_{\bar{\mathbf{u}},\mathbf{x}}\left[\left\|\nabla f_\mu(\mathbf{w}^{(t)})\right\|^2\right] + 3\mu^2 L^2(f)(\bar{n}^{(t)}+4)^3\right)$$

Choosing $h^{(t)} = \frac{1}{4(\bar{n}^{(t)}+4)L(f)}$, we have

$$\mathbb{E}_{\bar{\mathbf{u}},\mathbf{x}}\left[f_\mu(\mathbf{w}^{(t+1)})\right] \leq \mathbb{E}_{\bar{\mathbf{u}},\mathbf{x}}\left[f_\mu(\mathbf{w}^{(t)})\right] - \frac{1}{8(\bar{n}^{(t)}+4)L(f)}\mathbb{E}_{\bar{\mathbf{u}},\mathbf{x}}\left[\left\|\nabla f_\mu(\mathbf{w}^{(t)})\right\|^2\right] \tag{48}$$

$$+ \frac{3\mu^2}{32}L(f)(\bar{n}^{(t)}+4) \tag{49}$$

Recursively applying the inequalities above moving the index from $T + 1$ to 0, rearranging the terms and noting that $f_\mu^\star \leq f_\mu(w^{(t)})$ and $\hat{h} = \frac{1}{4(\hat{n}+4)L}$ where $L(f) \leq L$ for all $f(\mathbf{w}^{(t)})$ results in

$$\mathbb{E}_{\bar{u},x}\left[\left\|\nabla f_\mu(\mathbf{w}^{(T)})\right\|^2\right] \leq 8(\hat{n}+4)L\left[\frac{f_\mu(\mathbf{w}^{(0)}) - f_\mu^\star}{T+1} + \frac{3\mu^2}{32}L(\hat{n}+4)\right] \quad (50)$$

Thus, we obtain

$$\mathbb{E}_{\bar{u},x}\left[\left\|\nabla f(\mathbf{w}^{(T)})\right\|^2\right] \overset{(13)}{\leq} \frac{\mu^2 L^2}{2}(\hat{n}+4)^3 + 2\mathbb{E}_{\bar{u},x}\left[\left\|\nabla f_\mu(\mathbf{w}^{(T)})\right\|^2\right]$$

$$\overset{(52)}{\leq} 16(\hat{n}+4)L\frac{f_\mu(\mathbf{w}^{(0)}) - f_\mu^\star}{T+1} + \frac{\mu^2 L^2}{2}(\hat{n}+4)^2\left(\hat{n} + \frac{11}{2}\right) \quad (51)$$

Now we lower-bound the expected number of iteration. In order to get ϵ-accurate solution $\mathbb{E}_{\bar{u},x}\left[\left\|\nabla f(\mathbf{w}^{(T)})\right\|^2\right] \leq \epsilon^2$, we need to choose

$$\mu = \Omega\left(\frac{\epsilon}{\hat{n}^{3/2}L}\right)\cdots \quad (52)$$

so that the second term in the right hand side of (51) vanishes wrt. ϵ^2.

$$16(\hat{n}+4)L\frac{f_\mu(\mathbf{w}^{(0)}) - f_\mu^\star}{T+1} + \mathcal{O}(\mu^2 L^2\hat{n}^3) \overset{(52)}{=} 16(\hat{n}+4)L\frac{f_\mu(\mathbf{w}^{(0)}) - f_\mu^\star}{T+1} + \mathcal{O}(\epsilon^2)$$

$$\overset{!}{=} \mathcal{O}(\epsilon^2)$$

$$T = \mathcal{O}\left(\frac{\hat{n}L}{\epsilon^2}\right)$$

F   Miscellaneous

Lemma 4. *For any* $\mathbf{a}, \mathbf{b} \in \mathbb{R}^n$, *we have*

$$\|\mathbf{a}\|^2 \leq 2\|\mathbf{a} - \mathbf{b}\|^2 + 2\|\mathbf{b}\|^2 \quad (53)$$

Proof. For any $\mathbf{x}, \mathbf{y} \in \mathbb{R}^n$, it holds

$$\|\mathbf{x}+\mathbf{y}\|^2 + \|\mathbf{x}-\mathbf{y}\|^2 = \|\mathbf{x}\|^2 + 2\langle\mathbf{x},\mathbf{y}\rangle + \|\mathbf{y}\|^2 + \|\mathbf{x}\|^2 - 2\langle\mathbf{x},\mathbf{y}\rangle + \|\mathbf{y}\|^2$$

$$= 2\|\mathbf{x}\|^2 + 2\|\mathbf{y}\|^2.$$

Dropping either $\|\mathbf{x}+\mathbf{y}\|^2$ or $\|\mathbf{x}-\mathbf{y}\|^2$ on the left hand side, we get

$$\|\mathbf{x}\pm\mathbf{y}\|^2 \leq 2\|\mathbf{x}\|^2 + 2\|\mathbf{y}\|^2 \quad (54)$$

Substitute $\mathbf{x} \leftarrow \mathbf{a} - \mathbf{b}$ and $\mathbf{y} \leftarrow \mathbf{b}$ in $\|\mathbf{x}+\mathbf{y}\|^2 \leq 2\|\mathbf{x}\|^2 + 2\|\mathbf{y}\|^2$. Then we get

$$\|\mathbf{a}\|^2 \leq 2\|\mathbf{a} - \mathbf{b}\|^2 + 2\|\mathbf{b}\|^2.$$

Lemma 5. *Let be $t > 0$, $p \geq 2$, and $\tau \in (0,1)$. Then we have*

$$t^p e^{-\frac{\tau}{2}t^2} \leq \left(\frac{p}{\tau e}\right)^{\frac{p}{2}}. \tag{55}$$

Proof. Denote $\psi(t) := t^p e^{-\frac{\tau}{2}t^2}$ for some fixed $\tau \in (0,1)$. Find the point t s.t. $\psi'(t) = 0$.

$$\psi'(t) = pt^{p-1}e^{-\frac{\tau}{2}t^2} + t^p\left(-\tau t e^{-\frac{\tau}{2}t^2}\right) \overset{!}{=} 0$$

$$t = \sqrt{p/\tau} \qquad \text{since } t > 0, \tau \in (0,1)$$

$$\psi(\sqrt{p/\tau}) = (p/\tau)^{\frac{p}{2}} e^{-p/2}$$

$$= \left(\frac{p}{\tau e}\right)^{\frac{p}{2}}$$

Now we conduct the second derivative test.

$$\psi''(t) = \left(\tau^2 t^2 e^{-\frac{\tau}{2}t^2} - \tau e^{-\frac{\tau}{2}t^2}\right) t^p + (p-1)pe^{-\frac{\tau}{2}t^2} t^{p-2} - 2\tau p e^{-\frac{\tau}{2}t^2} t^p$$

$$= e^{-\frac{\tau}{2}t^2} \cdot t^{p-2} \left(\tau^2 t^4 - 2\tau p t^2 - \tau t^2 + p^2 - p\right)$$

$$\psi''(\sqrt{p/\tau}) = e^{-\frac{1}{2}\not{\frac{p}{\tau}}} \cdot \sqrt{\frac{p}{\tau}}^{\,p-2} \left(\not{\tau^2} \left(\frac{p}{\not{\tau}}\right)^2 - 2\not{\tau} p\frac{p}{\not{\tau}} - \not{\tau}\frac{p}{\not{\tau}} + p^2 - p\right)$$

$$= \underbrace{e^{-\frac{p}{2}}}_{>0} \underbrace{\left(\frac{p}{\tau}\right)^{\frac{p-2}{2}}}_{>0} \underbrace{\left(\not{p^2} - 2p^2 - p + \not{p^2} - p\right)}_{<0} < 0 \qquad \text{since } p \geq 2$$

Hence, we conclude that we have found a local maximum at $t = \sqrt{p/\tau}$, therefore $t^p e^{-\frac{\tau}{2}t^2} \leq \left(\frac{p}{\tau e}\right)^{\frac{p}{2}}$.

Lemma 6. *Let be $t > 0$, $p \geq 2$, $\tau \in (0,1)$, and $n > 0$. Then we have*

$$\left(\frac{p}{\tau e}\right)^{\frac{p}{2}} (1-\tau)^{-\frac{n}{2}} \leq (p+n)^{\frac{p}{2}} \tag{56}$$

Proof. Denote $\phi(\tau) := \left(\frac{p}{\tau e}\right)^{\frac{p}{2}} (1-\tau)^{-\frac{n}{2}}$ for some fixed $\tau \in (0,1)$. Find the point τ s.t. $\phi'(\tau) = 0$.

$$\phi'(\tau) = \frac{e^{-\frac{p}{2}}(1-\tau)^{-\frac{n}{2}-1}\left(\frac{p}{\tau}\right)^{\frac{p}{2}}(n\tau + p(\tau-1))}{2\tau} \overset{!}{=} 0$$

$$\tau = \frac{p}{n+p} \qquad \text{since } n+p \neq 0, np \neq 0$$

Let us check the sign of the second derivative of $\phi(\frac{p}{n+p})$.

$$\phi''(\tau) = \frac{1}{4(\tau-1)^2\tau^2} e^{-\frac{p}{2}} \left(\frac{p}{\tau}\right)^{\frac{p}{2}} (1-\tau)^{-\frac{n}{2}} \left(\tau^2((p+n)^2 + 2(p+n))\right.$$

$$\left. -2p\tau(p+n) - 4p\tau + 2p\right)$$

$$\phi''(\frac{p}{n+p}) = \underbrace{\frac{(n+p)^4}{4n^2p^2}}_{>0} \underbrace{e^{-\frac{p}{2}}}_{>0} \underbrace{(n+p)^{\frac{p}{2}}}_{>0} \underbrace{\left(\frac{n+p}{n}\right)^{\frac{n}{2}}}_{>0} \underbrace{\left(-p^2\left(1+\frac{2}{p+n}\right) + 2p\right)}_{<0} < 0$$

since $p \geq 2, n > 0$. Therefore, $\phi(\tau)$ takes a local maximum at $\tau = \frac{p}{n+p}$. Then we obtain

$$\phi(\frac{p}{n+p}) = e^{-\frac{p}{2}} (p+n)^{\frac{p}{2}} \left(\frac{n+p-p}{n+p}\right)^{-\frac{n}{2}} \leq (p+n)^{\frac{p}{2}}.$$

The last inequality follows because

$$e^{-\frac{p}{2}} \left(\frac{n}{n+p}\right)^{-\frac{n}{2}} \leq 1.$$

Lemma 7.

$$\nabla f(\mathbf{w}) = \mathbb{E}_{\mathbf{u}}\left[\langle \nabla f(\mathbf{w}), \mathbf{u}\rangle \mathbf{u}\right] \tag{57}$$

Proof. Let us denote the gradients $\nabla f(\mathbf{w})$ wrt. $\mathbf{w}$ as a column vector of derivatives of each component in $\mathbf{w}$.

$$\nabla f(\mathbf{w}) = \left[\frac{\partial f}{\partial w_1}, \frac{\partial f}{\partial w_2}, \cdots \frac{\partial f}{\partial w_n}\right]^{\mathsf{T}}$$

By definition of the scalar product, we have

$$\langle \nabla f(\mathbf{w}), \mathbf{u}\rangle = \sum_{j=1}^{n} \frac{\partial f}{\partial w_j} u_j \tag{58}$$

Then we get

$$\mathbb{E}_{\mathbf{u}}\left[\langle\nabla f(\mathbf{w}),\mathbf{u}\rangle\mathbf{u}\right] \stackrel{(58)}{=} \mathbb{E}_{\mathbf{u}}\left[\left(\sum_{j=1}^{n}\frac{\partial f}{\partial w_j}u_j\right)\mathbf{u}\right]$$

$$= \sum_{j=1}^{n}\mathbb{E}_{\mathbf{u}}\left[\left(\frac{\partial f}{\partial w_j}u_j\right)\mathbf{u}\right] \tag{59}$$

$$= \sum_{j=1}^{n}\frac{\partial f}{\partial w_j}\mathbb{E}_{\mathbf{u}}\left[u_j\cdot\mathbf{u}\right]$$

$$= \sum_{j=1}^{n}\frac{\partial f}{\partial w_j}\begin{bmatrix}\mathbb{E}\left[u_j\cdot u_1\right]\\ \mathbb{E}\left[u_j\cdot u_2\right]\\ \vdots\\ \mathbb{E}\left[u_j\cdot u_j\right]\\ \vdots\\ \mathbb{E}\left[u_j\cdot u_n\right]\end{bmatrix}$$

$$= \sum_{j=1}^{n}\frac{\partial f}{\partial w_j}\begin{bmatrix}\mathbb{E}\left[u_j\right]\cdot\mathbb{E}\left[u_1\right]\\ \mathbb{E}\left[u_j\right]\cdot\mathbb{E}\left[u_2\right]\\ \vdots\\ \mathbb{E}\left[u_j^2\right]\\ \vdots\\ \mathbb{E}\left[u_j\right]\cdot\mathbb{E}\left[u_n\right]\end{bmatrix} \tag{60}$$

We used the linearity of expectation in (59). The last equality (60) follows because each component u is independent. Since u_j was drawn from $\mathcal{N}(0,1)$, we know $\mathbb{E}[u_j]=0$, and $\mathbb{E}[(u_j-0)^2]=1$. Then we obtain

$$\mathbb{E}_{\mathbf{u}}\left[\langle\nabla f(\mathbf{w}),\mathbf{u}\rangle\mathbf{u}\right] \stackrel{(60)}{=} \sum_{j=1}^{n}\frac{\partial f}{\partial w_j}\begin{bmatrix}0\\ 0\\ \vdots\\ 1\\ \vdots\\ 0\end{bmatrix} = \begin{bmatrix}\frac{\partial f}{\partial w_1}\\ \frac{\partial f}{\partial w_2}\\ \vdots\\ \frac{\partial f}{\partial w_j}\\ \vdots\\ \frac{\partial f}{\partial w_n}\end{bmatrix} = \nabla f(\mathbf{w}) \tag{61}$$

G   Empirical Sparsity

Figures 7 and 8 illustrate our observation of empirical gradient sparsity across the entire learning process on both MNIST and CIFAR classification tasks. The histograms on the right side show that the majority of gradient coordinates have value zero throughout the learning process, while the values of the learned weight vector are centered around non-zero values, except obviously for pruning approaches.

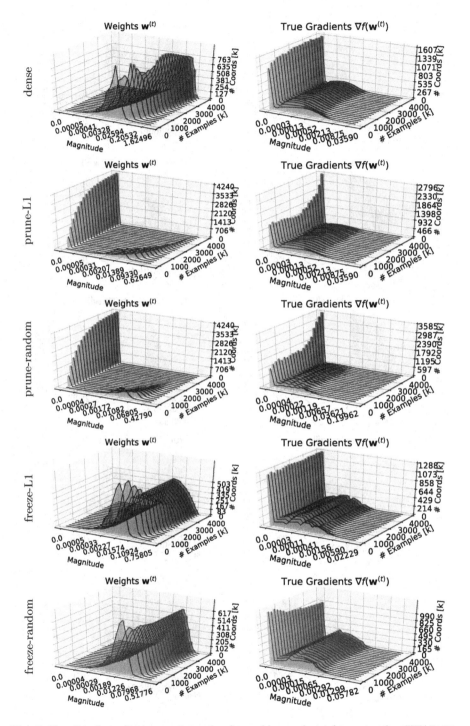

Fig. 7. Empirical gradient sparsity and values of learned weight vector for CIFAR-10.

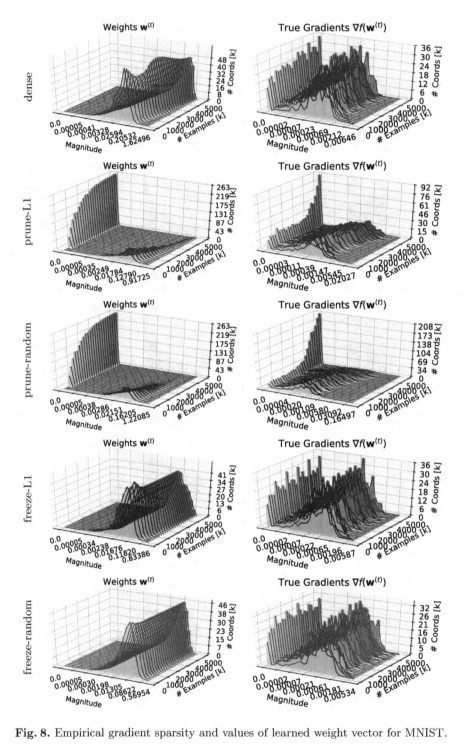

Fig. 8. Empirical gradient sparsity and values of learned weight vector for MNIST.

References

1. Agarwal, A., Dekel, O., Xiao, L.: Optimal algorithms for online convex optimization with multi-point bandit feedback. In: COLT (2010)
2. Balasubramanian, K., Ghadimi, S.: Zeroth-order nonconvex stochastic optimization: handling constraints, high-dimensionality and saddle-points. CoRR abs/1809.06474 (2018)
3. Chen, P.Y., Zhang, H., Sharma, Y., Yi, J., Hsieh, C.J.: Zoo: zeroth order optimization based black-box attacks to deep neural networks without training substitute models. In: AISec (2017)
4. Cheng, S., Dong, Y., Pang, T., Su, H., Zhu, J.: Improving black-box adversarial attacks with a transfer-based prior. In: NeurIPS (2019)
5. Duchi, J.C., Jordan, M.I., Wainwright, M.J., Wibisono, A.: Optimal rates for zero-order convex optimization: the power of two function evaluations. IEEE Trans. Inf. Theory **61**(5), 2788–2806 (2015)
6. Ebrahimi, S., Rohrbach, A., Darrell, T.: Gradient-free policy architecture search and adaptation. In: Proceedings of the Conference on Robot Learning (CoRL). Mountain View, CA, USA (2017)
7. Flaxman, A.D., Kalai, A.T., McMahan, H.B.: Online convex optimization in the bandit setting: gradient descent without a gradient. In: SODA (2005)
8. Frankle, J., Carbin, M.: The lottery ticket hypothesis: finding sparse, trainable neural networks. In: ICLR (2019)
9. Frankle, J., Dziugaite, G.K., Roy, D.M., Carbin, M.: Stabilizing the lottery ticket hypothesis. CoRR abs/1903.01611 (2019)
10. Fu, M.C.: Gradient estimation. In: Henderson, S., Nelson, B. (eds.) Handbook in Operations Research and Management Science, vol. 13, pp. 575–616. Elsevier (2006)
11. Ghadimi, S., Lan, G.: Stochastic first- and zeroth-order methods for nonconvex stochastic programming. SIAM J. Optim. **4**(23), 2342–2368 (2012)
12. Glorot, X., Bengio, Y.: Understanding the difficulty of training deep feedforward neural networks. PMLR **9**, 249–256 (2010)
13. Han, S., Pool, J., Tran, J., Dally, W.: Learning both weights and connections for efficient neural network. In: NIPS (2015)
14. Ioffe, S., Szegedy, C.: Batch normalization: accelerating deep network training by reducing internal covariate shift. In: ICML, pp. 448–456 (2015)
15. Keskar, N.S., Mudigere, D., Nocedal, J., Smelyanskiy, M., Tang, P.T.P.: On large-batch training for deep learning: generalization gap and sharp minima. In: ICLR (2017)
16. Kiefer, J., Wolfowitz, J.: Stochastic estimation of the maximum of a regression function. Ann. Math. Stat. **23**(3), 462–466 (1952)
17. Krizhevsky, A., Hinton, G.: Learning multiple layers of features from tiny images. Master's thesis, Department of Computer Science, University of Tront (2009)
18. Kushner, H.J., Yin, G.G.: Stochastic Approximation and Recursive Algorithms and Applications, 2nd edn. Springer, Boston (2003). https://doi.org/10.1007/978-1-4614-3223-4
19. LeCun, Y., Bottou, L., Bengio, Y., Haffner, P.: Gradient-based learning applied to document recognition. Proc. IEEE **86**(11), 2278–2324 (1998)
20. Liu, S., Kailkhura, B., Chen, P.Y., Ting, P., Chang, S., Amini, L.: Zeroth-order stochastic variance reduction for nonconvex optimization. In: Advances in Neural Information Processing Systems 31. Montreal, Canada (2018)

21. Mania, H., Guy, A., Recht, B.: Simple random search provides a competitive approach to reinforcement learning. In: NIPS (2018)
22. Nesterov, Y., Spokoiny, V.: Random gradient-free minimization of convex functions. Found. Comput. Math. **17**, 527–566 (2015)
23. Plappert, M., et al.: Parameter space noise for exploration. In: ICLR (2018)
24. Salimans, T., Ho, J., Chen, X., Sutskever, I.: Evolution strategies as a scalable alternative to reinforcement learning. CoRR abs/1703.03864 (2017)
25. Sehnke, F., Osendorfer, C., Rückstieß, T., Graves, A., Peters, J., Schmidhuber, J.: Parameter-exploring policy gradients. Neural Networks **23**(4), 551–559 (2010)
26. Shamir, O.: An optimal algorithm for bandit and zero-order convex optimization with two-point feedback. JMLR **18**, 1–11 (2017)
27. Sokolov, A., Hitschler, J., Riezler, S.: Sparse stochastic zeroth-order optimization with an application to bandit structured prediction. CoRR abs/1806.04458 (2018)
28. Spall, J.C.: Multivariate stochastic approximation using a simultaneous perturbation gradient approximation. IEEE Trans. Autom. Control **37**(3), 332–341 (1992)
29. Spall, J.C.: Introduction to Stochastic Search and Optimization: Estimation, Simulation, and Control. Wiley, Hoboken (2003)
30. Wang, Y., Du, S., Balakrishnan, S., Singh, A.: Stochastic zeroth-order optimization in high dimensions. In: AISTATS (2018)
31. Yue, Y., Joachims, T.: Interactively optimizing information retrieval systems as a dueling bandits problem. In: ICML (2009)

Learning Controllers for Adaptive Spreading of Carbon Fiber Tows

Julia Krützmann[1]([⊠])⬤, Alexander Schiendorfer[1]⬤, Sergej Beratz[2],
Judith Moosburger-Will[2], Wolfgang Reif[1]⬤, and Siegfried Horn[2]

[1] Institute for Software and Systems Engineering, University of Augsburg,
Augsburg, Germany
{julia.kruetzmann,alexander.schiendorfer,reif}@informatik.uni-augsburg.de
[2] Institute of Materials Resource Management, University of Augsburg,
Augsburg, Germany
{sergej.beratz,judith.moosburger,horn}@mrm.uni-augsburg.de

Abstract. Carbon fiber reinforced polymers (CFRP) are lightweight but strong composite materials designed to reduce the weight of aerospace or automotive components – contributing to reduced greenhouse gas emissions. A common manufacturing process for carbon fiber tapes consists of aligning tows (bundles of carbon fiber filaments) side by side to form tapes via a spreading machine. Tows are pulled across metallic spreading bars that are conventionally kept in a fixed position. That can lead to high variations in quality metrics such as tape width or height. Alternatively, one could try to control the spreading bars based on the incoming tows' profiles. We investigate whether a machine learning approach, consisting of a supervised process model trained on real data and a process control model to choose adequate spreading bar positions is able to improve the tape quality variations. Our results indicate promising tendencies for adaptive tow spreading.

1 Introduction to Spreading of Carbon Fiber Tows

Carbon fiber reinforced polymers (CFRP) are lightweight composite materials with extraordinary structural integrity. That makes them attractive for the construction of lighter aerospace and automotive parts (conventionally made from steel or aluminum) to reduce fuel consumption and $CO2$ emissions [5].

These composites are made from a polymer matrix that is reinforced with textiles containing carbon fibers, often in the form of long tapes (e.g., featuring a width between 25 and 50 mm) that are eventually layered to form a textile. Such a carbon fiber textile is composed of carbon fiber filaments that are grouped to form so-called *tows*. To align the tows side-by-side to form tapes, a spreading machine is commonly applied [9]: By pulling the tows across a number of metallic spreading bars at a constant velocity, overlapping tows are "entangled" and aligned properly, as Fig. 1 illustrates.

Conventionally, the positioning of the spreading bars is fixed for all profiles of the tows, leading to varying quality in the width and height of the outgoing

© Springer Nature Switzerland AG 2020
G. Nicosia et al. (Eds.): LOD 2020, LNCS 12566, pp. 65–77, 2020.
https://doi.org/10.1007/978-3-030-64580-9_6

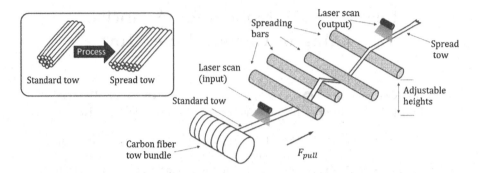

Fig. 1. Overview of tow spreading in the course of composite tape production.

tapes [4]. Especially for aerospace applications, only highest quality tows are suitable for the production of components resulting in a substantial amount of rejected parts. In this paper, *we investigate whether it is beneficial to adapt the spreading bar positions to the initial profiles* which could lead to a more efficient production of tapes. To do so, we first train a process model representing the forward behavior of the real system in a supervised fashion (cf. Fig. 2). It is based on experimental data obtained from a real spreading machine and thus alleviates the need to define an abstract mathematical model of the process, including numerically determining the process parameters. This model takes tow height profiles from a laser scanner as well as the spreading bar positions as inputs and predicts the resulting tow height profiles as output. We then apply Neuroevolution [3] to optimize for a process control model predicting the most suitable bar positions, given the scanner reading of the input tows and target tape widths (see Fig. 3). Our results show that, compared to the best (but fixed) spreading bar positioning, on average adaptive spreading can halve the deviations in the desired tape width.

Following a discussion of related work, we present the process model including the data acquisition in Sect. 2, present the Neuroevolution based process control model in Sect. 3, and conclude with experimental results in Sect. 4.

1.1   Related Work

Research related to our approach is divided into two major domains: one, the spreading process and advances in its automation and, two, machine learning techniques in similar industrial settings.

Several publications address different types of the spreading process in general [8,11] as well as the underlying acting forces [6]. Gizik et al. [4] compared the effects of spreading on various types of heavy tows and showed that deviations in width of the resulting tow can vary greatly. Appel et al. [1] aim to develop an active control of the spreading process and, as a preliminary result, have identified numerically the process parameters that have an immediate influence on the quality of the tow. They also generated regression models for different quality parameters that should build the base for a control unit. In contrast to our work, they focus on the mathematical relations of various process parameters – mainly spreading velocity, pre-tension force of the initial tow, and wrapping angle which is directly dependent on the vertical bar position – with the tow quality but do not consider variations in the input tows.

When applying techniques that learn from online experiences, such as Reinforcement Learning (RL), to real-world problems (e.g. industrial processes or in a medical context), sufficiently exploring the action space is often problematic. One common approach to reduce this issue is to pre-train RL models on simulations, before adapting them carefully to the real environment requiring less real data [12]. However, the development of an adequate simulation can be infeasible, e.g., if the effects of behavior or forces in physical processes are not fully known. Several variations of off-policy evaluation were suggested as a way to gain confidence in RL policies by evaluating them on pre-collected databases before applying them online [7,16]. Model-based reinforcement learning [10,15] reduces real-world interactions by using an internal world model to generate samples for planning ahead and evaluating the next steps before applying them. This model is built while performing actions in the environment and continuously updated.

Due to the lack of a feedback loop to perform actions on the real system, our approach can be understood as a combination of some of the core ideas of the RL techniques described. We use pre-collected data to develop an internal model that is used to evaluate the suitability of adaptive control in general and Neuroevolution as optimization strategy in particular. As we consider no sequences of actions but only evaluate single steps, it is more viable to sufficiently cover the action space with experiments in advance than for more complex reinforcement learning tasks.

2   The Process Model – A Supervised System Predictor

Since it is not feasible (not to mention highly inefficient) to randomly adjust spreading bar positions in the real world process and learn which of these lead to high quality tows (i.e., a process control policy), we first aspire to develop the process model that approximates a representation of the input-output behavior based on real data, before using it to train a second model that proposes suitable bar positions (the "process control model"). Fig. 2 shows the process model which is trained, in a supervised fashion, to map input tow height profiles to output tow height profiles based on two laser scanners.

2.1   Data Acquisition

In order to learn how tows are affected by different bar positions, a sufficiently large database is required. The data[1] we use to train the process model was

---

[1] Data available via https://doi.org/10.6084/m9.figshare.12213746.v1 Code is available here: https://github.com/isse-augsburg/adaptive-spreading.

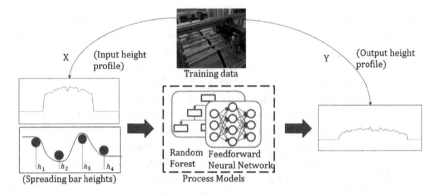

Fig. 2. The process model maps input height profiles and spreading bar heights to output height profiles. The model is trained from a real world tow spreading machine.

recorded using a system slightly different to the one shown in Fig. 1. It is composed of five metallic, non-rotating spreading bars through which a tow is pulled at a constant velocity. The spreading bars are placed horizontally and arranged alternately above and below the tow. A light section sensor is mounted above the first and last spreading bar to measure the height profile of the tape before and after the spreading process, respectively. To obtain a varied database, experiments were conducted with a total of fifteen different bar positions and twelve velocity settings for each bar setup. The height of the bars was altered within a range of 29 cm, while the pulling velocity was set to a value between 2.7 m/min and 15 m/min and the sensors took measurements at a frequency 500 Hz and a resolution of 800 pixels where one pixel covers about 0.1mm. Experiments were recorded over about 60 s each with a fixed bar position and velocity which results in raw data consisting of 687 files (one per 60 s of recording) with 15 million measuring points in total.

Due to the high measurement frequency and the absorption behavior of carbon fiber tows, the data contains substantial noise as well as artifacts. Additionally, the whole system was often adjusted manually to conduct various experiments leading to inaccuracies in the form of considerably differing value ranges of the measurements for different recordings. To facilitate the learning task, we therefore unify the data while maintaining relevant information such as irregularities in the profile. First, the measured profiles are smoothed by applying a Savitzky-Golay-Filter [13] and averaging sequential data points. Second, pixels beside the tow are set to zero as they hold no valuable information - serving as a form of supplementary sensor calibration. Moreover, since the two sensors are only synchronized to measure at the exact same time, the information which measurements belong to the same part of the tape before and after spreading is not immediately available in the data. That mapping is, however, crucial in order to learn the effect of bar positions on the tow, i.e., to obtain temporally aligned target values for the supervised training. Hence, in the last step, we calculate the time offset of the sensors for each experiment taking into account the velocity and the distance covered by the tow which depends directly on the bar positions.

These pre-processing steps lead to a reduced data set size of 650 files holding 800,899 samples. Table 1 shows the split of the data for training our models. When dividing the data, we ensure that sequential measurements of a single experiment are fully contained in the same set, avoiding the occurrence of highly similar data points in training and test set.

Analysis of the pre-processed data indicates that the average width of the tows is 212.9 pixels before spreading and 293.7 pixels after spreading with a standard deviation of 19.6 and 31.7 pixels respectively. Furthermore, there is no significant effect of the velocity on the spreading factor. Therefore, we focus on the measured height profiles and the bar positions as the main characteristics of the setup.

Table 1. Data set sizes – train/val/test-split

Data set	# Files	# Samples	% (Samples)
Training	416	511,727	63.9
Validation	104	138,500	17.3
Test	130	150,662	18.8
All	650	800,889	100

2.2 Models

The goal of the process model is to predict the tow height profile after spreading as accurately as possible based on the initial status, while also being robust against light noise (cf. Fig. 2). The input of the models, therefore, consists of 805 values (800 pixels as read from the laser scanner for tow profiles before spreading and current positions of the five bars), while the output has a dimension of 800. Besides these specifics, it is crucial that the model yields output that can be deemed realistic even for unknown inputs, i.e. bar setups previously not seen and new combinations of profiles and bar settings. For example, the tape predicted is expected to become wider when raising the middle bar. To reach a good solution, as part of this work, we compare two candidates: feedforward neural networks (FFNN) trained with backpropagation and random forests (RF) aggregating multiple multi-target decision tree regressors. Both approaches are generally suitable for the described supervised multivariate regression task, as they are capable of mapping non-linear dependencies while also providing robustness against noise which is essential in industrial applications.

3 A Process Control Model

Based on the process model, it is possible to train a control model that, by experimenting with the bar positions, learns which changes lead to a tow profile closer to the one desired (see Fig. 3). The process model fulfills a function partially comparable to a "world model" which is more commonly used in reinforcement learning. In detail, the workflow is such that the process control model receives a height profile and the current bar setting as input and yields a new setup which is optimized for a fixed target tow width. Based on this prediction of the process control model and the height profile, the process model then generates the resulting tow which, in turn, is used to determine the suitability of the suggested bar setting.

Since for a given input tow, current bar positions, and target tow width and height, the "target bar positions" are unknown (as they are not covered by data), supervised training is not applicable to optimize the process control model. Instead we choose a neuroevolutional approach based on a rather simplistic genetic algorithm. The applied algorithm as well as the terminology used in the following paragraph are based on [14].

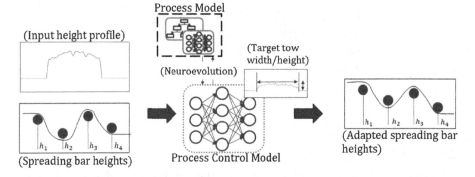

Fig. 3. The process control model that adaptively selects bar heights based on the input height profile, the current bar positions, and a fixed target width and height.

Neuroevolution Approach. For the implementation of the process control model, first, a population of neural networks is initialized with weights sampled from a scaled normal distribution. Second, the fitness of these "genotypes" is determined based on the training set (cf. Table 1) and the subset of the best performing neural networks is selected as parents for the next generation. As a third step, a neural network is randomly selected from the parents and mutated by adding normally distributed noise to its weights. These mutations and the parents that performed best form the population of the next generation. To perform the training of the controller, the second and third steps are repeated – leading to an "evolution" over multiple generations – until the performance of the fittest neural network does not improve on the validation set for multiple epochs (comparable to early stopping in gradient-based training). Finally, the fittest neural network of the last generation is chosen to be the process controller.

To determine the fitness (or reward) r of a single genotype, for each suggested action we generate the resulting tow profile p using the process model and compare its width and height to a designated target t. Additionally, we consider the distance of the height h of each of the b bars before and after the action, in order to favor solutions closer to the original position and, thus, reducing the required bar movement. Formally, r is defined as follows:

$$
r = -\left(k_{height} \cdot |height_t - height_p| + k_{width} \cdot |width_t - width_p| \right.
$$
$$
\left. + k_{distance} \cdot \sum_{i=1}^{b=5} |h_{\{current,i\}} - h_{\{p,i\}}| \right) \tag{1}
$$

where the hyperparameters k_{height}, k_{width} and $k_{distance}$ control the relative importance of each criterion. Each subterm should be minimized which is why r is negated, following the conventions of fitness-maximization in neuroevolution. For the following experiments, we set $k_{height} = 1.0$, $k_{width} = 4.0$ and $k_{distance} = 0.5$, mainly emphasizing the tow width. The accumulated fitness of a neural network is given by averaging the calculated rewards.

To obtain values in $[0, 1]$, indicating the relative bar positions, the output of the neural network is activated by the sigmoid function. Each of these values is transformed into its unique range specified by the data, meaning that the controller never suggests moving a bar below or above its lowest or highest position in the experiments with the real system.

Baseline: Fixed Bar Setup. To assess the benefits of an adaptive process controller in general, we also determine the fixed bar positions that perform best to achieve a given target width. The fixed setup is picked from a set of all the settings used to record the real data. This baseline is comparable to the current standard procedure for spreading carbon fiber tows where a fixed bar position is identified on the basis of experiments and expert knowledge independent of variations in the initial tow.

4 Evaluation

Since the mere error can be hard to interpret when evaluating solutions of regression tasks, we focus on the width of the predicted tows as the main criterion. For this purpose, we need, firstly, to clarify how the profile width is determined and, secondly, how we can derive a meaningful metric for the models. As described in Sect. 2.1, we pre-processed the data such that pixels beside the tape are set to zero. Accordingly, the most naive approach to calculate the tow width is to consider all positive values in a prediction as "tow" and to infer the width. Due to uncertainty in the developed models we see, however, softer edges in the predictions than in the real data, leading to overestimation of the width. As a dynamic threshold, we take the mean value of the predicted profile and consider the first and last pixels above this value as the tow edges. The derived width is set in relation to the width of the corresponding target. In order to factor in overestimations as well as underestimations, we calculate the quality metric m of a predicted profile p as follows:

$$m(p) = 1 + |1 - \frac{width_p}{width_{target}}| \tag{2}$$

Thus, the optimum score is 1, while it increases accordingly in case of deviations. For comparison, we take the geometric mean and the geometric standard deviation (due to m being a relative quantity) of the quality of all predictions on the same test set for each model.

Before applying this metric to evaluate the performance of the process control model, we put to test whether the underlying process model can be considered a sufficiently accurate representation of the real spreading process.

4.1 Process Model

As expected, keeping the dimension of the input data at 805 values is not feasible when training the random forest. Besides the well-known training runtime complexity of $O(n_{samples} \cdot n_{features} \cdot \log(n_{samples}))$, the size and, thus, the memory usage of the underlying decision trees increases immensely. To overcome this, using random subsets of the features, as described by [2], is a very applicable approach. Results from our experiments suggest that a subset size of 200 features is sufficient to achieve good scores, with no improvement in the quality of predictions when further increasing the input dimension. This seems reasonable, given the fact that the average tow width is slightly above 200 pixels, as mentioned in Sect. 2.1. Thus, about 600 values of a profile are mostly set to zero and can easily be dropped - with the caveat that such a model can presumably not be adapted to tows that are significantly wider.

By contrast, since the feedforward neural network does not suffer from runtime or space problems due to high dimensionality, we trained that model with the full input dimension to keep it as versatile as possible.

After optimizing the architectures of the feedforward neural network and the random forest, both approaches achieve similar m-scores (see Fig. 4) – with a slight advantage for the random forest. The average width of tows predicted on the test set matches the ground truth of the measured profiles to within two pixels, with the standard deviation of the predicted width being lower than the real one (cf. Fig. 4b) which can be accounted for by the fact that the models are a generalisation of the real process. In accordance with the geometric mean and standard deviation in Fig. 4a, the quality m_{rf} is in $[1.002, 1.081]$, while m_{ffnn} is in $[1.001, 1.082]$ which means the worst predictions are about 2.37 mm and 2.40 mm off while the mean prediction deviates about 1.18 mm and 1.19 mm for average tow width, respectively for the random forest and the neural network. Both models are equally likely to over- and underestimate the width. It is an open question if the predictions are sufficiently accurate to use the models in the real process, e.g. for quality monitoring, but we presume that they are a sufficiently approximate representation of the spreading behavior to help evaluate whether a control model as described in Sect. 3 can be beneficial.

Not only are the scores of both approaches highly similar but the predicted profiles are remarkably alike for the recorded bar positions (cf. Fig. 5). However, to decide which one of the two solutions is the more suitable backend for the controller, it is necessary to consider the behavior of the models for different bar settings as well. As there is no target profile available for random settings, the qualitative evaluation which prediction is closer to the real process is difficult. We, thus, examined empirically which model is more likely to show the expected behavior. Especially when altering the bar positions to more extreme heights (each within the ranges of the real experiments but new combinations thereof), the differences between the two models become apparent. At a certain point, the profiles predicted by the random forest do not get any wider. Fig. 6 shows that the random forest predicts narrower tows than the feedforward neural network for the same input data. We presume that, one, wide tows are less common in the original data, also indicated by the similarities in the predictions by the neural network, such as the elevation on the right side. And, two, the random forest tends to be a stronger generalisation of the process in general as its output is robust to unknown bar settings while the profiles of the neural networks become slightly noisier. Since we would expect a wider output for the maximum bar setup, we choose the FFNN as the underlying process model for the controller.

	FFNN	RF
Geo. Mean	1.0406	**1.0403**
Geo. Std.	1.0396	**1.0387**

(a) Geometric mean and geometric standard deviation of m.

	Ground Truth	FFNN	RF
Avg. Tow width	288.9	286.3	**287.3**
Std. Tow width	32.1	**27.5**	27.1

(b) Mean and standard deviation of the real and predicted tow width.

Fig. 4. Scores of the trained process models: feedforward neural (FFNN) and random forest (RF).

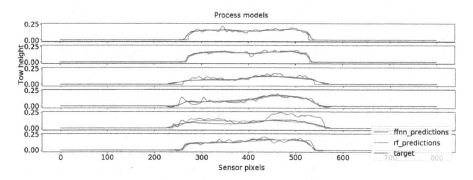

Fig. 5. Tow profiles after spreading as recorded during experiments and the according predictions by the neural network and random forest process model. Samples are taken at random.

To improve its robustness, we train it on additional synthetic data consisting of previously unseen bar positions that would not touch the tape and, therefore, lead to no differences in the profiles before and after spreading (i.e. the expected output is the same as the input profile), serving as a form of regularization.

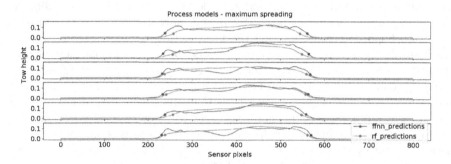

Fig. 6. Tow profiles predicted by the neural network and random forest, respectively, when setting the bars to extreme but realistic positions. This bar setup was not part of the recorded data. Samples are taken at random.

4.2   Process Control Model

When optimizing the hyperparameters of the neuroevolution approach, we found that the dimension of the neural networks is of secondary importance, with rather small neural networks with two hidden layers performing as well or slightly better than larger ones, whereas identifying a suitable population size and batch size has more impact. We observed the best performance with a population of 300 neural networks where the 15 fittest genotypes (on a rather small batch size of 2048 samples) would mutate and reproduce each generation.

Table 2. Geometric mean and geometric standard deviation of the absolute ratio of achieved and desired tape widths for fixed bar setups and the adaptive approach. Target width in pixels.

Target width	Fixed bar setup		Adaptive control	
	Geo. Mean	Geo. Std	Geo. Mean	Geo. Std
280	1.048	1.032	**1.023**	**1.023**
290	1.039	1.033	**1.028**	**1.025**
300	1.041	1.036	**1.023**	**1.022**
310	1.034	1.037	**1.022**	**1.022**
320	1.045	1.032	**1.021**	**1.026**
330	1.052	1.043	**1.028**	**1.030**

Overall, the evaluation of the process control model suggests that adaptive control is promising to generate high-quality tows and that neuroevolution may be a suitable approach to realize it. Examining the scores achieved, it is apparent that adjusting the bar positions consistently outperforms fixes settings (Table 2). In detail, developing a controller for the target tow width reduces the mean offset of the desired width by up to 47.9% in comparison to the fixed setup or down to 0.64 mm in general. Additionally, the standard deviation is lower as well. In fact, considering e.g. a target width of 28 mm, when using the adaptive controller, even the tow that differs most from the target width achieves a score of 1.046 while the overall geometric mean of the quality is 1.048 for the baseline fixed bar setup. For target widths greater 320 pixels, the quality declines for the two approaches which was to be expected. This can be attributed both to the fact that our process model is limited as less real data was available for this size and that this width might also not be achievable with the given bar settings. Interestingly, when visualising the bar positions proposed, we see that the control model tends to suggest solutions where mainly one bar (mostly the middle one) is adjusted depending on the input while the others are kept still. This finding is very promising, especially with respect to the time constraints that would apply when adapting the controller to the real process.

5   Conclusion

In this work, we presented a process controller for adaptive spreading of carbon fiber tows with an underlying process model, both implemented with neural networks. While the process model was trained conventionally using gradient descent, the process control model was optimized by applying a genetic algorithm. Regarding the process model, we showed that neural networks as well as random forests can adequately approximate a representation of the process behavior, even though they each carry certain shortcomings. Overall, we found a neural network to be more suitable for the application at hand as it yielded predictions for bar settings previously not seen that were closer to the expected behavior. In regards to the controller, adjusting the bar setup depending on the present tow profile is promising when compared to fixed setups. Our controller was able to reduce the deviations in tow width after spreading by half.

Future work may focus on two main issues: In the short term, the process controller as implemented is static in a sense, since it was trained to provide bar positions that result in tows with one given, fixed target width. Expanding its input to accept a desired tow width would lead to a more versatile solution. In addition, we consider implementing and evaluating other promising optimization strategies for the controller, as Bayesian optimization, for comparison. The long term goal is to replace the underlying process model with the real setup, which requires some changes on the controller. In particular, it is necessary to include time constraints of real world processes. Contrary to the process model, adjusting bar positions is not instantaneously possible and affects neighbouring parts of the tow, so it is mandatory to consider spatial dependencies. To reduce costly experiments with real carbon fiber tows, a cyclic approach is promising where findings from the real process are fed into the process model. In doing so, the tuning of the controller's hyperparameters is performed virtually, before evaluating the results on the real setup. If successfully applied to the real world system, this means that the quality of the tapes produced would be more consistent and, thus, the reject rate in aerospace production could be substantially reduced. The first iteration of this cycle, training a process model with real data and designing a suitable controller, was accomplished in this paper.

Acknowledgments. This research is partly funded by the Bavarian Ministry of Economic Affairs, Regional Development and Energy in the project LufPro.

References

1. Appel, L., Kerber, A., Abbas, B., Jeschke, S., Gries, T.: Determination of interactions between bar spreading process parameters and spreading quality for the development of an automated quality control of spread high modulus fiber tows. In: Proceeding of the 17th European Conference on Composite Materials, ECCM 2016, 17, pp. 26–30 (2016)
2. Breiman, L.: Random Forests. Machine Learning **45**, 5–32 (2001). https://doi.org/10.1023/A:1010933404324
3. Floreano, D., Dürr, P., Mattiussi, C.: Neuroevolution: from architectures to learning. Evol. Intel. **1**(1), 47–62 (2008). https://doi.org/10.1007/s12065-007-0002-4
4. Gizik, D., Metzner, C., Weimer, C., Middendorf, P.: Spreading of heavy tow carbon fibers for the use in aircraft structures. In: Proceeding of the 17th European Conference on Composite Materials, ECCM 2016, 17 (2016)
5. MacKenzie, H.D.J. et al.: On the Road toward 2050: Report Massachusetts Institute of Technology Potential for Substantial Reductions in Light-Duty Vehicle Energy Use and Greenhouse Gas Emissions. Massachusetts Institute of Technology (2015). https://energy.mit.edu/publication/on-the-road-toward-2050/
6. Irfan, M.S., et al.: Lateral spreading of a fiber bundle via mechanical means. J. Compos. Mater. **46**(3), 311–330 (2012)
7. Mandel, T., Liu, Y.E., Brunskill, E., Popovíc, Z.: Offline evaluation of online reinforcement learning algorithms. In: 30th AAAI Conference on Artificial Intelligence, AAAI 2016, pp. 1926–1933 (2016)
8. Marissen, R., van der Drift, L.T., Sterk, J.: Technology for rapid impregnation of fibre bundles with a molten thermoplastic polymer. Compos. Sci. Technol. **60**(10), 2029–2034 (2000)
9. Morgan, P.: Carbon Fibers and Their Composites. CRC Press, Boca Raton (2005)
10. Nagabandi, A., Kahn, G., Fearing, R.S., Levine, S.: Neural network dynamics for model-based deep reinforcement learning with model-free fine-tuning. In: 2018 IEEE International Conference on Robotics and Automation (ICRA), pp. 7559–7566 (2018)
11. Newell, J.A., Puzianowski, A.A.: Development of a pneumatic spreading system for kevlar-based sic-precursor carbon fibre tows. High Perform. Polym. **11**(2), 197–203 (1999)
12. Peng, X.B., Andrychowicz, M., Zaremba, W., Abbeel, P.: Sim-to-Real Transfer of Robotic Control with Dynamics Randomization. In: Proceedings - IEEE International Conference on Robotics and Automation, pp. 3803–3810 (October 2018)

13. Savitzky, A., Golay, M.J.E.: Smoothing and differentiation of data by simplified least squares procedures. Anal. Chem. **36**(8), 1627–1639 (1964)
14. Such, F.P., Madhavan, V., Conti, E., Lehman, J., Stanley, K.O., Clune, J.: Deep Neuroevolution: Genetic Algorithms Are a Competitive Alternative for Training Deep Neural Networks for Reinforcement Learning (2017). http://arxiv.org/abs/1712.06567
15. Sutton, R.S.: Dyna, an integrated architecture for learning, planning, and reacting. ACM SIGART Bull. **2**(4), 160–163 (1991)
16. Thomas, P.S., Theocharous, G., Ghavamzadeh, M.: High confidence off-policy evaluation. In: Proceedings of the National Conference on Artificial Intelligence **4**, 3000–3006 (2015)

Ensemble Kalman Filter Optimizing Deep Neural Networks: An Alternative Approach to Non-performing Gradient Descent

Alper Yegenoglu[1,2]([⊠]) [iD], Kai Krajsek[1] [iD], Sandra Diaz Pier[1] [iD],
and Michael Herty[2]

[1] SimLab Neuroscience, Jülich Supercomputing Centre (JSC), Institute for Advanced Simulation, JARA, Forschungszentrum Jülich GmbH, Jülich, Germany
a.yegenoglu@fz-juelich.de
[2] Institute of Geometry and Applied Mathematics, Department of Mathematics, RWTH Aachen University, Aachen, Germany

Abstract. The successful training of deep neural networks is dependent on initialization schemes and choice of activation functions. Non-optimally chosen parameter settings lead to the known problem of exploding or vanishing gradients. This issue occurs when gradient descent and backpropagation are applied. For this setting the Ensemble Kalman Filter (EnKF) can be used as an alternative optimizer when training neural networks. The EnKF does not require the explicit calculation of gradients or adjoints and we show this resolves the exploding and vanishing gradient problem. We analyze different parameter initializations, propose a dynamic change in ensembles and compare results to established methods.

Keywords: Deep Neural Networks · Kalman Filter · Activation Function · Vanishing Gradients · Initialization

1 Introduction

The performance of deep multilayered neural networks is very susceptible to the initialization of weights and the selected activation functions [10,31]. For example, non-optimally chosen parameters in the initialization stage may lead to a loss of input information in the feed-forward step or to the well known problem of exploding and vanishing gradients during the backpropgation phase [2,3,13, 29]. An example of this problem is illustrated in Fig. 1, which depicts the test error of a convolutional network trained for 10 epochs on the MNIST dataset (see Sect. 2.2 for details). The figure shows that, if the initialization is not done in an optimal way, the stochastic gradient descent (SGD) algorithm is not able to train the network. Further, training with ADAM [17] is heavily slowed down depending on the choice of the algorithmic parameters (c.f. Sects. 3.1 and 3.2).

© Springer Nature Switzerland AG 2020
G. Nicosia et al. (Eds.): LOD 2020, LNCS 12566, pp. 78–92, 2020.
https://doi.org/10.1007/978-3-030-64580-9_7

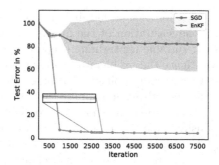

Fig. 1. Mean test error (dark line) of the network in % for 10 runs trained for one epoch using different optimizers. The shaded area is the standard deviation of the runs. The standard deviation of the normal distribution when initializing the weights is set as $\sigma = 1$. The test is performed on a separate test dataset while the training is ongoing.

The vanishing gradient problem has been explored analytically in detail [9,24, 26]. Although substantial work has been done regarding activation functions and initializations, most of the studies investigate the problems using gradient descent and backpropagation as optimization procedure. The aforementioned drawbacks are related to gradients obtained within deep neuronal networks [13].

Since, in general, optimal parameters can only be determined by successive testing, we are interested in alternative robust numerical methods for training deep neural networks. Hence, in this work we utilize a method that does not rely on gradient information (see Fig. 1). An advantage of this approach is that it also allows for non differentiable activation functions without introducing additional smoothing.

Dealing with Non Differentiability and the Vanishing Gradient Problem. Many of the recent approaches for training deep networks counteract against non differentiability by applying smoothing to the rectified linear unit (ReLU) [22], e.g. the exponential linear unit (ELU) [5]. This may also lead to numerical instability due to possible singular limits for small smoothing parameters. Other approaches have been suggested to circumvent the problem observed in Fig. 1. In particular, different initialization schemes like the Xavier [30] or Kaiming-initialization [11]. The latter incorporates the number of incoming or outgoing connections to or from the neurons of the next or previous layer, to stabilize the training by normalizing the weights. Sussillo and Abbott [28] showed that gradients scale only as the square root of the depth of the network. Based on this observation they proposed a heuristic for initializing networks with a procedure called Random Walk Initialization. They observe satisfactory performance under those modifications.

The role of the activation function and its properties have been investigated with regards to their influence on the training procedure. Lecun et al. [19] discuss that the logistic function will slow down training since it induces a slower convergence. Based on [19] Glorot et al. [30] investigate the selection of activation functions regarding the saturation of the gradients. Many studies remark the logistic function is not suitable for training deep networks [19,24,30].

In this paper we investigate the problem of training neural networks within ill-conditioned settings. Our starting point is that a possible explanation of the difficulties of the previous methods when training the networks could be related to the dependence on explicit gradient information. To diminish this dependence we explore the effects of using a particle based method, in particular, the Ensemble Kalman Filter (EnKF). The EnKF has been widely studied in engineering literature [1,7,15,27] as well as recently in the mathematical community [12,25]. We refer to [16] for a comprehensible introduction and [6,14] for more details on the method. However, it has only been recently shown that EnKF could also be applied to solve inverse problems and we formulate the training procedure as such a problem. The EnKF does not require to calculate gradients explicitly and omits the backpropagation step. Furthermore, this method does not require a particular form of the activation function. We follow the formulation of the EnKF proposed in [14]. In contrast to previous work having a similar setting such as in [8,18] we emphasize that the scope of our contribution does not focus on achieving high benchmark performance on the investigated datasets. Instead, our objective is to provide an analysis based on numerical solutions regarding the ability of the EnKF to cope with problems where gradient based algorithms fail, are very sensitive to initial conditions or provide a poor performance in general.

Among other results, we show that a classification problem can be solved using the EnKF as an optimizer, the logistic function as activation function and a normal distributed initialization of the weights. The method is shown to be independent to different realizations when initializing the weights. In contrast, the same network is not able to achieve sufficient performance in the task when optimized by SGD as indicated in Fig. 1. Since the EnKFdepends among other parameters on the chosen number of ensembles we propose a dynamic adaption of algorithmic parameters depending on the reached training accuracy indicated in Algorithm 1 and with corresponding results presented in Fig. 10.

2   Ensemble Kalman Filter Optimizing Neural Networks

The EnKF is a well-known iterative numerical method for nonlinear dynamic filtering problems under noise and has been applied more recently to inverse problems [25] as well. The parameter estimation problem is a particular inverse problem and we apply the EnKF here. When training a deep neural network only the evaluation of the feed-forward propagation is required, thus omitting the backpropagation. Further, the EnKF is easily parallelizable in contrast to gradient based approaches. First results on training deep neural networks and recurrent neural networks with the EnKF can be found in [18,21].

2.1   Description and Properties of the Ensemble Kalman Filter

Training a neural network to learn its weights and biases can be formulated as an optimization problem. Kovachki et al. [18] describe the training as a minimization problem for Φ given by

$$\Phi(\mathbf{u}, \mathbf{y}) = \|\mathcal{G}(\mathbf{u}) - \mathbf{y}\|_\Gamma^2 \tag{1}$$

where $\mathcal{G}(\mathbf{u})$ is the model output and $\mathbf{y}$ is the target or label to optimize for.

We introduce briefly the EnKF following the formulation from [14]:

$$\mathbf{u}_j^{n+1} = \mathbf{u}_j^n + \mathbf{C}(\mathbf{U}^n)\left(\mathbf{D}(\mathbf{U}^n) + \Gamma^{-1}\right)^{-1}\left(\mathbf{y} - \mathcal{G}(\mathbf{u}_j^n)\right) \tag{2}$$

where $\mathbf{U}^n = \{\mathbf{u}_j^n\}_{j=1}^J$ is the set of all ensembles, n is the iteration index, J is number of ensembles. In our setting, the ensemble $\mathbf{u}_j$ corresponds to the weights of the network (c.f. Sect. 2.2). Γ is the covariance matrix related to the measurement of noise, in our case we use Γ as an identity matrix multiplied with a small scalar, i.e. $\Gamma = \gamma\mathbf{I}$. The hyper-parameters J, γ, and n are the only hyper-parameters of the EnKF.

The matrices $\mathbf{C}(\mathbf{U}^n)$ and $\mathbf{D}(\mathbf{U}^n)$ are covariance matrices defined by:

$$\mathbf{C}(\mathbf{U}) = \frac{1}{J}\sum_{j=1}^J (\mathbf{u}_j - \overline{\mathbf{u}}) \otimes (\mathcal{G}(\mathbf{u}_j) - \overline{\mathcal{G}})^T,$$

$$\mathbf{D}(\mathbf{U}) = \frac{1}{J}\sum_{j=1}^J (\mathcal{G}(\mathbf{u}_j) - \overline{\mathcal{G}}) \otimes (\mathcal{G}(\mathbf{u}_j) - \overline{\mathcal{G}})^T, \tag{3}$$

$$\overline{\mathbf{u}} = \frac{1}{J}\sum_{j=1}^J \mathbf{u}_j, \quad \overline{\mathcal{G}} = \frac{1}{J}\sum_{j=1}^J \mathcal{G}(\mathbf{u}_j)$$

where $\otimes$ is the tensor-product. In [18] it was shown that under simplifed assumptions we have $\overline{\mathbf{u}}^\star = \arg\min(\Phi(\mathbf{u}, \mathbf{y}))$ and $\frac{1}{N}\sum_{j=1}^J \mathbf{u}_j^n \xrightarrow{n\to\infty} \overline{\mathbf{u}}^\star$.

2.2   Experimental Setup

In this section we present the network architecture and the training procedure for the Ensemble Kalman Filter and the backpropagation optimizers. All simulations were performed on a compute node with a NVIDIA Tesla K20c graphic cards, Intel i7-4770 CPU and Scientific Linux 7.4 as an operating system.

Network Architecture. The network (see also Fig. 2) we use in all of our experiments consists of two convolutional layers and a fully connected layer and the logistic function $(f(x) = 1/(1 + e^{-x}))$ applied on the output of the convolutional layers. The kernel size of the convolution for both layers is 5×5 with stride 1. Max pooling is applied on both convolutional layers with a kernel size of 2 and a stride of 2. The implementation is done with the PyTorch[1] [23] library.

---

[1] v.1.2.0.

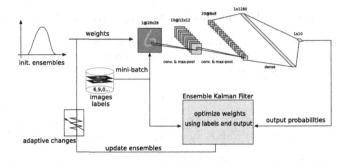

Fig. 2. Workflow depicting a training phase of a convolutional network optimized by an Ensemble Kalman Filter using adaptive choice of EnKF parameters.

Training with the Ensemble Kalman Filter. The training of the Convolutional Neural Network using the EnKF is shown in Fig. 2 as a workflow[2]. The experiments have been designed using the MNIST dataset [20]. In Sect. 3.3 we show classification results obtained on the letter dataset [4]. First we initialize the biases to be 0 and the weights $\mathbf{W} = (W_{i,k})_{i,k}$ for each layer as

$$W_{i,k} \sim \mathcal{N}(\mu, \sigma^2) \tag{4}$$

with $\mu = 0$, $\sigma \in [0.01, 10]$, where $\mathcal{N}$ is the normal distribution with mean μ and variance σ.

The initialization $n = 0$ and $\mathbf{u}_j^0$ is done J times, i.e. a single member j of the ensemble $\mathbf{u}_j$ corresponds to a random weight matrix $\mathbf{W}$ according to the normal distribution $\mathcal{N}(\mu, \sigma^2)$. This matrix is assigned to the ensemble $\mathbf{u}_j^0 := \mathbf{W} \in \mathbb{R}^{\text{layer} \times \text{weights}}$ for $j = 1, \ldots, J$. After that $\mathbf{u}_j^n$ for $n > 0$ is obtained by the iterative Formula (2). We choose $J = 5000$ ensembles as a basis. Different settings are discussed in Sect. 3. Additionally, the EnKF Formula (2) requires the model outputs $\mathcal{G}(\mathbf{u})$, i.e. the feed-forward network output. Thus, for J iterations we initialize the j-th ensemble as weights for the network and apply one classification step for a given mini-batch of size 64. The scaling factor γ is set constant for all experiments, $\gamma = 0.01$.

The output of the network and the targets are fed into the iterative EnKF update formula, where a new set of ensembles J is calculated according to Eq. (2). We observed that a repetitive presentation of the images before switching to new samples helped the network to converge faster and reach a better performance. We verified the number of repeated presentations required to reach a high accuracy on the training mini-batch was 8 repetitions. This number can be set less or omitted after 500 iterations when a high test accuracy is reached (see Fig. 8). Every 500th iterations, after the training is completed, we obtain the test errors. We initialize one network by setting the weights of the network with the mean vector $\bar{\mathbf{u}}$ of the ensembles. A test set of data images with corresponding targets

---

[2] Code can be found on GitHub: https://github.com/alperyeg/enkf-dnn-lod2020.

are forwarded as the input to the network and the classification error is evaluated. We also extended the setting with an adaptive mechanism to change the number of ensembles, iterations and repetitions depending on the test accuracy. For details we refer to Sect. 3.3.

Training with Backpropagation. Backpropagation is performed using SGD or ADAM on mini-batches of size 64. The training is completed using a similar strategy as described for the EnKF. The loss is calculated using the Cross Entropy Loss[3], which is a combination of the negative log-likelihood and log softmax applied on the model output $\mathbf{z}_j = \mathcal{G}(\mathbf{u}_j)$, i.e.

$$\sigma(\mathbf{z}_j) = -\log\left(\frac{e^{z_j}}{\sum_i e^{z_i}}\right). \tag{5}$$

Note that, if the network is optimized using SGD no learning effect or improvement can be observed for longer runs (c.f. Sect. 3.1).

3 Numerical Results

In this section we discuss how gradients and activation values vanish or saturate by initialization schemes described in Sect. 2.2 with focus on SGD and ADAM. Afterwards we show that different sigmoidal activation functions (Logistic function, ReLU, Tanh) and varying number of ensembles influence the performance of the EnKF on the basis of the test error. Furthermore, we present results obtained when training the network on the letters dataset [4].

3.1 Non Evolving Gradients and Activation Values with SGD

When SGD is applied as an optimizer in our experimental setting (see Sect. 2.2) the units saturate in the early training phase and they cannot recover, as it can be observed from Fig. 3 (left). Here, the evolution of the mean (dark blue, orange and green lines) and standard deviation (light blue, orange and green vertical bars) of the activation values obtained after the logistic function on the hidden layer are shown over all mini-batch iterations. For layer 1 and 2 the activation values saturate close to 200 iterations, with mean activation value of layer 1 of 0.4597, standard deviation of 0.2976 and a value for the mean activation of layer 2 of 0.3178 and 0.4624 standard deviation. The mean activation value in layer 3 (0.1) stays constant from the first iteration on and has a standard deviation of 0.2908. Saturation is clearly reflected in the gradients as depicted in Fig. 3 (right). The distribution shows that after 800 mini-batch iterations the distribution didn't change and the mean of the gradients is close to zero, confirming saturated gradients.

If the training is run for more epochs (e.g. 50 epochs) the network is still not able to learn. This is depicted in Fig. 4 (left) where the gradients stay at

---

[3] Following Pytorch's nomenclature.

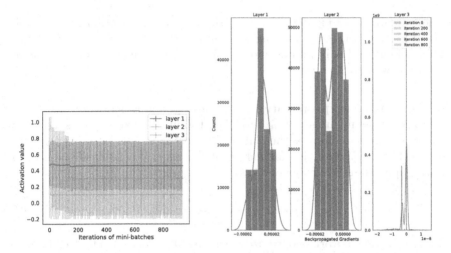

Fig. 3. Left: Mean and standard deviation of the activation values during training for the three hidden layers of the network. SGD is applied as optimizer. Note: only every 8th entry is shown. Right: normalized histogram depicting a distribution of the gradients with a mean value of 0 for all three layers. Iterations are one epoch run with a mini-batch size of 64.

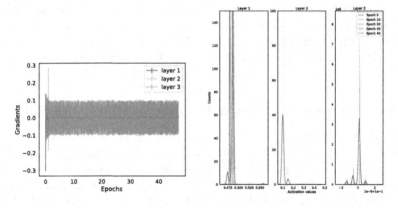

Fig. 4. Left: Mean (blue, orange and green lines) and standard deviation (blue, orange and green vertical bars) of the backpropagated gradients for all three layers for 50 epochs. Right: Mean distributions of activation values over 50 epochs for all three layers using SGD as optimizer. (Color figure online)

zero mean for all three layers. Only layer 3 has an oscillating standard deviation around the value 0.1. The same situation is presented for the activation functions as it can be seen in Fig. 4, in which the distribution of the mean activation is depicted. For layer 1 there is a very small shift of the mean from 0.475 towards 0.5, but the values of the other layers still do not change.

3.2   Slowly Evolving Gradients and Activation Values with ADAM

In our experiments we observe that ADAM is more robust when optimizing the network compared to SGD in terms of performance. When the weights are initialized with $\sigma = 1$ the network will slowly start to correctly classify the dataset. Within 10 epochs it reaches more than 95% of test accuracy (see also Fig. 5). When fixing $\sigma = 3$ a similar behavior to the SGD setting is observable in the first epochs (c.f. Fig. 5). When looking at the mean and standard deviation of the activation values within in the first epoch a fast saturation is not observed for layer 1 (c.f. Fig. 6 (left)). We see an increase of the mean value of the activation values from 0.49 up to 0.53 (overall mean value of 0.5124), with standard deviation of 0.2860. This indicates a slow saturation in the longer run, which can be verified by analyzing layer 2. Layer 2 also increases it's mean activation value (0.47) from the first iteration to the value 0.5, but stagnates at around the 400th iteration with a value of 0.5 (overall mean is 0.4957, standard deviation is 0.4965). The mean value of layer 3 stays constant at 0.1.

Figure 7 (left) shows the progress of the mean activation values over 50 epochs. An acceptable learning performance with regards to the test accuracy is achieved after 30 epochs. This is reflected especially in layer 1: here, we observe how the mean evolves over the epochs. While in epoch 1 the distribution is around 0.49, we can see after 40 epochs that the distribution evolves around 0.56. On the other hand the distribution of the activation values of layer 2 and 3 stay constant (around 0.5 and 0.0). The mean distribution of gradients obtained by backpropagation after one epoch is close to zero for all layers as one can observe in Fig. 6 (right). However, in the long run (i.e. after 30 epochs) the network starts to perform better (up to 84% accuracy). In Fig. 7 (left) the mean and standard deviations of the backpropagated gradients in layer 1 and layer 2 do not vary strongly up to epoch 10. Until then, the network shows no learning capability (test accuracy under 10%). Mean values in the first epochs for layer 1 are between -0.3 and 0.2 and standard deviations are between -0.8 and 0.6 showing a decreasing trend. For layer 2 mean values are between -0.1 and 0.1 and standard deviations are between -0.2 and 0.2. From epoch 10 up to epoch 28 both mean and standard deviation of layer 1 and 2 decrease and stay constant

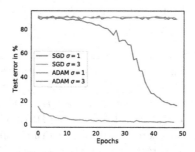

Fig. 5. Test error of the network trained for 50 epochs on the MNIST dataset. Different values for the standard deviation σ are used when initializing the weights.

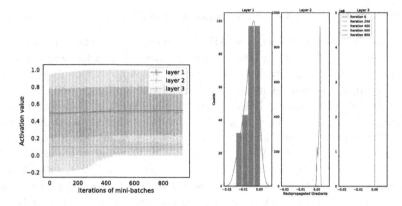

Fig. 6. Left: Mean (dark blue, orange and green lines) and standard deviation (light blue, orange and green vertical bars) of the activation values during training for the three hidden layers of the network in one epoch. ADAM is applied as optimizer. Right: Normalized histogram depicting a distribution of the gradients with a mean value of 0 for all three layers. Iterations are one epoch run with a mini-batch size of 64. (Color figure online)

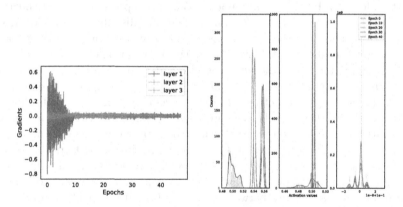

Fig. 7. Left: Mean (line) and standard deviation (vertical bar) of the backpropagated gradients for all three layers for 50 epochs. Right: Mean distributions of activation values over 50 epochs for all three layers using ADAM as optimizer.

close to zero for both values. After the 28th epoch the network starts to learn. An increase and a small variance in the mean and standard deviations of both layers 1 and 2 can be observed again (layer 1 means in $[-0.02, 0.02]$, standard deviations in $[-0.01, 0.06]$; layer 2 means in $[-0.01, 0.01]$, standard deviation in $[-0.01, 0.03]$). The mean activation value of layer 3 stays constant in all iterations at a very small value close to zero, in contrast to the standard deviation which starts in the first epoch at 0.16 and decreases to 0.01. We offer the following interpretation. First, this shows that a network can recover from the saturation of its gradients given enough time to train. Secondly, a good performance shows that the network converges to a suitable classification with high accuracy only for suitable initial settings.

3.3   Performance of the Ensemble Kalman Filter

Given the same setup using the Ensemble Kalman Filter enables the network to perform well (see Figs. 8 and 9). After the first epoch the network has a test error of 3.8% when classifying the MNIST dataset. In the following we analyze in detail the performance with varying number of ensembles and its sensitivity to different activation functions.

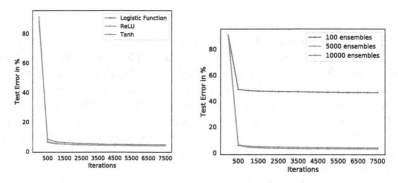

Fig. 8. Left: Test error during training on the MNIST dataset for different activation functions optimized by EnKF. Right: Test error during training on the MNIST dataset for different ensemble sizes within one epoch. Every 500th iteration is shown.

Different Activation Functions. We investigated the network training for different activation functions such as ReLU: $\text{ReLU}(x) = \max(0, x)$, Tanh: $\tanh(x) = \dfrac{e^x - e^{-x}}{e^x + e^{-x}}$ and a Logistic Function: $f(x) = \dfrac{1}{1 + e^{-x}}$. Figure 8 (left) depicts the test errors in % using different activation functions over iterations. The test errors do not change strongly when ReLU or Logistic Function as activation functions are applied. There is a small gap of 0.45% error after the 500th iteration (test errors with Logistic Function: 6.11%, ReLU: 6.56%). The test error after one epoch is 3.8% when Logistic Function is applied and 3.75% for ReLU. If Tanh is used as activation function the performance is slightly worse, in the 500th iteration the error is at 8.15%, while after 7500 iterations it is 4.51%. The discrepancy may be induced because of the range of the Tanh function which is in $[-1, 1]$, whereas the images are normalized between $[0, 1]$ (c.f. [19]). Due to the fast convergence the EnKF reaches a suitable accuracy already after 500 iterations. This behavior results in a slower improvement of the accuracy between the 500th and the 7500th iteration, approximately 3%.

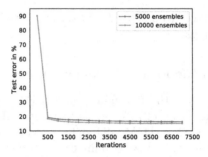

Fig. 9. Test error during training on the letters dataset for different ensemble sizes within one epoch. Every 500th iteration is shown.

EnKF-NN Training on Letters Dataset. Figure 9 presents the test error on the letters dataset using the EnKF optimizer. The parameter-setting is the same as in Sect. 2.2. We use 5000 and 10000 ensembles during the training. We observe a higher number of ensembles achieves a slightly better classification with a test error 15.1% within one epoch. With 5000 ensembles the test error is at 16.32%. We can conclude that a higher number of ensembles enables a better performance but is not necessarily an overall solution to increase the accuracy. To this point we didn't test if there is a smaller ensemble size than 10000 which achieves the same error rate.

A hyperparameter for the EnKF is the number of ensembles. The convergence behavior of the EnKF is strongly influenced by this value. The test error for a different number of ensembles is depicted in Fig. 8 (right). A small size of a few hundred ensembles is not sufficient to optimize the network to perform with a suitable accuracy. Here, the test error stagnates around 47% which is just above chance level when classifying the test dataset. In contrast, a higher number of ensembles, e.g. 5000 ensembles, is sufficient to achieve a good performance on the test set. The test error after 7500 iterations is at 3.8%. Having even more ensembles decreases the test error slightly, e.g. with 10000 ensembles the error drops to 3%.

Adaptive Changes. The results of the previous sections indicate a possible reduction of ensembles, repetitions and iterations. To achieve these reductions we implemented an adaptive mechanism. The main idea is based on comparing the actual with the previous test accuracy (t_n, t_{n-1}) and deciding to change if the number stay in given upper and lower boundaries (b_u, b_l). The algorithm is outlined in Algorithm 1.

Figure 10 depicts two runs of the EnKF optimization on the MNIST dataset with the adaptive change (orange graphs, "adaptive") and without (blue graphs, "fixed"). The fixed run follows the setting described in Sect. 2.2 and e.g. conducted for Fig. 8, the values of the parameters are fixed. The test error is shown on the left y-Axis and the number of ensembles and repetitions are on the right y-Axes. In every 500th iteration τ the test errors are evaluated and changes can

Algorithm 1. Algorithm for adaptive choice of parameters

Input: Lower bound b_l, upper bound b_u, error threshold ϵ, test error t_n, mini-batch size m, length of dataset D_s, update interval τ, ensemble update step v, error difference threshold κ

Output: Number of ensembles J, total iterations N, repetitions r

if $n \bmod \tau = 0$ and $n > 0$ **then**

    if $t_n < \epsilon$ and $t_n < t_{n-1}$ **then**

        if $J > b_l$ and $t_{n-1} - t_n \leq \kappa$ **then**

            $J = J - v$

            $N = \frac{D_s}{m} \cdot r$

            $r = r - 1$

        else if $n < b_u$ and $t_{n-1} - t_n > \kappa$ **then**

            $J = J + v$

            $N = \frac{D_s}{m} \cdot r$

            $r = r + 1$

be applied if $t_n < 10\%$. The test error of the fixed run is shown up to iteration 2500 as by then one epoch is reached with the adaptive setting. We observed that the adaptive changes (Fig. 10 left, orange line with dots) do not increase the test error while they reach the same accuracy as the fixed run (blue line with dots). Due to the fast convergence of the EnKF we were able decrease the repetitions r (Fig. 10 right, orange line with stars) by 1 in every 500th iteration. Additionally, we noticed that more repetitions do not increase the performance after a few iterations and after a satisfactory test accuracy is reached. After 1500 iterations the number of ensembles N (orange line with squares) are decreased by $v = 1000$. We were rather conservative by slowly increasing or decreasing all numbers. We decreased for instance the number of ensembles only by $v = 1000$ and not e.g. by half to ensure smaller changes and stability. The lower bound b_l is set to minimal number of 1000 ensembles and the upper bound b_u is set to a maximal number of 10000 ensembles. For the same reasons we required that

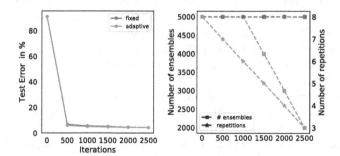

Fig. 10. Two runs on the MNIST dataset with (adaptive, orange graphs) and without (fixed, blue graphs) adaptive changes. In the adaptive setting, every 500 iteration the values of the parameters required by the EnKF are dynamically changed depending on the test errors. (Color figure online)

in order to decrease the values the test error has to be equal or less than $\kappa = 1$ between the actual and previous test errors. This restrictive behavior causes the slow decrease of the number of ensembles after 1500 iterations. Since there was no increase in the test error an increase of the values did not occur.

4   Conclusions and Outlook

Network performance is susceptible to weight initialization and proper selection of activation functions. Improperly selected parameters cause the vanishing gradient problem especially when training with gradient descent and backpropagation. We have shown in Sect. 3 that also simple deep networks are affected by this problem. For a better understanding we investigated how gradients and activation function values vanish or saturate by analysing their distributions, mean and standard deviation per layer over iterations and epochs for different parameters. Our setting focused on SGD and ADAM as optimizers, random normal distributed weight initializations and the logistic function as activation function. The network was not able to learn or only very slowly on the MNIST classification task. In our experiments ADAM was more robust in contrast to SGD and performed better for different parameter settings. As an alternative to backpropagation we analyzed the use of the Ensemble Kalman Filter (EnKF), which omits the calculation of gradients and only requires the feed-forward step in the training phase.

Given the same settings, the EnKF is able to optimize the network on the MNIST and letters datasets and achieve a performance above 96% and 85%. We investigated the activation values of different network layers by analyzing the distribution, mean and standard deviations over several iterations within one epoch. Further, we experimented with the number of ensembles and concluded that a larger ensemble size does not necessarily give a significant better performance as one could expect (Sect. 3.3). Additionally, we investigated how different initializations of the weights influence the overall learning performance. Our study didn't focus on reaching state of the art classification accuracy. Regarding solely the classification performance we refer to [8,18]. Due to the fast convergence of the EnKF the network was able to perform on the MNIST dataset already after 500 iterations. The fast convergence on the other hand may be problematic if the method is stuck in a local minimum and exploration of other possible minima is desirable. To counteract this issue it is possible to add noise to the ensembles in every update step as it was suggested in [18]. Another approach would be to adapt the scaling γ of the covariance matrix $\mathbf{\Gamma}$ dynamically (see Eq. 3). To dynamically change the number of iterations, repetitions and ensembles we extended the setting with an adaptive algorithm to modify parameters depending on the test errors.

Our approach provides an alternative to optimization of multilayered neural networks which can overcome the problems introduced by different activation functions and initialization settings. This opens new avenues in research of the benefits of using these activation functions and initializations. It also provides a basis for new research on gradient free alternatives to optimize learning in neural networks.

In training neural networks there is a large number of variables which can affect the results. In our work we have chosen a narrow set of parameters to explore, while leaving everything else fixed. This approach allows us to control the conclusions we draw from these experiments. We consider our work a first step into exploring setups where using the EnKF can be beneficial and maybe even a unique option to deal with specific problems of interest to the scientific community.

Acknowledgements. AY would like to thank Giuseppe Visconti and Wouter Kljin for fruitful discussions. Partially funded by the Helmholtz Association through the Helmholtz Portfolio Theme "Supercomputing and modeling for the Human Brain". This work was performed as part of the Helmholtz School for Data Science in Life, Earth and Energy (HDS-LEE). This project has received funding from the European Union's Horizon 2020 research and innovation programme under grant agreements No. 785907 (HBP SGA2) and No. 826421 (VirtualBrainCloud). It has also received funding from the CSD-SSD no. 20190612, PHD-PROGRAM-20170404, and DFG EXC-2023 Internet of Production - 390621612. This work is partly supported by the Helmholtz Association Initiative and Networking Fund under project number ZT-I-0003.

References

1. Aanonsen, S.I., Nævdal, G., Oliver, D.S., Reynolds, A.C., Vallès, B., et al.: The ensemble kalman filter in reservoir engineering-a review. Spe J. **14**(03), 393–412 (2009)
2. Bengio, Y., Frasconi, P., Simard, P.: The problem of learning long-term dependencies in recurrent networks. In: IEEE International Conference on Neural Networks, vol. 3, pp. 1183–1188 (March 1993). https://doi.org/10.1109/ICNN.1993.298725
3. Bengio, Y., Simard, P., Frasconi, P.: Learning long-term dependencies with gradient descent is difficult. IEEE Trans. Neural Networks **5**(2), 157–166 (1994). https://doi.org/10.1109/72.279181
4. Bulatov, Y.: notMNIST. Kaggle dataset (February 2018). https://www.kaggle.com/jwjohnson314/notmnist#notMNIST_large
5. Clevert, D.A., Unterthiner, T., Hochreiter, S.: Fast and accurate deep network learning by exponential linear units (elus). arXiv preprint arXiv:1511.07289 (2015)
6. Evensen, G.: Data Assimilation. Springer, Berlin (2009). https://doi.org/10.1007/978-3-642-03711-5
7. Evensen, G.: Sequential Data Assimilation, pp. 27–45. Springer, Berlin (2009). https://doi.org/10.1007/978-3-642-03711-5_4
8. Haber, E., Lucka, F., Ruthotto, L.: Never look back - A modified EnKF method and its application to the training of neural networks without back propagation (2018)
9. Hanin, B.: Which neural net architectures give rise to exploding and vanishing gradients? In: Advances in Neural Information Processing Systems, pp. 582–591 (2018)
10. Hayou, S., Doucet, A., Rousseau, J.: On the impact of the activation function on deep neural networks training. arXiv preprint arXiv:1902.06853 (2019)
11. He, K., Zhang, X., Ren, S., Sun, J.: Delving deep into rectifiers: Surpassing human-level performance on imagenet classification (2015)

12. Herty, M., Visconti, G.: Kinetic methods for inverse problems. Kinetic & Related Models vol. 12, pp. 1109 (2019) 19375093_2019_5_1109
13. Hochreiter, S., Bengio, Y., Frasconi, P., Schmidhuber, J., et al.: Gradient flow in recurrent nets: the difficulty of learning long-term dependencies (2001)
14. Iglesias, M.A., Law, K.J.H., Stuart, A.M.: Ensemble kalman methods for inverse problems. Inverse Prob. **29**(4), 045001 (2013). https://doi.org/10.1088/0266-5611/29/4/045001
15. Janjic, T., McLaughlin, D., Cohn, S.E., Verlaan, M.: Conservation of mass and preservation of positivity with ensemble-type kalman filter algorithms. Mon. Weather Rev. **142**(2), 755–773 (2014)
16. Katzfuss, M., Stroud, J.R., Wikle, C.K.: Understanding the ensemble kalman filter. Am. Stat. **70**(4), 350–357 (2016)
17. Kingma, D.P., Ba, J.: Adam: A method for stochastic optimization. arXiv preprint arXiv:1412.6980 (2014)
18. Kovachki, N.B., Stuart, A.M.: Ensemble kalman inversion: a derivative-free technique for machine learning tasks (2018)
19. Lecun, Y., Bottou, L., Orr, G.B., Müller, K.R.: Efficient backprop (1998)
20. LeCun, Y., Cortes, C., Burges, C.: Mnist handwritten digit database. AT&T Labs **2**, 18 (2010). http://yann.lecun.com/exdb/mnist
21. Mirikitani, D.T., Nikolaev, N.: Dynamic modeling with ensemble kalman filter trained recurrent neural networks. In: 2008 Seventh International Conference on Machine Learning and Applications, IEEE (2008). https://doi.org/10.1109/icmla.2008.79
22. Nair, V., Hinton, G.E.: Rectified linear units improve restricted boltzmann machines. In: Proceedings of the 27th International Conference on Machine Learning (ICML-10), pp. 807–814 (2010)
23. Paszke, A., et al.: Automatic differentiation in PyTorch. In: NeurIPS Autodiff Workshop (2017)
24. Pennington, J., Schoenholz, S., Ganguli, S.: Resurrecting the sigmoid in deep learning through dynamical isometry: theory and practice. In: Advances in Neural Information Processing Systems, pp. 4785–4795 (2017)
25. Schillings, C., Stuart, A.M.: Convergence analysis of ensemble Kalman inversion: the linear, noisy case. Appl. Anal. **97**(1), 107–123 (2018). https://doi.org/10.1080/00036811.2017.1386784
26. Schoenholz, S.S., Gilmer, J., Ganguli, S., Sohl-Dickstein, J.: Deep information propagation. arXiv preprint arXiv:1611.01232 (2016)
27. Schwenzer, M., Stemmler, S., Ay, M., Bergs, T., Abel, D.: Ensemble kalman filtering for force model identification in milling. Procedia CIRP **82**, 296–301 (2019)
28. Sussillo, D., Abbott, L.: Random walk initialization for training very deep feedforward networks. arXiv preprint arXiv:1412.6558 (2014)
29. Sutskever, I., Martens, J., Dahl, G., Hinton, G.: On the importance of initialization and momentum in deep learning. In: International conference on machine learning, pp. 1139–1147 (2013)
30. Xavier Glorot, Y.B.: Understanding the difficulty of training deep feedforward neural networks (2010)
31. Xie, D., Xiong, J., Pu, S.: All you need is beyond a good Init: Exploring better solution for training extremely deep convolutional neural networks with orthonormality and modulation. In: Proceedings of the IEEE Conference on Computer Vision and Pattern Recognition, pp. 6176–6185 (2017)

Effects of Random Seeds on the Accuracy of Convolutional Neural Networks

Christofer Fellicious$^{(\boxtimes)}$ ⓘ, Thomas Weissgerber ⓘ, and Michael Granitzer ⓘ

University of Passau, 94032 Passau, Germany
{christofer.fellicious,thomas.weissgerber}@uni-passau.de

Abstract. The domain of Artificial Neural Networks is growing very fast and new research ideas and results are published constantly. The advent of big data brought more complex network architectures and with that complexity, the training time often runs into multiple days. In such scenarios, a significant number of published papers only report the result on the basis of a single run. We explore the possibility whether such a reported result is robust enough and we investigate whether the variations in the result are significant.

In this paper, we explore the effects of random seeds and different optimizers on the accuracy of different network architectures.

Keywords: Neural networks · Random seeds

1 Introduction

Throughout the past two decades, the machine learning community has been growing by leaps and bounds. The drastic reduction in price to performance of raw computing power and the ability to create large datasets by crawling different platforms on the World Wide Web have contributed to a large number of rapid advancements in this field. Particularly Artificial Neural Networks (ANN) have been showing great promise in solving complex tasks such as image classification, text classification and autonomous driving. The core theory of Artificial Neural Networks reflects its biological counter part. Neurons relay information depending on an input and environmental parameters. In classification or regression tasks, a network makes decision based on its initialized weights. Depending on the initialization method this initially leads to random often imprecise predictions. To readjust the weights of the network it is fed with correct training data. Over the course of many iterations, the network learns to identify the features that are required for a particular task.

Nevertheless some properties of Artificial Neural Networks may hamper their usability. Artificial Neural Network may introduce a aspect of randomness or at least unpredictability in the workflow. The weights and biases in the network are generally initialized randomly which are drawn from distributions like the uniform distribution or normal distribution. Depending on a different initialization the same network architecture can procure different predictions even after

© Springer Nature Switzerland AG 2020
G. Nicosia et al. (Eds.): LOD 2020, LNCS 12566, pp. 93–102, 2020.
https://doi.org/10.1007/978-3-030-64580-9_8

a number of learning iterations. Another element for randomness can be found in the shuffling of training data. Data is generally shuffled before training and sometimes after every epoch of training. Due to this, releasing the initialization model parameters might not be sufficient to reproduce the exact machine learning model. These factors make Artificial Neural Networks more non-deterministic compared to other machine learning algorithms like Support Vector Machines or Decision Trees. The lack of determinism in Artificial Neural Networks can cause fluctuations in the accuracy between different experiments. Looking at the CIFAR-10 [8] leaderboard, architectures are tightly grouped especially around the 88%–93% accuracy range. There are twenty two published papers for the aforementioned accuracy values [1]. Some methods differ by less than 0.1% on the leaderboard. Nevertheless small improvements of scoring metrics are often reported. One has to ask themselves, if the network is better because of its architecture or might be better because it got random values that were conducive to reach the top spot. This pattern also repeats on the leaderboards of Kaggle[1] and of PapersWithCode[2].

Developing initialization strategies for neural networks yielding better end results and faster training times is a highly discussed topic. This includes papers and professional blogs like towards data science [5]. Nevertheless sources often fail to report details of their initialization strategy. They employ random initialization or do not specify whether they used a random seed for initialization of the weights and biases. This impedes an direct comparison of different experiments. In this paper, we compare different initialization seeds along with different optimizers to assess the impact of seed based randomness on produced metrics for each of them. For these means we kept hardware, software and input constant, varying only the initialization seed. All calculations where done on the same deterministic GPU setup. We then collate how different optimizers converge for the same random seed therefore initialization. The accuracy of the network is then compared over different initialization seeds and optimizers.

2   Background

During the training of an artificial neural network, some parameters introduce randomness into the network and this randomness can affect the outcome of the training. One of the main factors is the initialization strategy. There are different schemes for initialization of the parameters in a network such as Glorot [4] and He [6]. These initialization values can be drawn from a normal or uniform distribution. If a normal distribution is picked then the standard deviation plays a role on the probability of the values chosen.

There have been studies on different initialization strategies and its effects on the convergence of neural networks. Godoy [5] explains how different initialization strategies coupled with the different activation functions can lead to vanishing or exploding gradients. Ramos et Al. [14] and Kumar et al. [9]

---

[1] https://www.kaggle.com/c/cifar-10/leaderboard.

[2] https://paperswithcode.com/sota/image-classification-on-cifar-10.

for example look into how to initialize a neural network. Ramos et al. [14] did a quantitative experiment analysis of different initialization strategies on the MNIST dataset, mostly concentrating on the effect of different strategies on the same network architecture with random initialization rather than random seed based initialization. Kumar [9] explains why the Xavier initialization is bad for a network containing non-differential activation functions such as Rectified Linear Units(ReLU). This is significant as more and more networks use such functions as the activation function. While these papers show that choosing the right initialization strategy for the job is important, more often than not, they are used without a second thought and not reported in research publications. This is in our opinion one of the reasons for a current difficulty related to reproducibility and verifiability of scientific conclusions in the field of machine learning.

Another factor introducing randomness is the ordering of the input training data set. Can [2] has shown that as much as some changes in the ordering of the input data can cause significant changes in accuracy, while all the other parameters are kept constant. This experiment was conducted on two different architectures the ResNet-18 and the DenseNet. In the process of the experiment five different random seeds, two different learning rates and two different batch sizes for each network architecture were employed. Their experimental results via the ANOVA tests showed that the deviation of results were significant irrespective of the batch size, learning rate and the architecture. The conclusion of the paper is that researchers should provide the data ordering along with the results to improve the reproducibility of the experiment.

There are many different libraries for training neural networks such as PyTorch, TensorFlow and Keras. We need to look into how these libraries implement the different functions and whether they round off numbers in the GPU/CPU differently. These subtle changes could have an effect on how the optimizer traverses the loss landscape. Taking into account that computations additionally are based on a multitude of library and software versions, the experimental search space shouldn't just be constrained to the space defined by hyperparameter configurations.

Another aspect of variance might be found in the mode of network training itself. Normally, most networks are trained on GPU's because they are faster. Depending on the hardware results can differ and there is a multitude of different hardware available. Google for example has its own Tensor Processing Unit(TPU)s [7]. There are already observations reported on how TensorFlow computations on the GPU and may diverge from the results produced on a CPU [17]. Individual computations might only result in a deviation of 10^{-3} but such an discrepancy over thousands of computations multiplied by thousands of iterations could lead to significant differences. We also may need to look into the different CPU/GPU architectures. It has been reported that different GPUs display different results on pretrained models [3]. This is also another factor that affects reproducibility of experiments.

Musgrave et al. [10] shows that even though many papers claim significant advances in accuracy, a close inspection reveals that many accuracy advancements are marginal at best. They also "examine flaws in the current literature, including the problem of unfair comparisons, the bad practice of training with test set feedback, and the weaknesses of commonly used accuracy metrics". Musgrave et al. also points out flaws in existing literature such as unfair comparisons. They argue that for a new network to be better, as many parameters should be kept as constant as possible. This would help in making sure that it is the superior architecture of the new network and not a random hyperparameter that caused the increase in accuracy. They also point out that papers might omit small details that could have a huge impact on accuracy. Such omissions hurt the reproducibility of the experiment and may hinder the progress of science.

3   Experimental Setup

The experiment was run in a software reproducibility and tracking tool for machine learning known as PaDRe[3], which was developed in parallel with this work. The tool provided support for tracking the needed results such as individual prediction and automatically compute the metrics and store them in a standardized way. This experiment was a classification experiment and therefore the precision, recall, f1-score and accuracy metrics were computed. We only use the accuracy metric as that is the most commonly noted metric in most research papers for the CIFAR-10 dataset.

PaDRe was initially written for the scikit-learn [13] library. In order to incorporate neural networks into PaDRe, we needed to expand the functionality of PaDRe. PyTorch was chosen due to its easy integration into Pythonic nature and the popularity of the library [11,12].

The plugin supports all the layers, optimizers, transformations available with the PyTorch library. The user can create layers by giving the required layer a name, specifying its type and all the additional parameters that are required by the layer. This allows users to create networks via JSON files without any programming knowledge on the part of the user. The plugin fits within the PaDRe framework by extracting the different hyperparameters of the network such as the different layers, optimizers, loss functions, testing and training parameters, if they are present. The user also has additional options to set a random seed to make the network deterministic, use the GPU if it is available, test after every epoch of training and so on. Also, this plugin allowed us to run the required networks easily and also modify the networks without worrying about the underlying code change because the network creation complexity was taken care of by the Plugin itself. We then used PaDRe to extract the different hyperparameters of the experiment including the network architecture, the optimizer, the loss function, the training duration automatically and store them on to the disk. This automatic parameter extraction was helpful as it helped us not worry about whether the experiment was tracked and the changes in the parameters of the experiment. And once the experiment was done, PaDRe allowed us to easily access the metrics and use it for plotting the results.

---

[3] https://github.com/padre-lab-eu/pypadre.

For this experiment, we chose six different Convolutional Neural Network architectures. Four of these were based on the VGG network and two simple network architectures called Baseline-1 and Baseline-2. VGG architectures were chosen because they were linear architectures without any branching such as in Googlenet [16] and our PyTorch plugin only supported linear architectures. We need to get an overview on how the accuracy was affected with increased number of parameters and we also needed to analyze it against smaller networks having considerably lesser number of parameters in this case Baseline-1 and Baseline-2 which had 80k and 150k parameters approximately. The accuracy of the VGG network was around 90% which is a densely populated section in different CIFAR-10 leaderboards. In addition to this, the VGG networks are easier to train when compared with network architectures having much higher number of parameters. Thus VGG provided the best balance between performance, ease of training and being on one of the densely populated areas of the CIFAR-10 leaderboards.

The VGG inspired networks were named VGGScaled, VGG-11, VGG-13 and VGG-16. We also chose SGD [15], Nesterov Accelerated Gradient and the Adadelta optimizers for training each of these networks. When choosing the optimizers, Stochastic Gradient Descent was the primary choice for its simplicity and robustness. The Nesterov momentum was selected to investigate the effect of the future look ahead of momentum on the accuracy. Also, Nesterov momentum was similar to that of SGD and it is interesting to have an evaluation on the effect of the momentum on the final accuracy of the network. The previously chosen optimizers depended on the initial learning rate and also required to be tuned manually with learning rate schedulers to extract the best performance. Now, we needed an optimizer that did not depend on the learning rate. Adadelta was chosen as it did not need a learning rate term[4] and used a windowed approach when compared to Adagrad where the learning of the network slowed down greatly over time due to the accumulated gradients. This would help us compare how each optimizer performs with the same random seed based initialization of the parameters in the network. We also used three different initialization methods to assess the variance in the results for each combination of network architecture and optimizer. The three different initialization methods are

Sequential Initialization Initialized with values from 0 to 19
Pseudorandom Initialization Initialization with random values generated by the numpy library. The values were chosen by feeding numpy with the sequence 0–19 as the random seed and extracting a single integer value between 0 and 65535.
Uninitialized Initialized randomly by the library itself as this is the usual initialization method

Using this we can set the seed for the Pseudo Random Number Generator(PRNG) and this would help us to recreate the exact same values for the

---

[4] Set to 1.0.

parameters over and over again across different hardware, provided the PRNG does not change.

The dataset for this experiment was the CIFAR-10 dataset. CIFAR-10 offered the advantage that it was a standard dataset and it is small and easier to test out networks when compared to the Imagenet dataset. For this experiment, we kept the split of the training and testing datasets constant and in the same order across all the runs of different initialization, optimizer and network architecture. In addition to this, the input training dataset was kept static throughout the training process.

We also made the results of the network deterministic by keeping the experimental setup and environment static. We initialized the network with the same weights for a fixed random seed, using a predefined dataset splitting for training and not shuffling the dataset after each epoch. An identical hardware setup was used for all the experiment executions.

All the source code and the raw experimental data is available at Zenodo[5]. This includes the Jupyter notebook used to aggregate and plot the various results which is named as *metrics.ipynb*. The data used for training and testing is present within */data* with the images stored in *features.npy* and the corresponding labels in *label.npy*. The different JSON files used for creating the networks are stored with in */code*. This folder also includes the python directories used to execute the networks. The experimental output data is collected in */experiments*. The results are then further grouped by network architecture.

4 Results and Discussion

We executed each network architecture with three different optimizers and three different initialization methods. For each combination of network architecture and optimizer we had twenty different runs for each initialization method. This added up to 180 runs for each network architecture.

All these runs were aggregated and the mean and standard deviation of the accuracy values were computed. We are more interested in the variation of accuracies with respect to the initialization method and optimizer. Ideally, a robust network should have very little variance in the metrics which translates to better experiment reproducibility. If a network has a large variance, a user trying to reproduce the experiment might not be able to pin the variance to an error in the experimental setup or the natural variance inherent in the network itself.

This also translates to relying more on luck or random chance to beat other networks on the leaderboards.

The tabulated results are grouped by architecture and the minimum, maximum mean and standard deviations of accuracies are calculated grouped by initialization method. Table 1 shows the lowest accuracy, highest accuracy, highest mean accuracy and lowest standard deviation for a specific architecture.

---

[5] https://zenodo.org/record/3676968.

Table 1. Overview of accuracies over different network architectures

Architecture	Minimum	Maximum	μ	σ
Baseline 1	76.56	81.66	80.402	0.469
Baseline 2	78.80	82.80	80.899	0.334
VGGScaled	85.10	89.06	88.320	0.217
VGG-11	87.20	89.88	88.849	0.191
VGG-13	88.50	91.54	90.584	0.216
VGG-16	10.20	90.88	89.719	0.218

All the networks except the VGG-Scaled architecture reported their lowest standard deviations with SGD. This means that out of all the optimizers, SGD is the most robust but also requires fine tuning on the learning rate scheduling parameters. The best average results for the Baseline-1 architecture was also reported with SGD as shown in Fig. 1a and this was the only architecture that had the best results with SGD. For the Baseline-2 architecture, Nesterov optimizer performs similar to the Adadelta and with lesser variance as seen in Fig. 1b while for other architectures it performs similar to SGD.

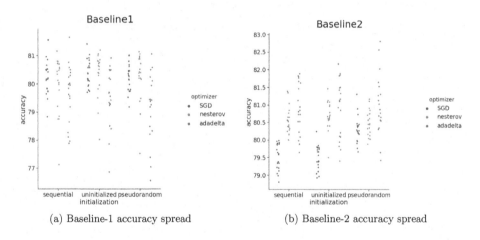

(a) Baseline-1 accuracy spread   (b) Baseline-2 accuracy spread

Fig. 1. Accuracies of baseline networks

From the graphs we see that the Adadelta optimizer outperforms all the other optimizers in almost all architectures tested here. This implies that Adadelta works quite well out of the box in training the parameters.

A user can simply do multiple runs on a network with Adadelta and end up with a good result without requiring any fine tuning. On the downside, Adadelta also has the highest variance among all the optimizers which makes multiple runs necessary for approximating the true capability of the neural network architecture. Even then looking at the Figs. 2b and 3a, we see that Adadelta is head and shoulders above the other two optimizers. An exemplary example would be the performance of the VGG-Scaled network with Adadelta as seen in Fig. 2a. Another interesting phenomenon was that the network failed to converge twice during training on the VGG-16 architecture when using initialized methods. These are marked as outliers in Fig. 3b. The VGG-16 architecture has similar accuracy measures to that of VGG-13, and we believe that the VGG-16 architecture will benefit from additional epochs of training because of its higher number of parameters.

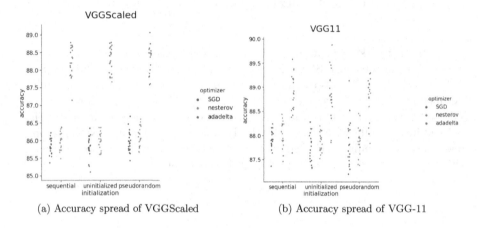

(a) Accuracy spread of VGGScaled   (b) Accuracy spread of VGG-11

Fig. 2. Accuracy of VGG-Scaled and VGG-11

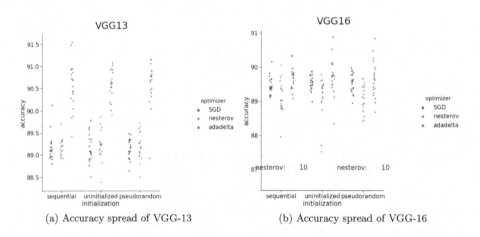

(a) Accuracy spread of VGG-13   (b) Accuracy spread of VGG-16

Fig. 3. Accuracy of VGG-13 and VGG-16

We understand that running the network multiple times is not always feasible especially if the network and dataset are quite large. But for openness and reproducibility of the data, we require more information about how the network was initialized. And in these cases we suggest storing the initial parameter values even before the first weight update takes place.

5   Conclusion

We have shown that irrespective of the optimizer or initialization method there is a significant variance in accuracy. We also see that the Adadelta optimizer exhibits the greatest variance among the optimizers we tested regardless of the network architecture. Nevertheless assessing if some optimizers might be more prone to variance compared to others when the initialization seed is changed needs further research. For the scope of the experiment it has to be acknowledged that the covered search space of random seeds was very small compared to the possible values. But even in this small selection of the search space, we could show that aggregated metrics vary in a considerable degree. Providing only a single run can therefore not be the foundation for robust conclusions. To allow for detailed experimentation the reproducibility of experiments has to be secured. A larger number of runs and a heuristic tailored for such a situation will help us gain better understanding on the performance and the robustness of a certain network. We also could not provide a single random seed that was able to provide consistently exceptional results over different optimizers. This indicates that the final result not only depends on the initialization values but also on how the weights are updated. Neural networks in their current state still continue to be a black box, while being used in a multitude of domains.Finding networks which improve on aggregated metrics for experimental setups might further practical applications, but a need to invest more effort into the exploration of what exactly makes a network perform well also exists. This can be summarized as a necessity for more openness, better documentation and extensive analysis of experimental setups, given that critical factors for the scores of aggregated metrics are sometimes not tracked.

6   Future Work

We have only considered the effects of the provided random seed on python environment level among the many different random factors that could influence the accuracy of a network. While we attempted to keep them as stable as possible for this experiment, we also need to look into the scope of their effect to enable a assessment for independent experiments. Such parameters can be for example the used hardware and software versions as well as hardware configuration parameters. Especially the versions of used drivers seem to impact the training of networks, when using Graphics Processing Units. Following this line of thought the fluctuation of accuracy with respect to training on CPU and a GPU needs to be inspected. For an extensive study all these parameters should be tracked and classified as well as analyzed on their properties. There might be random factors, which reinforce or dampen their effects on the final scoring metric. We also need a heuristic that will provide us with how many times a network should be executed for the results to be independent of the variations of the initial random seeds.

References

1. Benenson, R.: Classification datasets results. https://rodrigob.github.io/are_we_there_yet/build/classification_datasets_results.html. Accessed 08 Dec 2019
2. Can, E.F., Ezen-Can, A.: The effect of data ordering in image classification (2020)
3. Eengineer: gpgpu - tensorflow: different results on different gpus - stack overflow. https://stackoverflow.com/questions/46303920/tensorflow-different-results-on-different-gpus Accessed 29 Apr 2020
4. Glorot, X., Bengio, Y.: Understanding the difficulty of training deep feedforward neural networks. In: Proceedings of the Thirteenth International Conference on Artificial Intelligence and Statistics, pp. 249–256 (2010)
5. Godoy, D.: Hyper-parameters in action! part ii - weight initializers. https://towardsdatascience.com/hyper-parameters-in-action-part-ii-weight-initializers-35aee1a28404 Accessed 29 Apr 2020
6. He, K., Zhang, X., Ren, S., Sun, J.: Delving deep into rectifiers: Surpassing human-level performance on imagenet classification. In: Proceedings of the IEEE international conference on computer vision, pp. 1026–1034 (2015)
7. Jouppi, N.P., et al.: In-datacenter performance analysis of a tensor processing unit. In: Proceedings of the 44th Annual International Symposium on Computer Architecture, pp. 1–12 (2017)
8. Krizhevsky, A., Hinton, G., et al.: Learning multiple layers of features from tiny images. Technical report, Citeseer (2009)
9. Kumar, S.K.: On weight initialization in deep neural networks. arXiv preprint arXiv:1704.08863 (2017)
10. Musgrave, K., Belongie, S., Lim, S.N.: A metric learning reality check. arXiv preprint arXiv:2003.08505 (2020)
11. Paszke, A., et al.: Automatic differentiation in PyTorch (2017)
12. Paszke, A., et al.: Pytorch: An imperative style, high-performance deep learning library. In: Advances in Neural Information Processing Systems, pp. 8024–8035 (2019)
13. Pedregosa, F., et al.: Scikit-learn: machine learning in Python. J. Mach. Learn. Res. **12**, 2825–2830 (2011)
14. Ramos, E.Z., Nakakuni, M., Yfantis, E.: Quantitative measures to evaluate neural network weight initialization strategies. In: 2017 IEEE 7th Annual Computing and Communication Workshop and Conference (CCWC), pp. 1–7. IEEE (2017)
15. Ruder, S.: An overview of gradient descent optimization algorithms. arXiv preprint arXiv:1609.04747 (2016)
16. Szegedy, C., et al.: Going deeper with convolutions. In: Proceedings of the IEEE Conference on Computer Vision and Pattern Recognition, pp. 1–9 (2015)
17. zergylord: Inaccuracies in tf.reduce_sum · issue #5527 · tensorflow/tensorflow · github. https://github.com/tensorflow/tensorflow/issues/5527. Accessed 29 Apr 2020

Benchmarking Deep Learning Models for Driver Distraction Detection

Jimiama Mafeni Mase[1]([✉]), Peter Chapman[2], Grazziela P. Figueredo[1], and Mercedes Torres Torres[1]

[1] School of Computer Science, The University of Nottingham, Nottingham, UK
psxjmma@nottingham.ac.uk
[2] School of Psychology, The University of Nottingham, Nottingham, UK

Abstract. The World Health Organisation reports distracted driving as one of the main causes of road traffic accidents. Current studies to detect distraction postures focus on analysing image features. However, there is lack of a comprehensive evaluation of state-of-the-art deep learning techniques employed for driver distraction detection, which limits or misguides future research in this area. In this paper, we conduct an in depth review of deep learning methods used in driver distraction detection and benchmark these methods including other popular state-of-the-art CNN and RNN techniques. This will assist researchers to compare their novel deep learning methods with state-of-the-art models for driver distraction posture identification. We evaluate 10 state-of-the-art CNN and RNN methods using the average cross-entropy loss, accuracy, F1-score and training time on the American University in Cairo (AUC) Distracted Driver Dataset, which is the most comprehensive and detailed dataset on driver distraction to date. Results show that pre-trained InceptionV3 CNNs coupled with stacked Bidirectional Long Short Term Memory outperforms state-of-the-art CNN and RNN models with an average loss and F1-score of 0.292 and 93.1% respectively.

Keywords: Deep learning · Image classification · Driving distraction postures

1 Introduction

Distracted driving (such as daydreaming, cell phone usage and looking at something outside the car) accounts for a large proportion of road traffic fatalities worldwide. Researchers are exploring artificial intelligence to understand driving behaviours, find solutions to mitigate road traffic incidents and develop driver assistance systems [3,13,27,28]. However, the number of road traffic fatalities has been continuously increasing over the last few years [1].

J.M. Mase—The first author is supported by the Horizon Centre for Doctoral Training at the University of Nottingham (UKRI Grant No. EP/L015463/1) and by Microlise.

© Springer Nature Switzerland AG 2020
G. Nicosia et al. (Eds.): LOD 2020, LNCS 12566, pp. 103–117, 2020.
https://doi.org/10.1007/978-3-030-64580-9_9

The Center for Disease Control and Prevention (CDC) provides a broad definition of distracted driving by taking into account its visual, cognitive and manual causes [8]. *Visual distraction* refers to situations where drivers take their eyes off the road, usually due to roadside distractions. *Cognitive distractions* are caused by listening, conversing, and daydreaming. *Manual distractions* are mainly concerned with drivers' activities other than driving, such as reaching behind their seats and using cell phones. The majority of the above examples of distractions can easily be captured using video cameras, which are unobtrusive and can capture naturalistic driving behaviour.

With the dramatic increase in computational power, deep neural networks have demonstrated impressive performance in automatically extracting image features for computer vision tasks, such as image classification [39,43], object detection [24,31] and other image analysis tasks [25,26]. This has caused a shift in image analysis from hand-crafted feature learning (where features are manually derived using expert knowledge) to deep learning. Current studies have explored Convolutional Neural Networks (CNNs) to extract the spatial information from images [20,29,44] and Recurrent Neural Networks (RNNs) to extract spectral information [28] for the classification of driving postures, with promising results. However, there is lack of a comprehensive review and evaluation of these models for driver distraction detection, which limits or misguides future research in this area. Therefore, benchmarking state-of-the-art CNN and RNN models will produce fair performance rankings and assist future research in developing novel approaches for driver distraction detection.

In this paper, we present an in-depth review of the literature on driver distraction detection and benchmark the deep learning methods including other popular state-of-the-art CNN and RNN techniques on the American University in Cairo (AUC) Distracted Driver Dataset [2,12], which is the most comprehensive and detailed dataset for driver distraction detection. We compare the architectures using three adequate classification evaluation metrics including the average training time i.e., average cross-entropy loss, accuracy and F1-score.

This paper is organised as follows, in Sect. 2 we review the literature on driver distraction detection using traditional machine learning and deep learning techniques. Later in Sect. 3, we provide an overview of CNNs and RNNs architectures, and the classification evaluation metrics used to evaluated the methods. In Sect. 4, we introduce the publicly available driver distracted dataset, describe hyperparameter configuration and the evaluation protocol. In Sect. 5, the results are presented along with discussion, and Sect. 6 concludes the paper and establishes the opportunity for future work.

2 Literature Review

2.1 Traditional Machine Learning Approaches

Initial studies on driver distraction detection were based on handcrafted features and traditional machine learning methods, such as Support Vector Machine (SVM) classifiers. For example, Artan *et al.* [4] used SVMs for detecting driver cell phone usage using a Near Infrared (NIR) camera system directed at the vehicle's front windshield. Their methodology consisted of localisation of facial landmarks using Deformable Part Model(DPM) and the classification of the facial landmarks using Support Vector Machines (SVM). The authors captured 1500 front-view vehicle images from a public roadway, with 378 images of drivers using cell phone and 1122 images without using cell phone.

Similarly, Berri et al. [6] explored SVMs for detecting cell phone usage in frontal images of drivers. They used hand-crafted feature learning techniques to extract Haar-like-features for the identification of distraction. Their method was evaluated on a driver frontal image dataset consisting of 100 images of drivers using cell phones and 100 images with no phone. In additional, the authors used Genetic Algorithm (GA) for hyperparameter optimisation.

Later, Craye and Karray [11] explored hand-crafted features using AdaBoost and Hidden Markov models. Four feature sets were extracted from 8 drivers: 1) arm position, 2) face orientation, 3) action units and 4) gaze estimation and eye closure. The extracted features were concatenated into a seventeen-features vector for each image frame. Subsequently, five postures were analysed: making a phone call, drinking, sending an SMS, looking at an object inside the vehicle and driving normally.

The above studies use hand-crafted feature learning techniques, which are time consuming and require expert knowledge to manually extract the image features. In addition, the studies use very small number of images (less than 2000 images in both training and test sets) with limited number of drivers (less than 8 drivers) to evaluate their methods. The datasets also consist of a small variety of distracted driving postures (less than five distracted driving postures).

2.2   Deep Learning

Recently, deep learning methods have proven to outperform these traditional machine learning techniques when automatically detecting distracted postures. Kim et al. [20] proposed a method of detecting driver distraction using RestNet and MobileNet CNN models. However, their study only focused on two types of distraction: looking in-front and not looking in-front postures. Their results on training the models from scratch and using fine-tuned pre-trained models show that performance of fine-tuned models significantly outperforms training from scratch. In addition, the MobileNet CNN model outperforms the Inception RestNet for both accuracy and processing time. However, their dataset consists of images from only two drivers with only two distraction postures. which makes their models not generalisable.

Similarly, Yan et al. [44] examine CNNs to classify driver distraction postures. The authors use pre-trained CNNs and evaluate their model on three datasets: the *Southeast University Driving Posture* dataset, and two datasets developed by the authors called *Driving-Posture-atNight* and *Driving-Posture-inReal* datasets. Results show high classification accuracy with the three driving posture datasets which outperformed methods using hand-crafted features.

However, their datasets have only four distraction postures and are not publicly available for benchmarking.

Later, Majdi *et al.* [29] employ CNNs for detecting driver distraction postures. The authors adopted the U-Net CNN architecture for capturing context around the objects. Their model was trained on the American University in Cairo (AUC) Distracted Driver dataset. Their results show great improvement in accuracy compared to Support Vector Classifiers and other CNN architectures. Likewise, Eraqi *et al.* [12] propose a weighted ensemble of CNNs using four different CNN architectures (i.e AlexNet network [22], InceptionV3 [40] networks, ResNet networks [16], and VGG-16 networks [45]). The CNNs are trained on five different image sources of the AUC distracted driver dataset i.e. raw images, skin-segmented images, face images, hands images, and face and hands images. The results from the individual CNNs show the best accuracy when trained on the raw images. Subsequently, the predictions from the different CNNs are combined using a weighted Genetic Algorithm (GA) and the results from the fusion show improved accuracy compared to the independent CNNs and majority voting fusion.

Driving postures with similar spatial information are difficult to classify using CNNs. As a result, Mafeni *et al.* [28] propose a hybrid deep learning model consisting of pre-trained InceptionV3 CNNs to learn the spatial information in the images and stacked Bidirectional LSTMs to extract the spectral information amongst the feature maps from the pre-trained CNNs. The authors evaluate their model on the AUC Distracted Driver Dataset. Results show improved detection of similar driving postures (e.g., "reaching behind the vehicle" vs "turn and talk to a passenger" postures) with far more accuracy compared to CNN models. In addition, their approach beats state-of-the-art CNN models in average classification accuracy.

The studies reviewed above lack comprehensive evaluation of their methods with state-of-the-art CNN and RNN models in identifying distraction postures. In this paper, we address this limitation by benchmarking the deep learning methods reviewed in the literature including other popular state-of-the-art CNN and RNN models on the most comprehensive and detailed dataset on driver distraction (the AUC Distracted Driver Dataset) using adequate classification evaluation metrics.

3   Methodology

3.1   An Overview of CNNs and RNNs

Convolutional Neural Networks. CNNs [23,34] are neural networks consisting of filtering (or convolution), pooling and activation layers. The inputs go through the convolution layer, where they are filtered to produce stacked smaller dimensional features (feature maps). The stacked feature maps go through the pooling layer, which downsamples the input representations using a sample-based discretisation process, such as max pooling [30]. The activation layer later converts the stacked downsampled data into specific features depending on the activation function that is used (e.g. Rectified Linear Unit (ReLU) converts all negative values to zero and maintains all positive values). These filtering, pooling and activation layers allow CNNs to learn hierarchical discriminative features.

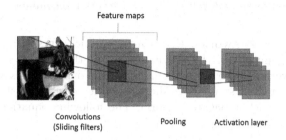

Fig. 1. A simple CNN architecture consisting of convolution, pooling and activation layers

Figure 1 presents a simple CNN architecture with one convolution, one pooling and one activation layer. Most state-of-the-art CNN models [40,42,45,46] consist of a concatenation of many of such layers with additional units for Batch normalisation and regularisation. For a more elaborate description of CNN architectures, please refer [32,33]. The studies (i.e., [32,33]) provide an in-depth review of the different state-of-the-art CNN architectures for image classification tasks, a review of classification loss functions (e.g., softmax loss) and optimisation techniques (e.g., Stochastic Gradient Descent). The state-of-the-art CNN approaches include AlexNet [22], VGG [37], Inception [41], ResNet [16] and DenseNet [19].

- **AlexNet:** AlexNet was introduced in 2012. The network consists of five convolutional layers and three fully connected layers. ReLU activation function is applied to the output feature maps of every convolutional layer. The last convolutional layer outputs 256*13*13 feature maps.
- **VGG:** VGG was introduced in 2014. The network is deeper than ALexNet, with thirteen convolutional layers and three fully connected layers. VGG uses smaller kernels compared to AlexNet but also uses ReLU non-linear activation functions.
- **Inception:** Inception was introduced in 2014 with its main motivation to deal with the uncertainty in choosing the kernel size of convolutional layers. The network uses multiple kernel sizes in each convolutional layer to complement the advantages of different kernel sizes while avoiding deeper architectures which can lead to overfitting.
- **ResNet:** ResNet was introduced in 2015 to handle the vanishing gradient problem [5]. The network solves this problem by introducing skip connections between layers. This allows the gradients to flow properly through the skip connections to any other earlier layer.

- **DenseNet:** DenseNet was introduced in 2017 as an upgrade of ResNet. In DenseNet, each layer connects to every other layer i.e., the input of any layer is the concatenated feature maps of all subsequent layers. This reduces the number of parameters compared to ResNet, however, very deep DenseNet variations have been recently developed that have very large number of parameters.

Recurrent Neural Networks. RNNs are neural networks with feedback loops that connect the output of the previous state to the current state [47]. This enables the network to learn temporal or contextual dependencies among sequential data. However, RNNs only have short term memory due to the issue of vanishing gradients (backpropagated errors). The following equations represent a typical RNN:

$$h_t = f_h(W_h X_t + U_h h_{t-1} + b_h)$$
$$y_t = f_o(W_o h_t + b_o)$$

Where X_t is the current input data at time t, h_t is the current hidden layer units obtained from the current input data and the previous hidden layer units(h_{t-1}) and y_t represents the output. W, U, and b represent the weight matrices and bias vectors which need to be learned during training, while f_h and f_o are the non-linear activation functions.

LSTMs [18] are an extension of RNN, which are capable of learning short and long-term dependencies in the data by introducing additionally gates to the memory cell; forget, input, and output gate layers. The input gate controls which state is updated. The forget gate controls how much information needs to be retained or forgotten, and the output gate decides which part of the cell state is outputted to the next LSTM unit. These gates control information flow into and out of the LSTM cell unit.

The interactions between the gate layers in the LSTM memory unit is given by the following equations:

$$f_t = \text{sigm}(W_f X_t + U_f h_{t-1} + b_f)$$
$$i_t = \text{sigm}(W_i X_t + U_i h_{t-1} + b_i)$$
$$o_t = \text{sigm}(W_o X_t + U_o h_{t-1} + b_o)$$
$$C_t = f_t \odot C_{t-1} + i_t \odot \tanh(W_c X_t + U_c h_{t-1} + b_c)$$
$$h_t = o_t \odot \tanh(C_t)$$

Where f_t is the forget gate's activation vector, i_t is the input or update gate's activation vector, and o_t is the output gate's activation vector. W, U, and b represent the weight matrices and bias vectors which need to be learned during training.

Stacked LSTMs were introduced by Graves *et al.* [15] for speech recognition. The model consists of multiple hidden states where each hidden layer contains multiple memory cells. This approach potentially allows the hidden state at each level to operate at different and deeper timescales. In addition, Schuster and Paliwal [36] developed the Bidirectional Recurrent Neural Network (BiRNN), which merges information from past and future states simultaneously. BiRNNs are now implemented using LSTMs (Bidirectional LSTM) and have produced state-of-the-art results in many popular classification tasks [14, 26] tasks.

3.2   Evaluation Metrics

Accuracy. Classification accuracy is the proportion of correct predictions among the total number of input samples. With reference to truth table values (i.e., True Positives (TP), True Negatives (TN), False Positives (FP) and False Negatives (FN)), the accuracy of a model can be represented as follows:

$$Accuracy = (TP + TN)/(TP + FP + FN + TN)$$

Accuracy is a valid evaluation metric for balanced datasets. However, the AUC test set is not well balanced with one posture having 346 images and another with 143 images, and a standard deviation of 57 images amongst the postures. Therefore, we also use the F1-score to evaluate model performance, which is a better evaluation metric for uneven class distributions.

F1-Score. F1 Score gives a better measure of incorrectly predicted instances i.e. False Positives and False Negatives. It is the weighted average of precision and recall. The F1 score of a model is expressed as follows:

$$Precision = TP/(TP + FP)$$
$$Recall = TP/(TP + FN)$$

$$F1 = 2 * (Precision * Recall)/(Precision + Recall)$$

Cross-Entropy Loss. Cross-entropy loss measures the performance of a classification model whose outputs are class probabilities between 0 and 1. It measures how much the predicted value varies from the actual label using the log of the predicted probability. During the learning process, the cross-entropy loss is minimised for greater prediction accuracy. The cross-entropy loss is expressed as follows:

$$CrossentropyLoss = -1/N \sum_{i=1}^{N} \sum_{j=1}^{M} y_{ij} * log(p_{ij})$$

Where N is the number of instances and M is the number of classes. y_{ij} is 1 if the instance i belongs to class j else 0 and p_{ij} is the prediction probability of instance i belonging to class j.

4   Experiments

In this section we introduce the AUC distracted driver dataset used to benchmark the deep learning models. We also describe the hyperparameter configurations of the models and the evaluation protocol.

4.1   The American University in Cairo Distracted Driver Dataset

The American University in Cairo (AUC) Distracted Driver dataset [2,12] is the largest and most comprehensive publicly available dataset for driver distraction identification. The dataset captures most real-world distracted driving postures (up to 10 postures): safe driving (c0), text right (c1), right phone usage (c2), text left (c3), left phone usage (c4), adjusting radio (c5), drinking (c6), reaching behind (c7), hair or makeup (c8), and talking to passenger (c9). The dataset was captured using an ASUS ZenPhone rear camera (Model Z00UD), and consists of 1080 * 1920 pixel images. The dataset contains images from 44 drivers. 38 drivers are used in the training set and 6 drivers in the test set. Table 1 shows the number of images in the training and test sets for each driving posture.

Table 1. Description of AUC Distracted driver dataset

Types of driving postures	Number of images in training set	Number of images in test set
C0	2,640	346
C1	1,505	213
C2	1062	194
C3	944	180
C4	1,150	170
C5	953	170
C6	933	143
C7	891	143
C8	898	146
C9	1,579	218

4.2   Hyper-parameter Configuration

Table 2 presents the hyper-parameter configuration of the CNN and RNN models. First, we initialise the parameters of the CNN models with the parameters obtained when they are trained on ImageNet dataset [35] (represented by Pretrained CNN = True in Table 2). Subsequently, we replace the number of outputs

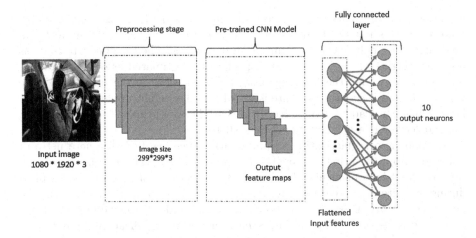

Fig. 2. CNN architecture for detecting driving distraction postures

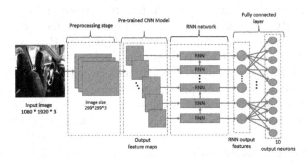

Fig. 3. CNN-RNN architecture for detecting driving distraction postures

in the fully connected layers of the CNN models with the number of distracted postures (10 neurons) as shown in Fig. 2. While for the CNN-RNN models, we remove the fully connected layers of the CNNs and use their output feature maps as inputs to the RNN networks as shown in Fig. 3. It is important to note that we use Inception-V3 as the CNN architecture for the CNN-RNN models because Inception-V3 showed remarkable performance compared to other state-of-the-art CNN models (Alexnet, Resnet and VGGnet) and had comparable performance with Densenet with far lower training time as shown in Fig. 4.

Neural networks are trained by optimising their objective functions. In this case, the objective function is the cross-entropy loss function described in Sect. 3.2. Most machine learning techniques use stochastic gradient descent [7] to optimise the loss function because of its capability to deal with very large datasets by randomly selecting samples (to form a batch of samples) that are used to estimate the gradient for each iteration. In this study, we evaluate our models using the simple stochastic gradient descent algorithm (referred to as SGD in pytorch) and Adam [21] (which is a variant of SGD). It is important to note that over the years, there have been improvements and variations of SGD, however, this is not the focus of this study and our evaluation framework can be extended to include other optimisation techniques. The learning rate controls how the weights are updated with respect to the estimated error. If the learning rate is very low, the learning process will be slow as the updates will be very small, and if the learning rate is very high, the weight updates will be very large which can lead to divergence. In this study, we optimise the models using popular learning rates between 0.1 and 0.0001 [38].

Dropout is a regularisation technique to reduce overfitting, where network nodes are dropped during training [17] to prevent co-adaptations of feature detectors. Dropout is implemented by setting the percentage of nodes to be dropped during each iteration (e.g., setting the dropout value to 0.5 means 50% of the nodes will be dropped). We use no dropout, 0.5 and 0.2 to evaluate the RNN architectures as these dropout values have proven to significantly improve performance [17]. It should be noted that we did not implement dropouts in the CNN architectures because we adapted already existing networks. Batch size defines the number of instances to be propagated before updating the model's parameters [10]. By considering the number of images in datasets (less than 15k images) and the depth of the networks (very deep networks), we choose small batch sizes of 8, 16, 32 and 64 to evaluate the models. The number of layers of the RNN networks are the number of RNN memory cells, and the hidden size is the number of hidden states. To prevent overfitting, we selected small number of layers between 1 and 4 and hidden states in the range 16 to 128 (increments of 16) to evaluate the RNN models. Lastly, the input features of the RNN architectures is chosen between 64 and 256 (increments of 64). We choose this range due to the size of the output feature maps of the CNNs, which is less than 256 in most of the CNN networks.

We split the AUC training set that consist of 38 drivers into a new training set with 35 drivers and a validation set with 3 drivers. Later, experiments are carried out by evaluating the validation loss of different values of each hyperparameter while keeping the others constant to obtain the optimal values in Table 2. We use early stopping to stop the learning process when the validation loss stops decreasing for 5 consecutive epochs. This also helps to avoid model overfitting. The AUC test set consisting of 6 drivers was used to benchmark the models. The experiments were executed on a graphics processing unit (GPU) using 4 CPU cores and 6GB RAM. Our code was implemented in Pytorch with an epoch size of 50 for each experiment.

4.3   Evaluation Protocol

After optimising the models, we evaluated their classification performance using the Average Cross-entropy Loss (ACL), Average Accuracy (AA), Average F1-score (AF) and the Average Training Time (ATT) across the different distracted driving postures after 5 runs.

Table 2. Hyper-parameter Configuration of CNN and RNN models

	VGG-19	Densenet-201	Resnet50	AlexNet	InceptionV3	Inception V3-RNN	Inception V3-LSTM	Inception V3-GRU	Inception V3-BiGRU	Inception V3-BiLSTM
Pretrained CNN	True	True	True	True	True	True	True	True	True	True
Re-trained layers	All	All	All	All	All	All	All	All	All	All
Batch size	32	32	32	32	32	16	16	16	8	8
Learning rate	0.0001	0.0001	0.0001	0.0001	0.0001	0.0001	0.0001	0.0001	0.0001	0.0001
Optimizer	Adam	Adam	Adam	Adam	Adam	Adam	Adam	Adam	Adam	Adam
Input features	/	/	/	/	/	256	256	256	64	64
Hidden size	/	/	/	/	/	64	64	64	128	128
Number of layers	/	/	/	/	/	1	1	1	1	1
Dropout	/	/	/	/	/	No	No	No	No	No

5   Results and Discussion

Table 3. ACL, AA, AF, ATT and the average accuracy of each posture after five runs performed by models on the AUC 'split-by-driver' Distracted Driver test dataset

	VGG-19	Densenet-201	Resnet50	AlexNet	Inception V3	Inception V3-RNN	Inception V3-LSTM	Inception V3-GRU	Inception V3-BiGRU	Inception V3-BiLSTM
ACL	0.531 ± 0.14	0.395 ± 0.07	0.442 ± 0.06	1.024 ± 0.16	0.442 ± 0.10	0.418 ± 0.06	0.375 ± 0.02	0.348 ± 0.01	0.336 ± 0.06	**0.292 ± 0.01**
AA	0.833 ± 0.06	0.890 ± 0.03	0.877 ± 0.02	0.738 ± 0.03	0.884 ± 0.01	0.884 ± 0.01	0.902 ± 0.01	0.903 ± 0.01	0.917 ± 0.01	**0.917 ± 0.003**
AF	0.835 ± 0.06	0.895 ± 0.02	0.882 ± 0.02	0.741 ± 0.04	0.890 ± 0.00	0.899 ± 0.02	0.906 ± 0.02	0.909 ± 0.01	0.922 ± 0.00	**0.931 ± 0.004**
ATT	214.1 ± 6.21	298.9 ± 47.7	209.1 ± 9.94	**196.0 ± 0.90**	221.5 ± 4.34	240.7 ± 5.12	264.8 ± 5.23	250.7 ± 20.9	277.4 ± 32.0	289.8 ± 4.65
C0	0.86 ± 0.11	0.89 ± 0.05	0.89 ± 0.06	0.78 ± 0.09	0.87 ± 0.08	0.94 ± 0.05	0.93 ± 0.03	0.91 ± 0.05	0.95 ± 0.01	**0.95 ± 0.00**
C1	**0.97 ± 0.01**	0.83 ± 0.06	0.89 ± 0.13	0.74 ± 0.09	0.83 ± 0.09	0.94 ± 0.08	0.89 ± 0.03	0.87 ± 0.06	0.90 ± 0.08	0.82 ± 0.05
C2	0.99 ± 0.01	**1.00 ± 0.00**	1.00 ± 0.00	0.83 ± 0.15	**1.00 ± 0.00**	0.99 ± 0.00	0.99 ± 0.00	**1.00 ± 0.00**	**1.00 ± 0.00**	0.98 ± 0.01
C3	0.76 ± 0.20	0.96 ± 0.01	0.92 ± 0.08	0.91 ± 0.12	0.97 ± 0.04	0.99 ± 0.01	0.96 ± 0.03	0.98 ± 0.01	0.99 ± 0.03	**1.00 ± 0.00**
C4	0.94 ± 0.05	0.99 ± 0.01	0.98 ± 0.01	0.82 ± 0.07	0.97 ± 0.03	**0.99 ± 0.00**	0.98 ± 0.01	**0.99 ± 0.00**	0.98 ± 0.01	0.98 ± 0.02
C5	0.99 ± 0.02	**1.00 ± 0.00**	1.00 ± 0.00	0.99 ± 0.00	0.99 ± 0.00	**1.00 ± 0.00**	1.00 ± 0.00	0.96 ± 0.07	**1.00 ± 0.00**	**1.00 ± 0.00**
C6	0.82 ± 0.20	0.98 ± 0.06	0.98 ± 0.02	0.69 ± 0.19	0.95 ± 0.04	0.92 ± 0.10	0.90 ± 0.09	**0.98 ± 0.01**	**0.98 ± 0.01**	0.95 ± 0.01
C7	0.59 ± 0.21	0.59 ± 0.16	0.61 ± 0.09	0.54 ± 0.23	0.62 ± 0.07	0.54 ± 0.10	0.57 ± 0.08	0.59 ± 0.05	0.66 ± 0.17	**0.73 ± 0.13**
C8	0.62 ± 0.17	0.81 ± 0.13	0.69 ± 0.14	0.53 ± 0.16	0.79 ± 0.09	0.77 ± 0.11	**0.86 ± 0.14**	0.78 ± 0.08	0.78 ± 0.10	0.73 ± 0.08
C9	0.85 ± 0.10	0.93 ± 0.03	0.87 ± 0.06	0.66 ± 0.07	0.95 ± 0.02	0.90 ± 0.15	0.93 ± 0.05	0.96 ± 0.01	0.92 ± 0.09	**0.97 ± 0.02**

We evaluate 10 deep learning methods using Average Cross-entropy Loss (ACL), Average Accuracy (AA), Average F1-score (AF) and the Average Training Time (ATT) on the AUC test set after 5 runs as shown in Table 3. These methods consist of AlexNet [46], Inception-V3 [40], ResNet-50 [42], VGG-19 networks [45],

DenseNet-201 [19], and Inception-V3 coupled with variations of RNN networks
(i.e., RNN [47], GRU [9], LSTM [18], BiGRU [26] and BiLSTM [26]). We choose
these models due to their remarkable performance in image classification [24, 39,
43] and driver distraction detection [20, 28, 29, 44].

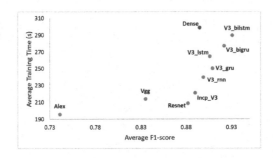

Fig. 4. F1-score vs training time of the models

InceptionV3-BiLSTM (CNN-BiLSTM) shows best performance with an average loss, average accuracy and average F1-score of 0.292 ± 0.01, 0.917 ± 0.003,
and 0.931 ± 0.004 respectively. While AlexNet has the least average training
time of 196.0 min. AlexNet is one of the first state-of-the-art deep networks in
image classification and is less computationally intensive compared to the other
models. However, there is generally a trade-off between accuracy and computation efficiency as deeper networks turn to produce more accurate results. This
is illustrated in Fig. 4, which presents a plot between the F1-scores and training
time of the models. In the plot, one can observe that the more accurate models
have higher training time.

CNN-BiLSTM also significantly outperforms the CNN architectures due to
its ability to extract spectral information. Likewise, it outperforms other unidi-
rectional RNN architectures due to the forward and backward processing that
captures extra spectral features. However, CNN-BiLSTM shows comparable performance with CNN-BiGRU as both architectures are capable of learning short
and long-term dependencies in the data. While CNN-BiGRU is computationally
more efficient than CNN-BiLSTM as shown by its lower training time in Fig. 4.
This is because GRUs use only two gates (reset and update gates) as opposed to
LSTMs, which use three gates. Therefore, GRUs may be a suitable model over
LSTMs for developing a real-time distraction detection system.

Furthermore, DenseNet CNN model outperforms other CNN architectures
due to its dense connections from any layer to all subsequent layers i.e. each
layer receives the feature maps of all preceding layers. This dense connections
significant increase training time as shown in Fig. 4, making Inception-V3 CNN
a better choice for real-time detection. Lastly, we observe that the most misclas-
sified posture is "reaching behind" (c7). This may be due to the posture having
the least number of training images (891 images) and its spatial similarities with

other postures such as "turning and talking to passenger" (c9) posture. CNN-BiLSTM, however, best identifies these postures compared to the other models.

6   Conclusions and Future Work

Distracted driving is one of the major causes of road traffic accidents worldwide. Therefore, monitoring and detecting driver distraction postures can help in the development of Advanced Driver-Assistance and alert systems to mitigate the problem. In this paper, we evaluated state-of-the-art CNN and RNN models that capture spatial and spectral features in images for the classification of driver distraction postures. We used three adequate classification evaluation metrics including the average training time to compare the models' performance on the most comprehensive, publicly-available distracted driver dataset. Results showed that Inception-V3 coupled with Bidirectional LSTMs outperformed other CNN and RNN architectures with an average F1-score of 93.1%.

For future work, we plan on acquiring video streams of driver distraction and benchmark these methods on the data to capture the temporal dynamics of naturalistic driving. In addition, we intend to extend the methods to include anomaly detection techniques, which are capable of identifying new types of distracted postures.

References

1. Global status report on road safety: World Health Organization; 2018. https://www.who.int/violence_injury_prevention/road_safety_status/2018/
2. Abouelnaga, Y., Eraqi, H.M., Moustafa, M.N.: Real-time distracted driver posture classification. arXiv preprint arXiv:1706.09498 (2017)
3. Agrawal, U., Mase, J.M., Figueredo, G.P., Wagner, C., Mesgarpour, M., John, R.I.: Towards real-time heavy goods vehicle driving behaviour classification in the united kingdom. In: 2019 IEEE Intelligent Transportation Systems Conference (ITSC), pp. 2330–2336. IEEE (2019)
4. Artan, Y., Bulan, O., Loce, R.P., Paul, P.: Driver cell phone usage detection from hov/hot nir images. In: Proceedings of the IEEE Conference on Computer Vision and Pattern Recognition Workshops, pp. 225–230 (2014)
5. Bengio, Y., Simard, P., Frasconi, P.: Learning long-term dependencies with gradient descent is difficult. IEEE Trans. Neural Networks 5(2), 157–166 (1994)
6. Berri, R.A., Silva, A.G., Parpinelli, R.S., Girardi, E., Arthur, R.: A pattern recognition system for detecting use of mobile phones while driving. In: 2014 International Conference on Computer Vision Theory and Applications (VISAPP), vol. 2, pp. 411–418. IEEE (2014)
7. Bottou, L.: Large-scale machine learning with stochastic gradient descent. In: Lechevallier, Y., Saporta, G. (eds.) Proceedings of COMPSTAT 2010, pp. 177–186. Springer, Berlin (2010)
8. CDC: Distracted driving (2016). www.cdc.gov/motorvehiclesafety/distracted driving/ Acessed 24 Jan 2020
9. Chung, J., Gulcehre, C., Cho, K., Bengio, Y.: Gated feedback recurrent neural networks. In: International Conference on Machine Learning, pp. 2067–2075 (2015)

10. Cotter, A., Shamir, O., Srebro, N., Sridharan, K.: Better mini-batch algorithms via accelerated gradient methods. In: Advances in Neural Information Processing Systems, pp. 1647–1655 (2011)

11. Craye, C., Karray, F.: Driver distraction detection and recognition using rgb-d sensor. arXiv preprint arXiv:1502.00250 (2015)

12. Eraqi, H.M., Abouelnaga, Y., Saad, M.H., Moustafa, M.N.: Driver distraction identification with an ensemble of convolutional neural networks. J. Adv. Transp. **2019**, (2019)

13. Figueredo, G.P., et al.: Identifying heavy goods vehicle driving styles in the United Kingdom. IEEE Trans. Intell. Transp. Syst. (2018). https://doi.org/10.1109/TITS.2018.2875343

14. Graves, A., Fernández, S., Schmidhuber, J.: Bidirectional LSTM networks for improved phoneme classification and recognition. In: Duch, W., Kacprzyk, J., Oja, E., Zadrożny, S. (eds.) ICANN 2005. LNCS, vol. 3697, pp. 799–804. Springer, Heidelberg (2005). https://doi.org/10.1007/11550907_126

15. Graves, A., Mohamed, A.R., Hinton, G.: Speech recognition with deep recurrent neural networks. In: 2013 IEEE International Conference on Acoustics, Speech and Signal Processing, pp. 6645–6649. IEEE (2013)

16. He, K., Zhang, X., Ren, S., Sun, J.: Deep residual learning for image recognition. In: Proceedings of the IEEE Conference on Computer Vision and Pattern Recognition, pp. 770–778 (2016)

17. Hinton, G.E., Srivastava, N., Krizhevsky, A., Sutskever, I., Salakhutdinov, R.R.: Improving neural networks by preventing co-adaptation of feature detectors. arXiv preprint arXiv:1207.0580 (2012)

18. Hochreiter, S., Schmidhuber, J.: Long short-term memory. Neural Comput. **9**(8), 1735–1780 (1997)

19. Huang, G., Liu, Z., Van Der Maaten, L., Weinberger, K.Q.: Densely connected convolutional networks. In: Proceedings of the IEEE Conference on Computer Vision and Pattern Recognition, pp. 4700–4708 (2017)

20. Kim, W., Choi, H.K., Jang, B.T., Lim, J.: Driver distraction detection using single convolutional neural network. In: 2017 International Conference on Information and Communication Technology Convergence (ICTC), pp. 1203–1205. IEEE (2017)

21. Kingma, D.P., Ba, J.: Adam: A method for stochastic optimization. arXiv preprint arXiv:1412.6980 (2014)

22. Krizhevsky, A., Sutskever, I., Hinton, G.E.: Imagenet classification with deep convolutional neural networks. In: Advances in Neural Information Processing Systems, pp. 1097–1105 (2012)

23. Lawrence, S., Giles, C.L., Tsoi, A.C., Back, A.D.: Face recognition: a convolutional neural-network approach. IEEE Trans. Neural Networks **8**(1), 98–113 (1997)

24. LeCun, Y., Bengio, Y., Hinton, G.: Deep learning. Nature **521**(7553), 436 (2015)

25. Litjens, G., et al.: A survey on deep learning in medical image analysis. Med. Image Anal. **42**, 60–88 (2017)

26. Liu, Q., Zhou, F., Hang, R., Yuan, X.: Bidirectional-convolutional lstm based spectral-spatial feature learning for hyperspectral image classification. Remote Sens. **9**(12), 1330 (2017)

27. Mafeni, J.M.: Capturing uncertainty in heavy goods vehicle driving behaviour. In: IEEE International Conference on Intelligent Transportation Systems, vol. 2020 (2020)

28. Mafeni, J.M., Chapman, P., Figueredo, G.P., Torres, M.T.: A hybrid deep learning approach for driver distraction detection (2020)

29. Majdi, M.S., Ram, S., Gill, J.T., Rodríguez, J.J.: Drive-net: Convolutional network for driver distraction detection. In: 2018 IEEE Southwest Symposium on Image Analysis and Interpretation (SSIAI), pp. 1–4. IEEE (2018)

30. Nagi, J., et al.: Max-pooling convolutional neural networks for vision-based hand gesture recognition. In: 2011 IEEE International Conference on Signal and Image Processing Applications (ICSIPA), pp. 342–347. IEEE (2011)

31. Ouyang, W., et al.: Deepid-net: deformable deep convolutional neural networks for object detection. In: Proceedings of the IEEE Conference on Computer Vision and Pattern Recognition, pp. 2403–2412 (2015)

32. Pak, M., Kim, S.: A review of deep learning in image recognition. In: 2017 4th International Conference on Computer Applications and Information Processing Technology (CAIPT), pp. 1–3. IEEE (2017)

33. Rawat, W., Wang, Z.: Deep convolutional neural networks for image classification: a comprehensive review. Neural Comput. **29**(9), 2352–2449 (2017)

34. Rengasamy, D., Morvan, H.P., Figueredo, G.P.: Deep learning approaches to aircraft maintenance, repair and overhaul: a review. In: 2018 21st International Conference on Intelligent Transportation Systems (ITSC), pp. 150–156. IEEE (2018)

35. Russakovsky, O., et al.: Imagenet large scale visual recognition challenge. Int. J. Comput. Vision **115**(3), 211–252 (2015)

36. Schuster, M., Paliwal, K.K.: Bidirectional recurrent neural networks. IEEE Trans. Signal Process. **45**(11), 2673–2681 (1997)

37. Simonyan, K., Zisserman, A.: Very deep convolutional networks for large-scale image recognition. arXiv preprint arXiv:1409.1556 (2014)

38. Smith, L.N.: Cyclical learning rates for training neural networks. In: 2017 IEEE Winter Conference on Applications of Computer Vision (WACV), pp. 464–472. IEEE (2017)

39. Sun, Y., Wang, X., Tang, X.: Deep learning face representation from predicting 10,000 classes. In: Proceedings of the IEEE Conference on Computer Vision and Pattern Recognition, pp. 1891–1898 (2014)

40. Szegedy, C., Ioffe, S., Vanhoucke, V., Alemi, A.A.: Inception-v4, inception-resnet and the impact of residual connections on learning. In: Thirty-First AAAI Conference on Artificial Intelligence (2017)

41. Szegedy, C., et al.: Going deeper with convolutions. In: Proceedings of the IEEE Conference on Computer Vision and Pattern Recognition, pp. 1–9 (2015)

42. Targ, S., Almeida, D., Lyman, K.: Resnet in resnet: Generalizing residual architectures. arXiv preprint arXiv:1603.08029 (2016)

43. Xiao, T., Xia, T., Yang, Y., Huang, C., Wang, X.: Learning from massive noisy labeled data for image classification. In: Proceedings of the IEEE Conference on Computer Vision and Pattern Recognition, pp. 2691–2699 (2015)

44. Yan, C., Coenen, F., Zhang, B.: Driving posture recognition by convolutional neural networks. IET Comput. Vis. **10**(2), 103–114 (2016)

45. Yu, W., Yang, K., Bai, Y., Xiao, T., Yao, H., Rui, Y.: Visualizing and comparing AlexNet and VGG using deconvolutional layers. In: Proceedings of the 33rd International Conference on Machine Learning (2016)

46. Yuan, Z.W., Zhang, J.: Feature extraction and image retrieval based on AlexNet. In: Eighth International Conference on Digital Image Processing, ICDIP 2016, vol. 10033, p. 100330E, International Society for Optics and Photonics (2016)

47. Zuo, Z., et al.: Convolutional recurrent neural networks: Learning spatial dependencies for image representation. In: Proceedings of the IEEE Conference on Computer Vision and Pattern Recognition Workshops, pp. 18–26 (2015)

Chronologically Guided Deep Network for Remaining Useful Life Estimation

Abhay Harpale[(⊠)]

GE Global Research, New York, USA
harpale@ge.com
https://www.cs.cmu.edu/~aharpale

Abstract. In this paper, we introduce a new chronological loss function for training models to predict remaining useful life (RUL) of industrial assets based on multivariate time-series observations. The chronological loss, an alternative to the more traditional mean-squared error (MSE) loss, incorporates a monotonicity constraint, an upper bound, and a lower bound on the RUL estimates at each time step. We also present a fully-convolutional network (FCN) as a superior competitor to the current state-of-the-art approaches that are based on LSTM. Our experiments on public benchmark datasets demonstrate that deep models trained using chronological loss outperform those trained using the traditional MSE loss. We also observe that the proposed FCN architecture out-performs LSTM-based predictive models for RUL estimation on most datasets in this study. Our experiments demonstrate the potential of the proposed models to assist in observing degradation trends. Finally, we derive a sensor-importance score from the trained FCN model to enable cost savings by minimizing the number of sensors that need to be placed for asset monitoring without sacrificing RUL estimation accuracy.

1 Introduction

Predicting the time to failure, or the remaining useful life (RUL), of an asset is an important challenge, with applications in the area of material science, biostatistics, structural engineering, and econometrics [13]. Accurate RUL estimation can lead to condition-based scheduled maintenance, spare parts provisioning, operational efficiency, improved profitability, and the prevention of catastrophic failure of the affected system and related components.

RUL estimation is a well-studied problem, with a history spanning at least half a century. There has been significant amount of work in studying the degradation profiles of assets. Describing formulae for rates of degradation requires significant understanding of the mechanics, the physics, and chemistry of deterioration. Such domain-specific models may be infeasible for novel asset types and complex materials such as composites.

This has led to increased interest in purely data-driven general approaches to RUL estimation [12]. In recent years, data-driven state-of-the-art approaches for RUL estimation have been developed using deep learning architectures [18].

© Springer Nature Switzerland AG 2020
G. Nicosia et al. (Eds.): LOD 2020, LNCS 12566, pp. 118–130, 2020.
https://doi.org/10.1007/978-3-030-64580-9_10

These approaches are modeled as regression problems. They are trained by minimizing the mean squared loss (MSE) of predicted RUL and ground truth RUL from training set of historical observations.

In this paper, we introduce a novel *chronological* loss that respects properties of RUL at various times in the history of the asset's life. Specifically, in addition to the MSE loss, we introduce monotonic decrease, a lower bound, and an upper bound on the RUL estimate at each point along the trajectory of the asset's life.

In our experiments on publicly available benchmark datasets, we observe superior predictive accuracy of models trained using this composite loss compared to training the same model using MSE. We also analyze the contributions of each component of the proposed chronological loss for enabling application-specific variants. We also notice that the sequence of RUL predictions trained using the proposed loss are better than MSE for degradation profiling.

Recurrent networks such as those based on long short-term memory (LSTM) [1,3,14,16,17] currently dominate other deep architectures for RUL estimation. We introduce fully convolutional networks (FCN) [9] as a strong competing architecture. In addition to getting superior performance on some datasets, we also present a strategy for analyzing the trained FCN to identify sensors that are indicative of degradation—a crucial information for system designers who wish to minimize the number of sensors that need to be placed on the assets being monitored, without comprising RUL estimation accuracy.

2   Problem Setting

Data-driven models for RUL estimation are trained with multivariate time-series of sensor observations from previously monitored units supplemented with a ground-truth RUL. In some cases, accurate RUL estimation for providing supervised examples can be invasive and destructive, rendering the asset useless after RUL measurement for supervision. We consider such a problem setting in this paper. Thus, for training the models presented in this paper, we only have the ground-truth RUL estimate for the last step, not necessarily end of life, in the sequence of observations in the life of the asset.

In this work, we denote multivariate time-series observations as a matrix $\mathbf{X}_i \in \mathbb{R}^{T_i \times D}$ for an entity i being monitored. Note that T_i is not necessarily the overall lifespan of the asset i. It is usually the case that $0 < T_i \leq L_i$, if L_i is the true lifespan of the asset. For each such matrix, as supervision in the training set, we also have the RUL $y_{i,T_i} \in \mathbb{R}$, such that $T_i + y_{i,T_i} = L_i$. As discussed earlier, note that the ground-truth RUL is not provided at each step in the time-series, only at the time T_i. Thus, if $\mathbb{A}$ denotes the set of all assets in the training dataset, then the training set for the predictive model is $\mathbb{D} = \{(\mathbf{X}_i, y_{i,T_i}) : \forall i \in \mathbb{A}\}$

For prediction, the model is presented a multivariate time-series of observations $\mathbf{X}_j \in \mathbb{R}^{t \times D}$ for an asset j at some time t in its life. The goal of the predictive model is to then provide expected RUL $\hat{y}_{j,t}$ at that time t, such that $t + \hat{y}_{j,t} = L_j$, where L_j is the true overall lifespan of the asset j.

2.1 Chronological Loss

Our loss function for fitting RUL estimation models has four components.

Estimation Error. The first component is the estimation error. This is the prediction error between the predicted RUL and the true RUL of an asset at the time of measurement. The prediction error is measured as the squared error between the predicted RUL and true RUL for the asset for the time t.

$$\ell_1(y_{i,t}, \hat{y}_{i,t}) = (\hat{y}_{i,t} - y_{i,t})^2 \tag{1}$$

Thus, this component of the loss is equivalent to the mean squared error loss commonly used for fitting regression models.

Monotonicity Constraint. In the absence of preventive maintenance, the RUL is a decreasing function of time. This means, the RUL at any time is always greater than the RUL at a later time. Thus, the following relationship always holds $y_{it} \geq y_{it'}, \forall t < t'$. In fact, for contiguous time steps, due to the elapsed time between them, it is always the case that $y_{i,t} - y_{i,t+1} \geq 1$ for all $0 < t < L_i$. We devise the second component of our chronological loss based on this observation.

$$\ell_2(\hat{y}_i) = \frac{1}{T_i} \sum_{t=1}^{T_i} (\max(0, \hat{y}_{i,t} - \hat{y}_{i,t-1} + 1))^2 \tag{2}$$

This loss component is inspired by the hinge loss to only penalize when the predicted RUL at a time-step is higher than that predicted at a previous time-step. Notice that ℓ_2 does not depend on the knowledge of true RUL. It ensures monotonic decrease of the RUL estimates at each time step.

Lower Bound. For an asset i with a lifespan L_i, at a time t before failure, the RUL is at least $L_i - t$. At least, because the asset may not yet be in the accelerated degradation phase, when the RUL drops more rapidly. Thus, it should be the case that $\hat{y}_{i,t} \geq L_i - t$ for all $0 < t < L_i$. This criterion takes into account the various rates of degradation, without assuming a particular parametric form for the rate of degradation. Thus, the third component of the chronological loss is defined as

$$\ell_3(\hat{y}_i, L_i) = \frac{1}{T_i} \sum_{t=1}^{T_i} (\max(0, L_i - t - \hat{y}_{it}))^2 . \tag{3}$$

Similar to ℓ_2, the third component of loss ℓ_3 is one-sided. It penalizes the predictions if they are lower than the lower bound. Notice that ℓ_3 requires the knowledge of L_i for each asset in the training set. In our problem setting the ground-truth RUL y_{i,T_i} is available at time T_i leading to an easy calculation of this particular loss. Thus, the overall lifespan of the asset can be calculated as $L_i = T_i + y_{i,T_i}$.

Upper Bound. Finally, asset types might have a maximum lifespan, as dictated by underlying materials or safety regulations. Such maximum lifespan can be used to upper bound each of the RUL estimates along the sequence. This leads us to the fourth component of the loss, ℓ_4, which depends on the knowledge of the maximum possible lifespan for the asset type, L_{max}.

$$\ell_4(\hat{y}_i, L_{max}) = \frac{1}{T_i} \sum_{t=1}^{T_i} (\max(0, \hat{y}_{i,t} - L_{max}))^2 \tag{4}$$

As in loss ℓ_2 and ℓ_3, this fourth component is also one-sided. It penalizes only if the predicted RUL at any time is higher than the maximum possible lifespan for that asset type.

We combine these four components as a simple sum to calculate the overall chronological loss to train our model.

$$\mathcal{L}(y_i, \hat{y}_i) = \ell_1 + \ell_2 + \ell_3 + \ell_4 \tag{5}$$

For a training mini-batch, we compute the mean of this composite loss over the batch.

3   Model Architectures

Note that the chronological loss does not depend on the specific model architecture. It can replace MSE as the optimization loss for any RUL estimation model that currently uses MSE for learning. In this section, we study two families of deep architectures that can benefit from the proposed chronological loss.

3.1   Fully Convolutional Network

Fully convolutional networks (FCN) [9], in addition to their efficacy in computer vision tasks, have been shown to be particularly effective for time-series classification problems [15]. To our knowledge, ours is the first implementation of FCN to the problem of RUL estimation. The FCN architecture used in this paper is shown in Fig. 1a.

Similar to prior successful work in applying FCN to time-series classification, the basic building block of our model consists of a convolution layer, followed by a batch normalization layer and then a Rectified Linear Unit (ReLU) as a nonlinear activation layer. We stack multiple such blocks to extract multi-scale information from the time-series signals.

The convolution layer applies the convolution operation to the input. In our implementation, the convolution filters on the first layer have the width D, the same as that of the number of sensors in the observations. Thus, the filters on the first level are expected to glean inter-sensor relationships and pass on that information to the next layer. The convolution filters in the remaining layers are one-dimensional with stride 1. They operate only along the time-dimension. The heights of the filters are dataset dependent and chosen using

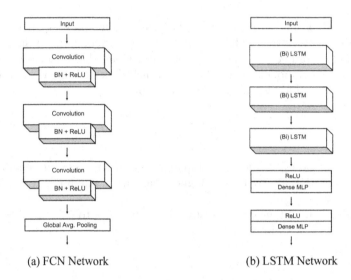

(a) FCN Network   (b) LSTM Network

Fig. 1. (a) FCN network (b) LSTM network

cross-validation. Batch Normalization [6] (BN) is a strategy to prevent internal covariate shift by normalizing layer inputs. The ReLU [10] is a non-negative element-wise maximum of its input that clips the negative portion to zero and retains the positive part.

The chronological loss requires that the model output a RUL at each time-step to be able to apply the monotonicity constraint and lower/upper bounds. To ensure that the convolution operation with one stride transforms the input to an output of the same length, we pad the input with zeros before applying the convolution operator. The global average pooling (GAP) layer then computes the average of the filters from the final convolutional block at each time-step resulting in the predictions at each time step $\hat{y}_{i,t}$ for all $0 < t \leq T_i$.

3.2 Recurrent Networks

Recurrent neural networks allow directed connections between nodes that form a temporal sequence. Long short-term memory (LSTM) [4], a recurrent neural network, is the current dominant approach for data-driven RUL estimation [3,16,17].

A particular LSTM variant, the bidirectional LSTM [2] connects two hidden layers of opposite directions to the output. This way, the output layer can get information from both the past (backwards direction) and the future (forward direction). For the problem of RUL estimation, the bidirectional design may seem like a bad choice given that it cannot have foresight into future observations to support the forward direction. However, note that such design is valuable for inferring the entire trajectory of RUL by being able to predict intermediate RUL values from the start of the sequence up to the present time. Owing to this modeling efficacy, it has been found to be superior to LSTMs for RUL estimation [1,14,19]. This design is particularly suitable for training with the chronological loss since the entire trajectory of RUL estimations can then be fitted more accurately.

In our implementation, depicted in Fig. 1b the LSTM network consists of multiple stacked layers of LSTM *cells*. Each layer transforms an input sequence to an output sequence of the same length. After passing through multiple such layers, these sequential outputs pass through a two densely connected linear layers, each with a ReLU activation at the input. This architecture is based on the current state-of-art bidirectional LSTM for RUL estimation [14].

4   Experiments

4.1   Comparative Baselines

Our goal is to develop a general domain-agnostic data-driven approach that can be the first line-of-attack for failure modes and degradation profiles that have not been well-understood to derive physical formulations of RUL estimate. To this end, we do not incorporate any prior domain-knowledge about the specific RUL task, features, or failure modes into our models or experiments. Hence, we do not compare our predictive model to domain-specific models that introduce prior knowledge of the task in their modeling.

We instead compare to current state of the art purely data-driven deep learning approaches. These include the currently popular recurrent networks such as LSTM and BiLSTM trained on the MSE loss, as described in the literature. We also compare the FCN architecture, proposed in this paper for the task of RUL estimation, using both the MSE as well as the chronological loss. We use the suffixes *+MSE* and *+Chrono* to denote the loss criterion used for training a particular model. For example, *LSTM+MSE* denotes an LSTM trained using the MSE loss and *LSTM+Chrono* denotes the LSTM trained using the chronological loss.

As a strawman baseline, we also include a *Random* prediction model that randomly predicts from the distribution of the observed lifespans of assets in the training set.

4.2   Datasets

NASA Commercial Modular Aero-Propulsion System Simulation dataset (C-MAPSS) [11] has been widely used as a publicly available benchmark for RUL estimation approaches. The corpus consists of 4 datasets with different faults and operating conditions, known as FD001 (I), FD002 (II), FD003 (III), and FD004 (IV). NASA has provided train/test splits for each of these datasets. The training data includes run-to-failure observations of various assets. The test data includes a truncated time-series for each asset some time before its failure along with the ground truth RUL at the time of truncation. Thus, the test data is more suitable for our problem setting and has been used for these experiments. This means, the training, validation, and test sets in our experiments were based on the portion of the data that NASA created for testing.

The specifics of the dataset are listed in Table 1.

Table 1. C-MAPSS datasets: Characteristics and distribution of lifespans

Data set	I	II	III	IV
Assets	100	259	100	248
Operating conditions	1	6	1	6
Fault modes	1	1	2	2
Max. lifespan	341	378	484	554
Min. lifespan	141	126	137	126
Median lifespan	199	204	222	235

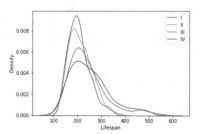

4.3   Experimental Setup

All the reported results have been averaged over a 10-fold cross-validation setup. About 10% of observations in each training fold were further retained for validation to discover the hyperparameters for better performance. For fitting the model parameters, we used ADAM [7] for the optimization routine. Suitable learning rate was discovered using grid-search in the range $[10^{-5}, 1.0]$ by evaluating the corresponding performance on the validation set. Each model was trained for 100 epochs with adaptive reduction of learning rate when performance plateaued [5]. Early stopping was applied if no reduction in the error was observed for 3 epochs. For the FCN architecture, the number of convolution filters were discovered using grid search in the range $[2, 256]$. The convolution kernel sizes were discovered from the range $[2, 32]$. For LSTM and Bi-LSTM baselines, the hidden dimensions were discovered from the range $[2, 256]$ and the number of layers from the range $[1, 3]$.

For reproducibility, the source code for the models and the experimentation platform have been made available. The datasets used in these experiments can be downloaded from the public repository[1].

4.4   Predictive Performance

In Table 2, we present the mean absolute error (MAE) of the predictive performance of the compared approaches. It can be observed that the models trained using chronological loss outperform those that are trained using the hitherto popular MSE loss. It can also be observed that even within the same model family,

---

[1] https://ti.arc.nasa.gov/tech/dash/groups/pcoe/prognostic-data-repository/.

for example BiLSTM, the performance is always better with the chronological loss (BiLSTM+Chrono) than with the MSE loss(BiLSTM+MSE).

Also note that in most cases, the FCN architecture is superior to the current state-of-the-art RUL estimation models, the LSTM and the BiLSTM, irrespective of the loss it is being trained on. In fact, the improvement from FCN+MSE to FCN+Chrono seems less when compared to the performance improvement from LSTM and BiLSTM based approaches to FCN-based approaches. Owing to this result we propose FCN, irrespective of the loss being minimized, as a strong baseline for any new RUL estimation approaches of the future.

Table 2. Predictive performance comparison (MAE). Top performance marked in bold text.

Data set	I	II	III	IV
Random	45.86	67.10	44.48	62.77
LSTM+MSE	38.95	45.67	40.39	46.72
LSTM+Chrono	35.73	43.16	36.44	46.45
BiLSTM+MSE	37.46	45.83	36.83	46.64
BiLSTM+Chrono	**15.93**	41.95	20.10	42.996
FCN+MSE	17.90	32.09	16.88	34.18
FCN+Chrono	17.63	**27.48**	**16.74**	**31.48**

4.5 Loss Analysis

Our proposed chronological loss consists of several components described in Eqs. 1, 2, 3, and 4. We investigate the contribution of each of these components to the overall performance of the model. To enable this study, we conduct the same experiments as in the previous section on the FCN+Chrono and BiLSTM+Chrono models, but with different combinations of the loss components. The standard MSE loss component ℓ_1 is always included otherwise the optimization criterion will be ill-formed with just bounds and monotonicity constraints and no ground-truth measurement to anchor to.

In Table 3, we present the MAE of the predictive performance when trained using these different combinations. The setting $\ell_1 + \ell_2 + \ell_3$ means that the loss component ℓ_4 was ignored in the calculation of the chronological loss. The BiLSTM+Chrono model makes the case for having all four loss components. Although that setting is not the best for FCN+Chrono, using all the four components does not lead to significantly inferior performance compared to the best performing FCN+Chrono variant on each dataset. Interestingly, inclusion of any combination of the loss components ℓ_2, ℓ_3, and ℓ_4 leads to improvements over the plain ℓ_1 loss. Particularly among the FCN+Chrono variants, the better performing configurations usually have the loss ℓ_3, the lower bound. The same cannot be said for BiLSTM+Chrono variants, where the performance improves with the inclusion of every loss component.

Table 3. Analysis of loss components

FCN+Chrono variants					BiLSTM+Chrono variants				
Data set	I	II	III	IV	Data set	I	II	III	IV
ℓ_1	15.65	35.52	17.04	35.14	ℓ_1	18.46	41.37	32.52	46.38
$\ell_1 + \ell_2$	14.69	33.16	16.91	32.95	$\ell_1 + \ell_2$	15.57	40.27	19.44	45.82
$\ell_1 + \ell_3$	15.31	33.53	**16.56**	32.06	$\ell_1 + \ell_3$	15.15	42.62	18.60	44.04
$\ell_1 + \ell_4$	15.38	33.58	17.07	33.03	$\ell_1 + \ell_4$	15.06	44.16	35.38	45.77
$\ell_1 + \ell_2 + \ell_3$	15.37	**31.02**	16.72	**31.72**	$\ell_1 + \ell_2 + \ell_3$	15.02	41.25	18.85	44.73
$\ell_1 + \ell_2 + \ell_4$	15.40	33.42	17.56	33.12	$\ell_1 + \ell_2 + \ell_4$	15.48	**38.00**	24.14	46.59
$\ell_1 + \ell_3 + \ell_4$	**14.21**	31.70	17.20	32.40	$\ell_1 + \ell_3 + \ell_4$	14.99	41.94	19.43	45.35
$\ell_1 + \ell_2 + \ell_3 + \ell_4$	14.26	31.02	16.78	32.04	$\ell_1 + \ell_2 + \ell_3 + \ell_4$	**14.98**	40.56	**17.77**	**43.09**

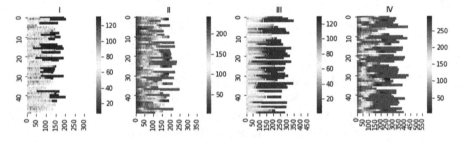

Fig. 2. Degradation trends of predicted RUL sequence by FCN+Chrono. Red indicates low RUL region, the likely phase of rapid degradation

4.6 Degradation Analysis

In Fig. 2, we examine the RUL trends of individual assets as reported by the FCN+Chrono approach. Each heatmap shows the predicted RUL trend of 50 randomly chosen assets from the corresponding dataset The assets are lined up along the vertical axis and predicted RUL along the horizontal axis at each time-step during the available monitoring data from that asset. Because FCN+Chrono penalizes non-monotonically decreasing RULs, we expected to see decreasing values of predicted RUL as time advances in the life of each asset. This is the case in the heatmaps for all the datasets. In all the datasets, the RUL decrease is very gradual till the asset reaches the region of low RUL, when the degradation is rapid. This aligns with the expectation of the designers of the dataset [11]. In fact, it is common practice in other models to assume a phase of near-constant RUL and then a phase of rapid decrease in the life of an asset after entering the failure mode [14]. We made no such assumption and let the model discover it automatically.

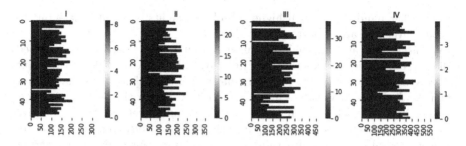

Fig. 3. Degradation trends of predicted RUL sequence by FCN+MSE

By observing the regions of low RUL in the life of each asset, exemplified by the dark red regions of the heatmaps in Fig. 2, analysts can identify the beginnings of rapid decline in the life of each asset. The knowledge of the time of failure initiation can then lead to better understanding of the root cause of such failures by investigating contemporaneous operating conditions, environmental variables, and controls. This deeper understanding of the failure modes can then help mitigate such failures from recurring on those or other similar assets.

Compare these heatmaps with those in Fig. 3 where we present the sequence of RUL predictions provided by the FCN+MSE model. Note that throughout the life of the asset, FCN+MSE always predicts the asset to have a low RUL. For example, the maximum predicted RUL, even at a 0-th time-step is seldom higher than 8 in the dataset I, although we can clearly see that most assets survive for at least 100 time-steps. We believe, this is the result of lack of lower bound on the sequence of estimates in the loss. No wonder the loss component ℓ_3 seemed important for the FCN architecture for the experiments presented in the Sect. 4.5.

4.7 Sensor Importance

Lifing analysts are interested in identifying sensors that are indicative of the degradation. This enables minimizing the number of sensors that need to monitor the assets, thereby saving production costs as well as ongoing maintenance costs from having to service such sensors. This task is usually supported by a separate step of feature selection so that the sensors deemed important can be provided as inputs to the predictive model. Current state-of-the-art deep learning models apply such feature selection steps separately before passing the data to the models [3,14]. This can be a sub-optimal approach because separate feature selection using other means may discard features that could have been potentially useful to the deep learning model. In our FCN model architecture, we instead investigate the filters in the first layer to understand the importance of various sensors.

Filter importance scoring has been studied as a means of gauging the utility of filters so that less useful filters can be pruned to reduce the parameter space. The sum of absolute values of a filter's weights has been shown to be a good approximate measure of a filter's importance in the convolutional network [8]. Intuitively, this means that filters with low absolute weights do not pass any meaningful information to the next layer and hence such filters can be removed.

On the other hand, those with higher absolute weights dominate the input to the next layer and are important for the task.

We use similar analogy to derive our feature importance score. As described earlier, in our model architecture, the first layer filters are as wide as the input feature space. Since we do not pad the input width-wise, this means that a filter combines all the input variables into a composite value for the next layer. Thus, the magnitude of each column of the first layer filters represents the relative *importance* of the corresponding input variable. So, to discover importance of a sensor, we sum the absolute values of the column representing that sensor across all the filters in the first layer. If the convolution kernel weights of the first layer are represented by the tensor $\mathbf{K} \in \mathbb{R}^{C_{in} \times H_{in} \times D}$, then we calculate the importance score of a feature $f_i \in \{f_1, \ldots, f_D\}$ as

$$\text{Importance}(f_i) = \sum_{c=1}^{C_{in}} \sum_{h=1}^{H_{in}} |\mathbf{K}_{c,h,i}| \tag{6}$$

In Table 4a, we present these importance scores for the operating conditions and sensors in the datasets. The sensor importance scores for each dataset have been scaled to the range $[0, 1]$. Higher scores imply a relatively more important sensor compared to the others for that particular dataset, while the black bands (lower importance scores) clearly mark the sensors that are not needed for RUL estimation.

To validate that we are indeed selecting important features using this score, we first identified the top 5 and worst 5 sensors by fitting FCN+Chrono on training set consisting of all the sensors. Then we conducted 10-fold experiments using only best 5 sensors and only worst 5 sensors as input to the model. The Table 4b list the MAE achieved by FCN+Chrono under such information scarcity due to limited sensors. It can be observed that the top sensors identified by our approach significantly outperform the bottom features in terms of the model performance, implying inherent feature selection by FCN+Chrono. This could result in cost savings in producing and monitoring such assets due to minimizing the number of sensors that need to be placed.

Table 4. Selecting important sensors

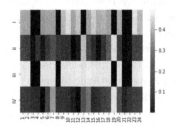

Data set	I	II	III	IV
Worst 5 sensors	44.29	47.73	47.82	54.96
Best 5 sensors	**15.25**	**30.42**	**17.52**	**38.34**

(b) Sensor selection-based predictive performance comparison

(a) Importance scores for sensors in the CMAPSS datasets

5   Conclusion and Future Work

We proposed a novel approach for remaining useful life estimation by introducing a custom chronological loss that respects the degradation profile of assets without explicitly assuming the nature of degradation. Our experiments on publicly available benchmark datasets demonstrate the superior performance of the proposed approach compared to state-of-the-art deep learning baselines for RUL estimation.

Our current approach is general and domain-agnostic. Incorporation of domain knowledge into the deep architecture and the chronological loss is likely to improve the performance of our model. There needs to be further exploration on the nature of such fusion of our idea with domain-specific customization and their impact on performance. In this work, we have not imparted the model any guidance about the possibility of multiple failure modes and multiple phases of degradation. In the future, we would like to investigate the impact of inclusion of such knowledge in the model.

References

1. Elsheikh, A., Yacout, S., Ouali, M.S.: Bidirectional handshaking lstm for remaining useful life prediction. Neurocomputing **323**, 148–156 (2019)
2. Graves, A., Fernández, S., Schmidhuber, J.: Bidirectional LSTM networks for improved phoneme classification and recognition. In: Duch, W., Kacprzyk, J., Oja, E., Zadrożny, S. (eds.) ICANN 2005. LNCS, vol. 3697, pp. 799–804. Springer, Heidelberg (2005). https://doi.org/10.1007/11550907_126
3. Gugulothu, N., TV, V., Malhotra, P., Vig, L., Agarwal, P., Shroff, G.: Predicting remaining useful life using time series embeddings based on recurrent neural networks. arXiv preprint arXiv:1709.01073 (2017)
4. Hochreiter, S., Schmidhuber, J.: Long short-term memory. Neural Comput. **9**(8), 1735–1780 (1997)
5. Hoffer, E., Hubara, I., Soudry, D.: Train longer, generalize better: closing the generalization gap in large batch training of neural networks. In: Advances in Neural Information Processing Systems, pp. 1731–1741 (2017)
6. Ioffe, S., Szegedy, C.: Batch normalization: Accelerating deep network training by reducing internal covariate shift. arXiv preprint arXiv:1502.03167 (2015)
7. Kingma, D.P., Ba, J.: Adam: a method for stochastic optimization. arXiv preprint arXiv:1412.6980 (2014)
8. Li, H., Kadav, A., Durdanovic, I., Samet, H., Graf, H.P.: Pruning filters for efficient convnets. arXiv preprint arXiv:1608.08710 (2016)
9. Long, J., Shelhamer, E., Darrell, T.: Fully convolutional networks for semantic segmentation. In: Proceedings of the IEEE Conference on Computer Vision and Pattern Recognition, pp. 3431–3440 (2015)
10. Nair, V., Hinton, G.E.: Rectified linear units improve restricted boltzmann machines. In: Proceedings of the 27th International Conference on Machine Learning (ICML-10), pp. 807–814 (2010)
11. Saxena, A., Goebel, K., Simon, D., Eklund, N.: Damage propagation modeling for aircraft engine run-to-failure simulation. In: Prognostics and Health Management, 2008. PHM 2008. International Conference on, pp. 1–9. IEEE (2008)

12. Schwabacher, M., Goebel, K.: A survey of artificial intelligence for prognostics. In: AAAI Fall Symposium: Artificial Intelligence for Prognostics, pp. 108–115 (2007)
13. Si, X.S., Wang, W., Hu, C.H., Zhou, D.H.: Remaining useful life estimation-a review on the statistical data driven approaches. Eur. J. Oper. Res. **213**(1), 1–14 (2011)
14. Wang, J., Wen, G., Yang, S., Liu, Y.: Remaining useful life estimation in prognostics using deep bidirectional lstm neural network. In: 2018 Prognostics and System Health Management Conference (PHM-Chongqing), pp. 1037–1042. IEEE (2018)
15. Wang, Z., Yan, W., Oates, T.: Time series classification from scratch with deep neural networks: a strong baseline. In: 2017 International Joint Conference on Neural Networks (IJCNN), pp. 1578–1585. IEEE (2017)
16. Wu, Y., Yuan, M., Dong, S., Lin, L., Liu, Y.: Remaining useful life estimation of engineered systems using vanilla lstm neural networks. Neurocomputing **275**, 167–179 (2018)
17. Zhang, Y., Xiong, R., He, H., Liu, Z.: A lstm-rnn method for the lithuim-ion battery remaining useful life prediction. In: 2017 Prognostics and System Health Management Conference (PHM-Harbin), pp. 1–4. IEEE (2017)
18. Zhao, G., Zhang, G., Ge, Q., Liu, X.: Research advances in fault diagnosis and prognostic based on deep learning. In: 2016 Prognostics and System Health Management Conference (PHM-Chengdu), pp. 1–6. IEEE (2016)
19. Zhao, R., Yan, R., Wang, J., Mao, K.: Learning to monitor machine health with convolutional bi-directional lstm networks. Sensors **17**(2), 273 (2017)

A Comparison of Machine Learning and Classical Demand Forecasting Methods: A Case Study of Ecuadorian Textile Industry

Leandro L. Lorente-Leyva[1]([✉]) [iD], M. M. E. Alemany[1] [iD],
Diego H. Peluffo-Ordóñez[2] [iD], and Israel D. Herrera-Granda[1] [iD]

[1] Universitat Politècnica de València, Camino de Vera S/N, 46022 València, Spain
lealo@doctor.upv.es
[2] Yachay Tech University, Hacienda San José, Urcuquí, Ecuador

Abstract. This document presents a comparison of demand forecasting methods, with the aim of improving demand forecasting and with it, the production planning system of Ecuadorian textile industry. These industries present problems in providing a reliable estimate of future demand due to recent changes in the Ecuadorian context. The impact on demand for textile products has been observed in variables such as sales prices and manufacturing costs, manufacturing gross domestic product and the unemployment rate. Being indicators that determine to a great extent, the quality and accuracy of the forecast, generating also, uncertainty scenarios. For this reason, the aim of this work is focused on the demand forecasting for textile products by comparing a set of classic methods such as ARIMA, STL Decomposition, Holt-Winters and machine learning, Artificial Neural Networks, Bayesian Networks, Random Forest, Support Vector Machine, taking into consideration all the above mentioned, as an essential input for the production planning and sales of the textile industries. And as a support, when developing strategies for demand management and medium-term decision making of this sector under study. Finally, the effectiveness of the methods is demonstrated by comparing them with different indicators that evaluate the forecast error, with the Multi-layer Neural Networks having the best results with the least error and the best performance.

Keywords: Demand forecasting methods · Textile industry · Machine learning · Classical methods · Forecast error

1 Introduction

All types of organizations require forecasts, since they are the starting point for many plans that, in the short, medium, and long term, are made with the aim of increasing competitiveness in global markets, which are totally changing and aggressive in terms of the strategies they adopt to achieve their survival. The aim of forecasts is to serve as a guide, for an organization, when estimating the behavior of future events; in production, they would trace the course of action of a production planner. It allows to determine the quantities of products to be satisfied in the coming months, the number of workers that

© Springer Nature Switzerland AG 2020
G. Nicosia et al. (Eds.): LOD 2020, LNCS 12566, pp. 131–142, 2020.
https://doi.org/10.1007/978-3-030-64580-9_11

would respond to fluctuations in demand or even to plan the size, location, etc., of new facilities. The aim is for the forecasts themselves to minimize the degree of uncertainty to which the input information to different processes is subject, so that decision-makers can generate effective results for the industry.

Today, manufacturing or service companies face the challenge of implementing new organizational and production techniques that allow them to compete in the marketplace. For this reason, several investigations have been developed on demand forecasting, to develop low-cost computational methods for dealing with forecast uncertainties [1]. In Ecuador, the textile industry is one of the sectors with the highest participation in manufacturing, representing 29% of the total of these companies. For these textile and clothing industries, effective planning will facilitate their survival in an increasingly demanding market [2].

Recent changes in the Ecuadorian context have shown the existence of variability in the tariff rates imposed on imports, both of finished products and raw materials. This has caused consumers to reduce purchase levels for certain products, because of price increases, causing uncertainty in the demand for these products. Influencing decisively during the forecast that the textile industry currently makes, variables such as sales prices of textile products, inflation, consumer price index, unemployment, market competition.

Currently, textile industry does not have the effective application of demand forecasting methods that provide solutions for decision making. That they consider all the frequent changes that occur in these companies, for the subsequent marketing and sales of the goods produced. That is why this work focuses on demand forecasting for textile products through a comparative analysis of a set of classical and machine learning methods, which provide an adequate development of demand forecasts.

The remaining of this manuscript is structured as follows: Sect. 2 sets out a review of the main related works to the subject of this paper. Section 3 describes the case study and Sect. 4 the demand forecasting methods applied. The results and comparison of forecasting errors are presented in Sect. 5. Finally, the conclusions are shown in Sect. 6.

2   Related Works

Forecasting is very useful, as it is used as input to a great number of key processes in companies. In 2015 Seifert et al. [3] study the variables that influence the effectiveness of demand forecasting in the fashion industry, presenting the implications for building decision support models. In this way the development of accurate forecasts also aids in decision making [4]. Prak and Teunter propose a general method for dealing with forecast uncertainty in inventory models, the estimation of which is applicable to any inventory and demand distribution model [5].

In 2017 Gaba et al. [6] considering some heuristics to combine forecast intervals and compare their performance, obtaining satisfactory results. Other researches [7] studies the problem of multiple product purchases, demand forecast updates and orders. Classification criteria for forecasting methods are also provided [8].

To make a forecast, it is necessary to have quantitative information on demand behavior over time, with analysis using classic statistical techniques such as ARIMA, Holt-Winters, among others, the most widely used to predict their behavior [4, 9].

For many years, this type of analysis has been carried out using linear methods that can be conveniently applied. However, the existence of uncertainty and non-linear relationships in the data largely limits their use. Therefore, it is necessary to use techniques capable of reflecting such behavior and this is where, lately, the use of artificial intelligence techniques is being applied in forecasting demand with much greater force. By the year 2018, in [10] propose a method that uses forecasting and data mining tools, applying them with a higher level of precision at the client level than other traditional methods. Subsequently, [11] examines the exponential smoothing model in the context of supply chain use and logistics forecasting, performing microeconomic time series forecasting.

The development of these innovative methods has led to satisfactory solutions in several areas, with the application of Artificial Neural Networks (ANN), Bayesian Networks (BNs), Naïve Bayes, Support Vector Machine (SVM), Random Forest, among other data mining and machine learning techniques that surpass in accuracy and performance the classic methods [12–14]. In recent years, these machine learning techniques have become very popular in time series forecasting in a large number of areas such as finance, power generation and water resources, among other application [15, 16]. Several authors have used different techniques to demand forecast [17], of the agricultural supply chain [18]. Forecasting urban water demand [19], and forecasting through BNs have also played a key role [20]. Gallego et al. [21] made a robust implementation using a Bayesian time series model to explain the relationship between advertising costs of a food franchise network. Also, hybridization methods and comparison of predictions based on ANN and classical statistical methods [22, 23].

From the literature review it appears that the applications of the different methods, mainly of machine learning and artificial intelligence, are the most used at present and that they have given the best results in the demand forecasts development, even more under uncertainty. However, the application of these forecasting methods in the textile industries is not very wide, even more so in small companies that handle information in an empirical way, where the planners experience is the main source for demand forecasting and production planning.

On this basis, this paper applies different methods of forecasting demand based on the need to make a forecast that contemplates the reality of the present scenarios, using historical series and comparing them to determine the most suitable for the textile sector. Each method is adapted according to the data and input variables, training process, to achieve the best results and reach the highest accuracy and quality in the demand forecast. The comparison between the quality of the applied methods is made based on the indicators mean absolute percentage error (MAPE) and root mean square error (RMSE), observing that the best method is ANN. Details about the case study and the demand forecasting methods are provided in the following sections.

3   Case Study Description

The textile industries studied use the initial inventory and sales forecast of a given product, a planner can calculate the amount of production needed per period to meet the expected demand of customers. This calculation becomes more complex when dealing with several products, where forecasting errors and capacity constraints can lead to uncertainty in the planning process. From this, problems are detected in these companies, where there is an inadequate or non-existent demand forecast, insufficient market analysis, and increased operating costs. As well as an inadequate production planning, with high levels of inventory with little rotation, all this influenced greatly by the appearance of an intermittent demand and uncertainty in the same one, affecting directly the levels of service and the strategy of orders and a suitable demand forecast that facilitates the decision making in the production planning of this sector in study.

The development of this research will be done by means of an analysis of methods used in the forecast of the demand. For this purpose, historical data will be taken from a textile industry, which already has a planning system. This will seek greater stability in production planning, generate alternatives and make companies in this important sector more profitable. An integrated approach will be proposed for demand forecasting in the medium term using this historical data, by comparing the performance of classical and machine learning methods. Where the factors to be considered are indicated, such as the identification and pre-processing of the information with which the time horizon of the forecast is analyzed.

The company under study in this paper is a medium manufacturer and distributor industry of textile products, which operates in northern Ecuador. Using a set of forecasting methods and data from historical series of Sublimated Uniforms, Sport T-shirts, Sublimated T-shirts and Polo T-shirts, defined consecutively as Stock Keeping Unit (SKU), with the objective of obtaining more solid predictions with less forecasting error and improve compliance in deliveries to the customer thus increasing their satisfaction. To this end, data is available for these textile products in demand between 2016 and 2019. Every day, sales orders, and orders for these 4 main products are recorded. Figure 1 clearly shows the behavior of the demand forecast in relation to previous years.

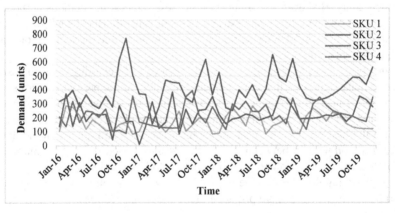

Fig. 1. Dataset textile industry.

The demand study seeks to analyze and predict its behavior in the future based on the analysis of a set of classic and machine learning methods, in addition to the diagnosis of the forecasting process that the industry initially carries out. Finally, a demand forecasting framework is designed for the company under study based on methods used through R programming language.

4 Demand Forecasting Methods

Different methods are used to develop demand forecasting, which can be adapted to specific data, considering all the variables involved in the process and the conditions of each production environment. In our case, the methods used only take time as an independent variable, with a period of analysis per year, according to the historical data provided by the case study industry.

In this study, several classical time series and causal relationship methods are applied to demand forecast, including Hold-Winters, STL Decomposition and ARIMA, which are some of the most used methods in this area under study. Methods based on artificial intelligence include BNs, ANN, SVM and Random Forest. The application of the above methods to the case study, as well as their quality, is briefly described below.

Classical Methods
These time-series based methods are relatively easy to apply and can generate accurate predictions for demand. They have become one of the most important tools during forecasting due to their wide range of applicability and flexibility of use. The ARIMA model [9] is a classical forecasting model that combines autoregressive (AR) and moving average (MA) components with additional differing time series (I).

One of the most widely used methods is the Hold-Winters, effectively adapting to changes and seasonal patterns. It allows accurate forecasting of periodic series with few training samples [24]. Represented by smoothing equations and for the forecast by Eq. (1).

$$\hat{X}_t(k) = (S_t + kT_t)I_{t-s+k} \tag{1}$$

S_t, T_t y I_t represent the smoothing, level, trend, and seasonality equations. Where the observed values are projected to obtain the forecasts $\hat{X}_t(k)$.

Also, the STL Decomposition method, a robust and accurate method that analyzes time series. It decomposes time series into trends, seasonality, cycles, and random variations.

Figure 2 shows the historical data of the last 4 years distributed in 12 months, on which the multiplicative decomposition is performed analyzing its trend, seasonality and randomness.

Naive Method is an estimation technique that only uses the previous year's actual data as a forecast for the next period, without adjusting it or trying to establish causal factors. The application of this method is shown in Fig. 3.

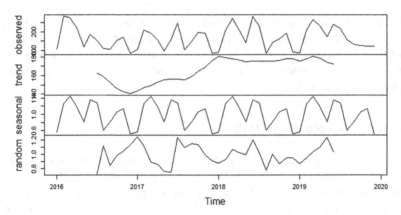

Fig. 2. Descomposition of multiplicative time series.

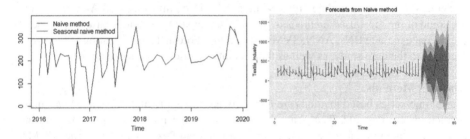

Fig. 3. Naive method application.

Machine Learning Methods

According to the related works, an exploration of machine learning techniques is conducted to demand forecasting. Based on [12], a typical specification of demand for product j of group h on the market m at time t would be:

$$Y_{jhmt} = f(X, D, p)'\beta + \zeta_{hm} + \eta_{mt} + \in_{jmt} \tag{2}$$

Where f generates interactions between observations (X), products (D) and prices (p). The dummy variables are represented by ζ_{hm}. Seasonality by the term η_{mt}, which varies according to a period of time.

Support vector machines (SVM) establish tolerance margins to minimize the error, looking for the optimal hyper plane as a decision function, maximizing the prediction margin [13], which using a linear kernel method would be represented as in Eq. (3).

$$y = b_0 \sum_{i=0}^{m} (\alpha_i - \alpha_i^*) \cdot (x_i, x) + b \tag{3}$$

During the training process, the radial base kernel coefficient and the regularization constant are set and updated using data during each iteration. The accuracy of the method, when implemented with the case study data is adjusted to 97% with a computational cost of less than one minute, as shown in Fig. 4.

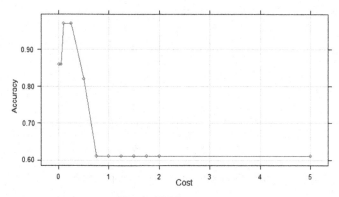

Fig. 4. SVM accuracy.

In Bayesian analysis, forecasts are based on the predictive distribution after the event [20]. It is insignificant to simulate from this, where the probability of subsequent inclusion is indicated for each predictor, as discussed in Eq. (4).

$$p(\tilde{y}|\mathbf{y}) = \int p(\tilde{y}|\phi)p(\phi|\mathbf{y})d\phi \tag{4}$$

Where $\tilde{y}$ is the set of values to predict, $p(\tilde{y}|\phi)$ the set of random extractions $p(\phi|\mathbf{y})$. The application of the method is presented below in Fig. 5, as a function of the distribution and demand forecasting.

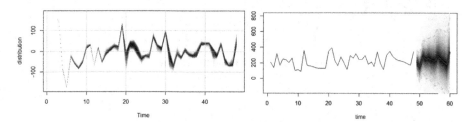

Fig. 5. Distribution and forecasting demand with BNs application.

The Random Forest algorithm has been extremely successful as a general-purpose classification and regression method [25]. It is proposed that by combining several random decision trees and aggregating their predictions by averaging, it has demonstrated excellent performance in scenarios where the number of variables is greater than the observations. The behavior of the relative error of the method when applying the combination of random trees for the prediction is presented in Fig. 6.

The applications of neural networks in demand forecasting have been significant, from the typical single-layer hidden power neural network, to multilayer, a MLP (multilayer perceptron) network specially designed for time series forecasting [23]. The function represented by a single-layer MLP with a single output is shown in Eq. (5).

$$f(Y, w) = \beta_0 + \sum_{h=1}^{H} \beta_h \, g\left(\gamma_{0i} + \sum_{i=1}^{I} \gamma_h \, Y_i\right) \tag{5}$$

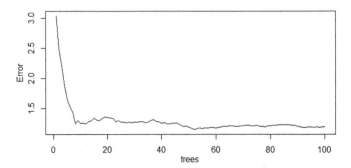

Fig. 6. Error with random forest application.

Where w is the weight of each neuron, β the output layers and γ the hidden. The variables β_0, γ_{0i} represent the biases of each neuron.

The configuration of the MLP neural network and the application result is shown in Fig. 7.

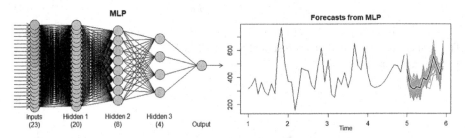

Fig. 7. MLP for demand forecasting.

5   Results and Discussion

The industry in the textile sector under study is located in the north of Ecuador. It is a small company with 25 workers. In order to apply the methods of demand forecasting, the historical demand from 2016 to 2019 of Sublimated Uniforms, Sport T-shirts, Sublimated T-shirts and Polo T-shirts, the main products of this Ecuadorian textile industry, are available. These products represent the largest quantity of orders and sales of the company, with a unit of SKU measurement and a period of 48 months. Based on the previous historical data, the methods described above are applied, obtaining the demand forecast for the next 12 months. Below, Fig. 8 shows the forecasts graphs obtained by applying each method.

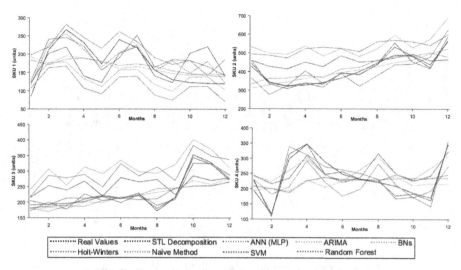

Fig. 8. Comparison of results between methods by SKU.

To evaluate the performance of the applied methods, error measurements comparing the difference between the forecast and the real value are used, such as mean absolute percentage error (MAPE) and root mean square error (RMSE), indicators of forecast accuracy widely used in time series analysis. Which can respectively expressed by Eqs. (6) and (7).

$$MAPE = \frac{1}{n}\sum_{t=1}^{n}\left|\frac{y_t - \hat{y}_t}{y_t}\right| \cdot 100\% \tag{6}$$

$$RMSE = \sqrt{\frac{1}{n}\sum_{t=1}^{n}(y_t - \hat{y}_t)^2} \tag{7}$$

In these equations the variables include the number of samples n, y_t is the real demand for textile products and $\hat{y}_t$ is the estimate of this. The ideal value for statistical metrics is zero, the closer they are, the greater the forecast accuracy, indicating a better performance of the method used. Figure 9 shows the MAPE and RMSE results for each method analyzed.

From the comparative analysis represented in the figure above, it can be determined that each product behaves differently, where the best results are obtained with ANN (MLP). With greater accuracy than the other methods and the least error in the forecast, the best performance is also obtained for each product analyzed. This value is significantly different from those of the other methods and even more so from the classical forecasting methods, where it demonstrates the quality and accuracy of the latter in forecasting demand and in comparison, with evaluation metrics. Figure 10 shows the behavior of MAPE for each model, where the SVM obtains less variability, that is, less dispersion of error on average, but by analyzing each SKU in particular, the ANN (MLP) develop better quality and precision in the results with respect to the other methods applied.

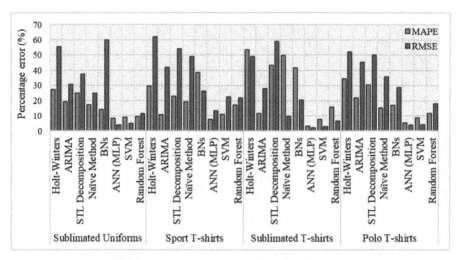

Fig. 9. Errors comparison of the applied forecasting methods.

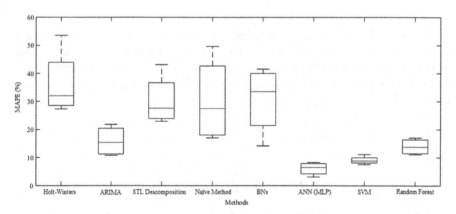

Fig. 10. MAPE analysis by each method.

By analyzing and comparing the demand forecast and the evaluation metrics used in the case study industry, it is identified that the most suitable method is ANN (MLP). Achieving an optimal network configuration through enhancement processes, to determine the nodes of each hidden layer: 22, 8, 4. Obtaining the best results of the forecast in the second iteration in the ANN (MLP) retraining, reaching the best values of the MAPE and RMSE indicators with a low computational cost. That, with respect to the other methods applied, they obtain indisputably the best results in terms of the performance indicators analyzed.

6   Conclusions

Accurate demand forecasting is essential to increase companies' competitiveness as multiple processes use it as input to make decisions. Ecuador's textile industry is no exception. This industry is characterized by high inventory levels with low turnover, increased operating costs, where the demand forecast is inadequate or non-existent, with uncertainty in the same, which is why it requires forecasting methods that consider all possible scenarios and provide feasible solutions for decision making.

In the present article we have applied more classic methods of demand forecasting like Holt-Winters, ARIMA, STL Decomposition, Naïve Method, and others of more actuality like Random Forest, ANN (MLP), SVM, BNs. The objective is to be able to compare the accuracy of these methods and determine the one that best suits the textile sector in Ecuador, with the ANN (MLP) having the best results in each product analyzed.

As a result of the analyses carried out in the study, it can be concluded that the use of machine learning methods for the forecasts development, when there is a large amount of data and uncertainty in them, achieve better results, which greatly influence the forecasts accuracy. A better quality forecasting will serve as a basis, for the use of resources and labor, for a better production planning of the Ecuadorian textile industry and in this way to improve the manufacture times, as well as, the delivery on time to the clients, increasing their satisfaction and with it, the sales.

As a future work, it is proposed to go deeper into the configuration of machine learning methods, analyzing in detail the accuracy and computational cost of the solutions during the demand forecasting development.

Acknowledgment. The authors are greatly grateful by the support given by the SDAS Research Group (https://sdas-group.com/).

References

1. Silva, P.C.L., Sadaei, H.J., Ballini, R., Guimaraes, F.G.: Probabilistic forecasting with fuzzy time series. IEEE Trans. Fuzzy Syst. (2019). https://doi.org/10.1109/TFUZZ.2019.2922152
2. Lorente-Leyva, L.L., et al.: Optimization of the master production scheduling in a textile industry using genetic algorithm. In: Pérez García, H., Sánchez González, L., Castejón Limas, M., Quintián Pardo, H., Corchado Rodríguez, E. (eds.) HAIS 2019. LNCS (LNAI), vol. 11734, pp. 674–685. Springer, Cham (2019). https://doi.org/10.1007/978-3-030-29859-3_57
3. Seifert, M., Siemsen, E., Hadida, A.L., Eisingerich, A.B.: Effective judgmental forecasting in the context of fashion products. J. Oper. Manag. **36**, 33–45 (2015). https://doi.org/10.1016/j.jom.2015.02.001
4. Tratar, L.F., Strmčnik, E.: Forecasting methods in engineering. IOP Conf. Ser. Mater. Sci. Eng. **657**, 012027 (2019). https://doi.org/10.1088/1757-899X/657/1/012027
5. Prak, D., Teunter, R.: A general method for addressing forecasting uncertainty in inventory models. Int. J. Forecast. **35**, 224–238 (2019). https://doi.org/10.1016/j.ijforecast.2017.11.004
6. Gaba, A., Tsetlin, I., Winkler, R.L.: Combining interval forecasts. Decis. Anal. **14**, 1–20 (2017). https://doi.org/10.1287/deca.2016.0340
7. Zhang, B., Duan, D., Ma, Y.: Multi-product expedited ordering with demand forecast updates. Int. J. Prod. Econ. **206**, 196–208 (2018). https://doi.org/10.1016/j.ijpe.2018.09.034
8. Januschowski, T., et al.: Criteria for classifying forecasting methods. Int. J. Forecast. **36**, 167–177 (2020). https://doi.org/10.1016/j.ijforecast.2019.05.008
9. Box, G.E., Jenkins, G.M., Reinsel, C., Ljung, M.: Time Series Analysis: Forecasting and Control, 5th edn. Wiley, Hoboken (2015)

10. Murray, P.W., Agard, B., Barajas, M.A.: Forecast of individual customer's demand from a large and noisy dataset. Comput. Ind. Eng. **118**, 33–43 (2018). https://doi.org/10.1016/j.cie.2018.02.007

11. Bruzda, J.: Quantile smoothing in supply chain and logistics forecasting. Int. J. Prod. Econ. **208**, 122–139 (2019). https://doi.org/10.1016/j.ijpe.2018.11.015

12. Bajari, P., Nekipelov, D., Ryan, S.P., Yang, M.: Machine learning methods for demand estimation. Am. Econ. Rev. **105**, 481–485 (2015). https://doi.org/10.1257/aer.p20151021

13. Villegas, M.A., Pedregal, D.J., Trapero, J.R.: A support vector machine for model selection in demand forecasting applications. Comput. Ind. Eng. **121**, 1–7 (2018). https://doi.org/10.1016/j.cie.2018.04.042

14. Herrera-Granda, I.D., et al.: Artificial neural networks for bottled water demand forecasting: a small business case study. In: Rojas, I., Joya, G., Catala, A. (eds.) IWANN 2019. LNCS, vol. 11507, pp. 362–373. Springer, Cham (2019). https://doi.org/10.1007/978-3-030-20518-8_31

15. Dudek, G.: Multilayer perceptron for short-term load forecasting: from global to local approach. Neural Comput. Appl. **32**(8), 3695–3707 (2019). https://doi.org/10.1007/s00521-019-04130-y

16. Salinas, D., Flunkert, V., Gasthaus, J., Januschowski, T.: DeepAR: probabilistic forecasting with autoregressive recurrent networks. Int. J. Forecast. (2019). https://doi.org/10.1016/j.ijforecast.2019.07.001

17. Weng, Y., Wang, X., Hua, J., Wang, H., Kang, M., Wang, F.Y.: Forecasting horticultural products price using ARIMA model and neural network based on a large-scale data set collected by web crawler. IEEE Trans. Comput. Soc. Syst. **6**, 547–553 (2019). https://doi.org/10.1109/TCSS.2019.2914499

18. Zhang, X., Zheng, Y., Wang, S.: A demand forecasting method based on stochastic frontier analysis and model average: an application in air travel demand forecasting. J. Syst. Sci. Complexity **32**(2), 615–633 (2019). https://doi.org/10.1007/s11424-018-7093-0

19. Lorente-Leyva, L.L., et al.: Artificial neural networks for urban water demand forecasting: a case study. J. Phys: Conf. Ser. **1284**(1), 012004 (2019). https://doi.org/10.1088/1742-6596/1284/1/012004

20. Scott, S.L., Varian, H.R.: Predicting the present with Bayesian structural time series. Int. J. Math. Model. Numer. Optim. **5**, 4–23 (2014). https://doi.org/10.1504/IJMMNO.2014.059942

21. Gallego, V., Suárez-García, P., Angulo, P., Gómez-Ullate, D.: Assessing the effect of advertising expenditures upon sales: a Bayesian structural time series model. Appl. Stoch. Model. Bus. Ind. **35**, 479–491 (2019). https://doi.org/10.1002/asmb.2460

22. Han, S., Ko, Y., Kim, J., Hong, T.: Housing market trend forecasts through statistical comparisons based on big data analytic methods. J. Manag. Eng. **34** (2018). https://doi.org/10.1061/(ASCE)ME.1943-5479.0000583

23. Lee, J.: A neural network method for nonlinear time series analysis. J. Time Ser. Econom. **11**, 1–18 (2019). https://doi.org/10.1515/jtse-2016-0011

24. Trull, O., García-Díaz, J.C., Troncoso, A.: Initialization methods for multiple seasonal holt-winters forecasting models. Mathematics **8**, 1–16 (2020). https://doi.org/10.3390/math8020268

25. Biau, G., Scornet, E.: A random forest guided tour. Test **25**(2), 197–227 (2016). https://doi.org/10.1007/s11749-016-0481-7

Automatic Curriculum Recommendation for Employees

Abhay Harpale[✉]

GE Global Research, Niskayuna, USA
harpale@ge.com
https://www.cs.cmu.edu/~aharpale

Abstract. We describe a platform for recommending training assets to employees based on interests, career objectives, organizational hierarchy, and stakeholder or peer feedback. The system integrates content-based and interested-based recommendations across multiple data-streams and interaction modalities to arrive at superior recommendations to those based on just content or interests. The training assets span a wide variety of content formats such as blog articles, podcasts, videos, books, and summaries with the added complexity of multiple content providers. The system incorporates a gamut of interactive information such as likes, ratings, comments, and social activity such as sharing to further personalize recommendations to the user. The system has been deployed in a large organization and is continuously improving from user feedback.

1 Introduction

Employers benefit when their employees are better trained for their roles and responsibilities. Consequently, human resource managers in big organizations develop platforms and strategies to enable continuous employee development and growth. Many organizations maintain or subscribe to large corpora of training assets such as blog articles, video tutorials, podcasts, books, summaries, best practice articles, and also a catalog of on-site seminars and workshops. Organizations can help employees in navigating such complex but important libraries by providing effective information retrieval systems [6] that direct employees to relevant training assets.

In this paper, we present a general platform for automatically recommending training assets from such diverse libraries to employees that are effective for their growth and development. The modules presented in this paper are built upon well-studied concepts in information retrieval and machine learning. At a very high-level, we use content-based matching to first identify relevant assets and then use a matrix factorization based collaborative filtering approach to personalize the ordering of those relevant assets for the user. The diversity of the content library and the variety of query formulations lead to enhancements of the common approaches to address this recommendation problem.

The recommendation platform described in this paper has been implemented and deployed in a large organization with over a quarter million employees. It has

© Springer Nature Switzerland AG 2020
G. Nicosia et al. (Eds.): LOD 2020, LNCS 12566, pp. 143–155, 2020.
https://doi.org/10.1007/978-3-030-64580-9_12

been well received by employees and their human resource managers. Monitoring of training activity demonstrated a three-fold increase in training completions and registrations compared to the existing system which was a textual search interface supported by Apache Lucene.[1] The platform has also received top awards at important fora in relevant categories[2].

2   Problem Setting

Our problem setting is depicted in Fig. 1. We detail the various challenges in this section.

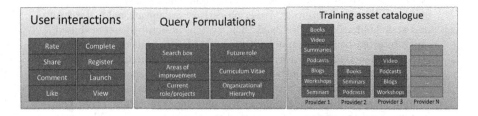

Fig. 1. Problem setting

Our catalog of training assets cover a wide variety of formats. There are multiple providers of training assets. The training assets are catalogued in enterprise-level databases with meta-data acquired from the providers. Typical fields of information associated with the training assets include title, description, categories, keywords, author, publisher, provider, and some unique identifier. Some asset types such as books and blog articles have significantly more textual content than other content such as videos and podcasts. Some assets are freely available to the employees. Others such as workshops and training seminars require organizational clearance, prerequisites, paid registration and may be undertaken at specific times of the year. The categories and keywords provided by individual providers do not conform to the same ontology or vocabulary. For example, one provider may assign the category label "communication skills", other may use the succinct "communication", while another may use more fine-grained categories such as "oral communication", "written communication", "technical communication" and "leadership communication". Some providers have manually

---

[1] https://lucene.apache.org/.

[2] The system titled "Enhancing Employee Development Through Personalized Learning Recommendations" was awarded the highest honors, the Gold awards, at the Brandon Hall Excellence in Technology Program in the categories "Best Advance in Performance Support Technology" and "Best Advance in Learning Management Technology" https://www.brandonhall.com/excellenceawards/excellence-technology.php?year=2018.

curated their content into categories, while others seem to have used automatic categorization leading to some evident mistakes in the labeling.

The catalog may be accessed by the employees in several ways and each of these channels formulate a different query to the recommender platform.

– There is a plain text search facility on the catalog website where the results are delivered by our platform. This search interface also allows the user to specifically apply filters on training format, provider, and other factors.
– The internal employee performance evaluation platform seamlessly accesses our system through an Application Programming Interface (API). When employees receive suggestions for improvement from their customers, peers, or managers, an adjoining panel displays training recommendations relevant to address those shortcomings. For example, if the feedback states "Please consider improving your written communication skills", the recommender system will present suitable trainings to address their written communication challenges.
– The internal career progression website displays future professional ladders for employees in their profile page. For example, a software engineer may be presented that they may aspire to become a senior software engineer or to consider a managerial track. By clicking on their chosen tracks employees can view the potential skills they may need to perform at the next level in their career. These skills are accompanied by a list of training assets delivered by an API that may help the employee progress in their career to the next role.
– Employees receive a regular email digest containing a list of recommendations for excelling in their current role. This is derived from their organizational information, past engagements, and internal resume.

The employees interact with the recommendations in multiple ways.

– Users may provide ratings on a scale from 1 to 5 with higher scores implying that the user found the content interesting.
– Users may share the training assets they come across with their colleagues.
– Users may comment on the training assets in free-form text. The comments may indicate user's happiness with the training asset, they may be a question, or they may also be a response to another question in the comments thread for that asset.
– A newer initiative introduced the *like* button which the users can utilize to express satisfaction with the training asset. It co-exists with the previous rating system, albeit conveys similar information. We noticed that some users tend to provide ratings as a fine-grained feedback while others like the simplicity of the like button.
– The system also tracks when users have registered for a training asset, when they initially launch it, whether they have viewed it before, and when they complete it by taking the necessary post-training assessments. Note that some of these interactions may not be available for all training assets. For example, external videos and blog-posts are not accompanied by assessments and the system does not perform eye-tracking to detection completion of those assets, but it may track when the user has launches the course material.

3   Design Considerations

This complex landscape of possibilities is further exacerbated by the requirement from the business managers to

- allow incremental updates. This is crucial to support employee turnover. It is also necessary because the organization periodically adds new content as well as content providers. Some subscriptions may also be terminated for various reasons.
- build a generalizable strategy that is uniform across asset types. So, at this time, we have not invested into automatically augmenting multimedia content by captioning video or transcribing audio.
- implement a modular system instead of an end-to-end system to support innovative integrations with other upcoming platforms. Thus, we have multiple components interacting through APIs instead of a state-of-the-art recommender systems which may be able to address content-based and interest-based recommendations simultaneously. As a result, the content-based recommendation module and interest-based personalization module are also being used independently to support other initiatives within the organization apart from curriculum recommendation.
- learn continuously from feedback. With so many different interaction types available, it is apt for the system to continually learn from feedback received from the employee for increased satisfaction in the future.

Based on these design guidelines, we split the platform into three high-level modules.

- Rule-based filter: Some training assets are designed to be inaccessible to certain employees. Others are available only during certain time of the year. Additional filters may be described by the business use-case or preset by the employee in their profile.
- Content-based matching module: Given a textual query, the content-based matching module discovers relevant training assets, a typical well-studied information retrieval task.
- Interest-based recommendation module: Given past interactions with the training assets, the interest-based recommendation module discovers the user's interests and assigns *interestingness* scores to the training assets.

These modules are applied in this sequence with the output of previous module being worked on by the next in the sequence. The rule-based filter module is a straightforward strict query and has been implemented for exact match filtering. The training assets that pass the rule-based filter are then evaluated by the content-based module for query-relevance. Training assets deemed relevant to the query are then passed to the interest-based personalization module. We will focus the rest of this section on the content-based and interest-based modules.

4 Content-Based Recommendations

A typical content-based retrieval approach involves running the query against an inverted index. Due to our problem setting and design considerations, we chose an alternative. In our setting the queries come from multiple different sources—the search box, the career development platform, and the peer or manager feedback portal. Each of these sources involve a different vocabulary and style of writing.

Queries that arrive from the search box are typical of information retrieval systems. They are short, contain discriminative terms, and are usually devoid of non-informative words that may be irrelevant to the search intent.

On the other hand, a query from the peer or manager feedback portal is longer and contained full formed sentences. The language was informal. Since it was a feedback there was also leading in pleasantries before the feedback provider suggested areas for improvement. Due to camaraderie among managers and their employees, the criticism was shrouded in the politeness. For example, "You have great communication skills. To progress in your career, consider improving your decision making skills". For this particular example, a direct content match will present both "communication" and "decision making" related training assets, but that is incorrect. In this context, we implemented a query formulation module to separate the criticism from the politeness. By training a sentence classifier based on support vector machines (SVM) [1], we were able to remove sentences that did not convey some area for improvement. The remaining sentences were then concatenated to form the query. Nevertheless, the problem of differing language models across query types persisted.

Similar challenges existed among the content and meta-data from multiple providers. The content spanned various modalities such as videos, books, podcasts, blog-posts, on-site workshops, and summaries. Even within the same modality, the content had been subscribed from various providers. For example, there were blog-posts and articles by employees within the organization as well content sourced from authoritative sources such as Harvard Business Review (HBR). Not only was the vocabulary across these providers different, the meta-data was also very different. Some publishers provided meticulously categorized content with abundant keywords, while others merely provided title, body, and in some cases, description. Content such as video and podcasts mostly included multimedia content which was sometimes tagged with keywords, but in most cases the only text included was the title. Treating all the content equally for the purpose of building an inverted index was ineffective because curated and verbose content was more likely to match queries. Moreover, we realized that employees preferred some modalities and sources over others and it was important to cater to that preference.

We addressed these challenges by treating each query source and each provider as a separate domain and implemented a similarity scoring mechanism to compare texts across domains. We describe this next.

4.1   Cross-Domain Similarity

Given that the content as well as queries were modeled as separate domains, we refer our content-based matching system the *Cross-domain similarity (CDS)* module. The overall schematic of the cross-domain similarity module is depicted in Fig. 2.

At a very high level, the cross-domain similarity module works by comparing the bag-of-words vector representation of the query with a bag-of-words representation of the training assets using weighted cosine similarity metric. Assets scoring high similarity scores are deemed relevant to the query. It is known as a cross-domain similarity module because the corpus statistics leading to the vectorization of the query and of the training asset are inferred separately.

Corpus Model. Longer queries with full-formed sentences also contained word repetitions across sentences. It was thus important to infer the corpus statistics of such long queries separately from other query types. To enable this, each domain was represented by its own *Corpus Model*. Corpus models are a mainstay of most information retrieval systems. In our case, the corpus model associated with each domain included the preprocessing pipeline associated with that domain and the inferred corpus statistics and customizations. The preprocessing pipeline incorporated the standard steps of tokenization and stemming. The corpus statistics included the book-keeping for frequency of word occurrence in individual documents as well as their prevalence across the particular domain. It was implemented to allow incremental learning of the corpus statistics to enable the ever-growing collection of assets and queries in their respective domain. Using our corpus model, we represented each document as a vector of term weights. The elements of the vector represent a score for every word in the document. For our model, we used the popular TF-IDF (Term Frequency -

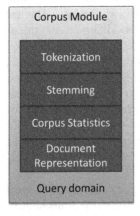
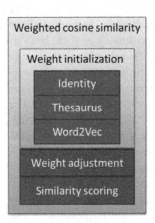
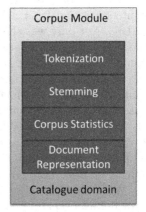

Fig. 2. Cross-domain similarity

Inverted Document Frequency) score, specifically the variant commonly known as ltc^3 [6].

4.2 Similarity Scoring

With the vectorized representation of both queries and training assets, we needed a similarity score to find training assets that matched the queries. One of the most popular similarity metric in information retrieval is the cosine similarity metric [6].

Cosine similarity of two n-dimensional real-valued vectors $\mathbf{x}$ and $\mathbf{y}$ is computed as the dot product of their unit vectors.

$$\cos(\mathbf{x}, \mathbf{y}) = \frac{\mathbf{x}\mathbf{y}^T}{\|\mathbf{x}\| \, \|\mathbf{y}\|} \tag{1}$$

But cosine similarity requires strict matching of terms. To enable true cross-domain similarity, we also wanted to leverage cross-term similarity. For example query might have the term "good" and the document might have the term "nice", which may be contextually similar. Moreover, if a certain term implies a different meaning across two domains, then we will need to respect that as well. So, we enhanced the standard cosine similarity metric with a weighted version which introduces term-level and cross-term weights for the vector terms.

We denote these weights by the weight matrix $\mathbf{W}$. In the weight matrix, each diagonal element, for example $\mathbf{W}_{i,i}, \forall i$ is the weight associated with the i-th term. Each non-diagonal element $\mathbf{W}_{i,j}$ is the cross-term weight between the i-th and the j-th term.

Equipped with the weight matrix, the weighted cosine similarity is computed as follows.

$$\mathrm{wcos}(\mathbf{x}, \mathbf{y}, \mathbf{W}) = \frac{\mathbf{x}\mathbf{W}\mathbf{W}^T\mathbf{y}^T}{\|\mathbf{x}\mathbf{W}\| \, \|\mathbf{y}\mathbf{W}\|} \tag{2}$$

It can be observed that if the weight matrix is an identity matrix, that is $\mathbf{W} = \mathbf{I}$, such that $\mathbf{W}_{i,i} = 1, \forall i$ and $\mathbf{W}_{i,j} = 0, \forall i \neq j$, then the weighted cosine similarity computes the standard cosine similarity.

Initializing Weights. If we initialize the weights to an identity matrix, $\mathbf{W} = \mathbf{I}$, we will be starting off with the standard cosine similarity metric. This is one possible approach. We propose two additional approaches. Our novel approach relies on the availability of a thesaurus, a listing of synonyms and antonyms of words in a language. Each word is deemed similar to itself, so the diagonal elements are 1. This means $\mathbf{W}_{i,i} = 1, \forall i$. All words that are listed as synonyms (similar in meaning) of a word get an equal fraction of the weight. Suppose a word i has n_i synonyms $\mathbb{S}_i = \{j_1, \ldots, j_{n_i}\}$, then the corresponding non-diagonal

---

[3] https://nlp.stanford.edu/IR-book/html/htmledition/document-and-query-weighting-schemes-1.html.

elements $\mathbf{W}_{i,j} = \frac{1}{n_i}, \forall j \in \mathbb{S}_i$. For our work, we utilized the Wordnet database [2] to discover synonyms of popular words. But even large lexicons like Wordnet were inadequate for us to cover all the terms in the corporate vocabulary. For such terms, we resorted to the identity weighting—1 weight for diagonal elements, 0 for the cross-term weights.

Continuously Learning Weights. Unlike standard cosine similarity, the weights in our weighted cosine similarity score can be adjusted to feedback. We implemented an automatic update to the weights using the feedback received from the end-user. If the user suggests that the search result is irrelevant or relevant to the query then the weights are adjusted to reflect that observation. We use the stochastic gradient descent (SGD) [8] approach to update the weights. The steps are outlined below.

When true cosine similarity between two vectors is provided, these steps will be followed to update the existing weights $\mathbf{W}$. The input to the update function will be $\mathbf{x}, \mathbf{y}, \mathbf{W}, c$. Here $\mathbf{x}$ and $\mathbf{y}$ are the vectors whose similarity we are computing. The learnt weights are denoted by $\mathbf{W}$. The desired cosine similarity between the two vectors, as provided by human annotator or some third party system is c.

- Step 1: Predict weighted cosine similarity

$$\hat{c} = \mathrm{wcos}(\mathbf{x}, \mathbf{y}, \mathbf{W}) \tag{3}$$

- Step 2: Compute error from true cosine similarity (as supplied by analyst)

$$\delta = c - \hat{c} \tag{4}$$

- Step 3: Compute gradient of the error with respect to $\mathbf{W}$.

$$\nabla(\delta) = 2 \frac{\mathbf{xWy}}{\|\mathbf{xW}\| \, \|\mathbf{yW}\|}$$
$$- 2 \frac{\cos(\mathbf{x}, \mathbf{y})}{\|\mathbf{xW}\| \, \|\mathbf{yW}\|} \left(\|\mathbf{yW}\| \frac{\mathbf{x} \odot \mathbf{x}}{\|\mathbf{xW}\|} + \|\mathbf{xW}\| \frac{\mathbf{y} \odot \mathbf{y}}{\|\mathbf{yW}\|} \right) \odot \mathbf{W}$$

- Step 4: Include the gradient of the regularization term for the weight matrix in the gradient

$$\nabla(\delta, \mathbf{W}) = -2\delta \nabla(\delta) + \lambda \mathbf{W} \tag{5}$$

- Step 5: Standard SGD Update with learning rate η:

$$\mathbf{W} = \mathbf{W} - \eta \nabla(\delta, \mathbf{W}) \tag{6}$$

- Step 6: Repeat steps 1–5 for all pairs of documents with known similarity scores in the corpus till desired accuracy is reached. In a deployed system, the steps 1–5 can be performed in the background when an annotator (typically a human analyst) suggests that two documents are similar or dissimilar. The system can then performs steps 1–5 to update the weights to satisfy the feedback received from the annotator.

5   Interaction-Based Personalization

In addition to content-based relevant training assets, we also wanted to personalize the recommendations to individual user tastes. To enable personalization [7], we leveraged information from the multiple ways the user had interacted with the system. These interactions were described earlier in Sect. 2. These interactions can be grouped into three types. Particularly, the platforms we supported enabled the user to perform the following interactions

- Numerical: This included the *ratings*-based interactions.
- Binary: This included the share, like, complete, register, launch, and view interaction types. In all these interactions, the user had either performed the said interaction or not. Compared to the numerical interaction type, these were implicit. There were numerous possibilities in making assumptions about user intent here. For example, if a user launched a course but did not complete it, does it mean she is dissatisfied with the content? We did not attempt to decipher this in the current version of the platform. Instead, we treated all these interactions are binary interactions in isolation with other interactions.
- Textual: Users could leave comments under each training asset. This was a textual field. We had two options here, we could treat it as a binary field just like "share" above—meaning if a user commented on an asset, they might like it. But on closer scrutiny, we found that many of the comments expressed negative sentiment about the user's dissatisfaction with the learning material. So, we first applied an off-the-shelf sentiment-classification algorithm on the comments to discover if the sentiment expressed is negative or positive and then appropriately converted this interaction to a numerical field by scaling the polarity score from the sentiment classifier from the range $[-1, 1]$ to the range $[1, 5]$.

Thus, we arrived at a numerical representation for each of the interaction types. We represented these as interaction matrices. In each of the interaction matrices, each row represented a user, each column represented a training asset, and the cell-value represented the interaction score for that user and training asset combination. For example, in the "rating" interaction matrix, each row represented a user, each column represented a training asset, and the cell-value represented the star rating provided by the user to the asset. Similarly, in the "comment" interaction matrix, cell-values represented the scaled polarity score and in the "share" interaction matrix, the cell-values represented whether the user had shared the particular training asset.

5.1   Collaborative Filtering

The interaction matrices are very sparse because users may have expressed their interest in a small subset of the assets through their interactions. The goal of collaborative filtering is to infer the interestingness of the remaining assets to the user.

Matrix factorization [5] are a popular and effective class of collaborative filtering [11] algorithms used in recommender systems. They work by decomposing the input user-item interaction matrix into two lower dimensional rectangular matrices, one associating the items to genres and the other representing user interests across the genres.

Given the multiple types of interaction types, we did not wish to contaminate the user-interest models resulting from each of the interaction types. Moreover, as a business use-case it was desirable to implement separate recommender modules based on each of the interaction types. For example, the organization may wish to increase sharing of content, so the model resulting from "share" interaction type could be leveraged, so that the system recommends assets the users are more likely to share.

To enable such possibilities, we factorized each of the interaction matrices separately. We implemented factorizing algorithm using the well-studied alternating least squares approach [4].

Alternating Least Squares. Consider an interaction matrix $\mathbf{R} \in \mathbb{R}^{U \times A}$ where U is the number of users and A is the number of training assets. The element $\mathbf{R}_{u,a}$ denotes the interestingness of the asset a to the user u, for example the value of $\mathbf{R}_{u,a}$ could be a rating on a scale 1 to 5. This is a scalar score and is also interchangeably denoted as $r_{u,a}$.

Let the interest profile of the user u be represented by a vector $\mathbf{x}_u \in \mathbb{R}^K$. Similarly, let the asset a be represented by a vector $\mathbf{y}_a \in \mathbb{R}^K$.

If the interest profile of the user closely aligns with the genre profile of the asset, then the rating will be higher, else it will be lower. Thus, if the user interest profile and item genre profile are available, we can predict the rating as follows

$$\hat{r}_{u,a} = \mathbf{x}_u^T \mathbf{y}_a \tag{7}$$

These vectors $\mathbf{x}_u$ and $\mathbf{y}_a$ need to be inferred based on the training set. This is achieved by minimizing the prediction error of the model above compared to the true rating, for the user-asset pairs with available ratings. The optimization problem minimizes the square of the prediction error $r_{u,a} - \hat{r}_{u,a}$ while restricting the profile vectors to have lower magnitudes by imposing an $L2$-regularization term for each of $\mathbf{x}$ and $\mathbf{y}$.

$$\underset{\mathbf{x},\mathbf{y}}{\text{minimize}} \sum_{r_{u,a} \text{ is known}} \left(r_{u,a} - \mathbf{x}_u^T \mathbf{y}_a\right)^2 + \lambda \left(\sum_u \|\mathbf{x}_u\|^2 + \sum_a \|\mathbf{y}_a\|^2\right)$$

The algorithm for solving this optimization problem is known as Alternating Least Squares (ALS) because the two unknowns $\mathbf{x}_u$ and $\mathbf{y}_a$ depend on each other and need to be solved in alternate steps.

Let $\mathbf{X}$ denote the matrix of all user factors such that $\mathbf{x}_u$ is the u-th row of the matrix. Similarly, let $\mathbf{Y}$ denote the matrix of all asset factors such that $\mathbf{y}_a$

is the a-th row of the matrix. The alternating steps for solving $\mathbf{x}_u$ and $\mathbf{y}_a$ are presented below.

First, holding $\mathbf{Y}$ fixed, we solve for user-profile vectors $\mathbf{x}_u$.

$$\mathbf{x}_u = \left(\mathbf{Y}^T\mathbf{Y} + \lambda\mathbf{I}\right)^{-1}\mathbf{Y}^T\mathbf{R}_u \tag{8}$$

Then keep $\mathbf{X}$ fixed, we solve for asset-profile vectors $\mathbf{y}_a$.

$$\mathbf{y}_a = \left(\mathbf{X}^T\mathbf{X} + \lambda\mathbf{I}\right)^{-1}\mathbf{X}^T\mathbf{R}_a \tag{9}$$

Alternating this way, the algorithm discovers user interest profiles and asset genre profiles. To ensure that we had non-negative interest profiles, at each iteration, we clipped negative values to zero, a common strategy in the projected gradient approaches.

Dealing with Binary Interactions. For binary interactions, such as those resulting in the "share" interaction type—whether a person has shared a particular asset or not, we modified the algorithm to introduce an $L1$ regularization term to introduce sparsity in the interest profile $\mathbf{X}$ and genre profile $\mathbf{Y}$. With the $L1$-regularization term, the appropriate update algorithm involved stochastic coordinate descent [12]. This differs from the update above by randomly suppressing weaker cell values in the $\mathbf{X}$ and $\mathbf{Y}$ matrices to zero in each iteration. It finally results in few dominant cell values in each row of $\mathbf{X}$ and $\mathbf{Y}$, a desirable sparse outcome.

5.2 Multi-interaction Model

We got separate user profile vectors as well as asset profile vectors for each of the interaction types. This helped us support individual recommendations for each of the interaction types. To provide overall recommendations, we first predicted the recommendation score using each interaction model separately. We then arrived at the overall personalized recommendation score using a linear combination of those scores. This allowed us to weigh scores arising from individual models according to prior beliefs about the accuracy and availability of the data in each of the interaction matrices. Thus, this was driven by the business managers' preferences in having more control over the weights in the linear combination of the recommendation scores arising out of the multiple interaction types.

5.3 Combining Scores of Content and Interest-Based Recommendations

We used a multi-step approach to incorporate personalized scores with the content-based scores. In the first step, the content-based scores were used to discover relevant training assets. Training assets that received content-based scores lower than a threshold were discarded from the search results.

In the next step, the relevant training assets discovered in the previous step were assigned a new score based on the linear combination of the content-based scores and recommendation scores. This enabled discovery of relevant content that also matched user interests.

5.4   Implementation

The system was developed as multiple interacting components. The frontend web interface was developed in Drupal[4] The backend databases were based in MySQL[5]. The algorithms for weighted cosine similarity, cross-domain similarity, alternating least squares for matrix factorization, and score combiners were developed in Python[6] using the popular Numpy[7] and Scipy[8] libraries for numerical computation [13]. For custom-derived word-embeddings we used Gensim [10]. Existing word-embeddings were used from the GloVe repository [9] Sentiment analysis on user-comments was performed using the TextBlob package[9] At this time, the weights in the multiple linear combinations are manually defined to support a particular business use-case and prior beliefs about the quality of various interaction types. As we gather more interaction from the platform, we will introduce ways to empirically assign such weights for a better optimized system.

6   Conclusion and Future Work

We developed a comprehensive curriculum recommendation system for employees in a large organization. So far, it has shown the promise of increased employee engagement with the learning material.

The modular nature of the platform has also enabled several other innovative integrations that we wish to investigate further. The cross-domain similarity module is being explored for its effectiveness in delivering suitable job openings to employees interested in changing their roles. It is also being investigated to suggest employees to staff a project based on the match between the project description and employee resume and experience.

The interest-based recommendation model is being used in other novel ways beyond asset recommendations. For example, we are exploring the possibility of using the employee's interest profile inferred by the matrix factorization to find similar other profiles within the organization, for the goal of succession planning or building communities for like-minded employees.

In the future, we wish to explore deep learning approaches to further strengthen some of the building blocks in our system. We enabled continuous

---

[4] https://www.drupal.org/.

[5] https://www.mysql.com/.

[6] https://www.python.org/.

[7] https://numpy.org/.

[8] https://www.scipy.org/.

[9] https://textblob.readthedocs.io/en/dev/quickstart.html#sentiment-analysis.

learning the weighted cosine similarity module and the interest-based recommendation module. We are currently monitoring the system to gauge its improvement over time from the accumulated feedback. This may lead to modifications or enhancements to our continuous learning strategy. We also intend to explore *active learning* [3] strategies to identify additional interests of the user.

References

1. Cortes, C., Vapnik, V.: Support-vector networks. Mach. Learn. **20**, 273–297 (1995). https://doi.org/10.1007/BF00994018
2. Fellbaum, C.: WordNet: An Electronic Lexical Database. Bradford Books, London (1998)
3. Harpale, A.S., Yang, Y.: Personalized active learning for collaborative filtering. In: Proceedings of the 31st Annual International ACM SIGIR Conference on Research and Development in Information Retrieval, SIGIR 2008, pp. 91–98. Association for Computing Machinery, New York (2008). https://doi.org/10.1145/1390334. 1390352
4. Hastie, T., Mazumder, R., Lee, J.D., Zadeh, R.: Matrix completion and low-rank SVD via fast alternating least squares. J. Mach. Learn. Res. **16**(1), 3367–3402 (2015)
5. Koren, Y., Bell, R., Volinsky, C.: Matrix factorization techniques for recommender systems (2009)
6. Manning, C.D., Raghavan, P., Schütze, H.: Introduction to Information Retrieval. Cambridge University Press, Cambridge (2008)
7. Micarelli, A., Gasparetti, F., Sciarrone, F., Gauch, S.: Personalized search on the world wide web. In: Brusilovsky, P., Kobsa, A., Nejdl, W. (eds.) The Adaptive Web. LNCS, vol. 4321, pp. 195–230. Springer, Heidelberg (2007). https://doi.org/ 10.1007/978-3-540-72079-9_6
8. Nocedal, J., Wright, S.J.: Numerical Optimization, 2nd edn. Springer, New York (2006). https://doi.org/10.1007/978-0-387-40065-5
9. Pennington, J., Socher, R., Manning, C.D.: Glove: global vectors for word representation. In: Empirical Methods in Natural Language Processing (EMNLP), pp. 1532–1543 (2014). http://www.aclweb.org/anthology/D14-1162
10. Řehůřek, R., Sojka, P.: Software framework for topic modelling with large corpora. In: Proceedings of the LREC 2010 Workshop on New Challenges for NLP Frameworks, pp. 45–50. ELRA, Valletta, Malta, May 2010. http://is.muni.cz/ publication/884893/en
11. Schafer, J.B., Frankowski, D., Herlocker, J., Sen, S.: Collaborative filtering recommender systems. In: Brusilovsky, P., Kobsa, A., Nejdl, W. (eds.) The Adaptive Web. LNCS, vol. 4321, pp. 291–324. Springer, Heidelberg (2007). https://doi.org/ 10.1007/978-3-540-72079-9_9
12. Shalev-shwartz, S., Tewari, A.: Stochastic methods for 1-regularized loss minimization. J. Mach. Learn. Res. **12**, 1865–1892 (2011)
13. Virtanen, P., et al.: SciPy 1.0-Fundamental Algorithms for Scientific Computing in Python. arXiv e-prints arXiv:1907.10121, July 2019

Target-Aware Prediction of Tool Usage in Sequential Repair Tasks

Nima Nabizadeh[1]([✉]), Martin Heckmann[2], and Dorothea Kolossa[1]

[1] Cognitive Signal Processing Group, Ruhr University, Bochum, Germany
{nima.nabizadeh,dorothea.kolossa}@rub.de
[2] Honda Research Institute Europe GmbH, Offenbach, Germany
martin.heckmann@honda-ri.de

Abstract. Repairing a device is usually a sequential task comprising several steps, each potentially involving a different tool. Learning the sequential pattern of tool usage would be helpful for various assistance scenarios, e.g. allowing a contextualized assistant to predict the next required tool in an unseen task. In this work, we examine the potential of this idea. We employ two prominent classes of sequence learning methods for modeling the tool usage, including Variable Order Markov Models (VMMs) and Recurrent Neural Networks (RNNs). We then extend these methods by also conditioning them on additional information from the repair manuals that represents the repair target. This information is present, for example, in the title of manuals. We investigate the effect of long-term dependencies and of target-awareness on the prediction performance and compare the methods in terms of accuracy and training time. The evaluation using an annotated dataset of repair manuals shows that both target-awareness and long-term dependencies have a substantial effect on the tool prediction. While the RNN has slightly more accurate predictions in most scenarios, the VMM has a lower training time and is beneficial when the prediction needs to be restricted with respect to the device category.

Keywords: Target-awareness · Sequence learning · Repair tasks · Variable Order Markov Model · Recurrent Neural Network

1 Introduction

An increasing demand exists for collaborative artificial agents that help humans to accomplish tasks and achieve common goals. Since performing many of these tasks requires interactions with objects in the environment, it is useful to equip the AI agent with knowledge about the possible object interactions. As an example, Whitney et al. [24] demonstrated that in a cooking task, modeling the sequence of mentioned ingredients in the recipes improves the performance of a robot when resolving the user's references to ingredients and helps in the disambiguation of the user's requests for such objects.

© Springer Nature Switzerland AG 2020
G. Nicosia et al. (Eds.): LOD 2020, LNCS 12566, pp. 156–168, 2020.
https://doi.org/10.1007/978-3-030-64580-9_13

Typically, repairing a device is a complex sequential task, in which each step has to be completed before the next one can be started. In carrying out such repairs, a user works on a series of components employing a sequence of tools. The order in which the tools are applied can be obtained from available repair manuals. Repair manuals for related device categories usually present similar patterns of tool use that correspond to the arrangement of components within the device. For example, most repairs of MacBook Pro devices include a first step of opening the lower case with a pentalobe screwdriver. The common next step is detaching the battery connections using a spudger, and then removing the logic board with a T6 Torx screwdriver. An AI agent capable of analyzing the repair manuals and modeling the tool usage could utilize this knowledge to predict the required tools during task execution, even if the exact ongoing task does not yet exist in the training data. For example, when a repair manual for replacing the hard drive on a 2017 Macbook Pro is not available, there may be manuals for other repair tasks on a Macbook Pro 2017, as well as manuals for hard drive replacement for other Macbook models. By leveraging this knowledge, an agent could disambiguate tool requests more easily and provide or suggest the needed tool to the user more quickly and proactively.

In this paper, we examine two well-known classes of sequence learning models for our purpose of predicting the required tools in unseen repair tasks: generative sequence modeling with a Variable Order Markov Model (VMM) and a discriminatively trained Recurrent Neural Network (RNN). We investigate a scenario, where the agent not only observes the sequence of used tools but is also informed about the task target. In this scenario, we assume that the user specifies the target, including information about the component under repair, e.g., "battery replacement," and the device category, e.g., "Macbook Pro Retina display." Based on the intuition that the training examples with more similar repair targets should provide more useful information for an accurate tool prediction, we propose methods for conditioning the predictor models on non-sequential information from repair manuals that represents the repair target.

The performance of the proposed methods is evaluated with a web-based dataset comprised of semi-structured repair manuals, in which the necessary tool for each repair step is annotated. We employ the titles of the repair manuals as a proxy to the full target of the repair tasks, containing the specific model of the device and the component that is to be repaired. Additionally, we utilize other information from the repair manuals, such as the hierarchy of the device categories, or the name of the component under repair. We are also concerned with the training time of the models because data available from the Internet can be updated frequently, and a model featuring a high training speed can leverage new instances of the source data faster.

2   Related Work

There have been many studies on obtaining and structuring knowledge from instructional data on the web, mainly in the domain of kitchen activities. The derived knowledge is sometimes common-sense knowledge about the individual objects that are involved in a task, e.g., "milk is in the refrigerator", or tool-action relations, e.g., "knives are used for cutting" [11,18]. However, the constructed knowledge-bases do not cover the temporal relation among the used objects in a task.

The temporal ordering of the task steps and their corresponding objects is taken into account in various studies on interpreting instructional texts and extracting a workflow from them [14,16,25,26]. However, in these studies the generalization of learned sequences to previously unseen tasks is not considered. Such generalization is achieved in [1], where authors proposed a graphical model to determine the likely temporal order of events, i.e., actions together with the objects that they operate on. In [5], the authors extracted step-level semantic frames of the tasks in the Wikihow website, and temporally arranged them by clustering the frames' embeddings. In [27], Zhou et al. utilized the task titles in addition to the descriptions of the steps for predicting the order of steps. Still, none of the mentioned works accounts for long-term dependencies among the steps in a task.

RNNs and VMMs have been employed for sequence learning considering long-term dependencies in various applications, such as recommendation systems [3,9,22], human activity recognition [6,17] or a wide range of natural language processing tasks, such as script learning [21]. Our work presents similar sequential models that are additionally conditioned on non-sequential information.

3 Models

In this section, we describe the proposed methods for modeling the dependencies among the used tools in various steps of the repair tasks along with the methods for conditioning the predictor models on the task target.

3.1 Problem Definition

The problem of tool prediction can be defined as follows: at each time step i, given a sequence of observed tools from k preceding steps $<t_{i-k}^{i-1}>$, we wish to predict the expected tool of the current step $(\hat{t}_i)$ from a finite set of tools in the training set (T). At each time step, the models produce a probability distribution over T, and the required tool is obtained by

$$\hat{t}_i = \operatorname*{argmax}_{t \in T} P(t_i = t | t_{i-k}^{i-1}), \tag{1}$$

The symbol in the first time step represents the start symbol and all sequences terminate in an end symbol, shared among all repair manuals.

3.2 Variable Order Markov Model

Variable order Markov models are an extension of first-order and fixed-order Markov models. In fixed-order Markov models, each random variable in a sequence depends on a specific number of preceding random variables. In VMMs, in contrast, the order of the model may vary in response to the statistics in the data. As a result of this, VMMs can be used to model sequential data of considerable complexity.

Begleiter et al. in [2] studied several prominent algorithms for the prediction of discrete sequences over a finite alphabet, using variable-order Markov models. Their results showed that in most cases the Prediction-by-Partial-Matching (PPM) algorithm performed best. In this paper, we use their implementation of the so-called method C (PPM-C)[1]. The value k in Eq. (1) can be interpreted as the upper bound for the maximal Markov order the model constructs and it limits the history to the k previous observations. When $k = 0$, it can be interpreted as a memoryless model which always predicts the most frequent tool in the training set, irrespective of any observation. For $0 < k \leq i$ the probability $P(t|t_{i-k}^{i-1})$ depends on the count of the sequence ($N(t_{i-k}^{i-1}, t)$) in the training set. If the count is zero, the model simply ignores the first observation in the sequence and recursively backs off to a lower order using an escape (E) probability:

$$P\left(t|t_{i-k}^{i-1}\right) = \begin{cases} P_k\left(t|t_{i-k}^{i-1}\right), & \text{if} <t_{i-k}^{i-1}, t> \text{exists in training set} \\ P_k\left(E|t_{i-k}^{i-1}\right) P\left(t|t_{i-k+1}^{i-1}\right), & \text{otherwise.} \end{cases} \qquad (2)$$

If Φ_k is the set of tools appearing after the history $<t_{i-k}^{i-1}>$ in the training set, the probability estimates based on the PPM-C method are:

$$P_k(t|t_{i-k}^{i-1}) = \frac{N(t_{i-k}^{i-1}, t)}{|\Phi_k| + \sum_{t' \in \Phi_k} N\left(t_{i-k}^{i-1}, t'\right)} \quad , \text{ if } t \in \Phi_k$$

$$P_k(E|t_{i-k}^{i-1}) = \frac{|\Phi_k|}{|\Phi_k| + \sum_{t' \in \Phi_k} N\left(t_{i-k}^{i-1}, t'\right)}$$

$$(3)$$

3.3 Recurrent Neural Network Model

Our RNN model consists of a Long-Short-Term Memory (LSTM) [10] layer, followed by a Fully Connected (FC) layer. The LSTM layer takes the tools encoded in 1-hot encoding from the input and transforms them into an embedding vector. The model estimates the probability of all possible tools in T, given the sequence of previous tools, by using a final soft-max layer that takes the 1-hot vector of the next tool as the ground truth during training. We use the cross-entropy loss

$$H(p, q) = - \sum_{x \in T} p(x) \log \left(q(x)\right) \qquad (4)$$

between the predicted p and target q probabilities of the tools using the Adam optimizer [13] with a learning rate of 0.001 over batches of 10 sequences. The

---

[1] We tried two other VMM algorithms using the SPMF open-source library, including compact prediction tree [8] and LZ78 [28], but they did not show a better performance than PPM and were therefore not considered in the later evaluation.

LSTM layer has 256 units attached to a fully connected hidden layer of size 256. During training, we also use dropout with a rate of 0.2. The training process benefits from an early stopping mechanism that watches the accuracy of the development set in each epoch and ends the training if the accuracy is not increased in 10 consecutive epochs.

3.4   Integrating the Task Target in a VMM

For integrating the task target in a VMM, we additionally condition the probability of observing a tool on the auxiliary information that represents the repair target (G).

Assume that the observed target of the task during test time is a string of words $G = w_1, w_2, .., w_m$. The probability of using tool t in the current step after observing the history of used tools and the task target can then be estimated via

$$P\left(t | t_{i-k}^{i-1}, G\right) = \frac{P(t | t_{i-k}^{i-1}) P(G | t_{i-k}^{i-1}, t)}{P(G | t_{i-k}^{i-1})} \propto P(t | t_{i-k}^{i-1}) P(G | t_{i-k}^{i-1}, t). \qquad (5)$$

The first term in Eq. (5), $P(t | t_{i-k}^{i-1})$, is derived similarly to the model without target awareness, using Eq. (3) independently of the task target. Similarly if $<t_{i-k}^{i-1}, t>$ does not exist in the training set, the model recursively ignores the earliest observation and repeats the process. The second term $P(G | t_{i-k}^{i-1}, t)$ is the probability of observing the target G given the history of tool usage and the tool candidate t. Let us assume that the sequence $<t_{i-k}^{i-1}, t>$ appears in one or more repair manuals in the training set. A set of all targets, e.g., titles of these manuals containing $<t_{i-k}^{i-1}, t>$, forms our reference for estimating $P(G | t_{i-k}^{i-1}, t)$. This set is denoted as D. We assume that the more similar G is to D, the higher the probability $P(G | t_{i-k}^{i-1}, t)$ should be. We use the Jaccard similarity between the sets of n-grams of various orders as the measure of text similarity between G and D. If G_n and D_n are the sets of all n-grams of order n in the strings of words in G and D respectively, the similarity score is: $\text{sim}(G_n, D_n) = |D_n \cap G_n| / |D_n \cup G_n|$. The final value is estimated by a weighted interpolation of the similarity scores of various n-gram orders:

$$P(G | t_{i-k}^{i-1}, t) \propto \epsilon + \sum_{n=1}^{m} \lambda(n) \text{sim}(D_n, G_n), \qquad (6)$$

where m is the total number of words in G, ϵ is a small smoothing constant to avoid zero values in case there is no common n-gram between G and D, and $\lambda(n)$ is the interpolation weight when $\sum_{n=1}^{m} \lambda(n) = 1$. The interpolation weight makes it easy to investigate the effect of various n-gram orders on the result. We examine three different approaches for λ, a constant weight ($\lambda_c(n) = 1/m$), a linear function that assigns higher weights to the higher-order n-gram similarities ($\lambda_l(n) = n / \sum_{j=1}^{m} j$), and a softmax function that exponentially grows with the order of n-gram similarities ($\lambda_s(n) = e^n / \sum_{j=1}^{m} e^j$).

3.5 Integrating the Task Target in an RNN Model

For conditioning RNN models on auxiliary features, several approaches have been proposed in the literature. In the *contextual recurrent networks*, see [7,15], a context vector is added to the input of the RNN cells at each time step. Unlike to our case, the context vector may change from one time step to the next. In contrast, in [12,23], the hidden state of the RNN is affected by an image embedder only at time zero for generating short image descriptions. For longer sequences, however, it may be difficult for the network to remember the auxiliary information until the end of the sequence. In our experiments, we achieved the best results when we included a representation of the repair target into the output of the LSTM hidden states in every time step, feeding into a fully connected layer. An advantage of this approach is that we do not need to re-train the LSTM weights, and the LSTM weights of the model in Sect. 3.3 can be used directly.

For generating a representation of the repair task, we employed another RNN with Gated Recurrent Unit (GRU) cells [4] of size 50, receiving the pretrained GloVe [20] word embeddings of the repair target description as input. The GRU cells receive the word embeddings and produce a representation of the repair target, which is concatenated to the representation of the sequence of the tools produced by the LSTM cells of the core tool prediction network. Figure 1 shows the architecture of this model.

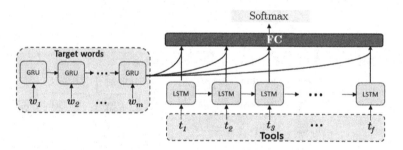

Fig. 1. Unrolled graph of the RNN model conditioned on the task target. The tools are encoded as 1-hot vectors and the target words are encoded with pre-trained GloVe embeddings.

4 Experiments

In this section, we describe the dataset we used for evaluating the performance of all considered sequence prediction models. First, we assess the models without an integrated target to analyze the effect of long-term dependencies, then we evaluate the performance of the new, target-aware predictors.

4.1 Dataset Description

We evaluate the methods of Sect. 3 on the human-annotated part of the MyFixit dataset[2]. MyFixit is a collection of repair manuals collected from the iFixit website [19]. The iFixit instructors split the manuals into their constituent steps, where at each step, the user typically works on one particular device component.

Since reassembly is often the reverse of disassembly, it is not included in the manuals of the iFixit website. Every step of manuals in the "Mac Laptop" category is annotated with the required tools of the step. In total, 1,497 manuals and 36,973 steps are annotated with 81 unique tools. Since our task is to predict the required tools, the steps that do not need a tool are not included in the manuals. On the other hand, more than one tool is used in 2.8% of the steps, which we split into multiple consecutive steps that only require one tool. Each repair manual has a title that usually contains detailed information about the device and the broken component, e.g., "MacBook Pro 15" Retina Display Mid 2014 Battery Replacement". We interpret the title of manuals as the full target of the task. Additionally, every manual has a hierarchical list of device *categories* and a *subject* field that refers to the component that needs to be repaired or replaced. We treat the name of device categories or the broken component as a limited target of the task. Figure 2 shows an instance of the device categories and components in MyFixit data. The name of the device categories in the data include the information from upper-level categories, for example, the complete name of "Late 2013" category in Fig. 2 is "MacBook Pro 15" Retina Display Late 2013" in the data.

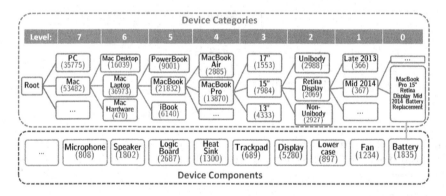

Fig. 2. Instances of the hierarchical categories of devices (above) and the component to be repaired (below) in the MyFixit dataset. The number in the parentheses shows the total number of steps within the device category or with same component to be repaired.

It can be seen in Fig. 2 that we identify each level of the device categories with a number, starting from zero on the leaf nodes that belong to the complete title

---

[2] https://github.com/rub-ksv/MyFixit-Dataset.

of the tasks. As the category level increases, we have a more general information about the device being repaired. One should note that the number of hierarchy levels is not the same for all the manuals; for example, MacBook Air devices do not have the level two in this figure relating to the Unibody/Retina display, and the level two for those devices corresponds to the level three in the figure, indicating the size of the screen.

4.2 Prediction Without Target-Awareness

In this section, we evaluate the models of Sect. 3.2 and 3.3 for the prediction of expected tools based only on the observation of previously applied tools. We use 10 folds of cross-validation; in each fold, we randomly split the data into 80% training and 20% test set. For the RNN-based model, we utilize 10% of the training set as a development set for hyper-parameter tuning and early stopping. In each cross-validation fold, the accuracy is calculated by counting the number of correct predictions divided by the total number of steps in the test set. By varying the maximum order of the VMM, i.e., k, we can scrutinize the effect of long-term dependencies on the prediction. When $k = $ max, the maximum order of the VMM is equal to the length of the longest sequence in the training set (61 on average). Training the RNN takes around 8 min on a single GTX TITAN X GPU and training the VMM on one core of a 2.10 GHz CPU with 11 MB cache takes 0.01 to 1.3 s depending on the maximum model order (k). Figure 3 shows the accuracy of the proposed models in percent. data$_N$ $N.txt Name x Accuracy st 1 VMM (k = 0) 131.30.2 VMM (k = 1) 248.50.4 VMM (k = 2) 353.30.5 VMM (k = 3) 459.10.4 VMM (k = 4) 563.10.3 VMM (k = 5) 666.10.3 VMM (k = 10) 773.50.4 VMM (k = max)$ 874.70.4 RNN 975.60.9.

Fig. 3. Accuracy (%) of tool prediction for VMM with different maximal Markov order (k) and RNN. The error bars show the standard deviation between the results of the 10 folds of cross validation.

As can be seen in Fig. 3, both approaches perform comparably well, where the RNN has a slightly higher accuracy, which is also accompanied by a somewhat higher standard deviation between the results of the 10 folds. The accuracy of the VMM with $k = 0$ reveals the percentage of the most frequent tool (the spudger) in the dataset. In contrast to Whitney et al., [24], who showed that increasing the order of the model to more than two, i.e., using *bi*-grams, does not have a considerable impact on deriving the correct ingredient, our result demonstrates that incorporating the longer history can significantly increase the accuracy of tool prediction. However, the effect of expanding the history to include more than 10 previous observations only leads to marginal improvements. Using an RNN model instead of a VMM results in a further small improvement. Yet this comes at the cost of considerably longer training times for the RNN.

4.3 Target-Aware Prediction

In the following, we assume that apart from the history of tool usage, the model has access to other pieces of information regarding the repair task. In one scenario, this information is the full task target, containing both the name of the device and the broken component (level 0 in Fig. 2). In another scenario, we assume that the information is limited either to the device category at various levels of generalization or to the name of the component being replaced or repaired.

Full Target: The models with target-awareness proposed in Sect. 3.4 and 3.5 are evaluated similarly to the model without target-awareness (Sect. 4.2) using 10 folds of cross-validation. Training the RNN model took around 4 min, while training and testing the VMM (k=max) conditioned on the task target took around 95 s for the constant and linear interpolation weight and approximately two minutes for the exponential weight. The computational hardware is the same as the ones in the previous section. Nearly one-third of this time was spent on the training phase when we build the transition matrix, and for each entry, a set of n-grams in the title of the manual is created. In the test phase, the similarity scores are computed in an online fashion. Table 1 shows the accuracy of the models.

Table 1. Accuracy (%) of the models without target-awareness, compared to the accuracy of the target-aware VMM with various interpolation weights (λ) of n-gram similarities, along with the RNN conditioned on the repair targets.

Target-awareness	Model	Accuracy	
		Mean	Std
Without	VMM	74.7	0.43
	RNN	75.6	0.88
With	VMM (Constant λ)	80.5	0.51
	VMM (Linear λ)	81.6	0.48
	VMM (Softmax λ)	83.4	0.36
	RNN	**86.3**	0.75

One can see that integrating target awareness improves the accuracy by 10.7% for RNN and 8.7% for VMM models. From Table 1 it can be seen that assigning larger weights to the higher n-gram similarities in VMM models yields a higher accuracy and lower standard deviation. The model based on an RNN gives the highest accuracy by producing a high-dimensional representation of the task target and the sequence of tools.

Limited Target: In this experiment, we investigate the cases with limited target knowledge, i.e., only knowing the general category or the broken component of the device. For the VMM, we compare two different strategies. In the first, similar to the models with full target-awareness, we condition the model's posterior probability on the sequence of words in the limited target, using Eq. (5) and the softmax interpolation weights. Since the similarity scores are calculated in an online manner, the model does not need extra training. In the second strategy, for each test instance, we train the VMM without an integrated target directly on the subset of the training data that shares the repair target. These subsets are denoted as "target sets". Because the training of the VMM, especially on small datasets, usually takes only a fraction of a second, this approach appears to be feasible in practice. If the target set of the category of a test manual is empty, i.e., there is no manual with the same category in the training set, the target set is constructed using a higher level category of the devices in the training set. For the components, we skip the target-awareness and use the complete training set if the target set becomes empty.

For the RNN model, we fine-tune the conditioned model (Sect. 3.5) on a training set with limited target information. We re-train the parameters of the GRU and fully connected layer on different variants of the training sets; in each one, the target of the manuals includes either the name of the component to be repaired or different levels of device categories. The models are evaluated on three levels of categories. The average numbers of manuals in the target set are 16.63, 72.59, and 149.99 for the category level 1, 2, and 3, respectively, and 22.16 for the device components. Table 2 represents the average accuracy of the suggested models for different levels of target-awareness.

Table 2. Average accuracy (%) of proposed VMM and RNN with full target-awareness (category level 0) and limited target-awareness (category 1, 2, 3 and device components).

Model	Category level				Component
	0	1	2	3	
VMM conditioned on target	83.4	82.6	80.4	78.1	76.2
VMM trained on target set	N/A	**83.3**	**81.1**	79.0	75.8
RNN conditioned on target	**86.3**	81.4	80.0	**79.3**	**77.5**

The result shows that the VMM trained on target sets performs slightly better when there is more explicit information about the category of the device, i.e., the category level is lower and there are fewer samples in the target set. For the most general category (level 3) and the device components, the RNN performs best.

5   Conclusion

In this paper, we investigate the use of sequence models for predicting tool usage from repair manuals. We introduce approaches to additionally condition the models on the repair task.

Our results show that the prediction of tool usage can be improved notably by considering the long-term dependencies among the repair steps. VMMs and RNNs are comparable for this task, while VMMs exhibit a considerably shorter training time and do not need the effort for hyper-parameter tuning that is required with neural network models. We have also shown that our approaches for including target awareness increase accuracy, and that this increase depends on the specificity of the provided repair target. Not surprisingly, the best performance is achieved when the target includes the exact model of the device and of the component that is to be replaced. For our targeted application domain, it could be beneficial to use both VMM and RNN models in parallel, as the VMM has the advantage that it can be trained on a smaller sub-set of the training set corresponding to the specified category of the device, while the RNN can offer performance benefits on larger data.

In this work, we have focused on task-conditional modeling of tool sequences, all the while assuming that the models can rely on a perfect observation of the ground-truth tool usage. In the future, we plan to explore how observation errors can influence the performance of each of the predictor models, with the goal of developing our methods towards real-life scenarios, where a repair assistant would only have imperfect data, e.g. from video recordings of the on-going repair process, on which to base its tool predictions and suggestions.

References

1. Abend, O., Cohen, S.B., Steedman, M.: Lexical event ordering with an edge-factored model. In: The 2015 Conference of the North American Chapter of the Association for Computational Linguistics: Human Language Technologies, pp. 1161–1171. ACL (2015)
2. Begleiter, R., El-Yaniv, R., Yona, G.: On prediction using variable order Markov models. J. Artif. Intell. Res. **22**, 385–421 (2004)
3. Cai, R., Wang, Q., Wang, C., Liu, X.: Learning to structure long-term dependence for sequential recommendation. arXiv preprint arXiv:2001.11369 (2020)
4. Cho, K., et al.: Learning phrase representations using RNN encoder-decoder for statistical machine translation. In: Proceedings of the 2014 Conference on Empirical Methods in Natural Language Processing, pp. 1724–1734. ACL (2014)

5. Chu, C.X., Tandon, N., Weikum, G.: Distilling task knowledge from how-to communities. In: Proceedings of the 26th International Conference on World Wide Web, pp. 805–814. ACM (2017)
6. Ding, W., Liu, K., Cheng, F., Zhang, J.: Learning hierarchical spatio-temporal pattern for human activity prediction. J. Vis. Commun. Image Represent. **35**, 103–111 (2016)
7. Ghosh, S., Vinyals, O., Strope, B., Roy, S., Dean, T., Heck, L.P.: Contextual LSTM (CLSTM) models for large scale NLP tasks. CoRR (2016)
8. Gueniche, T., Fournier-Viger, P., Tseng, V.S.: Compact prediction tree: a lossless model for accurate sequence prediction. In: Motoda, H., Wu, Z., Cao, L., Zaiane, O., Yao, M., Wang, W. (eds.) ADMA 2013. LNCS (LNAI), vol. 8347, pp. 177–188. Springer, Heidelberg (2013). https://doi.org/10.1007/978-3-642-53917-6_16
9. He, Q., et al.: Web query recommendation via sequential query prediction. In: 25th International Conference on Data Engineering, pp. 1443–1454. IEEE (2009)
10. Hochreiter, S., Schmidhuber, J.: Long short-term memory. Neural Comput. **9**(8), 1735–1780 (1997)
11. Kaiser, P., Lewis, M., Petrick, R.P., Asfour, T., Steedman, M.: Extracting common sense knowledge from text for robot planning. In: International Conference on Robotics and Automation (ICRA), pp. 3749–3756. IEEE (2014)
12. Karpathy, A., Fei-Fei, L.: Deep visual-semantic alignments for generating image descriptions. In: Proceedings of the IEEE Conference on Computer Vision and Pattern Recognition, pp. 3128–3137. IEEE (2015)
13. Kingma, D.P., Ba, J.: Adam: a method for stochastic optimization. In: 3rd International Conference on Learning Representations, ICLR (2015)
14. Maeta, H., Sasada, T., Mori, S.: A framework for procedural text understanding. In: Proceedings of the 14th International Conference on Parsing Technologies, IWPT, pp. 50–60. ACL (2015)
15. Mikolov, T., Zweig, G.: Context dependent recurrent neural network language model. In: Spoken Language Technology Workshop (SLT), pp. 234–239. IEEE (2012)
16. Mori, S., Maeta, H., Yamakata, Y., Sasada, T.: Flow graph corpus from recipe texts. In: Proceedings of the Ninth International Conference on Language Resources and Evaluation, pp. 2370–2377. ELRA (2014)
17. Murad, A., Pyun, J.: Deep recurrent neural networks for human activity recognition. Sensors **17**(11), 2556 (2017)
18. Nanba, H., Doi, Y., Tsujita, M., Takezawa, T., Sumiya, K.: Construction of a cooking ontology from cooking recipes and patents. In: Proceedings of the 2014 ACM International Joint Conference on Pervasive and Ubiquitous Computing: Adjunct Publication, pp. 507–516 (2014)
19. Nima Nabizadeh, Dorothea Kolossa, M.H.: MyFixit: an annotated dataset, annotation tool, and baseline methods for information extraction from repair manuals. In: Proceedings of Twelfth International Conference on Language Resources and Evaluation (2020)
20. Pennington, J., Socher, R., Manning, C.D.: Glove: global vectors for word representation. In: Empirical Methods in Natural Language Processing, pp. 1532–1543. ACL (2014)
21. Pichotta, K., Mooney, R.J.: Learning statistical scripts with LSTM recurrent neural networks. In: Schuurmans, D., Wellman, M.P. (eds.) Proceedings of the Thirtieth AAAI Conference on Artificial Intelligence, pp. 2800–2806. AAAI Press (2016)
22. Quadrana, M., Cremonesi, P., Jannach, D.: Sequence-aware recommender systems. ACM Comput. Surv. (CSUR) **51**(4), 1–36 (2018)

23. Vinyals, O., Toshev, A., Bengio, S., Erhan, D.: Show and tell: a neural image caption generator. In: IEEE Conference on Computer Vision and Pattern Recognition, CVPR, pp. 3156–3164 (2015)
24. Whitney, D., Eldon, M., Oberlin, J., Tellex, S.: Interpreting multimodal referring expressions in real time. In: International Conference on Robotics and Automation (ICRA), pp. 3331–3338. IEEE (2016)
25. Yamakata, Y., Imahori, S., Maeta, H., Mori, S.: A method for extracting major workflow composed of ingredients, tools, and actions from cooking procedural text. In: International Conference on Multimedia & Expo Workshops (ICMEW), pp. 1–6. IEEE (2016)
26. Zhang, Z., Webster, P., Uren, V.S., Varga, A., Ciravegna, F.: Automatically extracting procedural knowledge from instructional texts using natural language processing. In: Proceedings of the Eighth International Conference on Language Resources and Evaluation, pp. 520–527. ELRA (2012)
27. Zhou, Y., Shah, J.A., Schockaert, S.: Learning household task knowledge from WikiHow descriptions. arXiv preprint arXiv:1909.06414 (2019)
28. Ziv, J., Lempel, A.: Compression of individual sequences via variable-rate coding. IEEE Trans. Inf. Theory **24**(5), 530–536 (1978)

Safer Reinforcement Learning for Agents in Industrial Grid-Warehousing

Per-Arne Andersen$^{(\boxtimes)}$, Morten Goodwin, and Ole-Christoffer Granmo

Department of ICT, University of Agder, Grimstad, Norway
{per.andersen,morten.goodwin,ole.granmo}@uia.no

Abstract. In mission-critical, real-world environments, there is typically a low threshold for failure, which makes interaction with learning algorithms particularly challenging. Here, current state-of-the-art reinforcement learning algorithms struggle to learn optimal control policies safely. Loss of control follows, which could result in equipment breakages and even personal injuries.

On the other hand, a model-based reinforcement learning algorithm aims to encode environment transition dynamics into a predictive model. The transition dynamics define the mapping from one state to another, conditioned on an action. A sufficiently accurate predictive model should learn optimal behavior, also even in real environments.

The paper's heart is the introduction of the novel, Safer Dreaming Variational Autoencoder, which combines constrained criterion, external knowledge, and risk-directed exploration to learn good policies. Using model-based reinforcement learning, we show that the proposed method performs comparably to model-free algorithms safety constraints, with a substantially lower risk of entering catastrophic states.

Keywords: Deep reinforcement learning · Model-based reinforcement learning · Neural networks · Variational autoencoder · Markov Decision Processes · Exploration

1 Introduction

Reinforcement learning has recently demonstrated a high capacity to learn efficient strategies in environments where there are noisy or incomplete data [7]. We find achievements in many domains, such as robotics [12], wireless networking [20], and game-playing [13]. The common denominator between these domains is that they can be computer-simulated with significant resemblance to real-world environments, and therefore, let algorithms train at accelerated rates with strong safety guarantees.

The goal of reinforcement learning algorithms is to learn a behavioral policy that produces optimal actions based on sensory input and feedback from an environment. A policy is a parameterized model that is constructed in exact tabular form or using an approximation neural network with gradient descent

© Springer Nature Switzerland AG 2020
G. Nicosia et al. (Eds.): LOD 2020, LNCS 12566, pp. 169–180, 2020.
https://doi.org/10.1007/978-3-030-64580-9_14

[17]. The algorithm performs an iterative process of exploration (often through sampling), exploitation, and learning policy updates that moves the policy in the direction of better behavior. Exploration is commonly performed using a separate policy, such as random sampling from a Gaussian distribution. It is crucial that the algorithm balance exploration and exploitation with schemes such as ϵ-greedy so that the policy learns with the best possible data distribution.

The challenges towards achieving safer reinforcement learning are many. The main limitations of current state-of-the-art in this context are (1) It requires a tremendous amount of sampling to learn a good policy. (2) Stable and safe policies are challenging to achieve in dynamic environments. (3) Model-free exploration methods are not safe in mission-critical environments. (4) Reinforcement learning methods depend on negative feedback through trial and error and is therefore not applicable to mission-critical systems. Most reinforcement learning techniques are not designed for safe learning, and therefore, few solutions exist for mission-critical real-world environments.

Automated Storage and Retrieval Systems (ASRS) is a modern method of performing warehouse logistics where the system is partially or fully automated. In industry, including ASRS, it is common to rely on complex expert systems to perform tasks such as control, storage, retrieval, and scheduling. If-else statements and traditional pathfinding algorithms drive these tasks. The benefit of expert systems is that it is trivial to model operative safety bounds that limit the system from entering catastrophic states. The downside is that expert systems do not adapt to changes automatically, and requires extensive testing if altered. While it may be possible and perhaps trivial to construct safe routines with an expert system, it is inconceivable to expect optimal behavior due to the complexity of the environment. Reinforcement learning is possibly the most promising approach to solve these problems because it can generalize well across many domains [13] and is designed to work in noisy environments with partial state-space visibility [18].

This paper presents *The Safer Dreaming Varational Autoencoder* (DVAE-S), an algorithm for safe reinforcement learning in real-world environments. DVAE-S is based on previous work in [2], where a model-based approach was proposed for training fully off-line without direct access to the real environment. However, the algorithm suffers from making catastrophic actions, a common problem in reinforcement learning and, therefore, does not satisfy the safety requirements. The following research question was raised.

How to ensure that the agent acts within defined safety boundaries?

DVAE-S address this question by eliminating catastrophic states from the state-space and utilizing external knowledge from previously hand-crafted algorithms

The rest of the paper is structured as follows. Section 2 discusses related work to the proposed algorithm. Section 3 introduces preliminaries. Section 4 presents the DVAE-S algorithm for safe reinforcement learning. Section 5, presents an approach for safer learning of ASRS agents using the DVAE-S algorithm. Finally, Sect. 6 concludes the work and proposes a roadmap for future work in safe reinforcement learning.

2   Related Work

This section address related work to applied reinforcement learning in industry-near environments and on improving safety in reinforcement learning. Advancements in reinforcement learning have recently been more frequent and with substantial performance improvements in various domains [4]. Trial and error is central in standard reinforcement learning and therefore, little attention is given to safety during training.

Reinforcement learning is previously applied to industry-near environments, and perhaps the most widespread application is autonomous vehicles. The proposed method in this paper uses an auxiliary policy to label data for supervised training. With only 12 h of labeled data, [19] shows that it is possible to learn a performant policy using a direct perception approach with convolutional neural networks. This approach is much like a variational autoencoder that simplifies the perception of the world significantly. This simplifaction of the input significantly speeds up inference, which enables the system to issue control commands more frequently. Many other significant contributions in autonomous vehicle control directly relate to control in ASRS environments, such as [15].

2.1   Safe Reinforcement Learning

In the majority of established systems in the industry, an expert system acts as the controller for the environment. It is critical for safe and stable learning in real world-environments so that ongoing operations are not interrupted.

Similar to this work, [5] assumes a predictive model that learns the dynamics of the environment. The authors propose that the policy should be limited to a safe-zone, called the **R**egion **O**f **A**ttraction (ROA). Everything within the bounds of the ROA is "safe states" that the policy can visit, and during training, the ROA gradually expands by carefully exploring unknown states. The algorithm shrinks the ROA to ensure stability if the feedback signal indicates movement towards catastrophic states.

The proposed algorithm encodes observations as latent embeddings or vectors using a variational autoencoder (VAE), similar to the View model in [10]. The World Model approach defines three components. The *view* (VAE) encodes observations to a compact latent embedding. The *model* (MDM-RNN)[1] is the predictive model used to learn the world model. Finally, the *controller* (C) is an interoperability abstraction that enables model-free algorithms to interact with the world model.

3   Background

This work is modeled according to an **M**arkov **D**ecision **P**rocess (MDP) described formally by a tuple (S, A, T, R), where S is the state-space, A is the action-space, $T: S \times A \rightarrow S$ is the transition function and $R: S \times A \rightarrow \mathbb{R}$ is the reward function [17].

---

[1] Mixture Density Network combined with a Recurrent Neural Networks.

3.1 Safer Policy Updates

A policy π in reinforcement learning is a parametrized[2] model that maps (input) observations to (output) actions. The goal is to find an optimal policy $\pi^* = \arg\max_\pi V^\pi(s) \quad \forall s \in S$ where $\max_\pi V^\pi(s)$ denotes the highest possible state-value under policy π.

In traditional RL, the algorithm learns according to an *optimization criterion*. This optimization criterion varies with different algorithms but is commonly implemented to minimize time or to maximize reward. Any cost metric can be used and is defined by the algorithm designer. The *return maximization criterion* is one example where the agent seeks a policy that achieves the highest possible reward, where $R = \sum_{t=0}^{\infty} \gamma^t r_t$ denotes the *expected cumulative future discounted reward*. $\gamma \in [0, 1]$ is the *discount factor* that determines the importance of future rewards. The problem with the return maximization criterion is that it is **not sufficient** for environments with catastrophic states. We adopt the term *risk* from [11], which is a metric that explains the risk of doing actions using a policy π. There is several proposed criterion methods to improve safer policy updates, namely *Risk-Sensitive Criterion* [9], *Worst Case Criterion* [8], and *Constrained Criterion* [14].

This paper use the **Constrained Criterion**, where the objective is modified to maximize the return, similar to [5]. The difference from the traditional objective function is that it introduces additional constraints to the policy objective function. These constraints act as a lower or upper bound for the maximization of the return. The general form of the constrained criterion is defined

$$\max_{\pi \in \Pi} E_\pi(R) \text{ subject to } c_i \in C, c_i = \{h_i \gtrless \alpha_i\} \tag{1}$$

where c_i is the ith constraint in C that must be satisfied by the policy π, additionally $c_i = \{h_i \gtrless \alpha_i\}$ must specify a function related to the return h_i with a threshold value α_i.

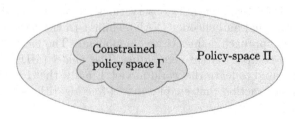

Fig. 1. The policy-space (blue) Π and the subset of policies (red) $\Gamma \subseteq \Pi$, where each policy $\pi \in \Gamma$ must satisfy the constraints $c_i \in C$. (Color figure online)

Figure 1 illustrates that a constrained policy-space eliminates a considerable portion of possible policies. The objective of the constraints is to maximize the

---

[2] All models in this paper are neural network approximations.

number of eliminated states that lead to catastrophic outcome and minimize elimination of safe states. The objective function is denoted $\max_{\pi \in \Gamma} E_\pi(R)$ given that only updates from the constrained subset of policies Γ is used. This increases the safety of policy updates far more than the traditional return maximization methods. Depending on the application, the constraints can be tuned using the threshold value α. Higher the value, the constraint is more permissive, and for lower values, more restrictive.

3.2 Safe Exploration

Policies with a constrained criterion do not guarantee safety in the short term. **Risk-directed Exploration** is a good approach to improve short term safety and uses risk to determine in which direction the agent should explore. There are many possible ways to define a risk metrics, but this paper uses *normalized expected return with weighted sum entropy* [6].

$$Risk(s, a) = \mathcal{R}(s, a) = wH(X) - (1 - w)\frac{\mathbb{E}[R(s, a, s')]}{\max_{a \in A} |\mathbb{E}[R(s, a, s')]|} \tag{2}$$

where the entropy is defined as

$$H(s, s) = \mathbb{E}[-\pi(a|s) \log \pi(a|s)]. \tag{3}$$

The resulting utility function denotes as follows:

$$Utility(s, a) = U(s, a) = p(1 - \mathcal{R}(s, a)) + (1 - p)\pi(s, a). \tag{4}$$

The calculated risk is multiplied with the action probability distribution for states in the MDP. The updated utility function ensures that sampling is performed in favor of safe and conservative actions, depending on the weight parameter w, and interpolation parameter p [6]. The risk-directed exploration plays well with introducing external knowledge from existing expert systems. The proposed method works well with external knowledge originating from expert systems already running in an environment. Initial knowledge significantly boosts the accuracy of risk and therefore improves the safety and performance of the agent. The downside of using external knowledge is that the predictive model becomes biased and may not learn well when entering unvisited areas of the state-space. However, the benefit is that large portions of the state-space are quickly labeled, which provides better safety guarantees during learning.

4   Safer Dreaming Variational Autoencoder

This work extends the *Dreaming Varational Autoencoder* (DVAE) from [2,3] to reduce the risk of entering catastrophic states in mission-critical systems. DVAE is a model-based reinforcement learning approach where the objective is two-fold.

1. The algorithm learns a predictive model of the environment. The goal of the predictive model is to capture the $(T \colon S \times A \to S)$ dynamics between states and to learn the $(R \colon S \times A \to \mathbb{R})$ reward function.
2. The DVAE algorithm uses the learned predictive model to train model-free algorithms, where the interaction during training is primarily with the predictive model.

Environment Isolation

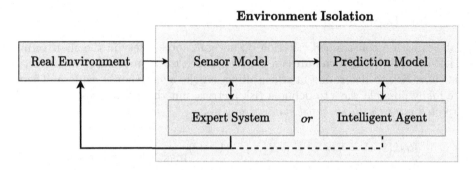

Fig. 2. The predictive model of DVAE-S learns the sensor model behavior. The intelligent agent uses the predictive model to train entirely off-line, without the risk of making mistakes in the real environment. After training, the algorithm performs inference in the real environment, with significantly less chance of making actions that lead to catastrophic states.

Figure 2 demonstrates the algorithm procedure in the following steps. (1) The predictive model observes and learns the real environment dynamics using a sensor model. The same sensor model is the interface that the expert system uses for making actions. (2) The intelligent agent (i.e., reinforcement learning agent) interacts with the predictive model and learns a policy. (3) When the intelligent agent is sufficiently trained, it replaces existing expert systems with comparable performance. (4) The intelligent agent can optionally train further in the real-world environment for improved performance, at the risk of making catastrophic actions.

4.1   Implementation

The DVAE-S algorithm is composed of the following three models: The **V**iew, **R**eason, and **C**ontrol (VRC).

The **view model** is responsible for transforming raw input data into a meaningful and compact feature embedding. DVAE-S uses a variational autoencoder primarily for this task, but there exist alternative methods, such as generative adversarial networks (GAN). It is possible to visualize the embedding $z_x \in \mathbb{Z}$ by manually altering input values. This is especially useful when predicting long

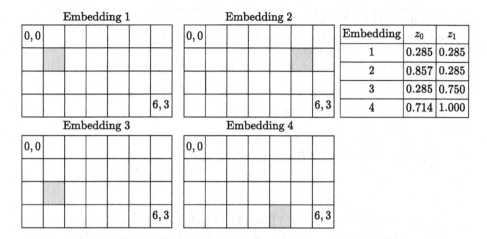

Fig. 3. Four distinct (table) embedding value sets and its (grid-world) decoded output. (Color figure online)

horizon futures. Figure 3 illustrates this with a (green) agent in an empty grid-world. The embedding layer consists of two neurons where the first and second neuron learns the vertical and horizontal location, respectively.

The **reason model** learns the time dependency between states, or in MDP terms, the transition function $T: S \times A \rightarrow S$. The reason model computes the future state embedding z_{t+1}^{π} based on a batch of previous embeddings from the view $Z_t^{\pi} = \{z_{t-n} \dots z_t\}^{\pi}$. The policy π denotes the behavior that DVAE-S follows. DVAE-S uses the *long short-term memory* (LSTM) architecture as it was found best learn the transition between states.

The **control model** is responsible for interaction with the environment. The control is the primary model for performing actions that are safe and progress the learning in the right direction. DVAE-S uses primarily Deep Q-Networks (DQN) and Proximal Policy Optimization (PPO) with **constrained optimization criterion** and the **risk-directed exploration** method. The input to the algorithm is a raw-state, commonly a high-dimensional data structure that is difficult to interpret for humans. The benefit of the DVAE-S architecture is that the reason model finds an embedding that is describes states with significantly fewer dimensions. The DVAE-S algorithm also enables initial training entirely off-line in a *dream* version of the real environment.

Algorithm 1. The DVAE-S algorithm

1: Initialize policy $\pi_\theta(s_t|a_t)$, predictive model $p_\psi(\hat{s}_{t+1}, \hat{r}, h_t|s_t, a_t^\pi)$
2: Let $Z = \{z_{t-n}^\pi \ldots z_t^\pi\}$, a vector of encoded states
3: Initialize encoder $ENC(z_t^\pi|s_t, a_t^\pi)$, temporal reasoner $TR(h_t^\pi|Z)$
4: **for** N epochs **do**
5:     $\mathcal{D}_{env} \leftarrow$ Collect samples from p_{env} under predefined policy π
6:     Train model p_ψ on data batch $\mathcal{D}_{env}$ via MLE

7: **for** M epochs **do**
8:     Sample initial state $s_0 \sim U(0,1)$ from $\mathcal{D}_{env}$
9:     Construct $\{\mathcal{D}_{p_\psi}|t < k, TR(h_t^{\pi_\theta}|ENC(z_t|s_t, a_t)^{\pi_\theta}, s_t = s_0)\}$
10:     Update policy π_θ using pairs of $(\hat{s}_t, a_t, \hat{r}_t, \hat{s}_{t+1})^{\pi_\theta}$ discard if $\pi_\theta \notin \Gamma$

Algorithm 1 works as follows. **(Line 1)** Initialize the control policy and the predictive model parameters. **(Line 2)** The Z variable denotes a finite set of sequential predictions from the view model (ENC) that captures time dependency between states from the reason model (TR). **(Line 5)** Collect samples from the real environment p_{env} under a predefined policy (i.e, expert system), see Fig. 2. **(Line 6)** The predictive model p_ψ trains using the collected data $\mathcal{D}_{env}$. The loss is calculated using maximum likelihood estimation (MLE) or mean squared error $MSE(p_\psi \| p_{env})$ until the loss is sufficiently low. **(Line 7)**Training of the control model now starts using the predictive model p_ψ in place of the real environment p_{env}. **(Line 8)** The first state s_0 is sampled from the real dataset $\mathcal{D}_{env}$ and the control policy makes action in the prediction model to create an artificial dataset of simulated states (sampling from the predictive model). **(Line 10)** The parametrized control policy is optimized using $(\hat{s}_t, a_t, \hat{r}_t, \hat{s}_{t+1})^{\pi_\theta \notin \Gamma}$ pairs under the constrained policy where "hat" denotes predicted values.

4.2 Exploration and Policy Update Constraints

There are significant improvements to the exploration and policy update for finding safe policies in DVAE-S. During sampling, the control model uses the *risk-directed exploration* bonus [6]. This bonus is added to the probability distribution over actions before sampling is performed, as described in Sect. 3. The **policy updates** are constrained to a set of criterion defined as follows. During the learning of the predictive model, feedback is received from the real-world environment. The model assumes that all actions that were not chosen by the agent are considered unsafe. This way, the algorithm gradually maps the unsafe policy space. It is important to note that this mapping does not influence the choices of the agent when learning the predictive model. When the agent revisits a state, the agent may select another action, and this will label the state as safe.

5 Results

The content of this section summarizes experiments with the following contributions. The ASRS-Lab and Deep RTS is introduced as the test environment for the experiments. The proposed method is then applied to Deep Q-Networks [13], which we refer to as DVAE-S-DQN, and Proximal Policy optimization, which we refer to as DVAE-S-PPO, [16] and tested in three environments, respectively, ASRS-Lab 11x11, ASRS-Lab 30x30, and DeepRTS Gold-Collect. Each experiment is averaged over 100 separate runs. Finally, the predictive model performance and safety performance is discussed with the resulting output from decoded embeddings.

The ASRS Lab[3] is a detailed simulator for research into autonomous agents in grid-warehousing [3]. The simulator's purpose is to create a framework that would efficiently represent a real-world environment without the risk of damaging costly equipment. Cost is perhaps the reason that safety in reinforcement learning is less prevalent than standard model-free approaches. There are many categories of ASRS systems in the real world, and the ASRS Lab environment supports a subset of existing setups. The ASRS-Lab features a shuttle-based, aisle-based, and grid-based warehouse simulation where this work focus on solving the grid-based architecture.

DeepRTS[4] is a lightweight game engine specifically designed for reinforcement learning research [1]. The motivation behind DeepRTS is to set up scenarios that this easily reproducible quickly. This paper focuses on solving the "DeepRTS Gold-Collect" environment where the goal is to accumulate as much gold as possible before a predefined timer runs out.

5.1 Predictive Model

The predictive model trains with a learning rate of 0.0003 using Adam optimizer. The observations originate from expert systems, a manhattan-distance agent in ASRS-Lab, and rule-based agent for DeepRTS Gold-Collect[5] Note that the experiments are performed after the safe policy pretraining.

Figure 4 shows the training loss for the predictive model for each environment. The second row is the decoded embeddings when predicting future states using the DVAE-S-DQN model. The model achieves this embedding and decoding quality using a learning rate of 0.002, batch size of 16 (previous states) with the adam optimizer. For the DVAE-S-DQN model, the performance gradually decreases for ASRS environments with similar results. The loss in the Deep RTS environment increased in early training but converged after approximately 500000 timesteps. The DVAE-S-PPO method cannot reproduce the DQN variant's performance because of divergence during agent training.

5.2 Failure Rate

The failure-rate experiment measures how often an agent makes actions that lead to catastrophic states in the environment where 1.0 consistently failure and

---

[3] https://github.com/cair/deep-warehouse.

[4] https://github.com/cair/deep-rts.

[5] The agent in DeepRTS walks to the nearest gold deposit when the inventory is empty. When inventory is full, it returns the gold to the base.

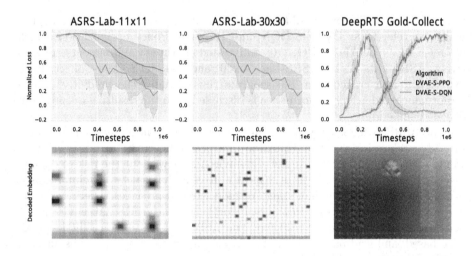

Fig. 4. Training loss of the DVAE-S predictive model including the resulting decoded latent-space variables for the DQN variant. The DVAE-S-DQN move towards convergence while DVAE-S-PPO fails to improve. The bottom row illustrates the quality of predicted future states after applying median filtering.

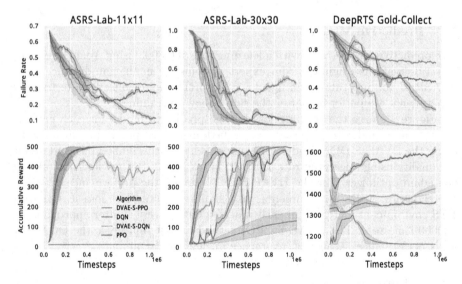

Fig. 5. A comparative study of DQN and PPO using DVAE-S. Each plot shows results in an environment, respectively, ASRS-Lab-11 x 11, ASRS-Lab-30 x 30, and DeepRTS Gold-Collect. The topmost row shows the failure rate of agents, and the bottom row illustrates the accumulated reward during the experiment. All plots have a duration of 1 million frames. The plots clearly show that DVAE-S-DQN performs comparably to standard DQN while having a significantly lower failure rate. On the contrary, DVAE-S-PPO performs worse than vanilla PPO.

0 is error-free behavior. The accumulated reward is recorded simultaneously to verify that the agent learns a better behavior. The parameters for risk-directed exploration is $w = 0.7$, and $p = 0.85$, and $\alpha = 0.4$ for all constraints.

Figure 5 shows that DQN agents using the DVAE-S algorithm significantly reduces the frequency of catastrophic states during a training period of 1 000 000 timesteps. The expert systems form the baseline performance of an accumulated reward of 300 in the ASRS-Lab environments and 1300 for the DeepRTS environment. The DVAE-S-DQN performs comparably to vanilla DQN, but the empirical evidence suggests that using constrained criterions and risk-directed exploration reduces the failure rate significantly. The PPO variant did not perform better, and it remains future work to investigate why it poses a challenge to combine direct-policy search algorithms and DVAE-S.

Setting the low performance of DVAE-S-PPO aside, the DQN variant shows impressive results where it performs significantly better than vanilla DQN and above the in the DeepRTS Gold-Collect environment. The performance of DVAE-S-DQN is marginally lower in the ASRS-Lab environment but at the benefit of increased safety, which is the primary concern addressed in this work.

6   Conclusion and Future Work

The results in this paper show that the combination of DVAE, constrained criterion, and risk-directed exploration is a promising approach for improving safety while maintaining comparable behavioral performance. The empirical evidence suggests a strong relationship between control performance and predictive model performance, where good policies lead to good predictive models and the contrary for bad policies.

By using the constrained criterion for policy updates and risk-directed exploration, the DVAE-S architecture significantly improves the safety of learned policy in some algorithms. The DVAE-S-DQN algorithm performs comparably to agents without safety constraints but has a significantly lower risk of entering catastrophic states. The DVAE-S architecture is a novel approach for reinforcement learning in industry-near environments.

This work's continued research is dedicated to better combine proximal policy optimization with DVAE-S for safe reinforcement learning. Temporal convolutional networks look promising for improving the performance of the predictive model. We also hope to experiment with the novel vector quantized variational autoencoders for better embedding high-dimensional states. While this paper contributes new findings in safe reinforcement learning, it is still room for significant improvements, in which we hope to contribute more in the future.

References

1. Andersen, P.-A., Goodwin, M., Granmo, O.-C.: Deep RTS: a game environment for deep reinforcement learning in real-time strategy games. In: 2018 IEEE Conference on Computational Intelligence and Games (CIG). IEEE (2018). https://doi.org/10.1109/CIG.2018.8490409

2. Andersen, P.-A., Goodwin, M., Granmo, O.-C.: The dreaming variational autoencoder for reinforcement learning environments. In: Bramer, M., Petridis, M. (eds.) SGAI 2018. LNCS (LNAI), vol. 11311, pp. 143–155. Springer, Cham (2018). https://doi.org/10.1007/978-3-030-04191-5_11

3. Andersen, P.-A., Goodwin, M., Granmo, O.-C.: Towards model-based reinforcement learning for industry-near environments. In: Bramer, M., Petridis, M. (eds.) SGAI 2019. LNCS (LNAI), vol. 11927, pp. 36–49. Springer, Cham (2019). https://doi.org/10.1007/978-3-030-34885-4_3

4. Arulkumaran, K., Deisenroth, M.P., Brundage, M., Bharath, A.A.: Deep reinforcement learning: a brief survey. IEEE Sig. Process. Mag. **34**, 26–38 (2017). https://doi.org/10.1109/MSP.2017.2743240

5. Berkenkamp, F., Turchetta, M., Schoellig, A.P., Krause, A.: Safe model-based reinforcement learning with stability guarantees. In: Advances in Neural Information Processing Systems 30, pp. 908–918. Curran Associates Inc., CA, USA (2017)

6. Edith, L.L., Melanie, C., Doina, P., Bohdana, R.: Risk-directed exploration in reinforcement learning. In: IJCAI (2005)

7. Fox, R., Pakman, A., Tishby, N.: Taming the noise in reinforcement learning via soft updates. In: Proceedings of the 32nd Conference on Uncertainty in Artificial Intelligence, UAI 2016, pp. 202–211. AUAI Press, Arlington (2016). https://doi.org/10.5555/3020948.3020970

8. Gaskett, C.: Reinforcement learning under circumstances beyond its control. In: Proceedings of the International Conference on Computational Intelligence for Modelling Control and Automation (2003). http://www.his.atr.co.jp/cgaskett/

9. Geibel, P., Wysotzki, F.: Risk-sensitive reinforcement learning applied to control under constraints. JAIR **24**, 81–108 (2005). https://doi.org/10.1613/jair.1666

10. Ha, D., Schmidhuber, J.: Recurrent world models facilitate policy evolution. In: Advances in Neural Information Processing Systems 31, pp. 2450–2462. Curran Associates Inc., Montréal, September 2018. https://arxiv.org/abs/1809.01999

11. Heger, M.: Consideration of risk in reinforcement learning. In: Proceedings of the 11th International Conference on Machine Learning, ICML 1994, pp. 105–111. Elsevier (1994). https://doi.org/10.1016/B978-1-55860-335-6.50021-0

12. Levine, S., Finn, C., Darrell, T., Abbeel, P.: End-to-end training of deep visuomotor policies. J. Mach. Learn. Res. **17**(1), 1334–1373 (2016)

13. Mnih, V., et al.: Playing Atari with deep reinforcement learning. Neural Information Processing Systems, December 2013. http://arxiv.org/abs/1312.5602

14. Moldovan, T.M., Abbeel, P.: Safe exploration in Markov decision processes. In: Proceedings of the 29th International Conference on Machine Learning (2012)

15. Santana, E., Hotz, G.: Learning a Driving Simulator. arxiv preprint arXiv:1608.01230 (2016)

16. Schulman, J., Wolski, F., Dhariwal, P., Radford, A., Klimov, O.: Proximal Policy Optimization Algorithms. arxiv preprint arXiv:1707.06347 (2017)

17. Sutton, R.S., Barto, A.G.: Reinforcement learning: An introduction, 2nd edn. A Bradford Book, Cambridge (2018). https://doi.org/10.5555/3312046

18. Wang, Y., He, H., Tan, X.: Robust Reinforcement Learning in POMDPs with Incomplete and Noisy Observations (2019)

19. Xiao, T., Kesineni, G.: Generative adversarial networks for model based reinforcement learning with tree search. Technical report, University of California, Berkeley (2016). http://tedxiao.me/pdf/gans_drl.pdf

20. Zhang, C., Patras, P., Haddadi, H.: Deep learning in mobile and wireless networking: a survey. IEEE Commun. Surv. Tutor. **21**(3), 2224–2287 (2019). https://doi.org/10.1109/COMST.2019.2904897

Coking Coal Railway Transportation Forecasting Using Ensembles of ElasticNet, LightGBM, and Facebook Prophet

Vladimir Soloviev[✉] [iD], Nikita Titov[iD], and Elena Smirnova[iD]

Financial University under the Government of the Russian Federation, 38 Shcherbakovskaya, Moscow 105187, Russia
{VSoloviev,NATitov,EKSmirnova}@fa.ru

Abstract. Based on monthly data, from January 2010 to December 2016, machine learning models were developed for coking coal railway transportation forecasting in two directions: export and domestic transportation. We built ensembles of ElasticNet, LightGBM, and Facebook Prophet models. The coal export transportation volumes are best predicted by an ensemble of ElasticNet and LightGBM models, giving the mean absolute percentage error at 10%. The best model for coking coal domestic transportation forecast is pure LightGBM, with the mean absolute percentage error at 6%.

Keywords: Russian Railways · Coking coal · Rail freight transportation forecasting · Time series · Machine learning · ElasticNet · LightGBM · Facebook Prophet

1 Introduction

In recent years, the balance of economic centers in the world is changing, and significant structural changes are taking place in the global economy. The role of regional economic unions is growing. New technologies of freight transportation are being introduced and widely spread. All these processes naturally affect the global coal market, which retains its great importance in the global energy balance and plays a vital role in the worldwide metallurgy, and therefore has a direct impact on all metal-intensive industries, from construction to engineering. Consequently, it is essential to be able to accurately predict the change in demand for coal, as well as for transportation services.

This paper aims to build machine learning models to predict the demand for coking coal freight transportation in Russian Railways.

The main challenge is to build adequate forecasts based on a very short time series of about a hundred records, in which a significant part of the values are zeros.

Six groups of features are identified that, in our opinion, most affect the transportation volume: GDP by country, coal prices, global coal imports, Russian coal exports, dollar exchange rate, coal production.

Based on the ensembling of ElasticNet, LightGBM, and Facebook Prophet, machine learning models were developed to estimate future export and domestic coking coal

© Springer Nature Switzerland AG 2020
G. Nicosia et al. (Eds.): LOD 2020, LNCS 12566, pp. 181–190, 2020.
https://doi.org/10.1007/978-3-030-64580-9_15

freight transportation efficiently, even trained on a limited dataset of about a hundred rows.

The remainder of the paper is structured as follows. Section 2 reviews related works on forecasting the demand for rail passenger and freight transportation using intelligent algorithms, Sect. 3 presents our research methods, including Russian Railways coking coal freight transportation dataset overview; ElasticNet, LightGBM, and Facebook Prophet ensembling approach; and forecasting quality methodology. Section 4 presents the results of coking coal export and domestic transportation forecasting using ensembles of ElasticNet, LightGBM, and Facebook Prophet. Finally, the results are discussed in Sect. 5.

2   Literature Review

Effective methods for forecasting the demand for rail passenger and freight transportation originate from Rao (1978), who proposed a model for forecasting railway transportation based on stochastic equations [1].

Since then, various techniques have been used to forecast railway transportation volumes, including:

- traditional ordinary least squares linear and non-linear regression, partial adjustment, time-varying parameter regression, and other classic econometric models [2–5];
- Holt-Winters, AR, VAR, ARIMA, ARIMAX, SARIMAX, and other time series models [6–13];
- support vector regression models [14–17];
- back propagation neural networks [18, 19], multi-layer perceptrons, multiple temporal units neural network, parallel ensemble neural network [20], generalized regression neural network [21–23], LSTM [24, 25], deep belief networks [26], and other deep neural networks [27–30].

These approaches work quite well when models are trained on sufficiently large datasets, typical, for example, for mass passenger traffic.

However, data on coal railway freight transportation are collected monthly, which raises the problem of training the model on a dataset containing about a hundred lines.

In recent years, effective methods for forecasting time series have appeared: ElasticNet [31], LightGBM [32], and Facebook Prophet [33].

It seems that these methods may improve the quality of coking coal railway transportation forecasting.

3   Research Methods

3.1   Russian Railways Coking Coal Freight Transportation Dataset

The considered dataset contains a total of only 114 records (January 2010–June 2019), with a significant part of the values represented by zeros. The dataset is published online by the authors [34].

The initial data are monthly values of the following features:

- actual coking coal transportation by Russian Railways;
- production volumes of steam and coking coal, steel, metallurgical coke from coal obtained by carbonization at high temperature, as well as electricity produced by thermal power plants;
- volumes of world trade in the coal of Russia with other countries;
- prices for coal from South Africa and Australia as substitutes;
- GDP volumes of individual countries;
- USD/RUB exchange rates.

As the labels, we considered export and domestic transportation of coking coal.

3.2 Ensembling of ElasticNet, LightGBM, and Facebook Prophet

ElasticNet [31] is a method that involves constructing a regression model using the least-squares method with a penalty in the form of the sum of the model coefficients squares and the sum of the model coefficients absolute values.

Thus, this technique summarizes the approaches of Ridge and Lasso regressions and allows us to overcome the disadvantages of the ordinary least squares method by regularization of parameters.

Namely, if the number of features is not much smaller than the number of observations, then there is a decrease in the forecasting accuracy due to increasing variance. If the number of regressors exceeds the number of observations, the least-squares estimates become inconsistent, as the forecast variance tends to infinity.

The ElasticNet allows to reduce the forecast variance and, accordingly, to improve the forecast accuracy due to slight bias.

LightGBM [32] is a method of constructing reinforced regression trees based on splitting the original dataset into more homogeneous subsets.

This method has several advantages compared to other tools for building boosted decision trees.

For example, it demonstrates the training speed several times faster, and with better accuracy, in comparison with similar boosted decision trees like XGBoost. It also does not require special processing of categorical variables and missing values.

Facebook Prophet [33] is a novel model proposed by Facebook engineers for time series prediction. Unlike ARIMA, ARCH, and other specialized tools, Facebook Prophet does not require in-depth knowledge and extensive experience with time series, and also does not require elaborate source data preparation.

This method is an additive regression model with some advantages. It can track annual seasonality using the Fourier series and distinguish weekly and daily seasonality using dummy variables. It also takes rare events (for example, holidays) into account.

The Prophet is resistant to jumps in trends (a trend is modeled by either a piecewise linear or a logistic function), small data gaps, and outliers.

It seems that these three methods can improve the forecasting quality for coking coal freight transportation.

To forecast export and domestic transportation of coking coal, we built optimal ensembles of ElacticNet, LightGBM, and Facebook Prophet as the final models.

3.3 Forecasting Quality Measurement

To assess the forecasting quality of machine learning models, the initial dataset was split into training and test subsets.

Since the source data are time series by nature, the split into two disjoint subsamples was performed as follows.

The training dataset included data from January 2010 to December 2016 and the test dataset - from January to December 2017.

Such split is typical when working with time series models and avoids "looking into the future by the model," which leads to inconsistent forecasts.

Data for 2018 and 2019 were not used since, for half the features, the history of observations in the existing data ends in 2017.

As the quality metrics for the models, we calculated *MAE, MSE, MAPE*, and R^2.

4 Results

4.1 Coking Coal Export Transportation Forecasting

First, three models were built based on the three methods described above: ElasticNet, LightGBM, and Facebook Prophet, using the following features:

- GDP volumes of countries;
- USD/RUB exchange rate;
- steam and coking coal prices;
- production indicators;
- coking coal export transportation for the previous month.

Based on the quality assessment of these models, it was decided to ensemble the LightGBM and ElasticNet models by calculating the weighted average forecast value with the following weights:

- 0.6 for the LightGBM model (with coal prices and production indicators as the features);
- 0.4 for the ElasticNet model (with the GDP of individual countries, coal prices, and production indicators as the features).

Figure 1 illustrates the result of coking coal export transportation forecasting by the resulting ensemble model, and Table 1 shows the results of hyperparameters tuning for the models.

Here and further in the figures, the test subsample is highlighted with a light green background, blue lines represent the true values of the target variable, and red lines represent the forecasts given by the models.

The following metrics characterize the quality of the resulting ensemble model on the test dataset:

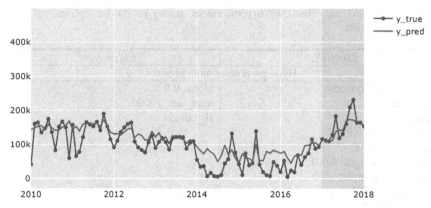

Fig. 1. Forecasting of coking coal export transportation. (Color figure online)

- $MAE = 17108.81613$;
- $MSE = 648819063,23557$;
- $R^2 = 0.55750$;
- $MAPE = 9.86880\%$.

The model showed a reasonably good forecasting quality, as evidenced by the value of the mean absolute percentage error of less than 10%.

Ensembling this model with Facebook Prophet did not improve the forecasting quality.

4.2 Coking Coal Domestic Transportation Forecasting

Based on the analysis of the source dataset, we used LightGBM to build the model with the following features:

- GDP volumes of countries;
- USD/RUB exchange rate;
- coking coal domestic transportation for the previous month.

The following optimal parameters of the LightGBM model were obtained:

- loss function: quantile;
- num_leaves: 6;
- max_depth: 2;
- learning_rate: 0.1;
- num_trees: 100;
- min_child_weight: 0.001;
- min_child_samples: 250;
- reg_alpha: 0.07;
- reg_lambda: 6;
- alpha: 0.6;

Table 1. LightGBM and ElasticNet hyperparameter tuning results for coking coal export transportation models.

LightGBM	ElasticNet
Hyperparameter tuning results	
loss function: quantile num_leaves: 8 max_depth: 3 learning_rate: 0.01 num_trees: 9500 min_child_weight: 0.001 min_child_samples: 250 reg_alpha: 0.001 reg_lambda: 0.01 alpha: 0.9 max_bin: 32 max_leaf_output: 0.1	alpha: 0.9 max_iter: 10000 l1_ratio: 1
Graph	
Quality metrics	
MAE: 20397.49666 MSE: 762840963.93090 R^2: 0.47974 MAPE: 12.19193%	MAE: 22524.12433 MSE: 797288893.85589 R^2: 0.45624 MAPE: 14.30514%

- max_bin: 128;
- max_leaf_output: 0.4.

Figure 2 presents the results of forecasting.

Assessment of forecasts on the test dataset showed the following values of quality metrics:

- *MAE*: 78818.00304;
- *MSE*: 11053153077.02472;
- R^2: 0.50022;
- *MAPE*: 6.12393%.

Thus, the model showed a reasonably good forecasting quality since the mean absolute percentage error was about 6%.

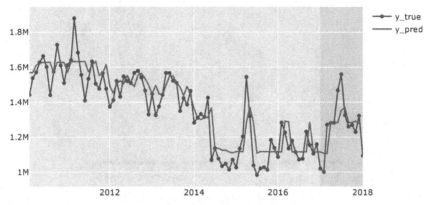

Fig. 2. Forecasting of coking coal domestic transportation.

Ensembling this model with ElasticNet and Facebook Prophet did not improve the quality of forecasting.

4.3 On Coking Coal Import and Transit Transportation Forecasting

Figure 3 shows that starting from March 2014 and ending in February 2017, the import of coking coal takes abnormally small, zero or close to zero, values that are uninformative for the model, and training an adequate model using such dataset is impossible.

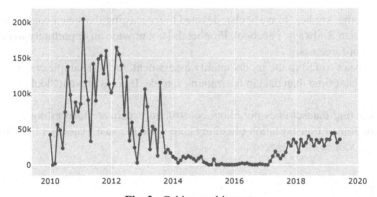

Fig. 3. Coking coal import.

In this regard, those anomalously small values must be deleted, which leads to the fact that the end of the training sample does not coincide with the beginning of the test sample (the interval is about three years). The number of examples in the training sample remains extremely small, containing only 50 values.

Figure 4 shows that coking coal transit is almost always equal to zero (only four values are nonzero).

Such datasets do not allow to train adequate machine learning models for coking coal import and transit transportation forecasting.

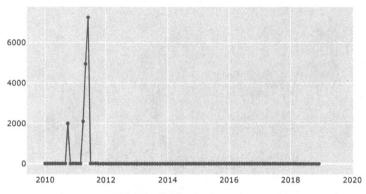

Fig. 4. Coking coal transit.

5   Discussion

Based on a limited set of data on the transportation of Russian Railways (just 114 records with many zero values) by ensembling the ElasticNet and LightGBM models, it was possible to build models for accurate short-term forecasting of coking coal export and domestic railway freight transportation.

As a result, the best forecasting model for coking coal export transportation is the ensemble of ElasticNet and LightGBM with MAPE at 10%, and the best forecasting model for coking coal domestic railway transportation is the pure LightGBM with MAPE at 6%.

Unlike other studies, in particular, devoted to forecasting the steam coal transportation by Russian Railways, Facebook Prophet did not provide an opportunity to improve the quality of forecasts.

In order not to fail in the model quality assessment, the test sample contained data related to a later time than data in the training sample. It helps prevent "looking into the future."

The existing dataset does not allow constructing adequate forecasting models for import and transit transportation of coking coal since the vast majority of label values are zeros or values close to zero.

References

1. Rao, P.S.: Forecasting the demand for railway freight services. J. Transp. Econ. Policy **12**(1), 7–26 (1978)
2. Doi, M., Allen, W.B.: A time series analysis of monthly ridership for an urban rail rapid transit line. Transportation **13**(3), 257–269 (1986). https://doi.org/10.1007/BF00148619
3. Profillidis, V.A., Botzoris, G.N.: Econometric models for the forecast of passenger demand in Greece. J. Stat. Manag. Syst. **9**(1), 37–54 (2006). https://doi.org/10.1080/09720510.2006. 10701192

4. Shen, S., Fowkes, T., Whiteing, T., Johnson, D.: Econometric modelling and forecasting of freight transport demand in Great Britain. In: Proceedings of the European Transport Conference, Noordwijkerhout, The Netherlands, 5 October 2009, pp. 1–21. Association for European Transport, Henley-in-Arden (2009). https://aetransport.org/public/downloads/GNK3F/3978-514ec5c9deed2.pdf. Accessed 09 May 2020

5. Odgers, J.F., Schijndel, A.V.: Forecasting annual train boardings in Melbourne using time series data. In: Proceedings of the 34th Australian Transport Research Forum (ATRF), Adelaide, Australia, 28–30 September 2011, pp. 1–20. Australian Transport Research Forum, Canberra (2011). https://www.researchgate.net/publication/241809181_Forecasting_annual_train_boardings_in_Melbourne_using_time_series_data. Accessed 09 May 2020

6. Babcock, M., Lu, X., Norton, J.: Time series forecasting of quarterly railroad grain carloadings. Transp. Res. Part E: Logist. Transp. Rev. **35**(1), 43–57 (1999). https://doi.org/10.1016/S1366-5545(98)00024-6

7. Kulshreshtha, M., Nag, B., Kulshreshtha, M.: A multivariate cointegrating vector auto regressive model of freight transport demand: evidence from Indian Railways. Transp. Res. Part A: Policy Pract. **35**, 29–45 (2001). https://ideas.repec.org/a/eee/transa/v35y2001i1p29-45.html. Accessed 09 May 2020

8. Smith, B.L., Williams, B.M., Oswald, R.K.: Comparison of parametric and nonparametric models for traffic flow forecasting. Transp. Res. Part C: Emerg. Technol. **10**(4), 303–321 (2002). https://doi.org/10.1016/S0968-090X(02)00009-8

9. Tong, M., Xue, H.: Highway traffic volume forecasting based on seasonal ARIMA model. J. Highway Transp. Res. Dev. **3**(2), 109–112 (2008). https://doi.org/10.1061/JHTRCQ.0000255

10. Cools, M., Moons, E., Wets, G.: Investigating the variability in daily traffic counts through use of ARIMAX and SARIMAX models: assessing the effect of holidays on two site locations. Transp. Res. Rec.: J. Transp. Res. Board **2136**, 57–66 (2009). https://doi.org/10.3141/2136-07

11. Guo, Y.N., Shi, X.P., Zhang, X.D.: A study of short term forecasting of the railway freight volume in China using ARIMA and Holt-Winters models. In: Proceedings of the 8th International Conference on Supply Chain Management and Information, SCMIS – 2010, Hong Kong, 6–9 October 2010, pp. 304–309. IEEE, Piscataway (2010). https://ieeexplore.ieee.org/document/5681738. Accessed 09 May 2020

12. Jiuran, H., Bingfeng, S.: The application of ARIMA-RBF model in urban rail traffic volume forecast. In: Proceedings of the 2nd International Conference on Computer Science and Electronic Engineering, ICCSEE – 2013, Dubai, UAE, 16–17 November 2013, pp. 1662–1665. Atlantis Press, Paris (2013). https://doi.org/10.2991/iccsee.2013.416

13. Wijeweera, A., To, H., Charles, M.B., Sloan, K.: A time series analysis of passenger rail demand in major Australian cities. Econ. Anal. Policy **44**(3), 301–309 (2014). https://doi.org/10.1016/j.eap.2014.08.003

14. Liu, Y., Lang, M.X.: Railway freight volume prediction based on support vector regression (SVR). Appl. Mech. Mater. **587–589**, 1954–1957 (2014). https://doi.org/10.4028/www.scientific.net/amm.587-589.1954

15. Plakandaras, V., Papadimitriou, Th., Gogas, P.: Forecasting transportation demand for the U.S. market. Transp. Res. Part A: Policy Pract. **126**, 195–214 (2019). https://doi.org/10.1016/j.tra.2019.06.008

16. Li, C., Wang, X., Cheng, Z., Bai, Y.: Forecasting bus passenger flows by using a clustering-based support vector regression approach. IEEE Access **8**, 19717–19725 (2020). https://doi.org/10.1109/ACCESS.2020.2967867

17. Shi, Z., Zhang, N., Schonfeld, P.M., Zhang, J.: Short-term metro passenger flow forecasting using ensemble-chaos support vector regression. Transportmetrica A: Transp. Sci. **156**(2), 194–212 (2020). https://doi.org/10.1080/23249935.2019.1692956

18. Sun, Y., Lang, M.X., Wang, D.Z., Liu, L.Y.: Prediction models for railway freight volume based on artificial neural networks. Appl. Mech. Mater. **543–547**, 2093–2098 (2014). https://doi.org/10.4028/www.scientific.net/amm.543-547.2093

19. Yongbin, X.Y., Xie, H., Wu, L.: Analysis and forecast of railway coal transportation volume based on BP neural network combined forecasting model. AIP Conf. Proc. **1967** (2018). Article 040052. https://doi.org/10.1063/1.5039126

20. Tsai, Ts.-Hs., Lee, Ch.-K., Wei, Ch.-H.: Neural network based temporal feature models for short-term railway passenger demand forecasting. Expert Syst. Appl. **36**(2), 3728–3736 (2009). https://doi.org/10.1016/j.eswa.2008.02.071

21. Hou, Z., Li, X.: Repeatability and similarity of freeway traffic flow and long-term prediction under big data. IEEE Trans. Intell. Transp. Syst. **17**(6), 1786–1796 (2016). https://doi.org/10.1109/TITS.2015.2511156

22. Guo, Z., Fu, J.-Y.: Prediction method of railway freight volume based on genetic algorithm improved general regression neural network. J. Intell. Syst. **27**(2), 291–302 (2018). https://doi.org/10.1515/jisys-2016-0172

23. Wang, P., Zhang, X., Han, B., Lang, M.: Prediction model for railway freight volume with GCA-genetic algorithm-generalized neural network: empirical analysis of China. Cluster Comput. **22**(2), 4239–4248 (2018). https://doi.org/10.1007/s10586-018-1794-y

24. Zhao, Z., Chen, W., Wu, X., Chen, P.C.Y., Liu, J.: LSTM network: a deep learning approach for short-term traffic forecast. IET Intell. Transp. Syst. **11**(2), 68–75 (2017). https://doi.org/10.1049/iet-its.2016.0208

25. Do, L.N.N., Vu, H.L., Vo, B.Q., Liu, Z., Phung, D.: An effective spatial-temporal attention based neural network for traffic flow prediction. Transp. Res. Part C: Emerg. Technol. **108**, 12–28 (2019). https://doi.org/10.1016/j.trc.2019.09.008

26. Zhai, H., Tian, R. Cui, L., Xu, X., Zhang, W.: A novel hierarchical hybrid model for short-term bus passenger flow forecasting. J. Adv. Transp. **2020** (2020). https://doi.org/10.1155/2020/7917353. Article 7917353

27. Xie, M.-Q., Li, X.-M., Zhou, W.-L., Fu, Y.-B.: Forecasting the short-term passenger flow on high-speed railway with neural networks. Comput. Intell. Neurosci. **2014** (2014). Article 375487. https://doi.org/10.1155/2014/375487

28. Licheng, Q., Wei, L., Wenjing, L., Dongfang, M., Yinhai, W.: Daily long-term traffic flow forecasting based on a deep neural network. Expert Syst. Appl. **121**, 304–312 (2019). https://doi.org/10.1016/j.eswa.2018.12.031

29. Gallo, M., De Luca, G., D'Acierno, L., Botte, M.: Artificial neural networks for forecasting passenger flows on metro lines. Sensors **19**(15), 3424–3436 (2019). https://doi.org/10.3390/s19153424

30. Yang, C., Li, X.: Research on railway freight volume prediction based on neural network. E3S Web Conf. **143** (2020). Article 01050. https://doi.org/10.1051/e3sconf/202014301050

31. Zou, H., Hastie, T.: Regularization and variable selection via the ElasticNet. J. Roy. Stat. Soc. B **67**(2), 301–320 (2005). https://doi.org/10.1111/j.1467-9868.2005.00503.x

32. Ke, G., et al.: LightGBM: a highly efficient gradient boosting decision tree. In: Proceedings of the 31st Conference on Neural Information Processing Systems, NIPS – 2017, Long Beach, CA, USA, 4–12 December 2017, pp. 1–9. Neural Information Processing Systems, San Diego (2017). https://papers.nips.cc/paper/6907-lightgbm-a-highly-efficient-gradient-boosting-decision-tree.pdf. Accessed 09 May 2020

33. Taylor, S.J., Letham, B.: Forecasting at scale. Am. Stat. **72**(1), 37–45 (2017). https://doi.org/10.1080/00031305.2017.1380080

34. Russian Railways coking coal freight transportation dataset. https://is.gd/cox_freight. Accessed 09 May 2020

Multi-parameter Regression of Photovoltaic Systems using Selection of Variables with the Method: Recursive Feature Elimination for Ridge, Lasso and Bayes

Jose Cruz$^{(\boxtimes)}$ ⓘ, Wilson Mamani$^{(\boxtimes)}$ ⓘ, Christian Romero$^{(\boxtimes)}$ ⓘ, and Ferdinand Pineda$^{(\boxtimes)}$ ⓘ

Universidad Nacional del Altiplano, Puno, Peru
{josecruz,ferpineda}@unap.edu.pe, wmamani@estudiante.unap.edu.pe, romeroc24@gmail.com

Abstract. The research focuses on the application of regularization techniques in a multiparameter linear regression model to predict the DC voltage levels of a photovoltaic system from 14 variables. Two predictions were made, in the first prediction, all the variables were taken, 14 independent variable and one dependent variable; Shrinkage Regularization types were applied, as a variable selection method. In the second prediction we propose the use of semiautomatic methods, we used Recursive Feature Elimination (RFE) as a variable selection method and to obtained results. We applied the following Shrinkage regularization methods: Lasso, Ridge and Bayesian Ridge.

The results were validated demonstrating: linearity, normality of error terms, non-self-correlation and homoscedasticity. In all cases the precision obtained is greater than 91.99%.

Keywords: Regularización shrinkage · Lasso · Ridge · Bayesian Ridge · RFE · Subset selection · Linear regression · Homoscedasticity · Auto correlation

1 Introduction

The generation of electrical energy from solar energy is a technology of great growth throughout the world. At the same time, Machine Learning techniques are used in all areas of knowledge. So the joint work of both technologies has been receiving attention and funds for their development. For example, the Advanced Research Projects-Energy Agency (ARPA-E) of the US Department of Energy. USA (DOE) announced $ 15 million in funding for 23 projects to accelerate the incorporation of machine learning and artificial intelligence into energy technology [1]. Within renewable energies, solar energy is one of the most widely used,

ⓒ Springer Nature Switzerland AG 2020
G. Nicosia et al. (Eds.): LOD 2020, LNCS 12566, pp. 191–202, 2020.
https://doi.org/10.1007/978-3-030-64580-9_16

taking into account its environmental impact compared to traditional forms of energy acquisition. Generation with photovoltaic panels from changing parameters, such as their non-linearity, levels of solar radiation during the day, ambient temperature and DC to AC converters, mean that the efficiency of this process does not be the optimal [3].

Several methods are used to predict the output voltage level of solar panels: Artificial neural networks such as multilayer perceptron with radial base function [4] transform, unsupervised machine learning methods, and support vector machine recognition [2], linear regression regularized with Lasso [5], the fusion of a basic time series model, data grouping, statistical model and learning automatic [6], deep neural networks. Likewise, analysis and review of the literature published in the Science Direct and IEEE databases since 1990, can be found in [7]. The regularization techniques, found within the "Selection of variables or predictors", are used in different fields where regressions are needed. [8] uses regularized logistic regression to classify cancers, [9] uses a weakly convex function as an approximation of pseudo rule 0; [10] uses a variation of the L1 regularization to improve high seismic resolution. In addition, [11] uses a modification of the L1 regularization to improve its results with data with outliers. [12] proposes a weighted crest regression method to overcome the collinearity problem in power systems. For cancer classification [13], use a new adaptive regression modeling strategy Similarly, [14] uses Lasso for model selection when the number of accessible measurements is much less than the space dimension of parameters. Bayesian regression is currently used in various fields, such as counting objects in images with a regression model [15], transfer decision algorithms based on machine learning [16], error correction at the level of software, [17]; Semi-supervised Bayesian principal component regression for nonlinear industrial processes [18]; in the same field [19] proposes a detection approach to implement virtual sensors; [20] proposes for the prediction of end points in the steel casting process. In this same industrial field [21] proposes a variational inference technique in incomplete industrial data sets. To linearize the energy flows in the transmission systems [22] it realizes a linearized model. [23] develops a convex optimization method to detect line faults. Similarly, [24] presents a regression model to predict the life of crack fatigue. [25] proposes a substitute modeling technique for electromagnetic dispersion analysis In regression models, a compromise must be made between the bias and the variance provided by the data to be predicted and the model performed. For this, the theory provides us with the following variable selection methods (Feature Selection): Subset selection, Shrinkage and Dimension reduction. The first identifies and selects among all available predictors those that are most related to the response variable. Shrinkage or Shrinkage fits the model, including all predictors, but including a method that forces the regression coefficient estimates to zero. While Dimension Reduction creates a small number of new variables from combinations of the original variables. Each of them has a subset of techniques such as for Subset selection: best subselection and stepwise selection (forward, backward and hybrid). For Shrinkage: Ridge, Lasso, and ElasticNet. For Dimension Reduction, we have Principal components, Partial Last Square and tSNE.

The research focuses on the first two methods of feature selection: Subset Selection and Shrinkage. Subset selection is the task of finding a small subset of the most informative elements in a basic set. In addition to helping reduce computational time and algorithm memory, due to working on a much smaller representative set, he has found numerous applications, including image and video summary, voice and document summary, grouping, feature selection, and models, sensor location, social media marketing, and product recommendation [26]. The recursive feature removal method (RFE) used works by recursively removing attributes and building a model on the remaining attributes. Use precision metrics to rank the feature based on importance. The RFE method takes the model to be used and the number of characteristics required as input. Then it gives the classification of all the variables, with 1 being the most important. It also provides support, True if it is a relevant feature and False if it is an irrelevant feature. Shrinkage fits the model by including all predictors, but by including a method that forces the regression coefficient estimates to tend or to be zero. Shrinkage has three methods: Lasso, Ridge, and ElasticNet, with some variations. Section 2 of the work describes the method used as well as the algorithms and formulas developed, Sect. 3 of the experimental results describe the data set, it is processing, experiments, and processed result, in Sect. 4 we validate the results meeting the linear regression criteria and finally, we present the conclusions.

2   Method

In this work, we use the recursive characteristics elimination method (RFE) that eliminates non-predominant variables, the algorithm of which is shown in Table 1.

Table 1. RFE algorithm

1.1 Train model using all predictors
1.2 Calculate model behavior
1.3 Calculate importance of the variables in Ranking
1.4 For each subset size Si, i = 1,... S do
1.4.1 Hold Yes for the most important variables
1.4.2 Optional: Pre-process data
1.4.3 Train model with training set using predictors. If
1.4.4 Calculate model behavior
1.4.5 Optional: recalculate ratings for each predictor
1.4.6 End
1.5 Calculate performance profile over. If
1.6 Determine appropriate number of predictors
1.7 Use the model corresponding to the optimum If

For Ridge the sum of squared errors for linear regression is defined by Eq. 1:

$$E = \sum_{i=1}^{N} (y_i - \hat{y}_i)^2 \tag{1}$$

Just as the data set we want to use to make machine learning models must follow the Gaussian distribution defined by its mean, μ and variance σ^2 and is represented by $N(\mu, \sigma^2)$, i.e., $X \sim N(\mu, \sigma^2)$ where X is the input matrix.

For any point x_i, the probability of x_i is given by Eq. 2.

$$P(x_i) = \frac{1}{2\pi\sigma^2} e^{-\frac{1}{2}\frac{(x_i - \mu)^2}{\sigma^2}} \tag{2}$$

The occurrence of each x_i is independent of the occurrence of another, the joint probability of each of them is given by Eq. 3:

$$p(x_1, x_2, ... x_N) = \prod_{i=1}^{N} \frac{1}{2\pi\sigma^2} e^{-\frac{1}{2}\frac{(x_i - \mu)^2}{\sigma^2}} \tag{3}$$

Furthermore, linear regression is the solution that gives the maximum likelihood to the line of best fit by Eq. 4:

$$P(X \mid \mu) = p(x_1, x_2, ... x_N) = \prod_{i=1}^{N} \frac{1}{2\pi\sigma^2} e^{-\frac{1}{2}\frac{(x_i - \mu)^2}{\sigma^2}} \tag{4}$$

Linear regression maximizes this function for the sake of finding the line of best fit. For this, we take the natural logarithm of the probability function (likelihood) (L), then differentiate and equal zero by Eq. 5.

$$ln(P(X \mid \mu)) = ln(p(x_1, x_2, ... x_N)) \tag{5}$$

$$= ln \prod_{i=1}^{N} \frac{1}{2\pi\sigma^2} e^{-\frac{1}{2}\frac{(x_i - \mu)^2}{\sigma^2}} = \sum_{i=1}^{N} ln\left(\frac{1}{2\pi\sigma^2} e^{-\frac{1}{2}\frac{(x_i - \mu)^2}{\sigma^2}}\right) \tag{6}$$

$$= \sum_{i=1}^{N} ln\left(\frac{1}{2\pi\sigma^2}\right) - \sum_{i=1}^{N} \mid \frac{1}{2}\frac{(x_i - \mu)^2}{\sigma^2} \tag{7}$$

$$\frac{\partial ln(P(X \mid \mu))}{\partial \mu} = \frac{\partial \sum_{i=1}^{N} ln\left(\frac{1}{2\pi\sigma^2}\right)}{\partial \mu} - \frac{\partial \sum_{i=1}^{N} \frac{1}{2}\frac{(x_i - \mu)^2}{\sigma^2}}{\partial \mu} \tag{8}$$

$$= 0 + \sum_{i=1}^{N} \frac{(x_i - \mu)}{\sigma^2} = \sum_{i=1}^{N} \frac{(x_i - \mu)}{\sigma^2} \tag{9}$$

$$\frac{\partial ln(P(X \mid \mu))}{\partial \mu} = \sum_{i=1}^{N} \frac{(x_i - \mu)}{\sigma^2} = 0 \Longrightarrow \mu = \frac{\sum_{i=1}^{N} x_i}{N} \tag{10}$$

We take into account here is that maximizing the probability function (likelihood) L is equivalent to minimizing the error function E. Furthermore, and it is Gaussian distributed with mean transposition (w) * X and variance σ^2 is show in Eq. 11.

$$y \sim N(\omega^T X, \sigma^2) \quad o \quad y = \omega^T X + \varepsilon \tag{11}$$

Where $\varepsilon \sim N(0, \sigma^2)$ ε is Gaussian distributed noise with zero mean and variance σ^2. This is equivalent to saying that in linear regression, the errors are Gaussian and the trend is linear. For new or outliers, the prediction would be less accurate for least squares, so we would use the L2 regularization method or Ridge regression. To do this, we modify the cost function and penalize large weights as follows by Eq. 12:

$$J_{RIDGE} = \sum_{i=1}^{N} (y_i - \hat{y}_i)^2 + \lambda |w|^2 \tag{12}$$

Where: $|w|^2 = w^T w = w_1^2 + w_2^2 + \cdots + w_D^2$
We, now have two probabilities:
Subsequent:

$$P(Y|X, w) = \prod_{i=1}^{N} \frac{1}{2\pi\sigma^2} exp(-\frac{1}{2\sigma^2}(y_n - w^T x_n)^2) \tag{13}$$

Prior:

$$P(w) = \frac{\lambda}{\sqrt{2\pi}} exp(-\frac{\lambda}{2} w^T w) \tag{14}$$

Then

$$exp(J) = \prod_{n=1}^{N} exp(-(y_n - w^T x_n)^2) exp(\lambda w^T w) \tag{15}$$

Applying Bayes: $J = (Y - Xw)(Y - Xw)^T + \lambda w^T w$

$$= Y^T T - 2Y^T Xw + w^T X^T Xw + \lambda w^T w \tag{16}$$

To minimize J, we use $\frac{\partial J}{\partial w}$ and set its value to 0. Therefore, $-2X^T + 2X^T Xw + 2\lambda w = 0$
So $(X^T X + \lambda I)w = X^T Y$ or $w = (X^T Y)$

This method encourages weights to be small since P (w) is a Gaussian centered around 0. The anterior value of w is called the MAP (maximize posterior) estimate of w.
In the same way for Lasso

$$J_{LASSO} = \sum_{n=1}^{N} (y_i - \hat{y}_i)^2 + \lambda ||w|| \tag{17}$$

Maximizing the likelihood

$$P(Y|X,w) = \prod_{n=1}^{N} \frac{1}{2\pi\sigma^2} exp(-\frac{1}{2\sigma^2}(y_n - w^T x_n)^2) \tag{18}$$

and prior (previous) is given by:

$$P(w) = \frac{\lambda}{2} exp(-\lambda|w|) \tag{19}$$

So that $J = (Y - X_w)^T(Y - X_w) + \lambda|w|$

y $\frac{\partial J}{\partial w} = -2X^T Y + 2X^T Y + 2X^T Xw + \lambda sign(w) = 0$

Where $sign(w) = 1$ If $x > 0$ and -1 if $x < 0$ and 0 if $x = 0$

3   Experimental Results

3.1   Description of the Data Set and Data Preprocessing

The data was collected in the city of Juliaca, province of San Román, department of Puno, whose coordinates are: 15° 29′ 27″ S and 70° 07′ 37″ O. The time period was April and August 2019. The data to be analyzed were: DC Voltage, AC Voltage, AC Current, Active Power, Apparent Power, Reactive Power, Frequency, Power Factor, Total Energy, Daily Energy, DC Voltage, DC Current, and DC Power. Those that were obtained through the StekaGrid 3010 Inverter that has an efficiency of 98.6%. The ambient and panel temperatures were read by the PT1000 sensors that are suitable for temperature-sensitive elements given their special sensitivity, precision, and reliability: And finally, the irradiance was obtained through a calibrated Atersa brand cell, whose output signal depends exclusively on solar radiation and not on temperature.

The amount of data is reduced from 331157 to 123120 because many of the values obtained are null (values read at night).

3.2   Experiments and Results

In this work, two groups of predictions were generated:

Prediction 1. All the variables were taken: 14 independent and one dependent (Tension DC) and the three types of Shrinkage Regularization: Lasso, Ridge and Bayesian Ridge were applied as the variable selection method.

The data set is divided into: training data 98496 (80%) and test data 24624 (20%), to obtain a seed with better performance to take the training and test data randomly; In order to obtain better results in the proposed methods, the value of the seed is 8083. We proceed to determine the Hyperparameters for Ridge and Lasso:

- LASSO(alpha = 0.01)
- RIDGE(alpha = 1.538).

For the models found, we determined R^2 and adjusted R^2, MAE, RMSE and Score as show as Table 2.

Prediction 2. We apply RFE inside Subset Selection as a variable selection method. To the obtained results we apply the following Shrinkage regularization methods: Lasso, Ridge and Bayesian Ridge.

The data set is divided into training data 98496 (80%) and test data 24624 (20%), the seed obtained for this case is 7780.

We apply the RFE algorithm, obtaining: ['Tension AC', 'Corriente AC', 'Potencia activa', 'Potencia aparente', 'Potencia reactiva', 'Frecuencia', 'Factor de potencia', 'Energia total', 'Energia diaria', 'Corriente DC', 'Potencia DC', 'Irradiancia', 'Temp modulo', 'Temp ambiente'] [True, True, True, True, True, True, True, False, True, True, True, True, True, True] So the variable eliminated by RFE is: Energáa total.

We proceed to determine the Hyperparameters for Ridge and Lasso:

- LASSO(alpha = 0.01)
- RIDGE(alpha = 1.538).

For the models found, we determined R^2 and adjusted R^2, MAE, RMSE and Score as show as Table 2.

The results table shows the values obtained for predictions 1 and 2:

Table 2. Results table

Parameters	No variable selection	Regularization Shinkage			Subset Selection and Regularization Shrinkage			
	OLS	Lasso	Ridge	Bayesian Ridge	REF-OLS	RFE-Lasso	RFE-Ridge	RFE-Bayesian Ridge
Mean absolute error R	4,8495397	4,85961896	4,84963005	4,84955257	4,90763389	4,91010124	4,90761474	4,90763040
Root mean square error R2	2,92351045	2,94801891	2,92355092	2,92351661	2,98428741	3,00560025	2,98424930	2,98428192
Determination coefficient	0,92255911	0,92223687	0,92255622	0,9225587	0,91997279	0,91989230	0,91997341	0,91997290
Adjusted coefficient of determination	0,9225503	0,92222802	0,92254741	0,92254989	0,91996433	0,91988384	0,91996496	0,91996445
Training time	0,037271	1,644358	0,01408	0,143112	0,033414	1,528573	0,012958	0,08514
Test time	0,000748	0,000736	0,000735	0,001128	0,000879	0,000876	0,000782	0,001317
Prediction	1				2			

4 Validation of Results

To validate the results provided by the model, we must review certain assumptions about linear regression. If they are not fulfilled, the interpretation of results will not be valid. The assumptions are:

4.1 Linearity

There must be a linear relationship between the independent variables and the dependent variable so that the model does not make inaccurate predictions. We will review to utilize a scatter diagram in which the values or points must be on or around the diagonal line of the diagram. We show the following linearity example for Bayesian Ridge prediction one and two in the Fig. 1. For the other models, the results are similar.

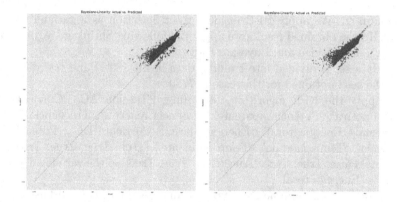

Fig. 1. Linearity - prediction 1 and 2, (a) correspond to model without RFE Bayessian Ridge, (b) correspond to model with RFE Bayessian Ridge

4.2   Normality of Error Terms

The error terms should be normally distributed, if this does not happen, the confidence intervals will increase. We will use the residual distribution diagram, in Fig. 2 we show the Normality of error terms for the first and second Bayesian Ridge predictions. For the other models, the results are similar.

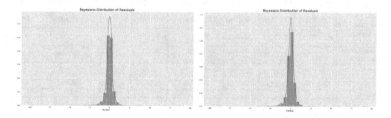

Fig. 2. Residual distribution - prediction 1 and 2, (a) correspond to model without RFE Bayessian Ridge, (b) correspond to model with RFE Bayessian Ridge

4.3   Non-multicollinearity Between Predictors - CORRELATION

The independent variables (predictors) should not be correlated with each other, as they would cause problems in the interpretation of the coefficients, as well as the error provided by each of them. To determine we will use a correlation heat map and the variance inflation factor (VIF) (Fig. 3).

Prediction 1 y 2:

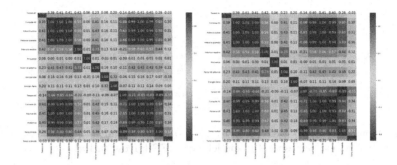

Fig. 3. Non-multicollinearity, (a) without RFE, (b) with RFE

4.4 No Autocorrelation of the Error Terms

Autocorrelation indicates that some information is missing that the model should capture. It would be represented by a systematic bias below or above the prediction. For this, we will use the Durbin-Watson test. Values of $1.5 < \times < 2.5$ assume no correlation. The results are show in Table 3.

Table 3. Durbin Watson Values

Models	D. W. Values	Correlation
OLS	1.996	NO
Lasso	1.995	NO
Ridge	1.995	NO
Bayesian Ridge	1.996	NO
RFE – OLS	1.995	NO
RFE - Lasso	1.995	NO
REF - Ridge	1.995	NO
REF – Bayesian Ridge	1.995	NO

4.5 Homocedasticity

The error produced by the model must always have the same variance, if this does not happen the model gives too much weight to a subset of the data, particularly when the error variance is greater. To detect the residuals that have a uniform variance we show the homoskedasticity for the first and second Bayesian Ridge predictions in Fig. 4. For the other models, the results are similar.

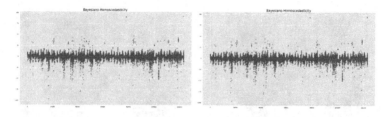

Fig. 4. Homocedasticity - prediction 1 and 2, (a) correspond to model without RFE Bayessian Ridge, (b) correspond to model with RFE Bayessian Ridge

5   Conclusion

In this work, we present two methods for the selection of variables in the realization of a multiparameter linear regression model to predict the output voltage levels of the photovoltaic system from 14 variables that model the system under study. We first used shrink regularization with Ridge, Lasso, and Bayesian Ridge. In the second, we perform a method crossing: first, we use within the Selection of subsets: the best recursive selection of subsectors (RFE) that eliminates an independent variable to simplify computational complexity; and to the results obtained, we again apply the other selection method: contraction regularization with its three submethods: Lasso, Ridge and Bayesian Ridge. From the results obtained, we observe that there is a better reduction in training time of 65.2% and 62.2% for the RFE-Ridge and Ridge methods, respectively. Likewise, the test time was reduced by 1.7% and 1.6% for Ridge and Lasso. In all cases the precision obtained is greater than 91.99%, obtaining an improvement in the adjusted determination coefficient of 0.29% and 0.28% for RFE-Ridge Bayesian and RFE-Lasso. Also, it was verified that the results obtained comply with the linear regression conditions: linearity, without multicollinearity, without autocorrelation and homoscedasticity. We can also indicate that external or climatic conditions are not the only factors that affect the efficiency of the photovoltaic system, but rather the intrinsic characteristics of the system such as: load, power, energy, and temperature. Based on this article, the following studies can be projected: use of nonlinear models, such as decision trees or neural networks; use of more data for the preprocessing phase; and use of imputation models for missing values.

References

1. Energy.gov: Department of energy announces 15 million for development of artificial intelligence and machine learning tools (2019)
2. Feng, C., Cui, M., Hodge, B., Lu, S., Hamann, H.F., Zhang, J.: Unsupervised clustering-based short-term solar forecasting. IEEE Trans. Sustain. Energy **10**(4), 2174–2185 (2019). https://doi.org/10.1109/TSTE.2018.2881531

3. Feshara, H.F., Ibrahim, A.M., El-Amary, N.H., Sharaf, S.M.: Performance evaluation of variable structure controller based on sliding mode technique for a grid-connected solar network. IEEE Access **7**, 84349–84359 (2019). https://doi.org/10.1109/ACCESS.2019.2924592

4. Abdullah, N.A., Koohi-Kamali, S., Rahim, N.A.: Forecasting of solar radiation in Malaysia using the artificial neural network and wavelet transform. In: 5th IET International Conference on Clean Energy and Technology (CEAT2018), pp. 1–8 (2018). https://doi.org/10.1049/cp.2018.1303

5. Tang, N., Mao, S., Wang, Y., Nelms, R.M.: Solar power generation forecasting With a LASSO-based approach. IEEE Internet Things J. **5**(2), 1090–1099 (2018). https://doi.org/10.1109/JIOT.2018.2812155

6. Wang, Y., Shen, Y., Mao, S., Chen, X., Zou, H.: LASSO and LSTM integrated temporal model for short-term solar intensity forecasting. IEEE Internet Things J. **6**(2), 2933–2944 (2019). https://doi.org/10.1109/JIOT.2018.2877510

7. Obando, E.D., Carvajal, S.X., Pineda Agudelo, J.: Solar radiation prediction using machine learning techniques: a review. IEEE Lat. Am. Trans. **17**(04), 684–697 (2019). https://doi.org/10.1109/TLA.2019.8891934

8. Yang, Z., Liang, Y., Zhang, H., Chai, H., Zhang, B., Peng, C.: Robust sparse logistic regression with the Lq $(0 < q < 1)$ regularization for feature selection using gene expression data. IEEE Access **6**, 68586–68595 (2018). https://doi.org/10.1109/ACCESS.2018.2880198

9. Shen, X., Gu, Y.: Nonconvex sparse logistic regression with weakly convex regularization. IEEE Trans. Signal Process. **66**(12), 3199–3211 (2018). https://doi.org/10.1109/TSP.2018.2824289

10. Li, F., Xie, R., Song, W., Chen, H.: Optimal seismic reflectivity inversion: data-driven lp loss-lq -regularization sparse regression. IEEE Geosci. Remote Sens. Lett. **16**(5), 806–810 (2019). https://doi.org/10.1109/LGRS.2018.2881102

11. Liu, J., Cosman, P.C., Rao, B.D.: robust linear regression via l0 regularization. IEEE Trans. Signal Process. **66**(3), 698–713 (2018). https://doi.org/10.1109/TSP.2017.2771720

12. Zhang, J., Wang, Z., Zheng, X., Guan, L., Chung, C.Y.: Locally weighted ridge regression for power system online sensitivity identification considering data collinearity. IEEE Trans. Power Syst. **33**(2), 1624–1634 (2018). https://doi.org/10.1109/TPWRS.2017.2733580

13. Park, H., Shiraishi, Y., Imoto, S., Miyano, S.: A novel adaptive penalized logistic regression for uncovering biomarker associated with anti-cancer drug sensitivity. IEEE/ACM Trans. Comput. Biol. Bioinf. **14**(4), 771–782 (2017). https://doi.org/10.1109/TCBB.2016.2561937

14. Owrang, A., Jansson, M.: A model selection criterion for high-dimensional linear regression. IEEE Trans. Signal Process. **66**(13), 3436–3446 (2018). https://doi.org/10.1109/TSP.2018.2821628

15. Jia, Y., Kwong, S., Wu, W., Wang, R., Gao, W.: Sparse Bayesian learning-based kernel poisson regression. IEEE Trans. Cybern. **49**(1), 56–68 (2019). https://doi.org/10.1109/TCYB.2017.2764099

16. Bang, J., Oh, S., Kang, K., Cho, Y.: A Bayesian regression based LTE-R handover decision algorithm for high-speed railway systems. IEEE Trans. Veh. Technol. **68**(10), 10160–10173 (2019). https://doi.org/10.1109/TVT.2019.2935165

17. Tatsumi, K., Matsuoka, T.: A software level calibration based on Bayesian regression for a successive stochastic approximation analog-to-digital converter system. IEEE Trans. Cybern. **49**(4), 1200–1211 (2019). https://doi.org/10.1109/TCYB.2018.2795238

18. Zhu, P., Liu, X., Wang, Y., Yang, X.: Mixture semisupervised Bayesian principal component regression for soft sensor modeling. IEEE Access **6**, 40909–40919 (2018). https://doi.org/10.1109/ACCESS.2018.2859366

19. Shao, W., Ge, Z., Yao, L., Song, Z.: Bayesian nonlinear Gaussian mixture regression and its application to virtual sensing for multimode industrial processes. IEEE Trans. Autom. Sci. Eng. **17**, 871–885 (2019). https://doi.org/10.1109/TASE.2019.2950716

20. Liu, C., Tang, L., Liu, J.: A stacked autoencoder with sparse Bayesian regression for end-point prediction problems in steelmaking process. IEEE Trans. Autom. Sci. Eng. **17**, 550–561 (2019). https://doi.org/10.1109/TASE.2019.2935314

21. Zhao, J., Chen, L., Pedrycz, W., Wang, W.: A novel semi-supervised sparse Bayesian regression based on variational inference for industrial datasets with incomplete outputs. IEEE Trans. Syst. Man Cybern.: Syst. **50**, 1–14 (2018). https://doi.org/10.1109/TSMC.2018.2864752

22. Liu, Y., Zhang, N., Wang, Y., Yang, J., Kang, C.: Data-driven power flow linearization: a regression approach. IEEE Trans. Smart Grid **10**(3), 2569–2580 (2019). https://doi.org/10.1109/TSG.2018.2805169

23. Soltan, S., Mittal, P., Poor, H.V.: Line failure detection after a cyber-physical attack on the grid using Bayesian regression. IEEE Trans. Power Syst. **34**(5), 3758–3768 (2019). https://doi.org/10.1109/TPWRS.2019.2910396

24. Shi, D., Ma, H.: A Bayesian inference method and its application in fatigue crack life prediction. IEEE Access **7**, 118381–118387 (2019). https://doi.org/10.1109/ACCESS.2019.2935404

25. Wang, K., Ding, D., Chen, R.: A surrogate modeling technique for electromagnetic scattering analysis of 3-D objects with varying shape. IEEE Antennas Wirel. Propag. Lett. **17**(8), 1524–1527 (2018). https://doi.org/10.1109/LAWP.2018.2852659

26. Elhamifar, E., De Paolis Kaluza, M.C.: Subset selection and summarization in sequential data. In: Guyon, I., et al. (eds.) Advances in Neural Information Processing Systems, vol. 30, pp. 1035–1045 (2017). http://papers.nips.cc/paper/6704-subset-selection-and-summarization-in-sequential-data.pdf

High-Dimensional Constrained Discrete Multi-objective Optimization Using Surrogates

Rommel G. Regis[(✉)] [iD]

Saint Joseph's University, Philadelphia, PA 19131, USA
rregis@sju.edu

Abstract. This paper presents a multi-objective optimization algorithm for computationally expensive high-dimensional problems with ordinal discrete variables and many black-box constraints. The algorithm builds and maintains multiple surrogates to approximate each of the black-box objective and constraint functions. The surrogates are used to identify promising sample points for the function evaluations from a large number of trial solutions in the neighborhood of a current nondominated point. The proposed method is implemented using radial basis function (RBF) surrogates and is applied to the large-scale 3-car Mazda benchmark problem involving 222 discrete variables, 54 black-box inequality constraints and 2 objective functions using a simulation budget of only 15 times the number of decision variables. The proposed algorithm substantially outperforms an NSGA-II implementation for discrete variables and yields a remarkable improvement over an initial feasible design on the Mazda benchmark given the limited computational budget.

Keywords: Multi-objective optimization · Discrete variables · Constrained optimization · High-dimensional optimization · Surrogate models · Radial basis functions

1 Introduction

Simulation-based multi-objective optimization problems involving multiple constraints are ubiquitous in engineering applications and many algorithms have been proposed for these problems. Among the most popular is NSGA-II [7] and its variants and extensions. In the computationally expensive setting where only a relatively limited number of function evaluations are feasible, many surrogate-based and surrogate-assisted multi-objective methods have been proposed for bound-constrained problems (e.g., [1,13,18,21,27]) and for problems with constraints (e.g., [5,9,12,17,24,25]). However, few surrogate-based multi-objective methods can be used for problems with discrete decision variables and even more scarce are methods that can be used for high-dimensional problems with hundreds of discrete variables and many constraints. In fact, relatively few surrogate approaches have been proposed even for computationally expensive single-objective optimization problems involving discrete variables [3]. One

© Springer Nature Switzerland AG 2020
G. Nicosia et al. (Eds.): LOD 2020, LNCS 12566, pp. 203–214, 2020.
https://doi.org/10.1007/978-3-030-64580-9_17

such surrogate-assisted multi-objective method that can handle discrete variables is the decomposition-based constrained multi-objective evolutionary algorithm CMOEA/D that uses an Extreme Learning Machine (ELM) by Ohtsuka et al. [19]. This algorithm was applied to the large-scale Mazda benchmark problem [14] for the design optimization of multiple car structures that involves 222 discrete variables. Another example is a variant of NSGA-II for constrained mixed-integer multi-objective optimization that uses radial basis function (RBF) networks for fitness approximation. This method was applied to a building design problem with 50 decision variables, of which 20 are discrete variables [4]. For surveys of multi-objective methods using surrogates, see [6] for evolutionary algorithms and [26] for non-nature inspired methods.

This paper presents the POSEIDON algorithm, which is a surrogate-based and non-population-based algorithm that can be used for high-dimensional multi-objective optimization with computationally expensive functions involving many ordinal discrete variables and constraints. POSEIDON stands for *Pareto Optimization using Surrogates for Expensive Inequality-constrained Discrete Ordinal and Nonlinear problems*. POSEIDON is implemented using RBF surrogates and applied to the 3-car Mazda benchmark problem that has 222 discrete decision variables, 54 black-box inequality constraints and 2 objective functions [14]. Numerical results given a limited computational budget show that two implementations of POSEIDON that use RBF surrogates substantially outperform NSGA-II that uses discrete variable encoding and yield remarkable improvements over an initial feasible design for the Mazda benchmark. Moreover, one big difference between POSEIDON with RBF surrogates on the Mazda benchmark and the ELM-assisted CMOEA/D in [19] is that POSEIDON builds and maintains an RBF model for each constraint individually, resulting in 54 RBF models for the constraints. In contrast, the ELM-assisted CMOEA/D aggregated the constraints for each type of car, resulting in only 3 constraints that are each modeled by an ELM. To the best of the author's knowledge, POSEIDON is among the first (if not the first) surrogate approach to a discrete constrained multi-objective problem of this size involving 222 ordinal discrete variables and 54 black-box inequality constraints that are each modeled by a surrogate.

2   An Algorithm for Constrained Discrete Multi-objective Optimization Using Surrogates

2.1   Algorithm Description

This paper proposes an algorithm for constrained multi-objective optimization problem with ordinal discrete variables of the general form:

$$\min \ F(x) = (f_1(x), \ldots, f_k(x))$$

s.t.

$$G(x) = (g_1(x), \ldots, g_m(x)) \leq 0$$
$$x^{(i)} \in D_i \subset \mathbb{R} \text{ with } |D_i| < \infty, \ i = 1, \ldots, d \tag{1}$$

Here, $x^{(i)}$ is the ith coordinate of $x \in \mathbb{R}^d$, which represents the ith variable, and D_i is the finite set consisting of the possible discrete settings of $x^{(i)}$. In practical applications, the elements of D_i are *not* necessarily integers. They could be fractional settings that are allowed for the ith variable. Moreover, the objective functions $f_i : \mathbb{R}^d \longrightarrow \mathbb{R}$, $i = 1, \ldots, k$ and constraint functions $g_j : \mathbb{R}^d \longrightarrow \mathbb{R}$, $j = 1, \ldots, m$ are black-box and might be obtained from an expensive, but deterministic simulation. The finite set $\prod_{i=1}^{d} D_i \subset \mathbb{R}^d$ is referred to as the *search space* for problem (1). Let $\ell_i = \min D_i$ and $u_i = \max D_i$ for $i = 1, \ldots, d$ be the lower and upper bounds of the variables. Note that the box region $[\ell, u] = \prod_{i=1}^{d} [\ell_i, u_i]$ encloses the search space of the problem. Throughout this paper, assume that one *simulation* at a given input vector $x \in \prod_{i=1}^{d} D_i$ yields the values of all the components of $F(x)$ and $G(x)$.

The proposed algorithm, called *POSEIDON (Pareto Optimization using Surrogates for Expensive Inequality-constrained Discrete Ordinal and Nonlinear problems)*, is an extension to problems with ordinal discrete variables of the Multi-Objective Constrained Stochastic RBF method [24], which was designed for continuous variables. The current focus is on constrained multi-objective problems where each decision variable takes on values from a finite number of possible discrete settings, which are not necessarily integers (e.g., $x^{(i)}$ takes on values from the set $\{1.2, 1.5, 2.3, 2.5\}$). However, we still assume that the objective and constraint functions can be treated as functions of the relaxed decision variables that vary on a continuous scale within the bounds. What is different is that the decision variables can only take values on an ordinal discrete scale due to practical limitations.

Below is a pseudo-code of the POSEIDON algorithm.

POSEIDON (Pareto Optimization using Surrogates for Expensive Inequality-constrained Discrete Ordinal and Nonlinear problems)

Inputs:

(i) Constrained discrete multi-objective optimization problem of the form (1)
(ii) Type of surrogates to use for the objective and constraints (e.g., RBF, Kriging/Gaussian Process, neural network or support vector regression models)
(iii) Initial points in the search space
(iv) Strategy for selecting a nondominated point that will be perturbed to generate trial points (Step 2(c) below)
(v) Mechanism for generating trial points in the neighborhood of a current nondominated point (Step 2(c) below)
(vi) Weights for the two criteria to choose simulation point from the promising trial points (Step 2(e))

Output: Nondominated set of points and corresponding objective vectors.

(1) *(Initial Simulations)* Perform simulations to obtain objective and constraint function values at initial set of points in the search space.
(2) *(Perform Iterations)* While the termination condition is *not* satisfied, do:

(a) *(Determine Nondominated Set)* Identify or update the set of nondominated points among those evaluated so far.

(b) *(Fit Surrogates)* Fit or update surrogate for each objective and each constraint.

(c) *(Generate Trial Points)* Select one of the nondominated points and generate a large number of trial points in some discrete neighborhood of this chosen nondominated point.

(d) *(Identify Promising Trial Points)* Evaluate the surrogates for the constraints at the trial points in Step 2(c) and identify trial points that are predicted to be feasible or those with the minimum number of predicted constraint violations.

(e) *(Select Simulation Point)* Evaluate the surrogates for the objectives at the promising trial points obtained in Step 2(d). Then, choose a trial point among those predicted to be nondominated according to two criteria: (i) minimum distance between the predicted objective vector of the trial point and the current nondominated objective vectors (in the objective space); and (ii) minimum distance between the trial point and the current nondominated points (in the decision space).

(f) *(Simulate)* Perform one simulation to obtain the objective and constraint function values at the promising trial point chosen in Step 2(e).

end.

(3) *(Return Nondominated Set)* Return set of nondominated points and corresponding nondominated objective vectors.

As with any surrogate-based method for computationally expensive optimization, POSEIDON begins by performing simulations to obtain the objective and constraint function values at an initial set of points in the search space in order to obtain information needed to fit the initial surrogate models. For now, assume that one of these initial points is feasible. However, the method can be extended to deal with the case where all initial points are infeasible by using a two-phase approach similar to that in [23]. After the objective and constraint function values are obtained at the initial points, the initial set of nondominated solutions are identified. Here, standard procedures for comparing feasible/infeasible solutions are used and a measure of constraint violation is used to compare infeasible solutions. Then, the initial surrogates are fit, one for each objective function and for each inequality constraint function. Next, one of the current nondominated points is chosen and perturbed many times to generate a large number of trial points in some neighborhood of the nondominated point. The nondominated point selected as the center of this neighborhood may be chosen uniformly at random among all current nondominated points (Random (RND) strategy) or it may be chosen to be the one whose objective vector is the most isolated (in terms of Euclidean distance) from the other nondominated objective vectors in the objective space (Objective Space Distance (OSD) strategy). These strategies for selecting the nondominated point that is perturbed to generate trial solutions were previously introduced in [24] in the context of surrogate-based continuous multi-objective optimization.

Various neighborhood structures can be considered, some can be designed to promote a focused local search while others can use a larger neighborhood that promotes global search. A neighbor trial point is obtained by perturbing some of the coordinates (values of the variables) of the chosen nondominated point x^*. That is, each coordinate is perturbed with a certain probability denoted by p_{pert}. A perturbation consists of stepping away from the current setting of a variable from a few discrete steps away either by increasing or decreasing its value. The perturbation for each discrete variable is controlled by a neighborhood depth parameter denoted by $depth_{\text{nbhd}}$, given as a percent, that indicates the fraction of the number of settings that the variable is allowed to increase or decrease. For example, suppose the current setting of variable $x^{(i)}$ in the chosen nondominated point is 0.7 and there are 10 possible settings of this variable, which are $\{0.1, 0.2, \ldots, 0.9, 1.0\}$. A depth parameter of 20% means that each discrete variable is allowed to take on up to 20% of the number of possible discrete settings for that variable above or below the current setting. Hence, variable $x^{(i)}$ will step away from 0.7 up to $0.2(10) = 2$ discrete steps away, and so, $x^{(i)}$ will take on the possible values $\{0.5, 0.6, 0.8, 0.9\}$. Note that if a variable is chosen to be perturbed, then its value needs to be changed.

From the collection of all trial points in a given iteration, the surrogate models for the constraint functions are used to identify the trial points that are predicted to be feasible, or if no such trial points are found, we identify the trial points that have the minimum number of predicted constraint violations. We refer to these as the *promising trial points*. Among these promising trial points, the surrogate models for the objective functions are used to identify the trial points that are predicted to be nondominated by the other promising trial points and by the current nondominated points. From this subset of trial points, we then select the trial point that is most promising according to a weighted combination of two criteria: (i) minimum distance between the predicted objective vector of the trial point and the current nondominated objective vectors (in the objective space); and (ii) minimum distance between the trial point and the current nondominated points (in the decision space). That is, we would like to select a trial point whose predicted objective vector is far away from the nondominated objective vectors to promote good spacing in the Pareto front approximation in the objective space. At the same time, we would like to select a trial point that is far away from the current nondominated points to promote exploration of the decision space. In setting the weights for these two criteria, we put more weight on the former than the latter. Once the most promising trial point is selected, a simulation is performed to obtain the objective and constraint values at that point. The process then iterates until the computational budget is exhausted.

2.2   Radial Basis Function Interpolation

Any type of surrogate can be used with the proposed method. Among the more popular surrogate modeling techniques in the context of expensive black-box optimization include Kriging or Gaussian process modeling [10], radial basis functions (RBF) (e.g., [1, 23]) and neural networks (e.g., [8]). Here, the RBF interpolation model in Powell [22] is used. This RBF model possesses desirable mathematical properties and differs from the more popular RBF network in the machine learning literature. In this RBF model, each data point is a center, the basis functions are not necessarily Gaussian, and training simply involves solving a linear system. The main advantage of this type of RBF model is that it has been successfully used for problems with hundreds of decision variables and many black-box constraints (e.g., see [2,23]) whereas it is well known that Kriging models have numerical issues and take an enormous amount of time to build and maintain in high dimensional problems with hundreds of variables.

Given n distinct points $x_1, \ldots, x_n \in \mathbb{R}^d$ and the function values $u(x_1), \ldots, u(x_n)$, where $u(x)$ is either an objective or constraint function, this RBF interpolation model has the form:

$$s(x) = \sum_{i=1}^{n} \lambda_i \phi(\|x - x_i\|) + p(x), \ x \in \mathbb{R}^d,$$

where $\| \cdot \|$ is the Euclidean norm, $\lambda_i \in \mathbb{R}$ for $i = 1, \ldots, n$, $p(x)$ is a linear polynomial in d variables, and ϕ has the *cubic* form: $\phi(r) = r^3$. The function ϕ can also be the thin plate spline ($\phi(r) = r^2 \log r$) and the Gaussian form ($\phi(r) = \exp(-\gamma r^2)$, where γ is a parameter). A cubic RBF model is used because of its simplicity and its success in prior RBF methods (e.g., [23,24]). Details of how to fit this model can be found in Powell [22] and also in Regis [23]).

3   Numerical Results on the Large-Scale Mazda Benchmark

The proposed POSEIDON algorithm is applied to the 3-car Mazda Benchmark problem [11,14,20] that was jointly developed by the Mazda Motor Corporation, Japan Aerospace Exploration Agency, and Tokyo University of Science. The multi-objective optimization problem involves 222 decision variables, 54 inequality constraints, and 2 objective functions and is among the largest benchmark of its kind in area of expensive black-box optimization. The problem is to determine the thickness of 222 structural parts that satisfy constraints such as collision safety performances with the simultaneous goal of minimizing the total weight of three types of cars (sport utility vehicle Mazda CX-5 (SUV), large vehicle Mazda 6 (LV), and small vehicle Mazda 3 (SV)) and maximizing the number of common gauge parts. For practical considerations, the thickness of a structural part can only take values on a finite set of discrete settings. Moreover, one of the objective functions (the number of common gauge parts) can only take on positive integer values even if the decision variables are allowed to take on values on a continuous scale. Simulations for the design optimization of car structures are computationally very expensive. However, the Mazda benchmark evaluates relatively quickly because the collision safety constraints are modeled by response surface approximations [14]. Details of the problem can be found at https://ladse.eng.isas.jaxa.jp/benchmark/index.html.

POSEIDON is implemented using RBF surrogates and two variants are considered. Both variants perform cycles of iterations where the perturbation probability p_{pert} and the neighborhood depth parameter $depth_{\text{nbhd}}$ vary within the cycle. Moreover, for both variants, the OSD strategy to select the nondominated point that is perturbed to generate the trial points is applied to an entire cycle of iterations followed by the RND strategy for the next cycle of iterations and alternates between these two strategies as the algorithm goes through multiple cycles of iterations. One variant, called *POSEIDON (RBF-local)*, is more focused on local search and uses a cycle of four iterations where the control parameters $(p_{\text{pert}}, depth_{\text{nbhd}})$ take on the values $\langle (0.5, 30\%), (0.1, 10\%), (0.05, 10\%), (0.01, 10\%) \rangle$. The other variant, *POSEIDON (RBF-global)*, performs a more global search and uses a cycle of five iterations where the control parameters $(p_{\text{pert}}, depth_{\text{nbhd}})$ take on the values $\langle (0.5, 50\%), (0.3, 50\%), (0.2, 10\%), (0.1, 10\%), (0.05, 10\%) \rangle$. Note that for a given perturbation probability p_{pert}, the number of variables perturbed follows a binomial distribution and the mean number of variables perturbed is $d \cdot p_{\text{pert}}$. In the case of the Mazda benchmark, when $p_{\text{pert}} = 0.01$, the mean number of variables perturbed is $222(0.01) = 2.22$, and when combined with a depth parameter of 10%, the generation of trial points is highly local. On the other hand, when $p_{\text{pert}} = 0.5$, the mean number of variables perturbed is $222(0.5) = 111$, and when combined with a depth parameter of 50%, the trial points generated promote global exploration.

The two variants of POSEIDON are compared with several variants of NSGA-II with discrete variable encoding as implemented in the NPGM software [15]. Discrete variable encoding is used because numerical results reported in [16] obtained better results for NSGA-II with this encoding than when using continuous variable encoding. In this study, the discrete settings for each variable are converted to the integers 0, 1, 2, up to the maximum number of possible discrete settings minus one. The population size is set to 300 as in [14]. Moreover, intermediate crossover is used since this is the only option supported by NPGM [15]. The NSGA-II variants differ in their crossover fraction, which is the fraction of the variables that are involved in crossover. In this investigation, the crossover fraction is set to 1.0, 0.75, 0.5 and 0.25.

To ensure a fair comparison, the initial population used in NSGA-II is also the initial set of points used for POSEIDON. This initial population is the same as the one used in [16] except that the initial feasible design was included as the first individual and the last individual was removed to keep the number of initial points to the population size of 300. The initial feasible design point is included since the current implementation of POSEIDON requires a feasible initial point. Besides, in a practical setting, one typically wishes to improve on a given feasible design point, which is the case for the Mazda benchmark.

All computations are performed in Matlab 9.4 on an Intel(R) Core(TM) i7-7700T CPU @ 2.90 GHz, 2904 Mhz, 4 Core(s), 8 Logical Processor(s) machine. Since the Mazda benchmark is released as a C++ source code, a Matlab interface was created to run the executable on the Matlab environment.

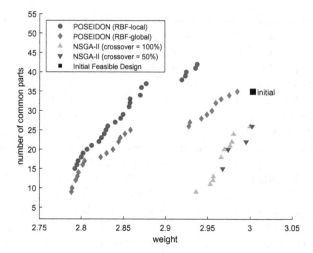

Fig. 1. Nondominated objective vectors obtained by two POSEIDON algorithms that use RBF surrogates and by NSGA-II with a population size of 300, discrete variable encoding and crossover fractions of 100% and 50%. The computational budget is $15d = 3330$ simulations for POSEIDON and 3600 simulations (12 generations) for NSGA-II.

Figure 1 shows the scatter plots of the nondominated objective vectors obtained by the POSEIDON and NSGA-II algorithms for the given computational budgets and starting with the same set of initial sample points of size 300 that includes the given initial feasible point. Recall that we wish to simultaneously minimize f_1 (the total weight of three types of cars) and maximize f_2 (the number of common gauge parts). The plots show that the POSEIDON algorithms obtained much better sets of nondominated objective vectors than the ones obtained by NSGA-II. In fact, the nondominated objective vectors obtained by each POSEIDON algorithm dominate all the nondominated objective vectors obtained by any of the NSGA-II algorithms with various crossover fractions. Moreover, the local variant of POSEIDON obtained a better set of nondominated points than the one obtained by the more global variant. It is also worth noting that the POSEIDON algorithms obtained nondominated objective vectors that substantially improve on the initial feasible design for the Mazda benchmark. For example, one of the nondominated objective vectors obtained by POSEIDON (RBF-local) is $[f_1, f_2] = [2.9372, 42]$, which is a remarkable improvement over the initial feasible objective vector of $[f_1, f_2] = [3.0028, 35]$ given the limited computational budget. On the other hand, none of the NSGA-II algorithms yielded such an improvement over the initial feasible design even with a somewhat larger computational budget.

A direct comparison between the earlier results on the 3-car benchmark problem from [14] and [19] is not possible because the current version of POSEIDON used the initial feasible design provided in [16]. Moreover, POSEIDON was only run for $15d = 3330$ simulations while NSGA-II in [14] was run for 30,000 simulations (100 generations with a population size of 300) and CMOEA/D with ELM was run for 100 generations (with the population size not reported). However, a visual comparison with the results in [14] and [19] suggests that the nondominated objective vectors obtained by POSEIDON in 3330 simulations dominate those obtained in [19] and are *not* dominated by those obtained in [14] with 30,000 simulations.

Table 1. Hypervolumes for one trial of POSEIDON algorithms (RBF-local and RBF-global) and mean hypervolumes for 10 trials of NSGA-II with various crossover fractions on the Mazda 3-Car Benchmark Problem. The number inside the parenthesis next to the mean is the standard error of the mean. The objective functions (weight and number of common parts) are normalized as specified in [16] and the reference point is $[1.1, 0]$.

Algorithm	Hypervolume
POSEIDON (RBF-local)	0.1524
POSEIDON (RBF-global)	0.1182
NSGA-II (crossover fraction = 100%)	0.0605 (0.0008)
NSGA-II (crossover fraction = 75%)	0.0549 (0.0006)
NSGA-II (crossover fraction = 50%)	0.0538 (0.0010)
NSGA-II (crossover fraction = 25%)	0.0579 (0.0010)

Table 1 shows the hypervolumes obtained by the various algorithms after normalizing the objective functions and using the reference point $[1.1, 0]$ as suggested in [16]. The normalized objectives are obtained by: $\widetilde{f_1} = f_1 - 2$ and $\widetilde{f_2} = f_2/74$. For the NSGA-II algorithms, the mean hypervolume and standard error for 10 trials are reported since NSGA-II runs relatively quickly. However, for the POSEIDON algorithms, the hypervolume for only one trial is reported because the algorithm incurs a large amount of computational overhead. It is clear from Table 1 that the hypervolumes obtained by each POSEIDON algorithm are much better than the hypervolumes obtained by the NSGA-II algorithms even as the computational budget for POSEIDON (3330 simulations) is less than that for the NSGA-II algorithms (3600 simulations).

Next, Fig. 2 shows the plot of the hypervolumes for the normalized nondominated objective vectors for one trial of the POSEIDON and NSGA-II algorithms vs the number of simulations. That is, for each algorithm, we calculate and plot the hypervolume of the normalized nondominated objective vectors obtained by the algorithm for each possible computational budget up to $15d = 3330$ simulations. It is also clear from Fig. 2 that the hypervolumes obtained by each POSEIDON algorithm are consistently much better than the hypervolumes obtained by the NSGA-II algorithms for various computational budgets up to 3330 simulations.

Finally, as with other surrogate-based algorithms, POSEIDON incurs a significant amount of computational overhead in building and maintaining the high-dimensional surrogate models for each of the 54 black-box constraints and in

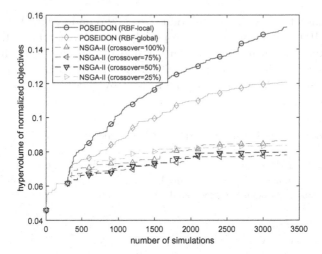

Fig. 2. Plot of hypervolumes for the normalized nondominated objective vectors for one trial of the POSEIDON and NSGA-II algorithms vs the number of simulations on the 3-car Mazda Benchmark Problem.

evaluating a large number of trial points throughout an entire run of the algorithm. In particular, POSEIDON (RBF-local) took about 15.5 h to run on the Mazda benchmark given a computational budget of $15d = 3330$ simulations. In contrast, NSGA-II (crossover fraction $= 100\%$) only took 18 min to run on the Mazda benchmark given a computational budget of 3600 simulations (12 generations). However, in the truly computationally expensive setting where one single simulation potentially takes many hours, this computational overhead is relatively small, and the use of a surrogate-based method is well justified.

4   Summary

This paper developed a surrogate-based algorithm for Constrained Discrete Multi-Objective Optimization called POSEIDON that can be used for high-dimensional problems involving hundreds of decision variables when the computational budget is relatively limited. Two variants of POSEIDON that utilize RBF surrogates are applied to the 3-car Mazda benchmark problem involving 222 discrete decision variables, representing the thickness of structural parts, 2 objective functions (total weight of three types of cars and number of common gauge parts), and 54 black-box inequality constraints such as collision safety performance constraints. The results show that when the computational budget is only about $15d = 3330$ simulations, the POSEIDON algorithms yielded much better nondominated objective vectors than NSGA-II that uses discrete variable encoding. Moreover, the local variant of the POSEIDON algorithm yielded a remarkable improvement over the initial feasible design provided for the benchmark problem while none of the NSGA-II algorithms yielded such an improvement for the given computational budget. These preliminary results represent substantial progress in pushing the capability of surrogate-based algorithms for solving practical problems in the sense that the proposed method is one of very few surrogate-based methods that has been applied to a constrained multi-objective optimization problem of this size involving hundreds of ordinal discrete variables and many constraints.

Acknowledgements. Thanks to Dr. Takehisa Kohira of the Mazda Corporation for clarifying the details of the Mazda benchmark. The Matlab NPGM code that implements NSGA-II with constraint handling is due to Song Lin. Thanks also to Yi Cao for his Matlab codes that find the set of nondominated solutions. Finally, thanks to Yohanes Bimo Dwianto for bringing the Mazda benchmark to my attention.

References

1. Akhtar, T., Shoemaker, C.A.: Multi objective optimization of computationally expensive multi-modal functions with RBF surrogates and multi-rule selection. J. Glob. Optim. **64**(1), 17–32 (2015). https://doi.org/10.1007/s10898-015-0270-y
2. Bagheri, S., Konen, W., Emmerich, M., Bäck, T.: Self-adjusting parameter control for surrogate-assisted constrained optimization under limited budgets. Appl. Soft Comput. **61**, 377–393 (2017)
3. Bartz-Beielstein, T., Zaefferer, M.: Model-based methods for continuous and discrete global optimization. Appl. Soft Comput. **55**, 154–167 (2017)
4. Brownlee, A.E.I., Wright, J.A.: Constrained, mixed-integer and multi-objective optimisation of building designs by NSGA-II with fitness approximation. Appl. Soft Comput. **33**, 114–126 (2015)
5. Chugh, T., Jin, Y., Miettinen, K., Hakanen, J., Sindhya, K.: A surrogate-assisted reference vector guided evolutionary algorithm for computationally expensive many-objective optimization. IEEE Trans. Evol. Comput. **22**(1), 129–142 (2018)
6. Chugh, T., Sindhya, K., Hakanen, J., Miettinen, K.: A survey on handling computationally expensive multiobjective optimization problems with evolutionary algorithms. Soft Comput. **23**(9), 3137–3166 (2019). https://doi.org/10.1007/s00500-017-2965-0
7. Deb, K., Agrawal, S., Pratap, A., Meyarivan, T.: A fast and elitist multi-objective genetic algorithm: NSGA-II. IEEE Trans. Evol. Comput. **6**(2), 182–197 (2002)
8. Dushatskiy, A., Mendrik, A.M., Alderliesten, T., Bosman, P.A.N.: Convolutional neural network surrogate-assisted GOMEA. In: Proceedings of the Genetic and Evolutionary Computation Conference, GECCO 2019, pp. 753–761 (2019)
9. Feliot, P., Bect, J., Vazquez, E.: A Bayesian approach to constrained single- and multi-objective optimization. J. Glob. Optim. **67**, 97–133 (2017)
10. Forrester, A.I.J., Sobester, A., Keane, A.J.: Engineering Design via Surrogate Modelling: A Practical Guide. Wiley, Hoboken (2008)
11. Fukumoto, H., Oyama, A.: Benchmarking multiobjective evolutionary algorithms and constraint handling techniques on a real-world car structure design optimization benchmark problem. In: Proceedings of the Genetic and Evolutionary Computation Conference Companion, GECCO 2018, pp. 177–178 (2018)
12. Habib, A., Singh, H.K., Chugh, T., Ray, T., Miettinen, K.: A multiple surrogate assisted decomposition based evolutionary algorithm for expensive multi/many-objective optimization. IEEE Trans. Evol. Comput. **23**(6), 1000–1014 (2019)

13. Knowles, J.: ParEGO: a hybrid algorithm with on-line landscape approximation for expensive multiobjective optimization problems. IEEE Trans. Evol. Comput. **10**(1), 50–66 (2006)
14. Kohira, T., Kemmotsu, H., Oyama, A., Tatsukawa, T.: Proposal of benchmark problem based on real-world car structure design optimization. In: Proceedings of the Genetic and Evolutionary Computation Conference Companion, GECCO 2018, pp. 183–184 (2018)
15. Lin, S.: NPGM - a NSGA program in Matlab, version 1.4 (2011). http://www.mathworks.com/matlabcentral/fileexchange/31166
16. Mazda Benchmark Problem. https://ladse.eng.isas.jaxa.jp/benchmark/index.html. Accessed 30 Apr 2020
17. Mlakar, M., Petelin, D., Tušar, T., Filipič, B.: GP-DEMO: differential evolution for multiobjective optimization based on gaussian process models. Eur. J. Oper. Res. **243**(2), 347–361 (2015)
18. Mueller, J.: SOCEMO: surrogate optimization of computationally expensive multiobjective problems. INFORMS J. Comput. **29**(4), 581–596 (2017)
19. Ohtsuka, H., Kaidan, M., Harada, T., Thawonmas, R.: Evolutionary algorithm using surrogate assisted model for simultaneous design optimization benchmark problem of multiple car structures. In: Proceedings of the Genetic and Evolutionary Computation Conference Companion, GECCO 2018, pp. 55–56 (2018)
20. Oyama, A., Kohira, T., Kemmotsu, H., Tatsukawa, T., Watanabe, T.: Simultaneous structure design optimization of multiple car models using the K computer. In: 2017 IEEE Symposium Series on Computational Intelligence (SSCI), Honolulu, HI, pp. 1–4 (2017)
21. Palar, P.S., Dwianto, Y.B., Zuhal, L.R., Tsuchiya, T.: Framework for robust optimization combining surrogate model, memetic algorithm, and uncertainty quantification. In: Tan, Y., Shi, Y., Niu, B. (eds.) Advances in Swarm Intelligence. ICSI 2016. Lecture Notes in Computer Science, vol. 9712. Springer, Cham (2016). https://doi.org/10.1007/978-3-319-41000-5_5
22. Powell, M.J.D.: The theory of radial basis function approximation in 1990. In: Light, W. (ed.) Advances in Numerical Analysis, Volume 2: Wavelets, Subdivision Algorithms and Radial Basis Functions, pp. 105–210. Oxford University Press, Oxford (1992)
23. Regis, R.G.: Constrained optimization by radial basis function interpolation for high-dimensional expensive black-box problems with infeasible initial points. Eng. Optim. **46**(2), 218–243 (2014)
24. Regis, R.G.: Multi-objective constrained black-box optimization using radial basis function surrogates. J. Comput. Sci. **16**, 140–155 (2016)
25. Singh, P., Couckuyt, I., Ferranti, F., Dhaene, T.: A constrained multi-objective surrogate-based optimization algorithm. In: 2014 IEEE Congress on Evolutionary Computation (CEC), Beijing, pp. 3080–3087 (2014)
26. Tabatabaei, M., Hakanen, J., Hartikainen, M., Miettinen, K., Sindhya, K.: A survey on handling computationally expensive multiobjective optimization problems using surrogates: non-nature inspired methods. Struct. Multidiscip. Optim. **52**(1), 1–25 (2015). https://doi.org/10.1007/s00158-015-1226-z
27. Yang, K., Li, L., Deutz, A., Back, T., Emmerich, M.: Preference-based multiobjective optimization using truncated expected hypervolume improvement. In: 2016 12th International Conference on Natural Computation, Fuzzy Systems and Knowledge Discovery (ICNC-FSKD), Changsha, pp. 276–281 (2016)

Exploring Gaps in DeepFool in Search of More Effective Adversarial Perturbations

Jon Vadillo[1]([⊠])(iD), Roberto Santana[1](iD), and Jose A. Lozano[1,2](iD)

[1] University of the Basque Country UPV/EHU, 20018 San Sebastian, Spain
{jon.vadillo,roberto.santana,ja.lozano}@ehu.eus
[2] Basque Center for Applied Mathematics (BCAM), 48009 Bilbao, Spain

Abstract. Adversarial examples are inputs subtly perturbed to produce a wrong prediction in machine learning models, while remaining perceptually similar to the original input. To find adversarial examples, some attack strategies rely on linear approximations of different properties of the models. This opens a number of questions related to the accuracy of such approximations. In this paper we focus on DeepFool, a state-of-the-art attack algorithm, which is based on efficiently approximating the decision space of the target classifier to find the minimal perturbation needed to fool the model. The objective of this paper is to analyze the feasibility of finding inaccuracies in the linear approximation of DeepFool, with the aim of studying whether they can be used to increase the effectiveness of the attack. We introduce two strategies to efficiently explore gaps in the approximation of the decision boundaries, and evaluate our approach in a speech command classification task.

Keywords: Adversarial examples · DeepFool · Robust machine learning

1 Introduction

The intriguing vulnerability of Deep Neural Networks (DNNs) to imperceptibly yet maliciously perturbed inputs, known in the literature as adversarial examples [24], has raised concerns regarding the robustness of these models in adversarial scenarios, and more particularly in security-critical applications.

While different hypotheses have been proposed in the literature to explain why DNNs are vulnerable to such imperceptible perturbations, most of them focus on the analysis of the decision spaces learnt by the classifiers [6,10,13, 19,24,26]. Furthermore, the underlying theoretical framework of different attack strategies relies directly on wisely exploiting different properties of such decision spaces [14–16], which is the case of DeepFool [16], a state-of-the-art algorithm based on linearly approximating the decision boundaries of the target classifier to efficiently approximate minimal perturbations capable of inducing a misclassification.

© Springer Nature Switzerland AG 2020
G. Nicosia et al. (Eds.): LOD 2020, LNCS 12566, pp. 215–227, 2020.
https://doi.org/10.1007/978-3-030-64580-9_18

Even outside the particular field of adversarial machine learning, the study of the decision boundaries of DNNs is currently a relevant yet understudied research topic [9,13,30], and advances in this direction are necessary to better understand the complex behaviour and decision making process in these models.

The objective of this paper is to study whether it is possible to exploit the inaccuracies in the linear approximation assumed in DeepFool in order to increase the effectiveness of the attack, while maintaining a minimal distortion. For instance, one could increase the number of incorrect output classes that can be produced in the model with a negligible overhead in the original algorithm, or find *shortcuts* to closer decision boundaries that are missed by DeepFool, reducing the amount of perturbation. From another point of view, the analysis of such inaccuracies in the linear approximation could also reveal interesting properties about the geometry of the decision space of the classifier, or provide a useful framework to study the proximity between the classes, which is directly related to the robustness of the classifiers to adversarial attacks.

For these purposes, in this paper we introduce two different methods for extending the DeepFool algorithm to efficiently explore inaccuracies in its linear approximation, and to study whether such inaccuracies can be exploited to generate more effective perturbations. Our experiments reveal that, although inaccuracies can be found in DeepFool, taking advantage of such inaccuracies does not result in a significant improvement over the original algorithm.

2   Related Work

The intriguing phenomenon of the vulnerability of DNNs to imperceptible adversarial perturbations was first reported by Szegedy et al. in [24]. Although multiple attack approaches [1–3,6,11,12,14,16,17,23], and defensive strategies [5–7,18,21,29] have been proposed for different tasks and domains, the explanation of these vulnerabilities, the connection between different attack strategies or the reason for common vulnerabilities on different models are still open questions.

Regarding the theoretical justification of adversarial examples, different hypotheses have been put forward. In [24], adversarial examples are attributed to the highly non-linearity nature of DNNs, causing dense low-probability *"pockets"* in the input space of the model, composed of inputs that, with very low probability, could be found by randomly sampling in the vicinity of a clean sample. Contrarily, in [6] it is stated that even linear models are also highly vulnerable to adversarial examples for high dimensional problems. In [22,25], the existence of adversarial examples is studied under the manifold hypothesis: the data lies on a low-dimensional manifold S embedded in a high-dimensional representation of the input space, and, due to the high dimensionality of the input data, samples close to S can be found outside the decision boundary of the corresponding class. In [8], however, a data-perspective explanation is provided, explaining adversarial examples as *non-robust* features of the input data, instead of flaws in the learnt representation of the DNNs.

In this work, we focus on the exploitability of the geometry of the decision regions learnt by the DNNs as a basis for studying the adversarial examples

[14–16]. In [16], the DeepFool algorithm was introduced, a state-of-the-art method for efficiently crafting adversarial examples, which is based on pushing an input to its closest decision boundary, approximated in the vicinity of the input according to an efficient linear approximation. A detailed explanation of this attack algorithm can be found in Sect. 3. In [14], this method has been extended in order to generate universal (input-agnostic) adversarial perturbations for the image classification task. In fact, the approach proposed in [14] is based on accumulating individual perturbations generated using the DeepFool algorithm. This strategy has also been used in the audio domain [27].

The theoretical framework introduced in [14] to generate universal perturbations is further developed in [15], where the authors study the relationship between the robustness of the models to universal adversarial perturbations and the geometry of their decision boundaries.

In this work, we intend to further exploit the decision boundaries of the target classifier to achieve more effective perturbations. In particular, we extend the DeepFool algorithm to expand its search space, and overcome the possible limitations that the assumed simplification of the decision boundaries may produce.

2.1 Technical Background

Let us consider a classification function $f : X \to Y$, being $X \subseteq \mathbb{R}^d$ the d-dimensional input space and Y a discrete output space of k classes, where $y_i \in \{y_1, \dots, y_k\}$ represents the i-th class. Let $x \in X$ be an input sample correctly classified by f. The objective of an adversarial attack is to produce a perturbed input x' which, being highly similar to x, produces a misclassification of f, that is, $f(x) \neq f(x')$. To ensure that x' is as similar as possible to x, in this work we require the adversarial example to satisfy $\varphi(x, x') \leq \varepsilon$, where $\varphi : \mathbb{R}^d \times \mathbb{R}^d \to \mathbb{R}$ represents a suitable distortion metric, typically an ℓ_p norm, and ε a maximum distortion threshold.

Depending on the malicious effect we want to produce on the classifier, we can consider different types of adversarial attacks. Given a clean input x, a *targeted* adversarial attack consists of perturbing x so that x' is wrongly classified as one particular target class $y_t \neq f(x)$. Contrarily, an *untargeted* adversarial attack consists of generating an adversarial example x' so that $f(x') \neq f(x)$, without any additional regard for the output class. A more comprehensive overview of the possible attack types can be found in [31].

3   Attack Algorithm

As previously mentioned, in this paper we use different variants of DeepFool [16], a state-of-the-art algorithm to generate adversarial attacks, initially introduced for images. Thus, in this section we provide a more detailed overview of this algorithm.

The objective of the attack is to find the minimal perturbation capable of sending an input sample x outside its decision region, by pushing it to the closest decision boundary. This can be seen as an optimization problem, in which we aim to find

$$r^* = \underset{r}{\arg\min}\, \varphi(x, x + r) \ \text{ s.t. } f(x + r) \neq f(x). \tag{1}$$

However, for high-dimensional non-linear decision spaces, which is the general case of DNNs, estimating the decision boundaries is a complex task, which makes this optimization intractable in practice. Due to this limitation, the DeepFool algorithm provides a strategy to approximate r^* by efficiently approximating the decision region of the model. This strategy consists of iteratively pushing an initial input x_0, of class $f(x_0) = y_c$, towards a linear approximation of the decision boundaries, based on the first-order derivatives in the vicinity of the input sample. This transforms the decision region into a polyhedron:

$$\tilde{\mathcal{R}}_i = \bigcap_{j=1}^{k} \{x : f_j(x_i) - f_c(x_i) + \nabla f_j(x_i)^\top x - \nabla f_c(x_i)^\top x \leq 0\}, \tag{2}$$

where x_i represents the input sample at step i and f_j represents the output of f corresponding to the class y_j. Thus, the task of determining the minimal distances to such boundaries is now tractable. Based on this approximation, the closest decision boundary, which will correspond to a class y_l, is determined as follows:

$$l = \underset{j \neq c}{\arg\min}\, \frac{|f'_j|}{||w'_j||_2}, \tag{3}$$

where $f'_j = f_j(x_i) - f_c(x_i)$, and $w'_j = \nabla f_j(x_i) - \nabla f_c(x_i)$ represents the direction of the perturbation. Finally, x_i is pushed towards the selected decision boundary according to the following rule:

$$r_i \leftarrow \frac{|f'_l|}{||w'_l||_2^2} w'_l \tag{4}$$

$$x_{i+1} \leftarrow x_i + r_i. \tag{5}$$

The algorithm stops when we finally produce a new incorrect class $f(x_i) \neq y_c$ in the target model.

As previously mentioned, we consider that the adversarial example should be restricted by a maximum amount of distortion ε in comparison to the original input, according to a suitable distortion metric: $\varphi(x, x') \leq \varepsilon$. Therefore, in this paper we assume that, if at any step of DeepFool the amount of distortion exceeds ε, then it is not possible to construct a *valid* adversarial example x' which is able to fool the classifier.

Finally, even if we mainly focus on untargeted attacks in this work, note that a targeted version of DeepFool can be obtained if the input is pushed towards the direction of the boundary corresponding to the target class $y_t \neq f(x_0)$ until

$f(x_i) = y_t$ is satisfied. This can be easily achieved by removing the criterion specified in Eq. (3), and using the following update rule:

$$r_i \leftarrow \frac{|f'_t|}{||w'_t||_2^2} w'_t \qquad (6)$$

$$x_{i+1} \leftarrow x_i + r_i. \qquad (7)$$

4   Exploiting Gaps in the Linear Approximation of the Decision Boundaries

DeepFool relies on the assumption that we can accurately approximate the boundaries of the decision region of the classifier in the proximity of a given input using a linear approach. Based on this assumption, at each step, the input is moved in a *greedy* way towards the (estimated) closest boundary, using the perturbation r_i (see Eqs. (3) and (4)). However, there could exist alternative perturbations that, with the same amount of distortion employed by DeepFool, can reach a different decision region, or even reach it with less distortion. Being able to find such *shortcuts* in an efficient way can increase the effectiveness of the attack, or reduce the required amount of perturbation to fool the model. Note that, if this is possible, it is because the linear approximation of the decision regions is not sufficiently accurate, and, therefore, we refer to them as *gaps* or *holes* in the algorithm. Based on this hypothesis, in this section we propose two different strategies to explore, during the original attack process, the existence of such gaps in the algorithm.

Let x_i be the input sample at step i, assuming that $f(x_i) = f(x_0) = y_c$. Let $\widehat{W} = \{\hat{w}_1, \ldots \hat{w}_{c-1}, \hat{w}_{c+1} \ldots, \hat{w}_k\}$ be the set of normalized direction vectors computed by DeepFool for each class (except y_c), where:

$$\hat{w}_j = \frac{w'_j}{||w'_j||_2}. \qquad (8)$$

It is worth mentioning that, following the criterion described in Eq. (3), the direction $\hat{w}_l$ is the one that will be followed by DeepFool according to its *greedy* criterion.

The first search strategy that we propose, named *Gap Finder Local Search* (GFLS), consists of exploring all the possible directions in $\widehat{W} - \{\hat{w}_l\}$ at each step, projected in the sphere of radius $||r_i||_2$ and centered at the current point x_i. That is, we consider a *gap* if, with a perturbation amount of $||r_i||_2$, we can change the output of the model by pushing x_i in any other direction $\hat{w}_j \in \widehat{W} - \{\hat{w}_l\}$:

$$\exists j \neq l : f(x_i + \hat{w}_j \cdot ||r_i||_2) \neq f(x_i). \qquad (9)$$

In the second strategy, instead of taking the *local* neighborhood of x_i as reference, we will take as reference the initial point x_0, and consider the accumulated perturbation $r_{tot} = (x_i - x_0) + r_i$ at the end of the step i. We will refer to this second strategy as *Gap Finder Global Search* (GFGS). This criterion is more suitable if we want to check alternative directions, while ensuring that the *total* perturbation amount that those directions produce is the same as the one produced by DeepFool. Thus, now the search space will be bounded to the surface of the sphere of radius $||r_{tot}||_2$ and centered at the starting point x_0:

$$\exists j \neq l : f\left(x_0 + r_{tot}^j \cdot \frac{||r_{tot}||_2}{||r_{tot}^j||_2}\right) \neq f(x_i) , \; r_{tot}^j = (x_i - x_0) + \frac{|f_j'|}{||w_j'||_2^2}w_j'. \quad (10)$$

An illustration of the two search strategies is provided in Fig. 1.

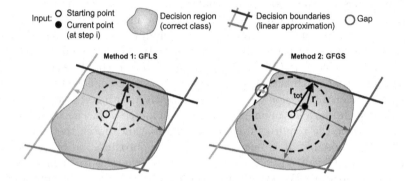

Fig. 1. Illustration of the two search methodologies employed to find gaps: *Gap Finder Local Search* (GFLS) and *Gap Finder Global Search* (GFGS).

Note that, in both cases, we assume that a gap is found if an alternative perturbation (that is, different to the one proposed by DeepFool) is capable of changing the output of the model, with no additional regard for the incorrect class that is produced. Thus, the proposed methodologies are particularly well suited for untargeted attacks, as in targeted attacks we focus on reaching a particular output rather than producing any possible incorrect class.

5 Evaluating Gaps in the Linear Approximation of the Decision Boundary

5.1 Case of Study

We will use the Speech Command Dataset [28], which consists of a set of 16-bit WAV audio files of 30 different spoken commands. The dataset can be downloaded from[1]. The duration of all the files is the same in all the audio clips, exactly 1 s. The sample-rate of the audio signals is 16 kHz, so every audio signal is composed of 16000 values.

---

[1] http://download.tensorflow.org/data/speech_commands_v0.02.tar.gz.

As in previous publications [1, 28], we focused on the following set of classes out of the 30 different classes in the original dataset: *Yes, No, Up, Down, Left, Right, On, Off, Stop*, and *Go*. In addition to this set, we will consider two special classes for a more realistic setup: *Unknown* and *Silence*. We used a Convolutional Neural Network as the classification model, based on the architecture proposed in [20], which is considered a benchmark in speech command classification [1, 4, 28, 32].

To explore the existence of gaps, we selected a set of $N = 500$ files per class from the training set, forming a total of 6000 inputs. For each audio, a default untargeted DeepFool attack is executed, and at each step of the process the two strategies proposed in Sect. 4 are applied to search for gaps in the vicinity of the inputs. We launched the experiment for the following maximum distortion thresholds, with which the ℓ_2 norm of the perturbations will be bounded: $\varepsilon = \{0.05, 0.1, 0.15\}$. Our implementation is available upon request.

5.2   Analysis of the Results

In this section we evaluate with which frequency *gaps* were found in the linear approximation of the decision boundary employed by DeepFool, as described in Sect. 4. We also analayze the gain in terms of effectiveness with respect to the original algorithm, or the insights that the study of these gaps provide regarding the structure of the decision space of the classifier.

First, we report the percentage of inputs for which, at any step of the algorithm, the introduced search methods found a *gap*. Table 1 contains the obtained percentages for different distortion thresholds. As can be seen, for the majority of the classes, the GFLS approach was capable of finding a gap for approximately 30% of the inputs samples. With the GFGS approach, however, we only found gaps for less than 10% of the inputs for most of the classes. This can reflect that the accumulated perturbation, after multiple steps, achieves a much better approximation of the optimal perturbation r^* than the individual perturbations applied at each step.

In spite of these positive results, if we analyze the iteration number in which those gaps were found, it can be seen that a very high percentage of gaps was found in the last iteration of the process, considering the last step the one in which $f(x_{i+1}) \neq f(x_i) = f(x_0)$ is satisfied. This information is displayed in Table 2. These percentages are particularly high for the GFLS approach, in which more than approximately 80% of the gaps were found in the last step, independently of the class. Although this is also true for some classes for the GFGS method, the percentages are considerably lower for the rest, around 40% in the lowest cases.

These results suggest, especially for the local strategy, that the gaps are mainly found in the proximity of the decision boundary. Therefore, we can hypothesize that, even through gaps, the input is being pushed towards the same decision boundary that is reached using the default untargeted attack. This hypothesis would also explain why the local search followed in GFLS achieved considerably more gaps than GFGS, as in the close proximity of the decision

Table 1. Percentage of inputs for which gaps were found in the linear approximation of the decision boundaries, during the attack process of DeepFool.

Class	GFLS			GFGS		
	ε			ε		
	0.05	0.10	0.15	0.05	0.10	0.15
Go	31.00	31.20	31.20	9.20	9.20	9.20
Stop	26.60	28.80	29.80	6.20	7.40	8.60
Off	30.80	31.40	31.40	21.40	22.00	22.20
On	31.80	32.00	32.00	7.40	7.80	7.80
Right	21.80	22.20	22.40	3.40	3.40	3.40
Left	27.60	28.60	28.80	5.60	6.00	6.40
Down	25.20	26.80	27.00	7.20	8.00	8.00
Up	29.80	30.40	30.60	7.00	7.40	7.40
No	27.20	27.60	27.60	6.40	6.60	6.80
Yes	26.80	28.00	28.40	11.60	12.60	14.20
Unknown	27.00	27.20	27.20	6.40	6.60	6.60
Silence	6.00	7.60	8.20	22.40	31.20	41.20

boundary there is a higher chance of surpassing the boundary by moving the sample to other directions. To validate this hypothesis, we computed the percentage of gaps that reach a different class than the default untargeted DeepFool algorithm. According to the results, for both methods, only for a percentage below 1% of the inputs is it possible to produce a wrong output class different to that which is produced by the default attack.

All these results might also suggest that the linear approximation of the decision boundaries employed by DeepFool is highly accurate in our case, and as a consequence, that the generated perturbations are close to the optimal ones. However, another explanation can be that, in the proximity of the natural input samples, there is generally a decision boundary for one class much closer than those corresponding to the rest of the classes. In fact, if we compute the frequency with which each input is classified as another class after being fooled by the original DeepFool algorithm, we can see that, for the majority of the classes, there are always one or two classes with higher frequency. This information is shown in Fig. 2, for the case of $\varepsilon = 0.15$, although the same pattern is given for the other distortion thresholds tried.

To continue with the analysis, in order to assess the distortion gain that the gaps suppose, if any, we also compared the distortion introduced by the gaps with the one introduced by the untargeted attacks. The comparison is made by considering those inputs for which a perturbation capable of fooling the model was found with both attack types. If more than one possible gap was found for one input, the one with the lowest perturbation has been considered. The results

Table 2. Percentage of gaps that were found in the last step of the process. We consider the last step the one in which, when we apply the last perturbation to the (still correctly classified) input, we reach a new decision region, that is, when $f(x_{i+1}) \neq f(x_i) = f(x_0)$.

Class	GFLS			GFGS		
	ε			ε		
	0.05	0.10	0.15	0.05	0.10	0.15
Go	80.65	80.77	80.77	50.00	50.00	50.00
Stop	87.97	88.89	87.92	67.74	72.97	76.74
Off	83.77	84.08	84.08	65.14	65.18	65.49
On	86.16	86.25	86.25	62.16	64.10	64.10
Right	89.91	90.09	90.18	82.35	82.35	82.35
Left	78.99	79.72	79.17	39.29	40.00	43.75
Down	83.33	83.58	83.70	55.56	55.00	55.00
Up	87.92	87.50	87.58	74.29	75.68	75.68
No	80.88	81.16	81.16	42.42	41.18	42.86
Yes	91.11	90.78	90.91	65.52	61.90	63.89
Unknown	81.48	81.62	81.62	40.62	42.42	42.42
Silence	80.00	78.95	76.19	67.26	71.07	73.93

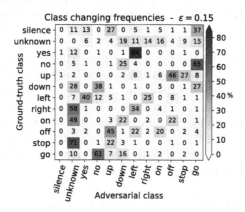

Fig. 2. Class changing frequencies caused by the default untargeted DeepFool algorithm. Each row in the matrix represents the percentage of cases changed from the corresponding ground-truth class to a different class.

are shown in Fig. 3, for the different distortion thresholds.[2] As can be seen, the gain is negligible, showing again that the found gaps are not very different to the perturbation provided by DeepFool, which reinforces our previous hypothesis.

---

[2] The outliers have been removed for the sake of visualization.

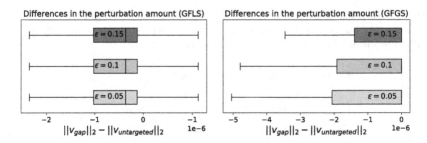

Fig. 3. Comparison of the ℓ_2 norm between the perturbations obtained using gaps, v_{gap}, and the perturbation obtained with the default untargeted attack, $v_{untargeted}$.

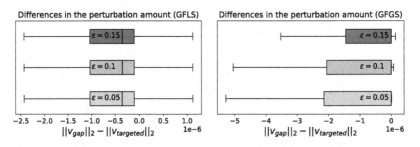

Fig. 4. Comparison of the ℓ_2 norm between the perturbations obtained using gaps, v_{gap}, and the perturbation obtained with the default targeted attack, $v_{targeted}$.

Finally, we also tested the hypothesis that, through the gaps, it could be possible to generate perturbations capable of fooling the model for those inputs for which, through the original algorithm, it would not be possible to find such a perturbation. However, according to the results obtained by the GFLS strategy, in no case we were able to fool the model using only the perturbations generated through gaps. For the GFGS strategy, the percentage of inputs for which this happens is below 1%. Thus, for almost all the inputs for which gaps were found, we could also find a perturbation capable of fooling the model using the default algorithm.

Similarly, if we compare the results with the targeted version of DeepFool, relying on gaps to find targeted attacks is much less effective than generating them by moving the input directly towards the decision boundary of the target class. Only in less than 1% of the cases were gaps capable of reaching a particular class that could not be reached through the targeted attack, according to both search strategies. Moreover, the difference in terms of distortion level is negligible in this case also, as is shown in Fig. 4.

6   Conclusion

In this paper, we have proposed two different strategies to search for inaccuracies in the linear approximation of the decision boundaries employed by the Deep-

Fool algorithm, a state-of-the-art adversarial attack, to assess whether they can increase the effectiveness of the attack, as well as a framework to study the properties of the decision spaces of machine learning classifiers. We experimentally tested our approach for a Convolutional Neural Network in the speech command classification task, an exemplary machine learning model in the audio domain.

The results obtained revealed that, in approximately 30% of the inputs evaluated, the approach introduced was capable of exploiting those inaccuracies to find alternative directions in which the model can be fooled, with a negligible overhead with respect to the original algorithm, although the gain in terms of effectiveness was not significant enough to consider it an improvement. In fact, we discovered that most of these gaps lead to perturbations highly similar to those found by the default algorithm. We intend to improve the results by extending the search strategies in future work.

As future research lines, we also believe that the comparison of the geometrical properties of the decision boundaries of different machine learning models can provide insightful contributions regarding their vulnerability to adversarial attacks. For this reason, we intend to extend our methodology to evaluate a wider range of decision models, including more complex structures, such as recurrent neural networks, as well as to other tasks and domains, for instance, speech transcription, computer vision tasks or natural language processing problems.

Acknowledgments. This work is supported by the Basque Government (BERC 2018-2021 and ELKARTEK programs, IT1244-19, and PRE_2019_1_0128 predoctoral grant), by the Spanish Ministry of Economy and Competitiveness MINECO (projects TIN2016-78365-R and PID2019-104966GB-I00) and by the Spanish Ministry of Science, Innovation and Universities (FPU19/03231 predoctoral grant). Jose A. Lozano acknowledges support by the Spanish Ministry of Science, Innovation and Universities through BCAM Severo Ochoa accreditation (SEV-2017-0718).

References

1. Alzantot, M., Balaji, B., Srivastava, M.: Did you hear that? Adversarial examples against automatic speech recognition. arXiv preprint arXiv:1801.00554 (2018)
2. Carlini, N., Wagner, D.: Towards evaluating the robustness of neural networks. In: 2017 IEEE Symposium on Security and Privacy (SP), pp. 39–57. IEEE (2017)
3. Cisse, M., Adi, Y., Neverova, N., Keshet, J.: Houdini: fooling deep structured prediction models. arXiv preprint arXiv:1707.05373 (2017)
4. Du, T., Ji, S., Li, J., Gu, Q., Wang, T., Beyah, R.: Sirenattack: generating adversarial audio for end-to-end acoustic systems. arXiv preprint arXiv:1901.07846 (2019)
5. Esmaeilpour, M., Cardinal, P., Lameiras Koerich, A.: A robust approach for securing audio classification against adversarial attacks. IEEE Trans. Inf. Forensics Secur. **15**, 2147–2159 (2020)
6. Goodfellow, I.J., Shlens, J., Szegedy, C.: Explaining and harnessing adversarial examples. arXiv preprint arXiv:1412.6572 (2014)
7. Grosse, K., Manoharan, P., Papernot, N., Backes, M., McDaniel, P.: On the (statistical) detection of adversarial examples. arXiv preprint arXiv:1702.06280 (2017)
8. Ilyas, A., Santurkar, S., Tsipras, D., Engstrom, L., Tran, B., Madry, A.: Adversarial examples are not bugs, they are features. arXiv preprint arXiv:1905.02175 (2019)

9. Karimi, H., Tang, J.: Decision boundary of deep neural networks: challenges and opportunities. In: Proceedings of the 13th International Conference on Web Search and Data Mining, pp. 919–920 (2020)

10. Khoury, M., Hadfield-Menell, D.: On the geometry of adversarial examples. arXiv preprint arXiv:1811.00525 (2018)

11. Kurakin, A., Goodfellow, I., Bengio, S.: Adversarial examples in the physical world. arXiv preprint arXiv:1607.02533 (2016)

12. Luo, B., Liu, Y., Wei, L., Xu, Q.: Towards imperceptible and robust adversarial example attacks against neural networks. In: Thirty-Second AAAI Conference on Artificial Intelligence (2018)

13. Mickisch, D., Assion, F., Greßner, F., Günther, W., Motta, M.: Understanding the decision boundary of deep neural networks: an empirical study. arXiv preprint arXiv:2002.01810 (2020)

14. Moosavi-Dezfooli, S.M., Fawzi, A., Fawzi, O., Frossard, P.: Universal adversarial perturbations. In: Proceedings of the IEEE Conference on Computer Vision and Pattern Recognition, pp. 1765–1773 (2017)

15. Moosavi-Dezfooli, S.M., Fawzi, A., Fawzi, O., Frossard, P., Soatto, S.: Analysis of universal adversarial perturbations. arXiv preprint arXiv:1705.09554 (2017)

16. Moosavi-Dezfooli, S.M., Fawzi, A., Frossard, P.: DeepFool: a simple and accurate method to fool deep neural networks. In: Proceedings of the IEEE Conference on Computer Vision and Pattern Recognition, pp. 2574–2582 (2016)

17. Papernot, N., McDaniel, P., Jha, S., Fredrikson, M., Celik, Z.B., Swami, A.: The limitations of deep learning in adversarial settings. In: 2016 IEEE European Symposium on Security and Privacy (EuroS&P), pp. 372–387. IEEE (2016)

18. Papernot, N., McDaniel, P., Wu, X., Jha, S., Swami, A.: Distillation as a defense to adversarial perturbations against deep neural networks. In: 2016 IEEE Symposium on Security and Privacy (SP), pp. 582–597. IEEE (2016)

19. Roos, M.J.: Utilizing a null class to restrict decision spaces and defend against neural network adversarial attacks. arXiv preprint arXiv:2002.10084 (2020)

20. Sainath, T.N., Parada, C.: Convolutional neural networks for small-footprint keyword spotting. In: Sixteenth Annual Conference of the International Speech Communication Association, pp. 1478–1482 (2015)

21. Samangouei, P., Kabkab, M., Chellappa, R.: Defense-GAN: protecting classifiers against adversarial attacks using generative models. arXiv preprint arXiv:1805.06605 (2018)

22. Stutz, D., Hein, M., Schiele, B.: Disentangling adversarial robustness and generalization. In: Proceedings of the IEEE Conference on Computer Vision and Pattern Recognition, pp. 6976–6987 (2019)

23. Su, J., Vargas, D.V., Sakurai, K.: One pixel attack for fooling deep neural networks. IEEE Trans. Evol. Comput. **23**(5), 828–841 (2019)

24. Szegedy, C., et al.: Intriguing properties of neural networks. In: 2nd International Conference on Learning Representations (2014)

25. Tanay, T., Griffin, L.: A boundary tilting perspective on the phenomenon of adversarial examples. arXiv preprint arXiv:1608.07690 (2016)

26. Tramèr, F., Papernot, N., Goodfellow, I., Boneh, D., McDaniel, P.: The space of transferable adversarial examples. arXiv preprint arXiv:1704.03453 (2017)

27. Vadillo, J., Santana, R.: Universal adversarial examples in speech command classification. arXiv preprint arXiv:1911.10182 (2019)

28. Warden, P.: Speech commands: a dataset for limited-vocabulary speech recognition. arXiv preprint arXiv:1804.03209 (2018)

29. Yang, Z., Li, B., Chen, P.Y., Song, D.: Characterizing audio adversarial examples using temporal dependency. arXiv preprint arXiv:1809.10875 (2018)
30. Yousefzadeh, R., O'Leary, D.P.: Investigating decision boundaries of trained neural networks. arXiv preprint arXiv:1908.02802 (2019)
31. Yuan, X., He, P., Zhu, Q., Li, X.: Adversarial examples: attacks and defenses for deep learning. IEEE Trans. Neural Netw. Learn. Syst. **30**(9), 2805–2824 (2019)
32. Zhang, Y., Suda, N., Lai, L., Chandra, V.: Hello edge: Keyword spotting on micro-controllers. arXiv preprint arXiv:1711.07128 (2017)

Lottery Ticket Hypothesis: Placing the k-orrect Bets

Abhinav Raj[1,2] and Subhankar Mishra[1,2(✉)] ⓘD

[1] National Institute of Science Education and Research, Jatani 752050, Odisha, India
{abhinav.raj,smishra}@niser.ac.in
[2] Homi Bhabha National Institute, Anushaktinagar, Mumbai 400094, India
https://www.niser.ac.in/

Abstract. Neural Network pruning has been one of the widely used techniques for reducing parameter count from an over-parametrized network. In the paper titled *Deconstructing Lottery Tickets*, the authors showed that pruning is a way of training, which we extend to show that pruning a well-trained network at initialization does not exhibit significant gains in accuracy. *Stabilizing the Lottery Ticket Hypothesis* motivates us to explore pruning after k^{th} epoch. We show that there exists a minimum value of k above which there is insignificant gain in accuracy, and the network enjoys a maximum level of pruning at this value of k while maintaining or increasing the accuracy of the original network. We test our claims on MNIST, CIFAR10, and CIFAR100 with small architectures such as lenet-300-100, Conv-2,4,6, and more extensive networks such as Resnet-20. We then discuss why pruning at initialization does not result in considerable benefits compared to pruning at k.

Keywords: Pruning · Lottery ticket · Machine learning · Neural networks

1 Introduction

The effectiveness of Neural Networks lies in the fact that they can generalize unreasonably well i.e., perform well on unseen data. This generalization ability is usually attributed to a large number of parameters (compared to the data) present in a generic neural network; a setting commonly referred to as over-parametrization.

1.1 Need for Over-Parametrization

It was believed that a large over-parametrized network is essential for training as it has high representational power and is less sensitive to hyperparameter selection, making the model more robust and easier to train [15,22]. Empirically, *a good rule of thumb for obtaining good generalization is to use the smallest network that still fits the data* [22]. However, practically finding such a network

© Springer Nature Switzerland AG 2020
G. Nicosia et al. (Eds.): LOD 2020, LNCS 12566, pp. 228–239, 2020.
https://doi.org/10.1007/978-3-030-64580-9_19

is complicated, and even if we know the optimal size of the network, it might be infeasible to actually train the network as the optimization technique might not be good enough to reach the global minima in the loss landscape. Also, in the training process, if we get a sub-optimal result, it would be hard to determine the root cause of the problem, whether it is the choice of hyperparameters or the network being too small to fit the data or possibly both.

1.2 Network Sparsifying Techniques

Neural Architecture Search (NAS) and Network Pruning have emerged as the two popular approaches for achieving optimal network structure. NAS aims to grow a small, sparse, more efficient network from scratch by performing a search in a given search space with search strategies such as evolutionary algorithms [1,7,8], bayesian optimization [2] or Reinforcement Learning [6]. Each model is then evaluated using Evaluation Strategies to select the best model. While NAS automates the process of finding optimal network structures, vast computational resources are required to achieve this (800 GPUs for three to four weeks [6]). While significant improvements have been made, NAS remains very computationally intensive.

Compared to that, network pruning aims to remove a redundant/useless set of weights from the large initial network. One again, the large initial size allows for a safe window for pruning the model from so that we can remove redundant weights without significantly hurting model performance. Pruning techniques as early as [19,20] measured the saliency of weights using the diagonal terms of the Hessian matrix. More recently, [21] did *weight magnitude pruning* where the weights with lowest magnitudes were removed. This approach has minimal computational overhead and can aggressively prune networks by removing up to 90% of weights without harming accuracy.

1.3 Lottery Tickets and Network Pruning

Contrary to the popular beliefs about over-parametrized networks, recent work by [18] showed that one could start from a small, sparse network and train it to an accuracy, matching that of the original network while having significantly less number of parameters. The technique they use is called *Iterative Magnitude Pruning* [21] where we: (1) train a network till convergence in T iterations (2) use the magnitude of final weights as pruning criteria to create a mask (3) Apply the mask to the trained network and fine-tune (4) Repeat till desired pruning level is met or in One-Shot.

In [18], in step 3, the mask is applied to the initial weights resulting in a sparse network, which they call *Lottery Tickets* and showed them to have better generalization capabilities such as lower minimum validation loss and higher accuracy, than the original network. They also take at most T iterations for reaching the stopping criteria of minimum validation loss. Later, [17] analysed the Lottery Tickets and concluded that pruning is a form of training. The weights pruned are the ones that were already headed towards zero so, setting them to zero in one-step accelerates the process of training.

1.4   Rewind to k

[16] came up with rewinding to k, a modification to the iterative pruning approach, which enabled them to find lottery tickets in larger networks and datasets. Instead of resetting the parameters back to values in initialization, they are set to weight values at some epoch k intermediate in training. The mask is applied at this point, and then a masked network is trained from here on. Note that, the whole network is rewound to k *including optimizer and scheduler states*.

2   Various Approaches to Pruning

In this section, we explore various settings in which we perform.

2.1   Pruning at Initialization

Following the claims of [17], we investigate the effect of training on pruning. In particular, we test if the benefits of pruning would be much less pronounced on networks trained with better optimizer(s) or better optimization techniques. We expect well-trained networks to be more resilient to drastic improvements in accuracy and validation loss. As we shall see, in most cases, accuracy only slightly increases with mild pruning, and any further prune always results in a reduction of accuracy; this reduction is directly proportional to the amount of pruning.

2.2   Pruning with Rewinding to k

Since rewinding to initialization does not show any increase (actually a decrease in accuracy), we take inspiration from [16] and try rewinding to k. Given a fixed pruning rate, we expect the accuracy to increase comparatively more, as instead of starting from a random set of weights, we are starting from a partially trained network, which is also precisely what we observe but with a small twist that there is a value of k at which the accuracy is maximum or close to maximum and after which the accuracy seems to decline.

2.3   Optimal Value of k

The value of k is optimal in the sense of accuracy and pruning; as at this value of k, the network enjoys a maximum level of pruning with a reasonably high amount of accuracy. Above this value of k, there is no significant increase in accuracy, and the maximum pruning is reduced. The effects of this optimal value of k are much more pronounced at partial training, where we do not adequately train the network before pruning it. We hypothesize that at lower values of k, the network is not stable enough to significantly increase the accuracy. While, at higher values of k, the network **does not get enough time to counter the effects of pruning**, i.e., after some weights are pruned, other weights do not get enough time to adjust their values in a way such that there is no reduction in overall accuracy.

2.4   Training in Constrained Setting

Here, we check the effects of incomplete training before pruning is done. We train-prune the network in half the number of full training epochs. With the optimal value of k, we have a simple method that allows us to produce sparse trained networks with slightly reduced accuracy relatively cheaply, even with constrained resources. A slightly higher than the usual value of momentum makes up a bit for reduced training time.

3   Our Work

Our basic intuition was that, since pruning is a form of training, the benefits of pruning should be significantly diminished on a well-trained network and vice versa. In particular, when using better training techniques such as **One Cycle Policy** [23], the increase in accuracy after pruning is far less. In particular, there is barely any increase in accuracy with masked-initialised network compared to the original network. Taking inspiration from [16], we then explore pruning with rewinding to k. We summarise our findings below:

- The network fares badly when pruning is done with rewinding to initial values given the network is well-trained, especially on a complex dataset.
- The existence of an optimal k such that rewinding to k, gives the most significant increase in accuracy compared to other values. The network enjoys maximum pruning at this value of k.
- This value of k depends primarily on the dataset along with the architecture and optimizer.
- We develop a faster approach to train a network using an iterative combination of partial training and pruning, which delivers us a trained-pruned network faster than any methods presently known and without a considerable drop in accuracy without extra computation.

4   Experimental Setup

Here, we present the setup that was used for performing the experiments.

4.1   Datasets and Architectures Used

Standard datasets such as MNIST [13], CIFAR10, CIFAR100 [14]. All standard architectures are used including LeNet [11], ResNet20 [24], and Conv-2,4,6 which are variants of VGG [12]. ResNet20 for pytorch was used as mentioned in [25].

4.2   Hyperparameter Selection

Here we use non-standard ways of training neural networks. AdamW [10] and SGD are used as optimizers, but instead of using standard training schedules, One Cycle Policy is used for training networks. Initially, the learning rate is chosen with the help of a learning rate test [9], and then we do a fine-tuned search by varying the learning rate linearly to find an optimal value. For training Resnet20 on CIFAR10, we use SGD with a cyclic learning rate with a maximum of 0.4, cyclic momentum of 0.85–0.95, and a weight decay of 10^{-4} as mentioned in [5]. For all other networks, the maximum learning rate used was 0.002 with similar values for cyclic momentum but with a weight decay of 0.01.

4.3   Pruning Method

We follow the pruning procedure used in [18]. In all the experiments, except one on ResNet20, the pruning rate is 20% for linear layers and 10% for convolutional layers. Pruning is done layer-wise, and only the layer's weights are pruned, not the bias. The last (linear) layer is pruned at half the rate at which the rest of the linear layers are pruned. For ResNet20, since there is only one linear layer, we do not prune it. All other convolutional layers are pruned at a rate of 20%.

The code is available at https://github.com/smlab-niser/korrectBets.

5   Results

5.1   Initialization Experiment

As we can see, there is a small initial increase in accuracy at mild pruning, after which there is a masked drop in accuracy, which significantly worsens with an increase in pruning level. This trend is consistent across all datasets with all the architectures, which are evident in Fig. 1.

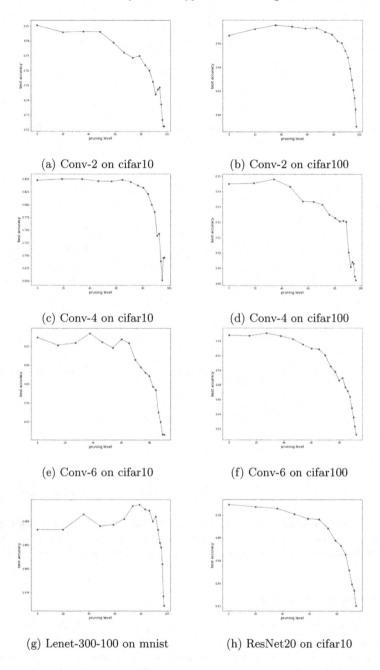

Fig. 1. This figure compares the accuracy of various architectures at initialization.

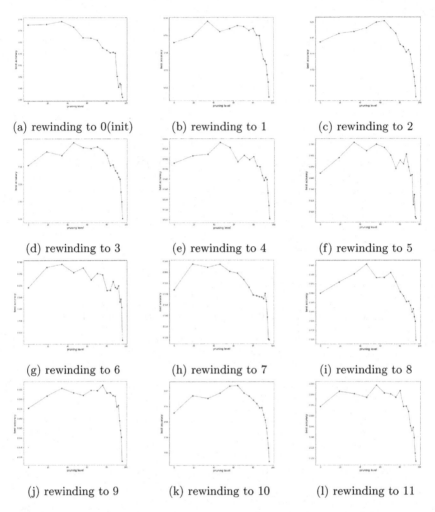

Fig. 2. This figure compares the accuracy of Conv-4 on cifar100 with rewinding to k, while the maximum accuracy is at $k = 10$, maximum pruning occurs at $k = 9$ with good enough accuracy. After $k = 10$, there is a drop in both accuracy and sparsity.

5.2 Rewind to k Experiment

In this section, we explore rewinding to $k(0 \leq k < T)$ where T is the total number of epochs the network is trained. For partially trained networks, this value of k is precise, and any deviation leads to a marked decrease in accuracy. In these conditions, $k = 1$, is always a clear winner, see Fig. 3. At $k = 0$ (at initialization), the network performs consistently worse with increased pruning, while at $k = 1$ there is a drastic increase in accuracy with increased pruning (to a maximum value). At any values higher than 1, a reduction in accuracy is visible as shown in Fig. 2. The results are summarized in Table 1 and 2. Also,

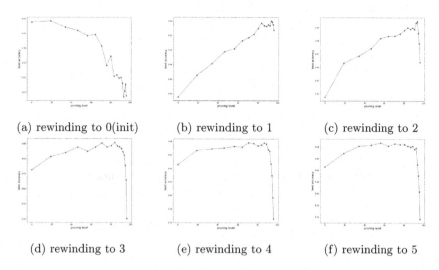

(a) rewinding to 0(init)   (b) rewinding to 1   (c) rewinding to 2

(d) rewinding to 3   (e) rewinding to 4   (f) rewinding to 5

Fig. 3. This figure compares the accuracy of Conv-4 on cifar100 with rewinding to k with constrained resources. Training was done only for 6 epochs compared to the standard 18 epochs. The best accuracy comes at $k = 1$ which reduces with an increase in the value of k.

Table 1. Overall summary of the results for Conv4, Conv4 with constrained resources (trained for third of total training time) and ResNet-20, all on CIFAR-100

	Max accuracy and pruning level for various values of k		
Value of k in each case	Conv4	Conv4 with constrained resources	ResNet20
0	54.6 at 40	44.9 at 19	90.0 at 0
1	55.8 at 37	54.0 at 82	91.7 at 61
2	56.0 at 65	49.4 at 90	91.8 at 43
3	56.2 at 45	48.1 at 82	91.6 at 65
4	56.3 at 45	47.9 at 83	91.5 at 61
5	56.0 at 38	48.2 at 60	91.0 at 45
6	55.8 at 37	-	-
7	56.4 at 20	-	-
8	56.5 at 45	-	-
9	56.6 at 78	-	-
10	57.2 at 62	-	91.5 at 78
15	-	-	91.4 at 44

note that we are able to find lottery tickets in larger networks such as ResNet20 using One Cycle without any changes to the training schedule or using global pruning as visible in Fig. 4.

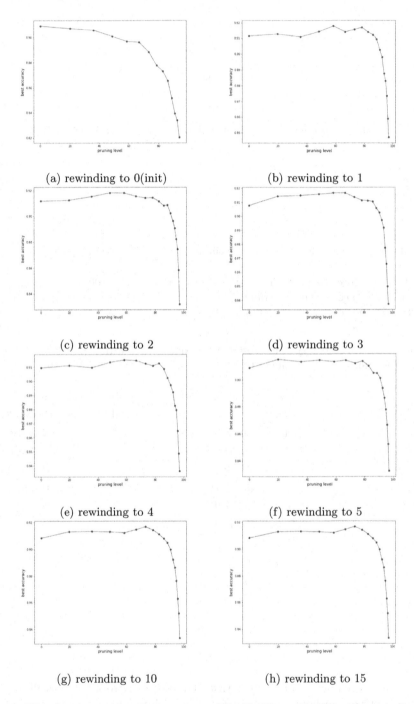

(a) rewinding to 0(init)   (b) rewinding to 1

(c) rewinding to 2   (d) rewinding to 3

(e) rewinding to 4   (f) rewinding to 5

(g) rewinding to 10   (h) rewinding to 15

Fig. 4. This figure compares the accuracy of ResNet20 on cifar100 with rewinding to k, here, we can see that there are no significant gains after rewinding to k with values as early as $k = 1$ and with a network sparsity of 80%.

Table 2. Overall summary of the results for ResNet-56 on CIFAR-10

Value of k for ResNet56	Max accuracy and pruning level for various values of k
0	94.0 at 48.5
1	94.2 at 73.4
2	94.1 at 66.9
3	94.0 at 58.7
8	94.1 at 58.7
15	94.1 at 66.9
20	93.9 at 66.9
30	94.2 at 58.7
40	93.8 at 35.8
60	94.0 at 35.8
100	94.5 at 78.6
130	94.4 at 78.6

6 Conclusion

We empirically showed that rewinding to initialization does not produce excellent results for well-trained networks. Hence, we turn our attention to rewinding at k in hopes of finding a better result which and indeed it yields. We further find the existence of k, which is optimal and allows for maximum pruning determined to be around 80%. The value of k, in general, is not very precise but typically occurs between a fourth to a third of the total number of training epochs. The effect of rewinding to this value of k is much more pronounced when we train the network partially, half to a third of the total training epochs.

7 Future Work

Since using better optimizer results in a better-trained network that resists pruning, one may hypothesize if a network can be trained to a degree such that even rewinding to k does no benefit. We can try new, better optimizers such as [4] which enable faster, better training of the neural network. We want further insights into how and why rewinding to k works so much better than initialization. One can try much bigger datasets such as ImageNet on which pruning would be significant as these pruned, sparse networks can be used on various other datasets [3].

References

1. Stanley, K.O., Clune, J., Lehman, J., Miikkulainen, R.: Designing neural networks through neuroevolution. Nat. Mach. Intell. **1**(1), 24–35 (2019)
2. Bergstra, J., Yamins, D., Cox, D.D.: Making a science of model search: hyperparameter optimization in hundreds of dimensions for vision architectures. Jmlr (2013)
3. Mehta, R.: Sparse transfer learning via winning lottery tickets. arXiv preprint arXiv:1905.07785 (2019)
4. Zhang, M., Lucas, J., Ba, J., Hinton, G.E.: Lookahead optimizer: k steps forward, 1 step back. In: Advances in Neural Information Processing Systems, pp. 9593–9604 (2019)
5. Smith, L.N.: A disciplined approach to neural network hyper-parameters: part 1-learning rate, batch size, momentum, and weight decay. arXiv preprint arXiv:1803.09820 (2018)
6. Zoph, B., Le, Q.V.: Neural architecture search with reinforcement learning. arXiv preprint arXiv:1611.01578 (2016)
7. Jozefowicz, R., Zaremba, W., Sutskever, I.: An empirical exploration of recurrent network architectures. In: International Conference on Machine Learning, pp. 2342–2350 (2015)
8. Stanley, K.O., Miikkulainen, R.: Evolving neural networks through augmenting topologies. Evol. Comput. **10**(2), 99–127 (2002)
9. Smith, L.N.: No more pesky learning rate guessing games. CoRR abs/1506.01186, 5 (2015)
10. Loshchilov, I., Hutter, F.: Decoupled weight decay regularization. arXiv preprint arXiv:1711.05101 (2017)
11. LeCun, Y., Bottou, L., Bengio, Y., Haffner, P.: Gradient-based learning applied to document recognition. Proc. IEEE **86**(11), 2278–2324 (1998)
12. Simonyan, K., Zisserman, A.: Very deep convolutional networks for large-scale image recognition. arXiv preprint arXiv:1409.1556 (2014)
13. LeCun, Y., Cortes, C., Burges, C.J.: MNIST handwritten digit database (2010)
14. Krizhevsky, A., Hinton, G., et al.: Learning multiple layers of features from tiny images. Citeseer (2009)
15. Liu, Z., Sun, M., Zhou, T., Huang, G., Darrell, T.: Rethinking the value of network pruning. arXiv preprint arXiv:1810.05270 (2018)
16. Frankle, J., Dziugaite, G.K., Roy, D.M., Carbin, M.: Stabilizing the lottery ticket hypothesis. arXiv preprint arXiv:1903.01611 (2019)
17. Zhou, H., Lan, J., Liu, R., Yosinski, J.: Deconstructing lottery tickets: zeros, signs, and the supermask. arXiv preprint arXiv:1905.01067 (2019)
18. Frankle, J., Carbin, M.: The lottery ticket hypothesis: finding sparse, trainable neural networks. arXiv preprint arXiv:1803.03635 (2018)
19. LeCun, Y., Denker, J.S., Solla, S.A.: Optimal brain damage. In: Advances in Neural Information Processing Systems, vol. 2, pp. 598–605. D.S. Touretzky, Morgan Kaufmann (1990)
20. Hassibi, B., Stork, D.G.: Second order derivatives for network pruning: optimal brain surgeon. In: Lippman, D.S., Moody, J.E., Touretzky, D.S.: Advances in Neural Information Processing Systems, vol. 5, pp. 164–171. Morgan Kaufmann (1993)
21. Han, S., Pool, J., Tran, J., Dally, W.: Learning both weights and connections for efficient neural network. In: Advances in Neural Information Processing Systems, pp. 1135–1143 (2015)

22. Reed, R.: Pruning algorithms-a survey. IEEE Trans. Neural Netw. **4**(5), 740–747 (1993)
23. Smith, L.N., Topin, N.: Super-convergence: very fast training of neural networks using large learning rates. In: Artificial Intelligence and Machine Learning for Multi-Domain Operations Applications, vol. 11006, p. 1100612. International Society for Optics and Photonics (2019)
24. He, K., Zhang, X., Ren, S., Sun, J.: Deep residual learning for image recognition. In: Proceedings of the IEEE Conference on Computer Vision and Pattern Recognition, pp. 770–778 (2016)
25. Idelbayev, Y.: Proper ResNet implementation for CIFAR10/CIFAR100 in PyTorch. https://github.com/akamaster/pytorch_resnet_cifar10

A Double-Dictionary Approach Learns Component Means and Variances for V1 Encoding

S. Hamid Mousavi[(⊠)], Jakob Drefs, and Jörg Lücke

Machine Learning Lab, Department of Medical Physics and Acoustics,
University of Oldenburg, Oldenburg, Germany
{hamid.mousavi,jakob.drefs,joerg.luecke}@uol.de

Abstract. Sparse coding (SC) is a standard approach to relate neural response properties of primary visual cortex (V1) to the statistical properties of images. SC models the dependency between latent and observed variables using one weight matrix that contains the latents' generative fields (GFs). Here, we present a novel SC model that couples latent and observed variables using two matrices: one matrix for component means and another for component variances. When training on natural image patches, we observe Gabor-like and globular GFs. Additionally, we obtain a second dictionary for each component's variances. The double-dictionary model is thus the first to capture first- and second-order statistics of natural image patches using a multiple-causes latent variable model. If response probabilities of V1 simple cells are not restricted to first order statistics, the investigated model is likely to be more closely aligned with neural responses than standard SCs or independent component analysis (ICA) models.

Keywords: Unsupervised learning · Maximal causes analysis · Non-linear sparse coding · Double dictionary learning · Variational EM

1 Introduction

The two well-known standard models of SC [9] and ICA (e.g., [4]) have been successfully used to link the response properties of simple cells in V1 to the view of sensory systems as optimal information encoders. Since they have been introduced, it has repeatedly been pointed out that generalizations of the original model assumptions are required. Such generalizations include, e.g., extensions of ICA which include an encoding of dependencies between latent activities, addition of intensity variables and latents, hierarchical features and many more (e.g. [5,6,13]). Also the encoding of variances of image components has been

---

S.H. Mousavi and J. Drefs—Authors contributed equally.

© Springer Nature Switzerland AG 2020
G. Nicosia et al. (Eds.): LOD 2020, LNCS 12566, pp. 240–244, 2020.
https://doi.org/10.1007/978-3-030-64580-9_20

observed to be important for image processing but has so far only been realized for mixture models (e.g., [14]).

In previous work [1], we have studied the impact of occlusion like non-linearities on predicted simple cell responses by exploiting a maximum superposition (in contrast to the linear superposition of the standard SCs). Here we step forward and ask if such established non-linear SC model can be generalized by coupling latents to observables using two matrices: one to model the means of the observable distribution and another to model the variances. Although such generalization imposes a few challenges on the SC model, we further show that a novel parameterization of mean and variance components enables us to obtain an efficient and applicable learning algorithm.

2   Sparse Coding with Mean and Variance Dictionaries

We build upon maximal causes analysis (MCA) models [1,7,10] with binary latents and Gaussian noise. Binary latents suggest themselves for probabilistic neural encoding [12], and the maximum non-linearity of MCA can be motivated by statistical properties of stimuli processed by primary visual cortex [1,10] and auditory cortex [11]. Concretely, we propose the following generative model:

$$p(s|\Theta) = \prod_{h=1}^{H} \text{Bernoulli}(s_h; \pi_h) \quad \text{with} \quad s_h \in \{0,1\}, \pi_h \in [0,1], \quad (1)$$

$$p(y|s,\Theta) = \prod_{d=1}^{D} \mathcal{N}(y_d; \bar{W}_d(s,\Theta), \bar{\Sigma}_d(s,\Theta)) \quad \text{with} \quad y_d \in \mathbb{R}. \quad (2)$$

Hidden units s_h are activated with probability π_h and given the latents, observables y_d are assumed to follow a Gaussian with mean and variance parameters given by:

$$\left.\begin{array}{l} \bar{W}_d(s,\Theta) = W_{dh(d,s,\Theta)} \\ \bar{\Sigma}_d(s,\Theta) = \Sigma_{dh(d,s,\Theta)} \end{array}\right\} h(d,s,\Theta) = \text{argmax}_h |s_h W_{dh}|, \quad h = 1, \ldots, H. \quad (3)$$

The weight W_{dh} denotes the mean associated with hidden unit s_h and observable y_d. Component means combine non-linearly ($\bar{W}_d(s,\Theta) = \max_h |s_h W_{dh}|$). Previous work [1,7,10] considered MCA models with a scalar parameter to model component variances ($\Sigma_{dh(d,s,\Theta)} = \sigma^2 \ \forall d = 1, \ldots, D$ and $h = 1, \ldots, H$). Here, we extend previous MCA models by introducing individual component variances which are learned alongside of component means: each hidden unit s_h is associated with a set of weights for component means $\{W_{dh}\}_{d=1,\ldots,D}$ *and* an additional set of weights for component variances $\{\Sigma_{dh}\}_{d=1,\ldots,D}$ (see Fig. 1). Component variances combine non-linearly according to Eq. (3). The proposed double-dictionary MCA model (1)–(3) is parameterized by $\Theta = (\pi, W, \Sigma)$ with $W = \{W_{dh}\}_{h=1,\ldots,H}^{d=1,\ldots,D}$ and $\Sigma = \{\Sigma_{dh}\}_{h=1,\ldots,H}^{d=1,\ldots,D}$ denoting the dictionaries of component means and variances and with $\pi = \{\pi_h\}_{h=1,\ldots,H}$.

Given a set of data points $\{y^{(n)}\}_{n=1,\ldots,N}$, we seek maximum likelihood parameter Θ^* based on an expectation maximization (EM) approach. Crucial for enabling effective and efficient learning are two recent results: first, novel truncated variational approaches [2] for highly efficient variational E-steps and

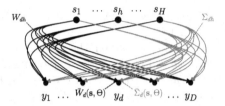

Fig. 1. Structure of the proposed double dictionary model. Hidden units s_h are coupled to observables y_d using weights W_{dh} and Σ_{dh} for component means and component variances, respectively. For a given latent vector $s = (s_1, ..., s_H)$, the value y_d of the observables is drawn from a Gaussian with mean and variance parameters $\bar{W}_d(s, \Theta)$ and $\bar{\Sigma}_d(s, \Theta)$, respectively (see text for details).

second, novel theoretical results to obtain parameter update equations (M-step) that were first more generally derived for the exponential family observables [8]. For the double-dictionary model (1)–(3), derived parameter update equations are given as follows (derivations not shown due to space limitations):

$$
W_{dh}^{new} = \frac{\sum_{n=1}^{N} <\mathcal{A}_{dh}(s, \Theta)>_{q^{(n)}} y_d^{(n)}}{\sum_{n=1}^{N} <\mathcal{A}_{dh}(s, \Theta)>_{q^{(n)}}}, \quad \mathcal{A}_{dh}(s, \Theta) := \begin{cases} 1 & h = h(d, s, \Theta) \\ 0 & \text{otherwise}, \end{cases}
$$

$$
\Sigma_{dh}^{new} = \frac{\sum_{n=1}^{N} <\mathcal{A}_{dh}(s, \Theta)>_{q^{(n)}} \left(y_d^{(n)} - W_{dh}\right)^2}{\sum_{n=1}^{N} <\mathcal{A}_{dh}(s, \Theta)>_{q^{(n)}}}, \quad \pi_h^{new} = \frac{1}{N} \sum_{n=1}^{N} <s_h>_{q^{(n)}}
$$

$$
\tag{4}
$$

where $<g(s)>_{q^{(n)}}$ represents the expected value of an arbitrary function $g(s)$ w.r.t. a variational distribution $q^{(n)}(s)$. To train the model at large scales, we use truncated variational distributions and apply the evolutionary EM (EEM) approach of [2] (also compare references therein for details).

3   Encoding of Natural Image Patches

Using E- and M-steps described above, we fitted the model (1)–(3) to ZCA-whitened image patches from a standard database [3]. After learning, we observed a large variety of generative GFs for the component means (including the familiar Gabor-like and globular fields) as well as a large variety of GFs for component variances (see Fig. 2). If we identify image patches that maximally activate a given latent, we can systematically change these patches while measuring changes of the latent's responses. By adding specific types of changes including addition of specific noise, the response properties of the latent changes depending on how it encodes component variances. For instance, noise added in proximity of a Gabor component or distant from it has a different effect for latents with individual variances compared to latents with the same variance for all pixels and all components (as used for standard SC). Such differences in responses between the models could be used to design stimuli that can detect potential variance encoding in V1 and other sensory areas.

A Means

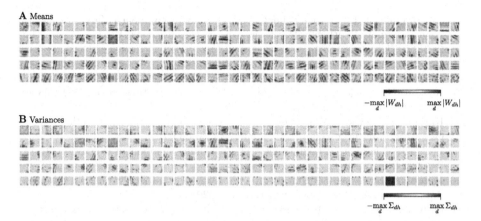

$$-\max_{d}|W_{dh}| \qquad \max_{d}|W_{dh}|$$

B Variances

$$-\max_{d}\Sigma_{dh} \qquad \max_{d}\Sigma_{dh}$$

Fig. 2. Dictionaries with component means and component variances learned from $N = 100,000$ ZCA-whitened image patches ($D = 12 \times 12$ pixels) using the proposed model. In total $H = 1,000$ dictionary elements were learned; we here show every fifth field, picked after sorting all fields according to their activation frequencies (frequencies decrease from left to right and top to bottom). For the experiment, we ran 300 EEM iterations using the "fitparents-randflip" variant of EEM and using $S = 60$ variational states (compare [2]).

4 Conclusion

We presented a double-dictionary SC model capable of learning component variances alongside of component means. For natural image patches, the here observed variety of variance GFs stands in contrast to standard SC which assumes a uniform variance with equal value for all components. A single variance parameter is statistically suboptimal which suggests that primary visual cortex may, likewise, represent component variances as well as component means. Using different alterations of image patches including different noise types, we pointed out how variance encoding can in principle be detected in novel *in vivo* experiments.

Acknowledgments. We acknowledge funding by the German Research Foundation (DFG) in project 390895286 (cluster of excellence H4a 2.0, EXC 2177/1) and the German Ministry of Research and Education (BMBF) project 05M2020 (SPAplus, TP 3).

References

1. Bornschein, J., Henniges, M., Lücke, J.: Are V1 simple cells optimized for visual occlusions? A comparative study. PLoS Comput. Biol. **9**(6), e1003062 (2013)
2. Guiraud, E., Drefs, J., Lücke, J.: Evolutionary expectation maximization. In: GECCO. ACM (2018)

3. van Hateren, J.H., van der Schaaf, A.: Independent component filters of natural images compared with simple cells in primary visual cortex. Proc. R. Soc. Lond. **265**, 359–366 (1998)

4. Hyvärinen, A., Oja, E.: A fast fixed-point algorithm for independent component analysis. Neural Comput. **9**(7), 1483–1492 (1997)

5. Karklin, Y., Lewicki, M.S.: Learning higher-order structures in natural images. Netw.: Comput. Neural Syst. **14**, 483–499 (2003)

6. Karklin, Y., Lewicki, M.S.: Emergence of complex cell properties by learning to generalize in natural scenes. Nature **457**, 83–86 (2009)

7. Lücke, J., Sahani, M.: Maximal causes for non-linear component extraction. JMLR **9**, 1227–1267 (2008)

8. Mousavi, S.H., Drefs, J., Hirschberger, F., Lücke, J.: Maximal causes for exponential family observables. arXiv preprint arXiv:2003.02214 (2020)

9. Olshausen, B.A., Field, D.J.: Sparse coding with an overcomplete basis set: a strategy employed by V1? Vis. Res. **37**, 3311–3325 (1997)

10. Puertas, J., Bornschein, J., Lücke, J.: The maximal causes of natural scenes are edge filters. In: Advances in Neural Information Processing Systems (2010)

11. Sheikh, A.S., et al.: STRFs in primary auditory cortex emerge from masking-based statistics of natural sounds. PLoS Comput. Biol. **15**(1), e1006595 (2019)

12. Shivkumar, S., Lange, R., Chattoraj, A., Haefner, R.: A probabilistic population code based on neural samples. Adv. Neural Inf. Process. Syst. **31**, 7070–7079 (2018)

13. Wainwright, M.J., Simoncelli, E.: Scale mixtures of Gaussians and the statistics of natural images. Adv. Neural Inf. Process. Syst. **12**, 855–861 (2000)

14. Zoran, D., Weiss, Y.: From learning models of natural image patches to whole image restoration. In: ICCV (2011)

A Forecasting Model to Predict the Demand of Roses in an Ecuadorian Small Business Under Uncertain Scenarios

Israel D. Herrera-Granda[1]([✉]) (iD), Leandro L. Lorente-Leyva[1] (iD),
Diego H. Peluffo-Ordóñez[2] (iD), and M. M. E. Alemany[1] (iD)

[1] Universitat Politècnica de València, Camino de Vera S/N 46022, València, Spain
ishergra@doctor.upv.es
[2] Yachay Tech University, Hacienda San José, Urcuquí, Ecuador

Abstract. Ecuador is worldwide considered as one of the main natural flower producers and exporters –being roses the most salient ones. Such a fact has naturally led the emergence of small and medium sized companies devoted to the production of quality roses in the Ecuadorian highlands, which intrinsically entails resource usage optimization. One of the first steps towards optimizing the use of resources is to forecast demand, since it enables a fair perspective of the future, in such a manner that the in-advance raw materials supply can be previewed against eventualities, resources usage can be properly planned, as well as the misuse can be avoided. Within this approach, the problem of forecasting the supply of roses was solved into two phases: the first phase consists of the macro-forecast of the total amount to be exported by the Ecuadorian flower sector by the year 2020, using multi-layer neural networks. In the second phase, the monthly demand for the main rose varieties offered by the study company was micro-forecasted by testing seven models. In addition, a Bayesian network model is designed, which takes into consideration macroeconomic aspects, the level of employability in Ecuador and weather-related aspects. This Bayesian network provided satisfactory results without the need for a large amount of historical data and at a low-computational cost.

Keywords: Bayesian-Networks-based forecasting · Demand forecast · Floriculture sector · Neural-Networks-based forecasting

1 Introduction

The floriculture sector represents one of the main income sources for the Ecuadorian agricultural sector [1]. In addition, given the geographic location and the goodness of the climate and soil, provides among its star products the high quality Ecuadorian rose, which is highly appreciated in countries like the United States, Russia, and countries of the European Union. Therefore, these figure as one of the main non-oil export products in the country, generating the creation of enterprises, employment and dynamization of the economy -especially at the Ecuador northern area.

© Springer Nature Switzerland AG 2020
G. Nicosia et al. (Eds.): LOD 2020, LNCS 12566, pp. 245–258, 2020.
https://doi.org/10.1007/978-3-030-64580-9_21

Nonetheless, companies dedicated to growing and processing quality roses daily face a wide variety of problems. Which in turn generate scenarios of uncertainty such as difficulty in accessing bank credit, the constant growth of inflation, the appearance of new local taxes and tariffs, the increase in the level of unemployment and the constant rotation of personnel during seasons of low demand.

Within this topic, it is also important to mention the difficulty of access to water resources, since the floriculture sector uses water in most of its processes and in the irrigation of its crops, which has often caused problems with the inhabitants and village communities who manage the irrigation canals on a community basis. Moreover, given the scarcity of access to water resources and the variety of climates in times of drought, and with the emergence of new enterprises also dedicated to this activity, the problem of access to water resources increases, and affects negatively manufacturing costs.

Despite the fact that the northern zone of Ecuador has adequate conditions in terms of soil and climate for the cultivation of high quality roses. It is important to mention that the climate is highly variable with respect to other areas of Ecuador. Specifically, two weather stations have been identified to be relatively-colder-temperature ones within the months of January to May, and the second cold season in the months of September to November, which are accompanied by increased rainfall. This variability in climate requires the use of greenhouses that seek to keep the climate of the crops stable [2, 3].

Climate variability and droughts encourage the appearance of pests, which are treated with pesticides. Within this aspect, it is also important to mention the negative impact that flower production has on the properties of the soil, since when it is overexploited it loses its chemical properties and it becomes necessary to use agricultural fertilizers, which in turn increase production costs and are a risk to human health. Therefore, all the above factors negatively affect the manufacturing costs and the quality of the final product.

In addition, we must be take into account the uncertainty of demand in the international target market comprising mainly the United States, Russia, and European Union countries. To aggravate the scenario of uncertainty in the Ecuadorian floricultural sector we can mention the variability in the international sales price, which constantly fluctuates within a constant game between offer and demand.

In an optimistic scenario, all Ecuadorian rose production would be sold and exported immediately after their production. Given the perishable nature of this product. However, unexpected events may occur such as the emergence of new rose producing countries or even pandemics or foreign trade restrictions on flower exports. It is important to note that local consumption of roses is negligible compared to exports, which may be due to the fact that international markets offer better prices for them [1].

Because of the above, there is a need to design a specific demand-forecasting model for the rose sector that considers previously mentioned variables. In addition, this model should be constantly updated to reduce uncertainty in flower production scenarios and make better demand decisions that optimize the entire Agricultural Supply Chain (ASC) and the appropriate use of water and soil for crops.

The rest of the manuscript is structured as follows: Sect. 2 shows a chronological review of the work done on the forecast of agricultural products under uncertainty scenarios. Section 3 establishes the forecasting methods used on the case study. Section 4 shows the results obtained by applying 7 different forecasting methods and a Bayesian network (BN) specially designed for the case study. Finally, the conclusions obtained by applying the different forecasting models to the case study are discussed.

1.1 Bibliographic Review

Forecasting methodologies have had various applications within agricultural sciences, for example, in [4] they developed a model to manage risks in cotton production, taking into account climate uncertainty. Similarly [5] they address the problem of estimating agricultural production costs under the effects of local climatic phenomena that are highly variable such as the "El Niño" phenomenon.

The optimal allocation of land to crops has been the subject of studies, in [6] they have developed a quantitative tool for making optimal decisions. However, for its correct execution, this tool depends on an initial process to forecasting its variables, especially the variable of climate in the crops.

In [7], forecasting tools are used for market analysis and final grain prices in Germany and Poland. In the same year [8], developed a model based on Bayesian networks to represent the irrigation needs of crop farms and their effects on water sources, thus it represents a valuable tool for making future decisions on the optimal use of irrigation water. Within this same in-focus [9] they address the uncertainty of the climate variable and how its short-term forecast can improve climate control in closed and automatically controlled environments.

Subsequently, [10] used forecasting tools and market analysis techniques to assess the availability of feed for poultry and livestock in the United States. As well as analyzing the effects of the income of emerging markets who start selling feed or their raw materials in that country and its effect on costs in agricultural production. Finally, they presented their results to the U.S. Congress with possible solutions to improve the outlook. In the same year, [11] they analyze the economic impacts caused by climate change on agriculture in the state of California in the United States, in which after considering some variables on uncertain scenarios they finally give a possible short-term forecast scenario.

In [12] they discusses the effect of combining a good climate forecast, simulation of future crop yields, and the judgement of local farmers to reduce the uncertainty in the future crop reality. The need to implement a model to make the decisions was eliminated because the farmers were able to make them correctly.

With the above background, the need to develop a medium and short term forecast model that is specific to the sector of quality rose production under scenarios of uncertainty becomes apparent. In this context, the use of models based on Bayesian networks, such as the one proposed by [13] seems to be appropriate. This model also incorporates a process of training and heuristic optimization of its objective indicators. Because of the application of this model, forecasts about the future state of a variable are obtained, assuming a range of probability that the forecast is correct. This type of forecast could benefit from data provided by climate agencies that constantly monitor and forecast the climate, as is the case with the project Improving Predictions Of UK Drought (IMPETUS) [14] which forecasts droughts in the United Kingdom.

Subsequent work also addressed the problem of forecasting the market for agricultural products, for example, [15] used hybrid methods between neural networks and diffuse systems for this purpose, while in [16] analyzed and forecasted feed and milk time series using three deterministic methods. In turn, [17] they predict the performance of pastures by means of machine learning methods, whose historical series were obtained through satellite images. On the other hand, [18] they forecast the energy demand of greenhouses by means of non-linear forecasting methods.

Within this scenario of modelling production systems in an uncertain context, [19] they analyses the impact of climate change on soybean and maize production. In [20] used Bayesian networks to forecast the yield of palm oil crops. Also [21] they use Bayesian networks to forecast relative humidity in cultivated coffee.

In 2019, we can cite the work of [22] who analyses the cereal production sector in Egypt and makes long-term forecasts of cereal production using deterministic methods. On the other hand, [23] they carry out experiments to measure the response of wheat farmers to inaccurate forecasts.

On the other hand, [24] they forecast prices and yields of cereal crops in Tanzania using stochastic and multi-variable methods. In [25] they use stochastic forecasting methods and Bayesian procedures to forecast pesticide concentrations in the crop soil in real time. Using a similar approach, [26] they model a Bayesian network that allows estimate the risks in the agricultural supply chain.

Within this context of uncertainty, it is important to cite the work of [27] who determine and evaluated the factors of uncertainty that affect the performance of an agricultural enterprise. That is, *for production*: agro-technical requirements, scientific results, sustainability of soil fertility; *for trade*: forecasting and scenario analysis and modeling; *for the financial area*: monitoring financial conditions and reserves to compensate for fluctuating market conditions.

As for forecasting methods, there are several classifications, but the one proposed by [28] is interesting, since it establishes two main forecasting methods: *deterministic models* are those in which a historical series and its future behavior can be adequately modeled through mathematical expression, without influencing these methods of unknown factors. On the other hand, there are *probabilistic or stochastic models* in which the future values of a time series only can be expressed as a probability of being correct.

Within this classification, we have included an additional section for *hybrid prediction models*, which would include those models that use both deterministic and probabilistic or stochastic models, as is the case in the field of machine learning. Note that both of the above models could be classified as quantitative. However, we could also add forecasting models that consider qualitative variables such as criteria, observation or empiricism [29].

After reviewing the literature, a lack of specific works for the floricultural sector can be noted. Therefore, within the present proposal a Bayesian (BN) network will be designed that is capable of capturing the most relevant aspects of the problems faced by the floricultural companies in Ecuador, which were described in the introduction section. In addition, the performance of the BN will be compared with other forecasting methods found in the literature, in which methods from the field of machine learning (ML) are also found. As described in the introduction section, most aspects that cause high uncertainty in the demand for roses involve stochastic processes that generate scenarios of future demand uncertainty. Therefore, the present proposed model for forecasting could be classified as a Stochastic Method. However, the proposed BN model is also tested against several deterministic and even hybrid models from the ML field such as neural networks.

This proposal provides a model to forecast the demand of the Ecuadorian floricultural sector, which has gone through a rigorous validation process and can be properly implemented in small and medium sized floricultural enterprises at a low computer cost.

Indeed, one can expect that if a good forecast is available initially, uncertainty in the future demand for products would be reduced, which would translate into producers who would be supplied in advance and in the right quantities. An adequate allocation of fields for the crop in which the soil is not overexploited and the desired rotation in the use of the crops are feasible, and working personnel could be planned, avoiding frequent dismissals and hiring in seasons of high demand, i.e., stabilization of personnel. The effect of a good demand forecast on the other components of the supply chain is also beneficial, as suppliers and their suppliers could be adequately supplied and take advantage of economies of scale. Likewise, distributors could plan the use of their facilities and could even start offering their products in foreign markets by obtaining better negotiations in terms of prices of the final product and taking advantage of foreign trade facilities in terms of the use of Incoterms, or lower cost modes and form of transportation.

Additionally, it is important to mention that although this study uses the forecasting model for Bayesian networks proposed by Scott et al. [13]. That work in its conclusions section promotes future research on a procedure to determine if certain predictors should be included in the model, and requests a research on a methodology to standardize the predictors included in the BN model.

Case Study. The study company is located in the north of Ecuador, specifically in the Pedro Moncayo village; this company consists of a small business dedicated to the production of high quality roses both natural and artificially dyed which are marketed mostly abroad. One of its main objectives is to optimize the management of the post-harvest process through a forecast of future demand, thus balancing its installed supply with his demand.

Currently, the rose market has experienced a wide growth, so there is a significant unsatisfied demand that the company could aim to meet through an accurate forecast.

The external variables (predictors) to be considered in this case are included in two large groups, financial and technical variables.

The financial variables: access to financing, inflation, finance and taxes, unemployment, high market competition, the selling price of the final product, and international demand.

Technical variables: access to water resources and droughts, climate variability, storms and pests, wear and tear on soil properties

2 Materials and Methods

This section describes the forecasting process for the case study, which has been divided into two phases. In the first phase, or macro-forecast: we used historical data of the monthly demand in tons for exported roses from Ecuador to all markets since the year 1990 to 2019. With this data, we forecasted the monthly demand for the year 2020, the idea of this first forecast is to have a general idea of the market for Ecuadorian roses. Which can help the company to determine the top percentage it can enter in the national market with its products, so that it can balance its offer with his demand, at a strategic level. According to the indicator Root of Medium Square Error (RMSE), the Multilayer perceptron (MLP) Neural Network method was the one that provided the best results.

In the second phase or micro-forecast, the quantities demanded of the four main varieties of roses offered by the study company are estimated. To do this, we used order data in monthly units from January 2016 to December 2019, in order to forecast the monthly units demanded for the months of 2020.

2.1   Phase 1: Macro-Forecast of Tons of Ecuadorian Roses to Be Exported

In this phase, we forecasted the tons to be exported from Ecuador to the various markets of the world through for the 12 months of the year 2020 using MLP. The historical database was obtained from the Ecuadorian Rose Growers Association (EXPOFLORES) [1], thus consolidating a database of tons exported from 1990 to 2019. This macro-forecast will serve as a reference framework for knowing the global rose market in which the study company can participate.

2.2   Phase 2: Micro-Forecast of Rose Varieties Offered

In this phase, six forecast models were tested using the RMSE: {HoltWinters (hw), Auto arima (Arima), Exponential smoothing with 2 different optimization criteria (ets1, ets2), Linear regression (tslm), Neural Network (nnetar), Multi Layer Perceptron (MLP)}. Note that the first five models can be classified as deterministic, while nnetar and MLP use neural networks, therefore, they could be considered as hybrid and ML models. The results of the micro-forecast phase are discussed in the results phase.

To complete this second phase, a stochastic model of the Bayesian network (BN) was designed. It considers the historical series for the year 2019 but also considers the macroeconomic aspects and level of employability in Ecuador as well as its climate-related aspects.

Design of a Bayesian Network (BN) for the Micro-Forecast of the Demand.   In order to create a forecast model based on Bayesian networks (BN) which is more accurate for the study company also it can consider more than the historical monthly demand D_{hist}. In addition, it needs to consider other monthly variables that reflect the problems of the Ecuadorian rose market, then we have designed a Bayesian network model (BN) implemented through the package of R bsts [13], which allows us to include additional variables (predictors) to the historical series provided by the company. In this BN model, we used a Markov Chain-Monte Carlo (MCMC) sampling algorithm to simulate the posterior distribution, which smoothed the predictions over a large number of potential models using the average of the Bayesian model [13]. Below in Fig. 1, the architecture of the BN is shown with its predictors that affect the model output.

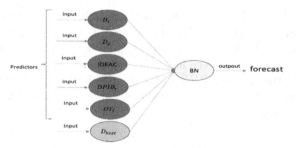

Fig. 1. BN architecture

The Climate Variability (D_c) represents quantitatively the percentage deviation from the average in the climate of the Ecuadorian mountain range. In which there are two markedly colder seasons within the months of January to May and the second cold season in the months of September to November and consequently in those months there was an increase of rainfalls and pests, which in turn, cause the need to use greenhouses. The ideal temperature for a rose to develop and produce normally in a greenhouse is between 18 and 21 °C [30]. Therefore, in the variable D_c the variation in temperature within the glasshouse is represented with respect to the average or ideal temperature, which should be within the aforementioned range, taking into account in this function that the minimum environmental temperature is 7 °C and the maximum is 21 °C [31]. The environmental temperature in the Ecuadorian sierra tends to be colder than inside the greenhouse, that is why in the case of this variable its decrease affects directly proportional in the model, that is to say its decrease affects negatively the crops [2, 3].

The *variable Employment deviation* in Ecuador (D_e) quantitatively represents the percentage deviation from the average in the gross employment positions available in Ecuador. Therefore, its increase positively affects the result of the model [32].

The *Index of Economic Activity* (IDEAC) describes the country's economic reality in terms of the amount of production, so its increase has a positive effect on the model's results [33]. The IDEAC measures the percentage deviations in its values with respect to its average. Therefore, it is not necessary to normalize this variable.

Likewise, the variable *Deviation of the Ecuadorian rose sector's Gross Domestic Product* ($DPIB_r$) measures the percentage deviation from the average contributed by the rose sector to Ecuador's gross domestic product, its increase positively affecting the model [1].

The *inflation rate variable* (DT_i) reflects the increase in the prices of goods and services in Ecuador; therefore, its increase negatively affects the results of the model [33].

Process for Normalization of Predictors for the BN. The aforementioned BN were used to forecast the monthly production quantities of the different rose varieties for the year 2020. In addition, it is important to mention that the scales used in each of these variables directly affect the outcome of the model, which is why for the present case all the predictors have been normalized to keep amplitude at the same scale of a fractional number between 0 and 100.

However, there are also variables that indirectly affect the model proportionally, so fractional values between 0 and −100 are included. That is, if a predictor increases in magnitude it similarly affected to the estimation of the predicted quantity in the BN, and in another case, it is said to be inversely proportional. For this reason, the inverse proportional variables affect the BN model in the opposite way and we decided to change its symbol to treat all the variables in a standard way. As shown in Fig. 2.

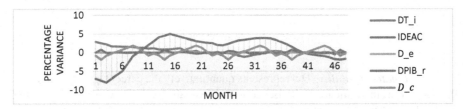

Fig. 2. BN normalized predictors

The formula used to normalize predictor values is described below. If x is a variable to be used as a predictor, then x_i will be its observations taken m times corresponding to the historical series D_{hist}

$$D(x_i) = \left[\frac{x_i - \bar{x}}{\bar{x}}\right] * 100, \ \forall \ i = 1, 2, \ldots m \tag{1}$$

3   Results

This section shows the results of the application of the forecasting models cited in the previous section, divided into two phases: the first corresponds to the macro-forecast of the demand for the entire Ecuadorian rose market for the year 2020 in tons. While the second phase corresponds to the micro-forecast of the four varieties of roses offered by the study company.

3.1   Results of the Macro-Forecast (Phase 1)

From the historical data of monthly orders of roses in tons from 1990 to 2019. We want to forecast the demand for the year 2020. For this purpose, a neuronal network type MLP has been used through Rstudio's NNFOR package, since the latter has demonstrated in many similar cases its efficiency at a low computational cost, capturing trends, seasonality, and other shorter patterns within the demand by means of its Iterative Neural Filter (INF)) [34, 35]. In this case, it was possible to reach the optimal configuration of the network using the pyramid rule and the optimal forecast result was reached in the third iteration of the re-training process.

The best result, specifically an RMSE of 1.364 was obtained with a manual parameter selection using the pyramid rule to determine the number of nodes in its three hidden layers: (22, 18, 15) and 40 repetitions and an automatic selection of frequency delays. After several experiments, we noted that the retraining processes do not substantially improve the MSE of the MLP model. However, the computational cost is high for each retraining, so the following shows the result of the optimal MLP network and the forecast of rose exports for the year 2020. As shown in Fig. 3.

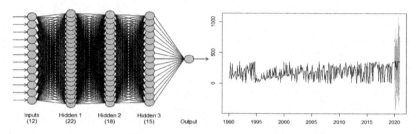

Fig. 3. Macro-forecast with MLP

3.2   Forecast of the Demand for the Different Varieties of Roses (Phase 2)

The above macro-forecast gave us a global picture of the rose market in 2020. Now we will work on forecasting the demand for roses for 2020 for each of the varieties of roses offered by the case study company. To do this, we used the historical monthly demands of the last three years for the four varieties of roses offered by the company, and then define a production plan to balance offer with demand in the company. The varieties of roses offered are FREEDOM, MONDIAL, PINK FLOYD and EXPLORER.

By means of the TSstudio package of the statistical language R, the NNFOR package, and the MSE indicator. It was possible to test simultaneously 7 forecast models: HoltWinters stats package (hw). Auto arima forecast package with autoregressive moving average model (Arima) [34]. Exponential smoothing state space model with error, trend and Seasonal, and 2 different optimization criteria (ets1, ets2). Neural Network time series forecast package(nnetar). Linear regression forecasting model using the tslm function (tslm). And Multi Layer Perceptron (MLP) without enhancement process, with only 1 hidden layer and without automatic delay selection [35]. A summary of the results obtained with the above-mentioned forecasting methods are shown in Fig. 4.

Forecasting Results with the BN Designed for the Case Study. A detail of the cumulative absolute errors (CAE) for each variable of each of the predictors and all of them together is shown in the left section of Fig. 5. It is possible to note that in all cases the CAE decreases by applying all the predictors simultaneously in comparison with incorporating these predictors individually. In this work, a code was developed that allows to include one by one of the predictors and finally, through the command compare.bsts [13], all the models have been graphed simultaneously. In addition, in the right section of the figure, the historical series and its forecast for the year 2020 have been included in the four types of roses offered. It can be noted that the historical demands have complex and stochastic structures.

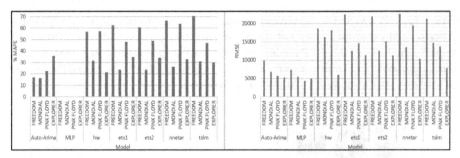

Fig. 4. Error comparison of the applied forecast models

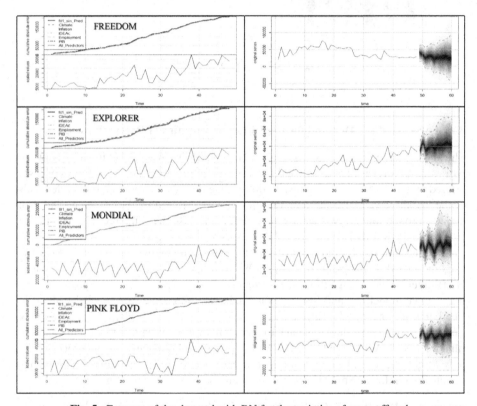

Fig. 5. Forecast of the demand with BN for the varieties of roses offered.

It is important to mention that in the BN forecast for the four varieties of roses offered by the company. We used as a measure of the BN error, the standard deviation of the one step forward errors in the prediction (Pre-diction.sd) for the training data. Although this indicator works in a similar way to the RMSE it is not comparable with the latter since its scales differ by the amount of MCMC iterations (burn) used in the BN. As shown in Table 1.

Table 1. Quantified errors of the proposed BN

Method	Varieties	Prediction.sd
BN	FREEDOM	12464,3400
	MONDIAL	7420,5040
	PINK FLOYD	5912,4030
	EXPLORER	5520,2260

4 Conclusions

This paper proposes a model for forecasting the demand of quality Ecuadorian roses based on Bayesian networks. Which considers five exogenous variables called predictors, which include percentage deviations from climate, employment, and other macroeconomic variables such as IDEAC, inflation and gross domestic product of the rose sector. These predictors influence positively or negatively on the results of the forecast, so it is important to establish a standardization process that detects in equal conditions the variations of each predictor in the analyzed time interval, so that finally these patterns can be projected into the future. The present work is an easy implementation with low cost for the small agricultural companies and that could be constantly refined. The aspect of the incorporation of new predictors is of importance for agricultural technical professionals that could suggest new agricultural variables to incorporate and even to incorporate quantitatively their technical judgment.

The design of the mentioned BN with the five predictors and the historical series of monthly data of the last four years showed acceptable values in the Standard Deviation of the Errors of a step forward in the prediction (Pre-diction.sd). However, it was not possible to compare this indicator against other forecasting methods reviewed in the literature. Since due to the Bayesian nature of the BN its error indicators are measured within probability intervals and also because its scales differ by the amount of MCMC iterations (burn) used in the BN. Specifically in traditional forecasting models indicators such as MSE, RMSE, MAPE, among others are used. While in the BN the CAE is used measuring the error in each iteration, and on average in the BN 200 to 500 iterations are used in the experiments carried out, taking advantage of the computational performance of the MCMC algorithm. For this reason, although the RMSE indicator and the CAE used in the BN measure the difference between the estimation model with respect to the historical data series, in the second indicator the error is measured many more times, which is why it would be incorrect to compare the RMSE and CAE values.

For the above reason it is recommended as future work to create new error indicators in the forecast model to allow a comparison between deterministic forecasting methods and stochastic models such as BN.

In the present work, only historical data and predictors corresponding to the months from January 2016 to December 2019 are used. However, in future works, we are aimed to include in the BN model a new variable that reflects the global risk of pandemics that affect negatively to the production of roses. As the case, we are currently facing with the

COVID-19 pandemic, which has not been considered in the present work and which, if it had been considered, would substantially improve the forecast for the 2020 year.

Acknowledgments. Authors of this publication acknowledge the contribution of the Project 691249, RUC-APS "Enhancing and implementing Knowledge based ICT solutions within high Risk and Uncertain Conditions for Agriculture Production Systems" (www.ruc-aps.eu), funded by the European Union under their funding scheme H2020-MSCA-RISE-2015. In addition, the authors are greatly grateful by the support given by the SDAS Research Group (www.sdas-group. com).

Annexes

A link has been included to access the historical series and the standardized predictors where each row has a monthly value. Therefore, the readers can replicate the experiment: https://utneduec-my.sharepoint.com/:f:/g/personal/idherrera_utn_edu_ec/Et2foq HulEZJpsnR8xVVxRMBZ5bGtgLvUq7dMNkmo1KyuA?e=En1Ia1.

References

1. Asociación de Productores y Exportadores de Flores: Inicio – Expoflores. https://expoflores. com/
2. Palacios, J., Rosero, D.: Análisis de las condiciones climáticas registradas en el Ecuador continental en el año 2013 y su impacto en el sector agrícola. Estud. e Investig. meteorológicas. Ina. Inst. Nac. Meteorol. e Hidrol. Ecuador, 28, p. (2014)
3. Hidalgo-Proaño, M.: Variabilidad climática interanual sobre el Ecuador asociada a ENOS. CienciAmérica **6**, 42–47 (2017)
4. Ritchie, J.W., Abawi, G.Y., Dutta, S.C., Harris, T.R., Bange, M.: Risk management strategies using seasonal climate forecasting in irrigated cotton production: a tale of stochastic dominance. Aust. J. Agric. Resour. Econ. **48**, 65–93 (2004). https://doi.org/10.1111/j.1467-8489. 2004.t01-1-00230.x
5. Letson, D., Podesta, G.P., Messina, C.D., Ferreyra, R.A.: The uncertain value of perfect ENSO phase forecasts: Stochastic agricultural prices and intra-phase climatic variations. Clim. Change **69**, 163–196 (2005). https://doi.org/10.1007/s10584-005-1814-9
6. Weber, E.U., Laciana, C., Bert, F., Letson, D.: Agricultural decision making in the argentine Pampas: Modeling the interaction between uncertain and complex environments and heterogeneous and complex decision makers (2008)
7. Loy, J.-P., Pieniadz, A.: Optimal grain marketing revisited a german and polish perspective. Outlook Agric. **38**, 47–54 (2009). https://doi.org/10.5367/000000009787762761
8. Wang, Q.J., Robertson, D.E., Haines, C.L.: A Bayesian network approach to knowledge integration and representation of farm irrigation: 1. Model development. WATER Resour. Res. 45 (2009). https://doi.org/10.1029/2006wr005419
9. Keesman, K.J., Doeswijk, T.: uncertainty analysis of weather controlled systems (2010). https://www.scopus.com/inward/record.uri?eid=2-s2.0-79960073961&doi=10.1007% 2F978-3-642-03735-1_12&partnerID=40&md5=210525584472097e996a9f124f96fddb
10. Schnepf, R.: U.S. livestock and poultry feed use and availability: background and emerging issues. In: Feed Market Dynamics and U.S. Livestock Implications. pp. 1–36. Nova Science Publishers, Inc., CRS, United States (2012)

11. Medellín-Azuara, J., Howitt, R.E., MacEwan, D.J., Lund, J.R.: Economic impacts of climate-related changes to California agriculture. Clim. Change **109**, 387–405 (2011). https://doi.org/10.1007/s10584-011-0314-3
12. McCown, R.L., Carberry, P.S., Dalgliesh, N.P., Foale, M.A., Hochman, Z.: Farmers use intuition to reinvent analytic decision support for managing seasonal climatic variability. Agric. Syst. **106**, 33–45 (2012). https://doi.org/10.1016/j.agsy.2011.10.005
13. Scott, S.L., Varian, H.R.: Predicting the present with bayesian structural time series. Available SSRN 2304426 (2013)
14. Prudhomme, C., Shaffrey, L., Woollings, T., Jackson, C., Fowler, H., Anderson, B.: IMPE-TUS: Improving predictions of drought for user decision-making. International Conference on Drought: Research and Science-Policy Interfacing, 2015. pp. 273–278. CRC Press/Balkema, Centre for Ecology and Hydrology, Wallingford, Oxfordshire, United Kingdom (2015)
15. Wiles, P., Enke, D.: A hybrid neuro-fuzzy model to forecast the Soybean complex. International Annual Conference of the American Society for Engineering Management 2015, ASEM 2015. pp. 1–5. American Society for Engineering Management, Missouri University of Science and Technology, Engineering Management and Systems Engineering Department, United States (2015)
16. Hansen, B.G., Li, Y.: An analysis of past world market prices of feed and milk and predictions for the future. Agribusiness **33**, 175–193 (2017). https://doi.org/10.1002/agr.21474
17. Johnson, M.D., Hsieh, W.W., Cannon, A.J., Davidson, A., Bedard, F.: Crop yield forecasting on the Canadian Prairies by remotely sensed vegetation indices and machine learning methods. Agric. For. Meteorol. **218**, 74–84 (2016). https://doi.org/10.1016/j.agrformet.2015.11.003
18. Chen, J., Yang, J., Zhao, J., Xu, F., Shen, Z., Zhang, L.: Energy demand forecasting of the greenhouses using nonlinear models based on model optimized prediction method. Neurocomputing **174**, 1087–1100 (2016). https://doi.org/10.1016/j.neucom.2015.09.105
19. Fodor, N., et al.: Integrating plant science and crop modeling: assessment of the impact of climate change on soybean and maize production. Plant Cell Physiol. **58**, 1833–1847 (2017). https://doi.org/10.1093/pcp/pcx141
20. Chapman, R., et al.: Using Bayesian networks to predict future yield functions with data from commercial oil palm plantations: a proof of concept analysis. Comput. Electron. Agric. **151**, 338–348 (2018). https://doi.org/10.1016/j.compag.2018.06.006
21. Lara-Estrada, L., Rasche, L., Sucar, L.E., Schneider, U.A.: Inferring Missing Climate Data for Agricultural Planning Using Bayesian Networks. LAND. 7 (2018). https://doi.org/10.3390/land7010004
22. Abdelaal, H.S.A., Thilmany, D.: Grains production prospects and long run food security in Egypt. Sustain. 11 (2019). https://doi.org/10.3390/su11164457
23. Kusunose, Y., Ma, L., Van Sanford, D.: User responses to imperfect forecasts: findings from an experiment with Kentucky wheat farmers. Weather. Clim. Soc. **11**, 791–808 (2019). https://doi.org/10.1175/wcas-d-18-0135.1
24. Kadigi, I.L., et al.: Forecasting yields, prices and net returns for main cereal crops in Tanzania as probability distributions: a multivariate empirical (MVE) approach. Agric. Syst. **180** (2020). https://doi.org/10.1016/j.agsy.2019.102693
25. McGrath, G., Rao, P.S.C., Mellander, P.-E., Kennedy, I., Rose, M., van Zwieten, L.: Real-time forecasting of pesticide concentrations in soil. Sci. Total Environ. **663**, 709–717 (2019). https://doi.org/10.1016/j.scitotenv.2019.01.401
26. Yang, B., Xie, L.: Bayesian network modelling for "direct farm" mode based agricultural supply chain risk. Ekoloji **28**, 2361–2368 (2019)

27. Zaporozhtseva, L.A., Sabetova, T. V, Yu Fedulova, I.: Assessment of the uncertainty factors in computer modelling of an agricultural company operation. International Conference on Information Technologies in Business and Industries, ITBI 2019. Institute of Physics Publishing, Voronezh State Agrarian University, Michurina Str. 30, Voronezh, 394087, Russian Federation (2019)

28. Box, G.E.P., Jenkins, G.M., Reinsel, G.C., Ljung, G.M.: Time series analysis: forecasting and control. Wiley (2015)

29. Hanke, J., Wichern, D.: Business forecast. Pearson Educación (2010)

30. Novagric: Invernaderos para Cultivo de Rosas. https://www.novagric.com/es/invernaderos-rosas

31. Weather Spark: Clima promedio en Quito, Ecuador, durante todo el año - Weather Spark. https://es.weatherspark.com/y/20030/Clima-promedio-en-Quito-Ecuador-durante-todo-el-año

32. Instituto Nacional de Estadísticas y Censos-INEC: Encuesta Nacional de Empleo, Desempleo y subempleo-ENEMDU. https://www.ecuadorencifras.gob.ec/empleo-diciembre-2019/

33. Central Bank of Ecuador: Central Bank of Ecuador. www.bce.fin.ec

34. Hyndman, R., Athnasopoulos, G.: Forecasting: Principles and Practice. OTexts, Australia (2018)

35. Herrera-Granda, I.D., et al.: Artificial neural networks for bottled water demand forecasting: a small business case study. In: Rojas, I., Joya, G.C.A. (eds.) International Work-Conference on Artificial Neural Networks, pp. 362–373. Springer, Canaria (2019)

Steplength and Mini-batch Size Selection in Stochastic Gradient Methods

Giorgia Franchini[1,2](✉) [iD], Valeria Ruggiero[3] [iD], and Luca Zanni[1] [iD]

[1] Department of Physics, Informatics and Mathematics,
UNIMORE, Modena, Italy
{giorgia.franchini,luca.zanni}@unimore.it
[2] Department of Mathematics and Computer Science,
UNIFE, Brussels, Belgium
[3] Department of Mathematics and Computer Science,
University of Ferrara, Brussels, Belgium
valeria.ruggiero@unife.it

Abstract. The steplength selection is a crucial issue for the effectiveness of the stochastic gradient methods for large-scale optimization problems arising in machine learning. In a recent paper, Bollapragada et al. [1] propose to include an adaptive subsampling strategy into a stochastic gradient scheme. We propose to combine this approach with a selection rule for the steplength, borrowed from the full-gradient scheme known as Limited Memory Steepest Descent (LMSD) method [4] and suitably tailored to the stochastic framework. This strategy, based on the Ritz-like values of a suitable matrix, enables to give a local estimate of the local Lipschitz constant of the gradient of the objective function, without introducing line-search techniques, while the possible increase of the subsample size used to compute the stochastic gradient enables to control the variance of this direction. An extensive numerical experimentation for convex and non-convex loss functions highlights that the new rule makes the tuning of the parameters less expensive than the selection of a suitable constant steplength in standard and mini-batch stochastic gradient methods. The proposed procedure has also been compared with the Momentum and ADAM methods.

Keywords: Stochastic gradient methods · Learning rate selection rule · Adaptive subsampling strategies · Reduction variance techniques

1 Introduction

The design of numerical optimization methods to identify the parameters of a system able to make decisions based on yet unseen data is a crucial topic in machine learning. Indeed, on the basis of currently available data or examples, these parameters are chosen to be optimal for a loss function [2], measuring some *cost* associated with the prediction of an event. The problem we consider is the unconstrained minimization of the sum of cost functions depending on a

© Springer Nature Switzerland AG 2020
G. Nicosia et al. (Eds.): LOD 2020, LNCS 12566, pp. 259–263, 2020.
https://doi.org/10.1007/978-3-030-64580-9_22

finite training set. For very large training set, the computation of the gradient of this sum is too expansive and it results unsuitable for an online training with a growing amount of samples. Stochastic Gradient (SG) method and its variants, as Momentum and ADAM [6], requiring only the gradient of one or few terms at each iteration, have been chosen as the main approaches for addressing this kind of problem. In this work, we propose a rule to select the steplength (learning rate) in SG approaches and evaluate its performance by numerical experimentation.

2   The Method: Steplength Selection via Ritz and Harmonic Ritz Values

Among the state-of-the-art steplength selection strategies for deterministic gradient methods, the LMSD rule proposed in [4] is one of the most effective ideas for capturing second-order information on the objective function.The idea proposed in this work is to adapt LMSD to the stochastic case by combining it with a technique for increasing the mini-batch size inspired by that presented in [1].

The Deterministic Framework. The LMSD rule provides the steplengths for performing groups of $m \geq 1$ iterations, where m is a small number (generally not larger than 7). After each group of m iterations, called sweep, a symmetric tridiagonal $m \times m$ matrix is defined by exploiting the gradients computed within the sweep. The m eigenvalues of the tridiagonal matrix, called λ_i, are interpreted as approximations of the eigenvalues of the Hessian of the objective function at the current iteration and their inverses define the m steplengths for the new sweep. The crucial point of this approach consists in building the tridiagonal matrix in an inexpensive way, starting from the information acquired in the last sweep. The strategy suggested by the LMSD method is suitable for the minimization of nonlinear and non-convex objective functions [4,7]. A variant of LMSD is based on the use of the harmonic Ritz values (see [3] for details).

Stochastic Framework. The main difference with respect to the deterministic case consists in building the tridiagonal matrix from a matrix whose columns are stochastic gradients instead of full gradients. Its eigenvalues are named Ritz-like values. With a similar strategy, we can also obtain the harmonic Ritz-like values. The reciprocals of the Ritz-like values (or the harmonic Ritz-like values), bounded within a prefixed interval $(\alpha_{min}, \alpha_{max}]$, are used as steplengths in the iterates of the next sweep. Furthermore, based on the analysis developed in [7], we adopt an alternation of the Ritz-like and harmonic Ritz-likes values. A first approach alternates the two steplength rules at each sweep (Alternate Ritz-like values or **A-R**); a second strategy consists in replacing the Ritz-like values with the harmonic Ritz-like values when the size of the current subsample is increased (Adaptive Alternation of Ritz-like values or **AA-R**) [5]. In both cases, the mini-batch size increasing strategy is based on the augmented inner product test proposed in [1] to control the variance of the descent direction.

3   Numerical Experiments

We consider the optimization problems arising in training binary and multi-labels classifiers on the well known data-set *MNIST* of size $n = 60000$. We consider two kinds of problems, the first relating to convex objective functions and the second involving a Convolutional Neural Network (CNN).

Convex Problems. We built linear classifiers corresponding to three different convex loss functions: Logistic Regression (LR), Square Loss (SL) and Smooth Hinge Loss (SH). We compare the effectiveness of the following schemes: 1) SG, Momentum and ADAM [2] with a fixed mini-batch size and a fixed steplength; 2) **A-R** and **AA-R** methods with the proposed adaptive selection of a suitable steplength. These numerical experiments are carried out in Matlab® on 1.6 GHz Intel Core i5 processor. We decided to conduct the experiments on a maximum of 10 epochs because this is sufficient to reach an accuracy comparable with other popular learning methodology on the MNIST data-set. The Fig. 1 shows the behaviour of the objective function optimality gap, in the case of LR with ADAM method and SH with Momentum method. In particular the dashed black line refers to **ADAM-mini** and **MOMENTUM-mini** (with best-tuned steplength), whereas the red and the blue lines are related to **A-R** and **AA-R** versions of the two methods respectively. These results, together with further numerical experiments, highlight that the results obtained with the **A-R** and **AA-R** methods are comparable with the ones obtained with the mini-batch standard versions equipped with the best-tuned steplength. Both **A-R** and **AA-R** methods have been tested on 4 different settings of $(\alpha_{min}, \alpha_{max}]$: 1) $(10^{-5}, 0.5]$, 2) $(10^{-6}, 0.5]$, 3) $(10^{-5}, 1]$, 4) $(10^{-6}, 1]$. It is important to remark that the adaptive steplength rules in **A-R** and **AA-R** methods seem to be slightly dependent on the values of α_{max} and α_{min}, making the choice of a suitable learning rate a less difficult task with respect to the selection of a good constant value in standard methods. For reasons of brevity we cannot report the final accuracies obtained by the methods on the testing set, but they differ at most to the third decimal digit. Finally, in Fig. 2, we compare the behaviour of **SG-mini** and **A-R** and **AA-R** methods with a steplength different from the best-tuned value. Figure 2 highlights that a too small fixed steplength in **SG-mini** produces a slow descent of the optimality gap; on the other hand, a steplength value larger than the best-tuned one can cause oscillating behavior of the optimality gap and, sometimes, it does not guarantee the convergence of **SG-mini** method. As regard **A-R** and **AA-R** methods, they appear less dependent on an optimal setting of the parameters and enable us to obtain smaller optimality gap values after the same number of epochs exploited by **SG-mini**. Furthermore, we observe that in **A-R** method, the optimality gap shows more stability with respect to the steplength bounds than **AA-R**. Nevertheless, **AA-R** method can produce a smaller optimality gap at the end of 10 epochs.

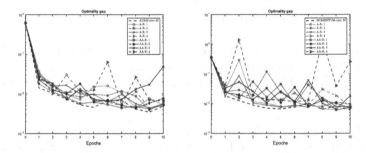

Fig. 1. Optimality gap in 10 epochs for **A-R** and **AA-R** methods: on the left panel, comparison with **ADAM-mini** on the LR loss function and, on the right panel, comparison with **MOMENTUM-mini** on the SH loss function. (Color figure online)

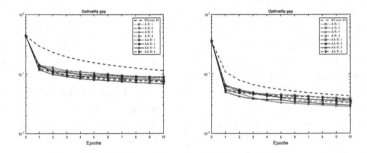

Fig. 2. Comparison of **A-R** and **AA-R** with **SG-mini** equipped with a steplength smaller than the best-tuned value; on the left panel, LR loss function and, on the right panel, SL loss function.

A Non-Convex Problem. The optimization problem arising in training a multi-class classifier for the MNIST data-set can be formulated as the minimization of a non-convex loss function obtained by a Convolutional Neural Network (CNN). We examine the effectiveness of **SG-mini** with fixed steplength and the corresponding **A-R** and **AA-R** versions. These numerical experiments are carried out in Matlab® on Intel(R) Core(TM) i7-6700 CPU @ 3.40 GHz with 8 CPUs. For the sake of brevity, let us mention just a few of the results after 5 epochs. We observe that the choice of the steplength is crucial for **SG-mini**. Indeed, after several trials, very expensive from the computational point of view, we found that the steplength 0.5 provides the highest accuracy, about equal to 0.97. With a smaller steplength we obtain very low accuracy. On the other hand, **A-R** and **AA-R** methods are slightly dependent on the interval $(\alpha_{min}, \alpha_{max}]$, with an accuracy that does not drop below 0.9. This makes the new methods more robust and less expensive in the steplength setting phase with respect to standard SG methods.

Conclusions. In this work the numerical experimentation highlighted that the proposed steplength selection rules enable to obtain an accuracy similar to the one obtained with SG, Momentum and ADAM mini-batch with fixed best-tuned steplength. Although also the new rules require to carefully select a thresholding range for the steplength, the proposed approach appears slightly dependent on the bounds imposed on the steplength, making the parameters setting less expensive with respect to the SG framework. In conclusion, the proposed technique provides a guidance on the learning rate selection and it allows to perform similarly to the SG approaches equipped with the best-tuned steplength.

References

1. Bollapragada, R., Byrd, R., Nocedal, J.: Adaptive sampling strategies for stochastic optimization. SIAM J. Optimiz. **28**(4), 3312–3343 (2018)
2. Bottou, L., Curtis, F.E., Nocedal, J.: Optimization methods for large-scale machine learning. SIAM Rev. **60**(2), 223–311 (2018)
3. Curtis, F.E., Guo, W.: Handling nonpositive curvature in a limited memory steepest descent method. IMA J. Numer. Anal. **36**(2), 717–742 (2016). https://doi.org/10.1093/imanum/drv034
4. Fletcher, R.: A limited memory steepest descent method. Math. Program. Ser. A **135**, 413–436 (2012). https://doi.org/10.1007/s10107-011-0479-6
5. Franchini, G., Ruggiero, V., Zanni, L.: On the steplength selection in stochastic gradient methods. In: Sergeyev, Y.D., Kvasov, D.E. (eds.) NUMTA 2019. LNCS, vol. 11973, pp. 186–197. Springer, Cham (2020). https://doi.org/10.1007/978-3-030-39081-5_17
6. Kingma, D.P., Ba, J.: Adam: a method for stochastic optimization (2014)
7. di Serafino, D., Ruggiero, V., Toraldo, G., Zanni, L.: On the steplength selection in gradient methods for unconstrained optimization. Appl. Math. Comput. **318**, 176–195 (2018)

Black Box Algorithm Selection
by Convolutional Neural Network

Yaodong He and Shiu Yin Yuen[(✉)]

City University of Hong Kong, Hong Kong, China
yaodonghe2-c@my.cityu.edu.hk, kelviny.ee@cityu.edu.hk

Abstract. The no free lunch theorems inform us that no algorithm can beat others on all types of problems. Different types of optimization problems need different optimization algorithms. To deal with this issue, researchers propose algorithm selection to suggest the best optimization algorithm from the algorithm set for a given unknown optimization problem. Deep learning, which has been shown to perform well on various classification and regression tasks, is applied to the algorithm selection problem in this paper. Our deep learning architecture is based on convolutional neural network and follows the main architecture of visual geometry group. This architecture has been applied to many different types of 2-D data. Moreover, we also propose a novel method to extract landscape information from the optimization problems and save the information as 2-D images.

Keywords: Black box optimization · Algorithm selection · Optimization problems · Deep learning · Convolutional neural network

1 Introduction

Many engineering problems can be treated as black box optimization problems [20,29]. The algebraic expression and gradient information of this type of problems are unknown. Thus, traditional numerical methods are not suitable for them. According to the number of objectives, black box optimization problems can be classified as single-objective and multi-objective problems. According to the type of variables, they can be classified as continuous and discrete problems. In this paper, without loss of generality, we only focus on continuous single-objective optimization problems.

Although a large number of optimization algorithms such as evolutionary algorithms have been invented, the no free lunch (NFL) theorems inform us that no algorithm can beat others on all optimization problems [25]. Thus, we still need different algorithms for different types of problems. To deal with this issue, the algorithm selection problem is proposed. Given an unknown optimization problem, the goal of algorithm selection is to predict the most suitable algorithm from a pre-defined algorithm set. Usually, the predictor is a well-trained classification model or a regression model. Using a classification model or a

© Springer Nature Switzerland AG 2020
G. Nicosia et al. (Eds.): LOD 2020, LNCS 12566, pp. 264–280, 2020.
https://doi.org/10.1007/978-3-030-64580-9_23

regression model depends on whether one treats the algorithm selection problem as a classification task or a regression task. In the paper, we consider it as a classification task.

Although many researches have been done on algorithm selection for constraint satisfaction problems, black box algorithm selection has attracted little attention until the past decade [9]. Most of the researches focus on defining suitable features or using reliable machine learning models. In recent years, deep neural network, also called deep learning [12], and its variants have won many competitions in various fields including computer vision, natural language processing, speech recognition, etc. However, only a few works focus on applying deep learning models to the algorithm selection problem. To fill the gap, we propose a novel approach to extract landscape information from optimization problems, and use a convolutional neural network (CNN), which is a variant of deep neural network, to understand the information. In this paper, we do not use any existing human-defined landscape features.

This paper has three main contributions: 1) We propose a novel method to extract landscape information from optimization problems and save the information as 2-D images. One can use this method to extract landscape information from optimization problems for their own deep learning architecture. 2) We apply a well-known structure, which is called Visual Geometry Group (VGG) [21] in the computer vision field, to algorithm selection. 3) We build an overall framework applying deep learning to algorithm selection problem and show its effectiveness.

The rest of this paper is organized as follows: We review and summarize related works in Sect. 2. The framework to extract landscape information and details of the applied convolutional neural network are introduced in Sect. 3. In Sect. 4, we conduct three experiments to investigate the performance of our approach. In the last section, we conclude this paper and discuss the future works.

2 Related Works

2.1 Portfolio Approaches and Ensemble Approaches

Besides algorithm selection approaches, combining and running multiple optimization algorithms together is another way to solve the issue mentioned in NFL theorems. This type of approaches are called portfolio approaches or ensemble approaches. So far, many portfolio approaches have been proposed. Peng et al. [19] propose a population based portfolio approach. In their approach, two parameters l and s are pre-defined. For each l generations, the worst s candidates generated by each algorithm are selected and replaced by the best s candidates generated by other algorithms. The budget is equally shared by the algorithms from the algorithm set. Vrugt et al. [24] propose an approach called multi-algorithm genetically adaptive method. At the end of each generation, the number of offsprings generated by each algorithm is modified according to its previous performance. Yuen et al. [28] propose a portfolio approach called MultiEA. At the end of each iteration, this approach will predict the fitness of each algorithm at the nearest common future point. The algorithm that has the best predicted fitness will be run for the next iteration.

Different from the approaches mentioned in the last paragraph, there is another type of approaches that combine the same algorithm with different parameter settings. Zhao et al. [30] propose a decomposition-based multiobjective evolutionary algorithm which uses ensemble of neighbourhood sizes. Wu et al. [27] combine multiple mutation strategies to improve the performance of differential evolution variants. Three different mutation strategies are applied. A detail survey paper about portfolio and ensemble approaches can be found in [26].

2.2   Exploratory Landscape Analysis Features and Algorithm Selection Approaches

In the past two decades, most researchers focus on solving the algorithm selection problem by conventional machine learning models. These models are not as complicated as deep neural network but require well-defined features, or else their performance will not be gratifying. These human-defined features for algorithm selection are called exploratory landscape analysis (ELA) features [18]. Now, a large number of ELA features have been proposed. Some examples are fitness distance correlation, probability of convexity, dispersion, entropy, etc. These features can describe landscape characteristics such as separability, smoothness, basins of attractions, ruggedness, etc. Kerschke et al. [8] release a user platform to calculate these ELA features. Researchers can apply these features to different machine learning models. Based on their own platform, Kerschke et al. [10] use various machine learning models including support vector machine, regression tree, random forest, extreme gradient boosting and multivariate adaptive regression splines for classification task, regression task, and pairwise regression task. Bischl et al. [1] treat algorithm selection as a regression task and propose a cost sensitive learning method to predict the performance of optimization algorithms on the black-box optimization benchmarking (BBOB) functions [3]. Muñoz et al. [16] propose an information content-based method for algorithm selection problem. The classification accuracy of their method which classifies the BBOB function into five groups is higher than 90%. In [17], Muñoz et al. use footprints to analyze the performance of algorithms on the BBOB functions. They cluster combinations of algorithms and problems into different regions. These regions can depict which algorithm is statistically better than others on which problem. The result can be used to motivate researchers to design new algorithms for unexplored regions. He et al. [6] propose a sequential approach. The approach incorporates a restart mechanism. At the beginning of each restart, a set of algorithm based ELA features [5] are extracted and used as the input to prediction models. The models will suggest an optimization algorithm for the next run. The optimization will not stop until the maximal number of evaluations are reached. A recently published survey paper about algorithm selection approaches can be found in [9].

2.3 Deep Learning and Its Application to Algorithm Selection Problem

Deep neural network, also called deep learning, has shown great capability of solving classification and regression tasks in various engineering fields. Compared with conventional machine learning models, it requires less human-defined features but many training instances. This method is good at learning from hidden information. However, only a few researches focus on applying deep learning to algorithm selection. In [15], Muñoz et al. try to predict the performance of CMA-ES by using deep neural network. Their network has two hidden layers and 10 trainable parameters for each hidden layer. A set of ELA features and algorithm parameters are concatenated as the input. However, the input size and the number of trainable parameters are small. The approach cannot completely show the capability of deep neural network. Loreggia et al. [13] use convolutional neural network to recommend optimization algorithms for the boolean satisfiability (SAT) problem and show appealing results. Since the SAT problems are described in text files, they convert these text files to images. The images are the input of a convolutional neural network with three convolutional layers and two fully connected layers. However, the authors did not propose an approach to solve the optimization problems that cannot be described in text files.

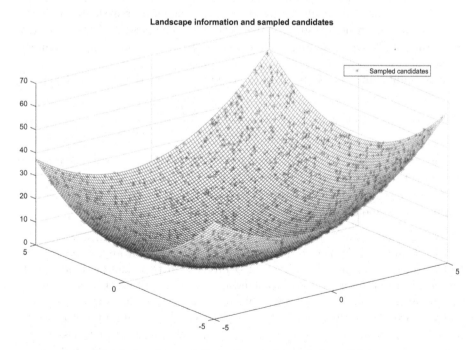

Fig. 1. Landscape information and sampled candidates.

3   Algorithm Selection by Convolutional Neural Network

In this section, we introduce our overall framework. In the first subsection, we introduce a novel method to extract landscape information from the optimization problems and save the information as images. In the second subsection, we introduce the architecture of our network.

3.1   Extract Information from Optimization Problems

The process of sampling landscape information and saving the information as images from optimization problems is described as follows:

1) We use scaling methods to make the search space of all training and testing instances the same. In the experimental section, we use the BBOB function instances which have an identical search space $[-5, 5]^D$ as our training and testing instances. Thus, they do not need scaling.
2) We uniformly generate a set of sample coordinates from the entire search space. The generated sample coordinates are recorded as a $N \times D$ matrix, where N is the number of sampled candidates and D is the problem dimension.
3) Given an optimization problem, according to the $N \times D$ matrix generated in 1), we sample a set of N candidates from the entire search space for the problem. The fitness results of the candidates compose a N-length fitness vector $\overrightarrow{Fit}$. The ith value of $\overrightarrow{Fit}$ records the ith fitness result of the sampled candidates.
4) The sampled fitness results are normalized to $[0, 1]$ by the following equation.

$$\overrightarrow{Fit}_{normalized} = \frac{\overrightarrow{Fit} - MIN(\overrightarrow{Fit})}{MAX(\overrightarrow{Fit}) - MIN(\overrightarrow{Fit})} \tag{1}$$

where MAX and MIN operations are to choose the maximal and minimal value of the given vector $\overrightarrow{Fit}$ respectively.
5) We resize the normalized fitness vector $\overrightarrow{Fit}_{normalized}$ to a $\sqrt{N} \times \sqrt{N}$ 2-D image by standard image resize operators.

We give an example to illustrate how an image is generated. Suppose the given optimization problem is a 2-D problem. Note that our approach can solve higher dimensional problems. Using 2-D problem in this example is for a better visualization. Figure 1 shows the landscape of the optimization problem and the sampled candidates. We uniformly sample 2025 candidates from the entire search space. The fitness results of the 2025 candidates compose a fitness vector. Figure 2 shows the fitness vector $\overrightarrow{Fit}$ and the normalized fitness vector $\overrightarrow{Fit}_{normalized}$. The normalized fitness vector is converted to a 45×45 image. The image is shown as a gray scale image in Fig. 3.

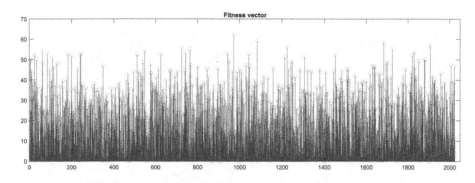

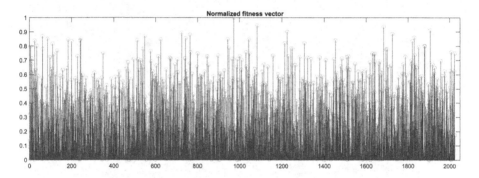

Fig. 2. Fitness vector and normalized fitness vector.

3.2   Convolutional Neural Network

Our CNN architecture follows the VGG architecture. Compared with other CNN architectures, VGG incorporates a smaller kernel size and a greater depth. In the experimental section, we investigate the problem with dimension $D = 2$ and $D = 10$. The sample size of the two types of problems should be different. We set the sample size of problem with $D = 2$ and $D = 10$ to 2025 and 10000 respectively. Since the input sizes of the two types of problems are different, two slightly different CNN architectures (a) and (b) are presented in this section.

The overall architecture consists of several groups of convolutional layers, several fully connected layers and a softmax layer. We use a softmax layer as the output layer because we treat the algorithm selection problem as a classification task, and the softmax layer is a commonly used output layer for the classification task. The number of outputs equals to the number of classes. We use rectified linear unit (ReLU) as the activation function and the max pooling filter to reduce the size of feature map.

Generated image

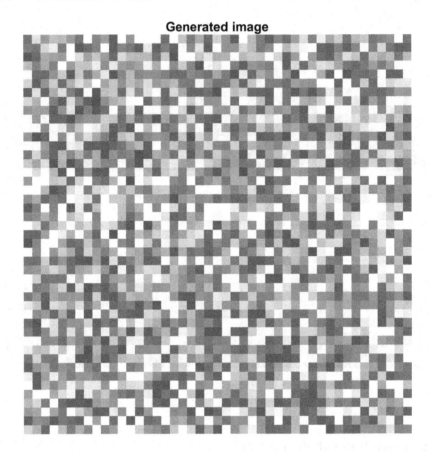

Fig. 3. Generated 2-D image.

We adopt ReLU at the end of each layer. ReLU is a commonly used activation function which retains the non-negative part and forces the negative part to be 0. It has been shown performing well on 2-D data [2]. The formula of ReLU can be described as $f(x) = MAX(0, x)$, where x is the input of ReLU and $f(x)$ is the output of ReLU.

We also adopt max pooling layers at the end of each group layer. These pooling layers are used to reduce the size of the output feature maps, which are the input of the next layers.

The overall framework is shown in Fig. 4. The detailed settings of the architecture is shown in Table 1. In this paper, we provide two convolutional architectures for two different types of input data. The architecture (a) for the instances with $D = 10$ consists of five groups of convolutional layers, three fully connected layers and a softmax layer. The convolutional layers are identical to those in VGG16. The architecture (b) for the instances with $D = 2$ consists of four groups of convolutional layers, three fully connected layers and a softmax layer.

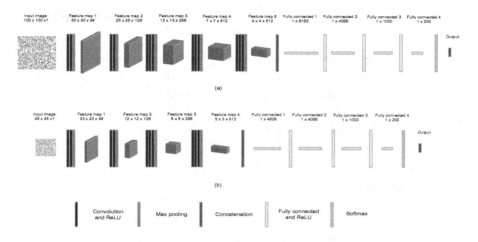

Fig. 4. The overall CNN architecture.

The four groups of convolutinal layers of the architecture (b) are identical to the first four groups of convolutional layers in the architecture (a).

The number of outputs of softmax layer equals to the number of classes. For example, if the task is to recommend a suitable algorithm for the given problem from the algorithm set consisting of three algorithms, the number of outputs of softmax layer in this task should be three. Softmax layer use cross entropy loss to evaluate the classification error. The formula of cross entropy loss for each instance is:

$$Loss = -\sum_{i=1}^{C} p_i \log(\hat{p}_i) \tag{2}$$

where p_i is the actual probability distribution of the ith class, and $\hat{p}_i$ is the predicted probability distribution of the ith class. C is the number of classes. Our CNN architecture is to minimize the loss value $Loss$.

4   Experimental Results

4.1   Data Description

In the following experiments, we use the BBOB functions to evaluate the performance of our approach. The BBOB functions can be classified into 24 problem classes. For each class, different problem instances can be generated by random transformation and rotation. The search space of the instances are $[-5,5]^D$, where D is the problem dimension. We investigate instances with $D = 2$ and $D = 10$ in the first experiment, and instances with $D = 10$ in the second and the last experiment.

Table 1. Parameter settings of the architectures. The convolutional group 5 is not included in the architecture for the instances with $D = 2$.

Group	Layers	Kernel size	Stride	Output channel
Group 1	conv3-64	3×3	1	64
	conv3-64	3×3	1	64
	max2	2×2	2	64
Group 2	conv3-128	3×3	1	128
	conv3-128	3×3	1	128
	max2	2×2	2	128
Group 3	conv3-256	3×3	1	256
	conv3-256	3×3	1	256
	conv3-256	3×3	1	256
	max2	2×2	2	128
Group 4	conv3-512	3×3	1	512
	conv3-512	3×3	1	512
	conv3-512	3×3	1	512
	max2	2×2	2	512
Group 5	conv3-512	3×3	1	512
	conv3-512	3×3	1	512
	conv3-512	3×3	1	512
	max2	2×2	2	512
	fc-4096	1×1	1	4096
	fc-1000	1×1	1	1000
	fc-200	1×1	1	200
	softmax	1×1	1	Number of classes

4.2 Training Details

We set the number of training epochs to 150, where one epoch denotes the network is trained by the entire training data once. We set the batch size, which is the number of instances for each training, to 60. We use the Adam optimizer [11] to optimize the neural network, and set the learning rate to 10^{-4}. The first momentum, the second momentum and epsilon of the Adam optimizer is set to 0.9, 0.999 and 10^{-8} respectively following their recommended settings. The CNN architecture is implemented by Python language with Tensorflow library on a Nvidia 1080Ti graphic card. At the end of each epoch, we use the validation instances to test the performance of our network. The network whose parameters perform the best (i.e., having the highest accuracy) on the validation instances will be saved. After the training, we use the testing instances to evaluate the performance of the well-trained network. To obtain reliable results, we repeat the training and testing process five independent times and record the mean of the prediction accuracy.

4.3   Classification Results of Classifying Problem Instances to Different Classes

In the first experiment, we investigate the classification capability of our approach on the BBOB function instances with $D = 2$ and $D = 10$. In the BBOB competition settings [3], the total number of evaluations for an optimization process is $10000 \times D$, where D is the problem dimension. In our experiment, the number of sampled candidates for the two kinds of instances are approximately 10% of the total budget. We sample $45 \times 45 = 2025$ candidates for instances with $D = 2$ which corresponds to 10.125% of the total budge, and sample $100 \times 100 = 10000$ candidates for instances with $D = 10$ which corresponds to 10% of the total budge. We investigate the instances with $D = 2$ because their landscape are similar to images. We investigate instances with $D = 10$ because this kind of instances are commonly investigated in many papers of algorithm selection field.

In this experiment, we evaluate the performance of using the deep neural network to classify the BBOB functions. For each of the 24 problem class, we generate 250 different instances. Thus, there are $250 \times 24 = 6000$ instances in total. We use 70% instances for training, 10% instances for validation, and 20% instances for testing. Given an unknown instance from the testing set, the classification task for our approach is to predict which problem class it belongs to. There are 24 possible prediction results in total. Thus, the baseline accuracy should be $1/24 = 4.17\%$. Table 2 shows the classification results for instances with $D = 2$ and $D = 10$.

Both results for the instances with $D = 2$ and $D = 10$ are much higher than the baseline accuracy. It indicates that our approach can classify the optimization functions into their corresponding classes. For the instances with $D = 2$, the average accuracy is 65.27%. For the instances with $D = 10$, the average accuracy is 57.58%. It is reasonable that the performance on the 2-D problems is better. According to [14], if the increasing of the sample size does not match the exponential growth of D, the sample strategy will suffer from a convergence behaviour. This convergence behaviour discussed in depth in [14]. However, in our experiment, the sample size increases linearly. The phenomenon also implies that the deep neural network may not work for very high dimensional optimization problems unless a large sample size is applied.

For the problem instances having salient characteristics such as Sphere function (which is f1 of the BBOB suite), our approach can identify them into correct classes with a very high accuracy. However, for the problem instances having no salient characteristics such as Ellipsoidal function (both f2 and f10 of the BBOB suite belong to Ellipsoidal function), our approach inevitably classifies many instances into incorrect classes. Misclassification happens because of the inherent similarity between functions rather than the limitations of our approach. This may not be a big problem as problem instances sharing similar characteristics may map to the same optimization algorithm.

Table 2. The accuracy of classifying the BBOB instances with $D = 2$ and $D = 10$ into 24 problem classes.

	$D = 2$	$D = 10$
f1	100.00%	100.00%
f2	65.20%	59.20%
f3	90.80%	82.40%
f4	51.20%	54.80%
f5	100.00%	87.20%
f6	66.80%	26.80%
f7	31.60%	37.60%
f8	93.20%	90.40%
f9	98.40%	96.40%
f10	18.40%	22.40%
f11	25.60%	31.60%
f12	44.40%	14.00%
f13	89.20%	59.20%
f14	33.60%	47.60%
f15	52.00%	50.00%
f16	56.80%	69.20%
f17	49.60%	21.20%
f18	45.60%	37.20%
f19	99.60%	76.00%
f20	98.00%	94.80%
f21	32.00%	59.20%
f22	38.40%	22.40%
f23	97.60%	96.80%
f24	88.40%	45.60%
Average	65.27%	57.58%

4.4 Classification Results of Algorithm Selection

In this experiment, we investigate the performance of our approach on recommending the optimization algorithms for the BBOB functions with $D = 10$. The algorithm set consists of three optimization algorithms. They are artificial bee colony (ABC) [7], covariance matrix adaptation evolution strategy (CMA-ES) [4] and linear population size reduction differential evolution (L-SHADE) [23]. The population sizes of ABC and CMA-ES are 125, 40 respectively. The initial population size of L-SHADE is set to 200. The settings follow their recommended settings.

Table 3. The accuracy of recommending the most suitable algorithms for the BBOB instances with $D = 10$.

	$D = 10$
f1	100.00%
f2	95.40%
f3	95.60%
f4	94.80%
f5	92.40%
f6	98.00%
f7	64.40%
f8	89.20%
f9	99.60%
f10	99.60%
f11	92.40%
f12	96.80%
f13	84.00%
f14	30.80%
f15	95.60%
f16	92.40%
f17	95.40%
f18	94.80%
f19	92.00%
f20	92.40%
f21	90.40%
f22	71.60%
f23	92.00%
f24	94.40%
Average	89.33%

Different from the first experiment, we do not test on the instances with $D = 2$ in this experiment. The reason is that the instances with $D = 2$ are very easy to solve. For most instances with $D = 2$, all the three algorithms can find the global optima. In this case, the algorithm selection problem is insignificant. We test on the instances with $D = 10$ only. Following the previous settings, we also sample 10000 candidates and convert the sampled vector to 100×100 image for each problem instance with $D = 10$.

For each of the 24 problem class, we also generate 250 different instances. Thus, there are 6000 instances in total. For each instance, we run the three optimization algorithms on it 51 independent times and select the best algorithm as the label. We still use 70% instances for training, 10% instances for validation,

Table 4. The rank results of our deep learning approach and single optimization algorithms. In this table, (+), (-) and (=) denote the corresponding algorithm is inferior to, superior to, and equally well to the deep learning approach on the corresponding problem respectively.

	Deep Learning	ABC	CMA-ES	L-SHADE
f1	1	1(=)	1(=)	1(=)
f2	1	1(=)	1(=)	1(=)
f3	1	1(=)	4(+)	3(+)
f4	1	1(=)	4(+)	3(+)
f5	1	1(=)	1(=)	1(=)
f6	1	4(+)	1(=)	1(=)
f7	1	4(+)	3(+)	1(=)
f8	1	4(+)	1(=)	1(=)
f9	1	4(+)	1(=)	1(=)
f10	1	4(+)	1(=)	1(=)
f11	1	4(+)	1(=)	1(=)
f12	1	4(+)	1(=)	3(+)
f13	1	4(+)	1(=)	3(+)
f14	1	4(+)	2(=)	1(=)
f15	2	4(+)	1(=)	3(+)
f16	2	3(+)	1(=)	4(+)
f17	1	4(+)	3(+)	2(=)
f18	3	4(+)	2(=)	1(=)
f19	2	4(+)	1(=)	3(+)
f20	3	1(-)	4(+)	2(-)
f21	3	1(=)	4(+)	2(=)
f22	3	1(=)	4(+)	2(=)
f23	3	2(=)	1(-)	4(+)
f24	2	4(+)	1(=)	3(+)
Average	1.583	2.875	1.875	2.000

	ABC	CMA-ES	L-SHADE
Inferior to the deep learning approach	15	7	9
Superior to the deep learning approach	1	1	1
Equally well to the deep learning approach	8	16	14

and 20% instances for testing. However, there are still some instances that more than one algorithm can find the global optima. In this experiment, we eliminate these instances from our dataset because these instances are not suitable for training and testing. Using instances with undetermined label will aggravate the performance of our neural network. There are 991 instances marked with undetermined labels from the training data, 149 instances from the validation data, and 292 instances from the testing data. After eliminating these unsuitable instances, there are 3209 instances used for training, 451 instances for validation, and 908 instances for testing.

There are three possible predicted outputs. Thus, the baseline accuracy in this experiment should be $1/3 = 33.33\%$. Following the previous settings, we also record the parameters of the networks that perform the best on the validation instances and use testing instances to evaluate the recorded network. The training and testing process are also repeated five times to obtain a reliable result.

Table 3 records the accuracy results. The average accuracy is 89.33%, which greatly outperforms the baseline accuracy. For most problem classes, the accuracy is more than 90%. It indicates that our approach can recommend the most suitable optimization algorithms for most of the optimization problem instances.

4.5   Comparisons of Fitness Results

In this experiment, we investigate the optimization capability of our approach. We treat our approach as a portfolio approach and compare it with the three single optimization algorithms (i.e., ABC, CMA-ES and L-SHADE). We also test on the instances with $D = 10$ only. For each instance, we use 10% budget to sample candidates (i.e., $100 \times 100 = 10000$ candidates are sampled) and generate an image as the input of deep learning. In the second experiment, we repeatedly train and test five times and generated five well-trained networks. In this experiment, we use the well-trained network having the median accuracy performance in the second experiment (the experiment in the Sect. 4.4) to recommend an optimization algorithm from the algorithm set for each testing instance. The testing instances are identical to the testing instances in the Sect. 4.4. We use the remaining 90% budget to run the recommended algorithm on the instance. The fitness results found by our approach are compared with the results obtained by the single optimization algorithms with 100% budget. Our approach and the three single optimization algorithms are run on each testing instance 51 independent times.

Table 4 shows the rank results. We use Kruskal-Wallis test and multiple comparison test with p-value = 0.05 to test the significance. Compared with the three single algorithms, our deep learning approach has the best average ranks. For most problems, the deep learning approach can find the best optimization algorithm. Note that the key of our approach is to select the most suitable algorithm from the algorithm set. Thus, it is reasonable that our approach does not beat all algorithms on all problems. The detailed results with averaged fitness results are shown in Table 5.

Table 5. The detailed average results of our deep learning approach and single optimization algorithms.

	Deep Learning	ABC	CMA-ES	L-SHADE
f1	0.00E+00	0.00E+00	0.00E+00	0.00E+00
f2	0.00E+00	0.00E+00	0.00E+00	0.00E+00
f3	0.00E+00	0.00E+00	4.88E+00	5.15E-01
f4	0.00E+00	0.00E+00	9.73E+00	1.42E+00
f5	0.00E+00	0.00E+00	0.00E+00	0.00E+00
f6	0.00E+00	2.42E-01	0.00E+00	0.00E+00
f7	0.00E+00	1.37E+00	5.15E-02	0.00E+00
f8	0.00E+00	4.71E-03	0.00E+00	0.00E+00
f9	0.00E+00	9.58E-01	0.00E+00	0.00E+00
f10	0.00E+00	1.43E+03	0.00E+00	0.00E+00
f11	0.00E+00	3.12E+01	0.00E+00	0.00E+00
f12	0.00E+00	1.21E+00	0.00E+00	4.03E-05
f13	0.00E+00	6.61E-01	0.00E+00	1.76E-08
f14	0.00E+00	1.80E-03	0.00E+00	0.00E+00
f15	3.65E+00	2.19E+01	2.96E+00	6.45E+00
f16	8.77E-01	1.45E+00	1.36E-01	2.53E+00
f17	1.79E-06	1.21E+00	8.00E-06	6.96E-06
f18	8.68E-04	3.13E+00	6.28E-04	3.33E-05
f19	4.29E-01	1.38E+00	1.19E-01	1.36E+00
f20	1.22E+00	1.91E-01	1.56E+00	5.01E-01
f21	1.68E-01	1.42E-02	2.54E+00	1.65E-01
f22	2.51E+00	2.45E-01	6.16E+00	2.06E+00
f23	9.18E-01	4.76E-01	2.43E-01	1.10E+00
f24	1.68E+01	2.71E+01	1.16E+01	2.66E+01

5   Conclusion and Future Works

Algorithm selection is an important topic since different types of optimization problems require different optimization algorithms. In recent years, researchers treat algorithm selection as a classification or regression task and propose many approaches for tackling it. On the other hand, deep neural network especially convolutional neural network has shown great capability of dealing with classification and regression tasks. However, only a few works focus on applying deep learning to algorithm selection, and little work has been done on applying deep learning to black box algorithm selection. In this paper, we propose an approach employing convolutional neural network architecture for algorithm selection.

We conduct three experiments to investigate the performance of our approach. In the first experiment, we show the classification capability of classifying the BBOB instances into their corresponding problem classes. In the second experiment, we show the capability of recommending optimization algorithms for given optimization problems. The accuracy is vastly better than the baseline accuracy. In the last experiment, we investigate the optimization capability of our approach. We use the well-trained convolutional neural network in the second experiment to recommend optimization algorithms, and use the predicted algorithms to optimize the given problems. The comparison results indicate that our deep learning approach can solve optimization problems effectively.

There are still lots of works that can be done in the future. In this paper, we only treat algorithm selection problem as a classification task, while treating it as a regression task is also reasonable. Moreover, our convolutional neural network follows the architecture of VGG, while other powerful architectures can also be used. The sampling strategy can also be improved. We uniformly sample candidates from the entire search space, while other sample strategies such as Latin hypercube sampling [22] can also be applied.

References

1. Bischl, B., Mersmann, O., Trautmann, H., Preuss, M.: Algorithm selection based on exploratory landscape analysis and cost-sensitive learning. In: Proceedings of the 14th Annual Conference on Genetic and Evolutionary Computation, pp. 313–320. ACM (2012)

2. Glorot, X., Bordes, A., Bengio, Y.: Deep sparse rectifier neural networks. In: Proceedings of the Fourteenth International Conference on Artificial Intelligence and Statistics, pp. 315–323 (2011)

3. Hansen, N., Auger, A., Ros, R., Finck, S., Pošík, P.: Comparing results of 31 algorithms from the black-box optimization benchmarking bbob-2009. In: Proceedings of the 12th Annual Conference Companion on Genetic and Evolutionary Computation, pp. 1689–1696. ACM (2010)

4. Hansen, N., Müller, S.D., Koumoutsakos, P.: Reducing the time complexity of the derandomized evolution strategy with covariance matrix adaptation (CMA-ES). Evol. Comput. 11(1), 1–18 (2003)

5. He, Y., Yuen, S.Y., Lou, Y.: Exploratory landscape analysis using algorithm based sampling. In: Proceedings of the Genetic and Evolutionary Computation Conference Companion, pp. 211–212. ACM (2018)

6. He, Y., Yuen, S.Y., Lou, Y., Zhang, X.: A sequential algorithm portfolio approach for black box optimization. Swarm Evol. Comput. 44, 559–570 (2019)

7. Karaboga, D., Basturk, B.: A powerful and efficient algorithm for numerical function optimization: artificial bee colony (abc) algorithm. J. Global Optimizat. 39(3), 459–471 (2007)

8. Kerschke, P., Dageforde, J.: flacco: Feature-based landscape analysis of continuous and constraint optimization problems. R-package Version 1 (2017)

9. Kerschke, P., Hoos, H.H., Neumann, F., Trautmann, H.: Automated algorithm selection: survey and perspectives. Evol. Comput. 27(1), 3–45 (2019)

10. Kerschke, P., Trautmann, H.: Automated algorithm selection on continuous black-box problems by combining exploratory landscape analysis and machine learning. Evol. Comput. **27**(1), 99–127 (2019)
11. Kingma, D.P., Ba, J.: Adam: A method for stochastic optimization. arXiv preprint arXiv:1412.6980 (2014)
12. LeCun, Y., Bengio, Y., Hinton, G.: Deep learning. Nature **521**(7553), 436–444 (2015)
13. Loreggia, A., Malitsky, Y., Samulowitz, H., Saraswat, V.: Deep learning for algorithm portfolios. In: Thirtieth AAAI Conference on Artificial Intelligence (2016)
14. Morgan, R., Gallagher, M.: Sampling techniques and distance metrics in high dimensional continuous landscape analysis: limitations and improvements. IEEE Trans. Evol. Comput. **18**(3), 456–461 (2013)
15. Muñoz, M.A., Kirley, M., Halgamuge, S.K.: A meta-learning prediction model of algorithm performance for continuous optimization problems. In: International Conference on Parallel Problem Solving from Nature, pp. 226–235. Springer (2012)
16. Muñoz, M.A., Kirley, M., Halgamuge, S.K.: Exploratory landscape analysis of continuous space optimization problems using information content. IEEE Trans. Evol. Comput. **19**(1), 74–87 (2014)
17. Muñoz, M.A., Smith-Miles, K.A.: Performance analysis of continuous black-box optimization algorithms via footprints in instance space. Evol. Comput. **25**(4), 529–554 (2017)
18. Muñoz, M.A., Sun, Y., Kirley, M., Halgamuge, S.K.: Algorithm selection for black-box continuous optimization problems: a survey on methods and challenges. Inf. Sci. **317**, 224–245 (2015)
19. Peng, F., Tang, K., Chen, G., Yao, X.: Population-based algorithm portfolios for numerical optimization. IEEE Trans. Evol. Comput. **14**(5), 782–800 (2010)
20. Roy, R., Hinduja, S., Teti, R.: Recent advances in engineering design optimisation: challenges and future trends. CIRP Annals **57**(2), 697–715 (2008)
21. Simonyan, K., Zisserman, A.: Very deep convolutional networks for large-scale image recognition. arXiv preprint arXiv:1409.1556 (2014)
22. Stein, M.: Large sample properties of simulations using latin hypercube sampling. Technometrics **29**(2), 143–151 (1987)
23. Tanabe, R., Fukunaga, A.S.: Improving the search performance of shade using linear population size reduction. In: 2014 IEEE Congress on Evolutionary Computation (CEC), pp. 1658–1665. IEEE (2014)
24. Vrugt, J.A., Robinson, B.A., Hyman, J.M.: Self-adaptive multimethod search for global optimization in real-parameter spaces. IEEE Trans. Evol. Comput. **13**(2), 243–259 (2008)
25. Wolpert, D.H., Macready, W.G.: No free lunch theorems for optimization. IEEE Trans. Evol. Comput. **1**(1), 67–82 (1997)
26. Wu, G., Mallipeddi, R., Suganthan, P.N.: Ensemble strategies for population-based optimization algorithms-a survey. Swarm Evol. Comput. **44**, 695–711 (2019)
27. Wu, G., Mallipeddi, R., Suganthan, P.N., Wang, R., Chen, H.: Differential evolution with multi-population based ensemble of mutation strategies. Inf. Sci. **329**, 329–345 (2016)
28. Yuen, S.Y., Chow, C.K., Zhang, X., Lou, Y.: Which algorithm should i choose: an evolutionary algorithm portfolio approach. Appl. Soft Comput. **40**, 654–673 (2016)
29. Zhang, X., Fong, K.F., Yuen, S.Y.: A novel artificial bee colony algorithm for hvac optimization problems. HVAC&R Res. **19**(6), 715–731 (2013)
30. Zhao, S.Z., Suganthan, P.N., Zhang, Q.: Decomposition-based multiobjective evolutionary algorithm with an ensemble of neighborhood sizes. IEEE Trans. Evol. Comput. **16**(3), 442–446 (2012)

A Unified Approach to Anomaly Detection

Richard Ball[(⊠)], Hennie Kruger, and Lynette Drevin

North-West University, Potchefstroom, South Africa
richardstevenball@gmail.com, {hennie.kruger,
lynette.drevin}@nwu.ac.za

Abstract. Anomalous actors are becoming increasingly sophisticated both in methodology and technical ability. Fortunately, the companies impacted by anomalous behaviour are also generating more data and insights on potential anomalous cases than ever before. In this paper, a unified approach to managing the complexity of constructing a useable anomaly detection system is presented. The unified approach is comprised of three algorithms, a Neural Architecture Search (NAS) implementation for autoencoders, an anomaly score threshold optimisation algorithm, and a Gaussian scaling function for anomaly scores. NAS is applied to a data set containing instances of credit card fraud. The NAS algorithm is used to simulate a population of 50 candidate deep learning architectures, with the best performing architecture being selected based on a balanced score, comprised of an average of the Area under the ROC curve, Average Precision and normalised Matthews Correlation Coefficient scores. The threshold optimisation algorithm is used to determine the appropriate threshold between the classes, for the purposes of producing the binary classification outcome of each architecture. The Gaussian scaling algorithm is applied to the raw anomaly scores of the optimal architecture into order to generate useable probability scores. Not only did the proposed unified approach simplify the process of selecting an optimal neural architecture whose output is interpretable by business practitioners and comparable with other probability score producing models, but it also contributes to anomaly detection in a transactional setting by eliminating subjective thresholds when classifying anomalous transactions.

Keywords: Anomaly detection · Artificial intelligence · Decision support systems · Neural networks · Evolutionary computations

1 Introduction

The aim of this paper is to report on the creation of a unified system for detecting anomalies in a transactional setting. Building a neural network-based anomaly detection capability not only requires an effective implementation of a chosen algorithm, but it should also include processes relating to the usability of the capability within an operational environment. Unsupervised learning techniques have proven to be effective for detecting anomalies [13] although their usage contains some uncertainties regarding the choice of an optimal number of clusters in the case of clustering [16], or the choice

© Springer Nature Switzerland AG 2020
G. Nicosia et al. (Eds.): LOD 2020, LNCS 12566, pp. 281–291, 2020.
https://doi.org/10.1007/978-3-030-64580-9_24

of a suitable threshold in the case of an autoencoder [10]. An approach is required that supports the experimental nature of unsupervised learning, but also caters for the structured needs of an organization.

The unified system begins with the development of a sparse autoencoder for detecting anomalies. The choice of structure for the sparse autoencoder architecture is justified by means of an implementation of neural architecture search. That is, multiple candidate architectures are simulated, with the simulated architecture that produces the best evaluation metric being selected. The determination of a suitable threshold for classifying anomalies is generally subjective, and this paper proposes an approach for finding the optimal threshold. Following the application of the optimal threshold, the sparse autoencoder produces an anomaly score per instance. These are determined by the mean square error, and are used to determine how anomalous an instance might be. The larger the anomaly score, the more likely that the instance is anomalous. A challenge with these anomaly scores however is that they are difficult to interpret, beyond the fact that higher means more anomalous [11]. A Gaussian scaling function is proposed in order to transform the anomaly scores into a distribution ranging from 0 to 1, a form which is both comparable across different models and understood universally. These transformed scores can therefore be considered the probability that the instance is anomalous in nature.

The motivation for the study is founded in the observation that anomaly detection projects often fail to be deployed in an operational setting due to issues around a lack of interpretability and effective evaluation [15]. The unified approach provides practitioners with a means of constructing a performant architecture, as evaluated against multiple preferred performance metrics for highly imbalanced classes, and producing an output in the form of a probability score.

The remainder of this paper is organised as follows: Sect. 2 provides some background on the use of unsupervised learning, more specifically autoencoders, for anomaly detection. Following this, Sect. 3 presents the data set used for experimental research. The implementation of neural architecture search, threshold optimisation, and Gaussian scaling are then illustrated in Sect. 4, followed by a discussion of the results of the study in Sect. 5. Finally, conclusions about the research study are made in Sect. 6.

2   Autoencoders for Anomaly Detection

For a typical anomaly detection problem, the instances of labelled records indicating anomalous behaviour are far outweighed by those labelled as non-anomalous. This presents a class imbalance problem i.e. one in which the majority class far outweighs the minority. A typical approach to resolving this problem is through the use of various sampling techniques including under sampling the majority or over sampling the minority classes [5].

A problem with under sampling is that potentially useful data may be discarded, whereas with over sampling the problem of overfitting arises through the use of copies (synthetic or otherwise) of existing data [17]. Another problem with using a supervised learning approach to solve the anomaly detection problem is that labels are known *a posteriori*, either through a customer compliant or chargeback from the bank [2].

In contrast to supervised learning, unsupervised learning techniques do not require knowledge of the labels, but rather seek to characterise the distributions of the data related to the problem space. An autoencoder is used for the anomaly detection component of the study. Autoencoders are unsupervised neural networks that are used to learn efficient data encodings and which have been used to detect instances of accounting anomalies [14]. Autoencoders can be described mathematically with lots of computational detail [12], although due to a page limit this detail has been omitted.

According to [8], an autoencoder is a neural network that is trained to try copy its input to its output. In terms of structure, an autoencoder has a hidden layer h that generates a function used to describe the model's input. It also contains two distinct parts, namely an encoder function $h = f(x)$ and a decoder function $r = g(h)$. The decoder function r attempts to produce a reconstruction of the input, hence it is termed reconstruction. It is not feasible nor useful for the autoencoder to simply learn to copy the input x perfectly. Instead the model is restricted in the sense that they can only copy input that resembles the training data, and that they are only allowed to copy approximately. Because of these restrictions, the model is likely to learn useful representations of the data.

According to [1] training an autoencoder is generally easier than training a deep neural network. Therefore autoencoders have been used as kinds of building blocks for deep neural networks, where each individual layer in the deep network can be represented by an autoencoder that can be trained in isolation. Figure 1 illustrates the general structure of an autoencoder.

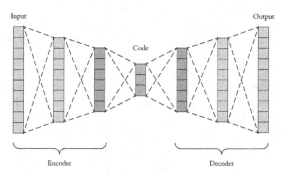

Fig. 1. An example of an autoencoder

3   Data

Data related to anomaly detection is notoriously difficult to obtain according to [13]. Organizations are traditionally somewhat conservative when it comes to publishing data related to cases of in-house anomalies. The reasons for this could be that exposing their internal anomalies might pose reputational risk, or the data may present an opportunity for prospective bad actors to capitalize on potential perceived gaps in the system.

The study was conducted using a publicly available anomaly data set that has been cited in the literature [5]. This particular data set's use is intended to illustrate the concepts for the purposes of this paper, but in follow-up studies the intention is to use real-world data. The public data set was sourced from Kaggle, a website dedicated to hosting machine learning competitions [9], and was collected by Worldline S.A., a European payments provider.

The data contains transactions made over a 2-day period from September 2013. This is a European user credit card data set, and contains over 285,000 transactions of which only 492 (or 0.172%) are confirmed as fraudulent. The data set is labelled and can therefore be used for either classification of anomaly detection purposes. Principal Component Analysis (PCA) has been performed on most of the features, with the resultant data set containing 28 unknown features. There is therefore no way in which to infer identities or additional user information from them. There are also two additional features that have not been processed via PCA, namely the time and amount of the transaction. The labels consist of 1 s and 0 s, with 1 indicating a confirmed case of fraud and 0 indicating a genuine transaction. Although the data used for this study is transactional in nature, the unified approach can be extended to any anomaly detection application.

4 Methodology

The autoencoder model discussed in this section follows from the discussion in Sect. 2. In this section Neural Architecture Search, threshold optimization, and Gaussian scaling are applied to the credit card fraud detection data set presented in Sect. 3.

Neural Architecture Search (NAS) is the process of automatically engineering neural network architectures. According to [6] there are three dimensions to NAS. They are defining the search space, the search strategy, and the performance estimation strategy. The search space relates to the neural architectures that will be simulated; the search strategy relates to how the search space is intended to be explored, while the performance estimation strategy relates to how the performance of each architecture is measured.

The following implementation of NAS simulates a population of candidate architectures. In general researchers typically use a single metric to evaluate the performance of an architecture. In this paper we propose an average of the Area Under the ROC Curve [3], the Average Precision [7], and a normalized Matthews Correlation Coefficient [4] to derive a balanced score that can be used to select the best performing architecture. Each candidate architecture is created by means of assigning it a random depth and a random number of nodes, with an upper bound set to the size of the input layer, for each layer of the corresponding depth. For the purposes of the study, each architecture has the same regularization constraint placed on it.

The search space for this research are chain-structured densely connected neural network architectures, created through the assignment of a randomly generated length and number of nodes per architectural element. The search strategy is deduced by simulating the search for a number of m iterations. Finally, the performance estimation strategy is defined by the balanced score mentioned above. This implementation is presented as Algorithm 1 below.

Algorithm 1 (Neural architecture search algorithm)

Input: number of n training epochs, intended population size m of generated architectures, size of the input layer l

Output: An array $S = \{s_1, ..., s_m\}$ containing the AUC scores of the m simulated architectures.

1 for $i = 1$ *to* m do
2   generate random depth d;
3   generate random architecture A, with a node upper bound l and length d;
4   for $j = 1$ *to* d do
5   if $A[j] < min$ then
6         $min \leftarrow A[j]$;
7   end
8   end
9   remove min from A;
10   encArchitecture $\leftarrow A$ sorted in descending order;
11   decArchitecture $\leftarrow A$ sorted in ascending order;
12   build input layer of size l;
13   for $k = 1$ to $d - 1$ do
14   build encoder layer with encArchitecture[k] nodes;
15   end
16   build bottleneck with min nodes;
17   for $n = 1$ to $d - 1$ do
18   build decoder layer with decArchitecture[n] nodes;
19   end
20   build output layer of size l;
21   train the autoencoder;
22   calculate the balanced score s_l using test data;
23   $S \leftarrow s_l$;
24 end
25 output S;

When computing the balanced score in Algorithm 1 (step 22), the AUROC and Average Precision metrics can be calculated directly from the raw anomaly score as produced by the autoencoder. For the Matthews Correlation Coefficient (MCC) however, a binary outcome is needed. That one may determine these binary outcomes, a threshold for the raw anomaly scores is required in order to split the scores into a binary classification outcome. A method for determining the optimal threshold for the MCC (or any other binary classification) is presented as Algorithm 2. The threshold optimization process is illustrated in Fig. 3, where it is apparent that the highest MCC score is achieved at a threshold of around 4. The threshold is represented by the gold line in Fig. 2 which attempts to best split the two classes, with the non-anomalous classes mainly occurring below the threshold due to their low reconstruction error. The converse is true for the anomalous class, where most have a high reconstruction error and are therefore above the threshold. Algorithm 2 effectively finds the best fit for the threshold, as determined by the MCC.

Algorithm 2 (Anomaly score threshold optimisation algorithm)

Input: $Y = \{y_1, \ldots, y_m\}$ is a linearly spaced vector, starting at zero, and extending to a value of n, containing m samples.

Output: An optimal threshold value $Y[index]$

1 for $i = 1$ to m do

2 read y_i from Y;

3 calculate the binary classification outcomes for y_1;

4 calculate the MCC score from the binary classification outcomes ;

5 populate array $R \leftarrow r_i$, the MCC value computed ;

6 end

7 $max \leftarrow R[0]$;

8 for $i = 1$ to m do

9 if $R[i] > max$ then

10   $max \leftarrow R[i]$;

11   $index \leftarrow i$;

12 end

13 end

14 output $Y[index]$;

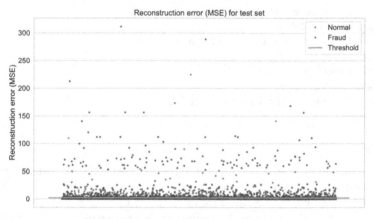

Fig. 2. Distribution of reconstruction errors

Following the implementation of Algorithms 1 and 2, a winning architecture is returned. An example of 50 simulated architectures with their corresponding balanced scores is presented in Fig. 4. The best performing architecture is the 8th architecture with a balanced score of 87.6%. This architecture has 10 hidden layers and 4687 learnable parameters. The distribution of the 50 balanced scores exhibits a reasonable amount of variance, which is expected considering the random nature of the architecture construction process. Due to the upper bound and depth constraints placed on the construction process, there are no exceedingly bad or failed architectures, with the lowest balanced score at 74.3%.

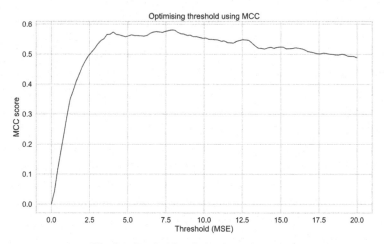

Fig. 3. Threshold optimization using MCC

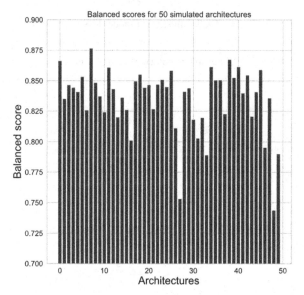

Fig. 4. Balanced scores for 50 simulated architectures

Figure 5 illustrates the model loss during training. In order to confirm that the model effectively discriminates between the two classes, the model was reshaped to include a bottleneck of 2 nodes, as presented in Fig. 6. It is clear that there is some degree of separation between the classes in the 2-dimensional latent space.

The output of the winning architecture is an array of reconstruction scores, one for each transaction. These reconstruction scores, as calculated by the MSE, are not entirely interpretable in their current form, as illustrated in Fig. 2. In order to determine the probability that the transaction is anomalous, a normalization approach needs to be

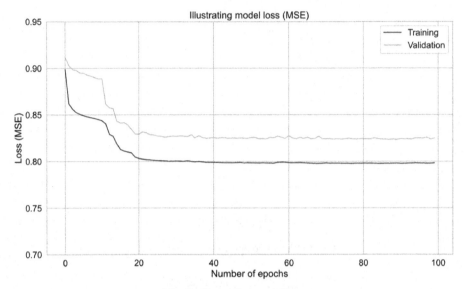

Fig. 5. Illustrating model loss

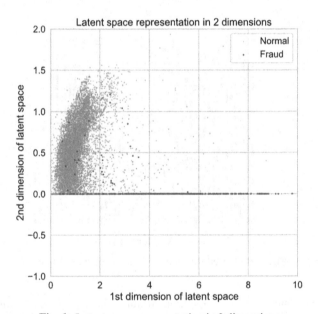

Fig. 6. Latent space representation in 2 dimensions

taken in order to scale the data to a range of [0; 1]. One approach is to apply linear scaling to the scores, although due to the nature of the distribution of the anomaly score, this approach does not effectively capture the separation between the classes as presented in Fig. 7.

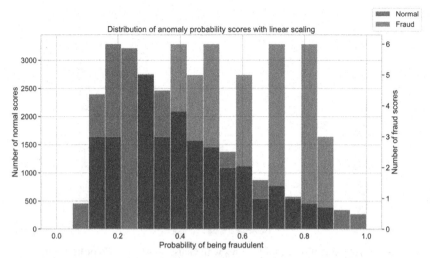

Fig. 7. Distribution of probability scores with linear scaling

An alternative approach, applying a Gaussian scaling function to the scores, is proposed in Algorithm 3. [11] propose that Gaussian scaling is a good general choice function for transforming raw anomaly scores. As is illustrated in Fig. 8, the resulting distribution of probability scores shows a much better separation between the classes. With the non-anomalous transactions in green using the left y-axis as its reference, and the anomalous transactions in red using the right y-axis as its reference, a superior split and one which captures the distribution of the underlying anomaly scores is presented. According to Fig. 8, the majority of the normal instances fall within the lowest score tier, whereas the majority of the anomalies fall within the 90% + tier. This is a favorable outcome for an operational function that makes use of probability scores such as these for their decision-making processes.

Algorithm 3 (Gaussian scaling of anomaly score algorithm)
Input: $S = \{s_1, \ldots, s_m\}$ is an array of anomaly scores, containing m samples.
Output: An array $U = \{u_1, \ldots, u_m\}$ containing the associated unified scores for m samples.
1 in one pass of array S
2   calculate μ_s; // mean
3   calculate σ_s; // standard deviation
4      $cdf = \frac{S(o) - \mu_s}{\sigma_s * \sqrt{2}}$; // calculate the cumulative distribution function of S
5   $erf(cdf)$; // transform the cdf with the Gaussian error function
6   normalise the transformation to the range [0; 1];
7 output U;

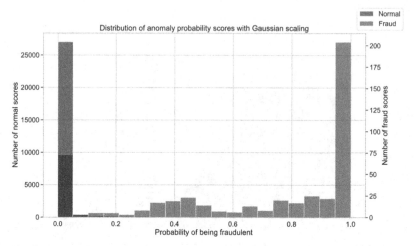

Fig. 8. Distribution of probability scores with Gaussian scaling (Color figure online)

5   Results and Discussion

Combining NAS and the combination of the 3 metrics for performance measurement provides a formal method for deciding on the appropriate architecture. The approach is straightforward to implement and understand, and the results show an increase in accuracy of around 15% when compared with a trial and error approach. The initial trial and error experiments are not reported in this paper however.

The determination of an optimized threshold is advantageous over the visual inspection approach often used in practice. While it may be possible to visually discriminate effectively, it is unlikely that this will yield a solution that is optimal according to a performance metric well suited to the distribution of the classes. Without the optimized threshold it is a subjective decision to decide which transactions are fraudulent. The optimized threshold approach therefore adds structure and consistency to the anomaly detection problem.

The Gaussian scaling function effectively transforms the raw anomaly scores into a distribution that is indicative of an anomaly detection problem. According to Fig. 7 using linear scaling yields a distribution that contains some level of discrimination between the classes, but as is shown by Fig. 8, Gaussian scaling helps to represent the scores in a manner that represents the discrimination between the classes evident in Fig. 2.

6   Conclusion

The purpose of this study was to report on the creation of a unified approach for detecting anomalies in a transactional setting. An example credit card fraud data set was used throughout the study to evaluate the unified approach.

A population of candidate models was simulated by means of NAS, with the optimal model being selected via the determination of a balanced score, namely an average of three effective evaluation metrics for anomaly detection problems. A threshold optimization algorithm was used to determine the binary classification outcomes within the NAS process. Finally, a Gaussian scaling function was applied to the resulting anomaly scores in order to transform them into a probability score. The resultant scores are easier to interpret than anomaly scores, and provide a more accurate representation of the distribution when compared with a linear scaling approach. Furthermore, the unified approach provides practitioners with a structured method for applying anomaly detection in an organizational context.

Future efforts in this space may be dedicated towards determining an upper bound on the anomaly scores, to investigate whether a better distribution of probability scores can be constructed. Other evolutionary approaches, such as genetic algorithms, may further enhance the NAS process.

References

1. Bengio, Y.: Learning deep architectures for AI, foundations and trends in machine. Learning **2**(1), 1–127 (2009)
2. Carcillo, F., Le Borgne, Y., Caelen, O., Kessaci, Y., Oble, F., Bontempi, G.: Combining unsupervised and supervised learning in credit card fraud detection. Inf. Sci. (2019). https://doi.org/10.1016/j.ins.2019.05.042
3. Chawla, N.V.: Data Mining for Imbalanced Datasets: An Overview, Data Mining and Knowledge Discovery Handbook, pp. 853–867. Springer, Boston (2005)
4. Chicco, D., Jurman, G.: The advantages of the Matthews correlation coefficient (MCC) over F1 score and accuracy in binary classication evaluation. BMC Genom. **21**(1), 6 (2020)
5. Dal Pozzolo, A.: Adaptive machine learning for credit card fraud detection, Ph.D. thesis, Université Libre De Bruxelles (2015)
6. Elsken, T.: Neural architecture search: a survey. J. Mach. Learn. Res. **20**, 1–21 (2019)
7. Fan, G., Zhu, M.: Detection of rare items with target. Stat. Interface **4**, 11–17 (2011)
8. Goodfellow, I., Bengio, Y., Courville, A.: Deep Learning, MIT Press (2016). www.deeplearningbook.org
9. Kaggle: Credit card fraud detection, Machine Learning Group – ULB (2017). www.kaggle.com/mlg-ulb/creditcardfraud
10. Khan, S., Taati, B.: Detecting unseen falls from wearable devices using channel-wise ensemble of autoencoders. Expert Syst. Appl. **87**, 280–290 (2017)
11. Kriegel, H., Kroger, P., Schubert, E., Zimek, A.: Interpreting and unifying outlier scores. In: 11th SIAM International Conference on Data Mining (SDM), Mesa, AZ, 2011 (2011)
12. Ng, A.: Sparse autoencoder, CS294A lecture notes, Stanford University, p. 1–19 (2011)
13. Phua, C., Lee, V., Smith, K., Gayler, R.: A comprehensive survey of data mining-based fraud detection research. In: Intelligent Computation Technology and Automation (ICICTA), pp. 50–53 (2010)
14. Schreyer, M., Sattarov, T., Schulze, C., Reimer, B., Borth, D.: Detection of accounting anomalies in the latent space using adversarial autoencoder neural networks. In: KDD-ADF, Association for Computing Machinery (2019)
15. Sommer, R., Paxson, V.: Outside the Closed World: on Using Machine Learning for Network Intrusion Detection. IEEE Symposium on Security and Privacy, Berkeley, CA (2010)
16. Sugar, C., James, G.: Finding the number of clusters in a data set. J. Am. Stat. Assoc. **98**(463), 750–763 (2003)
17. Weiss, G.M., McCarthy, K., Zabar, B.: Cost-sentitive learning vs sampling: which is best for handling unbalanced classes with unequal error costs. In: Proceedings of the 2007 International Conference on Data Mining (DMIN-07), CSREA Press, Las Vegas, NV (2007)

Skin Lesion Diagnosis with Imbalanced ECOC Ensembles

Sara Atito Ali Ahmed[1]([✉]) [iD], Berrin Yanikoglu[1] [iD], Cemre Zor[2] [iD],
Muhammad Awais[3] [iD], and Josef Kittler[3] [iD]

[1] Faculty of Engineering and Natural Sciences, Sabanci University, Istanbul, Turkey
sara.atito@gmail.com
[2] Centre for Medical Image Computing (CMIC), University College London,
London, UK
[3] Centre for Vision, Speech and Signal Processing (CVSSP), University of Surrey,
Guildford, UK

Abstract. Diagnosis of skin lesions is a challenging task due to the similarities between different lesion types, in terms of appearance, location, and size. We present a deep learning method for skin lesion classification by fine-tuning three pre-trained deep learning architectures (Xception, Inception-ResNet-V2, and NasNetLarge), using the training set provided by ISIC2019 organizers. We combine deep convolutional networks with the Error Correcting Output Codes (ECOC) framework to address the open set classification problem and to deal with the heavily imbalanced dataset of ISIC2019. Experimental results show that the proposed framework achieves promising performance that is comparable with the top results obtained in the ISIC2019 challenge leaderboard.

Keywords: Deep learning · Classification · Skin cancer · ECOC · Ensemble · Imbalanced dataset · Data augmentation

1 Introduction

Early detection of cancer is vital for successful treatment. Even though human experts can achieve a diagnostic accuracy of approximately 80% for classifying different skin cancer types [15], the number of dermatologists is insufficient when compared against the frequency of the disease occurrence [14]. On the other hand, the performance of computer-aided diagnostic tools has reached unprecedented levels in recent years. In particular, deep learning systems have reached a level of precision that is comparable to qualified dermatologists in classifying skin cancers [1,9].

As an effort to support clinical training and boost the performance of automated systems, the International Skin Imaging Collaboration (ISIC) has developed the ISIC Archive, an international repository of validated dermoscopic images, to enable the automated diagnosis of melanoma from dermoscopic images. The ISIC challenge has been organized annually since 2016, to measure performance of automated systems on this task [11].

© Springer Nature Switzerland AG 2020
G. Nicosia et al. (Eds.): LOD 2020, LNCS 12566, pp. 292–303, 2020.
https://doi.org/10.1007/978-3-030-64580-9_25

In previous years, the ISIC challenge aimed several sub-tasks related to the detection, segmentation and classification of skin lesions. This year, the ISIC2019 challenge focuses on lesion classification. The dataset contains 8 classes of skin cancer, while the evaluation is done as an open-set problem where images belonging to other cancer types need to be classified as the 9th class ("None of the others").

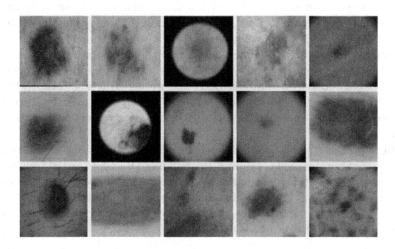

Fig. 1. Random samples of skin lesions from ISIC2019 Training set.

This paper presents the system we developed for the ISIC2019 challenge and details our findings. Our system is based on a set of various state-of-art Convolutional Neural Networks (CNNs), varying from each other in terms of architecture, pre-processing, training configurations, and data augmentation techniques. We combine deep convolutional networks with the Error Correcting Output Codes (ECOC) framework [8], to address the open set classification problem and to deal with the heavily imbalanced dataset. To the best of our knowledge, the proposed framework (ImbECOC) is novel and the end-to-end architecture has many advantages. Several task-specific data augmentation techniques are performed for further enhancement.

The rest of this paper is organized as follows. Section 2 describes the developed system based on the fine-tuning of Xception, Inception-ResNet-V2, and NasNetLarge models for skin lesion classification and the incorporation of ECOC framework. Description of the utilized dataset, data augmentation, and classifier fusion, along with the experiments and results are discussed in Sect. 3. The paper concludes in Sect. 4 with a summary and discussion of the utilized methods and obtained results.

2   Skin Lesion Classification with ECOC Ensemble

We adopt a deep learning approach, as used in many computer vision problems in recent years [2,21,23], with a special focus on two challenging aspects of the skin lesion classification problem: the open-set classification with the unknown class and the imbalance among the classes. Starting from multiple convolutional networks that are fine-tuned for this problem, we adopt the Error Correcting Output Codes (ECOC) approach, with the aim of addressing these two problems.

In our previous studies, we had tried training ECOC ensembles with deep features obtained from a convolutional network, as a quick and efficient way of obtaining deep learning ensembles, without much success in terms of accuracy [3]. In this work, the initial motivation was to use the decoding distance as a way to identify unknown samples. Later on, we added a new term in the loss function to address the class imbalance. Note that the end-to-end nature of the proposed ImbECOC model allows for the use of error correction together with a new loss function.

Our approach is composed of 3 elements: (i) fine-tuning deep learning models as the base of our ensemble approach (ii) building an ECOC model from each of these base networks (iii) fusing them by simple ensemble averaging. In Sect. 2.1, we describe the use of transfer learning to train our base convolutional neural networks, and in Sect. 2.2, we describe the use of ECOC codes to construct an ensemble from each trained base network.

2.1   Base Models

We use three of the top-ranked architectures trained on the ImageNet Large-Scale Visual Recognition Challenge (ILSVRC) in 2014 [18]: Xception [5], Inception-ResNet-V2 [19], and NasNetLarge [24].

The first network, Xception [5], is an extension and an extreme version of the Inception [20] architecture, which replaces the standard Inception modules with depth-wise separable convolutions. The network is 71 layers deep with only 22.9 million parameters. The image input size of the network is 299-by-299.

The second network, Inception-ResNet-V2 [19], is an advanced convolutional neural network that combines the inception module with ResNet [12] to increase the efficiency of the network. The network is 164 layers deep with only 55.9 million parameters. The image input size of the network is 299-by-299.

The third network we used, NasNetLarge [24], is built by training an architectural building block on a small dataset (e.g. CIFAR-10) and transfering it to a larger dataset by stacking together more copies of this block. Along with this new architecture, a new regularization technique called ScheduledDropPath is proposed that significantly improves the generalization of the proposed network. The image input size of the network is 331-by-331.

The models are pre-trained on the ILSVRC 2012 dataset with 1.2 million labeled images of 1,000 object classes and fine-tuned on the ISIC dataset. The distribution of the 8 given classes of the ISIC dataset is shown in Table 2. All training and testing were conducted on a Linux system with a Titan X Pascal

GPU and 12 GB of video memory. The network configurations are given in Table 1 and details of the different configurations are discussed in Sect. 3.2.

Table 1. Specifications of the trained CNN models.Specifications of the trained CNN models.

#	Model	Specifications		
		Batch Size	# Epochs	Loss Function
1	Xception	32	100	Cross Entropy
2	Xception	32	100	Focal Loss ($\gamma = 3$)
3	Inception-ResNet-V2	64	60	Cross Entropy
4	Inception-ResNet-V2	64	60	Focal Loss($\gamma = 3$)
5	Inception-ResNet-V2 (Shades of Gray)	64	60	Cross Entropy
6	NasNetLarge	20	25	Cross Entropy

Neural networks can have high variance due to the stochastic training approach: the models may find a different set of weights each time they are trained, which may produce different predictions. Furthermore, deep learning systems can overfit due to the large number of parameters, especially when dealing with small training data. Therefore, for the ISIC2019 challenge, we use an ensemble of 6 ImbECOC models that are obtained as described in Sect. 2.2.

2.2 The ImbECOC Framework

Ensemble learning techniques have been studied theoretically and experimentally, in the last 20 years [13], both in the context of obtaining strong classifiers from weak ones and also to improve performance even with highly accurate deep learning systems. They have been shown successfully in many machine learning and computer vision tasks [16,17,25].

In this work, we used an Error-Correcting Output Codes (ECOC) approach as the ensemble method to be used with convolutional neural networks.

In addition, we propose a novel approach to address the large imbalance in the ISIC dataset by changing the loss function of the ECOC ensemble,

Standard ECOC Framework: The main idea behind ECOC is to decompose the multi-class classification problem into several binary sub-problems. Standard ECOC has 3 stages: encoding, learning, and decoding.

In the encoding phase, given a multi-class classification problem with k classes, an encoding matrix $\mathbf{M}$, also called the *code matrix*, is often randomly generated to specify the binary subproblems for the ensemble. Each column M^j of the matrix indicates the desired output for the corresponding binary classifier of the ensemble. Each row M_i corresponds a unique codeword of length l for the corresponding class i, indicating the desired output for that class by each of the binary classifiers, also called the *base classifiers* in the ensemble.

In the learning phase, l independent binary classifiers are trained according to the given code matrix. Specifically, a base classifier $h_j, j = 1, \ldots l$ is learned according to the column M^j of the code matrix.

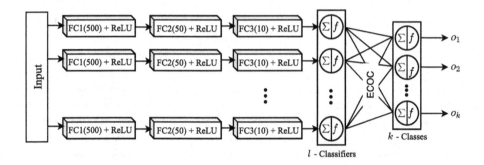

Fig. 2. ImbECOC architecture.

In decoding phase, a given test instance $\mathbf{x}$ is classified by each base classifier to obtain the output vector $\mathbf{Y} = [y_1, \ldots, y_l]$ where y_i is the hard or soft output of the classifier h_i for input $\mathbf{x}$. Then, the distance between $\mathbf{Y}$ and the *codeword* $\mathbf{M_i}$ of class c_i is computed, typically using the Hamming distance. The class $c*$ associated with the minimum distance is chosen as the predicted class, such that

$$c* = \arg \min_{i=1\ldots k} \text{dist}(\mathbf{Y}, \mathbf{M_i}) \tag{1}$$

ImbECOC Model: We use the ECOC framework in combination with the base networks described in 2.1, to obtain the ImbECOC model, which is designed to address the open-set classification problem and the class imbalance issue.

The input to ImbECOC consists of deep features extracted from the last pooling layer of a base network. Using this input, the base classifiers of the ECOC model are trained simultaneously, using a multi-task learning approach, within a single network. Finally, an output layer whose weights are obtained from the ECOC matrix, is added, so as to directly map the input to the classes. In this way, we obtain an end-to-end network that can be trained with a variety of loss functions. The network architecture is shown in Fig. 2.

The output o_i of ImbECOC, corresponding to class i is the dot product between the output $\mathbf{Y}$ of the base classifiers of the ECOC and the codeword $\mathbf{M_i}$ of class i. This output ranges in $[-l, l]$ and is maximum when the output of all l base classifiers match the desired codeword of the input. The output is minimum when the base classifiers' output are totally the opposite of the codeword.

The loss function for the correct class c is shown in Eq. 2, where x is the input; y_i is the output of the base classifier h_i; M(c,i) is the ECOC matrix element for the correct class and the ith base classifier and o_c is the output of the correct class:

$$Loss = \sum_{i=1}^{l} (y_i - M(c,i))^2 + (o_c - l)^2 \tag{2}$$

The first term penalizes difference between the prediction the base classifiers and the corresponding codeword entries for the correct class. The second term is used to maximize the output o_c corresponding to the correct class.

The standard ECOC aims to maximize the accuracy of the binary classification task regardless of the importance of the input sample. This makes standard ECOC inadequate to deal with imbalanced multi-class data. To address the imbalanced class issue, we added the normalized inverse frequency of each class, W_c, to the loss function as follows:

$$Loss = \sum_{i=1}^{l} (y_i - M(c,i))^2 + W_c \times (o_c - l)^2 \tag{3}$$

where W_c is the inverse frequency of class c. By incorporating W_c to the loss function, the optimal weights are obtained by minimizing a weighted loss in favor of the minority classes.

During testing, the posterior probability of each class i given the input x ($P(c_i|x)$) is directly obtained by normalizing the output of the ImbECOC network The normalization is done by omitting the negative values (least probable classes) and dividing by l, which is the maximum value that can be obtained, to normalize the probabilities to range between 0 and 1. Low confidence scores are used in detecting unknown samples, which was the original motivation for the use of ECOC in this work.

3 Experiments and Results

3.1 Dataset and Problem Definition

The training data of ISIC2019 includes skin lesion images from several datasets, such as: HAM10000 [22], BCN20000 [7], and MSK [6] datasets. The dataset consists of 25,331 images for training across 8 different categories, while the test set contains an additional outlier (unknown) class. The goal of ISIC2019 competition is to classify dermoscopic images among these nine categories, as shown in Table 2.

Two tasks are available for this competition: 1) classify dermoscopic images without meta-data, and 2) classify images with additional available meta-data. In this paper, we target the first task where only the provided images are used, without using the meta-data or any external dataset.

3.2 Base Networks

In order to construct an ensemble, we train several convolutional neural network models that are differentiated by varying the input channels, batch sizes, loss functions, and training durations.

Table 2. Distribution of the available ISIC2019 training images across the 8 given skin lesion categories.

Diagnostic	# Images	Ratio
Melanoma (MEL)	4522	17.85%
Melanocytic nevus (NV)	12875	50.83%
Basal cell carcinoma (BCC)	3323	13.12%
Actinic keratosis (AK)	867	3.42%
Benign keratosis (BKL)	2624	10.36%
Dermatofibroma (DF)	239	0.94%
Vascular lesion (VASC)	253	1%
Squamous cell carcinoma (SCC)	628	2.48%
None of the others (UNK)	0	0%

For the input of the models, we feed the employed $CNNs$ with the RGB images except for one model where Shades of Gray [10] color constancy method is applied to the images before training with Minkowski norm $p = 6$ as suggested in [4]. For the loss functions, we evaluated the cross entropy and focal loss. In particular, the focal loss function given in Eq. 4, is used to address the imbalance between classes.

$$FL(p, y) = -\sum_i \alpha_i y_i (1 - p_i)^\gamma \log(p_i) \tag{4}$$

where p_i and y_i are the prediction and the ground-truth of a given sample, respectively. The value of α is set to the inverse class frequency. Lastly, the value of γ is finetuned. The highest validation accuracy is obtained when γ is equal to 3 in all of our experiments. The training configurations that are used are summarized in Table 1.

RMSPROP optimizer is used for training all of the CNN models with an initial learning rate of 3e−3. The learning rates were selected based on the validation accuracy.

3.3   ImbECOC Model

The architecture of the ImbECOC model is shown in Fig. 2. It is trained on top of each CNN trained model, using the extracted features of the last pooling layer of that model as input.

Every branch in the model consists of 3 hidden layers with 500, 50, and 10 units. The first 2 hidden layers are followed by a dropout layer and a Rectified Linear Unit (ReLU) activation function. Third layer is followed by only ReLU activation layer. Output layer consists of l units and followed by a hyperbolic tangent (Tanh) layer.

We used the RMSPROP optimizer for training ImbECOC with an initial learning rate of 3e−4 and batch size of 512 samples.

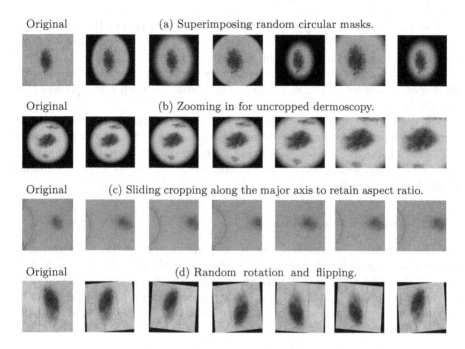

Fig. 3. Random augmented samples from ISIC2019 training set.

3.4 Data Augmentation

To build a powerful deep learning classifier, large and quality training dataset are essential. Data augmentation plays a crucial role in avoiding overfitting and increasing the number of training images specially when number of images of different categories varies widely.

In this work, we applied several commonly used data augmentation techniques during training, such as rotation $[-10$ to $10]$, x and y translation $[-5, 5]$, and vertical and horizontal flipping. Random augmented samples are shown in Fig. 3. Three task-specific data augmentation techniques are also proposed for further enhancement, as described below.

The ISIC2019 dataset consists of a large number of *uncropped* dermoscopy images, where black area surrounds the lesion. As one task specific data augmentation technique, we randomly superimposed circular masks to the images during training (top row of Fig. 3). Particularly, we multiplied a circular mask positioned in the center of the dermoscopy image with random radius followed by a Gaussian filter.

A second augmentation technique is used to eliminate the dark circle around the uncropped dermoscopy[1]. For this, several crops with random sizes are performed around the center of the image (second row from the top in Fig. 3).

---

[1] We consider a dermoscopy as uncropped if one-fourth the number of original pixels are dark.

Lastly, to retain the aspect ratio and avoid deformation of the dermoscopy, we feed the network with random square cropping along the major axis of the dermoscopy images (third row from the top in Fig. 3).

Random samples of the superimposed images, zooming in, and slide cropping strategies are shown in Fig. 3, where the first image in each row represents the original image.

All data augmentation were applied on the fly, which means that, at every iteration, different setting of augmentations are applied on top of the original batch of images. Finally, in testing time, we applied test time augmentation. Specifically, we applied 50 random augmentations similar to the techniques applied during training.

3.5   Results and Discussion

For the evaluation, augmented test images are passed to the 6 trained deep learning models and test features extracted from the last pooling layer from each model. These are then fed to the trained ImbECOC models. Score-level averaging is applied to combine the prediction scores assigned to each class for all the augmented patches, within a single ImbECOC model; and also globally, among the different ImbECOC models.

Table 3. Validation and testing accuracies across different models. The reported accuracy is the normalized multi-class accuracy.

Model	Validation Accuracy		Testing Accuracy	
	CNN	w/ ImbECOC	CNN	w/ ImbECOC
Model 1	0.674	0.808	0.480	0.497
Model 2	0.657	0.670	0.486	0.514
Model 3	0.809	0.798	0.498	0.499
Model 4	0.634	0.797	0.512	0.519
Model 5	0.715	0.744	0.475	0.495
Model 6	0.789	0.851	0.512	0.521
Ensemble	0.837	0.904	0.553	0.602

The maximum probability obtained by any of the k classes, $maxscore$, is calculated and used in rejecting a sample (to label it as UNK). The probability of the UNK class is set to $1 - maxscore$ if $maxscore$ is less than 0.3 (indicating that the agreement between the output vector and the closest codeword is less than 30%. If $maxscore$ is higher than 0.3, the sample is accepted to belong to one of the 8 classes and the probability of the UNK class is set to 0. The 0.3 threshold is found experimentally.

Following the aforementioned strategy, the validation and testing accuracies with and without ImbECOC are shown in Table 3. Considering these results, we see that simple ensemble averaging improves over the individual networks, obtaining 55.3% normalized multi-class accuracy and ImbECOC further improves the normalized multi-class accuracy by 4.9% points, reaching 60.2%.

The ROC curve (true positive rate against the false positive rate) of each lesion category, individually is shown in Fig. 4.

The system described in this paper has the 2nd highest ranked results in the live competition at the time of paper submission, *among teams that do not use external data*[2].

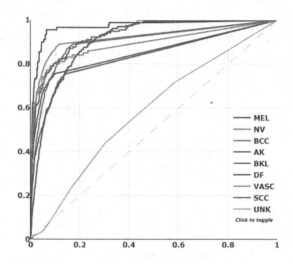

Fig. 4. ROC curve of the 9 skin lesion classes using deep CNNs ensembles and ImbECOC.

4   Conclusions

The core of our approach is based on an ensemble of deep networks, based on three separate pre-trained deep learning architectures (Xception, Inception-ResNet-V2, and NasNetLarge), trained using training images provided by the ISIC2019 organizers.

ImbECOC approach is presented to address the unknown class and the imbalanced dataset issues, as well as to improve the prediction accuracy. The presented system is 2nd from the top among groups not using external data, at the ISIC competition, at the time of the submission.

---

[2] https://challenge2019.isic-archive.com/live-leaderboard.html.

References

1. Alquran, H., et al.: The melanoma skin cancer detection and classification using support vector machine. In: 2017 IEEE Jordan Conference on Applied Electrical Engineering and Computing Technologies (AEECT), pp. 1–5. IEEE (2017)
2. Atito, S., Yanikoglu, B.A., Aptoula, E.: Plant identification with large number of classes: sabanciu-gebzetu system in plantclef 2017. In: Working Notes of CLEF 2017 - Conference and Labs of the Evaluation Forum, Dublin, Ireland, September 11–14, 2017. vol. 1866 (2017)
3. Atito, S., et al.: Plant identification with deep learning ensembles. In: Working Notes of CLEF 2018 - Conference and Labs of the Evaluation Forum, Avignon, France, 10–14 September 2018, vol. 2125 (2018)
4. Barata, C., Celebi, M.E., Marques, J.S.: Improving dermoscopy image classification using color constancy. IEEE J. Biomed. Health Inform. **19**(3), 1146–1152 (2014)
5. Chollet, F.: Xception: deep learning with depthwise separable convolutions. In: Proceedings of the IEEE Conference on Computer Vision and Pattern Recognition, pp. 1251–1258 (2017)
6. Codella, N.C.F., et al.: Skin lesion analysis toward melanoma detection. In: IEEE 15th International Symposium on Biomedical Imaging (ISBI), pp. 168–172 (2018)
7. Combalia, M., et al.: Bcn20000: Dermoscopic lesions in the wild (2019)
8. Dietterich, T.G., Bakiri, G.: Solving multiclass learning problems via error-correcting output codes. J. Artif. Intell. Res. **2**, 263–286 (1994)
9. Esteva, A., et al.: Dermatologist-level classification of skin cancer with deep neural networks. Nature **542**(7639), 115–118 (2017)
10. Finlayson, G.D., Trezzi, E.: Shades of gray and colour constancy. In: Color and Imaging Conference, vol. 2004, pp. 37–41. Society for Imaging Science and Technology (2004)
11. Gutman, D., et al.: Skin lesion analysis toward melanoma detection: a challenge at the international symposium on biomedical imaging (ISBI) 2016, hosted by the international skin imaging collaboration (ISIC). arXiv preprint arXiv:1605.01397 (2016)
12. He, K., Zhang, X., Ren, S., Sun, J.: Deep residual learning for image recognition. In: Proceedings of the IEEE Conference on Computer Vision and Pattern Recognition, pp. 770–778 (2016)
13. Kearns, M.J., Valiant, L.G.: Learning boolean formulae or finite automata is as hard as factoring (1988)
14. Kimball, A.: The US dermatology workforce: a specialty remains in shortage. J. Am. Acad. Dermatol. **59**, 741–745 (2008)
15. Kittler, H., Pehamberger, H., Wolff, K., Binder, M.: Diagnostic accuracy of dermoscopy. The Lancet Oncol. **3**, 159–165 (2002)
16. Liu, M., Zhang, D., Chen, S., Xue, H.: Joint binary classifier learning for ecoc-based multi-class classification. IEEE Trans. Pattern Anal. Mach. Intell. **38**, 2335–2341 (2016)
17. Qin, J., Liu, L., Shao, L., Shen, F., Ni, B., Chen, J., Wang, Y.: Zero-shot action recognition with error-correcting output codes. In: The IEEE Conference on Computer Vision and Pattern Recognition (CVPR), July 2017
18. Russakovsky, O., et al.: ImageNet large scale visual recognition challenge. Int. J. Comput. Vis. **115**(3), 211–252 (2015). https://doi.org/10.1007/s11263-015-0816-y
19. Szegedy, C., Ioffe, S., Vanhoucke, V., Alemi, A.A.: Inception-v4, inception-resnet and the impact of residual connections on learning. In: Thirty-First AAAI Conference on Artificial Intelligence (2017)

20. Szegedy, C., et al.: Going deeper with convolutions. In: Proceedings of the IEEE Conference on Computer Vision and Pattern Recognition, pp. 1–9 (2015)
21. Tan, C., Sun, F., Kong, T., Zhang, W., Yang, C., Liu, C.: A Survey on Deep Transfer Learning: 27th International Conference on Artificial Neural Networks, Rhodes, Greece, October 4–7, 2018, Proceedings, Part III, pp. 270–279 (2018)
22. Tschandl, P., Rosendahl, C., Kittler, H.: The ham10000 dataset, a large collection of multi-source dermatoscopic images of common pigmented skin lesions. Scientific Data **5**, 180161 (2018)
23. Ud Din, I., Rodrigues, J., Islam, N.: A novel deep learning based framework for the detection and classification of breast cancer using transfer learning. Pattern Recogn. Lett. **125** (2019)
24. Zoph, B., Vasudevan, V., Shlens, J., Le, Q.V.: Learning transferable architectures for scalable image recognition. In: Proceedings of the IEEE conference on Computer Vision and Pattern Recognition, pp. 8697–8710 (2018)
25. Zor, C., Windeatt, T., Yanikoglu, B.A.: Bias-variance analysis of ECOC and bagging using neural nets. In: Ensembles in Machine Learning Applications (2011)

State Representation Learning
from Demonstration

Astrid Merckling$^{(\boxtimes)}$, Alexandre Coninx, Loic Cressot, Stephane Doncieux,
and Nicolas Perrin

Sorbonne University, CNRS, Institute for Intelligent Systems and Robotics (ISIR),
Paris, France
{astrid.merckling,alexandre.coninx,stephane.doncieux,
nicolas.perrin}@sorbonne-universite.fr

Abstract. Robots could learn their own state and universe representation from perception, experience, and observations without supervision. This desirable goal is the main focus of our field of interest, State Representation Learning (SRL). Indeed, a compact representation of such a state is beneficial to help robots grasp onto their environment for interacting. The properties of this representation have a strong impact on the adaptive capability of the agent. Our approach deals with imitation learning from demonstration towards a shared representation across multiple tasks in the same environment. Our imitation learning strategy relies on a multi-head neural network starting from a shared state representation feeding a task-specific agent. As expected, generalization demands tasks diversity during training for better transfer learning effects. Our experimental setup proves favorable comparison with other SRL strategies and shows more efficient end-to-end Reinforcement Learning (RL) in our case than with independently learned tasks.

Keywords: State Representation Learning · Imitation learning from demonstration · Deep reinforcement learning

1 Introduction

Recent RL achievements might be attributed to a combination of (i) a dramatic increase of computational power, (ii) the remarkable rise of deep neural networks in many machine learning fields including robotics which takes advantage of the simple idea that training with quantity and diversity helps. The core idea of this work consists of leveraging task-agnostic knowledge learned from several task-specific agents performing various instances of a task.

Learning is supposed to provide animals and robots with the ability to adapt to their environment. RL algorithms define a theoretical framework that is efficient on robots [16] and can explain observed animal behaviors [23]. These algorithms build policies that associate an action to a state to maximize a reward. The state determines what an agent knows about itself and its environment. A

© Springer Nature Switzerland AG 2020
G. Nicosia et al. (Eds.): LOD 2020, LNCS 12566, pp. 304–315, 2020.
https://doi.org/10.1007/978-3-030-64580-9_26

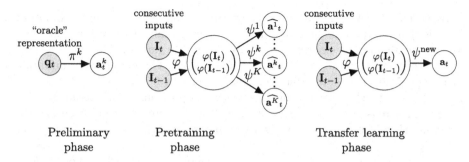

Preliminary   Pretraining   Transfer learning
phase   phase   phase

Fig. 1. Our State Representation Learning from Demonstration method (SRLfD) consists of three phases. **(a)** Preliminary phase: learning of K target policies π^k solving multiple tasks in the same environment with "oracle" representations as inputs. **(b)** Pretraining phase: learning of one shared representation function φ and K specific heads ψ^k for each task indexed by k with high-dimensional observations (e.g. raw images) as inputs. This phase's goal is to imitate the decisions $\mathbf{a}_t^k$ from the previous phase. Each head ψ^k defines a sub-network that contains the parameters $\boldsymbol{\theta}_\varphi$ of φ and the parameters $\boldsymbol{\theta}_{\psi_k}$ of ψ^k. The set of all network's parameters is $\boldsymbol{\theta} = \{\boldsymbol{\theta}_\varphi, \boldsymbol{\theta}_{\psi_1}, \ldots, \boldsymbol{\theta}_{\psi_K}\}$. **(c)** Transfer learning phase: the now trained network φ provides representations for the new learning process ψ^{new} on a new task.

large state space – raw sensor values, for instance – may contain the relevant information but would require a too large exploration to build an efficient policy. Well-thought feature engineering can often solve this issue and make the difference between the failure or success of a learning process. In their review of representation learning, Bengio et al. [2] formulate the hypothesis that the most relevant pieces of information contained in the data can be more or less entangled and hidden in different representations. If a representation is adequate, functions that map inputs to desired outputs are somewhat less complex and thus easier to construct via learning. However, a frequent issue is that these adequate representations may be task-specific and difficult to design, and this is true in particular when the raw data consists of images, i.e. 2D arrays of pixels. One of the objectives of deep learning methods is to automatize feature extraction to make learning algorithms effective even on raw data. By composing multiple non-linear transformations, the neural networks on which these methods rely are capable of progressively creating more abstract and useful representations of the data in their successive layers.

SRL is especially meaningful for transfer learning. Indeed, completely unsupervised state representations are unlikely to succeed as all information from sensors is not always useful for operating an agent in RL (this is still true even for simpler image classification settings [24] in which representation learning is getting increasing attention). In the context of RL, we believe SRL towards transfer learning is a natural avenue of research because physical laws like 3D orientation, gravity, motion, etc. should not be relearned every time we tackle a new task from the same environment. Thus transferring a form of shared knowledge between many tasks seem reasonable and promising. The intuition behind our work is that many tasks operated in the same environment share some common knowledge about that environment. This is why learning all these tasks with a shared representation at the same time is beneficial. The literature in imitation learning [17,20] has shown that demonstrations can be very valuable to learn new policies. To the best of our knowledge, no previous work has focused on constructing reusable state representations from raw inputs solely from demonstrations, therefore, here we investigate the potential of this approach for transfer learning.

In this paper, we are interested in solving robotic control tasks via end-to-end RL, using raw images as inputs, without having access to any other sensor, which means in particular that the robot configuration is a priori unknown (in the illustrative example we consider for experimental evaluation, we do not give access to the joint angles and angular velocities of a 2D robot arm). We assume that at all times the consecutive high-dimensional observations $(\mathbf{I}_{t-1}, \mathbf{I}_t)$ (images) contain enough information to know the robot configuration $\mathbf{q}_t$ and that the controller only needs to rely on this configuration to choose actions. Intuitively, the representation $\mathbf{q}_t$, which we will call *ground truth representation* in the remainder of the paper, could probably be a much better input for an RL algorithm than the raw images, but without prior knowledge, it is not easy to get $\mathbf{q}_t$ from $\mathbf{I}_t$. In robotics, SRL [18] aims at constructing a mapping from high-dimensional observations to lower-dimensional and denser representations that, similarly to $\mathbf{q}_t$, can be advantageously used instead of $\mathbf{I}_t$ to form the inputs of a policy.

Our proposed experimental setup consists in three different training processes. Preliminary phase (Fig. 1a): we place in *laboratory conditions*, where the availability of sensory inputs (what we call "oracle" representation) makes it relatively easy to train K target policies π^k solving different instances of a task. It could be human agents. For the sake of the experiences, we trained almost fake tasks. Pretraining phase (Fig. 1b): we derive a state representation that can be relied on to reproduce any of these policies. We do so via imitation learning on a multi-head neural network consisting of a first part that outputs a common state representation $(\varphi(\mathbf{I}_{t-1}^k), \varphi(\mathbf{I}_t^k))$ used as input to K heads ψ^k trained to predict actions $\mathbf{a}_t^k$ executed by the target policies π^k from the previous phase. Transfer learning phase (Fig. 1c): we use the previously trained representation $(\varphi(\mathbf{I}_{t-1}^k), \varphi(\mathbf{I}_t^k))$ as input of a new learning process ψ^{new} to solve a new task in the same environment. This method, which we call State Representation Learning from Demonstration (SRLfD), is presented in more detail in Sect. 3, after an overview of the existing related work in Sect. 2. We show that using state representations instead of raw images can significantly accelerate RL (using the popular SAC algorithm [9]). When the state representation is chosen to be low-dimensional, the speed up brought by our method is greater than the one resulting from state representations obtained with deep autoencoders, or with principal component analysis (PCA).

2   Related Work

SRL for control is the idea of extracting from the sensory stream the information that is relevant to control the robot and its environment and representing it in a way that is suited to drive robot actions. It has been subject to a lot of recent attention [18]. It was proposed as a way to overcome the curse of dimensionality, to speed up and improve RL [4], to achieve transfer learning, to ignore distractors, and to make artificial autonomous agents more transparent and explainable.

Since the curse of dimensionality is a major concern, many state representation techniques are based on dimension reduction [2] and classic unsupervised learning techniques such as principal component analysis (PCA) [6] or its nonlinear version, the autoencoder [11]. Those techniques allow to compress the observation in a compact latent space, from which it can be reconstructed with minimal error. Further developments led to variational autoencoders (VAE) [15]. However, the goal of those methods is to model the observation data; they do not take actions or rewards into account, and the representation they learn is optimized to minimize a reconstruction loss, not to extract the most relevant information for control. In particular, their behavior is independent of actions, rewards, or the temporal structure of the observations, and they cannot discriminate distractors.

To overcome this limitation, a common approach to state representation is to couple an autoencoder to a forward model predicting the future state [26]. A different approach to state representation is to forego observation reconstruction and to learn a representation satisfying some physics-based priors like temporal coherence, causality, and repeatability [12] or controllability [13]. Those methods have been shown to learn representations able to speed up RL, but this improvement is contingent on the careful choice and weighting of the priors suited to the task and environment.

Learning state representations from demonstrations of multiple policies solving different tasks instances, as we propose, also has some similarities with multi-task and transfer learning [25]. Multi-task learning aims to learn several similar but distinct tasks simultaneously to accelerate the learning or improve the quality of the learned policies, while transfer learning strives to exploit the knowledge of how to solve a given task to then improve the learning of a second task. Not all multi-task and transfer learning works rely on explicitly building a common representation, but some do, either by using a shared representation during multiple task learning [21] or by distilling a generic representation from task-specific features [22]. The common representation can then be used to learn new tasks. All those techniques do allow to build state representations in a sample-efficient way, but they rely on exploration of the task space and on-policy data collection, whereas we focus on situations where only a limited number of preexisting demonstrations from a given set of existing policies are available.

In another perspective, the learning from demonstration literature typically focuses on learning from a few examples and generalizing from those demonstrations, for example by learning a parameterized policy using control-theoretic methods [20] or RL-based approaches [17]. Although those methods typically assume prior knowledge of a compact representation of the robot and environment, some of them directly learn and generalize from visual input [7] and do learn a state representation. However, the goal is not to reuse that representation to learn new skills but to produce end-to-end visuomotor policies generalizing the demonstrated behaviors in a given task space.

3   Learning a State Representation from Demonstration

3.1   Demonstrations

Let us clarify the hierarchy of the objects that we manipulate and introduce our notations. This work focuses on simultaneously learning K tasks[1] $(T^1, T^2, \ldots, T^K)$ sharing a common state representation function φ and with K task-specific heads for decision $(\psi_1, \psi_2, \ldots, \psi_K)$ (see Fig. 1b). For each task T^k, the algorithm has seen demonstrations in a form of paths $P_1^k, P_2^k, \ldots, P_P^k$ from a random position initialization to the same objective corresponding to the task k generated by running the target policy π^k of the preliminary phase (see Fig. 1a). More specifically, during a path P_p^k, an agent is shown a demonstration of $(\mathbf{I}_{t-1}^{k,p}, \mathbf{I}_t^{k,p}, \mathbf{a}_t^{k,p})$ from which it can build its own universe-specific representation. Here, $\mathbf{I}_{t-1}^{k,p}$ and $\mathbf{I}_t^{k,p}$ are consecutive high-dimensional observations, and $\mathbf{a}_t^{k,p}$ is a real-valued vector corresponding to the action executed right after the observation $\mathbf{I}_t^{k,p}$ was made.

Algorithm 1. SRLfD algorithm

1: **Input:** A set of instances of tasks T^k, $k \in [\![1, K]\!]$, and for each of them a set of paths P_p^k, $p \in [\![1, P]\!]$ of maximum length T
2: **Initialization:** A randomly initialized neural network following the architecture described in Fig.1b with weights $\boldsymbol{\theta} = \{\boldsymbol{\theta}_\varphi, \boldsymbol{\theta}_{\psi_1}, \ldots, \boldsymbol{\theta}_{\psi_K}\}$
3: **while** $\boldsymbol{\theta}$ has not converged **do**
4:    Sample uniformly a task T^k
5:    Compute

$$\mathcal{L} \leftarrow \frac{1}{PT} \sum_{p=1}^{P} \sum_{t=1}^{T} \|\psi^k \left(\varphi(\mathbf{I}_{t-1}^{k,p}), \varphi(\mathbf{I}_t^{k,p}) \right) - \mathbf{a}_t^{k,p}\|_2^2$$

6:    Perform a gradient descent step on $\mathcal{L}$ w.r.t. $\boldsymbol{\theta}$
7: **end while**

---

[1] Roughly, different tasks refer to objectives of different natures, while different instances of a task refer to a difference of parameters in the task. For example, reaching various locations with a robotic arm is considered as different instances of the same reaching task.

3.2   Imitation Learning from Demonstration

Following the architecture described in Fig. 1b, we use a state representation neural network φ that maps high-dimensional observations $\mathbf{I}_t^{k,p}$ to a smaller real-valued vector $\varphi(\mathbf{I}_t^{k,p})$. This network φ is applied to consecutive observations $\mathbf{I}_{t-1}^{k,p}$ and $\mathbf{I}_t^{k,p}$, and the concatenated outputs $(\varphi(\mathbf{I}_{t-1}^{k,p}), \varphi(\mathbf{I}_t^{k,p}))$ is the state representation. This state representation $(\varphi(\mathbf{I}_{t-1}^{k,p}), \varphi(\mathbf{I}_t^{k,p}))$ is sent to the ψ^k network, where ψ^k is one of the K independent heads of our neural network architecture, corresponding to the current task: ψ^1, ψ^2, ..., ψ^K are head networks with similar structure but different weights, each one corresponding to a different task. Each head has continuous outputs with the same number of dimensions as the action space of the robot. We denote by $\psi^k(\varphi(\mathbf{I}_{t-1}^{k,p}), \varphi(\mathbf{I}_t^{k,p}))$ the output of the k-th head of the network on the input $(\mathbf{I}_{t-1}^{k,p}, \mathbf{I}_t^{k,p})$. We train the global network to imitate all the target policies via supervised learning in the following way. Our goal is to minimize the quantities $\|\psi^k(\varphi(\mathbf{I}_{t-1}^{k,p}), \varphi(\mathbf{I}_t^{k,p})) - \mathbf{a}_t^{k,p}\|_2^2$ that measure how well the target policies are imitated. We want to solve the optimization problem

$$\mathcal{L} = \frac{1}{PT} \sum_{p=1}^{P} \sum_{t=1}^{T} \|\psi^k\left(\varphi(\mathbf{I}_{t-1}^{k,p}), \varphi(\mathbf{I}_t^{k,p})\right) - \mathbf{a}_t^{k,p}\|_2^2 \tag{1}$$

$$\arg\min_{\theta} \mathcal{L}(\boldsymbol{\theta})$$

for $k \in [\![1, K]\!]$. We give an equal importance to all target policies by uniformly sampling $k \in [\![1, K]\!]$, and performing a gradient descent step on $\mathcal{L}$ to adjust $\boldsymbol{\theta}$. Algorithm 1 describes this procedure.

The network of SRLfD is trained to reproduce the demonstrations, but without direct access to the structured state. Each imitation can only be successful if the required information about the robot configuration is extracted by the state representation $(\varphi(\mathbf{I}_{t-1}^{k,p}), \varphi(\mathbf{I}_t^{k,p}))$. However, a single task may not require the knowledge of the full robot state. Hence, we cannot be sure that reproducing only one instance of a task would yield a good state representation. By learning a common representation for various instances of tasks, we increase the probability that the learned representation is general and complete. It can then be used as a convenient input for downstream learning tasks, in particular end-to-end RL.

4   Experimental Setup

This section describes the details of our experimental setup (see Fig. 1), consisting of the preliminary phase that creates the target policies π^k, the pretraining phase that builds the state representation function φ, and the transfer learning phase that is chosen here as a challenging RL training from raw pixels.

Environment. We consider a simulated 2D robotic arm with 2 torque-controlled joints, as shown in Fig. 2. We adapted the *Reacher* environment from the OpenAI Gym [3] benchmark on continuous control tasks with PyBullet [5]. An instance

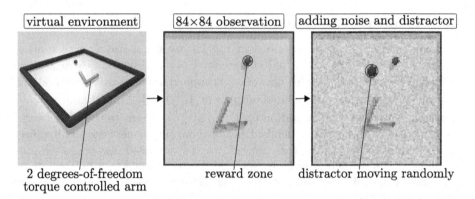

| virtual environment | 84×84 observation | adding noise and distractor |

2 degrees-of-freedom torque controlled arm   reward zone   distractor moving randomly

Fig. 2. The *Reacher* environment, with a reward of 1 when the end-effector reaches a position close to the goal, and 0 otherwise. For more challenging inputs (on the right), we add Gaussian noise with zero mean and standard deviation 10, and a distractor is added in the environment with random initial position and velocity.

of this task is parameterized by the position of a goal that the end-effector of the robot must reach within some margin of error (and in limited time). We use as raw inputs grayscale images of 84×84 pixels. As the heart of our work concerns state representation extraction, we have focused on making perception challenging, by adding in some cases randomly moving distractor and Gaussian noise, as shown in Fig. 2. We believe that the complexity of the control part (i.e. the complexity of the tasks) is less important to validate our method, as it depends more on the performance of the RL algorithm. To solve even just the simple reaching task, the information of the robot arm position, for instance, is required and needs to be extracted from the image for the RL algorithm to converge. Indeed our results show that this is the case when SRLfD compressed state representation is used as input of SAC [9].

Generating Demonstrations. For simplicity, the preliminary phase of training K target policies π^k (see Fig. 1a) is done by running the SAC [9] with HER [1] RL algorithms. Here, the "oracle" representations used in inputs are the *ground truth representations* $\mathbf{q}_t$, where $\mathbf{q}$ is the configuration of the robot arm, represented as a vector of size four: the two torques angles and velocities. HER also exploits the cartesian coordinates of the goal position. It returns a parameterized policy capable of producing reaching motions to any goal position. Note that the purpose of our method is to generate state representations from (possibly noisy) inputs that are hard to exploit (such as raw images), so only the preliminary phase has access to the "oracle" representation. For the pretraining phase of SRLfD (see Fig. 1b) these target policies π^k generated $K = 16$ different instances of the reaching task, with each of them 238 paths for training and 60 paths for validation of maximal length 50, computed from various initial positions.

Baseline Methods. We compare state representations obtained with our method (SRLfD) to five other types of state representations. The ground truth representation of the robot configuration is represented as a vector of size four: the two torques angles and velocities. With raw pixels, the network has the same structure (it receives $(\mathbf{I}_{t-1}, \mathbf{I}_t)$ in input) with the same dimensionality reduction after φ as other methods, but all of its parameters are trained end-to-end via RL. We compare also with three other trained or fixed representations: PCA, VAE [15], and a random network representation [8] where instead of training the first part φ of the network (as in the "raw pixels" case), its parameters are simply fixed to random values drawn from a normal distribution of zero mean and standard deviation 0.02. Note that the representations learned with baseline methods replace φ in the architecture of Fig. 1c, so the full state representation $(\varphi(\mathbf{I}_{t-1}), \varphi(\mathbf{I}_t))$ has twice as many dimensions as the number of dimensions of the latent space. All baselines are learned on the same demonstration data as SRLfD method, and share the same neural network structure for φ whose output size is 8 or 24^2.

Implementation Details. For SRLfD network architecture (adapted from the one used in [19]), φ (see Fig. 1) sends its 84×84 input to a succession of three convolution layers. The first one convolves 32 8×8 filters with stride four. The second layer convolves 64 4×4 filters with stride two. The third hidden layer convolves 32 3×3 filters with stride one. It ends with a fully connected layer with half as many output units as the chosen state representation dimension (because state representations have the form $(\varphi(\mathbf{I}_{t-1}), \varphi(\mathbf{I}_t))$. The heads ψ^k take as input the state representation and are composed of three fully connected layers, the two first ones of size 256 and the last one of size two, which corresponds to the size of the action vectors (one torque per joint). The Rectified linear units (ReLU) is used for the activation functions between hidden layers.

For SAC network architecture we choose a policy network that has the same structure as the heads ψ^k used for imitation learning, also identical to the original SAC implementation [9], and use the other default hyper-parameters.

We use ADAM [14] with a learning rate of 0.0001 to train the neural network φ, and 0.001 to train all the heads ψ^k. For SAC training, we use early stopping of patience 20 that corresponds to how many epochs we want to wait after the last time the validation loss improved before breaking the training loop (to avoid overfitting).

5   Results

For the transfer learning phase (see Fig. 1c) we consider a whole end-to-end RL training, which is challenging despite the simplicity of the task considered in this work. When high dimension observations are mapped to a lower dimensional space before feeding the RL a lot of information is compressed and valuable information for robotics control may be lost. We conducted a quantitative evaluation

---

[2] The number of 24 dimensions has been selected empirically (not very large but leading to good RL results).

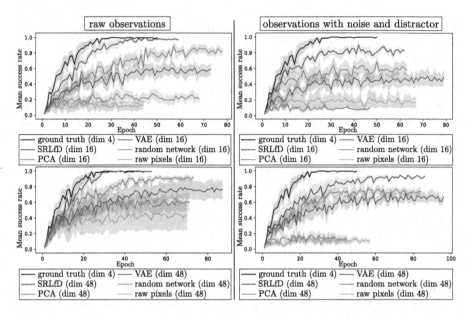

Fig. 3. Success rate learning curves (mean and half standard deviation in shaded areas for 8 runs; the higher, the better) with SAC algorithm with a random position of the target goal for each run, based on various state representations. The indicated dimensions for methods SRLfD, PCA, VAE, random network and raw pixels correspond to the size of the state representation $(\varphi(\mathbf{I}_{t-1}), \varphi(\mathbf{I}_t))$. The use of our SRLfD state representation (red) in SAC outperforms all the other baselines, except the case in which SAC is given a direct access to the ground truth representation. (Color figure online)

of the transferability of the state representations generated with our method used as inputs for an RL algorithm (SAC [9]) on a new instance of the reaching task chosen randomly, and compare the success rates to the ones obtained with state representations originating from other methods. The performance of a policy is measured as the probability to reach the reward (i.e. the goal) from a random initial configuration in 50 time steps or less. We expect that better representations yield faster learning progress (on average).

Table 1 displays the success rates corresponding to the end of the success rate learning curves of Fig. 3 obtained with SAC and different state representations in four different contexts: with a representation of either 16 or 48 dimensions (except for the ground truth representation, of size 4), and on "clean" i.e. raw observations (Fig. 2, in the middle) or observations with noise and a randomly moving distractor (Fig. 2, on the right). Not surprisingly, the best results are obtained with the ground truth representation, but we see that out of the five other state representations, only SRLfD, PCA, and VAE representations can be successfully used by SAC to solve reaching tasks when noise and a distractor are added to the inputs. SAC fails to train efficiently the large neural network

Table 1. Success rates (mean ± standard deviation over 8 runs; the higher, the better) corresponding to the end of the curves of Fig. 3.

Method	Raw observations	With noise and distractor
Ground Truth (dim 4)	0.995 ± 0.0054	0.995 ± 0.0054
SRLfD (dim 48)	**0.992 ± 0.0080**	**0.928 ± 0.029**
SRLfD (dim 16)	**0.980 ± 0.022**	**0.833 ± 0.082**
PCA (dim 48)	0.908 ± 0.072	0.725 ± 0.10
PCA (dim 16)	0.832 ± 0.13	0.591 ± 0.18
VAE (dim 48)	0.749 ± 0.27	0.650 ± 0.13
VAE (dim 16)	0.574 ± 0.16	0.448 ± 0.18
Raw Pixels (dim 48)	0.365 ± 0.39	0.106 ± 0.063
Raw Pixels (dim 16)	0.118 ± 0.13	0.149 ± 0.14
Random Network (dim 48)	0.552 ± 0.21	0.143 ± 0.14
Random Network (dim 16)	0.239 ± 0.13	0.0863 ± 0.037

that takes raw pixels in input, whether its representation is of size 16 or 48, and using fixed random weight for the first part of this network (random network representation) is not a viable option either.

The results show that with fewer dimensions our method (SRLfD) leads to better RL performances than with methods of the observation compression type (PCA and VAE). We assume that observation compression with restricted bottleneck size is disturbed by the reconstruction observation objective concerning the robotics control (and dramatically more on the challenging observations). This explains why PCA and VAE tend to need additional dimensions compared to the minimal number of dimensions of our robotics task (four dimensions: the two torques angles and velocities). This makes clear that with a carefully chosen unsupervised learning objective such as the one used for SRLfD, it is possible to compress in a minimal number of dimensions only robotics information.

Another surprising observation is that PCA outperforms VAE in our study. By design, the VAE is trained to encode and decode with as few errors as possible, and it can generally do so better than PCA by exploiting the non-linearities of neural networks. However, it does not put constraints on the features of the latent space, whereas PCA builds independent features. The goal of good input reconstruction may not be ideal when the actual true goal is to perform RL well. Besides that, although it is not clear that uncorrelated variables in input lead to better RL or control, the difference between PCA and VAE regarding the decorrelation of features may explain in part the better performances observed with PCA. Of course, much more work and carefully designed experiments would be necessary to confirm and explain these observations, for instance using variants of VAE such as β-VAE [10] that favor disentanglement of the latent features, but this is not the purpose of this paper and we leave it as potential future work.

6   Conclusion

We presented a method (SRLfD) for learning state representations from demonstration, more specifically from runs of target policies on different instances of a task. Our results indicate that the state representations obtained can advantageously replace raw pixel inputs to learn policies on new task instances via end-to-end RL. By simultaneously learning an end-to-end technique for several tasks sharing *common useful knowledge* intuitively, the proposed method forces the state representation to be general, provided that the tasks are diverse. Moreover, since the representation is trained together with heads that imitate the target policies, we believe that it is likely to be more appropriate for control than other types of representations (for instance ones that primarily aim at enabling a good reconstruction of the raw inputs). The experimental results we obtained tend to confirm that belief, as SRLfD state representations were more efficiently exploited by the SAC RL algorithm than several other types of state representations.

Acknowledgments. This article has been supported within the Labex SMART supported by French state funds managed by the ANR within the Investissements d'Avenir program under references ANR-11-LABX-65 and ANR-18-CE33-0005 HUSKI. We gratefully acknowledge the support of NVIDIA Corporation with the donation of the Titan Xp GPU used for this research.

References

1. Andrychowicz, M., et al.: Hindsight experience replay. In: Advances in Neural Information Processing Systems, pp. 5048–5058 (2017)
2. Bengio, Y., Courville, A., Vincent, P.: Representation learning: a review and new perspectives. IEEE Trans. Pattern Anal. Mach. Intell. **35**(8), 1798–1828 (2013)
3. Brockman, G., et al.: OpenAI Gym. arXiv preprint arXiv:1606.01540 (2016)
4. de Bruin, T., Kober, J., Tuyls, K., Babuška, R.: Integrating state representation learning into deep reinforcement learning. IEEE Robot. Autom. Lett. **3**(3), 1394–1401 (2018)
5. Coumans, E., Bai, Y., Hsu, J.: Pybullet physics engine (2018)
6. Curran, W., Brys, T., Taylor, M., Smart, W.: Using PCA to efficiently represent state spaces. arXiv preprint arXiv:1505.00322 (2015)
7. Finn, C., Yu, T., Zhang, T., Abbeel, P., Levine, S.: One-shot visual imitation learning via meta-learning. arXiv preprint arXiv:1709.04905 (2017)
8. Gaier, A., Ha, D.: Weight agnostic neural networks. In: Advances in Neural Information Processing Systems, pp. 5364–5378 (2019)
9. Haarnoja, T., Zhou, A., Abbeel, P., Levine, S.: Soft actor-critic: off-policy maximum entropy deep reinforcement learning with a stochastic actor. arXiv preprint arXiv:1801.01290 (2018)
10. Higgins, I., et al..: beta-VAE: learning basic visual concepts with a constrained variational framework. In: International Conference on Learning Representations (2017)
11. Hinton, G.E., Salakhutdinov, R.R.: Reducing the dimensionality of data with neural networks. Science **313**(5786), 504–507 (2006)

12. Jonschkowski, R., Brock, O.: Learning state representations with robotic priors. Auton. Robots **39**(3), 407–428 (2015). https://doi.org/10.1007/s10514-015-9459-7
13. Jonschkowski, R., Hafner, R., Scholz, J., Riedmiller, M.: PVEs: position-velocity encoders for unsupervised learning of structured state representations. arXiv preprint arXiv:1705.09805 (2017)
14. Kingma, D.P., Ba, J.: ADAM: a method for stochastic optimization. arXiv preprint arXiv:1412.6980 (2014)
15. Kingma, D.P., Welling, M.: Auto-encoding variational bayes. arXiv preprint arXiv:1312.6114 (2013)
16. Kober, J., Bagnell, J.A., Peters, J.: Reinforcement learning in robotics: a survey. Int. J. Robot. Res. **32**(11), 1238–1274 (2013)
17. Kober, J., Wilhelm, A., Oztop, E., Peters, J.: Reinforcement learning to adjust parametrized motor primitives to new situations. Auton. Robots **33**(4), 361–379 (2012). https://doi.org/10.1007/s10514-012-9290-3
18. Lesort, T., Díaz-Rodríguez, N., Goudou, J.F., Filliat, D.: State representation learning for control: an overview. Neural Netw. **108**, 379–392 (2018)
19. Mnih, V., et al.: Playing Atari with deep reinforcement learning. arXiv preprint arXiv:1312.5602 (2013)
20. Pastor, P., Hoffmann, H., Asfour, T., Schaal, S.: Learning and generalization of motor skills by learning from demonstration. In: 2009 IEEE International Conference on Robotics and Automation, ICRA 2009, pp. 763–768. IEEE (2009)
21. Pinto, L., Gupta, A.: Learning to push by grasping: using multiple tasks for effective learning. In: 2017 IEEE International Conference on Robotics and Automation (ICRA), pp. 2161–2168. IEEE (2017)
22. Rusu, A.A., et al.: Policy distillation. arXiv preprint arXiv:1511.06295 (2015)
23. Schultz, W., Dayan, P., Montague, P.R.: A neural substrate of prediction and reward. Science **275**(5306), 1593–1599 (1997)
24. Shin, H.C., et al.: Deep convolutional neural networks for computer-aided detection: CNN architectures, dataset characteristics and transfer learning. IEEE Trans. Med. Imaging **35**(5), 1285–1298 (2016)
25. Taylor, M.E., Stone, P.: Transfer learning for reinforcement learning domains: a survey. J. Mach. Learn. Res. **10**(Jul), 1633–1685 (2009)
26. Watter, M., Springenberg, J., Boedecker, J., Riedmiller, M.: Embed to control: a locally linear latent dynamics model for control from raw images. In: Advances in Neural Information Processing Systems, pp. 2746–2754 (2015)

A Deep Learning Based Fault Detection Method for Rocket Launcher Electrical System

Huanghua Li[1], Zhidong Deng[1(✉)], Jianxin Zhang[2], Zhen Zhang[2],
Xiaozhao Wang[2], Yongbao Li[2], Feng Li[2], and Lizhong Xie[2]

[1] Institute for Artificial Intelligence, Beijing National Research Center
for Information Science and Technology, Department of Computer Science and
Technology, Tsinghua University, Beijing 100084, China
`michael@tsinghua.edu.cn`
[2] Institute of Rocket Launcher, Hubei Jiangshan Heavy Industry Co. Ltd.,
Xiangyang 441005, China

Abstract. This paper proposes a fault detection method for a rocket launcher electrical system by using 1D convolutional neural network. Compared with the method based on analysis of mechanism model and the method based on knowledge, this end-to-end data-driven fault detection method, which only relies on the rich data generated during the running of the system, has the ability of automatic extraction of hierarchical features. The experimental results show that the 1D convolutional neural network designed in this paper achieves the accuracy of 98.66% in the practical fault detection for a rocket electrical system, which is improved by 29% and 13% higher than the traditional fully connected shallow neural network and support vector machine, respectively, which further verifies the feasibility and effectiveness of data-driven deep learning method in fault detection applications.

Keywords: Rocket launcher · Fault detection · Deep learning · Electrical system

1 Introduction

Fault detection method has extremely important value both in theory and application. It also plays a significant role in system operation and maintenance. For safety-critical system, it inevitably causes unpredictable serious consequences, once there occurs a fault and it is unable to be detected, isolated, and handled in time. For example, the air crash of Malaysia Airlines in 2014 has a great impact on society. Reliable fault detection of complex industrial system not only ensures the safety of production, but also increase the maintainability of products, and reduce economic losses.

Fault detection methods can be roughly divided into three categories: 1) model-based method, 2) knowledge-based, and 3) data-driven one. The model-based method is suitable for the system that can be precisely modeled, that is,

© Springer Nature Switzerland AG 2020
G. Nicosia et al. (Eds.): LOD 2020, LNCS 12566, pp. 316–325, 2020.
https://doi.org/10.1007/978-3-030-64580-9_27

the whole process can be expressed by using an accurate mathematical model. This method requires the designer to fully understand the mechanism of the system. For example, Moseler [1] implemented model-based fault detection of a brushless DC motor. The knowledge-based one can be used for systems that are hard to model. One of the most representative methods include experts system and fuzzy inference system. For example, Visinsky [2] designed an experts system framework for robot fault detection and tolerance. Although both model-based and knowledge-based methods are applicable to systems that have small amount of I/O data and simple state, they are not well-suited for complex systems that have multivariate associations and massive data. In fact, it is quite difficult for the model-based method to derive an accurate mathematical model to comply with all mechanism details of complex industrial system. Moreover, the knowledge-based method requires the designer not only to have rich industrial domain knowledge, but also to have long-term accumulated experience of designing model, which is too high for the designer.

With the rapid development of digital technology and industrial Internet, especially the continuous progress of large industrial systems, various deployed sensors can collect a large number of data in the process of system operation. In theory, as long as data-mining is done, we can get a reasonable representation of fault features, so as to conduct fault detection and analysis. Such methods are generally called data-driven ones. One of their advantages is that they no longer depend on both the domain knowledge and experience of the designer, and do not need to consider the complexity of the system. Since 2012, supported by big data and strong computing power, a new generation of artificial intelligence, in which deep learning is generally viewed as the core, has been pushed to an unprecedented height. At the same time, artificial intelligence has demonstrate performance in a wide range of fields, including computer vision, speech recognition, natural language processing, and big data analysis.

The research work of this paper mainly focuses on the fault detection of the electrical system of a rocket launcher. With the development of automation, informatization, and intelligence of modern rocket launcher, its electrical system is becoming more and more complex. The limitations of traditional fault detection methods, which highly depend on experience and knowledge, are increasingly prominent. In modern battlefield, the response speed and maneuverability of equipment are required to be higher and higher. In the course of rocket service, the failure ratio of electrical system is the highest, and the severity of the failure is also the highest, which greatly affects the mission completion. Therefore, it is of great significance to investigate the intelligent fault detection method of rocket launcher electrical system. In this paper, a fault detection method based on deep convolutional neural network is proposed for a specific rocket launcher electrical system. The experimental results show that 98.66% of the fault detection accuracy can be obtained by using the method proposed in this paper. Compared with the traditional fully connected shallow neural network and support vector machine, the corresponding performance is improved by 29% and 13%, respectively.

2   Related Work

2.1   Deep Learning

Deep learning model has strong ability in feature representation and modeling. It can automatically learn the layer-wise feature representation so as to conduct hierarchical abstraction and description of objects. In 1986, Rumelhart et al. [3] proposed the error back propagation (BP) algorithm of multi-layer forward perceptron. BP algorithm uses labels in training dataset and supervised learning to adjust weights of neural network. Based on such BP algorithms, deep learning can find complex structures underlying big data. It can be employed to transform original data into high-level abstract expression through large amount of simple nonlinear neurons [4]. Therefore, computer can learn hierarchical features of patterns automatically and avoid a series of challenging problems caused by hand-crafted features. In 2006, Hinton et al. [5,6] first proposed the deep learning, which increasingly attracted great attention and achieved remarkable success in a wide range of fields such as initial breakthrough in speech recognition [7]. In a large-scale image classification problem, Krizhevsky et al. [8] constructed a deep convolutional neural network and made significant achievements in 2012. As for object detection tasks, deep learning methods [9–11] have been proven to greatly exceed the performance of traditional ones.

In general, fully connected neural networks are unable to perfectly deal with local correlation in images. For a specific problem, they have relatively too many connection weights to be trained, probably leading to poor generalization ability. Inspired by neurobiological model of mammals' brain visual pathway [12], convolutional neural networks (CNNs) proposed by LeCun et al. [13] uses convolution kernels with few shared connection weights to have feature extractions from either raw images or feature maps. Meanwhile, it can effectively retain local correlation information of original data with not too many weights to be trained. In recent years, various deep CNN models have been extensively applied in computer vision tasks including image classification, visual object detection, and action prediction, which constantly refreshes SOTA performances [8,14–16].

2.2   Deep Learning Based Fault Detection Method

Traditional data-driven methods of fault detection and diagnosis often need to conduct preprocessing of signals and perform feature engineering. After effective feature extractions are done, classical machine learning models, e.g.., fully connected shallow neural network, support vector machine, and naive Bayesian network, are usually employed as classifiers [22–24]. One of the disadvantages of this method is that designers require to have rich domain experience of how to design or choose excellent features. The quality of feature selection will directly affect accuracy of models. Considering amazing ability of deep CNN models in automatic feature extraction, deep learning based fault detection and diagnosis not only avoid any tedious and inefficient feature selection, but also considerably surpass performance of traditional data-driven models like statistical analysis.

In addition to image processing, deep learning has also made a lot of progress in 1D signal processing such as ECG and EEG [26,27]. In the field of fault detection and diagnosis, Guo et al. [17] constructed a 1D deep CNN with hierarchical architecture for bearing fault diagnosis. Abdeljaber et al. [18] used 1D CNN to conduct a real-time detection of structural damages based on vibration. In [19], Ince et al. carried out a real-time fault detection of motors using 1D CNN.

3 Method

3.1 Problem Description

In this paper, we focus on the fault detection of the electrical system of a rocket launcher. Figure 1 shows the overall architecture of the rocket launcher fault detection system. It can be seen from the figure that the electrical system of the rocket launcher is equipped with 104 related sensors. Through monitoring the level of checkpoints, it can be used to determine whether the system is in a fault status. If there are any malfunction, then further locate and repair the fault. The fault types of the system are described in the form of hierarchical structure, which are divided into level II fault, level III fault and level IV fault from top to bottom. The level IV failures comprise 57 categories, including output faults of generator, KM0 damages of contactor, F1 damages of breaker, and damages of pump boot modules. Actually, we must take advantage of multiple consecutive sensors to jointly judge the occurrence of faults.

As shown in Fig. 1, we take all the level data collected by 104 monitoring sensors as the input of our AI-based fault detection model, and the corresponding outputs contain 57 fault labels in level IV. On such sampling data, our fault detection model is trained and tested. Essentially, this can be regarded as a data-driven binary classification problems with multiple inputs and multiple outputs.

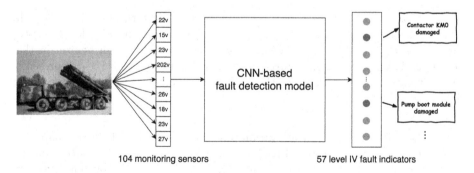

Fig. 1. General architecture of the fault detection for a rocket launcher electrical system.

Loss. For the above binary classification task, we define the corresponding loss function by the binary cross entropy for multiple outputs,

$$L = -\frac{1}{N} \sum_{i=1}^{N} \sum_{j=1}^{D} y_{ij} \log(p_{ij}) + (1 - y_{ij}) \log(1 - p_{ij}) \tag{1}$$

where N is the number of samples, D is the output dimension, and p_{ij} is the predicted output of the input vector x_i through the fault detection model. It satisfies that $p_{ij} \in (0, 1)$, and the classification label $y_{ij} \in \{0, 1\}$.

Evaluation. We use the average precision of each fault classification as performance evaluation.

$$\text{AP} = \frac{1}{ND} \sum_{i=1}^{N} \sum_{j=1}^{D} c(p_{ij}, y_{ij}) \tag{2a}$$

where

$$c(p_{ij}, y_{ij}) = \begin{cases} 1, & p_{ij} \geq 0.5 \text{ and } y_{ij} = 1 \\ 1, & p_{ij} < 0.5 \text{ and } y_{ij} = 0 \\ 0, & \text{otherwise} \end{cases} \tag{2b}$$

3.2 A DL-Based Fault Detection Method for Rocket Launcher Electrical System

Similar to typical deep CNN, the core of 1D CNN lies in its 1D convolution layer. In each 1D convolution layer, the formulation of forward calculation is given as follows:

$$x_k^l = b_k^l + \sum_{i=1}^{N_{l-1}} \text{conv1D}(w_{ik}^{l-1}, p_i^{l-1}) \tag{3}$$

where x_k^l and b_k^l denote the input and the bias of the k-th neuron in the l-th layer, respectively, p_i^{l-1} indicates the output of the i-th neuron in the $(l-1)$-th layer, and w_{ik}^{l-1} stands for the connection weight from the i-th neuron in the $(l-1)$-th layer to the k-th neuron in the l-th layer. The conv1D($\cdot$) represents 1D convolution operation. Thus, the output p_k^l is calculated by the input x_k^l as follows:

$$p_k^l = f(x_k^l) \tag{4}$$

where the $f(\cdot)$ is a nonlinear activation function. We use ReLU in this paper.

During training, the error of back propagation comes from the output of the last MLP layer. Let $l = 1$ be the input layer and $l = L$ the output layer. In the multi-input-multi-output binary classification problem of this paper, we let N_L denote the total number of output classes. For an input vector x^1, the label vector of the correct classification is $[y_1, y_2, ..., y_{N_L}]$, and the network's output vector is $[p_1^L, p_2^L, ..., p_{N_L}^L]$. According to Eq. (1), the output loss of binary cross entropy can be given below:

$$E_p = -\sum_{j=1}^{N_L} y_j \log(p_j^L) + (1 - y_j)\log(1 - p_j^L) \tag{5}$$

Thus, based on the chain rule, we can calculate the gradient to the parameters of the convolution kernel, and then update the connection weight w_{ik}^l and the bias b_k^l. For the detailed derivation of the back propagation process, see reference [20].

Considering that each input sample is a vector, the local correlation is just the local information between adjacent components. It is because a fault may be jointly determined by a consecutively deployed sensors. Therefore, we employ 1D convolutional neural network as model, which can provide this kind of local perception field and then effectively capture local correlation.

The architecture of the 1D deep CNN designed in this paper is given as follows.

- (Conv1D 128*3 + BatchNormalization + ReLU) * 2
- (Conv1D 64*3 + BatchNormalization + ReLU) * 2
- (Conv1D 32*3 + BatchNormalization + ReLU) * 2
- (Conv1D 16*3 + BatchNormalization + ReLU) * 2
- Flatten
- FullyConnected 1024
- ReLU
- FullyConnected 57
- Sigmoid

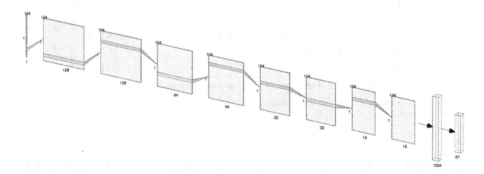

Fig. 2. The architecture of our 1D deep convolutional neural network.

Obviously, the size of all 1D convolution kernels is taken as 3, the stride is set as 1, and the number of convolution kernels in each convolution layer is chosen as 128, 128, 64, 64, 32, 32, 16 and 16, respectively. A batch normalization layer [21] is connected behind each convolution layer to reduce overfitting. After that, it passes through a ReLU layer. In order to complete the fault classification, the network finally uses a two-layer MLP as classifier. As the final results are 57 binary outputs, the sigmoid function is selected as the nonlinear activation function of the last layer. The overall network architecture is shown in Fig. 2.

4 Experimental Results

4.1 Dataset Setup

This paper works for the fault detection of a rocket launcher electrical system. There are 104 monitoring sensors in the system. Each sensor monitors the level of the corresponding checkpoint. The fault detection model eventually outputs 57 kinds of level IV faults, some of which are shown in Table 1.

Table 1. A partial fault list

Fault code	Level II fault	Level III fault	Level IV fault
F1	Distribution system failure	Generator failed	Generator damaged
F2		Power grid connection failed	Contactor KM0 damaged
F3			Breaker F1 damaged
F4			Contactor KM1 damaged
F5		Power voltage instability	External power damaged
F6	Driving system failure	Pump motor damaged	Pump boot module damaged
F7	Control system failure	Interlock control failure	Interface circuit failure

It should be pointed out that whether each fault occurs or not may be determined jointly by multiple consecutive sensors. Some criteria are listed in Table 2.

Table 2. Sensors involved in the fault detection (partial)

Fault code	Level IV fault	Level III fault	Involved sensors
F1	Generator damaged	Generator failed	A1–A4
F2	Contactor KM0 damaged	Power grid connection failed	A5–A8
F3	Breaker F1 damaged		A9
F4	Contactor KM1 damaged		A10–A13
F5	External power damaged	Power voltage instability	A14
F6	Pump boot module damaged	Pump motor damaged	A12
F7	Interface circuit failure	Interlock control failure	A15–A93

The checkpoint levels collected by the sensors are adopted as the input of the fault detection model. Each input sample is a 104-dimensional vector, and the value of each dimension component is in the interval $[0, 220]$, which corresponds to the 0-220V level of each monitoring sensor. Each output sample is a 57-dimensional vector, where the value of each dimension component is taken as 1 or 0, indicating whether the corresponding fault will occur or not.

In our experiment, the complete dataset consists of 20,000 input-output samples collected from maintenance record of faulty vehicles. We randomly divide it into three parts: the training dataset with 15,000 samples, the validation dataset with 2,500 ones and the test dataset with 2,500 ones. We ensure that there is no significant difference in the data distribution among the three datasets.

4.2 Results and Performance

In the phase of training, we use the above loss function and evaluation criteria. We use the Adam optimizer in training. The mini-batch size is set to 128. We select the initial learning rate as 0.001, which is reduced by half after every 10 epochs. We trained 45 epochs in total, with the loss and the AP curve shown in Figs. 3.a and 3.b, respectively.

As shown in Figs. 3.a and 3.b, the model converges after about 3,000 iterations. After 4,000 iterations, the loss and the AP curve become basically stable, and there is no large fluctuation.

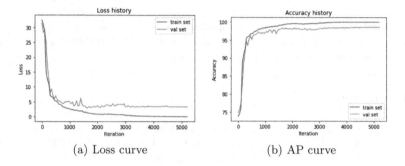

(a) Loss curve   (b) AP curve

Fig. 3. Training process

Table 3. The fault detection performance with different models

Fault detection model	AP (100%)
Fully connected SNN [3]	76.28 ± 1.73
SVM [25]	87.23 ± 0.84
Our 1D CNN	**98.66 ± 0.52**

After completing training, we use the test dataset to evaluate performance of our fault detection model, and the average precision after 10 run independent trials are given in Table 3. Here we also compare our method with two typical end-to-end machine learning models. We use a fully connected network with 4 layers and SVM as comparison. It is easy to see that in the fault detection task investigated in this paper, the performance of our 1D deep CNN model is significantly better than other two ones.

5 Conclusion

Based on the deep convolutional neural network, we present a fault detection method for the rocket launcher electrical system. With the help of the deep learning ability to automatically extract hierarchical features from big data, we achieve an end-to-end model that can detect 57 kinds of faults with high accuracy. Compared with methods based on mechanism model and knowledge, the above data-driven method does not rely too much on the domain knowledge and experts experience, only on the rich data naturally generated during the running of the system. The experimental results show that the 1D deep convolutional neural network designed in this paper yields the fault detection accuracy of 98.66% in the practical rocket launcher electrical system, which is 29% and 13% higher than traditional fully connected shallow neural network and support vector machine, respectively. It further verifies the feasibility and effectiveness of data-driven deep learning method in practical fault detection applications.

Acknowledgment. This work was supported in part by Joint Fund of NORINCO Group of China for Advanced Research under Grant No. 6141B012102 and by a grant from the Institute Guo Qiang, Tsinghua University.

References

1. Moseler, O., Isermann, R.: Application of model-based fault detection to a brushless DC motor. IEEE Trans. Industr. Electron. **47**(5), 1015–1020 (2000)
2. Visinsky, M.L., Cavallaro, J.R., Walker, I.D.: Expert system framework for fault detection and fault tolerance in robotics. Comput. Electr. Eng. **20**(5), 421–435 (1994)
3. Rumelhart, D.E., Hinton, G.E., Williams, R.J.: Learning representations by back-propagating errors. Nature **323**(6088), 533–536 (1986)
4. LeCun, Y., Bengio, Y., Hinton, G.: Deep learning. Nature **521**(7553), 436–444 (2015)
5. Hinton, G.E., Salakhutdinov, R.R.: Reducing the dimensionality of data with neural networks. Science **313**(5786), 504–507 (2006)
6. Hinton, G.E., Osindero, S., Teh, Y.W.: A fast learning algorithm for deep belief nets. Neural Comput. **18**(7), 1527–1554 (2006)
7. Hinton, G., et al.: Deep neural networks for acoustic modeling in speech recognition: the shared views of four research groups. IEEE Signal Process. Mag. **29**(6), 82–97 (2012)
8. Krizhevsky, A., Sutskever, I., Hinton, G.E.: Imagenet classification with deep convolutional neural networks. In: Proceedings Advances in Neural Information Processing Systems, pp. 1097–1105 (2012)
9. Ren, S.Q., He, K.M., Girshick, R., Sun, J.: Faster R-CNN: towards real-time object detection with region proposal networks. IEEE Trans. Pattern Anal. Mach. Intell. **39**(6), 1137–1149 (2017)
10. Girshick, R., Donahue, J., Darrell, T., Malik, J.: Rich feature hierarchies for accurate object detection and semantic segmentation. In: Proceedings of the 2014 IEEE Conference on Computer Vision and Pattern Recognition (CVPR), pp. 580–587. Columbus, Ohio, USA. IEEE (2014)
11. Girshick, R.: Fast R-CNN. In: Proceedings of the 2015 IEEE International Conference on Computer Vision (ICCV), Santiago, Chile, pp. 1440–1448. IEEE (2015)
12. Hubel, D.H., Wiesel, T.N.: Receptive fields binocular interaction and functional architecture in the cat's visual cortex. J. Physiol. **160**(1), 106–154 (1962)

13. LeCun, Y., Bottou, L., Bengio, Y., Haffner, P.: Gradient-based learning applied to document recognition. Proc. IEEE **86**(11), 2278–2324 (1998)
14. Simonyan, K., Zisserman, A.: Very deep convolutional networks for large-scale image recognition. CoRR, vol. abs/1409.1556 (2014). http://arxiv.org/abs/1409.1556
15. Szegedy, C., et al.: Going deeper with convolutions. In: Proceedings Conference Computer Vision Pattern Recognition, pp. 1–9 (2015)
16. He, K., Zhang, X., Ren, S., et al.: Deep residual learning for image recognition. In: Proceedings of the IEEE Conference on Computer Vision and Pattern Recognition, pp. 770–778 (2016)
17. Guo, X., Chen, L., Shen, C.: Hierarchical adaptive deep convolution neural network and its application to bearing fault diagnosis. In: Measurement: Journal of the International Measurement Confederation, vol. 93, pp. 490–502 (2016)
18. Abdeljaber, O., Avci, O., Kiranyaz, S., Gabbouj, M., Inman, D.J.: Real-time vibration-based structural damage detection using one-dimensional convolutional neural networks. J. Sound Vib. **388**, 154–170 (2017)
19. Ince, T., Kiranyaz, S., Eren, L., Askar, M., Gabbouj, M.: Real-time motor fault detection by 1-D convolutional neural networks. IEEE Trans. Industr. Electron. **63**(11), 7067–7075 (2016)
20. Kiranyaz, S., Ince, T., Abdeljaber, O., et al.: 1-D convolutional neural networks for signal processing applications. In: ICASSP 2019–2019 IEEE International Conference on Acoustics, Speech and Signal Processing (ICASSP), pp. 8360–8364. IEEE (2019)
21. Ioffe, S., Szegedy, C.: Batch normalization: Accelerating deep network training by reducing internal covariate shift. arXiv preprint arXiv:1502.03167 (2015)
22. Yang, Y., Yu, D., Cheng, J.: A fault diagnosis approach for roller bearing based on IMF envelope spectrum and SVM. Measurement **40**(9–10), 943–950 (2007)
23. Yang, J., Zhang, Y., Zhu, Y.: Intelligent fault diagnosis of rolling element bearing based on SVMs and fractal dimension. Mech. Syst. Signal Process. **21**(5), 2012–2024 (2007)
24. Yu, Y., Junsheng, C.: A roller bearing fault diagnosis method based on EMD energy entropy and ANN. J. Sound Vib. **294**(1–2), 269–277 (2006)
25. Cortes, C., Vapnik, V.: Support-vector networks. Mach. Learn. **20**(3), 273–297 (1995)
26. Kiranyaz, S., Ince, T., Gabbouj, M.: Real-time patient-specific ECG classification by 1-D convolutional neural networks. IEEE Trans. Biomed. Eng. **63**(3), 664–675 (2015)
27. Acharya, U.R., Oh, S.L., Hagiwara, Y., et al.: Deep convolutional neural network for the automated detection and diagnosis of seizure using EEG signals. Comput. Biol. Med. **100**, 270–278 (2018)

Heuristic Search in LegalTech: Dynamic Allocation of Legal Cases to Legal Staff

Mayowa Ayodele$^{(\boxtimes)}$ ⓘ, Richard Allmendinger ⓘ, and K. Nadia Papamichail ⓘ

The University of Manchester, Manchester, UK
{mayowa.ayodele,richard.allmendinger,n.papamichail}@manchester.ac.uk
http://www.manchester.ac.uk

Abstract. We investigate an allocation problem inspired by the process of assigning legal cases (matters) to staff in law firms. Addressing this problem is important as it can prevent issues around unbalanced workloads and over-recruitment, thus decreasing costs. This initial study on the topic frames the problem as a combinatorial dynamic single-objective problem (minimising tardiness) with constraints modelling staff-client relationships, staff capacities, and earliest start dates of matters. The paper motivates the allocation problem and puts it in context with the literature. Further contributions include: (i) a formal problem definition, (ii) the proposal and validation of a feature-rich problem generator to create realistic test cases, (iii) an initial analysis of the performance of selected heuristics (a greedy approach, a nature-inspired approach, and random search) on different test instances, and finally (iv) a discussion on directions for future research.

Keywords: Genetic Algorithm · Heuristic search · LegalTech · Resource allocation · MRCPSP · Scheduling

1 Introduction

The term *LegalTech* describes technology and software used in the legal sector [8]. An area of LegalTech that has been of interest in recent years and predicted to be of significant relevance, e.g. due to productivity gains, trust and safety improvements, and cost reductions, is the use of artificial intelligence [13]. Application areas include predicting legal outcomes, e-discovery, automated legal reasoning and work allocation [7,8,13].

This study is making a contribution to the area of work allocation. In particular, we consider the problem of efficiently assigning legal cases to legal staff (working e.g. in a law firm). These legal cases are often referred to as *matters* and staff are referred to as *fee-earners*. The process of allocating matters is an important task for law firms [7]. Traditionally, law firms work on matters that are charged by the hour. In more recent years, law firms have had to consider alternative billing approaches; one such approach is fixed-fee billing [11]. An example of this is when law firms are instructed by insurance companies to represent them

© Springer Nature Switzerland AG 2020
G. Nicosia et al. (Eds.): LOD 2020, LNCS 12566, pp. 326–338, 2020.
https://doi.org/10.1007/978-3-030-64580-9_28

on insurance claims. This category of work is often large in quantity, low value and of fixed-fee nature. These characteristics make it imperative for law firms to work efficiently, for example, to avoid spending more than the fee earned and achieve a high throughput of matters. The objectives highlighted in this article include maximising utilisation, preventing over-allocation, and optimally dealing with peaks and troughs. In reality, clients expect law firms to acknowledge and provide an action plan on a received matter. Doing this as early as possible gives a law firm a competitive advantage. We therefore use the worst case tardiness to ensure work is started on every matter as early as possible. In this paper, the term tardiness refers to the amount of delay there is before work starts on a given matter. Assigning matters to the most suitable and available fee-earners as early as possible ensures that matters are being processed swiftly without overloading a fee-earner with work. We also investigate the impact on workload distribution across the workforce (though this is not considered as a criteria by the optimiser in this study). Although law firms are looking into automating this process, the truth is that workload allocation is often still done manually [7].

To this end, we model the workload allocation problem considered (we will refer to it as the *Matter Allocation Problem* (MAP)) as a combinatorial dynamic and constrained single-objective problem. The constraints relate to aspects such as adhering to existing relationships between fee-earners and clients and maximum capacities of fee-earners. The dynamic aspect of the problem relates to the fact that matters need to be allocated as soon as they arrive while respecting previous allocation commitments (this property makes the MAP a type of time-linkage problem [3,17]).

The next section puts the problem into context with existing work in the area. Section 3 then provides a formal definition of the problem. Section 4 introduces three methods to tackle the problem including a greedy approach, a nature-inspired approach (a variant of a genetic algorithm to be precise), and random search, which will serve as a baseline method. The experimental study, presented in Sect. 5, covers three aspects: (i) the introduction of a problem generator to create realistic test problem instances, (ii) summary of algorithm and problem parameter settings, and (iii) validation of the generator and the algorithms on test problems of varying complexity. Finally, conclusions and further work are presented in Sect. 6.

2   Background

The MAP considered in this study shares similarities with some existing problem types such as Job-Shop Scheduling Problem (JSP) [9], Generalised Assignment Problem (GAP) [19], Resource Allocation Problem (RAP) [18] and Multi-Mode Resource Constrained Project Scheduling Problem (MRCPSP) [22].

JSP entails assigning a set of jobs to a set of machines. A job needs to be processed by more than one machine and each machine can only process one job at a time. The MAP shares some similarities with the flexible JSP [9] (which entails selecting one of a set of suitable machines) and dynamic JSP [15] (which considers the concept of old and new jobs). In the MAP, however, each matter needs to be processed by only one fee-earner. Also, a fee-earner can work on more than one matter at a time. GAP [19] describes the process of assigning a set of tasks to agents such that costs are minimised. A task needs to be completed by one of many agents that have limited capacities and varying costs. MAP is also constrained by limited capacities but matters can only be assigned to fee-earners that have a relationship with the client. Also, the aim of the MAP is to minimise tardiness. Although tardiness and earliness is an important objective of the earliness-tardiness problem [14], the problem focuses on completing jobs, just in time. For the MAP, one of the factors clients use to measure performance of law firms is tardiness and the best case scenario is for jobs to be performed on the day they are released.

RAP [18] is another well-studied allocation problem in the literature. It entails allocating resources to tasks or projects such that a business objective is satisfied. Although the RAP has been defined in many ways in previous work, it has been studied within the same context as the well-defined Resource Constrained Project Scheduling Problem (RCPSP) [1]. The MRCPSP [22] (a generalisation of the RCPSP) shares more similarities with the MAP. MRCPSP consists of two sub-problems: *activity scheduling* and *mode assignment*. *Activity scheduling* is used to depict the order in which the activities of a project should be executed. *Mode assignment* is used to represent the mode in which each activity is performed [6]. The objective of the MRCPSP is to minimise the makespan. The similarities between MAP and MRCPSP are (1) Multi-Component Problem i.e. the need to determine the ordering of matters is similar to *activity scheduling* while allocating matters to fee-earners is similar to *mode assignment* (2) Varying Duration i.e. the lifecycle (time to process) of a matter depends on the assigned fee-earner as the duration of an activity is determined by the mode of execution in the MRCPSP and (3) Resource Constraints i.e. a fee-earner has a maximum number of matters (called capacity) they can work on simultaneously. This is similar to the renewable resource constraint of the MRCPSP. However, the MAP is not constrained by precedence or non-renewable resource constraints like the MRCPSP. Also, we use an objective function that focuses on minimising tardiness. Note, in order to model the actual real-world problem at hand, it is not unusual to modify an existing basic problem type (as we do it with MRCPSP). Adjustments to an existing problem were reported, for example, for a construction project scheduling problem [5], allocation of computational resources [21], and truck and trailer scheduling problem [20]. In [20], the problem was formulated as a Vehicle Routing Problem (VRP), and, to suit the real-world constraints, the VRP was reformulated to handle additional constraints, was dynamic and considered heterogeneous vehicles.

To formally define the MAP, we draw on inspiration from the definition of the MRCPSP but add to the problem a number of constraints: Law firms can receive new matters on a daily basis meaning that a MAP considers the day of instruction as the earliest start day of a matter. In addition, some fee-earners will be busy working on previously allocated matters when new ones arrive (exhibiting a time-linkage property [3]), which needs to be taken into account by the optimiser. Furthermore, matters may originate from different clients. Clients will often prefer to engage with fee-earners they have previously worked with. In this paper, we implement an additional constraint where fee-earners can only work with clients they have existing relationship with.

Nature-inspired optimisation methods, such as evolutionary algorithms, are well suited to tackle MRCPSP and related problems due to the various complexities exhibited by these problems [2,6,10,22,23]. Consequently, one of the approaches we consider in this study is an evolutionary algorithm; more precisely, it is a variant of a Genetic Algorithm (GA). The experimental study will compare this approach against a greedy approach and a random search.

3   The Matter Allocation Problem (MAP)

The MAP consists of n matters, $j = 1, \ldots, n$, with each matter j originating from a client $k = 1, \ldots, q$, as indicated by the given binary matrix, $e_{jk} \in \{0, 1\}$. A matter has to be processed by one of the m fee-earners (legal staff), $i = 1, \ldots, m$ but only fee-earners i that have a relationship with the client k from which the matter originates are eligible to process the matter (Eq. (2)); relationships are specified by $a_{ik} \in \{0, 1\}$. Each fee-earner is continuously available from time zero onward and can handle at most b_i matters at any day d (Eq. (3)). Each matter j is available for processing from its release date r_j but the actual start date, S_j, of the processing may be delayed (Eq. (4)), e.g. if no eligible fee-earners has capacity. The processing of a matter takes an uninterrupted duration (lifecycle) of p_{ik}, and depends on (i) the fee-earner i that is processing the matter and (ii) the client k that has issued the matter. Since preemption is not allowed, the completion time of a matter j, originating from client k and carried out by fee-earner i is $C_j = S_j + p_{ik}$. The lifecycles simulate different skill levels of legal staff and levels of relationships.

Formally, we can define the MAP as:

$$\text{Minimise } T_{max} = \max_{\substack{i=1,\ldots,m; \\ j=1,\ldots,n}} (S_j - r_j) x_{ij} \tag{1}$$

$$\text{subject to } \sum_{i=1}^{n} \sum_{k=1}^{q} a_{ik} e_{jk} x_{ij} = 1, \quad j = 1, \ldots, n \tag{2}$$

$$\sum_{j=1}^{m} \delta_{dj} x_{ij} \leq b_i, \ \delta_{dj} = \begin{cases} 1, & \text{if } S_j \leq d \leq C_j, \\ 0, & \text{otherwise} \end{cases}, d = 0, \ldots; \quad i = 1, \ldots, m \tag{3}$$

$$S_j \geq r_j, \quad j = 1, \ldots, n \tag{4}$$

$$x_{ij} \in \{0, 1\}, \ S_j \in \mathbb{N}_0, \quad i = 1, \ldots, m; \ j = 1, \ldots, n \tag{5}$$

The objective is to find a schedule to minimize maximum tardiness T_{max}, that is, the maximal delay before any of the matters is being processed. For this objective, the optimizer will need to decide (i) which of the fee-earners i is processing which matter j, defined by the binary decision variables $x_{ij} \in \{0,1\}$, and (ii) the processing start time $S_j \in \mathbb{N}_0$ of a matter j (Eq. (5)).

This paper considers the **dynamic** version of the MAP problem, i.e. where matters arrive over time, the data of matters (such as the release dates, which client is releasing the matter, and the number of matters to be scheduled) are unknown until they arrive and decisions are made after the arrival of the matters.

Illustrative Example. The problem configuration below presents an example of a MAP consisting of 11 matters, $J_1, \ldots, J_{11}$, originating from 3 clients, and 3 fee-earners $M_1 \ldots, M_3$. Matters $J_1, \ldots, J_3$ have previously been assigned to fee-earners and thus cannot be changed anymore. We assume that matters $J_4, \ldots, J_{11}$ have arrived from all three clients on the same current day, $d = 3$, and need to be assigned to fee-earners. The question marks (?) indicated the decision variables that need to be set by the optimizer.

Figure 1 (left plot) shows an example of a solution to this problem. In this solution, fee-earner M_1 is assigned matters J_8, J_{11} and J_5, fee-earner M_2 is assigned matters J_7, J_6 and J_{10} and fee-earner M_3 to assigned matters J_9 and J_4. Start times are assigned as soon as there is capacity in the order $J_8 \rightarrow J_7 \rightarrow J_{11} \rightarrow J_9 \rightarrow J_6 \rightarrow J_5 \rightarrow J_{10} \rightarrow J_4$. The resulting schedule is presented in Fig. 1 (right plot) showing start and finish times of each matter. We see that the matter J_2 has the highest tardiness of $T_{max} = 7$ days.

$$A = \begin{pmatrix} 1 & 0 & 1 \\ 0 & 1 & 1 \\ 1 & 1 & 0 \end{pmatrix}, \ P = \begin{pmatrix} 2 & 0 & 5 \\ 0 & 3 & 4 \\ 3 & 2 & 0 \end{pmatrix}, \ B = \begin{pmatrix} 2 \\ 2 \\ 1 \end{pmatrix}, \ X = \begin{pmatrix} 1 & 1 & 0 & ? & ? & 0 & ? & ? & ? & 0 & ? \\ 0 & 0 & 0 & 0 & 0 & ? & ? & ? & 0 & ? & ? \\ 0 & 0 & 1 & ? & ? & ? & 0 & 0 & ? & ? & 0 \end{pmatrix}$$

$$E^T = \begin{pmatrix} 0 & 0 & 1 & 1 & 1 & 0 & 0 & 0 & 1 & 0 & 0 \\ 0 & 0 & 0 & 0 & 0 & 1 & 0 & 0 & 0 & 1 & 0 \\ 1 & 1 & 0 & 0 & 0 & 0 & 1 & 1 & 0 & 0 & 1 \end{pmatrix}, \quad \begin{array}{l} R^T = \begin{pmatrix} 0 & 2 & 1 & 3 & 3 & 3 & 3 & 3 & 3 & 3 \end{pmatrix}, \\ \\ S^T = \begin{pmatrix} 0 & 2 & 1 & ? & ? & ? & ? & ? & ? & ? \end{pmatrix}. \end{array}$$

Ordering of Matters	8	7	11	9	6	5	10	4
Fee-earner Allocation	1	2	1	3	2	1	2	3

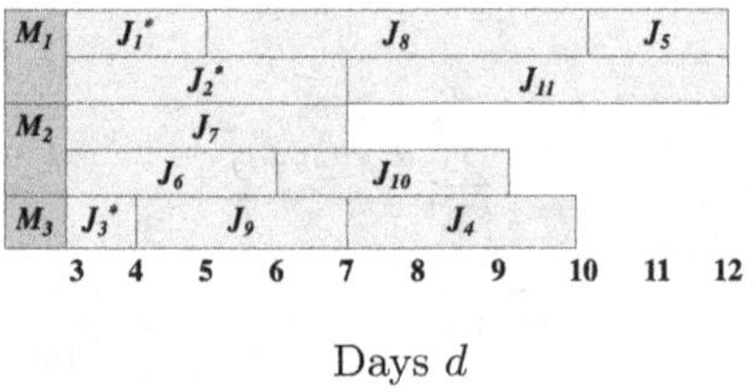

Days d

Fig. 1. Illustration of a sample solution string (left) and resulting schedule (right). The matters J_1^*, J_2^*, J_3^* have an asterisk to indicate that they have been allocated previously.

4   Solution Approach

4.1   Representation

Previous research on MRCPSP has used Activity List (AL) or Random Key (RK) representations to determine the ordering of activities, and a Mode List (ML) or Mode Vector (MV) to represent the assignment of modes [22].

In AL, the representation is a permutation $(x_1, ..., x_n)$ where x_i is the ith activity to be scheduled. With RK, x_i is a priority value, a matter i with the lowest value x_i is scheduled first. However, ML and MV are both integer representations $(y_1, ..., y_n)$. MV is used in combination with AL [22]. For MV, a value y_i represents the mode of activity x_i in an AL. In ML however, a value y_i represents the mode of activity i. In [12,16], a combination of AL and MV was used leading to competitive results. In this paper, the ordering of matters is represented as AL, while the allocation of fee-earners to matters is encoded as MV. The length of both representations are exactly the number of matters to be scheduled. For example, AL $(2, 1, 3)$ indicates that matter J_2 will be executed first, then J_1, and then J_3, while MV $(5, 4, 1)$ indicates that matters J_2, J_1 and J_3 will be executed by fee-earners M_5, M_4 and M_1, respectively.

4.2   Algorithms

This section presents a description of algorithms[1] used in this study.

Random Search. When executed (i.e. on a daily basis), this approach generates and evaluates D random solutions, i.e. matter orderings and fee-earner assignments. The solution with the smallest maximum tardiness is then selected. We shall refer to D as the *budget* (also known as the number of objective function evaluations available for the search).

Greedy Search. The greedy approach uses a *first in first out* approach to determine the ordering of the matters (as matters may arrive at different times of a day). To generate the fee-earner allocation, the approach computes the tardiness (if the matter would be assigned to that fee-earner) for each fee-earner that could work on a matter and then assigns the matter to the fee-earner with the smallest tardiness. This is repeated for all matters.

Genetic Algorithm (GA). The GA employs one-point crossover as proposed in [12], and also illustrated in Fig. 2 (left plot). An offspring is created by copying the ordering of matters and the associated fee-earners (before a randomly selected crossover point) from the first parent. The remaining matters and the associated fee-earners are then filled in order of the second parent.

---

[1] Implementations are presented in https://github.com/mayoayodele/MAPSolver.

Two mutation operators applied independently as illustrated in Fig. 2 (right plot): The *swap* mutation operator exchanges the positions of two randomly selected matter/fee-earner combinations. Once it has iterated over all matters and mutations applied or not (depending on the mutation probability), the *fee-earner change* mutation operator is applied, again independently to each matter (using the same mutation probability). The *fee-earner change* mutation operator selects a matter-fee-earner combination at random, and replaces the allocated fee-earner with another randomly selected and eligible fee-earner. To ensure eligibility (i.e. satisfaction of the client relationship constraint), we only select from a pool of other fee-earners that are permitted to work for the client associated with the matter. Both mutation operators are similar to the ones presented in [12], the difference is that our operators ensure that the resulting solution is feasible. We use a binary tournament selection with replacement for parental selection. For environmental selection, we select the best solutions from the combined pool of parent and offspring population. Compared to standard generational selection, this strategy provided quicker convergence without sacrificing performance significantly.

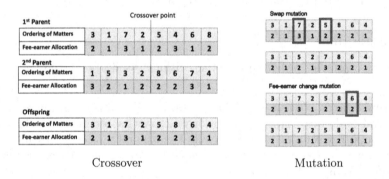

Crossover   Mutation

Fig. 2. Crossover (left) and mutation operator (right) for the MAP problem.

5   Experimental Setup

This section (i) introduces a feature-rich problem generator to create test problem instances of varying complexity and (ii) provides an overview of the problem and algorithm parameter settings used in the experimental study.

5.1   Problem Generator

The problem generator introduced here is inspired by real MAPs as faced by the authors observed in law firms.[2] The controllable problem features include

---

[2] Code of the generator is available at https://github.com/mayoayodele/MatterAllocat ionProblemGenerator.

the number of clients, total number of fee-earners, capacity per fee-earner (minimum and maximum), lifecycles (minimum and maximum), number of matters per client (minimum and maximum) and initial occupation level of fee-earners (minimum and maximum), which is the fraction of fee-earner's capacity that is dedicated to previously assigned matters; this is only relevant when the first batch of new matters is released. For features with a minimum and maximum value, the values set the boundaries within which the generator would select a number at random when generating a test problem instance, while the other features would be considered as constants.

The problem generator is similar to other test problems that involve a certain degree of randomness, such as NK landscapes [4]. A typical use case for such a generator is to have it create a set of test problem instances that share the same complexities. A search algorithm would then be applied over all these instances to give the algorithm designer an idea about the suitable of the algorithm for that problem class. We will adopt the same procedure here.

5.2   Algorithms and Problem Parameter Settings

We use the problem generator to create four problem classes, which we term as Small, Medium, Large and Actual, indicating the problem size. Each problem simulates a 28-day period. The problem class, Actual, is the actual problem we have had at hand when endeavouring on this research, while the other three problem classes can be seen as variations of it. Table 1 shows the problem feature values associated with the four problem classes. The experimental study will investigate the sensitivity of the various features on algorithm performance. For each problem class we generated 20 test instances, and then run each algorithm (except greedy search as it is deterministic) for 20 runs on each instance.[3] For each test instance, we then compute the average across the single best solutions found in each run, and it is these averages (of which we have 20, one per instance) that we show in the boxplots in the next section.

The GA parameters used in this study are: binary tournament selection, 1-point crossover with a crossover probability of 1, and two mutation operators, swap and fee-earner change, use a mutation rate of 1/#matters arrived on the day. The population size and number of evaluations (per day) are respectively set to 50 and 1500 for both algorithms. For the greedy method, although only one solution is being evolved, an exhaustive search is done on all possible fee-earners for each matter. This is however less than 1500 fitness evaluations.

5.3   Experimental Results

This section presents the results of running the GA using the parameters presented in Sect. 5.2 on the generated problem instances. This is compared with random search and greedy search.

---

[3] The test instances can be found at https://github.com/mayoayodele/MatterAllocat ionProblemGenerator/tree/master/Data.

Table 1. Different features that can be tuned by the proposed problem generator, and feature values of the four problem classes considered in this work.

Parameters	Actual	Small	Medium	Large
Number of clients	30	10	20	30
Number of fee-earners	100	40	70	100
Number of fee-earners per client [min, max]	[1, 27]	[1, 9]	[10, 18]	[19, 27]
Capacity per fee-earners [min, max]	[1, 45]	[1, 15]	[16, 30]	[31, 45]
Lifecycle [min, max]	[11, 250]	[11, 100]	[101, 190]	[191, 280]
Number of matters per client [min, max]	[1, 96]	[1, 32]	[33, 64]	[65, 96]
Initial occupation level of fee-earners [min, max]	[0, 1]	[0, 0.33]	[0.34, 0.66]	[0.67, 1]

Problem Generator: Sensitivity of Features. A sensitivity analysis was carried out to determine the sensitivity of each feature. For this, we have used the Medium problem class (see Table 1) as the baseline, and then set each feature in turn to the value of the Small and Large problem class. Figure 3 shows the results of the resulting tornado plot as obtained by the GA and greedy search (run on the same test instances). Results generated across all 20 instances of each problem category (feature) are presented in decreasing order of worst median tardiness. The different order of features (y-axis) in both plots indicates that the sensitivity in performance obtained by a GA and greedy search varies across the features.

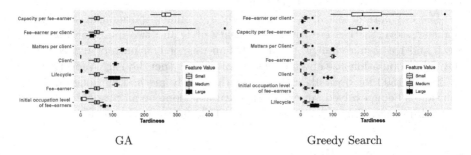

GA   Greedy Search

Fig. 3. Tornado plots measuring the sensitivity of features in the generator against algorithm performance of a GA (left) and greedy search (right). In both plots, the features are ranked in descending order of the median tardiness obtained for the most difficult feature value.

We assume that the lower the average tardiness, the simpler the problem for the optimizer. Decreasing the number of fee-earners per client, capacity per fee-earner and number of fee-earners make the problem more constrained on resources and therefore more difficult. However, decreasing the number of matters per client, number of clients, initial occupation level of fee-earners and lifecycle simplifies the problem. This leads to less work for fee-earners and therefore allows for more flexibility in schedules.

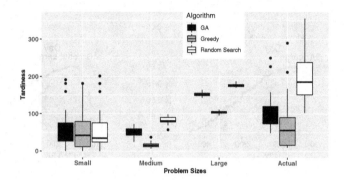

Fig. 4. Comparing schedule approaches

Maximum Utilisation: Comparing GA, Greedy Search and Random Search. In Fig. 4, the performance of all three search methods is similar on the Small problem class. This is because the search space is smaller and therefore all methods can perform well on it making it harder to discriminate between solution approaches. However, results produced by the GA and greedy search are better than random search on the Medium, Large and Actual problem class. In particular, greedy search produced the most promising results, and is statistically better than the GA on most problem instances.

To ensure that there are no over-allocation and that the fee-earners are optimally utilised, we asses the utilisation score of all fee-earners on each day of the resulting schedules. Utilisation is the fraction of a fee-earner's capacity that is used; a fraction of 1 indicates full utilisation and 0 no utilisation. We selected problem instances 5 and 15 from the Small and Large problem class, respectively, to compare the schedule produced by greedy search with the best schedule produced by the GA for these instances.

While there are no over-allocation of fee-earners in the results, as shown in Fig. 5, the median utilisation of fee-earners are closer to 1 for both algorithms during peak days (around day 28 when the last batch of matters were received). This gives an indication that over 50% of fee-earners are utilised to near full capacity. Although results produced by the GA on the small instance indicate that at least one fee-earner has nothing to do on each day of the project, this is not the case for the larger problem instances. Furthermore, the utilisation is generally more balanced for larger problem instances.

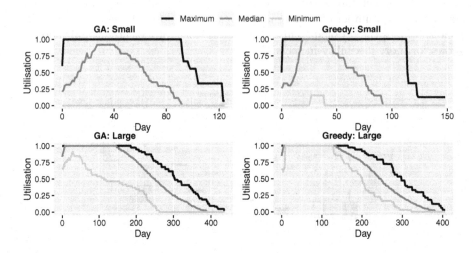

Fig. 5. Comparing Utilisation of Fee-earners on a Small (top) and Large problem Instance (bottom) as obtained by a GA (left) and greedy search (right).

6   Conclusion and Future Work

In this study, a matter allocation problem representing a real problem in the legal sector was motivated, formally defined and investigated experimentally. The problem has been linked to related problems from the literature, such as the MRCPSP. Further contributions of the work included (i) the proposal and validation of a feature-rich problem generator to create realistic test cases as well as an (ii) initial analysis of the performance of selected heuristics (a greedy approach, a nature-inspired approach, and random search) on a range of test instances. The GA and greedy search presented better results when compared to random search. The conclusion of this initial study is to suggest the adoption of greedy search to allocate matters to fee-earners upon arrival of a matter.

This initial study can be extended in a number of ways: One avenue is related to algorithm development. For example, one could explore the combination of nature-inspired search methods with problem-specific (greedy) heuristics. Replacing the representation with one that is less likely to feature redundant solutions is a promising approach too. Another avenue is to make the proposed MAP problem even more realistic by, for example, allowing preemption and reallocation of matters to fee-earners, and/or assigning priorities to matters. Moreover, one may want to add additional objectives to the problem, such as optimizing workload distribution to ensure all fee-earners are evenly loaded with work. Augmenting uncertainty to the problem, for example, in the lifecycles of matters and start times when matters are being processed, is of relevance to the legal sector too.

References

1. Alcaraz, J., Maroto, C.: A robust genetic algorithm for resource allocation in project scheduling. Ann. Oper. Res. **102**(1–4), 83–109 (2001)
2. Alcaraz, J., Maroto, C., Ruiz, R.: Solving the multi-mode resource-constrained project scheduling problem with genetic algorithms. J. Oper. Res. Soc. **54**(6), 614–626 (2003)
3. Allmendinger, R., Knowles, J.: On handling ephemeral resource constraints in evolutionary search. Evol. Comput. **21**(3), 497–531 (2013)
4. Altenberg, L.: NK-fitness landscapes. In: Bäck, Th., F.D.M.Z. (ed.) The Handbook of Evolutionary Computation. Oxford University Press, Oxford (1997)
5. Ayodele, M.: Effective and efficient estimation of distribution algorithms for permutation and scheduling problems. Ph.D. thesis, Robert Gordon University (2018)
6. Ayodele, M., McCall, J., Regnier-Coudert, O.: BPGA-EDA for the multi-mode resource constrained project scheduling problem. In: 2016 IEEE Congress on Evolutionary Computation (CEC), pp. 3417–3424. IEEE (2016)
7. Brent, R.: Smarter resourcing, September 2018. https://www.briefing.co.uk/wp-content/uploads/2018/09/Briefing-Sept18-Resourcing-supp-DIGI-final.pdf
8. Bues, M.-M., Matthaei, E.: LegalTech on the rise: technology changes legal work behaviours, but does not replace its profession. In: Jacob, K., Schindler, D., Strathausen, R. (eds.) Liquid Legal. MP, pp. 89–109. Springer, Cham (2017). https://doi.org/10.1007/978-3-319-45868-7_7
9. Chaudhry, I.A., Khan, A.A.: A research survey: review of flexible job shop scheduling techniques. Int. Trans. Oper. Res. **23**(3), 551–591 (2016)
10. Elloumi, S., Fortemps, P.: A hybrid rank-based evolutionary algorithm applied to multi-mode resource-constrained project scheduling problem. Eur. J. Oper. Res. **205**(1), 31–41 (2010)
11. Emons, W.: Legal fees and lawyers' compensation. In: Parisi, F. (ed.) The Oxford Handbook of Law and Economics, chap. 12, pp. 247–258. Oxford University Press, Oxford (2017)
12. Farshidi, S., Ziarati, K.: A bi-population genetic algorithm with two novel greedy mode selection methods for MRCPSP. Adv. Comput. Sci. Int. J. **5**(4), 66–77 (2016)
13. Haggerty, J.: Disruptive technology in the legal profession, January 2019. https://www2.deloitte.com/uk/en/pages/financial-advisory/articles/the-case-for-disruptive-technology-in-the-legal-profession.html
14. Kramer, A., Subramanian, A.: A unified heuristic and an annotated bibliography for a large class of earliness-tardiness scheduling problems. J. Sched. **22**(1), 21–57 (2019)
15. Kundakcı, N., Kulak, O.: Hybrid genetic algorithms for minimizing makespan in dynamic job shop scheduling problem. Comput. Ind. Eng. **96**, 31–51 (2016)
16. Lova, A., Tormos, P., Cervantes, M., Barber, F.: An efficient hybrid genetic algorithm for scheduling projects with resource constraints and multiple execution modes. Int. J. Prod. Econ. **117**(2), 302–316 (2009)
17. Nguyen, T.T., Yao, X.: Dynamic time-linkage problems revisited. In: Giacobini, M., et al. (eds.) EvoWorkshops 2009. LNCS, vol. 5484, pp. 735–744. Springer, Heidelberg (2009). https://doi.org/10.1007/978-3-642-01129-0_83

18. Osman, M., Abo-Sinna, M.A., Mousa, A.: An effective genetic algorithm approach to multiobjective resource allocation problems (MORAPs). Appl. Math. Comput. **163**(2), 755–768 (2005)

19. Özbakir, L., Baykasoğlu, A., Tapkan, P.: Bees algorithm for generalized assignment problem. Appl. Math. Comput. **215**(11), 3782–3795 (2010)

20. Regnier-Coudert, O., McCall, J., Ayodele, M., Anderson, S.: Truck and trailer scheduling in a real world, dynamic and heterogeneous context. Transp. Res. Part E: Logist. Transp. Rev. **93**, 389–408 (2016)

21. Tofan, S., Allmendinger, R., Zanda, M., Stephens, O.: Heuristic allocation of computational resources. In: Proceedings of the Genetic and Evolutionary Computation Conference, pp. 1256–1263 (2017)

22. Van Peteghem, V., Vanhoucke, M.: An experimental investigation of metaheuristics for the multi-mode resource-constrained project scheduling problem on new dataset instances. Eur. J. Oper. Res. **235**(1), 62–72 (2014)

23. Wang, L., Fang, C.: An effective estimation of distribution algorithm for the multi-mode resource-constrained project scheduling problem. Comput. Oper. Res. **39**(2), 449–460 (2012)

Unsupervisedly Learned Representations
– Should the Quest Be Over?

Daniel N. Nissani (Nissensohn)$^{(\boxtimes)}$

Tel Aviv, Israel
dnissani@post.bgu.ac.il

Abstract. After four decades of research there still exists a Classification accuracy gap of about 20% between our best Unsupervisedly Learned Representations methods and the accuracy rates achieved by intelligent animals. It thus may well be that we are looking in the wrong direction. A possible solution to this puzzle is presented. We demonstrate that Reinforcement Learning can learn representations which achieve the same accuracy as that of animals. Our main modest contribution lies in the observations that: a. when applied to a real world environment Reinforcement Learning does not require labels, and thus may be legitimately considered as Unsupervised Learning, and b. in contrast, when Reinforcement Learning is applied in a simulated environment it does inherently require labels and should thus be generally be considered as Supervised Learning. The corollary of these observations is that further search for Unsupervised Learning competitive paradigms which may be trained in simulated environments may be futile.

Keywords: Unsupervised learning · Representations learning · Features learning · Features extraction

1 Introduction

This is neither a "new results" nor a "new method" paper, rather a "quo vadis" call to our Unsupervised Learning research community.

Corroborated by observations, it is widely believed that the animal world, humans included, learn rich world representations by some Unsupervised Learning scheme. Yet, in spite of at least four decades of research in this area, there still exists a Classification accuracy gap of about 20% between our best Unsupervised Learning methods and the corresponding skills exhibited by humans. We deduce this from the facts that a. humans perform very close to our state-of-the-art Supervised Learning methods (a.k.a. deep neural networks) in e.g. challenging Classification tasks (Russakovsky et al. 2015); and b. an accuracy gap of about 20% exists between these Supervised Learning methods and our best Unsupervisedly Learning schemes.

We provide next a brief survey of related work and published evidence for this second assertion above.

As hinted above, the search for Unsupervisedly Learned Representations has been strongly inspired by observation of the animal world[1] and further motivated by the high

---

[1] See e.g. Yann LeCun Interview, IEEE Spectrum, Feb. 2015: "…The bottom line is that the brain is much better than our model at doing unsupervised learning…".

© Springer Nature Switzerland AG 2020
G. Nicosia et al. (Eds.): LOD 2020, LNCS 12566, pp. 339–349, 2020.
https://doi.org/10.1007/978-3-030-64580-9_29

labeling costs of data sets of increasing size and complexity. The hope has been to be able to map sensorial input spaces into feature or representation spaces so that the mapped class conditioned distributions achieve better inter-class linear separability than that of the input sensorial space (linear separability has been the goal since the last layer of a neural net may be viewed as a linear classifier).

We will exclude from our survey clustering methods (like k-Means, spectral clustering and the like) since they operate on the input space itself and do not execute any input to feature space mapping; and will also exclude handcrafted or engineered so called feature extractors (e.g. Lowe 1999) that while mapping an input to a representation space, their map is fixed and not trainable.

The vast majority of this effort has been focused in the area of Autoencoders, originally proposed by (Sanger 1988): neural nets which attempt to encode an input into a more useful representation and then decode this representation back into the same input. In many cases the unsupervised training of such a network is executed by means of standard backpropagation techniques (Bishop 2006) where the input reconstruction error is regularized by some intuitive criterion such as 'sparsifying' (Olshausen and Field 1996; Ranzato et al. 2007), 'contractive' (Rifai et al. 2011), 'denoising' (Vincent et al. 2010; Bengio et al. 2011). In other cases reconstruction is implemented by stochastic methods (Hinton et al. 1995; Salakhutdinov and Hinton 2007). Typically in these works several single layer Autoencoders are sequentially trained and layer-wise stacked. Such stacking can be viewed as an input to representation space mapping, and could in principle have been evaluated in conjunction with a classifier layer (e.g. SVM, linear discriminant, etc.); no such evaluation has been apparently reported. Instead, after training, Autoencoder stacks have been generally used to provide parameters initialization of (supervisedly trained) deep neural nets. These have been shown to exhibit faster convergence vs. random weights initialization (Bengio et al. 2011; Bengio 2012; Arora et al. 2014; Erhan et al. 2010). For a more comprehensive review the reader is referred to (Bengio and Courville 2014).

Another, smaller but more recent and relevant corpus of works, use an auxiliary task in order to unsupervisedly train a deep neural net and achieve such a representation mapping. These include context prediction (Doersch et al. 2016; Pathak et al. 2016), puzzle solving (Noroozi and Favaro 2016), video patches tracking (Wang and Gupta 2015), split image prediction (Zhang and Isola 2017), and Generative Adversarial Networks (Radford et al. 2016; Donahue et al. 2017). The Classification accuracies of these unsupervisedly learned representations (evaluated in conjunction with a classifier) are within a few percent from one another and about 20% worse than a fully supervisedly trained similar architecture net: e.g. 32.2% vs. 53.5% (Donahue et al. 2017) and 38.1% vs. 57.1% (Noroozi and Favaro 2016), unsupervised vs. supervised respectively, all Top-1 on ImageNet dataset.

Since, as we have seen, Unsupervised Learning in the animal (explicitly – human) world performs similarly as well as our artificial Supervised Learning does, and this last outperforms our artificial Unsupervised Learning by about 20%, we conclude by transitivity that, in spite of decades of research, the accuracy gap between the two Unsupervised Learning schemes, the 'natural' and the 'artificial', remains about 20% large, as asserted above.

We may have been looking at the wrong direction.

As we will demonstrate in the sequel Reinforcement Learning (RL) methods can learn representations which when applied to a linear classifier achieve Classification accuracy competitive with that of similar architecture Supervised Learning neural net classifiers.

Furthermore and more importantly, simple gedanken-experiments will lead us to the conclusion that if a RL scheme is applied in *real-world environments* (e.g.in the animal world) then labels of no sort are involved in the process, and it thus may be legitimately considered an Unsupervised Learning scheme; in contrast, application of RL methods in the context of *simulated or symbolic processing environments* (as is usually the case both during research and in business applications) will necessarily require labels for learning and is thus doomed to remain, with some exceptions, a Supervised Learning scheme[2].

In Sect. 2 we will provide a concise introduction to RL concepts. In Sect. 3 we bring up a simple demonstration of representation learning by a RL method, which achieves competitive Classification accuracy relative to state of the art neural nets of similar architecture. We discuss our results and their implications in Sect. 4, and conclude in Sect. 5.

2   The Reinforcement Learning Framework in a Nutshell

RL has reached outstanding achievements during the last few decades, including learning to play at champion level games like Backgammon (Tesauro 1994), highly publicized Chess and Go, and standing behind much of autonomous vehicles technology advances.

It is our goal in this brief Section to promote a connection, from Pattern Recognition community perspective, with the Reinforcement Learning methodology. There seems to exist, surprisingly, a relative disconnect that lasted for decades. For an excellent text on the RL subject we refer to (Sutton and Barto 2018). Readers knowledgeable of RL may directly skip to the next Section.

The basic RL framework consists of an Agent and an Environment. The Agent observes the Environment States (sometimes called Observations), and executes Actions according to some Policy. These Actions, in general, affect the Environment, modifying its State, and generating a Reward (which can also be null or negative). The new State and Reward are observed by the Agent and this in turn generates a new Action. The loop may go on forever, or till the end of the current episode, which may then be restarted. Actions may have non-immediate effects and Rewards may be delayed. A possible goal of the Agent is to maximize some, usually probabilistic, function of its cumulative Reward by means of modifying its Policy. The model is typically time discrete, stochastic, and usually assumed Markov, that is the next Environment State depends on the present State and Action only, and not in the further past. Formally stated, we have

$$G_t = R_{t+1} + \gamma R_{t+2} + \gamma^2 R_{t+3} + \ldots\ldots = R_{t+1} + \gamma G_{t+1} \qquad (1)$$

where G_t is the cumulative Reward (usually called the Return), R_{t+i}, $i = 1,2,\ldots$ are instantaneous Rewards, all referring to a time t, and $0 \leq \gamma \leq 1$ is a discount factor which accounts

---

[2] So that, like Moses on Mount Nebo, we may be able to see Promised Land but will never reach it (Deuteronomy 34:4).

for the relative significance we assign to future Rewards (and may also guarantee convergence in case of infinite Reward series). The recursive form at the RHS of (1) is important as we will momentarily see. R_{t+i}, S_t, A_t (as well as G_t) are all random variables governed by the Environment model distribution p(s', r | s, a) (shorthand for $\Pr\{S_{t+1} = s', R_{t+1} = r \mid S_t = s, A_t = a\}$; we will apply similar abbreviations for other distributions), the Policy distribution $\pi(a \mid s)$ and the States distribution $\mu(s)$. The sum (1) may have a finite number of terms or not depending on whether the model is episodic or not. To assess how good a State is we may use its Value, which is defined as the mean of the Return assuming we start from this specified State and act according to a fixed Policy π

$$V(S_t = s) = E[G_t \mid S_t = s] = E[R_{t+1} + \gamma G_{t+1} \mid S_t = s] \tag{2}$$

where E[.] is the expectation operator, and where the right side equality is based on the recursive form of (1). For a non-specified State realization s, the Value $V(S_t)$ is a function of the r.v. S_t and is thus a r.v. too. Assuming the model is indeed Markov we may derive from (2) to the important Bellman equation[3]

$$V(S_t = s) = E[R_{t+1} + \gamma V(S_{t+1}) \mid S_t = s] \tag{3}$$

Many (but not all) of the methods used by RL are based on the Bellman equation. If the involved distributions p and π are explicitly known then (3) may be directly evaluated. More commonly we may apply the law of large numbers, and iteratively approximate (3) by sampling $V(S_t)$ and $V(S_{t+1})$ and calculating

$$V(S_t = s)_{new} = V(S_t = s)_{old} + \alpha \left[R_{t+1} + \gamma V\left(S_{t+1} = s'\right)_{old} - V(S_t = s)_{old} \right] \tag{4}$$

which is a slight generalization of the moving average formula, and where α is an update parameter. When (4) converges the quantity in square parenthesis (appropriately called 'the error') vanishes so that it may become a legitimate goal in a minimization problem. If the Value is approximated by some neural network parameterized by W_t^V, as is many times the case, we can iteratively update W_t^V by means of stochastic gradient ascent

$$W_{t+1}^V = W_t^V + \eta^V [R_{t+1} + \gamma V\left(S_{t+1}; W_t^V\right) - V\left(S_t; W_t^V\right)] \nabla V\left(S_t; W_t^V\right) \tag{5}$$

where η^V is the learning step parameter, where the gradient is calculated w.r.t. W_t^V and where the factor within square parenthesis may be recognized as a parameterized version of the error in (4) above. Note that the gradient operator in (5) applies solely to $V(S_t; W_t^V)$ and not to $V(S_{t+1}; W_t^V)$ which also depends on W_t^V: this scheme constitutes a common and convenient formal abuse in RL methodology and is known as semi-gradient.

3 Learning Representations by Means of Reinforcement Learning

Forcing an Agent to learn a task, for which success the ability to correctly discriminate between different classes of objects is required, should end up with the Agent possessing sufficiently rich internal representations of these classes. After the task training is completed the resultant representations (more precisely the mapping which generates them) can be attached to a linear classifier (e.g. single layer perceptron, SVM, etc.) and used as a feature extractor.

---

[3] The reader may refer to https://stats.stackexchange.com/questions/243384/deriving-bellmans-equation-in-reinforcement-learning for a neat proof of the Bellman equation by Jie Shi.

We demonstrate this idea by means of a simple example using the MNIST dataset. To make matters simpler we will define the Agent task itself as that of a classifier, that is it should predict which class (digit in our case) is being presented to it by the Environment (any other task requiring class discrimination will make it). The Agent will be rewarded by a small Reward each time it succeeds; by a bigger Reward after N consecutive successes, after which the episode ends and a new episode is launched; and penalized by a negative Reward upon each error.

RL comprises today a rich arsenal of methods by which we can approach this problem. We choose here the so called Policy Gradient scheme (Sutton et al. 2000; Sutton and Barto 2018) where the Value, approximated by a neural network and the Policy, approximated by another, are concurrently trained so as to optimize a suitable goal. We define our goal here to be the maximization of the mean, over the state distribution $\mu(s)$, of the Value $V(S_t; W_t^V)$. This goal is translated into approximating the Value function by (5) above for all States, and learning the optimal Policy $\pi(a \mid s; W_t^\pi)$ by

$$W_{t+1}^\pi = W_t^\pi + \eta^\pi [R_{t+1} + \gamma V\left(S_{t+1}; W_t^V\right) - V\left(S_t; W_t^V\right)] \nabla \log \pi\left(A_t \mid S_t; W_t^\pi\right) \tag{6}$$

where η^π is the Policy learning step parameter and where the gradient is taken w.r.t. the parameters W_t^π of the Policy network (the derivation of (6) above is non-trivial and may be of interest, see e.g. Sutton and Barto 2018 for details). Both neural nets are concurrently trained by backpropagation.

In practice it is convenient to let all but the last layer of the neural nets to share the same parameters so we set [784 300] as dimensions for the shared layers ($784 = 28^2$ is the dimension of an MNIST digit vector) and 1 and 10 dimensions for the Value and Policy networks last layer respectively (all excluding a bias element). Please refer to Fig. 1 below. We used Leaky ReLU activation functions for both nets hidden layers and for the Value net last layer, and softmax for the Policy net last layer. The softmax output of the Policy net, expressing the probability of choosing each of the possible Actions,

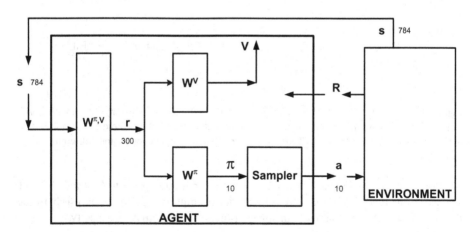

Fig. 1. Simple demonstration of Representations Learning by means of Reinforcement Learning. All but last layers of $\mathbf{W}^\pi$ and $\mathbf{W}^V$ are shared and denoted as $\mathbf{W}^{\pi,V}$. The last $\mathbf{W}^\pi$ layer serves also as classifier layer, and $\mathbf{r}$ is the learned Representation.

was sampled to generate the class prediction. We assigned a non-Terminal Reward (= +1) to each successful prediction, a Terminal Reward (= +10) after N (=5) consecutive successes, and a negative Reward (= −5) after each error.

We set $\eta^\pi = \eta^V = 1e{-}3$ and $\gamma = 0.9^4$. We trained the system permutating the Training set every epoch until it achieved 0.4% Error rate on the Training set (see Fig. 2 below) after 880 epochs. We then measured the Classification Error rate with the MNIST Test set, with frozen parameters: 1.98%. For reference purpose we trained with the same MNIST dataset a neural net classifier of same dimensionality ([784 300 10]) as our Policy network above: it achieved 0.4% Training set Error rate after 42 epochs (annotated for reference; and 0% Training set Error rate after 53 epochs), and a Test set Error rate of 2.09%, measured after attaining 0% Training set Error rate, negligibly inferior to our RL scheme above.

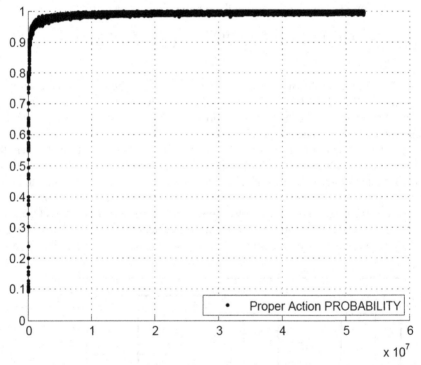

Fig. 2. Training set Pr{Proper Action}, equivalent to (1 − Pr{Training Error}), starts with a brief slightly descending segment (unseen) due to errors penalty, followed by a rapid uprise, and ending in slow final convergence to 0.4% Training set Error rate after 880 epochs, and still going.

Note that in our demonstrations above we are not attempting to beat any MNIST Test Error rate record; our goal here is to provide a simple comparative analysis between RL and a 'standard' neural net with similar function approximation capacity.

---

[4] Matlab code of this simple setup will be available upon request from the Author.

The learning steps in both demonstration setups were approximately set at the maximum values that lead to robust convergence. We observe in passing a large gap in learning rate (880 vs. 42 epochs till 0.4% Error rate as stated above) between both classifiers; it is premature to state whether this reflects a fundamental difference between the schemes, or not; and this falls beyond the scope of this paper.

Before we proceed it may be of interest to qualitatively probe whether learning representations under a less demanding task may still provide full inter-class discrimination, or just the minimal required to fulfill the task. For this purpose we trained a RL Policy Gradient network similar to the above to solve the task of 'Picking' a presented MNIST digit if 'odd' and 'Passing' it if 'even'. After training to high accuracy w.r.t. this task (about 99%), the first layer of the network, mapping the learned representations (of dimension 300), was frozen, and adjoined to a linear classifier, which was trained and tested for digits ('0' to '9') Classification accuracy.

As might be expected accuracy was poor (about 72%), reflecting the fact that the learned representations, during the RL training session, are able to almost perfectly discriminate between 'odd' and 'even' digits but, apparently, not much more. This failure may well also explain failures of the fore mentioned Unsupervised Learning "auxiliary tasks" methods (e.g. Donahue et al. 2017; etc.); these develop internal representations fit for the job at hand (e.g. puzzle solving, etc.) but not much more and certainly not for full inter-class discrimination.

We shall next see what are the circumstances under which RL may be considered a legitimate Unsupervised (or, alternatively, Supervised) Learning paradigm.

4   Unsupervised or Supervised Learning?

The following discussion might appear to some readers to be of philosophical nature (no offense implied), or alternatively, trivial. Nevertheless, it is important; and deeper than what it may appear.

It might be claimed by the astute reader that while the Agent in the demonstration of the last Section evidently does not need labels during its learning process (Rewards and States are its only inputs), the Environment does need. It has to have access to MNIST digits labels in order to 'decide' which type of Reward to assign (e.g. positive for correct prediction, negative for error, etc.). And so we may not consider the process as a whole, at system level, to be of Unsupervised Learning nature. This of course is true.

To crystallize we conduct 2 simple thought experiments:

Consider a *real world* Environment with a *real* Cat playing the Agent; by loose analogy we may call this an "in vivo" experiment. The *abstract entity* which we call Environment may present to the Cat a *real* Dog from which the Cat should escape to get a Reward (*real* safety in this case); or it may present it a *real* Mouse which the Cat should persecute in order to get a Reward (*real* fun in this case). Failing to act appropriately may result in a negative Reward (a *real* bite in case of the Dog or *real* boredom in case of the Mouse). After some reflection it should be easy to see that no decision from the part of the abstract Environment is required in order to assign a Reward: these are given, well, 'naturally'. And as a result, no label is involved in the process. The RL paradigm in this case should be considered a legitimate Unsupervised Learning scheme.

In sharp contrast to this we now consider a similar situation, but with a *simulated* Environment and a *simulated* Agent; by the same analogy this may now be called an "in vitro" scenario. The *real entity* which we call Environment (now a piece of computer code) may present to the Cat a *simulated* Dog (for example a photograph of a Dog) from which the Cat has to 'escape' in order to get a *simulated* Reward (such as a positive valued scalar), etc. The *simulated* 'escape' Action (assuming this is the selected Action) is presented to the *simulated* Environment in the form of some set of bits (say, or some other symbolic form). The Environment then, has to use the fore mentioned symbol representing the Action and the *label* of the presented State ('Dog'), which was provided by some genie, and use them combined to produce the appropriate Reward (e.g.by means of a look-up table). In the absence of a State *label* the process gets stuck[5]. We have no choice but to consider this RL case as a Supervised Learning scheme.

The 2 experiments above present extreme cases for the sake of clarity. In practice we may also envision scenarios where a *real* Agent acts upon a *simulated* Environment, or vice versa. The key point in any scenario is how the Reward assignment is produced by the Environment: whether 'naturally' (no labels required, Unsupervised Learning) or 'artificially' by means of some symbol processing (labels required, Supervised Learning).

5 Discussion and Conclusions

We have presented a comprehensive body of previous work. This has lead us to deduce that there exists today, after decades of research, a performance gap of about 20% between the recognition accuracy of our man-devised Unsupervised Learning methods and that of Unsupervised Learning as exhibited in the animal (specifically human) world. Animal Unsupervised Learning achievements naturally serve as our source of inspiration (just as flying birds inspired men to invent ways to fly), so that it was and is reasonable for us to expect that a competitive Unsupervised Learning scheme indeed exists; we just apparently happen to not have uncovered it yet. We proposed herein a possible resolution to this apparent puzzle.

As demonstrated above by a simple setup, Reinforcement Learning may force an Agent to learn rich enough representations to allow it to discriminate amongst classes as much as needed to successfully fulfill the task at hand. We have reasoned that such RL schemes, when implemented in a real world Environment should be considered to be of true Unsupervised Learning nature, slightly refining our conventional taxonomy (we may have been all along victims of our own terminology) which exclusively divides paradigms into *either* SL *or* UL *or* RL. RL boasts recognition competitive performance, as we explicitly illustrated in our MNIST Toy case above and as it was implicitly demonstrated by a myriad of RL experiments (e.g. Hessel et al. 2018). *Hence, RL may actually be that missing piece of our puzzle.*

---

[5] We could consider in principle the case wherein the Environment predicts by itself a label from the Dog's photograph; this however requires the Environment to be a trained classifier, which enters us into an endless loop.

As we indicated RL however, has the strange and unique peculiarity to metamorphose its nature when embodied in simulated Environments, and transform, with some exceptions, into a scheme of (not less) true Supervised Learning character.

This peculiarity might have been the source of our prolonged confusion. Indeed it is possible a result of this that RL has been usually considered in literature as a class of its own (i.e. neither Supervised nor Unsupervised) instigating further exploration efforts, to no avail so far.

Why all this matters? Because if RL turns out indeed to be the Unsupervised Learning scheme which the (real) animal world employs, then further search for the Holy Grail, a learning method with competitive accuracy which unsupervisedly learns representations in simulated environments, may be futile.

This is not to say that man-made systems of all kinds are always and inherently unable to competitively learn unsupervisedly. We may cite here 2 examples:

a. Robots acting in real world Environments (as might usually be the case). Such a Robot may e.g. learn by RL that when short of charge it should seek for a recharge station, its Reward being a real recharge. No labels are involved in this process.
b. Simulated Environments, but where the required labels are not provided by 'some genie', as mentioned above. In such cases the labels themselves are used in order to generate the associated objects; this may occur for example in systems (e.g. video games) where a label (e.g. 'Dog') is used to draw a dog caricature with some randomness in it, or to sample a realistic dog image from some distribution (e.g. by means of a generative network like GAN, etc.). Labels in such cases are associated with their respective objects by design (not externally provided), and it seems reasonable to consider these cases to be of unsupervised learning nature as well.

Of course one or both of our conjectures above may turn out false: animal brains, our source of inspiration, may be unsupervisedly learning by a completely different scheme than RL which we may try, once discovered, to imitate; and/or there might exist ways of rich representations learning other than RL, which may remain unsupervised even in simulated environments. Time will tell: Reinforcement Learning in Animals, the Neuroscience of Reinforcement Learning, and Unsupervised Learning of Representations are all active areas of research.

References

Arora, S., Bhaskara, A., Ge, R., Ma, T.: Provable bounds for learning some deep representations. In: ICML 2014 (2014)

Bengio, Y., et al.: Deep learners benefit more from out-of-distribution examples. In: AISTATS 2011 (2011)

Bengio, Y., Courville, A.: Representation learning: a review and new perspectives. arXiv:1206. 5538 (2014)

Bengio, Y., Lamblin, P., Popovici, D., Larochelle, H.: Greedy layer-wise training of deep networks. In: NIPS 2007 (2007)

Bengio, Y.: Deep Learning of Representations for Unsupervised and Transfer Learning, JMLR 2012

Bishop, C.M.: Pattern Recognition and Machine Learning. Springer, New York (2006)

Coates, A., Lee, H., Ng, A.Y.: An analysis of single-layer networks in unsupervised feature learning. In: AISTATS 2011 (2011)

Doersch, C., Gupta, A., Efros, A.A.: Unsupervised visual representation learning by context prediction. arXiv:1505.05192 (2016)

Donahue, J., Krahenbuhl, P., Darrell, T.: Adversarial feature learning. arXiv:1605.09782 (2017)

Erhan, D., Bengio, Y., Courville, A., Manzagol, P., Vincent, P.: Why does unsupervised pre-training help deep learning? J. Mach. Learn. Res. **11**, 625–660 (2010)

Hessel, M.: Rainbow: combining improvements in deep reinforcement learning. In: AAAI 2018 (2018)

Hinton, G.E., Dayan, P., Frey, B.J., Neal, R.M.: The wake-sleep algorithm for unsupervised neural networks. Science **268**(5214), 1158–1161 (1995)

Larochelle, H., Bengio, Y., Louradour, J., Lamblin, P.: Exploring strategies for training deep neural networks. J. Mach. Learn. Res. **1**, 1–40 (2009)

Lowe, D.G.: Object recognition from local scale-invariant features. In: Proceedings of the International Conference on Computer Vision (1999)

Mesnil, G., et al.: Unsupervised and transfer learning challenge: a deep learning approach. In: JMLR 2011 (2011)

Noroozi, Mehdi, Favaro, Paolo: Unsupervised learning of visual representations by solving jigsaw puzzles. In: Leibe, Bastian, Matas, Jiri, Sebe, Nicu, Welling, Max (eds.) ECCV 2016. LNCS, vol. 9910, pp. 69–84. Springer, Cham (2016). https://doi.org/10.1007/978-3-319-46466-4_5

Olshausen, B.A., Field, D.J.: Emergence of simple-cell receptive field properties by learning a sparse code for natural images. Nature **381**(6583), 607–9 (1996)

Pathak, D., Krahenbuhl, P., Donahue, J., Darrell, T.: Context encoders: feature learning by inpainting. In: IEEE CVPR 2016 (2016)

Radford, A., Metz, L., Chintala, S.: Unsupervised representation learning with deep convolutional generative adversarial network. In: ICLR 2016 (2016)

Ranzato, M.A., Poultney, C., Chopra, S., LeCun, Y.: Efficient learning of sparse representations with an energy-based model. In: NIPS 2006 (2006)

Ranzato, M.A., Huang, F., Boureau, Y., LeCun, Y.: Unsupervised learning of invariant feature hierarchies with applications to object recognition. In: IEEE CVPR 2007 (2007)

Rifai, S., Vincent, P., Muller, X., Glorot, X., Bengio, Y.: Contractive auto-encoders: explicit invariance during feature extraction. In: ICML 2011 (2011)

Russakovsky, Olga., et al.: ImageNet large scale visual recognition challenge. Int. J. Comput. Vis. **115**(3), 211–252 (2015). https://doi.org/10.1007/s11263-015-0816-y

Salakhutdinov, R., Hinton, G.: Learning a nonlinear embedding by preserving class neighborhood structure. J. Machi. Learn. Res. (2007)

Sanger, T.D.: An optimality principle for unsupervised learning. In: NIPS 1988 (1988)

Sutton, R.S., Barto, A.G.: Reinforcement Learning: An Introduction. MIT Press, Cambridge (2018)

Sutton, R.S., McAllester, D., Singh, S., Mansour, Y.: Policy gradient methods for reinforcement learning with function approximation. In: NIPS 2000 (2000)

Tesauro, G.: TD-gammon, a self-teaching backgammon program, achieves master-level play. Neural Comput. **6**, 215–219 (1994)

Vincent, P., Larochelle, H., Lajoie, I., Bengio, Y., Manzagol, P.A.: Stacked denoising autoencoders: learning useful representations in a deep network with a local denoising criterion. J. Mach. Learn. Res. (2010)

Wang, X., Gupta, A.: Unsupervised learning of visual representations using videos. In: IEEE ICCV 2015 (2015)

Xie, S., Girshick, R., Dollar, P., Tu, Z., He, K.: Aggregated residual transformations for deep neural networks. In; CVPR 2017 (2017)

Zhang, R., Isola, P., Efros, A.A.: Split-Brain autoencoders: unsupervised learning by cross-channel prediction. In: CVPR 2017 (2017)

Bayesian Optimization with Local Search

Yuzhou Gao[1], Tengchao Yu[1], and Jinglai Li[2(✉)]

[1] Shanghai Jiao Tong University, Shanghai 200240, China
[2] University of Birmingham, Edgbaston, Birmingham B15 2TT, UK
`j.li.10@bham.ac.uk`

Abstract. Global optimization finds applications in a wide range of real world problems. The multi-start methods are a popular class of global optimization techniques, which are based on the idea of conducting local searches at multiple starting points. In this work we propose a new multi-start algorithm where the starting points are determined in a Bayesian optimization framework. Specifically, the method can be understood as to construct a new function by conducting local searches of the original objective function, where the new function attains the same global optima as the original one. Bayesian optimization is then applied to find the global optima of the new local search defined function.

Keywords: Bayesian optimisation · Global optimisation · Multistart method

1 Introduction

Global optimization (GO) is a subject of tremendous potential applications, and has been an active research topic since. There are several difficulties associated with solving a global optimization problem: the objective function may be expensive to evaluate and/or subject to random noise, it may be a black-box model and the gradient information is not available, and the problem may admit a very large number of local minima, etc. In this work we focus on the last issue: namely, in many practical global optimization problems, it is often possible to find a local minimum efficiently, especially when the gradient information of the objective function is available, while the main challenge is to escape from a local minimum and find the global solution. Many metaheuristic GO methods, such simulated annealing [11] and genetic algorithm [6], can avoid being trapped by a local minimal, but these methods do not take advantage of the property that a local problem can be quite efficiently solved, which makes them less efficient in the type of problems mentioned above.

A more effective strategy for solving such problems is to combine global and local searches, and the multi-start (MS) algorithms [12] have become a very popular class of methods along this line. Loosely speaking the MS algorithms attempt to find a global solution by performing local optimization from multiple starting points. Compared to search based global optimization algorithms, the

© Springer Nature Switzerland AG 2020
G. Nicosia et al. (Eds.): LOD 2020, LNCS 12566, pp. 350–361, 2020.
https://doi.org/10.1007/978-3-030-64580-9_30

(MS) methods is particularly suitable for problems where a local optimization can be performed efficiently. The most popular MS methods include the clustering [8,18] and the Multi Level Single Linkage (MLSL) [14] methods and the OptQuest/NLP algorithm [19]. More recently, new MS algorithms have been proposed and applied to machine learning problems [5,10]. One of the most important issues in a MS algorithm is how to determine the initial points, i.e. the points to start a local search (LS) from. Most MS algorithms determine the initial points sequentially, which in each step requires to find the next initial point based on the current information. We shall adopt this setup in this work and so the question we want to address in the present work is *how to determine the next "best" initial point given the information at the current step.*

The main idea presented in this work is to sequentially determine the starting points in a Bayesian optimization (BO) [13,16,17] framework. The standard BO algorithm is designed to solve a global optimization problem directly without using LS: it uses a Bayesian framework and an experimental design strategy to search for the global minimizers. The BO algorithms have found success in many practical GO problems, especially for those expensive and noisy objective functions [2]. Nevertheless, the BO methods do not take advantage of efficient local solvers even when that is possible. In this work, instead of applying BO directly to the global optimization problem, we propose to use it to identify starting points for the local solvers in a MS formulation. Within the BO framework, we can determine the starting points using a rigorous and effective experimental design approach.

An alternative view of the proposed method is that we define a new function by solving a local optimization problem of the original objective function. By design the newly defined function is discrete-valued and has the same global optimizers as the original objective function. And we then perform BO to find the global minima for the new function. From this perspective, the method can be understood as to pair the BO method with a local solver, and we reinstate that the method requires that the local problems can be solved efficiently. For example, in many statistical learning problems with large amounts of data, a noisy estimate of the gradients can be computed more efficiently than the evaluation of the objective function [1], and it follows that a local solution can be obtained at a reasonable computational cost.

The rest of the work is organized as follows. In Sect. 2 we introduce the MS algorithms for GO problems, and present our BO based method to identify the starting points. In Sect. 3 we provide several examples to demonstrate the performance of the proposed method. Finally Sect. 4 offers some closing remarks.

2   Bayesian Optimization with Local Search

2.1   Generic Multi-start Algorithms

Suppose that we want to solve a bound constrained optimization problem:

$$\min_{\mathbf{x} \in \Omega} f(\mathbf{x}), \tag{1}$$

where Ω is a compact subspace of R^n. In general, the problem may admit multiple local minimizers and we want to find the global solution of it. As has been mentioned earlier, the MS algorithms are a class of GO methods for problems where LS can be conducted efficiently. The MS iteration consists of two steps: a global step where an initial point is generated, and a local step which performs a local search from the generated initial point. A pseudocode of the generic MS algorithm is given in Algorithm 1. It can be seen here that one of the key issues of the MS algorithm is how to generate the starting point in each iteration. A variety of methods have been proposed to choose the starting points, and they are usually designed for different type of problems. For example, certain methods such as [19] assume that the evaluation of the objective function is much less computationally expensive than the local searches, and as a result they try to reduce the number of local searches at the price of conducting a rather large number of function evaluations in the state space. On the other hand, in another class of problems, a satisfactory local solution may be obtained at a reasonable computational cost, and as will be discussed later we shall use the BO algorithm to determine the initial points. For this purpose, we next give a brief overview of BO.

Algorithm 1. A generic MS algorithm

1: set $i = 0$;
2: **while** Stopping criteria are not satisfied **do**
3:     $n = n + 1$;
4:     generate a new initial point $\mathbf{x}_n$ based on some prescribed rules;
5:     perform a LS from $\mathbf{x}_n$ and store the obtained local minimal value;
6: **end while**
7: **return** the smallest local minimum value found.

2.2   Bayesian Optimization

The Bayesian optimization (BO) is very popular global optimization method, which treats the objective function as a blackbox. Simply put, BO involves the use of a probabilistic model that defines a distribution over objective function. In practice the probabilistic model is usually constructed with the Gaussian Process (GP) regression: namely the function $f(\mathbf{x})$ is assumed to be a Gaussian process defined on Ω, the objective function is queried at certain locations, and the distribution of the function value at any location $\mathbf{x}$, conditional on the observations, which is Gaussian, can be explicitly computed from the Bayesian formula. Please see Appendix A for a brief description of the GP construction. Based on the current GP model of $f(\mathbf{x})$ the next point to query is determined in an experimental design formulation. Usually the point to query is determined by maximizing an acquisition function $\alpha(\mathbf{x}, \hat{f})$ where $\hat{f}$ is the GP model of f, which is designed based on the exploration and the exploitation purposes of the algorithm. Commonly used acquisition functions include the Expected Improvement, the Probability of Improvement, and the Upper Confidence Bound, and interested readers may consult [17] for detailed discussions and comparisons of these acquisition functions. We describe the standard version of BO in Algorithm 2.

Algorithm 2. The BO algorithm

1: generate a number of points $\{x_1, ..., x_{N_0}\}$ in Ω.
2: evaluate $y_n = f(\mathbf{x}_n)$ for $n = 1 : N_0$;
3: let $D_{N_0} = \{(\mathbf{x}_n, y_n)\}_{n=1}^{N_0}$;
4: construct a GP model from D_{N_0}, denoted as $\hat{f}_{N_0}$;
5: $n = N_0$;
6: **while** stopping criteria are not satisfied **do**
7:     $\mathbf{x}_{n+1} = \arg\max \alpha(\mathbf{x}; \hat{f}_n)$
8:     $y_{n+1} = f(\mathbf{x}_{n+1})$;
9:     augment data $D_{n+1} = D_n \cup \{(\mathbf{x}_{n+1}, y_{n+1})\}$
10:    update GP model obtaining $\hat{f}_{n+1}$;
11:    $n = n + 1$;
12: **end while**
13: **return** $y_{\min} = \min\{y_n\}_{n=1}^N$;

2.3 The BO with LS Algorithm

Now we present our method that integrate MS and BO. The idea behind the method is rather simple: we perform BO for a new function which has the same global minima as the original function $f(\mathbf{x})$. The new function is defined via conducting local search of $f(\mathbf{x})$. Specifically suppose we have local solver $\mathcal{L}$ defined as,

$$\mathbf{x}^* = \mathcal{L}(f(\cdot), \mathbf{x}), \tag{2}$$

where $f(\cdot)$ is the objective function, $\mathbf{x}$ is the initial point of the local search, and $\mathbf{x}^*$ is the obtained local minimal point. $\mathcal{L}$ can represent any local optimization approach, with or without gradient, and we require that for any given initial point $\mathbf{x}^*$, the solver $\mathcal{L}$ will return a unique local minimum $\mathbf{x}^*$. Using both $\mathcal{L}$ and F, we can define a new function

$$y = F_{\mathcal{L}}(\mathbf{x}) = f(\mathbf{x}^*), \tag{3}$$

where $\mathbf{x}^*$ is the output of Eq. (2) with objective function f and initial point $\mathbf{x}$. That is, the new function $F_{\mathcal{L}}$ takes a starting point $\mathbf{x}$ as its input, and returns the local minimal value of f found by the local solver $\mathcal{L}$ as its output. It should be clear that $F_{\mathcal{L}}$ is a well-defined function on R^n, which has the same global minima as function $f(\mathbf{x})$. Moreover, suppose that $f(\mathbf{x})$ only has a finite number of local minima, and $F_{\mathcal{L}}(\mathbf{x})$ is discrete-valued. Please see Fig. 1 for a schematic illustration of the new function defined by LS and its GP approximation. Next we apply standard BO algorithm to the newly constructed function $F_{\mathcal{L}}(\mathbf{x})$, and the global solution of $F_{\mathcal{L}}$ found by BO is regarded as the global solution of $f(\mathbf{x})$. We refer to the proposed algorithm as BO with LS (BOwLS) and we provide the complete procedure of it in Algorithm 3. We reinstate that, as one can see from the algorithm, BOwLS is essentially a MS scheme, which uses the BO experimental design criterion to determine the next starting point. When desired, multiple starting points can also be determined in the BO framework, and we refer to the aforementioned BO references for details of this matter.

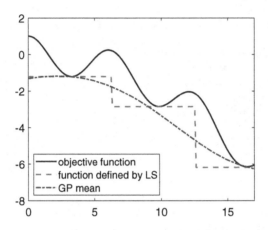

Fig. 1. A schematic illustration of the BOwLS algorithm: the solid line is the original objective function, the dashed line is the function defined by LS, and the dashed-dotted line is the GP regression of the LS defined function.

3   Numerical Examples

In this section, we provide several mathematical and practical examples to demonstrate the performance of the proposed method. In each example, we solve the GO problem with three methods: MLSL, the efficient multi-start (EMS) in [5] and the BOwLS method proposed in this work.

3.1   Mathematical Test Functions

We first consider six mathematical examples that are commonly used as the benchmarks for GO algorithms, selected from [4]. The objective functions, the domains and the global optimal solutions of these functions are provided in Appendix B. As is mentioned earlier, we solve these problems with MLSL, EMS and BOwLS methods, and, since all these algorithms are subject to certain randomness, we repeat the numerical experiments for 50 times. The local search is conducted with the conjugate gradient method using the SciPy package [7]. In these examples we shall assume that evaluating the objective function and its

Algorithm 3. The BOwLS algorithm

1: let $D_0 = \emptyset$;
2: **for** $n = 1{:}N_0$ **do**
3:     solve $[y^*, \mathbf{x}^*] = \mathcal{L}(f(\mathbf{x}), \mathbf{x}_n)$;
4:     let $y_n = y^*$;
5:     augment data $Dn = D_{n-1} \cup \{(\mathbf{x}_n, y_n)\}$
6: **end for**
7: construct a GP model from D_{N_0}, denoted as $\hat{f}_{N_0}$;
8: $n = N_0$;
9: **while** stopping criteria are not satisfied **do**
10:     $\mathbf{x}_{n+1} = \arg \max \alpha(\mathbf{x}; \hat{f}_n)$
11:     solve $[y^*, \mathbf{x}^*] = \mathcal{L}(f(\mathbf{x}), \mathbf{x}_{n+1})$;
12:     let $y_{n+1} = y^*$;
13:     augment data $D_{n+1} = D_n \cup \{(\mathbf{x}_{n+1}, y_{n+1})\}$;
14:     update GP model obtaining $\hat{f}_{n+1}$;
15:     $n = n + 1$;
16: **end while**
17: **return** $y_{\min} = \min\{y_i\}_{i=1}^N$

---

gradient is of similar computational cost, and so we measure the total computational cost by summation of the number of function evaluations and that of the gradient evaluations. For test purpose, we set the stopping criterion to be that the number of function/gradient combined evaluations exceeds 10,000. In our tests, we have found that all the three methods can reach the actual global optima within the stopping criterion in the first five functions. In Figs. 2 we compare the average numbers of the combined evaluations (and their standard deviations) to reach the global optimal value for all the three methods in the first five test functions. As we can see that in all these five test functions except the example (Price), the proposed BOwLS algorithm requires the least computational cost to research the global minima. We also note that it seems that EMS requires significantly more combined evaluations than the other two methods, and we believe that the reason is that EMS is particularly designed for problems with a very large number of local minima, and these test functions are not in that case. On the other hand, the last example (Ackley) is considerably more complicated, and so in our numerical tests all three methods have trials that can not reach the global minimum within the prescribed cost limit. To compare the performance of the methods, we plot the minimal function value obtained against the number of function/gradient combined evaluations in Fig. 3. First the plots show that, in this example, the EMS method performs better than MLSL in both 2-D and 4-D cases, due to the fact that this function is subject to more local minima. More importantly, as one can see, in both cases BOwLS performs considerably better than both EMS and MLSL, and the advantage is more substantial in the 4-D case, suggesting that BOwLS may become more useful for complex objective functions.

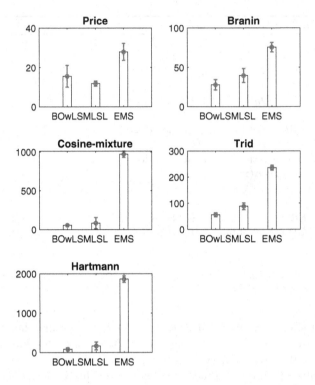

Fig. 2. The average numbers of the combined evaluations to reach the global optima for all the three methods (the error bars indicate the standard deviations) in the first five functions.

3.2   Logistic Regression

Finally consider a Logistic regression example. Logistic regression is a common tool for binary regression (or classification). Specifically suppose that we have binary regression problem where the output takes values at $y = 0$ or $y = 1$, and the probability that $y = 1$ is assumed to be the form of,

$$h_{\mathbf{w}}(\mathbf{x}) = \frac{1}{1 + \exp(-\sum_{i=1}^{m} x_i w_i - w_0)}, \tag{4}$$

where $\mathbf{x} = (x_1, ..., x_m)$ are the predictors and $\mathbf{w} = (w_0, ...w_m)$ are the coefficients to be determined from data. The cost function for the logistic regression is taken to be

$$C(h_{\mathbf{w}}(\mathbf{x}), y) = \begin{cases} -\log(h_{\mathbf{w}}(\mathbf{x})) & \text{if } y = 1 \\ -\log(1 - h_{\mathbf{w}}(\mathbf{x})) & \text{if } y = 0 \end{cases}.$$

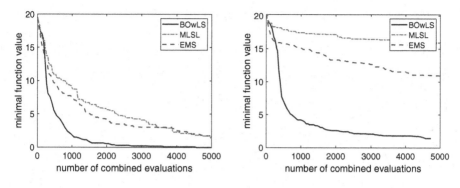

Fig. 3. The minimal function value plotted agains the number of function/gradient combined evaluations, for the 2-D and 4-D Ackley example.

Suppose that we have a training set $\{(\mathbf{x}_i, y_i)\}_{i=1}^{n}$, and we then determine the parameters $\mathbf{w}$ by solving the following optimization problem,

$$\min_{\mathbf{w} \in W} \sum_{i=1}^{n} C(h_{\mathbf{w}}(\mathbf{x}_i), y_i). \tag{5}$$

where W is the domain of $\mathbf{w}$.

In this example we apply the Logistic regression to the Pima Indians Diabetes dataset [15], the goal of which is to diagnose whether a patient has diabetes based on 8 diagnostic measures provided in the data set. The data set contains 768 instances and we split it into a training set of 691 instances and a test set of 77 ones. We solve the result optimization problem (5) with the three GO algorithms, and we repeat the computations for 100 times as before. The minimal function value averaged over the 100 trials is plotted against the number of combined evaluations in Fig. 4 (left). In this example, the EMS method actual performs better than MLSL, while BOwLS has the best performance measured by the number of the function/gradient combined evaluations. Moreover, in the BOwLS method, we expect that as the iteration approaches to the global optimum, the resulting Logistic model should become better and better. To show this, we plot in Fig. 4 (right) the prediction accuracy of the resulting model as a function of the BO iterations (which is also the number of LS), in six randomly selected trials out of 100. The figure shows that the prediction accuracy varies (overall increases) as the number of LS increases, which is a good evidence that the objective function in this example admits multiple local optima and the global optimum is needed for the optimal prediction accuracy.

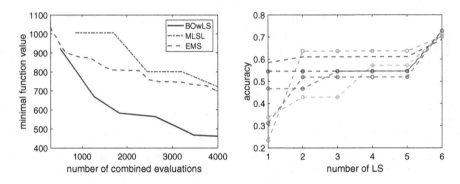

Fig. 4. The minimal function value plotted against the number of function/gradient combined evaluations, for the Logistic regression example.

4   Conclusions

In summary, we have presented a MS algorithm where the starting points of local searches are determined by a BO framework. A main advantage of the method is that the BO framework allows one to sequentially determine the next starting points in a rigorous and effective experimental design formation. With several numerical examples, we demonstrate that the proposed BOwLS method has highly competitive performance against many commonly used MS algorithms. A major limitation of BOwLS is that, as it is based on the BO framework, it may have difficulty in dealing with very high dimensional problems. We note however that a number of dimension reduction based approaches [3,9] have been proposed to enable BO for high dimensional problems, and we hope that these approaches can be extended to BOwLS as well. In addition, another problem that we plan to work on in the future to combine the BOwLS framework with the stochastic gradient descent type of algorithms to develop efficient GO algorithms for statistical learning problems.

Acknowledgement. This work was partially supported by the NSFC under grant number 11301337.

A   Construction of the GP Model

Given the dataset $D = \{(\mathbf{x}_j, y_j)\}_{j=1}^n$, the GP regression performs a nonparametric regression in a Bayesian framework [20]. The main idea of the GP method is to assume that the data points and the new point $(\mathbf{x}, y)$ are from a Gaussian Process defined on R^{n_x}, whose mean is $\mu(\mathbf{x})$ and covariance kernel is $k(\mathbf{x}, \mathbf{x}')$. Under the GP model, one can obtain directly the conditional distribution $\pi(y|\mathbf{x}, D)$ that is Gaussian: $\pi(y|\mathbf{x}) = \mathcal{N}(\mu_{\text{GP}}, \sigma_{\text{GP}}^2)$, where the posterior mean and variance are,

$$\mu_{\text{GP}}(\mathbf{x}) = \mu(\mathbf{x}) + k(\mathbf{x}, \mathbf{X})(k(\mathbf{X}, \mathbf{X}) + \sigma_n^2 I)^{-1}(\mathbf{y} - \mu(\mathbf{x})),$$
$$\sigma_{\text{GP}}^2 = k(\mathbf{x}, \mathbf{x}) - k(\mathbf{x}, \mathbf{X})(k(\mathbf{X}, \mathbf{X}) + \sigma_n^2 I)^{-1}k(\mathbf{X}, \mathbf{x}).$$

Here $\mathbf{y}^* = [y_1, \ldots, y_n]$, $\mathbf{X} = [\mathbf{x}_1, \ldots, \mathbf{x}_n]$, σ_n^2 is the variance of observation noise, I is an identity matrix, and the notation $k(\mathbf{A}, \mathbf{B})$ denotes the matrix of the covariance evaluated at all pairs of points in set $\mathbf{A}$ and in set $\mathbf{B}$ using the kernel function $k(\cdot, \cdot)$. In particular, if the data points are generated according to an underlying function $f(x)$ (which is the objective function in the BO setting), the distribution $\pi(y|\mathbf{x})$ then provides a probabilistic characterization of the function $f(\mathbf{x})$ which can be used to predict the function value of $f(\mathbf{x})$ as well as quantify the uncertainty in the prediction. In Sect. 2, we refer to this probabilistic characterization, i.e., the Gaussian distribution $\pi(y|\mathbf{x})$ as $\hat{f}$. There are a lot of technical issues of the GP construction, such as how to choose the kernel functions and determine the hyperparameters, are left out of this paper, and for more details of the method, we refer the readers to [20].

B   The Mathematical Test Functions

The test functions used in Sect. 3.1 are:

Price (2-D):
$$f(x) = 1 + \sin^2(x_1) + \sin^2(x_2) - 0.1e^{-x_1^2 - x_2^2}.$$

Branin (1-D):
$$f(x) = (-1.275\frac{x_1^2}{\pi^2} + 5\frac{x_1}{\pi} + x_2 - 6)^2 + (10 - \frac{5}{4\pi})\cos(x_1) + 10.$$

Cosine-mixture (4-D):
$$f(\mathbf{x}) = -0.1\sum_{i=1}^{4}\cos(5\pi x_i) - \sum_{i=1}^{4}x_i^2.$$

Trid (6-D):
$$f_{Trid}(\mathbf{x}) = \sum_{i=1}^{6}(x_i - 1)^2 - \sum_{i=2}^{6}x_i x_{i-1}.$$

Hartmann (6-D):
$$f(\mathbf{x}) = -\sum_{i=1}^{4}c_i \exp\left(-\sum_{j=1}^{6}a_{ij}(x_j - p_{ij})^2\right),$$

where

$$a = \begin{pmatrix} 10.0 & 3.0 & 17.0 & 3.50 & 1.70 & 8.0 \\ 0.05 & 10.0 & 17.0 & 0.10 & 8.00 & 14.0 \\ 3.0 & 3.50 & 1.70 & 10.0 & 17.0 & 8.0 \\ 17.0 & 8.0 & 0.05 & 10.0 & 0.10 & 14.0 \end{pmatrix}, \quad c = \begin{pmatrix} 1.0 \\ 1.2 \\ 3.0 \\ 3.2 \end{pmatrix},$$

$$p = \begin{pmatrix} 0.1312 & 0.1696 & 0.5569 & 0.0124 & 0.8283 & 0.5886 \\ 0.2329 & 0.4135 & 0.8307 & 0.3736 & 0.1004 & 0.9991 \\ 0.2348 & 0.1451 & 0.3522 & 0.2883 & 0.3047 & 0.6650 \\ 0.4047 & 0.8828 & 0.8732 & 0.5743 & 0.1091 & 0.0381 \end{pmatrix}.$$

Ackley (n-D):

$$f(x) = -20e^{-0.2\sqrt{\frac{1}{n}\sum_{i=1}^{n} x_i^2}} - e^{\frac{1}{n}\sum_{i=1}^{n} \cos(2\pi x_i)} + 20 + e,$$

where n is taken to be 2 and 4 respectively.

The domains and global minimal values of these functions are shown in Table 1.

Table 1. The domains and the global minimal values of the test functions.

Functions	Domain	Minimal value
Price	$[-10, 10]^2$	-3
Branin	$[-5, 10] \times [0, 15]$	0.397
Cosine-mixture 4d	$[-1, 1]^4$	-0.252
Trid	$[-20, 20]^6$	-50
Hartmann	$[0, 1]^6$	-3.323
Ackley	$[-32.768, 32.768]^n$	0

References

1. Bottou, L.: Large-scale machine learning with stochastic gradient descent. In: Lechevallier, Y., Saporta, G. (eds) Proceedings of COMPSTAT'2010, pp. 177–186. Physica-Verlag HD, Heidelberg (2010). https://doi.org/10.1007/978-3-7908-2604-3_16
2. Brochu, E., Cora, V.M., De Freitas, N.: A tutorial on Bayesian optimization of expensive cost functions, with application to active user modeling and hierarchical reinforcement learning. arXiv preprint arXiv:1012.2599 (2010)
3. Djolonga, J., Krause, A., Cevher, V.: High-dimensional Gaussian process bandits. In: Advances in Neural Information Processing Systems, pp. 1025–1033 (2013)
4. Gavana, A.: Global optimization benchmarks and AMPGO (2005). Accessed 30 Sept 2019
5. György, A., Kocsis, L.: Efficient multi-start strategies for local search algorithms. J. Artif. Intell. Res. **41**, 407–444 (2011)
6. Holland, J.H., et al.: Adaptation in Natural and Artificial Systems: An Introductory Analysis with Applications to Biology, Control, and Artificial Intelligence. MIT Press, Cambridge (1992)
7. Jones, E., Oliphant, T., Peterson, P., et al.: SciPy: open source scientific tools for Python (2001-). http://www.scipy.org/
8. Kan, A.R., Timmer, G.T.: Stochastic global optimization methods part I: clustering methods. Math. Program. **39**(1), 27–56 (1987). https://doi.org/10.1007/BF02592070
9. Kandasamy, K., Schneider, J., Póczos, B.: High dimensional Bayesian optimisation and bandits via additive models. In: International Conference on Machine Learning, pp. 295–304 (2015)

10. Kawaguchi, K., Maruyama, Y., Zheng, X.: Global continuous optimization with error bound and fast convergence. J. Artif. Intell. Res. **56**, 153–195 (2016)
11. Kirkpatrick, S., Gelatt, C.D., Vecchi, M.P.: Optimization by simulated annealing. Science **220**(4598), 671–680 (1983)
12. Martí, R., Lozano, J.A., Mendiburu, A., Hernando, L.: Multi-start methods. In: Handbook of Heuristics, pp. 1–21 (2016)
13. Mockus, J.: Bayesian Approach to Global Optimization: Theory and Applications, vol. 37. Springer, Dordrecht (2012). https://doi.org/10.1007/978-94-009-0909-0
14. Rinnooy Kan, A.H., Timmer, G.: Stochastic global optimization methods part II: multi level methods. Math. Program. **39**(1), 57–78 (1987)
15. Rossi, R.A., Ahmed, N.K.: The network data repository with interactive graph analytics and visualization. In: AAAI (2015). http://networkrepository.com
16. Shahriari, B., Swersky, K., Wang, Z., Adams, R.P., De Freitas, N.: Taking the human out of the loop: a review of Bayesian optimization. Proc. IEEE **104**(1), 148–175 (2015)
17. Snoek, J., Larochelle, H., Adams, R.P.: Practical Bayesian optimization of machine learning algorithms. In: Advances in Neural Information Processing Systems, pp. 2951–2959 (2012)
18. Tu, W., Mayne, R.: Studies of multi-start clustering for global optimization. Int. J. Numer. Methods Eng. **53**(9), 2239–2252 (2002)
19. Ugray, Z., Lasdon, L., Plummer, J., Glover, F., Kelly, J., Martí, R.: Scatter search and local NLP solvers: a multistart framework for global optimization. INFORMS J. Comput. **19**(3), 328–340 (2007)
20. Williams, C.K., Rasmussen, C.E.: Gaussian Processes for Machine Learning. MIT Press, Cambridge (2006)

Sparsity Meets Robustness: Channel Pruning for the Feynman-Kac Formalism Principled Robust Deep Neural Nets

Thu Dinh[1](✉), Bao Wang[2], Andrea Bertozzi[2], Stanley Osher[2], and Jack Xin[1]

[1] University of California, Irvine, USA
thud2@uci.edu, jxin@math.uci.edu
[2] University of California, Los Angeles, USA
wangbaonj@gmail.com, {bertozzi,sjo}@math.ucla.edu

Abstract. Deep neural nets (DNNs) compression is crucial for adaptation to mobile devices. Though many successful algorithms exist to compress naturally trained DNNs, developing efficient and stable compression algorithms for robustly trained DNNs remains widely open. In this paper, we focus on a co-design of efficient DNN compression algorithms and sparse neural architectures for robust and accurate deep learning. Such a co-design enables us to advance the goal of accommodating both sparsity and robustness. With this objective in mind, we leverage the relaxed augmented Lagrangian based algorithms to prune the weights of adversarially trained DNNs, at both structured and unstructured levels. Using a Feynman-Kac formalism principled robust and sparse DNNs, we can at least double the channel sparsity of the adversarially trained ResNet20 for CIFAR10 classification, meanwhile, improve the natural accuracy by 8.69% and the robust accuracy under the benchmark 20 iterations of IFGSM attack by 5.42%.

1 Introduction

Robust deep neural nets (DNNs) compression is a fundamental problem for secure AI applications in resource-constrained environments such as biometric verification and facial login on mobile devices, and computer vision tasks for the internet of things (IoT) [6,24,37]. Though compression and robustness have been separately addressed in recent years, much less is studied when both players are present and must be satisfied.

To date, many successful techniques have been developed to compress naturally trained DNNs, including neural architecture re-design or searching [18,43], pruning including structured (weights sparsification) [15,30] and unstructured (channel-, filter-, layer-wise sparsification) [17,36], quantization [7,39,44], low-rank approximation [8], knowledge distillation [27], and many more [1,22].

The adversarially trained (AT) DNN is more robust than the naturally trained (NT) DNN to adversarial attacks [2,23]. However, adversarial training (denoted as AT if no ambiguity arises, and the same for NT) also dramatically

© Springer Nature Switzerland AG 2020
G. Nicosia et al. (Eds.): LOD 2020, LNCS 12566, pp. 362–381, 2020.
https://doi.org/10.1007/978-3-030-64580-9_31

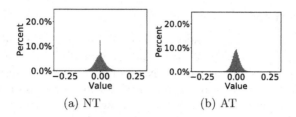

Fig. 1. Histograms of the ResNet20's weights.

reduces the sparsity of the trained DNN's weights. As shown in Fig. 1, start from the same default initialization in PyTorch, the NT ResNet20's weights are much sparser than that of the AT counterpart, for instance, the percent of weights that have magnitude less than 10^{-3} for NT and AT ResNet20 are 8.66% and 3.64% (averaged over 10 trials), resp. This observation motivates us to consider the following two questions:

- 1. *Can we re-design the neural architecture with minimal change on top of the existing one such that the new DNN has sparser weights and better robustness and accuracy than the existing one?*
- 2. *Can we develop efficient compression algorithms to compress the AT DNNs with minimal robustness and accuracy degradations?*

We note that under the AT, the recently proposed Feynman-Kac formalism principled ResNet ensemble [32] has much sparser weights than the standard ResNet, which gives a natural answer to the first question above. To answer the second question, we leverage state-of-the-art relaxed augmented Lagrangian based sparsification algorithms [9,36] to perform both structured and unstructured pruning for the AT DNNs. We focus on unstructured and channel pruning in this work.

1.1   Notation

Throughout this paper we use bold upper-case letters $\mathbf{A}$, $\mathbf{B}$ to denote matrices, bold lower-case letters $\boldsymbol{x}$, $\boldsymbol{y}$ to denote vectors, and lower case letters x, y and α, β to denote scalars. For vector $\boldsymbol{x} = (x_1, \ldots, x_d)^\top$, we use $\|\boldsymbol{x}\| = \|\boldsymbol{x}\|_2 = \sqrt{x_1^2 + \cdots + x_d^2}$ to represent its ℓ_2-norm; $\|\boldsymbol{x}\|_1 = \sum_{i=1}^d |x_i|$ to represent its ℓ_1-norm; and $\|\boldsymbol{x}\|_0 = \sum_{i=1}^d \chi_{\{x_i \neq 0\}}$ to represent its ℓ_0-norm. For a function $f : \mathbb{R}^d \to \mathbb{R}$, we use $\nabla f(\cdot)$ to denote its gradient. Generally, $\boldsymbol{w}^t$ represents the set of all parameters of the network being discussed at iteration t, e.g. $\boldsymbol{w}^t = (\boldsymbol{w}_1^t, \boldsymbol{w}_2^t, \ldots, \boldsymbol{w}_M^t)$, where $\boldsymbol{w}_j^t$ is the weight on the jth layer of the network at the t-th iteration. Similarly, $\boldsymbol{u}^t = (\boldsymbol{u}_1^t, \boldsymbol{u}_2^t, \ldots, \boldsymbol{u}_M^t)$ is a set of weights with the same dimension as $\boldsymbol{w}^t$, whose value depends on $\boldsymbol{w}^t$ and will be defined below. We use $\mathcal{N}(\mathbf{0}, \mathbf{I}_{d \times d})$ to represent the d-dimensional Gaussian, and use notation $O(\cdot)$ to hide only absolute constants which do not depend on any problem parameter.

1.2   Organization

This paper is organized in the following way: In Sect. 2, we list the most related work to this paper. In Sect. 3, we show that the weights of the recently proposed Feynman-Kac formalism principled ResNet ensemble are much sparser than that of the baseline ResNet, providing greater efficiency for compression. In Sect. 4, we present relaxed augmented Lagrangian-based algorithms along with theoretical analysis for both unstructured and channeling pruning of AT DNNs. The numerical results are presented in Sect. 5, followed by concluding remarks. Technical proofs and more related results are provided in the appendix.

2   Related Work

Compression of AT DNNs: [13] considered a low-rank form of the DNN weight matrix with ℓ_0 constraints on the matrix factors in the AT setting. Their training algorithm is a projected gradient descent (PGD) [23] based on the worst adversary. In their paper, the sparsity in matrix factors are unstructured and require additional memory.

Sparsity and Robustness: [14] shows that there is a relationship between the sparsity of weights in the DNN and its adversarial robustness. They showed that under certain conditions, sparsity can improve the DNN's adversarial robustness. The connection between sparsity and robustness has also been studied recently by [28,38], and et al. In our paper, we focus on designing efficient pruning algorithms integrated with sparse neural architectures to advance DNNs' sparsity, accuracy, and robustness.

Feynman-Kac Formalism Principled Robust DNNs: Neural ordinary differential equations (ODEs) [5] are a class of DNNs that use an ODE to describe the data flow of each input data. Instead of focusing on modeling the data flow of each individual input data, [21,31,32] use a transport equation (TE) to model the flow for the whole input distribution. In particular, from the TE viewpoint, [32] modeled training ResNet [16] as finding the optimal control of the following TE

$$\begin{cases} \frac{\partial u}{\partial t}(\boldsymbol{x},t) + G(\boldsymbol{x},\boldsymbol{w}(t)) \cdot \nabla u(\boldsymbol{x},t) = 0, & \boldsymbol{x} \in \mathbb{R}^d, \\ u(\boldsymbol{x},1) = g(\boldsymbol{x}), & \boldsymbol{x} \in \mathbb{R}^d, \\ u(\boldsymbol{x}_i,0) = y_i, & \boldsymbol{x}_i \in T, \text{ with } T \text{ being the training set.} \end{cases} \tag{1}$$

where $G(\boldsymbol{x},\boldsymbol{w}(t))$ encodes the architecture and weights of the underlying ResNet, $u(\boldsymbol{x},0)$ serves as the classifier, $g(\boldsymbol{x})$ is the output activation of ResNet, and y_i is the label of $\boldsymbol{x}_i$.

[32] interpreted adversarial vulnerability of ResNet as arising from the irregularity of $u(\boldsymbol{x},0)$ of the above TE. To enhance $u(\boldsymbol{x},0)$'s regularity, they added a diffusion term, $\frac{1}{2}\sigma^2 \Delta u(\boldsymbol{x},t)$, to the governing equation of (1) which resulting in the convection-diffusion equation (CDE). By the Feynman-Kac formula, $u(\boldsymbol{x},0)$ of the CDE can be approximated by the following two steps:

– Modify ResNet by injecting Gaussian noise to each residual mapping.
– Average the output of n jointly trained modified ResNets, and denote it as
  En_nResNet.

[32] have noticed that EnResNet can improve both natural and robust accuracies of the AT DNNs. In this work, we leverage the sparsity advantage of EnResNet to push the sparsity limit of the AT DNNs.

3   Regularity and Sparsity of the Feynman-Kac Formalism Principled Robust DNNs' Weights

From a partial differential equation (PDE) viewpoint, a diffusion term to the governing Eq. (1) not only smooths $u(\boldsymbol{x}, 0)$, but can also enhance regularity of the velocity field $G(\boldsymbol{x}, \boldsymbol{w}(t))$ [19]. As a DNN counterpart, we expect that when we plot the weights of EnResNet and ResNet at a randomly select layer, the pattern of the former one will look smoother than the latter one. To validate this, we follow the same AT with the same parameters as that used in [32] to train $\text{En}_5\text{ResNet20}$ and ResNet20, resp. After the above two robust models are trained, we randomly select and plot the weights of a convolutional layer of ResNet20 whose shape is $64 \times 64 \times 3 \times 3$ and plot the weights at the same layer of the first ResNet20 in $\text{En}_5\text{ResNet20}$. As shown in Fig. 2 (a) and (b), most of $\text{En}_5\text{ResNet20}$'s weights are close to 0 and they are more regularly distributed in the sense that the neighboring weights are closer to each other than ResNet20's weights. The complete visualization of this randomly selected layer's weights is shown in the appendix. As shown in Fig. 2 (c) and (d), the weights of $\text{En}_5\text{ResNet20}$ are more concentrated at zero than that of ResNet20, and most of the $\text{En}_5\text{ResNet20}$'s weights are close to zero.

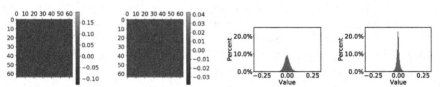

(a) ResNet20 (AT) (b) En ResNet20 (AT) (c) ResNet20 (AT) (d) En ResNet20 (AT)

Fig. 2. (a) and (b): weights visualization; (c) and (d): histogram of weights.

4   Unstructured and Channel Pruning with AT

4.1   Algorithms

In this subsection, we introduce relaxed, augmented Lagrangian-based, pruning algorithms to sparsify the AT DNNs. The algorithms of interest are the Relaxed Variable-Splitting Method (RVSM) [9] for weight pruning (Algorithm 1), and its variation, the Relaxed Group-wise Splitting Method (RGSM) [36] for channel pruning (Algorithm 2).

Our approach is to apply the RVSM/RGSM algorithm together with robust PGD training to train and sparsify the model from scratch. Namely, at each iteration, we apply a PGD attack to generate adversarial images x', which are then used in the forward-propagation process to generate predictions y'. The back-propagation process will then compute the appropriate loss function and apply RVSM/RGSM to update the model. Previous work on RVSM mainly focused on a one-hidden layer setting; In this paper, we extend this result to the general setting. To the best of our knowledge, this is the first result that uses RVSM/RGSM in an adversarial training scenario.

To explain our choice of algorithm, we discuss a classical algorithm to promote sparsity of the target weights, the alternating direction of multiplier method (ADMM) [3,10]. In ADMM, instead of minimizing the original loss function $f(w)$, we seek to minimize the ℓ_1 regularized loss function, $f(w) + \lambda\|u\|_1$, by considering the following augmented Lagrangian

$$\mathcal{L}(w, u, z) = f(w) + \lambda\|u\|_1 + \langle z, w - u \rangle + \frac{\beta}{2}\|w - u\|^2, \quad \lambda, \beta \geq 0. \quad (2)$$

which can be easily solved by applying the following iterations

$$\begin{cases} w^{t+1} \leftarrow \arg\min_w \mathcal{L}_\beta(w, u^t, z^t) \\ u^{t+1} \leftarrow \arg\min_u \mathcal{L}_\beta(w^{t+1}, u, z^t) \\ z^{t+1} \leftarrow z^t + \beta(w^{t+1} - u^{t+1}) \end{cases} \quad (3)$$

Although widely used in practice, ADMM has several drawbacks when it is used to regularize DNN's weights. First, one can improve the sparsity of the final learned weights by replacing $\|u\|_1$ with $\|u\|_0$; but $\|\cdot\|_0$ is not differentiable, thus current theory of optimization does not apply [34]. Second, the update $w^{t+1} \leftarrow \arg\min_w \mathcal{L}_\beta(w, u^t, z^t)$ is not a reasonable step in practice, as one has to fully know how $f(w)$ behaves. In most ADMM adaptation on DNN, this step is replaced by a simple gradient descent. Third, the Lagrange multiplier term, $\langle z, w - u \rangle$, seeks to close the gap between w^t and u^t, and this in turn reduces sparsity of u^t.

The RVSM we will implement is a relaxation of ADMM. RVSM drops the Lagrangian multiplier, and replaces $\lambda\|u\|_1$ with $\lambda\|u\|_0$, and resulting in the following relaxed augmented Lagrangian

$$\mathcal{L}_\beta(w, u) = f(w) + \lambda\|u\|_0 + \frac{\beta}{2}\|w - u\|^2. \quad (4)$$

The above relaxed augmented Lagrangian can be solved efficiently by the iteration in Algorithm 1. RVSM can resolve all the three issues associated with ADMM listed above in training robust DNNs with sparse weights: First, by removing the linear term $\langle z, w - u \rangle$, one has a closed form formula for the update of u^t without requiring $\|u\|_0$ to be differentiable. Explicitly, $u^t = H_{\sqrt{2\lambda/\beta}}(w^t) =$

$(w_1^t \chi_{\{|w_1| > \sqrt{2\lambda/\beta}\}}, ..., w_d^t \chi_{\{|w_1| > \sqrt{2\lambda/\beta}\}})$, where $H_\alpha(\cdot)$ is the hard-thresholding operator with parameter α. Second, the update of $\boldsymbol{w}^t$ is a gradient descent step itself, so the theoretical guarantees will not deviate from practice. Third, without the Lagrange multiplier term $\boldsymbol{z}^t$, there will be a gap between $\boldsymbol{w}^t$ and $\boldsymbol{u}^t$ at the limit (finally trained DNNs). However, the limit of $\boldsymbol{u}^t$ is much sparser than that in the case of ADMM. At the end of each training epoch, we replace $\boldsymbol{w}^t$ by $\boldsymbol{u}^t$ for the validation process. Numerical results in Sect. 5 will show that the AT DNN with parameters $\boldsymbol{u}^t$ usually outperforms the traditional ADMM in both accuracy and robustness.

Algorithm 1. RVSM

Input: $\eta, \beta, \lambda, max_{epoch}, max_{batch}$
Initialization: $\boldsymbol{w}^0$
Define: $\boldsymbol{u}^0 = H_{\sqrt{2\lambda/\beta}}(\boldsymbol{w}^0)$
for $t = 0, 1, 2, ..., max_{epoch}$ **do**
  for $batch = 1, 2, ..., max_{batch}$ **do**
    $\boldsymbol{w}^{t+1} \leftarrow \boldsymbol{w}^t - \eta \nabla f(\boldsymbol{w}^t) - \eta\beta(\boldsymbol{w}^t - \boldsymbol{u}^t)$
    $\boldsymbol{u}^{t+1} \leftarrow \arg\min_u \mathcal{L}_\beta(\boldsymbol{u}, \boldsymbol{w}^t) = H_{\sqrt{2\lambda/\beta}}(\boldsymbol{w}^t)$
  end for
end for

Algorithm 2. RGSM

Input: $\eta, \beta, \lambda_1, \lambda_2, max_{epoch}, max_{batch}$
Objective: $\tilde{f}(\boldsymbol{w}) = f(\boldsymbol{w}) + \lambda_2 \|\boldsymbol{w}\|_{GL}$
Initialization: Initialize $\boldsymbol{w}^0$, define $\boldsymbol{u}^0$
for $g = 1, 2, ..., G$ **do**
  $\boldsymbol{u}_g^0 = Prox_{\lambda_1}(\boldsymbol{w}_g^0)$
end for
for $t = 0, 1, 2, ..., max_{epoch}$ **do**
  for $batch = 1, 2, ..., max_{batch}$ **do**
    $\boldsymbol{w}^{t+1} = \boldsymbol{w}^t - \eta \nabla \tilde{f}(\boldsymbol{w}^t) - \eta\beta(\boldsymbol{w}^t - \boldsymbol{u}^t)$
    for $g = 1, 2, ..., G$ **do**
      $\boldsymbol{u}_g^{t+1} = Prox_{\lambda_1}(\boldsymbol{w}_g^t)$
    end for
  end for
end for

RGSM is a method that generalizes RVSM to structured pruning, in particular, channel pruning. Let $\boldsymbol{w} = \{\boldsymbol{w}_1, ..., \boldsymbol{w}_g, ..., \boldsymbol{w}_G\}$ be the grouped weights of convolutional layers of a DNN, where G is the total number of groups. Let I_g

be the indices of $\boldsymbol{w}$ in group g. The group Lasso (GLasso) penalty and group-ℓ_0 penalty [40] are defined as

$$\|\boldsymbol{w}\|_{GL} := \sum_{g=1}^{G} \|\boldsymbol{w}_g\|_2, \qquad \|\boldsymbol{w}\|_{G\ell_0} := \sum_{g=1}^{G} 1_{\|\boldsymbol{w}_g\|_2 \neq 0} \qquad (5)$$

and the corresponding Proximal (projection) operators are

$$\mathrm{Prox}_{GL,\lambda}(\boldsymbol{w}_g) := \mathrm{sgn}(\boldsymbol{w}_g) \max(\|\boldsymbol{w}_g\|_2 - \lambda, 0), \quad \mathrm{Prox}_{G\ell_0,\lambda}(\boldsymbol{w}_g) := \boldsymbol{w}_g 1_{\|\boldsymbol{w}_g\|_2 \neq \sqrt{2\lambda}}$$
$$(6)$$

where $\mathrm{sgn}(\boldsymbol{w}_g) := \boldsymbol{w}_g/\|\boldsymbol{w}_g\|_2$. The RGSM method is described in Algorithm 2, which improves on adding group Lasso penalty directly in the objective function [35] for natural DNN training [36].

4.2   Theoretical Guarantees

We propose a convergence analysis of the RVSM algorithm to minimize the Lagrangian (4). Consider the following empirical adversarial risk minimization (EARM)

$$\min_{f \in \mathcal{H}} \frac{1}{n} \sum_{i=1}^{n} \max_{\|\boldsymbol{x}_i' - \boldsymbol{x}_i\|_\infty \leq \epsilon} L(F(\boldsymbol{x}_i', \boldsymbol{w}), y_i) \qquad (7)$$

where the classifier $F(\cdot, \boldsymbol{w})$ is a function in the hypothesis class $\mathcal{H}$, e.g. ResNet and its ensembles, parametrized by $\boldsymbol{w}$. Here, $L(F(\boldsymbol{x}_i, \boldsymbol{w}), y_i)$ is the appropriate loss function associated with F on the data-label pair $(\boldsymbol{x}_i, y_i)$, e.g. cross-entropy for classification and root mean square error for regression problem. Since our model is trained using PGD AT, let

$$f(\boldsymbol{w}) = \mathbb{E}_{(x,y)\sim\mathcal{D}}[\max_{\boldsymbol{x}'} L(F(\boldsymbol{x}', \boldsymbol{w}), y)] \qquad (8)$$

where $\boldsymbol{x}'$ is obtained by applying the PGD attack to the clean data $\boldsymbol{x}$ [11,23, 25,32]. In a nutshell, $f(\boldsymbol{w})$ is the population adversarial loss of the network parameterized by $\boldsymbol{w} = (\boldsymbol{w}_1, \boldsymbol{w}_2, ..., \boldsymbol{w}_M)$. Before proceeding, we first make the following assumption:

Assumption 1. *Let $\boldsymbol{w}_1, \boldsymbol{w}_2, ..., \boldsymbol{w}_M$ be the weights in the M layers of the given DNN, then there exists a positive constant L such that for all t,*

$$\|\nabla f(\cdot, \boldsymbol{w}_j^{t+1}, \cdot) - \nabla f(\cdot, \boldsymbol{w}_j^t, \cdot)\| \leq L\|\boldsymbol{w}_j^{t+1} - \boldsymbol{w}_j^t\|, \ for \ j = 1, 2, \cdots, M. \quad (9)$$

Assumption 1 is a weaker version of that made by [29,33], in which the empirical adversarial loss function is smooth in both the input $\boldsymbol{x}$ and the parameters $\boldsymbol{w}$. Here we only require the population adversarial loss f to be smooth in each layer of the DNN in the region of iterations. An important consequence of Assumption 1 is

$$f(\cdot, \boldsymbol{w}_j^{t+1}, \cdot) - f(\cdot, \boldsymbol{w}_j^t, \cdot) \leq \langle \nabla f(\cdot, \boldsymbol{w}_j^t, \cdot), (0, ..., \boldsymbol{w}_j^{t+1} - \boldsymbol{w}_j^t, 0, ...) \rangle + \frac{L}{2}\|\boldsymbol{w}_j^{t+1} - \boldsymbol{w}_j^t\|^2$$
$$(10)$$

Theorem 1. *Under the Assumption 1, suppose also that the RVSM algorithm is initiated with a small stepsize η such that $\eta < \frac{2}{\beta+L}$. Then the Lagrangian $\mathcal{L}_\beta(\boldsymbol{w}^t, \boldsymbol{u}^t)$ decreases monotonically and converges sub-sequentially to a limit point $(\bar{\boldsymbol{w}}, \bar{\boldsymbol{u}})$.*

The proof of Theorem 1 is provided in the Appendix. From the descent property of $\mathcal{L}_\beta(\boldsymbol{w}^t, \boldsymbol{u}^t)$, classical results from optimization [26] can be used to show that after $T = O(1/\epsilon^2)$ iterations, we have $\nabla_{\boldsymbol{w}^t} \mathcal{L}_\beta(\boldsymbol{w}^t, \boldsymbol{u}^t) = O(\epsilon)$, for some $t \in (0, T]$. The term $\|\boldsymbol{u}\|_0$ promotes sparsity and $\frac{\beta}{2}\|\boldsymbol{w} - \boldsymbol{u}\|^2$ helps keep $\boldsymbol{w}$ close to $\boldsymbol{u}$. Since $\boldsymbol{u} = H_{\sqrt{2\lambda/\beta}}(\boldsymbol{w})$, it follows that $\bar{\boldsymbol{w}}$ will have lots of very small (and thus negligible) components. This explains the sparsity in the limit $\bar{\boldsymbol{u}}$.

5   Numerical Results

In this section, we verify the following advantages of the proposed algorithms:

- RVSM/RGSM is efficient for unstructured/channel-wise pruning for the AT DNNs.
- After pruning by RVSM and RGSM, EnResNet's weights are significantly sparser than the baseline ResNet's, and more accurate in classifying both natural and adversarial images.

These two merits lead to the fact that a synergistic integration of RVSM/RGSM with the Feynman-Kac formula principled EnResNet enables sparsity to meet robustness.

We perform AT by PGD integrated with RVSM, RGSM, or other sparsification algorithms on-the-fly. For all the experiments below, we run 200 epochs of the PGD (10 iterations of the iterative fast gradient sign method (IFGSM[10]) with $\alpha = 2/255$ and $\epsilon = 8/255$, and an initial random perturbation of magnitude ϵ). The initial learning rate of 0.1 decays by a factor of 10 at the 80th, 120th, and 160th epochs, and the RVSM/RGSM/ADMM sparsification takes place in the back-propagation stage. We split the training data into 45K/5K for training and validation, and the model with the best validation accuracy is used for testing. We test the trained models on the clean images and attack them by FGSM, IFGSM[20], and C&W with the same parameters as that used in [23,32,41]. We denote the accuracy on the clean images and under the FGSM, IFGSM[20] [12], C&W [4], and NAttack [20] [1] attacks as A_1, A_2, A_3, A_4, and A_5, resp. A brief introduction of these attacks is available in the appendix. We use both sparsity and channel sparsity to measure the performance of the pruning algorithms, where the sparsity is defined to be the percentage of zero weights; the channel sparsity is the percentage of channels whose weights' ℓ_2 norm is less than $1E - 15$.

---

[1] For NAttack, we use the default parameters in https://github.com/cmhcbb/attackbox.

5.1   Model Compression for at ResNet and EnResNets

First, we show that RVSM is efficient to sparsify ResNet and EnResNet. Table 1 shows the accuracies of ResNet20 and En_2ResNet20 under the unstructured sparsification with different sparsity controlling parameter β. We see that after the unstructured pruning by RVSM, En_2ResNet20 has much sparser weights than ResNet20. Moreover, the sparsified En_2ResNet20 is remarkably more accurate and robust than ResNet20. For instance, when $\beta = 0.5$, En_2ResNet20's weights are 16.42% sparser than ResNet20's (56.34% vs. 39.92%). Meanwhile, En_2ResNet20 boost the natural and robust accuracies of ResNet20 from 74.08%, 50.64%, 46.67%, and 57.24% to 78.47%, 56.13%, 49.54%, and 65.57%, resp. We perform a few independent trials, and the random effects is small.

Table 1. Accuracy and sparsity of ResNet20 and En_2ResNet20 under different attacks and β, with $\lambda = 1E - 6$. (Unit: %, n/a: do not perform sparsification. Same for all the following tables.)

	ResNet20					En_2ResNet20				
β	A_1	A_2	A_3	A_4	Sparsity	A_1	A_2	A_3	A_4	Sparsity
n/a	76.07	51.24	47.25	59.30	0	80.34	57.11	50.02	66.77	0
0.01	70.26	46.68	43.79	55.59	80.91	72.81	51.98	46.62	63.10	89.86
0.1	73.45	49.48	45.79	57.72	56.88	77.78	55.48	49.26	65.56	70.55
0.5	74.08	50.64	46.67	57.24	39.92	78.47	56.13	49.54	65.57	56.34

Second, we verify the effectiveness of RGSM in channel pruning. We lists the accuracy and channel sparsity of ResNet20, En_2ResNet20, and En_5ResNet20 in Table 2. Without any sparsification, En_2ResNet20 improves the four type of accuracies by 4.27% (76.07% vs. 80.34%), 5.87% (51.24% vs. 57.11%), 2.77% (47.25% vs. 50.02%), and 7.47% (59.30% vs. 66.77%), resp. When we set $\beta = 1$, $\lambda_1 = 5e - 2$, and $\lambda_2 = 1e - 5$, after channel pruning both natural and robust accuracies of ResNet20 and En_2ResNet20 remain close to the unsparsified models, but En_2ResNet20's weights are 33.48% (41.48% vs. 8%) sparser than that of ResNet20's. When we increase the channel sparsity level by increasing λ_1 to $1e - 1$, both the accuracy and channel sparsity gaps between ResNet20 and En_2ResNet20 are enlarged. En_5ResNet20 can future improve both natural and robust accuracies on top of En_2ResNet20. For instance, at ~55% (53.36% vs. 56.74%) channel sparsity, En_5ResNet20 can improve the four types of accuracy of En_2ResNet20 by 4.66% (80.53% vs. 75.87%), 2.73% (57.38% vs. 54.65%), 2.86% (50.63% vs. 47.77%), and 1.11% (66.52% vs. 65.41%), resp.

Table 2. Accuracy and sparsity of different EnResNet20. (Ch. Sp.: Channel Sparsity)

Net	β	λ_1	λ_2	A_1	A_2	A_3	A_4	A_5	Ch. Sp.
ResNet20	n/a	n/a	n/a	76.07	51.24	47.25	59.30	45.88	0
	1	5.E−02	1.E−05	75.91	51.52	47.14	58.77	45.02	8.00
	1	1.E−01	1.E−05	71.84	48.23	45.21	57.09	43.84	25.33
En$_2$ResNet20	n/a	n/a	n/a	80.34	57.11	50.02	66.77	49.35	0
	1	5.E−02	1.E−05	78.28	56.53	49.58	66.56	49.11	41.48
	1	1.E−01	1.E−05	75.87	54.65	47.77	65.41	46.77	56.74
En$_5$ResNet20	n/a	n/a	n/a	81.41	58.21	51.60	66.48	50.21	0
	1	1.E−02	1.E−05	81.46	58.34	51.35	66.84	50.07	19.76
	1	2.E−02	1.E−05	80.53	57.38	50.63	66.52	48.23	53.36

Third, we show that an ensemble of small ResNets via the Feynman-Kac formalism performs better than a larger ResNet of roughly the same size in accuracy, robustness, and sparsity. We AT En$_2$ResNet20 ($\sim$0.54M parameters) and ResNet38 ($\sim$0.56M parameters) with and without channel pruning. As shown in Table 3, under different sets of parameters, after RGSM pruning, En$_2$ResNet20 always has much more channel sparsity than ResNet38, also much more accurate and robust. For instance, when we set $\beta = 1$, $\lambda_1 = 5e - 2$, and $\lambda_2 = 1e - 5$, the AT ResNet38 and En$_2$ResNet20 with channel pruning have channel sparsity 17.67% and 41.48%, resp. Meanwhile, En$_2$ResNet20 outperforms ResNet38 in the four types of accuracy by 0.36% (78.28% vs. 77.92%), 3.02% (56.53% vs. 53.51%), 0.23% (49.58% vs. 49.35%), and 6.34% (66.56% vs. 60.32%), resp. When we increase λ_1, the channel sparsity of two nets increase. As shown in Fig. 3, En$_2$ResNet20's channel sparsity growth much faster than ResNet38's, and we plot the corresponding four types of accuracies of the channel sparsified nets in Fig. 4.

Table 3. Performance of En$_2$ResNet20 and ResNet38 under RVSM.

Net	β	λ_1	λ_2	A_1	A_2	A_3	A_4	Ch. Sp.
En$_2$ResNet20	n/a	n/a	n/a	**80.34**	**57.11**	**50.02**	**66.77**	0
ResNet38	n/a	n/a	n/a	78.03	54.09	49.81	61.72	0
En$_2$ResNet20	1	5.E−02	1.E−05	**78.28**	**56.53**	**49.58**	**66.56**	**41.48**
ResNet38	1	5.E−02	1.E−05	77.92	53.51	49.35	60.32	17.67
En$_2$ResNet20	1	1.E−01	1.E−05	**76.30**	**54.65**	**47.77**	**65.41**	**56.74**
ResNet38	1	1.E−01	1.E−05	72.95	49.78	46.48	57.92	43.80

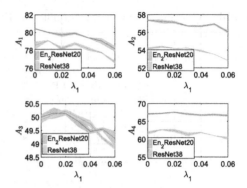

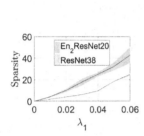

Fig. 3. Sparsity of $En_2ResNet20$ and ResNet38 under different λ_1. (5 runs)

Fig. 4. Accuracy of $En_2ResNet20$ and ResNet38 under different λ_1. (5 runs)

Table 4. Contrasting ADMM versus RVSM for the AT ResNet20.

	Unstructured pruning					Channel pruning				
	A_1	A_2	A_3	A_4	Sp.	A_1	A_2	A_3	A_4	Ch. Sp.
RVSM	70.26	46.68	43.79	55.59	**80.91**	**71.84**	**48.23**	**45.21**	**57.09**	**25.33**
ADMM	**71.55**	**47.37**	**44.30**	**55.79**	10.92	63.99	42.06	39.75	51.90	4.44

5.2 RVSM/RGSM Versus ADMM

In this subsection, we will compare RVSM, RGSM, and ADMM [42][2] for unstructured and channel pruning for the AT ResNet20, and we will show that RVSM and RGSM iterations can promote much higher sparsity with less natural and robust accuracies degradations than ADMM. We list both natural/robust accuracies and sparsities of ResNet20 after ADMM, RVSM, and RGSM pruning in Table 4. For unstructured pruning, ADMM retains slightly better natural ($\sim$1.3%) and robust ($\sim$0.7%, $\sim$0.5%, and 0.2% under FGSM, IFGSM^{20}, and C&W attacks) accuracies. However, RVSM gives much better sparsity (80.91% vs. 10.89%). In the channel pruning scenario, RVSM significantly outperforms ADMM in all criterion including natural and robust accuracies and channel sparsity, as the accuracy gets improved by at least 5.19% and boost the channel sparsity from 4.44% to 25.33%. Part of the reason for ADMM's inefficiency in sparsifying DNN's weights is due to the fact that the ADMM iterations try to close the gap between the weights $\boldsymbol{w}^t$ and the auxiliary variables $\boldsymbol{u}^t$, so the final result has a lot of weights with small magnitude, but not small enough to be regarded as zero (having norm less than 1e−15). The RVSM does not seek to close this gap, instead it replaces the weight $\boldsymbol{w}^t$ by $\boldsymbol{u}^t$, which is sparse, after each epoch. This results in a much sparser final result, as shown in Fig. 5: ADMM does result in a lot of channels with small norms; but to completely prune these

---

[2] We use the code from: github.com/KaiqiZhang/admm-pruning.

off, RVSM does a better job. Here, the channel norm is defined to be the ℓ_2 norm of the weights in each channel of the DNN [35].

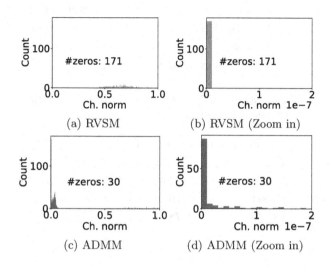

Fig. 5. Channel norms of the AT ResNet20 under RVSM and ADMM.

5.3 Beyond ResNet Ensemble and Beyond CIFAR10

5.4 Beyond ResNet Ensemble

In this part, we show that the idea of average over noise injected ResNet generalizes to other types of DNNs, that is, the natural and robust accuracies of DNNs can be improved by averaging over noise injections. As a proof of concept, we remove all the skipping connections in ResNet20 and En$_2$ResNet20, in this case, the model no longer involves a TE and CDE. As shown in Table 5, removing the skip connection degrades the performance significantly, especially for En$_2$ResNet20 (all the four accuracies of the AT En$_2$ResNet20 reduces more than the AT ResNet20's). These results confirm that skip connection is crucial, without it the TE model assumption breaks down, and the utility of EnResNets reduces. However, the ensemble of noise injected DNNs still improves the performance remarkably.

Next, let us consider the sparsity, robustness, and accuracies of ResNet20 and En$_2$ResNet20 without any skip connection, when they are AT using weights sparsification by either RVSM or RGSM. For RVSM pruning, we set $\beta = 5E-2$ and $\lambda = 1E-6$; for RGSM channel pruning, we set $\beta = 5E-2$, $\lambda_1 = 5E-2$, and $\lambda_2 = 1E-5$. We list the corresponding results in Table 6. We see that under both RVSM and RGSM pruning, En$_2$ResNet20 is remarkably more accurate on both clean and adversarial images, and significantly sparser than ResNet20. When we compare the results in Table 6 with that in Tables 1 and 2, we conclude that once we remove the skip connections, the sparsity, robustness, and accuracy degrades dramatically.

374 T. Dinh et al.

Table 5. Accuracies of ResNet20 and En_2ResNet20 with and without skip connections.

Net	Type	A_1	A_2	A_3	A_4
ResNet20	Base Net	76.07	51.24	47.25	59.30
	Without skip connection	75.45	51.03	47.22	58.44
En_2ResNet20	Base net	80.34	57.11	50.02	66.77
	Without skip connection	79.12	55.76	49.92	66.26

Table 6. Sparsity and accuracies of ResNet20 and En_2ResNet20 without skip connection under different pruning algorithms.

Net	Pruning algorithm	A_1	A_2	A_3	A_4	Sp	Ch. Sp
ResNet20	RVSM	73.72	50.46	46.98	58.28	0.05	0.15
(no skip connection)	RGSM	74.63	50.44	46.86	58.05	1.64	9.04
En_2ResNet20	RVSM	76.95	55.17	49.28	58.35	10.87	9.48
(no skip connection)	RGSM	78.51	56.55	49.71	67.08	7.95	15.48

5.5 Beyond CIFAR10

Besides CIFAR10, we further show the advantage of EnResNet + RVSM/RGSM in compressing and improving accuracy/robustness of the AT DNNs for CIFAR100 classification. We list the natural and robust accuracies and channel sparsities of the AT ResNet20 and En_2ResNet20 with different RGSM parameters (n/a stands for do not perform channel pruning) in Table 7. For $\lambda_1 = 0.05$, RGSM almost preserves the performance of the DNNs without channel pruning, while improving channel sparsity by 7.11% for ResNet20, and 16.89% for En_2ResNet20. As we increase λ_1 to 0.1, the channel sparsity becomes 18.37% for ResNet20 and 39.23% for En_2ResNet20. Without any channel pruning, En_2ResNet20 improves natural accuracy by 4.66% (50.68% vs. 46.02%), and robust accuracies by 5.25% (30.2% vs. 24.77%), 3.02% (26.25% vs. 23.23%), and 7.64% (40.06% vs. 32.42%), resp., under the FGSM, IFGSM[20], and C&W attacks. Even in very high channel sparsity scenario ($\lambda_1 = 0.05$), En_2ResNet20 still dramatically increase A_1, A_2, A_3, and A_4 by 2.90%, 4.31%, 1.89%, and 5.86%, resp. These results are similar to the one obtained on the CIFAR10 in Table 2. These results further confirm that RGSM together with the Feynman-Kac formalism principled ResNets ensemble can significantly improve both natural/robust accuracy and sparsity on top of the baseline ResNets.

Table 7. Accuracy and sparsity of different Ensembles of ResNet20's on the CIFAR100.

Net	β	λ_1	λ_2	A_1	A_2	A_3	A_4	Ch. Sp.
ResNet20	n/a	n/a	n/a	46.02	24.77	23.23	32.42	0
	1	5.E−02	1.E−05	45.74	25.34	23.55	33.53	7.11
	1	1.E−01	1.E−05	44.34	24.46	23.12	32.38	18.37
En$_2$ResNet20	n/a	n/a	n/a	50.68	30.2	26.25	40.06	0
	1	5.E−02	1.E−05	50.56	30.33	26.23	39.85	16.89
	1	1.E−01	1.E−05	47.24	28.77	25.01	38.24	39.23

6   Concluding Remarks

The Feynman-Kac formalism principled AT EnResNet's weights are much sparser than the baseline ResNet's. Together with the relaxed augmented Lagrangian based unstructured/channel pruning algorithms, we can compress the AT DNNs much more efficiently, meanwhile significantly improves both natural and robust accuracies of the compressed model. As future directions, we propose to quantize EnResNets and to integrate neural ODE into our framework.

Acknowledgements. The work was partially supported by NSF grants IIS-1632935 (TD and JX), DMS-1737770 (AB and SO), ATD-1924935 (BW), and DMS-1924548 (TD and JX); ONR Grant N00014181257 (BW and SO); AFOSR MURI FA9550-18-1-0502 (BW and SO); and Google Research Grant W0KPVLW22T86GB8M (TD).

A   Proof of Theorem 1

By the arg min update of u^t, $\mathcal{L}_\beta(w^{t+1}, u^{t+1}) \leq \mathcal{L}_\beta(w^{t+1}, u^t)$. It remains to show $\mathcal{L}_\beta(w^{t+1}, u^t) \leq \mathcal{L}_\beta(w^t, u^t)$. To this end, we show

$$\mathcal{L}_\beta(w_1^{t+1}, w_2^t, ..., w_M^t, u^t) \leq \mathcal{L}_\beta(w_1^t, w_2^t, ..., w_M^t, u^t)$$

the conclusion then follows by a repeated argument. Notice the only change occurs in the first layer w_1^t. For simplicity of notation, only for the following chain of inequality, let $w^t = (w_1^t, ..., w_M^t)$ and $w^{t+1} = (w_1^{t+1}, w_2^t, ..., w_M^t)$. Then for a fixed $u := u^t$, we have

$$\mathcal{L}_\beta(\boldsymbol{w}^{t+1}, \boldsymbol{u}) - \mathcal{L}_\beta(\boldsymbol{w}^t, \boldsymbol{u})$$

$$= f(\boldsymbol{w}^{t+1}) - f(\boldsymbol{w}^t) + \frac{\beta}{2}\left(\|\boldsymbol{w}^{t+1} - \boldsymbol{u}\|^2 - \|\boldsymbol{w}^t - \boldsymbol{u}\|^2\right)$$

$$\leq \langle \nabla f(\boldsymbol{w}^t), \boldsymbol{w}^{t+1} - \boldsymbol{w}^t \rangle + \frac{L}{2}\|\boldsymbol{w}^{t+1} - \boldsymbol{w}^t\|^2 + \frac{\beta}{2}\left(\|\boldsymbol{w}^{t+1} - \boldsymbol{u}\|^2 - \|\boldsymbol{w}^t - \boldsymbol{u}\|^2\right)$$

$$= \frac{1}{\eta}\langle \boldsymbol{w}^t - \boldsymbol{w}^{t+1}, \boldsymbol{w}^{t+1} - \boldsymbol{w}^t \rangle - \beta\langle \boldsymbol{w}^t - \boldsymbol{u}, \boldsymbol{w}^{t+1} - \boldsymbol{w}^t \rangle$$

$$+ \frac{L}{2}\|\boldsymbol{w}^{t+1} - \boldsymbol{w}^t\|^2 + \frac{\beta}{2}\left(\|\boldsymbol{w}^{t+1} - \boldsymbol{u}\|^2 - \|\boldsymbol{w}^t - \boldsymbol{u}\|^2\right)$$

$$= \frac{1}{\eta}\langle \boldsymbol{w}^t - \boldsymbol{w}^{t+1}, \boldsymbol{w}^{t+1} - \boldsymbol{w}^t \rangle + \left(\frac{L}{2} + \frac{\beta}{2}\right)\|\boldsymbol{w}^{t+1} - \boldsymbol{w}^t\|^2$$

$$+ \frac{\beta}{2}\|\boldsymbol{w}^{t+1} - \boldsymbol{u}\|^2 - \frac{\beta}{2}\|\boldsymbol{w}^t - \boldsymbol{u}\|^2 - \beta\langle \boldsymbol{w}^t - \boldsymbol{u}, \boldsymbol{w}^{t+1} - \boldsymbol{w}^t \rangle - \frac{\beta}{2}\|\boldsymbol{w}^{t+1} - \boldsymbol{w}^t\|^2$$

$$= \left(\frac{L}{2} + \frac{\beta}{2} - \frac{1}{\eta}\right)\|\boldsymbol{w}^{t+1} - \boldsymbol{w}^t\|^2$$

Thus, when $\eta \leq \frac{2}{\beta+L}$, we have $\mathcal{L}_\beta(\boldsymbol{w}^{t+1}, \boldsymbol{u}) \leq \mathcal{L}_\beta(\boldsymbol{w}^t, \boldsymbol{u})$. Apply the above argument repeatedly, we arrive at

$$\mathcal{L}_\beta(\boldsymbol{w}^{t+1}, \boldsymbol{u}^{t+1}) \leq \mathcal{L}_\beta(\boldsymbol{w}^{t+1}, \boldsymbol{u}^t)$$
$$\leq \mathcal{L}_\beta(\boldsymbol{w}_1^{t+1}, \boldsymbol{w}_2^{t+1}, ..., \boldsymbol{w}_{M-1}^{t+1}, \boldsymbol{w}_M^t, \boldsymbol{u}^t)$$
$$\leq ...$$
$$\leq \mathcal{L}_\beta(\boldsymbol{w}^t, \boldsymbol{u}^t)$$

This implies $\mathcal{L}_\beta(\boldsymbol{w}^t, \boldsymbol{u}^t)$ decreases monotonically. Since $\mathcal{L}_\beta(\boldsymbol{w}^t, \boldsymbol{u}^t) \geq 0$, $(\boldsymbol{w}^t, \boldsymbol{u}^t)$ must converge sub-sequentially to a limit point $(\bar{\boldsymbol{w}}, \bar{\boldsymbol{u}})$. This completes the proof.

B   Adversarial Attacks Used in This Work

We focus on the ℓ_∞ norm based untargeted attack. For a given image-label pair $\{\boldsymbol{x}, y\}$, a given ML model $F(\boldsymbol{x}, \boldsymbol{w})$, and the associated loss $L(\boldsymbol{x}, y) := L(F(\boldsymbol{x}, \boldsymbol{w}), y)$:

- Fast gradient sign method (FGSM) searches an adversarial, $\boldsymbol{x}'$, within an ℓ_∞-ball as
$$\boldsymbol{x}' = \boldsymbol{x} + \epsilon \cdot \text{sign}\left(\nabla_{\boldsymbol{x}} L(\boldsymbol{x}, y)\right).$$

- Iterative FGSM (IFGSM^M) [12] iterates FGSM and clip the range as
$$\boldsymbol{x}^{(m)} = \text{Clip}_{\boldsymbol{x}, \epsilon}\left\{\boldsymbol{x}^{(m-1)} + \alpha \cdot \text{sign}\left(\nabla_{\boldsymbol{x}^{(m-1)}} L(\boldsymbol{x}^{(m-1)}, y)\right)\right\}$$
with $\boldsymbol{x}^{(0)} = \boldsymbol{x}$, $m = 1, \cdots, M$.

- C&W attack [4] searches the minimal perturbation (δ) attack as
$$\min_\delta \|\delta\|_\infty, \quad \text{subject to} \quad F(\boldsymbol{w}, \boldsymbol{x} + \delta) = t, \ \boldsymbol{x} + \delta \in [0, 1]^d, \ \text{for } \forall t \neq y.$$

- NAttack [20] is an effective gradient-free attack.

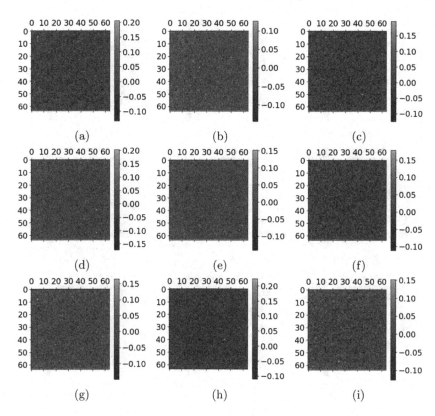

Fig. 6. Weights of a randomly selected convolutional layer of the PGD AT ResNet20.

C   More Visualizations of the DNNs' Weights

In Sect. 3, we showed some visualization results for part of the weights of a randomly selected convolutional layer of the AT ResNet20 and En$_5$ResNet20. The complete visualization results of this selected layer are shown in Figs. 6 and 7, resp., for ResNet20 and En$_5$ResNet20. These plots further verifies that:

- The magnitude of the weights of the adversarially trained En$_5$ResNet20 is significantly smaller than that of the robustly trained ResNet20.
- The overall pattern of the weights of the adversarially trained En$_5$ResNet20 is more regular than that of the robustly trained ResNet20.

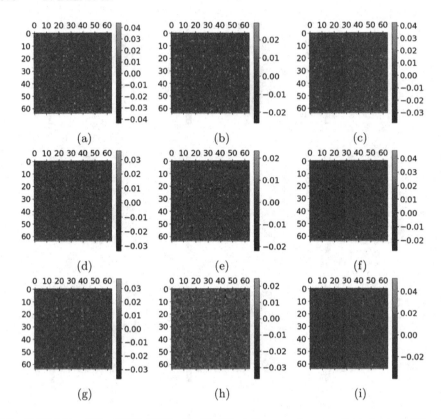

Fig. 7. Weights of the PGD AT En$_5$ResNet20 at the same layer as that shown in Fig. 6.

References

1. Alvarez, J.M., Salzmann, M.: Learning the number of neurons in deep networks. In: Advances in Neural Information Processing Systems, pp. 2270–2278 (2016)
2. Athalye, A., Carlini, N., Wagner, D.: Obfuscated gradients give a false sense of security: circumventing defenses to adversarial examples. arXiv preprint arXiv:1802.00420 (2018)
3. Boyd, S., Parikh, N., Chu, E., Peleato, B., Eckstein, J., et al.: Distributed optimization and statistical learning via the alternating direction method of multipliers. Found. Trends® Mach. Learn. **3**(1), 1–122 (2011)
4. Carlini, N., Wagner, D.: Towards evaluating the robustness of neural networks. In: IEEE European Symposium on Security and Privacy, pp. 39–57 (2016)
5. Chen, T., Rubanova, Y., Bettencourt, J., Duvenaud, D.: Neural ordinary differential equations. In: Advances in Neural Information Processing Systems, pp. 6571–6583 (2018)
6. Cheng, Y., Wang, D., Zhou, P., Zhang, T.: A survey of model compression and acceleration for deep neural networks. arXiv preprint arXiv:1710.09282 (2017)
7. Courbariaux, M., Hubara, I., Soudry, D., El-Yaniv, R., Bengio, Y.: BinaryNet: training deep neural networks with weights and activations constrained to +1 or −1. ArXiv abs/1602.02830 (2016)

8. Denil, M., Shakibi, B., Dinh, L., Ranzato, M., de Freitas, N.: Predicting parameters in deep learning. In: Proceedings of the 26th International Conference on Neural Information Processing Systems - Volume 2, NIPS 2013, pp. 2148–2156. Curran Associates Inc., USA (2013). http://dl.acm.org/citation.cfm?id=2999792.2999852

9. Dinh, T., Xin, J.: Convergence of a relaxed variable splitting method for learning sparse neural networks via ℓ_1, ℓ_0, and transformed-ℓ_1 penalties. arXiv preprint arXiv:1812.05719 (2018)

10. Goldstein, T., Osher, S.: The split Bregman method for L1-regularized problems. SIAM J. Imaging Sci. **2**(2), 323–343 (2009)

11. Goodfellow, I., Shlens, J., Szegedy, C.: Explaining and harnessing adversarial examples. CoRR abs/1412.6572 (2014)

12. Goodfellow, I.J., Shlens, J., Szegedy, C.: Explaining and harnessing adversarial examples. arXiv preprint arXiv:1412.6275 (2014)

13. Gui, S., Wang, H., Yu, C., Yang, H., Wang, Z., Liu, J.: Adversarially trained model compression: when robustness meets efficiency. arXiv preprint arXiv:1902.03538 (2019)

14. Guo, Y., Zhang, C., Zhang, C., Chen, Y.: Sparse DNNs with improved adversarial robustness. In: Advances in Neural Information Processing Systems, pp. 242–251 (2018)

15. Han, S., Pool, J., Tran, J., Dally, W.: Learning both weights and connections for efficient neural network. In: Advances in Neural Information Processing Systems, pp. 1135–1143 (2015)

16. He, K., Zhang, X., Ren, S., Sun, J.: Deep residual learning for image recognition. In: Proceedings of the IEEE Conference on Computer Vision and Pattern Recognition, pp. 770–778 (2016)

17. He, Y., Zhang, X., Sun, J.: Channel pruning for accelerating very deep neural networks. In: The IEEE International Conference on Computer Vision (ICCV), October 2017

18. Howard, A., et al.: MobileNets: efficient convolutional neural networks for mobile vision applications. arXiv preprint arXiv:1704.04861 (2017)

19. Ladyženskaja, O., Solonnikov, V., Ural'ceva, N.: Linear and quasi-linear equations of parabolic type, vol. 23. American Mathematical Society (1988)

20. Li, Y., Li, L., Wang, L., Zhang, T., Gong, B.: NATTACK: learning the distributions of adversarial examples for an improved black-box attack on deep neural networks. arXiv preprint arXiv:1905.00441 (2019)

21. Li, Z., Shi, Z.: Deep residual learning and PDEs on manifold. arXiv preprint arXiv:1708.05115 (2017)

22. Liu, B., Wang, M., Foroosh, H., Tappen, M., Pensky, M.: Sparse convolutional neural networks. In: Proceedings of the IEEE Conference on Computer Vision and Pattern Recognition, pp. 806–814 (2015)

23. Madry, A., Makelov, A., Schmidt, L., Tsipras, D., Vladu, A.: Towards deep learning models resistant to adversarial attacks. In: International Conference on Learning Representations (2018). https://openreview.net/forum?id=rJzIBfZAb

24. Mohammadi, M., Al-Fuqaha, A., Sorour, S., Guizani, M.: Deep learning for IoT big data and streaming analytics: a survey. IEEE Commun. Surv. Tutor. **20**(4), 2923–2960 (2018)

25. Na, T., Ko, J., Mukhopadhyay, S.: Cascade adversarial machine learning regularized with a unified embedding. In: International Conference on Learning Representations (2018). https://openreview.net/forum?id=HyRVBzap-

26. Nesterov, Y.: Introductory Lectures on Convex Optimization: A Basic Course, 1st edn. Springer, Boston (2014). https://doi.org/10.1007/978-1-4419-8853-9
27. Polino, A., Pascanu, R., Alistarh, D.: Model compression via distillation and quantization. arXiv preprint arXiv:1802.05668 (2018)
28. Rakin, A., He, Z., Yang, L., Wang, Y., Wang, L., Fan, D.: Robust sparse regularization: simultaneously optimizing neural network robustness and compactness. arXiv preprint arXiv:1905.13074 (2019)
29. Sinha, A., Namkoong, H., Duchi, J.: Certifiable distributional robustness with principled adversarial training. In: International Conference on Learning Representations (2018). https://openreview.net/forum?id=Hk6kPgZA-
30. Srinivas, S., Babu, R.V.: Data-free parameter pruning for deep neural networks. arXiv preprint arXiv:1507.06149 (2015)
31. Wang, B., Luo, X., Li, Z., Zhu, W., Shi, Z., Osher, S.: Deep neural nets with interpolating function as output activation. In: Advances in Neural Information Processing Systems, pp. 743–753 (2018)
32. Wang, B., Yuan, B., Shi, Z., Osher, S.: ResNet ensemble via the Feynman-Kac formalism to improve natural and robust accuracies. In: Advances in Neural Information Processing Systems (2019)
33. Wang, Y., Ma, X., Bailey, J., Yi, J., Zhou, B., Gu, Q.: On the convergence and robustness of adversarial training. In: Chaudhuri, K., Salakhutdinov, R. (eds.) Proceedings of the 36th International Conference on Machine Learning. Proceedings of Machine Learning Research, vol. 97, pp. 6586–6595. PMLR, Long Beach, 09–15 June 2019. http://proceedings.mlr.press/v97/wang19i.html
34. Wang, Yu., Yin, W., Zeng, J.: Global convergence of ADMM in nonconvex nonsmooth optimization. J. Sci. Comput. **78**(1), 29–63 (2018). https://doi.org/10.1007/s10915-018-0757-z
35. Wen, W., Wu, C., Wang, Y., Chen, Y., Li, H.: Learning structured sparsity in deep neural networks. In: Advances in Neural Information Processing Systems, pp. 2074–2082 (2016)
36. Yang, B., Lyu, J., Zhang, S., Qi, Y.Y., Xin, J.: Channel pruning for deep neural networks via a relaxed group-wise splitting method. In: Proceedings of 2nd International Conference on AI for Industries (AI4I), Laguna Hills, CA (2019)
37. Yao, S., Zhao, Y., Zhang, A., Su, L., Abdelzaher, T.: DeepIoT: compressing deep neural network structures for sensing systems with a compressor-critic framework. In: Proceedings of the 15th ACM Conference on Embedded Network Sensor Systems, p. 4. ACM (2017)
38. Ye, S., et al.: Second rethinking of network pruning in the adversarial setting. arXiv preprint arXiv:1903.12561 (2019)
39. Yin, P., Zhang, S., Lyu, J., Osher, S., Qi, Y., Xin, J.: Blended coarse gradient descent for full quantization of deep neural networks. Res. Math. Sci. **6**(1), 14 (2019). https://doi.org/10.1007/s40687-018-0177-6
40. Yuan, M., Lin, Y.: Model selection and estimation in regression with grouped variables. J. Roy. Stat. Soc. Ser. B **68**(1), 49–67 (2007)
41. Zhang, H., Yu, Y., Jiao, J., Xing, E., Ghaoui, L., Jordan, M.: Theoretically principled trade-off between robustness and accuracy. arXiv preprint arXiv:1901.08573 (2019)

42. Zhang, T., et al.: A systematic DNN weight pruning framework using alternating direction method of multipliers. arXiv preprint 1804.03294, July 2018. https://arxiv.org/abs/1804.03294
43. Zhang, X., Zhou, X., Lin, M., Sun, J.: ShuffleNet: an extremely efficient convolutional neural network for mobile devices. In: Proceedings of the IEEE Conference on Computer Vision and Pattern Recognition, pp. 6848–6856 (2018)
44. Zhou, A., Yao, A., Guo, Y., Xu, L., Chen, Y.: Incremental network quantization: Towards lossless CNNs with low-precision weights. arXiv preprint arXiv:1702.03044 (2017)

Limits of Transfer Learning

Jake Williams[1], Abel Tadesse[2], Tyler Sam[1], Huey Sun[3],
and George D. Montañez[1(✉)]

[1] Harvey Mudd College, Claremont, CA 91711, USA
{jwilliams,gmontanez}@hmc.edu
[2] Claremont McKenna College, Claremont, CA 91711, USA
[3] Pomona College, Claremont, CA 91711, USA

Abstract. Transfer learning involves taking information and insight from one problem domain and applying it to a new problem domain. Although widely used in practice, theory for transfer learning remains less well-developed. To address this, we prove several novel results related to transfer learning, showing the need to carefully select which sets of information to transfer and the need for dependence between transferred information and target problems. Furthermore, we prove how the degree of probabilistic change in an algorithm using transfer learning places an upper bound on the amount of improvement possible. These results build on the algorithmic search framework for machine learning, allowing the results to apply to a wide range of learning problems using transfer.

Keywords: Transfer learning · Algorithmic search framework · Affinity

1 Introduction

Transfer learning is a type of machine learning where insight gained from solving one problem is applied to solve a separate, but related problem [7]. Currently an exciting new frontier in machine learning, transfer learning has diverse practical application in a number of fields, from training self-driving cars [1], where model parameters are learned in simulated environments and transferred to real-life contexts, to audio transcription [11], where patterns learned from common accents are applied to learn less common accents. Despite its potential for use in industry, little is known about the theoretical guarantees and limitations of transfer learning.

To analyze transfer learning, we need a way to talk about the breadth of possible problems we can transfer from and to under a unified formalism. One such approach is the reduction of various machine learning problems (such as regression and classification) to a type of search, using the method of the algorithmic search framework [4,5]. This reduction allows for the simultaneous analysis of a host of different problems, as results proven within the framework can be applied to any of the problems cast into it. In this work, we show how transfer learning can fit within the framework, and define affinity as a measure of the extent

© Springer Nature Switzerland AG 2020
G. Nicosia et al. (Eds.): LOD 2020, LNCS 12566, pp. 382–393, 2020.
https://doi.org/10.1007/978-3-030-64580-9_32

to which information learned from solving one problem is applicable to another. Under this definition, we prove a number of useful theorems that connect affinity with the probability of success of transfer learning. We conclude our work with applied examples and suggest an experimental heuristic to determine conditions under which transfer learning is likely to succeed.

2 Distinctions from Prior Work

Previous work within the algorithmic search framework has focused on bias [3,6], a measure of the extent to which a distribution of information resources is predisposed towards a fixed target. The case of transfer learning carries additional complexity as the recipient problem can use not only its native information resource, but the learned information passed from the source as well. Thus, affinity serves as an analogue to bias which expresses this nuance, and enables us to prove a variety of interesting bounds for transfer learning.

3 Background

3.1 Transfer Learning

Definition of Transfer Learning. Transfer learning can be defined by two machine learning problems [7], a source problem and a recipient problem. Each of these is defined by two parts, a domain and a task. The domain is defined by the feature space, $\mathcal{X}$, the label space, $\mathcal{Y}$, and the data, $D = \{(x_i, y_i), \ldots, (x_n, y)\}$, where $x_i \in \mathcal{X}$ and $y_i \in \mathcal{Y}$. The task is defined by an objective function $P_f(Y|X)$, which is a conditional distribution over the label space, conditioned on an element of the feature space. In other words, it tells us the probability that a given label is correct for a particular input. A machine learning problem is "solved" by an algorithm $\mathcal{A}$, which takes in the domain and outputs a function $P_\mathcal{A}(Y|X)$. The success of an algorithm is its ability to learn the objective function as its output. Learning and optimization algorithms use a loss function $\mathcal{L}(\cdot)$ to evaluate an output function to decide if it is worthy of outputting. Such algorithms can be viewed as black-box search algorithms [4], where the particular algorithm determines the behavior of the black box. For transfer learning under this view, the output is defined as the final element sampled in the search.

Types of Transfer Learning. Pan and Yang separated transfer learning into four categories based on the type of information passed between domains [7]:

- *Instance transfer*: Supplementing the target domain data with a subset of data from the source domain.
- *Feature-representation transfer*: Using a feature-representation of inputs that is learned in the source domain to minimize differences between the source and target domains and reduce generalization error in the target task.

- *Parameter transfer.* Passing a subset of the parameters of a model from the source domain to the target domain to improve the starting point in the target domain.
- *Relational-knowledge transfer.* Learning a relation between knowledge in the source domain to pass to the target domain, especially when either or both do not follow i.i.d. assumptions.

3.2   The Search Framework

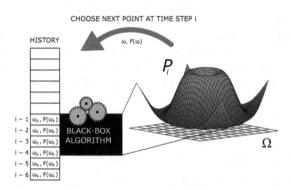

Fig. 1. Black-box search algorithm. We add evaluated queries to the history according to the distribution iteratively. Reproduced from [4].

To analyze transfer learning from a theoretical perspective, we take inspiration from previous work that views machine learning as a type of search. Montañez casts machine learning problems, including Vapnik's general learning problem (covering regression, classification, and density estimation) into an algorithmic search framework [5]. For example, classification is seen as a search through all possible labelings of the data, and clustering as a search through all possible ways to cluster the data [5]. This framework provides a common structure which we can use to analyze different machine learning problems, as each of them can be seen as a search problem with a defined set of components. Furthermore, any result we prove about search problems applies to all machine learning problems we can represent within the framework.

Within the algorithmic search framework, the three components of a search problem are the search space Ω, target set T, and external information resource F. The search space, which is finite and discrete due to the finite precision representation of numbers on computers, is the set of elements to be examined. The target set is a nonempty subset of Ω that contains the elements we wish to find. Finally, the external information resource is used to evaluate the elements of the search space. Usually, the target set and external information resource are related, as the external information resource guides the search to the target [4].

In this framework, an iterative algorithm searches for an element in the target set, depicted in Fig. 1. The algorithm is viewed as a black box that produces a probability distribution over the search space from the search history. At each step, an element is sampled from Ω according to the most recent probability distribution. The external information resource is then used to evaluate the queried element, and the element and its evaluation are added to the search history. Thus, the search history is the collection of all points sampled and all information gleaned from the information resource during the course of the search. Finally, the algorithm creates a new probability distribution according to its internal mechanisms. Abstracting the creation of the probability distribution allows the search framework to work with many different search algorithms [5].

3.3 Decomposable Probability-of-Success Metrics

Working within the same algorithmic search framework [4–6], to measure the performance of search and learning algorithms, Sam et al. [8] defined decomposable probability-of-success metrics as

$$\phi(t, f) = \mathbf{t}^\top \mathbf{P}_{\phi,f} = P_\phi(X \in t | f)$$

where $\mathbf{P}_{\phi,f}$ is not a function of target set t (with corresponding target function $\mathbf{t}$), being conditionally independent of it given information resource f. They note that one can view $\mathbf{t}^\top \mathbf{P}_{\phi,f}$ as an expectation over the probability of successfully querying an element from the target set at each step according to an arbitrary distribution. In the case of transfer learning, the distribution we choose should place most or all of its weight on the last or last couple of steps – since we transfer knowledge from the source problem's model after training, we care about our success at the last few steps when we're done training, rather than the first few.

3.4 Casting Transfer Learning into the Search Framework

Let $\mathcal{A}$ denote a fixed learning algorithm. We cast the source, which consists of $\mathcal{X}_s, \mathcal{Y}_s, D_s$, and $P_{f,s}(Y|X)$, into the algorithmic search framework as

1. $\Omega = \text{range}(\mathcal{A})$;
2. $T = \{P(Y|X) \in \text{range}(\mathcal{A}) \mid \Xi(P, P_{f,s}) < \epsilon\}$;
3. $F = \{D_s, \mathcal{L}_s\}$;
4. $F(\emptyset) = \emptyset$; and
5. $F(\omega_i) = \mathcal{L}_s(\omega_i)$.

where ω_i is the ith queried point in the search process, $\mathcal{L}_s$ the loss function for the source, and Ξ_s is an error functional on conditional distribution P and the optimal conditional distribution $P_{f,s}$.

Generally, any information from the source can be encoded in a binary string, so we represent the knowledge transferred as a finite length binary string. Let this string be $L = \{0, 1\}^n$. Thus, we cast the recipient, which consists of $\mathcal{X}_r, \mathcal{Y}_r, D_r$, and $P_{f,r}(Y|X)$, into the search framework as

1. $\Omega = \text{range}(\mathcal{A})$;
2. $T = \{P(Y|X) \in \text{range}(\mathcal{A}) \mid \Xi(P, P_{f,r}) < \epsilon\}$;
3. $F = \{D_r, \mathcal{L}_r\}$;
4. $F(\emptyset) = L$; and
5. $F(\omega_i) = \mathcal{L}_r(\omega_i)$.

where $\mathcal{L}_r$ is a loss function, and Ξ_t is an error functional on conditional distribution P and the optimal conditional distribution $P_{f,r}$.

4   Preliminaries

4.1   Affinity

In a transfer learning problem, we want to know how the source problem can improve the recipient problem, which happens by leveraging the source problem's information resource. So, we can think about how the bias of the recipient is changed by the learned knowledge that the source passes along. Recall that the bias is defined by a distribution over possible information resources [6]. However, we know that the information resource will contain f_{R_o}, the original information resource from the recipient problem. Our distribution over information resources will therefore take that into account, and only care about the learned knowledge being passed along from the source.

To quantify this, we let $\mathcal{D}_L$ be the distribution placed over L, the possible learning resources, by the source. We can use it to make statements similar to bias in traditional machine learning by defining a property called *affinity*.

Consider a transfer learning problem with a fixed k-hot target vector $\mathbf{t}$, fixed recipient information resource f_{R_o}, and a distribution $\mathcal{D}_L$ over a collection of possible learning resources, with $L \sim \mathcal{D}_L$. The **affinity** between the distribution and the recipient problem is defined as

$$\begin{aligned}
\text{Affin}(\mathcal{D}_L, \mathbf{t}, f_{R_o}) &= \mathbb{E}_{\mathcal{D}_L}[\mathbf{t}^\top \mathbf{P}_{\phi, f_{R_o}+L}] - \mathbf{t}^\top \mathbf{P}_{\phi, f_{R_o}} \\
&= \mathbf{t}^\top \mathbb{E}_{\mathcal{D}_L}[\mathbf{P}_{\phi, f_{R_o}+L}] - \phi(t, f_{R_o}) \\
&= \mathbf{t}^\top \int_{\mathcal{B}} \mathbf{P}_{\phi, f_{R_o}+L} \mathcal{D}_L(l) \mathrm{d}l - \phi(t, f_{R_o}).
\end{aligned}$$

Affinity can be interpreted as the expected increase or decrease in performance on the recipient problem when using a learning resource sampled from a set according to a given distribution.

Using affinity, we seek to prove bounds similar to existing bounds about bias, such as the Famine of Favorable Targets and Famine of Favorable Information Resources [6]. We present our results next, with proofs available online (arXiv).

5   Theoretical Results

We begin by showing that affinity is a conserved quantity, implying that positive affinity towards one target is offset by negative affinity towards other targets.

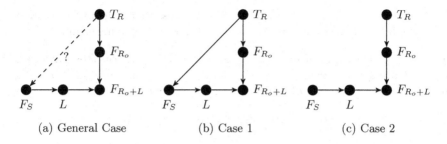

(a) General Case   (b) Case 1   (c) Case 2

Fig. 2. Dependence structure for transfer learning.

Theorem 1 (Conservation of Affinity). *For any arbitrary distribution $\mathcal{D}$ and any f_{R_o},*

$$\sum_{\mathbf{t}} \mathrm{Affin}(\mathcal{D}, \mathbf{t}, f_{R_o}) = 0.$$

This result agrees with other no free lunch [12] and conservation of information results [2,3,9], showing that trade-offs must always be made in learning.

Assuming the dependence structure of Fig. 2, we next bound the mutual information between our updated information resource and the recipient target in terms of the source and recipient information resources.

Theorem 2 (Transfer Learning under Dependence). *Define*

$$\phi_{TL} := \mathbb{E}_{T_R, F_{R+L}}[\phi(T_R, F_{R+L})] = \Pr(\omega \in T_R; \mathcal{A})$$

as the probability of success for transfer learning. Then,

$$\phi_{TL} \leq \frac{I(F_S; T_R) + I(F_R; T_R) + D(P_{T_R} \| \mathcal{U}_{T_R}) + 1}{I_\Omega}$$

where $I_\Omega = -\log |T_R|/|\Omega|$ (T_R being of fixed size), $D(P_{T_R} \| \mathcal{U}_{T_R})$ is the Kullback-Leibler divergence between the marginal distribution on T_R and the uniform distribution on T_R, and $I(F;T)$ is the mutual information.

This theorem upper bounds the probability of successful transfer (ϕ_{TL}) to show that transfer learning can't help us more than our information resources allow. This point is determined by $I(F_S; T_R)$, the amount of mutual information between the source's information resource and the recipient's target, by $I(F_R; T_R)$ the amount of mutual information between the recipient's information resource and the recipient's target, and how much P_{T_R} (the distribution over the recipient's target) 'diverges' from the uniform distribution over the recipient's target, $\mathcal{U}_{T_R}$. This makes sense in that

– the more dependent F_S and T_R, the more useful we expect the source's information resource to be in searching for T_R, in which case q_{TL} can take on larger values.

– the more P_{T_R} diverges from $\mathcal{U}_{T_R}$, the less helpless we are against the randomness (since the uniform distribution maximizes entropic uncertainty).

Theorem 3 (Famine of Favorable Learned Information Resources). *Let $\mathcal{B}$ be a finite set of learning resources and let $t \subseteq \Omega$ be an arbitrary fixed k-size target set. Given a recipient problem (Ω, t, f_{R_o}), define*

$$\mathcal{B}_{\phi_{\min}} = \{l \in \mathcal{B} \mid \phi(t, f_{R_o+l}) \geq \phi_{\min}\},$$

where $\phi(t, f_{R_o+l})$ is the decomposable probability-of-success metric for algorithm $\mathcal{A}$ on search problem (Ω, t, f_{R_o+l}) and $\phi_{\min} \in (0,1]$ represents the minimally acceptable probability of success under ϕ. Then,

$$\frac{|\mathcal{B}_{\phi_{\min}}|}{|\mathcal{B}|} \leq \frac{\phi(t, f_{R_o}) + \mathrm{Affin}(\mathcal{U}[\mathcal{B}], \mathbf{t}, f_{R_o})}{\phi_{\min}}$$

where $\phi(t, f_{R_o})$ is the decomposable probability-of-success metric with the recipient's original information resource.

Theorem 3 demonstrates the proportion of $\phi_{\min}$-favorable information resources for transfer learning is bounded by the degree of success without transfer, along with the affinity (average performance improvement) of the set of resources as a whole. Highly favorable transferable resources are rare for difficult tasks, within any neutral set of resources lacking high affinity. Unless a set of information resources is curated towards a specific transfer task by having high affinity towards it, the set will not and cannot contain a large proportion of highly favorable elements.

Theorem 4 (Futility of Affinity-Free Search). *For any fixed algorithm $\mathcal{A}$, fixed recipient problem (Ω, t, f_{R_o}), where $t \subseteq \Omega$ with a corresponding target function $\mathbf{t}$, and distribution over information resources $\mathcal{D}_L$, if $\mathrm{Affin}(\mathcal{D}_L, \mathbf{t}, f_{R_o}) = 0$, then*

$$\Pr(\omega \in t; \mathcal{A}_L) = \phi(t, f_{R_o})$$

where $\Pr(\omega \in t; \mathcal{A}_L)$ represents the expected decomposable probability of successfully sampling an element of t using $\mathcal{A}$ with transfer, marginalized over learning resources $L \sim \mathcal{D}_L$, and $\phi(t, f_{R_o})$ is the probability of success without L under the given decomposable metric.

Theorem 4 tells us that transfer learning only helps in the case that we have a favorable distribution on learning resources, tuned to the specific problem at hand. Given a distribution *not* tuned in favor of our specific problem, we can perform no better than if we had not used transfer learning. **This proves that transfer learning is not inherently beneficial in and of itself,** unless it is accompanied by a favorably tuned distribution over resources to be transferred. A natural question is how rare such favorably tuned distributions are, which we next consider in Theorem 5.

Theorem 5 (Famine of Favorable Affinity Distributions). *Given a fixed target function* $\mathbf{t}$ *and a finite set of learned information resources* $\mathcal{B}$, *let*

$$\mathcal{P} = \{\mathcal{D} \mid \mathcal{D} \in \mathbb{R}^{|\mathcal{B}|}, \sum_{l \in \mathcal{B}} \mathcal{D}(l) = 1\}$$

be the set of all discrete $|\mathcal{B}|$*-dimensional simplex vectors. Then,*

$$\frac{\mu(\mathcal{G}_{\mathbf{t},\phi_{\min}})}{\mu(\mathcal{P})} \leq \frac{\phi(\mathbf{t}, f_{R_o}) + \mathrm{Affin}(\mathcal{U}[\mathcal{B}], \mathbf{t}, f_{R_o})}{\phi_{\min}}$$

where $\mathcal{G}_{\mathbf{t},\phi_{\min}} = \{\mathcal{D} \mid \mathcal{D} \in \mathcal{P}, \mathrm{Affin}(\mathcal{D}, \mathbf{t}, f_{R_o}) \geq \phi_{\min}\}$ *and* μ *is Lebesgue measure.*

We find that highly favorable distributions are quite rare for problems that are difficult without transfer learning, unless we restrict ourselves to distributions over sets of highly favorable learning resources. (Clearly, finding a favorable distribution over a set of good options is not a difficult problem.) Additionally, note that we have recovered the same bound as in Theorem 3.

Theorem 6 (Success Difference from Distribution Divergence). *Given the performance of a search algorithm on the recipient problem in the transfer learning case,* ϕ_{TL}, *and without the learning resource,* ϕ_{NoTL}, *we can upperbound the absolute difference as*

$$|\phi_{TL} - \phi_{NoTL}| \leq |T|\sqrt{\frac{1}{2}D_{KL}(\mathbf{P}_{TL}||\mathbf{P}_{NoTL})}.$$

This result shows that unless using the learning resource significantly changes the resulting distribution over the search space, the change in performance from transfer learning will be minimal.

6 Examples and Applications

6.1 Examples

We can use examples to evaluate our theoretical results. To demonstrate how Theorem 2 can apply to an actual case of machine learning, we can construct a pair of machine learning problems in such a way that we can properly quantify each of the terms in the inequality, allowing us to show how the probability of successful search is directly affected by the transfer of knowledge from the source problem.

Let Ω be a 16×16 grid and $|T| = k = 1$. In this case, we know that the target set is a single cell in the grid, so choosing a target set is equivalent to choosing a cell in the grid. Let the distribution on target sets P_T be uniformly random across the grid. For simplicity, we will assume that there is no information about the target set in the information resource, and that any information will have to come via transfer from the source problem. Thus, $I(F_R; T_R) = 0$.

First, suppose that we provide no information through transfer, meaning that a learning algorithm can do no better than randomly guessing. The probability of successful search will be $1/256$. We can calculate the bound from our theorem using the known quantities:

- $I(F_S; T_R) = 0$;
- $I(F_R; T_R) = 0$;
- $H(T) = 8$ (as it takes 4 bits to specify a row and 4 bits to specify a column)
- $D(P_{T_R} \| \mathcal{U}_{T_R}) = \log_2 \binom{256}{1} - H(T) = 8 - 8 = 0$;
- $I_\Omega = -\log 1/256 = 8$;

Thus, we upper bound the probability of successful search at $1/8$.

Now, suppose that we had an algorithm which had been trained to learn which half, the top or bottom, our target set was in. This is a relatively easier task, and would be ideal for transfer learning. Under these circumstances, the actual probability of successful search doubles to $1/128$. We can examine the effect that this transfer of knowledge has on our probability of success.

- $I(F_S; T_R) = H(T_R) - H(T_R|F_S) = 8 - 7 = 1$;
- $I(F_R; T_R) = 0$;
- $H(T) = 8$;
- $D(P_{T_R} \| \mathcal{U}_{T_R}) = 0$;
- $I_\Omega = 8$;

The only change is in the mutual information between the recipient target set and the source information resource, which was able to perfectly identify which half the target set was in. This brings the probability of successful search to $1/4$, exactly twice as high as without transfer learning.

This result is encouraging, because it demonstrates that the upper bound for transfer learning under dependence is able to reflect changes in the use of transfer learning and their effects. The upper bound being twice as high when the probability of success is doubled is good. However, the bound is very loose. In both cases, the bound is 32 times as large as the actual probability of success. Tightening the bound may be possible; however, as seen in this example, the bound we have can already serve a practical purpose.

6.2   Transferability Heuristic

Our theoretical results suggest that we cannot expect transfer learning to be successful without careful selection of transferred information. Thus, it is imperative to identify instances in which transferred resources will raise the probability of success. In this section, we explore a simple heuristic indicating conditions in which transfer learning may be successful, motivated by our theorems. Theorem 2 shows that source information resources with strong dependence on the recipient target can raise the upper bound on performance. Thus, given a source problem and a recipient problem, our heuristic uses the success of an algorithm on the recipient problem after training solely on the source problem and

not the recipient problem as a way of assessing potential for successful transfer. Using a classification task, we test whether this heuristic reliably identifies cases where transfer learning works well.

We focused on two similar image classification problems, classifying tigers versus wolves[1] (TvW) and classifying cats versus dogs[2] (CvD). Due to the parallels in these two problem, we expect that a model trained for one task will be able to help us with the other. In our experiment, we used a generic deep convolutional neural network image classification model (VGG16 [10], using Keras[3]) to evaluate the aforementioned heuristic to see whether it correlates with any benefit in transfer learning. The table below contains our results:

Run	Source Problem	Source Testing Accuracy	Recipient Problem	Additional Training	Recipient Testing Accuracy
1	CvD	84.8%	TvW	N	74.24%
2	CvD	84.8%	TvW	Y	95.35%
3	TvW	92.16%	CvD	N	48.36%
4	TvW	92.16%	CvD	Y	82.44%

The `Source Problem` column denotes the problem we are transferring from, and the `Recipient Problem` column denotes the problem we are transferring to. The `Source Testing Accuracy` column contains the image classification model's testing accuracy on the source problem after training on its dataset, using a disjoint test dataset. The `Additional Training` column indicates whether we did any additional training before testing the model's accuracy on the recipient problem's dataset—N indicates no training, which means that the entry in the `Recipient Testing Accuracy` column contains the results of the heuristic, while Y indicates an additional training phase, which means that the corresponding entry in the `Recipient Testing Accuracy` column contains the experimental performance of transfer learning. In each run we start by training our model on the source problem.

Consider Runs 1 and 2. Run 1 is the heuristic run for the CvD → TvW transfer learning problem. When we apply the trained CvD model to the TvW problem without retraining, we get a testing accuracy of 74.24%. This result is promising, as it's significantly above a random fair coin flip, indicating that our CvD model has learned something about the difference between cats and dogs that can be weakly generalized to other images of feline and canine animals. Looking at Run 2, we see that taking our model and training additionally on the TvW dataset yields a transfer learning testing accuracy of 95.35%, which is higher than the testing accuracy when we train our model solely on TvW

---

[1] http://image-net.org/challenges/LSVRC/2014/browse-synsets.

[2] https://www.kaggle.com/c/dogs-vs-cats-redux-kernels-edition/data.

[3] https://keras.io/applications/#vgg16.

(92.16%). This is an example where transfer learning improves our model's success, suggesting that the pre-training step is helping our algorithm generalize.

When we look at Runs 3 and 4, we see the other side of the picture. The heuristic for the TvW → CvD transfer learning problem in Run 3 is a miserable 48.36%, which is roughly how well we would do randomly flipping a fair coin. It's important to note that this heuristic is not symmetric, which is to be expected—for example, if the TvW model is learning based on the background of the images and not the animals themselves, we would expect a poor application to the CvD problem regardless of how well the CvD model can apply to the TvD problem. Looking at Run 4, the transfer learning testing accuracy is 82.44%, which is below the testing accuracy when we train solely on the CvD dataset (84.8%). This offers some preliminary support for our heuristic—when the success of the heuristic is closer to random, it may be the case that pre-training not only fails to benefit the algorithm, but can even hurt performance.

Let us consider what insights we can gain from the above results regarding our heuristic. A high value means that the algorithm trained on the source problem is able to perform well on the recipient problem, which indicates that the algorithm is able to identify and discriminate between salient features of the recipient problem. Thus, when we transfer what it learns (e.g., the model weights), we expect to see a boost in performance. Conversely, a low value (around 50%, since any much lower would allow us to simply flip the labels to obtain a good classifier) indicates that the algorithm is unable to learn features useful for the recipient problem, so we would expect transfer to be unsuccessful. It's important to note that this heuristic is heavily algorithm independent, which is not the case for our theoretical results—problems with a large degree of latent similarity can receive poor values by our heuristic if the algorithm struggles to learn the underlying features of the problem.

These results offer preliminary support for the suggested heuristic, which was proposed to identify information resources that would be suitable for transfer learning. More research is needed to explore how well it works in practice on a wide variety of problems, which we leave for future work.

7   Conclusion

Transfer learning is a type of machine learning that involves a source and recipient problem, where information learned by solving the source problem is used to benefit the process of solving the recipient problem. A popular and potentially lucrative avenue of application is in transferring knowledge from data-rich problems to more niche, difficult problems that suffer from a lack of clean and dependable data. To analyze the bounds of transfer learning, applicable to a large diversity of source/recipient problem pairs, we cast transfer learning into the algorithmic search framework, and define affinity as the degree to which learned information is predisposed towards the recipient problem's target. In our work, we characterize various properties of affinity, show why affinity is essential for the success of transfer learning, and prove results connecting the probability of success of transfer learning to elements of the search framework.

Additionally, we introduce a heuristic to evaluate the likelihood of success of transfer, namely, the success of the source algorithm applied directly to the recipient problem without additional training. Our results show that the heuristic holds promise as a way of identifying potentially transferable information resources, and offers additional interpretability regarding the similarity between the source and recipient problems.

Much work remains to be done to develop theory for transfer learning. Through the results presented here, we learn that there are limits to when transfer learning can be successful, and gain some insight into what powers successful transfer between problems.

References

1. Choi, D., An, T.-H., Ahn, K., Choi, J.: Driving experience transfer method for end-to-end control of self-driving cars. arXiv preprint arXiv:1809.01822 (2018)
2. Dembski, W.A., Marks II, R.J.: Conservation of information in search: measuring the cost of success. IEEE Trans. Syst. Man Cybern.-Part A: Syst. Hum. **39**(5), 1051–1061 (2009)
3. Lauw, J., Macias, D., Trikha, A., Vendemiatti, J., Montanez, G.D.: The bias-expressivity trade-off. In: Ana, P.R., Luc, S., van den Herik, H.J. (eds.) Proceedings of the 12th International Conference on Agents and Artificial Intelligence, vol. 2, pp. 141–150. SCITEPRESS (2020)
4. George, D.M.: The famine of forte: few search problems greatly favor your algorithm. In: 2017 IEEE International Conference on Systems, Man, and Cybernetics (SMC), pp. 477–482. IEEE (2017a)
5. George, D.M.: Why machine learning works. Ph.D. thesis, Carnegie Mellon University (2017b)
6. Montañez, G.D., Hayase, J., Lauw, J., Macias, D., Trikha, A., Vendemiatti, J.: The futility of bias-free learning and search. In: Liu, J., Bailey, J. (eds.) AI 2019. LNCS (LNAI), vol. 11919, pp. 277–288. Springer, Cham (2019). https://doi.org/10.1007/978-3-030-35288-2_23
7. Pan, S.J., Yang, Q.: A survey on transfer learning. IEEE Trans. Knowl. Data Eng. **22**(10), 1345–1359 (2009)
8. Sam, T., Williams, J., Tadesse, A., Sun, H., Montanez, G.D.: Decomposable probability-of-success metrics in algorithmic search. In: Rocha, A.P., et al. (eds.) Proceedings of the 12th International Conference on Agents and Artificial Intelligence, vol. 2, pp. 785–792. SCITEPRESS (2020)
9. Schaffer, C.: A conservation law for generalization performance. In: Machine Learning Proceedings 1994, no. 1, pp. 259–265 (1994)
10. Simonyan, K., Zisserman, A.: Very deep convolutional networks for large-scale image recognition. arXiv preprint arXiv:1409.1556 (2014)
11. Wang, D., Zheng, T.F.: Transfer learning for speech and language processing. In: 2015 Asia-Pacific Signal and Information Processing Association Annual Summit and Conference (APSIPA), pp. 1225–1237. IEEE (2015)
12. Wolpert, D.H., Macready, W.G.: No free lunch theorems for optimization. IEEE Trans. Evol. Comput. **1**, 67–82 (1997)

Learning Objective Boundaries
for Constraint Optimization Problems

Helge Spieker[✉] and Arnaud Gotlieb

Simula Research Laboratory, P.O. Box 134, 1325 Lysaker, Norway
{helge,arnaud}@simula.no

Abstract. Constraint Optimization Problems (COP) are often considered without sufficient knowledge on the boundaries of the objective
variable to optimize. When available, tight boundaries are helpful to
prune the search space or estimate problem characteristics. Finding
close boundaries, that correctly under- and overestimate the optimum,
is almost impossible without actually solving the COP. This paper introduces Bion, a novel approach for boundary estimation by learning from
previously solved instances of the COP. Based on supervised machine
learning, Bion is problem-specific and solver-independent and can be
applied to any COP which is repeatedly solved with different data inputs.
An experimental evaluation over seven realistic COPs shows that an estimation model can be trained to prune the objective variables' domains
by over 80%. By evaluating the estimated boundaries with various COP
solvers, we find that Bion improves the solving process for some problems,
although the effect of closer bounds is generally problem-dependent.

Keywords: Machine learning · Constraint Optimization · Objective
boundaries

1 Introduction

Many scheduling or planning problems involve the exact optimization of some
variable (e.g., timespan), that depends on decision variables, constrained by a
set of combinatorial relations. These problems, called Constraint Optimization
Problems (COP), are notoriously difficult to solve [21]. They are often addressed
with systematic tree-search, such as branch-and-bound, where parts of the search
space with worse cost than the current best solution are pruned. In Constraint
Programming, these systematic techniques work without prior knowledge and
are steered by the constraint model. Unfortunately, the worst-case computational
cost to fully explore the search space is exponential in the worst case and the
performance of the solver depends on efficient domain pruning [33].

COP solving with branch-and-bound is often considered without sufficient
knowledge on boundaries of the objective variable [29]. When available, these
boundaries can support the solver in discovering near-optimal solutions early
during search and thus can reduce the computational effort [4,18,21]. In addition,

© Springer Nature Switzerland AG 2020
G. Nicosia et al. (Eds.): LOD 2020, LNCS 12566, pp. 394–408, 2020.
https://doi.org/10.1007/978-3-030-64580-9_33

providing tight estimates of the objective variable is useful user-feedback to estimate interesting features of COP [13]. Unfortunately, finding close under- and over estimations of the objective variable is still an open problem [19] and almost impossible without actually running the solver with a good heuristic. Domain boundaries are therefore usually obtained through problem-specific heuristics [4,18,24], which requires a deep understanding of the COP. Finding a generic method for closer domain estimation would allow many COP instances to be solved more efficiently.

This paper introduces Bion, a new method combining logic-driven constraint optimization and supervised machine learning (ML), for solving COP. Using the known results of already-solved instances of a COP, a trained data-driven ML estimation model predicts boundaries for the objective variable of a new instance. These boundaries are then exploited by a COP solver to prune the search space. In simpler words, for a given COP, an estimation model is trained once with already solved instances of the same problem. For a new instance, the trained model is exploited to estimate close boundaries of the objective variable. Using the estimated boundaries, additional domain constraints are added to the COP model and used by the solver to prune the search space at low cost. Note however that ML methods can only approximate the optimum and are therefore not a full alternative. To eliminate the inherent risk of misestimations, which can cause unsatisfiability, the ML model is trained with an asymmetric loss function, adjusted training labels, and other counter-measures. As a result, Bion is an exact method for solving COP and can complement advantageously any existing COP solver. Interestingly, the main computational cost of Bion lies in the training part, but the estimation cost for a new input is low.

Besides the general ability to estimate close objective boundaries, we explore with Bion how useful these boundaries are to prune the search space and to improve the COP solving process. Our results show that boundary estimation can generally improve solver performance, even though there are dependencies on the right combination of solver and COP model for best use of the reduced domains. The main contributions of this paper are threefold:

1. We introduce Bion, a new exact method combining ML and traditional COP solving to estimate close boundaries of the objective variable and to exploit these boundaries for boosting the solving process. Bion can be advantageously applied to any COP which is repeatedly solved with different data inputs. To the best of our knowledge, this is the first time a problem- and solver-independent ML method using historical data is proposed.
2. We discuss training techniques to avoid misestimations, compare various ML models such as gradient tree boosting, support vector machine and neural networks, with symmetric and asymmetric loss functions. A contribution lies in the dedicated feature selection and user-parameterized label shift, a new method to train these models on COP characteristics.
3. We evaluate Bion's ability to prune objective domains, as well as the impact of estimated and manually determined boundaries on solver performance with seven COPs.

2 Related Work

Many exact solvers include heuristics to initial feasible solutions that can be used for bounding the search, using for example, linear relaxations [21]. Others rely on branching heuristics and constraint propagation to find close bounds early [33]. Recent works have also considered including the objective variable as part of the search and branch heuristic [16,31]. By exploiting a trained ML model, our approach Bion adds an additional bounding step before the solver execution but it does not replace the solvers' bounding mechanisms. Hence, it complements these approaches by starting with a smaller search space.

The combination of ML and exact solvers has been previously explored from different angles [13,25]. One angle is the usage of ML for solver configuration and algorithm selection, for example by selecting and configuring a search strategy [3,10,27], deciding when to run heuristics [23] or lazy learning [17], or to efficiently orchestrate a solver portfolio [2,36]. In [7], Cappart *et al.* propose the integration of reinforcement learning into the construction of decision diagrams to derive bounds for solutions of optimization problems. A framework for mutual integration of CP and ML was presented, the Inductive Constraint Programming (ICP) loop [6], which is also applicable to Bion. Lombardi *et al.* present a general framework for embedding ML-trained models in optimization techniques, called empirical model learning [26]. This approach deploys trained ML models directly in the COP as additional global constraints, which is a promising approach, especially when the learning model works directly on the model input. In our case, the feature set uses external instance descriptions and works on different input sizes, which is a different type of integration. The coupling of data-driven and exact methods differs from the work on learning combinatorial optimization purely from training [5,12,14], without using any constraint solver. As the complexity of a full solution is much higher than estimating the objective value, these methods require considerably more training data and computational resources, and are not yet competitive in practice.

Previous work also exists on predicting instance characteristics of COPs. These characteristics include runtime prediction, e.g., for the traveling salesperson problem [22,28] or combining multiple heuristics into a single admissible A* heuristic [34]. However, these methods are tailored to a single problem and rely on problem-specific features for the statistical model, which makes them effective for their specific use-case. They require substantial analytic effort per problem. In contrast, our approach is both solver- and problem-independent.

3 Background

This section introduces constraint optimization in the context of Constraint Programming over Finite Domains [33] and necessities of supervised ML.

We define a Constraint Optimization Problem (COP) as a triple $\langle \mathcal{X}, \mathcal{C}, f_\mathbf{z} \rangle$ where $\mathcal{X} = \{\mathbf{x}_1, \ldots, \mathbf{x}_n\}$ is a set of variables, $\mathcal{C}$ is a set of constraints $\mathcal{C} = \{c_1, \ldots, c_m\}$ and $f_\mathbf{z}$ is an objective function with value $\mathbf{z}$ to optimize. Each variable $\mathbf{x}_i$, also called decision variable, is associated with a finite domain $\mathcal{D}(\mathbf{x}_i)$, representing all possible values. A constraint c_i is defined over a subset of r_i variables, represented by all allowed r_i-tuples of the corresponding relation.

A (feasible) solution φ of the COP is an assignment of each variable to a single value from its domain, such that all constraints are satisfied. Each solution corresponds to an objective value $\mathbf{z}_\varphi = f_\mathbf{z}(\varphi)$. The goal of solving the COP is to find at least one φ^*, called an optimal solution, such that $f_\mathbf{z}(\varphi^*) = \mathbf{z}_{opt}$ is optimal. Depending on the problem formulation, $f_\mathbf{z}$ has either to be minimized or maximized by finding solutions with a smaller respectively larger objective value $\mathbf{z}$. We use $\underline{\mathbf{x}}$ and $\overline{\mathbf{x}}$ to refer to the lower and upper domain boundaries of a variable $\mathbf{x}$, and the notation $\underline{\mathbf{x}}..\overline{\mathbf{x}}$ to denote the domain, i.e. the integer set $\{\, n \mid \underline{\mathbf{x}} \leq n \leq \overline{\mathbf{x}} \,\}$.

A COP $\langle \mathcal{X}, \mathcal{C}, f_\mathbf{z} \rangle$ is *satisfiable* iff it has at least one feasible solution. The instance is *unsatisfiable* iff it has no feasible solution. A COP is said to be *solved* iff at least one of its feasible solution is proved optimal.

A ML model can be trained for the regression task to approximate the objective function of a COP. In a typical regression task, a continuous value $\hat{y}$ is predicted for a given input vector $\mathbf{x}$: $f(\mathbf{x}) = \hat{y}$, e.g., for predicting closer domain boundaries. The model can be trained through supervised learning, that is, by examples of input vectors $\mathbf{x}$ and true outputs y: $\{(\mathbf{x_1}, y_1), (\mathbf{x_2}, y_2), \ldots, (\mathbf{x_m}, y_m)\}$. During successive iterations, the model parameters are adjusted with the help of a loss function, that evaluates the approximation of training examples [20].

We now introduce the concept of estimated boundaries, which refers to providing close lower and upper bounds for the optimal value $\mathbf{z}_{opt}$. An *estimation* is a domain $\hat{\underline{\mathbf{z}}}..\hat{\overline{\mathbf{z}}}$ which defines boundaries for the domain of $f_\mathbf{z}$. The domain boundaries are predicted by supervised ML, that is, $\hat{\overline{\mathbf{z}}} = f(\mathbf{x})$, $\hat{\underline{\mathbf{z}}} = f(\mathbf{x})$. An estimation $\hat{\underline{\mathbf{z}}}..\hat{\overline{\mathbf{z}}}$ is *admissible* iff $\mathbf{z}_{opt} \in \hat{\underline{\mathbf{z}}}..\hat{\overline{\mathbf{z}}}$. Otherwise, the estimation is *inadmissible*.

We further classify the two domain boundaries as *cutting* and *limiting* boundaries in relation to their effect on the solver's search process. Depending on whether the COP is a minimization or maximization problem, these terms refer to different domain boundaries. The *cutting boundary* is the domain boundary that reduces the number of reachable solutions. For minimization, this is the upper domain boundary $\overline{\mathbf{z}}$; for maximization, the lower domain boundary $\underline{\mathbf{z}}$. Similarly, the *limiting boundary* is the domain boundary that does not reduce the number of reachable solutions, but only reduces the search space to be explored. For minimization, this is the lower domain boundary $\underline{\mathbf{z}}$; for maximization, the upper domain boundary $\overline{\mathbf{z}}$.

4   Learning to Estimate Boundaries

In this section, we explain the initial part of our method Bion, which is to train a ML model to estimate close objective domain boundaries. This includes a discussion on the features to describe COPs and their instances, and how to train the model such that the estimated boundaries do not render the problem unsatisfiable or exclude the optimal objective value.

Training an estimator model has to be performed only once per COP. From each instance of an existing dataset, a feature vector is extracted in order to train an estimator model. This model predicts both lower and upper boundaries for the objective variable of each instance. When the estimated boundaries are used to support COP solving, they are embedded into the COP, either through additional hard constraints or with an augmented search strategy, which we will discuss in Sect. 4.3.

The training set is constructed such that the label y of each COP instance, i.e., the true optimal objective value z, is scaled by the original objective domain boundaries into $[0, 1]$, with $\underline{z} \triangleq 0$ and $\overline{z} \triangleq 1$. After estimating the boundaries, the model output is scaled back from $[0, 1]$ to the original domain. This scaling allows the model to relate the approximated objective function to the initial domain. This is useful both if the given boundaries tend to systematically under- and overestimate the optimal objective, and as it steers the estimations to be within the original boundaries and therefore improve admissibility. Furthermore, some ML models, such as those based on neural networks, benefit from having an output within a specified range for their estimation performance.

4.1   Instance Representation

The estimator ML model expects a fixed-size numeric feature vectors as its input. This feature vector is calculated from the COP model and the instance by analyzing its values and the model structure. As the presented method is problem-independent, a generic set of features is chosen, and problem-specific information is gathered by exploiting the variables and structure of the COP instance. This means, the set of features to describe a COP instance can be calculated for any instance of any COP model, which makes the method easily transferable without having to identify domain-dependent, specific features. Still, it is possible to extend the feature set by problem- or model-specific features to further improve the model performance.

At the same time, by requirement of the used ML methods, the size of the feature vector is fixed for all inputs of an estimator, i.e. for one COP model and all its possible instances. However, in practice, the size of each COP instance varies and the instance cannot directly function as a feature vector. For example, the number of locations varies in routing problems, or the number of tasks varies in scheduling problems, with each having a couple of additional attributes per location or task.

We construct the generic feature vector from two main sets of features. The first set of features focuses on the instance parameters and their values, i.e. these features directly encode the input to the COP model. The second set of features stem from the combination of COP model and instance and describe the resulting constraint system.

From the description of each decision variable of the COP, a first set of features is constructed. Thereby, each COP uses a problem-specific feature vector, depending on the number and the type of decision variables, constructed from problem-independent characteristics. Each decision variable of the COP instance is processed and described individually. Finite domains variables and constants are directly used without further processing. Data structures for collections of values, such as lists or sets, are described by 9 measures from descriptive statistics. Multidimensional and nested data structures are aggregated to a single, fixed-size dimension by the sum of the features for each nested dimension. The 9 statistical measures are: 1) The number of values in the collection and their 2) minimum and 3) maximum; 4) standard deviation and 5) interquartile range for dispersion; 6) mean and 7) median for central tendency; 8) skew and 9) kurtosis to describe the distribution's shape.

Additionally, the second set of features characterize the complete constraint system of COP model and instance. These features analyze the number of variables, constraints, etc. for the COP [1] and have originally been developed to efficiently schedule a portfolio of solvers [2]. Some of these features are without relevant information, because they can be constant for all instances of a COP, e.g. the number of global constraints in the model. Although these features are less descriptive than the first set, we observed a small accuracy improvement from including them.

Finally, when the feature vectors for all instances in the training set have been constructed, a final step is to remove features with a low variance, that add little or no information, to reduce the number of inputs and model complexity.

4.2   Eliminating Inadmissible Estimations

The boundaries of a domain regulate which optimum values a variable can take. For the objective variable, this means which objective values can be reached. By further limiting the domain through an external estimation, the desired effect is to prune unnecessary values, such as low-quality objective values. This pruning focuses the search on the high-quality region of the objective space.

The trained estimator only approximates an instance's objective value, but does not calculate it precisely, i.e., even after training estimation errors occur. These errors are usually approximately evenly distributed between under- and over estimations, because many applications make no difference in the error type. However, in the boundary estimation case, the type of error is crucial. If the estimator underestimates, resp. overestimates, the objective value in a minimization, resp. maximization COP, all solutions, including the optima, are excluded from the resulting objective domain. On the other hand, errors shifting the boundary away from the optimum only have the penalty of a larger resulting domain, but still find high-quality solutions, including the optima.

We consider three techniques to avoid inadmissible estimations. Two of these techniques, label shift and asymmetric losses, are applied during the estimator's training phase. The third affects the integration of estimated boundaries during constraint solving and will be discussed in Sect. 4.3.

Adjusting Training Labels. As inadmissible estimations are costly, but some estimation error is acceptable, the first technique *label shift* changes the training

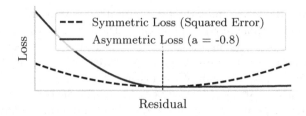

Fig. 1. Symmetric and asymmetric loss functions. The asymmetric loss assigns a higher loss to a negative residuals, but lower loss to overestimations.

label, i.e. the objective value of the sample instances, by a small margin, such that a small estimation error still leads to an admissible estimation. The estimator is thereby trained to always under- or overestimate the objective value. Label shift is similar to the prediction shift concept from [40], but is specifically designed for boundary estimation and the application on COP models.

Formally, label shift is defined as:

$$y' = y + \lambda \left(\overline{z} - y \right) \quad \textit{(Overestimation)}$$
$$y' = y - \lambda \left(y - \underline{z} \right) \quad \textit{(Underestimation)}$$

with adjustment factor $\lambda \in [0, 1)$. The configuration parameter λ steers the label shift margin, with a smaller λ being closer to the true label y. Therefore, setting λ is a trade-off between close estimations and their feasibility.

Training with Asymmetric Loss Functions. ML models are usually trained with symmetric loss functions, that do not differentiate between positive and negative errors. An asymmetric loss function, on the other hand, assigns higher loss values for either under- or overestimations, which penalizes certain errors stronger than others. Figure 1 shows an example of quadratic symmetric and asymmetric loss functions and the difference in penalization.

Shifted Squared Error Loss is an imbalanced variant of squared error loss. Formally speaking, the shifted squared error loss is defined as

$$L(r) = r^2 \cdot (sgn(r) + \alpha)^2 \text{ with absolute error } r = \hat{y} - y$$

where $\hat{y}$ is the estimated value and y is the true target value. The parameter α shifts the penalization towards under- or overestimation and influences the magnitude of the penalty.

4.3   Estimated Boundaries During Search

One potential application of estimated objective boundaries is their usage to improve the COP solving process. Using Bion to solve a COP consists of the following steps: 1) (Initially) Train an estimator model for the COP; 2) Extract a feature vector from each COP instance; 3) Estimate both a lower and an upper objective boundaries; 4) Update the COP with estimated boundaries; and 5) Solve the updated COP with the solver.

The boundaries provided by the estimator can be embedded as hard constraints on the objective variable, i.e., by adding $z \in \hat{\underline{z}} \ldots \hat{\overline{z}}$. The induced overhead is negligible, but dealing with misestimations requires additional control. If all feasible solutions are excluded, because the cutting bound is wrongly estimated, the instance is rendered unsatisfiable. This issue is handled by reverting to the original domain. If only optimal solutions are excluded, because the limiting bound is wrongly estimated, then only non-optimal solutions can be returned and this stays impossible to notice. This issue cannot be detected in a single optimization run of the solver. However, in practical cases where the goal is to find good-enough solutions early rather than finding truly-proven optima, it can be an acceptable risk to come-up with an good approximation of the optimal solutions only. In conclusion, hard boundary constraints are especially suited for cases where a high confidence in the quality of the estimator has been gained, and the occurrence of inadmissible estimations is unlikely.

5   Experimental Evaluation

We evaluate our method in three experiments, which focus 1) on the impact of label shift and asymmetric loss functions for training the estimator, 2) on the estimators' performance to bound the objective domain, and 3) on the impact of estimated boundaries on solver performance.

We selected the seven COP having the most instances from the MiniZinc benchmark repository[1], some of which are deployed in MiniZinc challenges [38]. These seven COPs are MRCPSP (11182 instances), RCPSP (2904 inst.), Bin Packing (500 inst.), Cutting Stock (121 inst.), Jobshop (74 inst.), VRP (74 inst.), and Open Stacks (50 inst.). Considering training sets of different sizes, from 50 to over 11,000 instances, is relevant to understand scenarios that can benefit from boundary estimation.

We consider four ML models to estimate boundaries: neural networks (NN), gradient tree boosting (GTB), support vector machines (SVM) and linear regression (LR). NN and GTB come in two variants, using either symmetric (GTB_s, NN_s) or asymmetric (GTB_a, NN_a) loss functions. All models are used with their default hyperparameters as defined by the libraries. The NN is a feed-forward neural network with 5 layers of 64 hidden nodes, a larger NN did not show to improve the results. Bion is implemented in Python, using scikit-learn [32] for SVM and LR. To support the implementation of asymmetric loss functions, NNs are based on Keras [9], and GTB on XGBoost [8].

Our experiments are focused towards the general effectiveness of Bion over a range of problems. Therefore, we used the default parameters and did not perform parameter tuning, although it could improve the performance. As loss

---

[1] github.com/MiniZinc/minizinc-benchmarks.

factors for the asymmetric loss functions, we set $a = -1$ for GTB_a and $a = -0.8$ for NN_a, where a smaller a caused problems during training.

Table 1. Reduction in objective domain through estimated boundaries (in %). *Gap*: Domain size between cutting boundary and optimum $((1 - (|\hat{\bar{z}} - z_{opt}|/|\bar{z} - z_{opt}|)) * 100)$. *Size*: Ratio between new and initial domain size $((1 - (|\hat{\bar{z}} - \hat{z}|/|\bar{z} - \underline{z}|)) * 100)$. Cells show the median and the median absolute deviation (MAD): No superscript indicator $\leq 5 \leq {}^{+} \leq 10 < {}^{*} \leq 20 < {}^{**} \leq 30 < {}^{***}$.

	GTB_a		GTB_s		LR		NN_a		NN_s		SVM	
	Gap	Size	Gap	Size	Gap	Size	Gap	Size	Gap	Size	Gap	Size
Bin Packing	68	65	60	58	48	48	**78***	**68***	50	48	15	18
Cutting Stock	64*	66	58*	59	48*	49	41***	**71**$^{+}$	48*	49	29*	17
Jobshop	69	69	60	60	50	50	**87**	**81**	50	48	19	20
MRCPSP	64	61	60	59	49	49	**80**	**76**	49	49	13	19
Open Stacks	64*	**60**$^{+}$	59*	53	43**	43	56**	33*	47*	42^{+}	15^{+}	15
RCPSP	65	64	60	60	50	50	**80**$^{+}$	**76**$^{+}$	50	50	13	20
VRP	70	70	60	60	50	50	**89**	**88**	50	50	0	0

We trained the ML models on commodity hardware without GPU acceleration, and the training time took less than 5 s per model, except for MRCPSP with up to 6 minutes with the NN model. The estimation performance of the ML models is measured via repeated 10-fold validation. The training set is split randomly into 10 folds, of which 9 folds are used to train the model and 1 fold for evaluation. This step is repeated 10 times, so each part is used for evaluation once. We report the median results over 10 runs.

5.1   Avoiding Inadmissible Estimations

The first experiment focuses on label shift, a technique to avoid inadmissible estimations. Label shift changes the training label of the ML model towards under- or overestimation based on an adjustment factor λ. Setting λ is a trade-off between close boundaries and inadmissible estimations. For the experiment, we considered 11λ values between 0 and 0.8 for every estimator variant.

All models benefit from training with label shift, but the adjustment factor for best performance varies. The asymmetric models NN_a and GTB_a only require small $\lambda = 0.1$, respectively 0.3, to maximize the number of admissible estimations. For LR ($\lambda = 0.5$), SVM (0.8), NN_s (0.5), and GTB_s (0.4), label shift is necessary to reach a large amount of admissible estimations. Here, the optimal λ is approximately 0.5, which shifts the label in the middle between optimum and domain boundary. As symmetric models do not distinguish between under- and overestimation error, this confirms the trade-off characteristic of λ. These results underline the benefit of training with both label shift and asymmetric loss functions for avoiding inadmissible estimations.

5.2   Estimating Tighter Domain Boundaries

We analyze here the capability of each model to estimate tight domain boundaries, as compared to the original domains of the COP. As evaluation metrics, the size of the estimated domain is compared to the original domain size. Furthermore, the distance between cutting boundary and optimal objective value is compared between the estimated and original domain. A closer gap between cutting bound and objective value leads to a relatively better first solution when using the estimations and is therefore of practical interest. Table 1 shows the estimation performance per problem and estimator. The results show that asymmetric models are able to estimate closer boundaries than symmetric models. For each model, the estimation performance is consistent over all problems.

First, we look at the share of admissible estimations. Most models achieve 100% admissible estimations in all problems. Exceptions exist for Cutting Stock (GTB$_a$, GTB$_s$, LR: 91%, SVM: 50%) and RCPSP (NN$_s$, SVM: 83%, all other models: $\geq$98%). In general, NN$_a$ has the highest number of admissible estimations, followed by GTB$_a$. The largest reduction is achieved by NN$_a$, making it the overall best performing model. GTB$_a$ is also capable to consistently reduce the domain size by over 60%, but not as much as NN$_a$. Cutting Stock and Open Stacks are difficult problems for most models, as indicated by the deviations in the results. LR and NN$_s$ reduce the domain size by approximately 50%, when the label shift adjustment factor λ is 0.5, as selected in the previous experiment.

Conclusively, these results show that Bion has an excellent ability to derive general estimation models from the extracted instance features. The estimators reduce substantially the domains and provide tight boundaries.

5.3   Effects on Solver Performance

Our third experiment investigated the effect of objective boundaries on the solver performance. The setup for the experiments is as follows. For each COP, 30 instances were randomly selected. Each instance was run in four configurations, using: 1) The original COP model without any modification; 2) The COP model with added upper and lower boundary constraints, estimated by Bion with NN$_a$; 3) The COP model with only an upper boundary constraint, estimated by Bion; 4) The COP model with a user-fixed upper boundary, set up on the middle between the true optimum and the first found solution when using no boundary constraints (z_{first}): $\bar{z} = z_{opt} + \lfloor (z_{first} - z_{opt})/2 \rfloor$. This model is a baseline to evaluate whether there is any benefit in solving the COP with additional boundary constraints.

We selected three distinct State-of-the-Art COP solvers, among those which have the highest rank in MiniZinc challenges: Chuffed (as distributed with MiniZinc 2.1.7) [11], Gecode 6.0.1 [35], and Google OR-Tools 6.8. All runs were performed with a 4-h timeout on a single-core of an Intel E5-2670 with 2.6 GHz.

Three metrics were used for evaluation (all in %), each comparing a run with added boundary constraint to the original COP model. The *Equivalent Solution Time* compares the time taken by the original model to reach a solution of similar

Table 2. Effect of boundaries on solver performance (in %). *Fixed*: Upper boundary set to middle between optimum and first found solution of unbounded run. *Upper*: Upper boundary set to estimated boundary. *Both*: Upper and lower boundary set to estimated boundaries. Results are averaged over 30 instances, lower values are better.

	Equiv. solution time			Quality of first			Time to completion		
	Fixed	Upper	Both	Fixed	Upper	Both	Fixed	Upper	Both
Bin Packing	−9.4	36.1	13.0	−37.9	−57.7	−57.7	36.1	2140.7	2364.5
Jobshop	−96.5	−96.4	−96.6	−38.1	−60.0	−60.0	−27.6	−53.6	−42.5
MRCPSP	0.0	0.0	0.0	−10.8	−0.4	0.0	1.2	0.3	−3.4
Open Stacks	−1.3	−1.3	−0.9	−24.0	−13.2	−13.2	2.0	−0.4	2.9
RCPSP	−3.2	197.4	25.3	−3.3	0.0	0.0	−4.2	0.0	−4.2
VRP	0.4	0.0	0.0	-23.5	0.0	0.0	2.0	7.0	7.0
(a) Chuffed									
Bin Packing	53.5	0.3	−0.6	−4.7	0.0	0.0	−10.3	−4.0	−13.0
Cutting Stock	5627.0	7.3	−29.5	−8.5	−5.5	−2.6	–	–	–
Jobshop	189.3	−6.4	37.4	−10.9	6.1	6.1	0.0	0.0	0.0
MRCPSP	0.0	0.0	23.6	−10.8	−0.4	−0.2	1.3	0.0	4.0
Open Stacks	0.0	−1.5	0.0	−24.0	−12.8	−12.8	8.9	6.4	6.8
RCPSP	−17.2	56.8	−14.4	−2.8	0.0	0.0	−11.8	12.0	−9.4
VRP	0.0	0.0	0.0	−21.0	0.0	0.0	−19.0	−18.0	−8.0
(b) Gecode									
Bin Packing	−22.7	35.0	39.2	−37.4	−57.0	−57.2	104.4	170.0	172.4
Jobshop	1.1	0.0	0.0	−16.5	−0.8	−0.8	0.0	0.0	0.0
MRCPSP	−3.2	−3.0	45.3	−10.8	−0.4	0.0	−2.4	−2.1	1.2
Open Stacks	−5.0	−2.6	−3.1	−24.0	−13.2	−13.2	6.3	−1.2	2.3
RCPSP	0.0	147.2	30.4	−3.3	0.0	0.0	−6.6	27.0	7.8
VRP	−95.3	0.0	0.0	−38.2	0.0	0.0	32.0	−3.0	−5.0
(c) OR-Tools									

or better quality than the first found solution when using Bion. It is calculated as $(t_{Bounds} - t_{Original})/t_{Original} * 100$. The *Quality of First* compares the quality of the first found solutions with and without boundary constraints and is calculated as $(1 - z_{Bounds}/z_{Original}) * 100$. The *Time to Completion* relates the times until the search completed and the optimal solution is found. It is calculated in the same way as the Equivalent Solution Time.

The results are shown in Table 2, listed per solver and problem. The results for the Cutting Stock problem for Chuffed and OR-Tools are not given, because none of the configurations, including the original COP, found a solution for more than one instance. Gecode found at least one solution for 26 of 30 instances. We obtain mixed results for the different solvers and problems, which indicates that benefiting from objective boundaries is both problem- and solver-specific. This holds true both for the boundaries determined by boundary estimation (columns *Upper* and *Both*) and the user-fixed boundary (column *Fixed*).

The general intuition, also confirmed by the literature, is that in many cases a reduced solution space allows more efficient search and for several COPs, this is confirmed. An interpretation for why the boundary constraints in some cases hinder effective search, compared to the original COP, is that the solvers can apply different improvement techniques for domain pruning or search once an initial solution is found. The best results are obtained for solving Jobshop with Chuffed, where the constraints improve both the time to find a good initial solution and the time until the search is completed. Whether both an upper and lower boundary constraint can be useful is visible for the combination of Gecode and RCPSP. Here, posting only the upper boundary constraint is not beneficial for the Equivalent Solution Time, but with both upper and lower boundary Gecode is 14% faster than without any boundaries. A similar behaviour shows for Chuffed and RCPSP regarding Time to Completion, where only the upper boundary has no effect, but posting both bounds reduces the total solving time by 4%. At the same time, we observe that posting both upper and lower boundaries, even though they reduce the original domain boundaries, does not always help the solver, such as for example in the combination of Chuffed and Jobshop. This can come from the efficiency of search heuristics, which can sometimes reach better initial solutions than those obtained with Bion in some cases.

In conclusion, our method Bion can generally produce objective variable boundaries which are helpful to improve the solver performance. Still, it is open to understand which combination of solvers, search heuristics and COP model benefits the most from strongly reduced domains. To the best of our knowledge, no clear answer is yet available in the literature. From the comparison with user-fixed boundaries that are known to reduce the solution space, we observe that the estimated boundaries with Bion are competitive and provide a similar behaviour. This makes Bion a promising approach in many contexts where COP are solved without any prior knowledge on initial boundaries.

6   Conclusion

This paper presents Bion, a boundary estimation method for constraint optimization problems (COP), based on machine learning (ML). A supervised ML-model is trained with solved instances of a given COP, in order to estimate close boundaries of the objective variable. We utilize two training techniques, namely, asymmetric loss functions and label shift, and another counter-measure to adjust automatically the training labels, and discard any wrong estimation. Bion is lightweight and both solver- and problem-independent. Our experimental results, obtained on 7 realistic COPs, show that already a small set of instances is sufficient to train an estimator model to reduce the domain size by over 80%.

Solving practical assignment, planning or scheduling problems often requires to repeatedly solve the same COP with different inputs [15,30,37,39]. Our approach is especially well-suited for those scenarios, where training data can be collected from previous iterations and historical data.

In future work, we plan to explore in depth the actual effects of objective boundaries on constraint solver performances with the goal to better predict in which scenarios Bion can be the most beneficial. Another research lead we intend to follow is to integrate Bion with search heuristics that explicitly focus on evaluating the objective variable [16,31].

References

1. Amadini, R., Gabbrielli, M., Mauro, J.: An enhanced features extractor for a portfolio of constraint solvers. In: Symposium on Applied Computing, pp. 1357–1359 (2014). https://doi.org/10.1145/2554850.2555114
2. Amadini, R., Gabbrielli, M., Mauro, J.: A multicore tool for constraint solving. In: International Joint Conference on Artificial Intelligence, pp. 232–238 (2015)
3. Arbelaez, A., Hamadi, Y., Sebag, M.: Continuous search in constraint programming. In: International Conference on Tools with Artificial Intelligence, pp. 53–60 (2010). https://doi.org/10.1109/ICTAI.2010.17
4. Beck, J.C., Feng, T.K., Watson, J.P.: Combining constraint programming and local search for job-shop scheduling. INFORMS J. Comput. **23**(1), 1–14 (2011). https://doi.org/10.1287/ijoc.1100.0388
5. Bello, I., Pham, H., Le, Q.V., Norouzi, M., Bengio, S.: Neural combinatorial optimization. In: International Conference on Learning Representation (2017). https://doi.org/10.1146/annurev.cellbio.15.1.81
6. Bessiere, C., et al.: The inductive constraint programming loop. IEEE Intell. Syst. **32**(5), 44–52 (2017). https://doi.org/10.1109/MIS.2017.3711637
7. Cappart, Q., Goutierre, E., Bergman, D., Rousseau, L.M.: Improving optimization bounds using machine learning: decision diagrams meet deep reinforcement learning. In: AAAI (2019)
8. Chen, T., Guestrin, C.: XGBoost. In: International Conference on Knowledge Discovery and Data Mining, pp. 785–794 (2016). https://doi.org/10.1145/2939672.2939785
9. Chollet, F., et al.: Keras (2015)
10. Chu, G., Stuckey, P.J.: learning value heuristics for constraint programming. In: Michel, L. (ed.) CPAIOR 2015. LNCS, vol. 9075, pp. 108–123. Springer, Cham (2015). https://doi.org/10.1007/978-3-319-18008-3_8
11. Chu, G., Stuckey, P.J., Schutt, A., Ehlers, T., Gange, G., Francis, K.: Chuffed, a lazy clause generation solver (2016)
12. Dai, H., Khalil, E., Zhang, Y., Dilkina, B., Song, L.: Learning combinatorial optimization algorithms over graphs. In: Advances in Neural Information Processing Systems, pp. 6351–6361 (2017)
13. De Raedt, L., Nijssen, S., O'Sullivan, B., Van Hentenryck, P.: Constraint programming meets machine learning and data mining. Dagstuhl Rep. **1**(5), 61–83 (2011). https://doi.org/10.4230/DagRep.1.5.61
14. Deudon, M., Cournut, P., Lacoste, A., Adulyasak, Y., Rousseau, L.-M.: Learning heuristics for the TSP by policy gradient. In: van Hoeve, W.-J. (ed.) CPAIOR 2018. LNCS, vol. 10848, pp. 170–181. Springer, Cham (2018). https://doi.org/10.1007/978-3-319-93031-2_12
15. Ernst, A.T., Jiang, H., Krishnamoorthy, M., Sier, D.: Staff scheduling and rostering: a review of applications, methods and models. Eur. J. Oper. Res. **153**(1), 3–27 (2004). https://doi.org/10.1016/S0377-2217(03)00095-X
16. Fages, J.G., Homme, C.P.: Making the first solution good! In: International Conference on Tools with Artificial Intelligence (ICTAI) (2017)

17. Gent, I.P., et al.: Learning when to use lazy learning in constraint solving. In: European Conference on Artificial Intelligence (ECAI), pp. 873–878 (2010). https://doi.org/10.3233/978-1-60750-606-5-873

18. Gualandi, S., Malucelli, F.: Exact solution of graph coloring problems via constraint programming and column generation. INFORMS J. Comput. **24**(1), 81–100 (2012). https://doi.org/10.1287/ijoc.1100.0436

19. Hà, M.H., Quimper, C.-G., Rousseau, L.-M.: General bounding mechanism for constraint programs. In: Pesant, G. (ed.) CP 2015. LNCS, vol. 9255, pp. 158–172. Springer, Cham (2015). https://doi.org/10.1007/978-3-319-23219-5_12

20. Hastie, T., Tibshirani, R., Friedman, J.: The Elements of Statistical Learning. SSS. Springer, New York (2009). https://doi.org/10.1007/978-0-387-84858-7

21. Hooker, J.N.: Integrated Methods for Optimization. International Series in Operations Research & Management Science, vol. 170, 2nd edn. Springer, Heidelberg (2012). https://doi.org/10.1007/978-1-4614-1900-6

22. Hoos, H., Stützle, T.: On the empirical scaling of run-time for finding optimal solutions to the traveling salesman problem. Eur. J. Oper. Res. **238**(1), 87–94 (2014). https://doi.org/10.1016/j.ejor.2014.03.042

23. Khalil, E.B., Dilkina, B., Nemhauser, G.L., Ahmed, S., Shao, Y.: Learning to run heuristics in tree search. In: International Joint Conference on Artificial Intelligence, pp. 659–666 (2017). https://doi.org/10.24963/ijcai.2017/92

24. Liu, C., Aleman, D.M., Beck, J.C.: Modelling and solving the senior transportation problem. In: van Hoeve, W.-J. (ed.) CPAIOR 2018. LNCS, vol. 10848, pp. 412–428. Springer, Cham (2018). https://doi.org/10.1007/978-3-319-93031-2_30

25. Lombardi, M., Milano, M.: Boosting combinatorial problem modeling with machine learning. In: International Joint Conference on Artificial Intelligence (2018). https://doi.org/10.24963/ijcai.2018/772

26. Lombardi, M., Milano, M., Bartolini, A.: Empirical decision model learning. Artif. Intell. **244**, 343–367 (2017). https://doi.org/10.1016/j.artint.2016.01.005

27. Loth, M., Sebag, M., Hamadi, Y., Schoenauer, M.: Bandit-based search for constraint programming. In: Schulte, C. (ed.) CP 2013. LNCS, vol. 8124, pp. 464–480. Springer, Heidelberg (2013). https://doi.org/10.1007/978-3-642-40627-0_36

28. Mersmann, O., Bischl, B., Trautmann, H., Wagner, M., Bossek, J., Neumann, F.: A novel feature-based approach to characterize algorithm performance for the traveling salesperson problem. Ann. Math. Artif. Intell. **69**(2), 151–182 (2013). https://doi.org/10.1007/s10472-013-9341-2

29. Milano, M., Wallace, M.: Integrating operations research in constraint programming. 4OR **4**(3), 175–219 (2006). https://doi.org/10.1007/s10288-006-0019-z

30. Mossige, M., Gotlieb, A., Spieker, H., Meling, H., Carlsson, M.: Time-aware test case execution scheduling for cyber-physical systems. In: Beck, J.C. (ed.) CP 2017. LNCS, vol. 10416, pp. 387–404. Springer, Cham (2017). https://doi.org/10.1007/978-3-319-66158-2_25

31. Palmieri, A., Perez, G.: Objective as a feature for robust search strategies. In: Hooker, J. (ed.) CP 2018. LNCS, vol. 11008, pp. 328–344. Springer, Cham (2018). https://doi.org/10.1007/978-3-319-98334-9_22

32. Pedregosa, F., et al.: Scikit-learn: machine learning in Python. J. Mach. Learn. Res. **12**, 2825–2830 (2011)

33. Rossi, F., Beek, P.V., Walsh, T.: Handbook of Constraint Programming (Foundations of Artificial Intelligence). Elsevier, Amsterdam (2006)

34. Samadi, M., Felner, A., Schaeffer, J.: Learning from multiple heuristics. In: AAAI Conference on Artificial Intelligence (2008)

35. Schulte, C., Tack, G., Lagerkvist, M.Z.: Modeling and programming with Gecode (2018)
36. Seipp, J., Sievers, S., Helmert, M., Hutter, F.: Automatic configuration of sequential planning portfolios. In: AAAI (2015)
37. Spieker, H., Gotlieb, A., Mossige, M.: rotational diversity in multi-cycle assignment problems. In: AAAI, pp. 7724–7731 (2019)
38. Stuckey, P.J., Feydy, T., Schutt, A., Tack, G., Fischer, J.: The MiniZinc challenge 2008–2013. AI Mag. **35**(2), 55–60 (2014)
39. Szeredi, R., Schutt, A.: Modelling and solving multi-mode resource-constrained project scheduling. In: Rueher, M. (ed.) CP 2016. LNCS, vol. 9892, pp. 483–492. Springer, Cham (2016). https://doi.org/10.1007/978-3-319-44953-1_31
40. Tolstikov, A., Janssen, F., Fürnkranz, J.: Evaluation of different heuristics for accommodating asymmetric loss functions in regression. In: Yamamoto, A., Kida, T., Uno, T., Kuboyama, T. (eds.) DS 2017. LNCS (LNAI), vol. 10558, pp. 67–81. Springer, Cham (2017). https://doi.org/10.1007/978-3-319-67786-6_5

Hierarchical Representation and Graph Convolutional Networks for the Prediction of Protein–Protein Interaction Sites

Michela Quadrini[1]([⊠]) [iD], Sebastian Daberdaku[1,2] [iD], and Carlo Ferrari[1] [iD]

[1] Department of Information Engineering, University of Padova,
Via Gradenigo 6/A, 35131 Padova (PD), Italy
{michela.quadrini,carlo.ferrari}@unipd.it
[2] Sorint.Tek, Sorint.LAB Group, Via Giovanni Savelli 102, 35129 Padova (PD), Italy
sebastian.daberdaku@latek.it

Abstract. Proteins carry out a broad range of functions in living organisms usually by interacting with other molecules. Protein–protein interaction (PPI) is an important base for understanding disease mechanisms and for deciphering rational drug design. The identification of protein interactions using experimental methods is expensive and time-consuming. Therefore, efficient computational methods to predict PPIs are of great value to biologists.

This work focuses on predicting protein interfaces and investigates the effect of different molecular representations in the prediction of such sites. We introduce a molecular representation according to its hierarchical structure. Therefore, proteins are abstracted in terms of spatial and sequential neighboring among amino acid pairs, while we use a deep learning framework, Graph Convolutional Networks, for data training. We tested the framework on two classes of proteins, Antibody–Antigen and Antigen–Bound Antibody, extracted from the Protein–Protein Docking Benchmark 5.0. The obtained results in terms of the area under the ROC curve (AU-ROC) on these classes are remarkable.

Keywords: Protein–protein interaction · Spatial neighboring · Sequential neighboring · Protein hierarchical structure · Graph convolutional networks

1 Introduction

Proteins consist of one or more amino acid sequences, which fold back on themselves determining complex three-dimensional conformations. The proteins structure organization can be described using a three levels hierarchy as follows: primary structure, secondary structure, tertiary structure. This hierarchical view

© Springer Nature Switzerland AG 2020
G. Nicosia et al. (Eds.): LOD 2020, LNCS 12566, pp. 409–420, 2020.
https://doi.org/10.1007/978-3-030-64580-9_34

is useful to describe the main forces involved in the stabilization of protein structures and, at the same time, provides a useful way to identify the molecular information, which is closely related to the biological function. The molecular conformations are the main predictors of the proteins' biological functions. Proteins carry out a broad range of functions, among which structural support, signal transmission, immune defense, transport, storage, biochemical reaction catalysis, and motility processes, usually by interacting with other molecules, among which other proteins, RNAs or DNAs [1]. The interactions between two proteins, referred to as protein–protein interactions (PPIs), determine the metabolic and signaling pathways [12]. Dysfunction or malfunction of pathways and alterations in protein interactions have been shown to be the cause of several diseases such as neurodegenerative disorders [22] and cancer [14].

The identification of PPI sites can help understand how a protein performs its biological functions and design new antibacterial drugs [8]. However, biological experimental methods such as NMR and X-ray crystallography are labor-intensive, time-consuming, and have high costs. Therefore, efficient computational methods to predict PPIs are of great value to biologists. Such methods can be divided, according to the protein representation, into sequence-based and structure-based. The former represent proteins using their residue sequence, while the latter model them using their three-dimensional (3D) structure. In such representations, different molecular information are codified. As a consequence, the results of a computation method vary in protein representation and the molecular representation is one of the challenges of molecular active site predictions. Although structure-based methods perform better than sequence-based ones, the latter allow the prediction of PPI sites when the structural information of the interacting proteins is not available. Several structure-based methods take advantage of molecular surface representations for describing protein structure, for instance by using Zernike descriptors or geometric invariant fingerprint (GIF) descriptors to identify possible binding sites [3,5,24]. However, a majority of computational methods employ machine learning algorithms for both groups, including support vector machines, neural networks, Bayesian networks, naive Bayes classifier, and random forests. As a recent development of neural networks, deep learning is a rapidly growing branch of machine learning that has also been used to predict PPI sites with convolutional neural networks [21,25], and Graph Convolutional Networks [6,7]. In these studies, feature selection is a crucial aspect. The commonly used features are evolutionary information, structural information (such as the secondary structure), and physico-chemical and biophysical features (such as the accessible surface area).

This work focuses on predicting protein interfaces starting from their given experimentally determined 3D structures by using a deep learning framework, namely Graph Convolutional Networks (GCNs) [13], for data training. The predictions are carried out on the single proteins without any knowledge of the potential binding partner. To represent the data, we introduce a molecular abstraction according to its hierarchical structure, considering *spatial* and *sequential neighboring* relationships among amino acids: two residues are sequential neighboring if they are consecutive in the sequence, while they are spatial neighboring if they are at a distance less than a fixed threshold. Taking into account these two relationships, we define a *hierarchical representation* considering three different thresholds (4.5 Å, 6 Å, and 10 Å) for the experiments.

In addition to representing the 3D data, another crucial aspect in the deep learning framework is the selection of features, as mentioned earlier. In the proposed framework, we used the accessible surface area and a set of eight high-quality amino acid indices of physico-chemical and biochemical properties extracted from the AAindex1 dataset, a database of numerical indices representing various physico-chemical and biochemical properties of residues and residue pairs [11]. We tested the framework on two classes of proteins, Antibody–Antigen and Antigen–Bound Antibody, extracted from the Protein–Protein Docking Benchmark 5.0 [20]. We are selected these two classes since antibodies are capable of specifically recognizing and binding to several antigens, and such characteristic makes them the most valuable category of biopharmaceuticals for both diagnostic and therapeutic applications.

To evaluate the model performance on the two data sets, we consider only the area under the receiver operating characteristic curve (AU-ROC) since the recognition of PPIs interface sites is an extremely imbalanced classification problem. This can lead to classifiers that tend to label all the samples as belonging to the majority class, thus trivially obtaining a high accuracy measure. The obtained results of AU-ROC on the data set are remarkable. Moreover, to investigate the effect of different representations in the prediction of PPI sites, we applied the framework using hierarchical protein representations, contact mapping, and, finally, only the residue sequence.

The paper is organized as follows. In Sect. 2, we describe different representations of proteins and the input features that we use in the Graph Convolutional Networks model. In Sect. 3, we describe the used dataset and the results obtained with the model, described in Sect. 2.3. The paper ends with some conclusions and future perspective, Sect. 4.

2   Materials and Methods

The prediction of PPI sites can be modeled as a classification of nodes in a graph and it can be formulated as a semi-supervised classification problem. We use a deep learning framework, GCNs, to address the problem, by representing each protein as a graph whose nodes are the amino acids and edges represent *spatial* and *sequential neighboring* relationships among the considered pairs of amino acids. The objective of the classification task is to assign a label, either 1 (interface) or 0 (no-interface), to each residue.

2.1   Representations

Protein representation is one of the challenges of protein prediction and comparison methods. In the literature, the residue sequence is usually represented as a string, while the 3D structure is codified in terms of the Protein Data Bank (PDB) file [2]. To abstract the 3D configuration, some representations have been introduced in the literature, among which lattices and map contacts. A map contact is a two-dimensional matrix that codifies the distance between all possible amino acid pairs of a 3D protein structure. Let i and j be two residues of a given protein, the $m_{i,j}$ element of the matrix is equal to 1 if the two residues are closer than a predetermined threshold, and 0 otherwise. Various contact definitions have been used: the distance between the alpha carbon (C_α) atoms with threshold 6–12 Å, the distance between the beta carbon (C_β) atoms with threshold 6–12 Å (C_α is used for Glycine), and the distance between the side-chain centers of mass. In this work, we consider the distances between C_α atoms.

Taking into account the hierarchical structure of each protein, which is the result of the folding process of the amino acid sequence, we introduce the concepts of *spatial* and *sequential neighboring* relationships among pairs of amino acids. Therefore, a residue pair is sequential neighboring if the two amino acids are consecutive in the sequence, while being spatial neighboring if their distance is less than a fixed threshold. Starting from the PDB file of each protein, we define the *hierarchical representation* matrix A by quantifying the spatial and sequential neighboring relations. Let i and j be two residues of a given protein, the $a_{i,j}$ element of the matrix is 1 if they are sequential residues in the chain or is equal to $1/(1+x)$ if x, the distance between the two C_α atoms, is less than a predetermined threshold, and 0 otherwise. This definition can be summarized as follows:

$$a_{i,j} = \begin{cases} 1 \; if \; i \; and \; j \; are \; sequential \; neighboring \; residues, \\ \frac{1}{1+x} \; if \; i \; and \; j \; are \; spatial \; neighboring \; residues, \\ 0 \; otherwise. \end{cases} \qquad (1)$$

As a consequence, the elements of the hierarchical representation matrix are values between 0 and 1. We consider three thresholds, 4.5 Å, 6 Å, and 10 Å. The proposed hierarchical representation allows us to interpret proteins as weighted undirected graphs with the residues as graph nodes, and A as the corresponding adjacency matrix. Moreover, we observe that such a representation can also be used to represent a protein whose 3D structure is unknown by codifying only the sequential neighboring. Such an aspect is crucial, for example, when a protein is engineered and the whole structure is still unknown.

To study the effects of various molecular representations on the prediction of PPI sites, in this work we consider the hierarchical representation, the contact map and only the residue sequence.

2.2   Input Features

In this work, we consider the accessible surface area for each amino acid and some physico-chemical and biochemical properties of residues published in the AAindex [11]. AAindex is a database that contains numerical indices representing various physico-chemical and biochemical properties of residues and residue pairs published in the literature. Each amino acid index is a set of 20 numerical values representing any of the different physico-chemical and biological properties of each amino acid: the AAindex1 section of the database is a collection of 566 such indices. By using a consensus fuzzy clustering method on all available indices in the AAindex1, Saha *et al.* [18] identified three high-quality subsets (HQIs) of all available indices, namely HQI8, HQI24, and HQI40. In this work, we use the features of the HQI8 amino acid index set, reported in Table 1, that are identified as the medoids (centres) of 8 clusters obtained by using the correlation coefficient between indices as a distance measure.

Associated with such physico-chemical and biochemical features, we introduce the relative solvent accessible surface area of the residue as computed by the DSSP program [19] and a one-hot encoding of the amino acid type (d = 22, 20 for the regular amino acids plus one for ASX (used when impossible to distinguish between Aspartate or Asparagine) and one for GLX (used when impossible to distinguish between Glutamate or Glutamine)).

Table 1. HQI8 indices.

Entry name	Description
BLAM930101	Alpha helix propensity of position 44 in T4 lysozyme
BIOV880101	Information value for accessibility; average fraction 35%
MAXF760101	Normalized frequency of alpha-helix
TSAJ990101	Volumes including the crystallographic waters using the ProtOr
NAKH920108	AA composition of MEM of multi-spanning proteins
CEDJ970104	Composition of amino acids in intracellular proteins (percent)
LIFS790101	Conformational preference for all beta-strands
MIYS990104	Optimized relative partition energies - method C

2.3   Graph Convolutional Networks

Graph Convolutional Networks, proposed by Kipf and Welling [13], are a neural network architecture that operates on graphs.

Let P be a protein represented by a weighted undirected graph $\mathcal{G}(V, E, w)$, where $V = \{v_1, v_2, \ldots v_n\}$ is the set of n nodes/residues, $E = \{e_1, e_2, \ldots, e_m\} \subset V \times V$ is the set of m edges/residue relationships, and $w : E \longrightarrow [0, 1]$ are the edge weights that quantify the relationships between pairs of residues. The graph can be represented by its adjacency matrix A, whose element $a_{i,j} = a_{j,i}$ is $\in (0, 1]$ if there is an edge between v_i and v_j, and 0 otherwise. Let m be the number of input features for each node. For the current graph $\mathcal{G}$ we define the input feature matrix $X \in \mathbb{R}^{n \times m}$, where each row represents the values of the m features of the corresponding node/residue. Finally, let $L \in \{0, 1\}^n$ be the vector of residue labels for the current protein.

The purpose of a GCN is to learn a function of the features on the graph $\mathcal{G}$ to correctly determine the labels on each node. The GCN model takes as input the adjacency matrix A of $\mathcal{G}$ and the input feature matrix X. Every neural network layer can be written as a non-linear function (propagation rule)

$$Z^{(h+1)} = f(Z^{(h)}, A), \quad h = 1, \ldots, H, \tag{2}$$

where $Z^{(0)} = X$ is the input feature matrix. Each layer $Z^{(h)}$ corresponds to a feature matrix where each row is a feature representation of a node. At each layer, these features are aggregated to form the next layer's features using the propagation rule f. The propagation rule used in this framework is

$$f(Z^{(h)}, A) = \sigma \left(\hat{D}^{-\frac{1}{2}} \hat{A} \hat{D}^{-\frac{1}{2}} Z^{(h)} W^{(h)} \right), \tag{3}$$

with $\hat{A} = A + I$, where I is the identity matrix, and $\hat{D}$ is the diagonal node degree of $\hat{A}$. $\hat{D}$ is a diagonal matrix, with element $d_{i,i}$ equal the number of incident edges of node v_i incremented by one. $W^{(h)}$ is the weight matrix for layer h, and σ is a non-linear activation function. In this work, we use the Rectified Linear Unit (ReLU) function as activation function.

The feature aggregation for each node can be calculated in vector form using the following equation:

$$g_{v_i}^{(h+1)} = \sigma \left(\sum_j \frac{1}{c_{ij}} g_{v_j}^{(l)} W^{(l)} \right) \tag{4}$$

where j iterates over the neighboring nodes of v_i, and c_{ij} is a normalization constant obtained from the adjacency matrix to account for the degree difference between v_i and v_j.

3   Experiments

3.1   Dataset

The Protein–Protein Docking Benchmark 5.0 (DB5) was used as a dataset in this work since it represents the standard benchmark dataset for assessing docking and interface prediction methods [20]. The benchmark consists of 230 non-redundant, high-quality structures of protein-protein complexes, and selected subset of structures from the Protein Data Bank (PDB). For each complex, DB5 includes both bound and unbound forms of each protein in the complex. The complexes are divided into 8 different classes: Antibody–Antigen (A), Antigen–Bound Antibody (AB), Enzyme–Inhibitor (EI), Enzyme–Substrate (ES), Enzyme complex with a regulatory or accessory chain, Others, G-protein containing (OG), Others, Receptor containing (OR), and Others, miscellaneous (OX).

In this preliminary study, we used only proteins of classes A and AB. To investigate the effect of different representations for different protein complex classes, we considered the two classes separately. For each class, we separated the receptor proteins from the ligand ones. Moreover, we applied our framework on the bound and unbound versions of the proteins' 3D structures. To easily compare our approach with other methods in the literature, we split the data into training and test sets according to [4]. Differently from this work, we further split the training set into two parts, training set, and validation set, as shown in Table 2. The residues of a given protein were labeled as part of the PPI interface if they had at least one heavy (non-hydrogen) atom within 5 Å from any heavy atom of the other protein (the same threshold used in [4]).

Table 2. The table gives the PDB code and chain ID of each protein used in this study (the PDB code in parentheses identifies the corresponding bound complex in the DB5 database).

Dataset	Training set	Validation set	Test set
A_r	2FAT.HL (2FD6)	1AY1.HL (1BGX)	1FGN.LH (1AHW)
	2I24.N (2I25)	1BVL.BA (1BVK)	1DQQ.CD (1DQJ)
	3EO0.AB (3EO1)	4GXV.HL (4GXU)	1QBL.HL (1WEJ)
	3G6A.LH (3G6D)		1GIG.LH (2VIS)
	3HMW.LH (3HMX)		2VXU.HL (2VXT)
	3L7E.LH (3L5W)		3RVT.CD (3RVW)
	3MXV.LH (3MXW)		
	3V6F.AB (3V6Z)		
A_r	1A43 (1E6J)	1TAQ.A (1BGX)	1HRC (1WEJ)
	1YWH.A (2FD6)	3LZT (1BVK)	2VIU.ACE (2VIS)
	1IK0.A (3G6D)	3F5V.A (3RVW)	1J0S.A (2VXT)
	1F45.AB (3HMX)		1QM1.A (2W9E)
	3M1N.A (3MXW)		1TGJ.AB (3EO1)
	3KXS.F (3V6Z)		3F74.A (3EOA)
	1DOL.A (4DN4)		2FK0.ABCDEF (4FQI)
	4I1B.A (4G6J)		
	1RUZ.HIJKLM (4GXU)		
AB_r	1K4C.AB (1K4C)	1BJ1.HL (1BJ1)	1IQD.AB (1IQD)
	1KXQ.H (1KXQ)	1FSK.BC (1FSK)	1NCA.HL (1NCA)
	2JEL.HL (2JEL)	1I9R.HL (1I9R)	1NSN.HL (1NSN)
	1QFW.HL (9QFW)		1QFW.IM (1QFW)
			2HMI.CD (2HMI)
AB_l	7NN9 (1NCA)	2VPF.GH (1BJ1)	1ALY.ABC (1I9R)
	1HRP.AB (1QFW)	1BV1 (1FSK)	1JVM.ABCD (1K4C)
	1S6P.AB (2HMI)	1D7P.M (1IQD)	1PPI (1KXQ)
	1POH (2JEL)		1KDC (1NSN)

3.2   Implementation and Results

The framework was implemented in TensorFlow 2.0 [9]. To test our framework, we used Stochastic Gradient Descent as optimization algorithm, a learning rate of 0.1, a hidden layer with 24 features and a dropout value of 0.5, similarly to the values proposed in the literature. The code used in this manuscript are available from the corresponding author upon reasonable request.

Any given set of proteins can be interpreted as a single graph composed by the union of the graphs representing each protein. This is achieved by concatenating the respective feature matrices and by building a sparse block-diagonal adjacency matrix where each block corresponds to the adjacency matrix of a given protein in the set (Fig. 1).

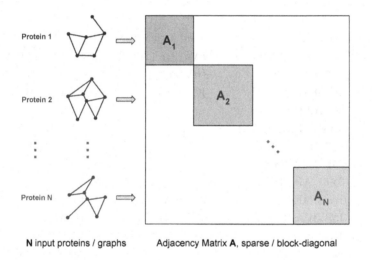

Fig. 1. Pooled adjacency graph construction for a given set of proteins.

The composite adjacency and feature matrices, can be logically split into training, validation and testing sets using the corresponding row indices as boundaries. Finally, by providing only the labels of the nodes belonging to the training set as an input to the GCN model, we can classify the ones in the validation and test sets in a semi-supervised fashion (Fig. 2).

The network was trained until either the performance on the validation set stopped improving or the maximum number of epochs was reached (2000). Such a limit is determined by empirical observation during the experimental face. We applied the framework on the previously described representations, i.e., the contact map, the residue sequence, and the hierarchical representation, considering distance thresholds of 4.5Å, 6 Å and 10 Å. The performance results, evaluated in terms of AU-ROC, for the proposed methodology on the test set are presented in Tables 3 and 4.

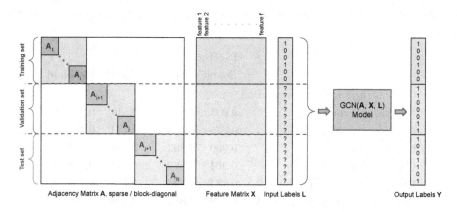

Fig. 2. PPI interface residue classification with a semi-supervised GCN framework.

From the analysis of these tables, we observe that the best classification performance in terms of AU-ROC is obtained with the hierarchical sequence map representation for both considered classes of molecules (both bound and unbound) with a threshold equal to 10 Å, except for ligands of A class. However, the table also shows that the best performance is obtained for the receptors, while the results are comparable for bound and unbound classes. The experiments were trained using 32 parallel threads on a HPC Server with eight 12-Core Intel Xeon Gold 5118 CPUs @2.30 GHz and 1.5 TB RAM running Fedora Linux 25.

Table 3. Classification results (AU-ROC) on the test set for the proteins of class A of DB5.

	Receptors		Ligands	
	Bound	Unbound	Bound	Unbound
GCN method with representation				
Contact map 4.5 Å	0.957	0.959	0.743	0.755
Contact map 6 Å	0.961	0.959	**0.757**	**0.775**
Contact map 10 Å	0.953	0.952	0.737	0.760
Hierarchical representation 4.5 Å	0.938	0.949	0.704	0.725
Hierarchical representation 6 Å	0.951	0.954	0.701	0.738
Hierarchical representation 10 Å	**0.963**	**0.962**	0.729	0.755
Other methods				
Daberdaku *et al.*	0.954	0.942	0.589	0.595
SPPIDER	0.773	0.754	0.630	0.575
NPS-HomPPI	0.796	0.780	0.610	0.626
PrISE	0.770	0.758	0.622	0.569

Table 4. Classification results (AU-ROC) on the test set for the proteins of class AB of DB5.

	Receptors		Ligands	
	Bound	Unbound	Bound	Unbound
GCN method with representation				
Residue sequence	0.694	0.684	0.540	0.559
Contact map 4.5 Å	0.875	0.875	0.720	0.739
Contact map 6 Å	0.900	0.892	0.741	0.798
Contact map 10 Å	0.904	0.899	0.711	0.778
Hierarchical representation 4.5 Å	0.835	0.823	0.718	0.703
Hierarchical representation 6 Å	0.864	0.853	0.739	0.739
Hierarchical representation 10 Å	**0.905**	**0.903**	**0.749**	**0.800**
Other methods				
Daberdaku et al.	0.813	0.840	0.599	0.729
SPPIDER	0.757	0.783	0.573	0.556
NPS-HomPPI	0.701	0.698	0.675	0.713
PrISE	0.776	0.789	0.683	0.649

Thanks to the appropriate division of molecules in testing, training and validation, we can compare our results with the ones obtained in [4]. The proposed methodology was also compared with two state-of-the-art homology-based PPI interface prediction algorithms: NPS-HomPPI [23] and PrISE [10], and with the well-known structure-based approach SPPIDER [15,16]. We observe that, when considering the hierarchical representation, the proposed methodology outperforms the competitor predictors in both the bound and unbound versions of the considered protein classes.

4   Conclusions and Future Work

In this work, we have focused on the prediction of protein interfaces starting from their given experimentally determined 3D structures by using a deep learning framework – Graph Convolutional Networks – for data training. We proposed a molecular abstraction according to the hierarchical structure of proteins, introducing the concepts of spatial and sequential neighboring relationships between amino acid pairs. Such a representation can also be used to represent a protein whose 3D structure is unknown, by codifying only the sequential neighboring. Such an aspect is crucial, for example, when a protein is engineered and the whole structure is still unknown. To test our approach, we have applied the framework using the hierarchical structure representation with different thresholds (4.5 Å, 6 Å and 10 Å) for the distances between C_α atoms. Moreover, we applied the framework using other representations, namely the contact map and the residue sequence, to investigate the effects of these representations on the PPI site prediction for different protein complex classes. The proposed method outperformed several competitor approaches when representing the proteins both as contact map and as hierarchical sequence map.

As a future work, we plan to apply our framework on the all classes of DB5 using various optimization algorithms, with various learning rates, and dropout values. Moreover, we intend to investigate how increasing the number of hidden layers will affect the classification performance. Feature selection is another important aspect to investigate since it represents a crucial step in any deep learning framework. Our approach achieves better classification results for receptors than ligands, so we plan to investigate different sets of features for the various protein classes. We will focus on structural features, such as the protein secondary structure, by further exploring the RNA-based methodology introduced in [17]. Although binding partner specificity was not explored in this work, we plan to address this issue in our future research.

Funding. This research has been partially supported by the University of Padua project BIRD189710/18 "Reliable identification of the PPI interface in multiunit protein complexes".

References

1. Berggård, T., Linse, S., James, P.: Methods for the detection and analysis of protein-protein interactions. Proteomics **7**(16), 2833–2842 (2007)
2. Berman, H.M., et al.: The protein data bank. Nucleic Acids Res. **28**(1), 235–242 (2000). https://doi.org/10.1093/nar/28.1.235
3. Daberdaku, S.: Structure-based antibody paratope prediction with 3D Zernike descriptors and SVM. In: Raposo, M., Ribeiro, P., Sério, S., Staiano, A., Ciaramella, A. (eds.) CIBB 2018. LNCS, vol. 11925, pp. 27–49. Springer, Cham (2020). https://doi.org/10.1007/978-3-030-34585-3_4
4. Daberdaku, S., Ferrari, C.: Exploring the potential of 3D Zernike descriptors and SVM for protein-protein interface prediction. BMC Bioinform. **19**(1), 35 (2018)
5. Daberdaku, S., Ferrari, C.: Antibody interface prediction with 3D Zernike descriptors and SVM. Bioinformatics **35**(11), 1870–1876 (2019)
6. Eyuboglu, E.S., Freeman, P.B.: Disease protein prediction with graph convolutional networks. Genetics **5**, 101–113 (2004)
7. Fout, A., Byrd, J., Shariat, B., Ben-Hur, A.: Protein interface prediction using graph convolutional networks. In: Advances in Neural Information Processing Systems, pp. 6530–6539 (2017)
8. Fry, D.C.: Protein-protein interactions as targets for small molecule drug discovery. Peptide Sci.: Orig. Res. Biomol. **84**(6), 535–552 (2006)
9. Girija, S.S.: TensorFlow: large-scale machine learning on heterogeneous distributed systems. Software available from tensorflow.org, vol. 39 (2016)
10. Jordan, R.A., Yasser, E.M., Dobbs, D., Honavar, V.: Predicting protein-protein interface residues using local surface structural similarity. BMC Bioinform. **13**(1), 41 (2012)

11. Kawashima, S., Pokarowski, P., Pokarowska, M., Kolinski, A., Katayama, T., Kanehisa, M.: AAindex: amino acid index database, progress report 2008. Nucleic Acids Res. **36**(suppl–1), D202–D205 (2007)

12. Keskin, O., Tuncbag, N., Gursoy, A.: Predicting protein-protein interactions from the molecular to the proteome level. Chem. Rev. **116**(8), 4884–4909 (2016)

13. Kipf, T.N., Welling, M.: Semi-supervised classification with graph convolutional networks. In: International Conference on Learning Representations (ICLR) (2017)

14. Liyasova, M.S., Ma, K., Lipkowitz, S.: Molecular pathways: CBL proteins in tumorigenesis and antitumor immunity-opportunities for cancer treatment. Clin. Cancer Res. **21**(8), 1789–1794 (2015)

15. Porollo, A., Meller, J.: Prediction-based fingerprints of protein-protein interactions. Proteins: Struct. Funct. Bioinform. **66**(3), 630–645 (2007)

16. Porollo, A., Meller, J., Cai, W., Hong, H.: Computational methods for prediction of protein-protein interaction sites. Protein-Protein Interact.-Comput. Exp. Tools **472**, 3–26 (2012)

17. Quadrini., M., Merelli., E., Piergallini., R.: Loop grammars to identify RNA structural patterns. In: Proceedings of the 12th International Joint Conference on Biomedical Engineering Systems and Technologies, Bioinformatics, vol. 3, pp. 302–309. SciTePress (2019)

18. Saha, I., Maulik, U., Bandyopadhyay, S., Plewczynski, D.: Fuzzy clustering of physicochemical and biochemical properties of amino acids. Amino Acids **43**(2), 583–594 (2012)

19. Touw, W.G., et al.: A series of PDB-related databanks for everyday needs. Nucleic Acids Res. **43**(D1), D364–D368 (2015)

20. Vreven, T., et al.: Updates to the integrated protein-protein interaction benchmarks: docking benchmark version 5 and affinity benchmark version 2. J. Mol. Biol. **427**(19), 3031–3041 (2015)

21. Xie, Z., Deng, X., Shu, K.: Prediction of protein-protein interaction sites using convolutional neural network and improved data sets. Int. J. Mol. Sci. **21**(2), 467 (2020)

22. Xu, W., et al.: Amyloid precursor protein-mediated endocytic pathway disruption induces axonal dysfunction and neurodegeneration. J. Clin. Investig. **126**(5), 1815–1833 (2016)

23. Xue, L.C., Dobbs, D., Honavar, V.: HomPPI: a class of sequence homology based protein-protein interface prediction methods. BMC Bioinform. **12**(1), 244 (2011)

24. Yin, S., Proctor, E.A., Lugovskoy, A.A., Dokholyan, N.V.: Fast screening of protein surfaces using geometric invariant fingerprints. Proc. Natl. Acad. Sci. **106**(39), 16622–16626 (2009)

25. Zeng, M., Zhang, F., Wu, F.X., Li, Y., Wang, J., Li, M.: Protein-protein interaction site prediction through combining local and global features with deep neural networks. Bioinformatics **36**(4), 1114–1120 (2020)

Brain-Inspired Spike Timing Model of Dynamic Visual Information Perception and Decision Making with STDP and Reinforcement Learning

Petia Koprinkova-Hristova[1]([✉])[iD] and Nadejda Bocheva[2][iD]

[1] Institute of Information and Communication Technologies,
Bulgarian Academy of Sciences, Sofia, Bulgaria
pkoprinkova@bas.bg
[2] Institute of Neurobiology, Bulgarian Academy of Sciences, Sofia, Bulgaria
nadya@percept.bas.bg

Abstract. The paper presents a brain-inspired spike timing neural network model of dynamic visual information processing and decision making implemented in the NEST simulator. It consists of multiple layers with functionality corresponding to the main visual information processing structures up to the areas responsible for decision making based on accumulated sensory evidence as well as the basal ganglia that modulate its response due to the feedback from the environment. The model has rich feedforward and feedback connectivity based on the knowledge about involved brain structures and their connections. The introduced spike timing-dependent plasticity and dopamine-dependent synapses allowed for its adaptation to external reinforcement signal. Simulations with specific visual stimuli and external reinforcement signal demonstrated that our model is able to change its decision via the considered as biologically plausible reinforcement learning.

Keywords: Reinforcement learning · Brain-inspired modeling · Visual perception · Decision making

1 Introduction

The pioneering work of Pavlov on conditional and unconditional reflexes of living creatures gave rise to the development of the theory of reinforcement learning. The seminal work of [2] and their actor-critic architecture that learns from a punish/reward feedback from the environment is probably the first simplified model of reinforcement learning in the brain. Parallel to the research based on behavioral experiments neurologists tried to discover the brain counterparts of sensory information processing and reinforcement learning.

This work is supported by the Bulgarian Science Fund project No DN02/3/2016 "Modelling of voluntary saccadic eye movements during decision making".

© Springer Nature Switzerland AG 2020
G. Nicosia et al. (Eds.): LOD 2020, LNCS 12566, pp. 421–435, 2020.
https://doi.org/10.1007/978-3-030-64580-9_35

The systematic investigations of the visual system of the human brain accumulated rich knowledge about its hierarchical processing structures starting from the eyes (light sensors) through the optic nerve to the visual cortex. It has been shown that brain structures like lateral intraparietal (LIP) area accumulate evidence based on the sensory information in support of various alternative decisions [29]. Many models assume that a choice of action is made when the accumulated evidence for it reaches a predefined threshold level modulated by another structure in the brain - the basal ganglia (e.g. [13]). Other studies (e.g. [6]) imply that the basal ganglia could also modify the rate of sensory evidence accumulation. Existing evidence (e.g. [1,9,15]) suggests also a significant role of the basal ganglia on learning by trial and error to acquire a reward, i.e. reinforcement learning. The role of the basal ganglia in reinforcement learning is related to the differential responses of the dopaminergic neurons in one structure of the basal ganglia (substantia nigra compacta) to unexpected and predicted rewards and to the omission of an expected reward. Recently, several modeling attempts (e.g. [4,7]) try to integrate these two functions of the basal ganglia – in decision-making processes and reinforcement learning in a common framework starting from the cortical input.

In contrast to these models, we developed an integrated spike timing neural network model that includes all major structures related to dynamic visual information processing (the sensory information) starting from the retinal input and up to decision-making areas [19]. The parallel structure of the model was adopted from [14] while the basal ganglia connectivity was adopted from [21]. While in [21] the model consisted of cellular network structures whose neurons are modeled by firing rate equations, in our model we have spike timing neurons organized in layers with similar connectivity. In [21] dopamine signal is calculated as temporal difference error that was directly exploited to adjust dopamine-dependent synapses. In our model, we have an additional layer of neurons (SNc) whose spiking activity is considered as equivalent to dopamine release. In [21] the layer responsible for visual information processing and generating sensory input to the basal ganglia is simplified and in [14] it is completely missing while our model includes multiple layers corresponding to hierarchical brain structures performing dynamic visual information processing. In contrast to [14] where lateral connections within layers are limited, our model has much more elaborated connectivity similar to that proposed in [21]. In [18] we also enhanced the feedback connectivity between sensory information processing layers and introduced spike timing-dependent plasticity (STDP) in some synapses to allow for their adaptation by external teaching signal. These characteristics of our model make it more realistic and provide greater opportunities for understanding the process of learning and decision making in the human brain. The model was implemented in NEST 2.12.0 simulator [22].

The present paper summarizes the model structure and connectivity and presents simulation results with specific visual stimuli and external reinforcement signal demonstrating its ability to change generated decision accordingly. We also investigated changes in STDP connections weights depending on the training signal whose parameters were determined by reaction times of humans from three age groups obtained in behavior experiments with similar stimuli as an attempt to simulate age-related changes in the brain.

The paper is organized as follows: Sect. 2 briefly describes the model structure; next a simulation experiment with moving dot stimulus and external reinforcement signal was presented; the paper finishes with results discussion and conclusions.

2   Model Structure

Based on the available data about human brain structures playing role in visual motion information processing and decision making, as well as their connectivity, the proposed in [19] hierarchical model consists of two basic substructures: related to visual information perception and sensory-based decision making and the basal ganglia that modulates the perceptual decision via external reinforcement. Each layer consists of neurons positioned in a regular two-dimensional grid. The receptive field (area of neurons from a given layer that is connected to a given neuron from the same or neighbor layer) of each neuron depends on the function of the layer it belongs to as well as on its position inside the layer. The neurons' dynamical models as well as intra- and inter-layer connectivity are described in detail in our previous works [17,19]. Here we briefly remind its structure and explain the novel feedback connections included.

The structure of perceptual layers up to the LIP area involved in the sensory-based decision making reported in [16,17,24] is shown on Fig. 1. Like in [20] we have two layers of retinal cells and their corresponding LGN neurons, having identical positions of "on-center off-surround" (ON) and "off-center on-surround" (OFF) cells placed in reverse order. The structure of the thalamic relay was adopted from [11] as shown in the lower right corner of Fig. 1. The structure of the next (V1) layer is as in [20]. In consists of four groups of neurons – two exciting and two inhibiting populations connected via corresponding excitatory and inhibitory intra-layer (lateral) connections as shown in the upper right inset of Fig. 1. The 2D maps containing the neurons' orientations and phases at the corresponding 2D grid of the V1 layer have typical for the mammalian brain "pinwheel-structure" generated by the approach from [28] (details are reported in [24]). These neurons are orientation sensitive and have elongated receptive fields defined by a Gabor probability function with orientation and phase parameters like in [12]. The next (MT) layer has a structure identical to the V1 layer. The lateral connections are designed in the same way while the connections from V1 cells depend on the angle between orientation preferences of each pair of cells according to [8]. The following Medial Superior Temporal Area (MST) was modeled like in [23] by two layers sensitive to expansion and contraction movement patterns (that occur during the self-motion of the observer) having a circular shape of on-center Gaussian receptive fields. Both layers have intra- and interlayer excitatory/inhibitory recurrent connections between cells having similar/different sensitivity (see left inset of Fig. 1 and more details in [17]). LIP area is the last layer of the perceptual part of the model that is responsible for making decisions based on accumulated sensory evidence. Since our model aims to decide whether the expansion center of moving dot stimulus is left or right from the stimulus center, in [17] we proposed a task-dependent design of excitatory/inhibitory connections as shown in Fig. 1. As in [18] the model has a complete set of feedback/forward connections between every two consecutive layers. The synapses denoted by dashed lines have spike-timing dependent plasticity to allow for their adaptation in dependence on a training signal.

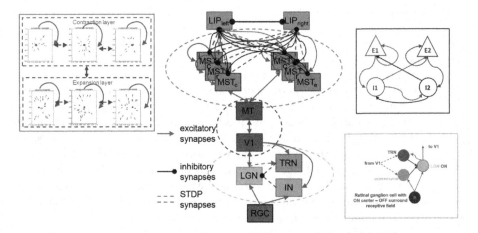

Fig. 1. Model of dynamic sensory information processing. Each box represents a two-dimensional layer of neurons. The color of connections corresponds to their type, i.e. red for excitatory and blue for inhibitory ones. Solid lines denote constant synapses while the dashed ones - those with spike-timing dependent plasticity (STDP). Rectangles in yellow, blue, and green represent detailed structures of the layers within corresponding dashed ovals. (Color figure online)

The Basal ganglia (BG) model is shown in Fig. 2. Like in [21], our model incorporates layers of Striatum, Globus Pallidus externa (GPe), Subthalamic Nucleus (STN), and Substantia Nigra pars reticulata (SNr). However, it consists of two parallel structures, receiving inputs from the left and right decision selecting LIP areas respectively. These two channels (left and right) are connected via mutually inhibiting connections through their GPe areas like in [14]. Additionally, in contrast to [21], our model has a complete 2D layer Substantia Nigra pars compacta (SNc) producing dopamine neuromodulator and dopamine-dependent synapses. The Striatum is divided into two sub-areas depending on the type of dopamine receptors they express (D1 and D2 in Fig. 2). Both are modeled as 2D layers of neurons whose lateral connections have short-range excitation and long-range inhibition characteristics like in [21]. The GPe/STN pairs consist of a 2D grid of pairs of neurons connected one-to-one via excitatory and inhibitory connections as shown in Fig. 2. The GPe layer has also lateral connections having a negative center and positive surround shape as in [21]. The structure receives inhibitory input from the second part of the Striatum (D2) via GPe and sends its output through STN via dopamine-dependent synapses to SNr (so-called indirect pathway from Striatum to the BG output layer). SNr was modelled by a 2D layer having short-range excitatory and long-range inhibitory lateral connections like both Striatum layers. Its input comes from both the D1 layer of the Striatum (direct pathway) and GPe/STN structure (indirect pathway) via dopamine-dependent synapses. SNr generates BG output to the motor-reaction controlling structure Superior colliculus (SC). Finally, the motor controlling structure SC was modeled by a 2D layer of neurons receiving inputs directly from the LIP area (decision according to accumulated sensory information) as well as from the external reinforcement modulated output of BG (via SNr). The overall model connectivity is also enhanced by excitatory feedback connections from SC to their corresponding D1 and D2 areas of the Stratum as well as to LIP areas following recently reported findings [25, 33, 34]. Moreover we introduced mutually inhibiting connections between the two SC groups. SNc is considered as the brain area producing the neuromodulator dopamine in dependence on external motivation (reinforcement) input signal. The input to SNc, coming from the D1 area of the Striatum, was considered as the value function estimation like in [21]. Thus to produce the dopamine proportional to the temporal difference error $\delta(t)$ at the output of SNc, we set its inputs to be the value function for two consecutive time steps and the reinforcement signal as follows:

$$\delta(t) = SNc = F(r(t) + \gamma D1(t+1) - D1(t)) \tag{1}$$

The discount factor γ was set to 0.9. The reinforcement signals on both left and right parts of the model r_{left} and r_{right} respectively are generated as the difference between SC left and right reactions and the external (teaching) input currents (I_{left} and I_{right} respectively). They affect both dopamine-producing areas (SNc left and right).

Following the commonly accepted models from [30], the reaction of retinal ganglion cells to luminosity changes was simulated by a spatio-tempoal filter whose spatial component has circular shape modeled by a difference of two Gaussians and a temporal component with bi-phasic profile determined by the difference of two Gamma functions. It generates a continuous signal by convolution of the spatio-temporal kernel with the visual stimuli (images falling on the retina).

The LGN neurons were simulated by the following model equations whose parameters were determined from in-vivo experiments as in [5]:

$$C\frac{dV}{dt} = -G_L(V - V_L) - G_E(V - V_E) - G_I(V - V_I) - G_A(V - V_A) \tag{2}$$

$$G_X = \sum_j g_X(t - t_j)H(t - t_j) \tag{3}$$

$$g_X(t) = \bar{g_X}\frac{t}{\tau_X}\exp-\frac{t - \tau_X}{\tau_X} \tag{4}$$

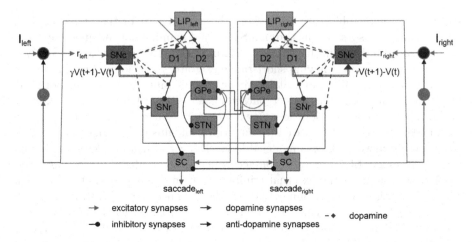

Fig. 2. Basal ganglia structure (in blue). It receives inputs from the decision-making area based on sensory information (LIP) as well as from the dopamine-releasing area (SNc) and generates activity biasing saccades generation via SC.

Here C is capacity of the neuron cell membrane, V is membrane potential, G_L is leakage conductance, G_E and G_I are the conductances of total excitatory and inhibitory synaptic inputs respectively, G_A is the conductance of the potassium-mediated afterhyperpolarization (AHP) channel and V_L, V_E, V_I and V_A are the corresponding reversal potentials. The time-dependent conductances (denoted here by G_X where X stands for E, I, and A respectively) are described by Eqs. (3) and (4), where t_j denotes the time of the event (release of a neurotransmitter into the corresponding synapse), τ_X is the duration of the event, H is Heviside step function and the alpha function g_X is defined by Eq. (4) to achieve its maximum $\bar{g}_X$ at the moment $t_j = \tau_X$.

The rest of neurons in our model were simulated according to the following equations proposed in [31]:

$$C\frac{dV}{dt} = -G_{rest}\left(V\left(t\right) - V_{rest}\right) + I_{syn} \tag{5}$$

$$I_{syn}\left(i\right) = \sum_j A_{ij}y_{ij}\left(t\right) \tag{6}$$

Here C is the capacity of the neuron cell membrane, V is membrane potential, G_{rest} is the membrane conductance at resting state V_{rest} and I_{syn} is the synaptic current that is modeled as the sum of postsynaptic currents from all neurons j connected to a given neuron i according to Eq. (6). The parameter A_{ij} determines the absolute strength of the synaptic connection. The factor y_{ij} describes the contribution to a synaptic current of neuron i of the postsynaptic currents from neurons j that is determined by a dynamic system of equations from [31].

We have two types of dynamic synapses: with spike timing-dependent plasticity (STDP) as in [27] and with dopamine-dependent plasticity as in [26]. Since our model includes also anti-dopamine synapses (from LIP to D2 sub-area of the Striatum) whose dynamics has to be opposite to that of dopaminergic ones, we've modified the model from [26] by converting the amplitudes A_+ and A_- of the dopamine eligibility trace dynamics from positive to negative.

Details about the parameters of all neuron models and dynamic synapses can be found in the references mentioned above as well as in NEST simulator documentation.

3   Simulation Results and Discussion

The model was tested with expansion dot patterns used in a behavioral experiment [17] with a focus of expansion to the left. These stimuli resemble the optic flow occurring during the self-motion of an observer in direction straight ahead with gaze shifted to the left. The adjustable parameters in the presented simulation are the strengths of the dopamine-dependent synapses that vary depending on spiking activity of both SNc layers as well as STDP synapses of the visual perception sub-structure (delineated by dashed lines in Fig. 1). Based on the estimated mean reaction times of the three age groups from [3] we've created training generating currents I_{left} and I_{right} for the left and right SNc neurons respectively as follows:

$$I_{left/right} = A_{left/right}/(1 + \exp(k_{left/right}t)) \qquad (7)$$

Amplitudes $A_{left/right}$ define the maximal input currents (in pA) while the value of parameter $k_{left/right}$ determines settling time of the exponent that corresponds to the mean reaction time determined from experiments for each age group and experimental conditions. The sign of $k_{left/right}$ determines whether the exponent will increase (promoted reaction) or decrease (suppressed reaction) in time. To evaluate the ability of the model to represent the modulatory effect of the basal ganglia on the motor response selection, we compared the activity in the brain structures involved in the decision making and motor response preparation without reinforcement signal (Fig. 3) to the case when the reinforcement signal corresponds to the perceptual decision (Figs. 4, 5 and 6) and when the reinforcement signal favors a decision opposite to the perceptual one (Figs. 7, 8 and 9). The figures represent the change in the activity of the brain structures for three age groups: young (20–34 years of age), middle-aged (36–52 years of age) and elderly (aged 57–84). For brevity, in the sequel, we will label the structures related to motor selection and preparation to the left and the right as left and right structures. The results in Fig. 3 imply that without a reinforcement signal the motor response selected by the model corresponds to the perceptual decision in LIP, while the combined activity of the basal ganglia and LIP leads to the suppression of the motor response generation to the opposite (right) direction. The results imply that the temporal evolution of the activity in the structures responsive for an expansion center to the left and the right complement each other, i.e. an increase in activity in the left structures leads to a decrease in the corresponding right structures and vice versa. The only exception of this is for the GPe and STN where both of these structures are active together.

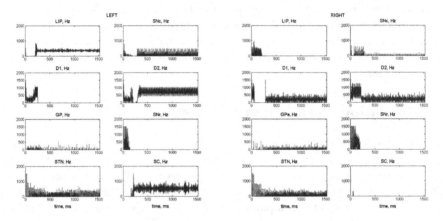

Fig. 3. LIP, BG, and SC reactions without external reinforcement - based only on the perceptual decision.

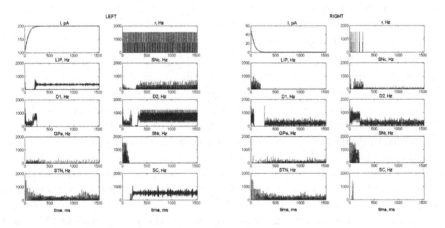

Fig. 4. LIP, BG, and SC reactions to reinforcement signal corresponding to the perceptual decision for the group of young test subjects.

Similarly, when a reinforcement signal confirming the perceptual decision is applied, the activity in SC related to the left response is facilitated, while the activity in the SC related to the right response is suppressed. Also, as in the case without a reinforcement signal, the activity in the left and right structures complement each other i.e. their activity changes in opposite directions. There are, however, several changes in the spike trains generated in the different brain

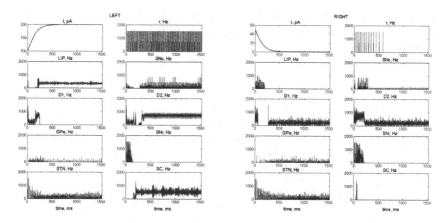

Fig. 5. LIP, BG, and SC reactions to reinforcement signal corresponding to the perceptual decision for the group of middle age test subjects.

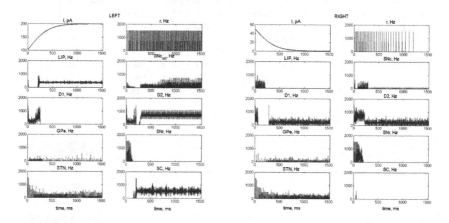

Fig. 6. LIP, BG, and SC reactions to reinforcement signal corresponding to the perceptual decision for the group of elderly test subjects.

structures like a tendency for an increase in the spike amplitudes with time in SNc, a trend for the higher amplitude of oscillations in D2 for the left response and a higher initial amplitude in D2 for the right response.

When the reinforcement signal is opposite to the perceptual decision, there are qualitative changes in the activity in both left and right structures leading to a reverse reaction in SC - the right one generates a response, while the left one is suppressed. This reversal is achieved by a similar activity in D1 and D2 in the left structures and increased activity in the right D2. These changes modulate the responses in GPe and STN, as well as alternate the activity in SNc.

The comparison of the data in Tables 1 and 2 shows significant differences in the mean spiking frequencies between the case without a reinforcement signal and the case when the reinforcement signal differs from the perceptual decision

Table 1. Mean and variance of spiking frequencies in the left and right channels of the model in case of reinforcement signal corresponding to the perceptual decision (left) in comparison with zero reinforcement.

Channel:	Left							
	Mean				Variance E^4			
Layer	Young	Middle	Elderly	Zero r	Young	Middle	Elderly	Zero r
LIP	314.9453	315.1627	313.7911	313.6112	1.828	2.002	2.019	2.023
SNc	81.1466	80.9068	80.9467	89.1813	2.002	2.308	2.631	1.373
D1	63.4782	64.7574	64.8373	64.6774	3.567	3.682	4.112	3.816
D2	606.7208	602.9633	622.9102	616.9541	14.860	7.357	8.517	8.303
GPe	14.1107	14.1907	13.8309	14.1507	0.360	0.375	0.374	0.385
SNr	24.1441	24.0642	23.7444	23.9442	2.616	2.611	2.581	2.569
STN	103.8517	103.7318	103.8517	103.9316	1.999	1.920	2.129	2.293
SC	461.1666	461.0867	459.2579	459.0880	4.536	5.791	6.028	5.825
Channel:	Right							
	Mean				Variance E^4			
Layer	Young	Middle	Elderly	zero r	Young	Middle	Elderly	zero r
LIP	21.4259	21.3335	21.0861	21.1386	1.070	1.071	1.012	0.976
SNc	22.1854	22.1454	22.0655	23.1847	0.716	0.866	0.601	0.655
D1	191.6339	191.6739	190.9943	190.2748	3.557	3.581	3.593	3.488
D2	270.6220	270.7019	271.2915	272.8605	5.640	5.615	6.011	6.302
GPe	28.3414	28.3414	28.7011	28.2614	0.592	0.611	0.598	0.563
SNr	41.2129	41.6526	40.5333	41.4927	3.882	3.863	3.669	3.711
STN	102.5625	102.6525	102.5725	102.7324	2.229	2.353	2.192	2.241
SC	3.0280	2.9880	1.6289	1.4291	0.388	0.383	0.185	0.166

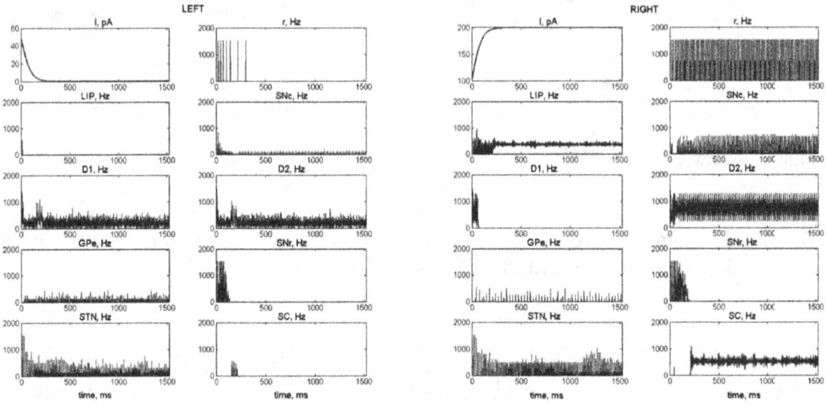

Fig. 7. LIP, BG, and SC reactions to reinforcement signal opposite to the perceptual decision for the group of young test subjects.

in both the left and right structures and negligible differences in these parameters when the reinforcement signal and the perceptual decision coincide. The thorough examination of the activity in the modeled brain structures related to the perceptual decision and motor response selection shows differences between the three age groups. Changes occur predominantly in the variance of the spiking

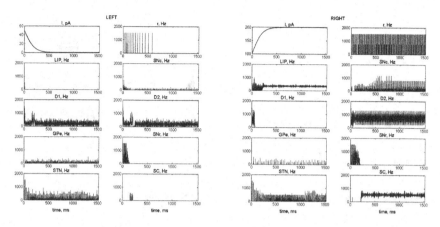

Fig. 8. LIP, BG, and SC reactions to reinforcement signal opposite to the perceptual decision for the group of middle age test subjects.

Table 2. Mean and variance of spiking frequencies in the left and right channels of the model in case of reinforcement signal opposite to the perceptual decision (left) in comparison with zero reinforcement.

Channel:	Left							
	Mean				VarianceE^4			
Layer	Young	Middle	Elderly	zero r	Young	Middle	Elderly	zero r
LIP	1.6514	1.7114	1.6439	313.6112	0.182	0.183	0.182	2.023
SNc	11.2726	11.1527	11.2726	89.1813	0.299	0.285	0.271	1.373
D1	218.2164	220.4150	216.9373	64.6774	2.741	2.754	2.487	3.816
D2	215.3783	210.1417	215.4583	616.9541	2.469	2.656	2.534	8.303
GPe	31.0196	31.0196	30.8997	14.1507	0.566	0.563	0.564	0.385
SNr	23.5045	23.5045	23.5045	23.9442	2.527	2.515	2.526	2.569
STN	102.6525	102.6525	102.8124	103.9316	2.3085	2.622	2.434	2.293
SC	4.9967	5.3964	4.5070	459.0880	0.307	0.319	0.287	5.825
Channel:	Right							
	Mean				Variance E^4			
Layer	Young	Middle	Elderly	zero r	Young	Middle	Elderly	zero r
LIP	331.9116	331.3295	332.7711	21.1386	1.366	1.346	1.316	0.976
SNc	91.4598	91.5398	91.5398	23.1847	3.791	3.870	6.133	0.655
D1	24.3040	24.3040	24.3040	190.2748	2.146	2.146	2.146	3.488
D2	717.1282	717.1282	717.1282	272.8605	10.623	10.623	10.623	6.302
GPe	12.4718	12.1120	12.1920	28.2614	0.446	0.446	0.439	0.563
SNr	40.4934	40.6932	40.3734	41.4927	3.634	3.689	3.705	3.711
STN	103.8917	103.8917	103.8917	102.7324	3.781	3.604	3.721	2.241
SC	453.9713	452.4823	456.1799	1.4291	4.312	4.203	4.024	0.166

frequencies in the different structures and to a lesser degree in the mean spiking frequencies (Tables 1 and 2).

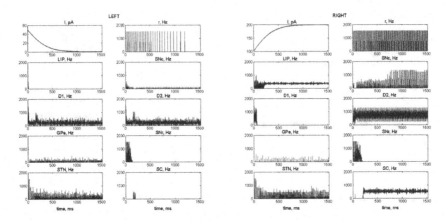

Fig. 9. LIP, BG, and SC reactions to reinforcement signal opposite to the perceptual decision for the group of elderly test subjects.

Figures 4, 5, 6, 7, 8 and 9 show also changes in the temporal evolution of the spike trains depending on the age of the participants and the type of the reinforcement signal. For example, when the perceptual decision and reinforcement signal coincide, in most of the modeled structures the variability in the spiking frequencies is highest in the left and lowest – in the right structures for the elderly. These differences and the interplay in the activity in the left and right brain structures can explain why in the output of the model the activity in the left SC shows the highest variability for the elderly. When the perceptual decision and the reinforcement signal differ, the motor response selected in the right SC has lower amplitude in the elderly that might be related to the lower suppression in the left structures due to the lesser variability in the activity. Changes in the evolution of the spike trains with age also occur in different brain structures. The most apparent change is for the activity in D2 for the left response. The amplitude of the oscillations in the spike frequencies is highest for the young and lowest – for the middle group, but the frequency of these oscillations is lowest for the elderly. These differences are reflected in the highest variance of the spike frequencies for the young group and the lowest – for the middle group, while the mean values for the elderly exceed that of the younger group. The temporal evolution of spiking frequencies in SNc also differs between the age groups showing an initial increase in their amplitude with dissimilar progression and duration. The rise is most apparent and with the longest length for the elderly. Hence, the simulation results show that age affects mostly the activity of D2 receptors and SNc in left and right structures leading to a change in the mean and variance of the responses in SC that depend on the type of the reinforcement signal.

In summary, for all age groups the response in SC is determined by the reinforcement signal – to the left when the reinforcement signal tolerates the response to the left and coincides with the perceptual decision and to the right when the reinforcement signal is opposite to the perceptual decision and tolerates the right response. However, the activity in the different brain structures related to the perceptual decision and motor response selection varies differently for the different age groups. Our model shows a behavior dependent on the correspondence between the perceptual decision and reinforcement signal by a balance between the activity in the left and the right structures involved in the processes of decision making and motor response selection and in this way it can inhibit the unwanted motor response.

4 Conclusions

The model, presented here incorporates all basic structures in the human brain responsible for decision making based on dynamic visual information in tasks with eye movement response starting from the visual information encoding, pre-processing, information extraction and accumulation and saccade generation biased by subcortical structures (BG) in the presence of external reinforcement. It can change its behavior depending on the reinforcement signal and the induced age differences.

The adjustment of the model parameters in the dynamic (dopamine and STDP) synapses by feeding reinforcement signal reflecting specific characteristics of the human performance provide further insight into the complicate interactions between different brain structures and their modification in the process of learning and acting.

Our future aim is to perform in-sillico modeling of brain lesions or other degenerative brain processes by simulations of the model with induced deterioration in any of its layers. The obtained results could support early and noninvasive diagnosis of some deceases of the human brain whose first sign might be violations from normal performance in visual tasks.

References

1. Barto, A.G.: Adaptive Critics and the Basal Ganglia. in Models of Information Processing in the Basal Ganglia. In: Houk, J.C., Davis, J.L., Beiser, DG., (eds.) MIT Press, Cambridge, pp. 215–232 (1995)
2. Barto, A.G., Sutton, R.S., Anderson, C.W.: Neuronlike adaptive elements that can solve difficult learning control problems. IEEE Trans. Syst. Man Cybern. B Cybern. **13**(5), 834–846 (1983)
3. Bocheva, N., Genova, B., Stefanova, M.: Drift diffusion modeling of response time in heading estimation based on motion and form cues. Int. J. Biol. Biomed. Eng. **12**, 75–83 (2018)
4. Bogacz, R., Larsen, T.: Integration of reinforcement learning and optimal decision-making theories of the basal ganglia. Neural Comput. **23**(4), 817–851 (2011)
5. Casti, A., Hayot, F., Xiao, Y., Kaplan, E.: A simple model of retina-LGN transmission. J. Comput. Neurosci. **24**, 235–252 (2008)
6. Dunovan, K., Lynch, B., Molesworth, T., Verstynen, T.: Competing basal-ganglia pathways determine the difference between stopping and deciding not to go. ELife, pp. 1–24 (2015) https://doi.org/10.7554/eLife.08723

7. Dunovan, K., Verstynen, T.: Believer-Skeptic meets Actor-Critic : Rethinking the role of basal ganglia pathways during decision-making and reinforcement learning. Frontiers Neurosci. **10**(March), 1–15 (2016). https://doi.org/10.1101/037085

8. Escobar, M.-J., Masson, G.S., Vieville, T., Kornprobst, P.: Action recognition using a bio-inspired feedforward spiking network. Int. J. Comput. Vis. **82**, 284–301 (2009)

9. Frank, M.J., Seeberger, L.C., O'Reilly, R.C.: By carrot or by stick: cognitive reinforcement learning in parkinsonism. Science **306**(5703), 1940–1943 (2004). https://doi.org/10.1126/science.1102941

10. Fregnac, Y., Bathellier, B.: Cortical correlates of low-level perception: from neural circuits to percepts. Neuron **88**, 110–126 (2015)

11. Ghodratia, M., Khaligh-Razavic, S.-M., Lehky, S.R.: Towards building a more complex view of the lateral geniculate nucleus: recent advances in understanding its role. Prog. Neurobiol. **156**, 214–255 (2017)

12. Gleeson, P., Martinez, R., Davison, A.: Network models of V1. http://www.opensourcebrain.org/projects/111 2016)

13. Herz, D.M., Zavala, B.A., Bogacz, R., Brown, P.: Neural Correlates of Decision Thresholds in the Human Subthalamic Nucleus. Current Biology **1–5**, (2016). https://doi.org/10.1016/j.cub.2016.01.051

14. Igarashi, J., Shounob, O., Fukai, T., Tsujino, H.: Real-time simulation of a spiking neural network model of the basal ganglia circuitry using general purpose computing on graphics processing units. Neural Networks **24**, 950–960 (2011)

15. Joel, D., Niv, Y., Ruppin, E.: Actor-critic models of the basal ganglia: new anatomical and computational perspectives. Neural Netw. **15**, 535–547 (2002)

16. Koprinkova-Hristova, P., Bocheva, N., Nedelcheva, S.: Investigation of feedback connections effect of a spike timing neural network model of early visual system. In: Innovations in Intelligent Systems and Applications (INISTA), 3–5 July 2018, Thessaloniki, Greece, (2018) https://doi.org/10.1109/INISTA.2018.8466292

17. Koprinkova-Hristova, P., Bocheva, N., Nedelcheva, S., Stefanova, M.: A model of self-motion perception developed in NEST. Frontiers Comput. Neurosci. (2019). https://doi.org/10.3389/fncom.2019.00020

18. Koprinkova-Hristova, P., et al.: STDP plasticity in TRN within hierarchical spike timing model of visual information processing, AIAI 2020. IFIP AICT **583**, 1–12 (2020)

19. Koprinkova-Hristova, P., Bocheva, N.: Spike Timing Neural Model of Eye Movement Motor Response with Reinforcement Learning, BG SIAM 2018 (accepted paper for LNCS volume, to appear), (2020)

20. Kremkow, J., et al.: Push-Pull Receptive Field Organization and Synaptic Depression: Mechanisms for Reliably Encoding Naturalistic Stimuli in V1. Frontiers in Neural Circuits (2016). https://doi.org/10.3389/fncir.2016.00037

21. Krishnan, R., Ratnadurai, S., Subramanian, D., Chakravarthy, V.S., Rengaswamyd, M.: Modeling the role of basal ganglia in saccade generation: Is the indirect pathway the explorer? Neural Networks **24**, 801–813 (2011)

22. Kunkel, S. et al.: NEST 2.12.0. Zenodo. (2017) https://doi.org/10.5281/zenodo.259534

23. Layton, O.W., Fajen, B.R.: Possible role for recurrent interactions between expansion and contraction cells in MSTd during self-motion perception in dynamic environments. J. Vis. **17**(5), 1–21 (2017)

24. Nedelcheva, S., Koprinkova-Hristova, P.: Orientation selectivity tuning of a spike timing neural network model of the first layer of the human visual cortex. In: Georgiev, K., Todorov, M., Georgiev, I. (eds.) BGSIAM 2017. SCI, vol. 793, pp. 291–303. Springer, Cham (2019). https://doi.org/10.1007/978-3-319-97277-0_24

25. Plotkin, J.L., Goldberg, L.A.: Thinking Outside the Box (and Arrow): Current Themes in Striatal Dysfunction in Movement Disorders, The Neuroscientist. https://doi.org/10.1177/1073858418807887. (2018)
26. Potjans, W., Morrison, A., Diesmann, M.: Enabling functional neural circuit simulations with distributed computing of neuromodulated plasticity. Front. Comp. Neurosci. **4**, (2010). https://doi.org/10.3389/fncom.2010.00141many
27. Rubin, J., Lee, D.D., Sompolinsky, H.: Equilibrium properties of temporally asymmetric Hebbian plasticity. Phys. Rev. Lett. **86**(2), 364–367 (2001)
28. Sadeh, S., Rotter, S.: Statistics and geometry of orientation selectivity in primary visual cortex. Biol. Cybern. **108**, 631–653 (2014)
29. Shadlen, M.N., Newsome, W.T.: Motion perception: seeing and deciding. Proc. Natl. Acad. Sci. USA **93**(2), 628–633 (1996)
30. Troyer, T.W., Krukowski, A.E., Priebe, N.J., Miller, K.D.: Contrast invariant orientation tuning in cat visual cortex: thalamocortical input tuning and correlation-based intracortical connectivity. J. Neurosci. **18**, 5908–5927 (1998)
31. Tsodyks, M., Uziel, A., Markram, H.: Synchrony generation in recurrent networks with frequency-dependent synapses. J. Neurosci. 20 RC50, 1–5 (2000)
32. Webb, B.S., Ledgeway, T.Y., McGraw, P.V.: Relating spatial and temporal orientation pooling to population decoding solutions in human vision. Vis. Res. **50**, 2274–2283 (2010)
33. Wei, W., Rubin, J.E., Wang, X.-J.: Role of the indirect pathway of the basal ganglia in perceptual decision making. J. Neurosci. **35**(9), 4052–4064 (2015)
34. Yan, H., Wang, J.: Quantification of motor network dynamics in Parkinson's disease by means of landscape and flux theory. PLoS ONE **12**(3), e0174364 (2017)

Automatic Classification of Low-Angle Fuze-Quick Craters Using Deep Learning

Sani Aji[1,3], Poom Kumam[1,2]([✉]), Punnarai Siricharoen[4],
and Ali Maina Bukar[5]

[1] KMUTTFixed Point Research Laboratory, Room SCL 802 Fixed Point Laboratory,
Science Laboratory Building, Department of Mathematics, Faculty of Science,
King Mongkut's University of Technology Thonburi (KMUTT),
126 Pracha-Uthit Road, Bang Mod, Thrung Khru, Bangkok 10140, Thailand
ajysani@yahoo.com, poom.kum@kmutt.ac.th

[2] KMUTT-Fixed Point Theory and Applications Research Group,
Theoretical and Computational Science Center (TaCS), Science Laboratory Building,
Faculty of Science, King Mongkut's University of Technology Thonburi (KMUTT),
126 Pracha-Uthit Road, Bang Mod, Thrung Khru, Bangkok 10140, Thailand

[3] Department of Mathematics, Faculty of Science,
Gombe State University, Gombe, Nigeria

[4] Department of Mathematics, Faculty of Science,
King Mongkut's University of Technology Thonburi (KMUTT),
126 Pracha-Uthit Road, Bang Mod, Thrung Khru, Bangkok 10140, Thailand
punnarai.sir@kmutt.ac.th

[5] Centre for Visual Computing, Faculty of Engineering and Informatics,
University of Bradford, Bradford, West Yorkshire, UK
alimainabukar@gmail.com

Abstract. Crater Analysis is a technique used for precautionary and investigative purpose, especially in confirming suspected locations of hostile fire and detecting the presence of enemy ammunition. Specifically, investigators have manually examined terrain disturbances due to blast craters, which are evidently caused by artillery fire. Although this manual process has been helpful in the identification of possible origins of fire, it is tedious and prone to errors. With the recent advancements in the fields of computer vision and machine learning (ML), this work investigates the automation of low-angle fuze-quick craters (LaFCs) classification using satellite imagery of the Ukranian border obtained from google

The authors acknowledge the financial support provided by the Petchra Pra Jom Klao Doctoral Scholarship for Ph.D. program of King Mongkut's University of Technology Thonburi (KMUTT) Thailand and Center of Excellence in Theoretical and Computational Science (TaCS-CoE), KMUTT. The first author was supported by the "Petchra Pra Jom Klao Ph.D. Research Scholarship from King Mongkut's University of Technology Thonburi (Grant No. 19/2562). The project was also supported by Center of Excellence in Theoretical and Computational Science (TaCS-CoE) Center under Computational and Applied Science for Smart Innovation research Cluster (CLASSIC), Faculty of Science, KMUTT. The authors would also like to aknowledge Rudiment (a research and development organization) for providing us with the dataset.

© Springer Nature Switzerland AG 2020
G. Nicosia et al. (Eds.): LOD 2020, LNCS 12566, pp. 436–447, 2020.
https://doi.org/10.1007/978-3-030-64580-9_36

earth. Due to limited volume of available satellite images (of LaFCs), we hereby propose the use of off-the-shelf ImageNet pretrained models (Resnet50, Resnet101 and ResNet152) for the automatic classification of the LaFCs. Thus, learned filters from these deep neural networks are used to extract features from the satellite images. These features are then fed into support vector machine (SVM) for classification. Promising results are obtained from our experiments, showing that it is possible to use ML to identify LaFCs. To the best of our knowledge, this is the first time classification of these type of craters has been investigated.

Keywords: Artillery craters · Classification · Deep learning · Residual networks · SVM · Transfer learning

1 Introduction

The Ukranian-Russian 2014 border conflict has drawn the attention of people around the world. It can be recalled that in late 2014, Ukranian armed forces were attacked by artillery fire resulting in the loss of lives as well as injuring many soldiers at the Ukranian border. Although there was an accusation that these attacks originated from Russia, however, the Russian officials denied this, and stated that no artillery attack was sent to Ukrain from the Russian territory [1]. According to a manual crater analysis conducted by Bellingcat [1], their investigation team claimed that the artillery attacks against the Ukranian armed forces were from the Russian territory. We would like to make it clear that our work is not aimed at confirming this allegation, rather, we simply aim to automate the process of classifying LaFCs using more efficient and robust ML approach.

Crater Analysis is one crucial and reliable way used for investigative purpose, specifically, that is, in the identification of suspected areas of hostile fire, confirming the presence of other artilleries, their directions as well as detecting new ammunition. Depending on their shapes, time taken before explosion and other features, artillery craters are categorized into different types including the LaFCs, low-angle fuze-delay craters, mortar craters and rocket craters among others. Different craters have different methods of identification, therefore, first key step in crater analysis is identifying its type.

Using satellite image analysis by Bellingcat research team, two types of craters were identified in the Ukranian-Russian attack. The LaFCs and the high-angle shell craters. LaFCs are artillery craters characterized by their movement in low angle and their explosion immediately they touch the ground. These craters have a distinctive arrowhead shape pointing to the direction of the weapon [19].

There are two manual methods involved in identifying these craters as well as measuring the direction from which they are fired. These methods are; the fuze-furrow and center of crater method, and side spray method, both of which involve using compass and stakes manually. However, the manual ways are prone to errors, consume time, and cause delay, which results in loosing information because craters exposed to weather and personnel deteriorate rapidly and lose their values as source of information. Thus, there is need for automating these processes to have the advantage of not only saving the time and energy involved in the manual methods, but also, it will yield more accurate results in tracing the direction of trajectories of these craters.

Motivated by the manual crater analysis of the Bellingcat, and the distinctive features of the LaFCs, this paper focuses on leveraging ML algorithms, specifically deep neural networks for automatic classification of the LaFCs with the Ukranian-Russian border as a case study.

2   Related Work

Looking at the importance of crater analysis in providing past and present information for investigative purpose, and the recent success of ML and deep learning algorithms in object classification and detection, a lot of approaches have been developed by many researchers to study, classify and detect different types of craters for different purposes.

Wetzler et al. [24] used ML techniques to detect small impact craters on planetary images. They applied different supervised learning algorithms including ensemble methods, SVM and continuously scalable template models (CSTM) in detecting craters from the ground-truth images. The detectors obtained in their work were tested on Viking Orbiter images of mars with 1000 craters. The result obtained using the SVM was approximately 60% which they claimed, proved to be good as compared to other approaches, however, this result does not quite reach the human level.

Ding et al. [8] also proposed an automatic crater classification approach by combining supervised learning classification techniques with transfer learning. They firstly used the concept of mathematical morphology to identify the regions that contain craters, used image texture feature extraction with the combination of supervised ensemble learning methods to classify craters and non-craters, and then combined transfer learning with boosting to improve the performance of their proposed method. Their method identifies small impact craters with high accuracy and is applicable in planetary research.

Recently, convolutional neural networks (CNNs) have been used in classifying different types of craters. It was shown in [7] that CNNs have advanced the performance of crater detection. Brenner et al. [5] also developed an automated approach for detecting bomb craters using aerial images from world war II. In their work, a DenseNet CNN architecture was first used for binary classification on image patches and the patches were then extracted for further processing. Their method considered bomb craters and is applicable in construction projects for inspection and risk estimation of exploded bombs.

Emami et al. [9] extended their earlier work in [10] by utilising transfer learning to classify lunar craters. Their results indicate that neural networks trained via transfer learning approach can be used to classify craters and non craters with high accuracy while requiring less training time and data.

Despite numerous works done in crater classification, none of these works has specifically looked at LaFCs with a view to eventual prediction of trajectories. While other researchers aimed at identifying craters for different reasons, this work mainly focuses on automating the classification of LaFCs to make crater analysis easy, accuarate and reliable for military precaution and investigations. In summary, this paper utilizes the dataset of the Ukranian border, and the use of deep transfer learning to automatically classify LaFCs. To our knowledge, this is the first study that focuses specifically on these special types of craters. Should the proposed approach prove promising, we shall then look at the problem of crater localisation, segmentation and mapping of trajectory to the gun.

2.1 Deep Convolutional Neural Networks

Deep convolutional neural networks (DCNNs) have widely been used recently in image classification and object recognition tasks due to their exceptionally good performance. It is worth mentioning that deep neural networks have made a significant improvement to the state-of-the art in image recognition, speech recognition, object detection [11,13,18] and medical image analysis [6,28].

Neural networks normally work by receiving an input through the input layer and pass it to some hidden layers serving as classifiers, then finally to an output layer which provides the classification result [14]. CNNs varies slightly as different inputs share weights instead of each of them to have a single weight as in traditional neural networks [17]. Therefore, CNNs use hierarchical layers to identifying the low level to high level features, thus having the ability to learn sophisticated abstractions that are used to make intelligent decisions. State of the art CNNs learn millions of parameters, and that is why they have been able to achieve very high accuracies while solving many problems. However, with all these breakthroughs, DCNNs performance depends heavily on the availability of huge training dataset, the more data given to these models during training, the more accuracy in the prediction result. In a situation where there is a limited amount of data available, data augmentation and transfer learning are used to overcome this problem. Data augmentation is a method used to increase diversity of the dataset, by performing image transformations, including but not limited to flipping, rotating, and zooming the image.

In transfer learning, weights of networks trained on large dataset are used in solving a different task i.e. different from the original task. More formally, given an initial domain I_d and an initial task I_t, and another new domain N_d with a new different task N_t, the idea of transfer learning is to learn the conditional probability $P(Y_T|X_T)$ in N_d using the information gained from I_d and I_t where $I_d \neq N_d$ or $I_t \neq N_t$ [20].

Several works have shown transfer learning to work efficiently despite training on less data. Additionally, the model seems to train at high speed. To conduct transfer learning, researchers normally use models trained on the ImageNet dataset (ILSVRC) [21] as the base model. ImageNet is a large dataset of annotated images developed specifically for computer vision research. The dataset consist of more than 14 million images with more than 21 thousand classes.

Researchers used subsets of the Imagenet dataset for the annual ILSVRC competition using around 1 million images with 1000 categories. Some of the most popular models trained on Imagenet includes AlexNet [16], VGG-16 [22], GoogleNet [23] and one of the numerous variants of ResNet [13] among others. Recently, a lot of researchers have used transfer learning for many classification problems by extracting features from images using these pretrained models and subsequently, the extracted features are used to train ML algorithms such as SVM for classification. An example of this where SVM is trained for classification on breast cancer after feature extraction is given by Araujo in [4] where 77.8% classification accuracy was recorded. Abubakar et al. [3] also used two pretrained models (VGG16 and VGG19) for feature extraction and then trained a linear kernel SVM for human skin burn classification and good classification accuracy was obtained.

Residual Networks (ResNets). Although deep neural networks have achieved extremely high accuracy in image recognition and classification, training deeper networks suffers from the vanishing gradient and degradation problems [12, 25]. That is information tends to vanish as it passes from one layer to another before it gets the end of the network. This resulted in poor performance of the networks. These problems were solved by the ResNets architectures [13] developed by microsoft research team in 2015. These architectures use a residual networks that are easier to optimize, making a shortcut connection by skipping one or more convolutional layers, taking the input of the previous layers and adding it to the result of the next skipped layers generating input for the remaining layers [13]. Different ResNets have different number of layers from 18, 34, 50, 101 and 152 thus, ResNet18, ResNet34, ResNet50, ResNet101, and ResNet152 respectively. Unlike plain networks in which going very deep results in poor performance, ResNet architectures gain more better accuracy by increasing the depth, thus producing better results than less deeper ones. Hence, In this paper, we employ three deep ResNet architectures. The ResNet50, ResNet101, and ResNet152.

3   Methodology

Since the artillery crater classification problem considered in this paper has limited amount of data, training the neural network from scratch will most likely lead to overfitting. Looking at the ILSVRC [21] also dealt with classifying images into different categories, and we also aimed at image classification but with limited amount of data, we decided to employ the concept of transfer learning by using three ResNet architectures (ResNet50, ResNet101, and ResNet152) for feature extraction and then, subsequently, a support vector machine is trained for classification. Since the three pretrained models considered already performed well on the ImageNet dataset having 1000 output classes, it is our hope that utilizing them for feature extraction may result in a good crater classification accuracy.

3.1   Data Collection

In this work, the crater dataset is collected from [2]. The dataset consist of satellite images collected via google earth from the territory of Ukrain and its border with Russia between July and September, 2014. At the moment, we are able to curate 120 sample craters and 120 non-craters, 240 images in total. The non craters are also satellite images consisting of potholes, decayed grasses and other environmental objects such as stumps, small holes, specifically, we would like to state that the non craters are challenging data samples that are even hard for human observer to distinguish them from craters. Samples of craters and non craters are given in Fig. 2 and 3 respectively. On the other hand, the LaFCs are described as an arrow shaped with four recognizable effects including some side spray areas that projects diagonally from the center of the crater. Due to these their distinct features, the direction from which the artillery was fired can be determined. The figure below shows the shape and the parts of these craters (Fig. 1).

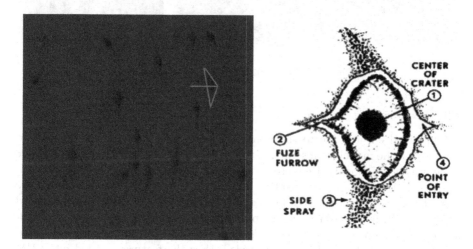

Fig. 1. Shape and parts of low-angle fuze quick craters

3.2   Data Preprocessing

After collecting our images, we cropped them into pieces so that each crater will be in a single image. We performed data augmentation in order to increase the amount of our dataset, and augmentation is done during training in order to avoid testing on an augmented data. Ten folds cross validation is used in our experiment, thus, at each time of training, one fold consisting of 24 images is held out of the total 240 images, and the remaining 216 are used for training. During training, the 216 images are augmented by $-10°$ rotation, $+10°$ rotation, flipped left to right and random cropping. Therefore, after data augmentation, 864 samples were added to the initial 216 training sets. The Fig. 2 and 3 below show the sample of craters and non-craters from the datasets.

Fig. 2. Sample craters from the dataset

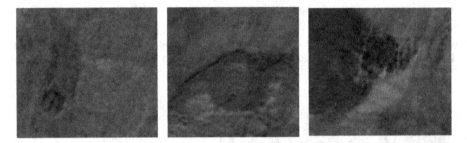

Fig. 3. Sample non-raters from the dataset

3.3 Feature Extraction

As explained earlier, due to the limited amount of crater dataset, we decided not to train the model from scratch. Instead, transfer learning is employed where three pretrained models (ResNet50, ResNet101 and ResNet152) are used for feature extraction. For all the three ResNet models compared in the experiment, we extract features by passing the images through the networks during the forward pass, the output of the last average pooling layer, which is a vector of 2048 values are what we used as features for training the SVM. That is, fully connected layers of each of the models are discarded, thus the features are fed into SVM for training. It is evident from the literature mentioned that this way proved to be effective when the available dataset is not in abundance.

3.4 Classification

After using the pretrained models for feature extraction, SVM is subsequently trained for the classification. SVM is a well known ML algorithm developed for classification problems, later on, extended for solving regression problems as well [15]. Given a training set (x_i, y_i), $x_i \in \mathbb{R}^d$, $y_i \in \{+1, -1\}$, y_i represents labels as $+1$ or -1, in our case, meaning crater or non crater respectively. A classifier is learned such that

$$y_i = \begin{cases} 1, & f(x_i) \geq 0, \\ -1, & f(x_i) < 0. \end{cases} \tag{1}$$

SVM aimed at finding the optimal separating hyperplane which separates the two classes, and there might be more than one hyperplane that can do that, however, the main goal in SVM is to find the best one. Thus, the problem becomes an optimization problem, and the best hyperplane is obtained by solving an objective function, with some kernel functions. For z_1, z_2, ... z_n support vectors, the formulation of the SVM in kernels is given by:

$$f(x) = \sum_{i=1}^{n} w_i K(z_i, x) + b \tag{2}$$

where w_1, w_2, ... w_n are the weights and b is a bias parameter. In this paper, a linear kernel SVM is used and the kernel is given as an inner product:

$$K(x_i, x_j) = x_i^T x_j + c. \tag{3}$$

The regularization parameter c which tells the SVM how much to avoid misclassification was set to 1.0. Therefore, the features extracted from the images using the pretrained models are then fed for training.

4   Experiments and Results

In this section, we present the results of the experiments performed using our proposed approach. The experiments were done in two steps, the first one is without any data augmentation and for the second one, data augmentation was carried out before feature extraction and classification.

We use k-fold cross validation technique with $k = 10$. Hence, our dataset is divided into 10 equal folds, $k_1, k_2, \cdots, k_{10}$. In each iteration, nine folds are used for training, and the one that was held out was used for testing. Thus, after ten iterations, all ten folds will have been used for training as well as testing the algorithm. The experiments were conducted using executable MATLAB file based on MatConvNet deep learning framework. This, we have made available to the public, and can be accessed on github https://github.com/sani-aji/link-to-drive/ commit/eaac3454c7da8a14dae59bb9cfa49e0f685bc813. The results of the experiments are given below. Table 1 is without data augmentation, and in Table 2, data augmentation was performed.

Table 1. Classification accuracy without data augmentation

CNN models	Accuracy (%)
ResNet50	96.25
ResNet101	96.25
ResNet152	96.67

Table 2. Classification accuracy with data augmentation

CNN models	Accuracy (%)
ResNet50	98.58
ResNet101	98.58
ResNet152	98.67

From Table 1 and 2 above, it can be seen that in both the experiments, high classification accuracy was achieved, thus an evidence that the extracted features are rich, hence, posessing discriminative power. It can be further observed from Table 2 that the results obtained after data augmentation surpassed the one without augmentation. This is not surprising, since the size of training data plays a vital role in ML; the more the training data, the better the performance.

The fact that the accuracies of the models on our dataset are closely together, we have taken a step further to run a non parametric statistical test in order to check the significance of the obtained results. We used a Wilcoxon test [26, 27], the non parametric version of student T test that is used to compare a paired data samples such as data in which each sample is independent but comes from the same distribution, the likes of when different algorithms are evaluated on same training and test data as in our case. We started by comparing ResNet50 against ResNet101, and then ResNet152. The p-values obtained are all greater than the significance level ($\alpha = 0.05$) and so, the null hypothesis supposing that the performance is the same was not rejected, and thus, no significant difference between the performance of these models on our dataset is observed, showing that all the three models used for feature extraction achieved good accuracy.

In order to further evaluate the classification performance, the precision and recall metrics were also calculated from the well known confusion matrix. The precision is simply the proportion of true positives (craters predicted as craters) in all the images that have been predicted to be craters. On the other hand, recall also known as sensitivity is the proportion of the identified positives (craters) from the images that are positively craters. These metrics were reported in the table below.

Table 3. Precision and recall

CNN models	Precision (%)	Recall (%)
ResNet50	97.55	99.67
ResNet101	97.55	99.67
ResNet152	97.71	99.67

From Table 3, it is clear that SVM performed very well when trained on features extracted using these pretrained models. High precision and recall were obtained by the SVM classifier. All the architectures performed very well, which can be attributed to their ability to extract discriminatory features.

Despite been a simple linear model, the SVM utilised for classification did very well in separating the two classes. This can be as a result of the efficiency in the feature extraction technique. Indeed using off-the-shelf pretrained deep models for feature extraction has proven to be a powerful approach especially in our case where we had small amount of training data.

5 Conclusions

In this paper, automatic classification of low-angle fuze-quick craters (LaFCs) was proposed and evaluated. To the best of our knowledge, this is the first work that focused specifically on these type of craters. Interestingly, they are special due to their distinctive signature when they hit the ground, which makes it possible for human experts to trace the direction of fire. Specifically, we utilized three pretrained residual neural network models (ResNet50, ResNet101 and ResNet152) for feature extraction, thereafter the extracted features were used to train a linear kernel SVM for binary classification. Subsequently, performance of the proposed algorithms were evaluated after training with and without data augmentation. Results obtained from the six experiments we conducted revealed that data augmentation improved the performance of the algorithm, it was also observed from the statistical test conducted that there is no significant difference between the performance of these pretrained models on the dataset. Thus, in the future, we can stick to ResNet50 since it is computationally less expensive compare to the other deeper variants. Having achieved promising classification accuracies between 97% and 99%, we therefore believe that our automatic classification approach can be used to identify LaFCs from satellite images. We are optimistic that this work will serve as an alternative to the manual crater identification process, which is both tedious and prone to errors. We foresee this as a tool that will help in human right investigations. For future research, we plan to investigate automatic crater detection as well as segmentation with a view to mapping of the crater's trajectory, to eventually identify the direction of fire. We would like to point out here that, despite citing the work conducted by Bellingcat, we are in no way saying we agree with the allegation they put forward. As scientific researchers, this work was done solely to investigate the possibility of automating the tedious, an error-prone process of crater analysis.

Conflicts of Interest. The authors declare no conflict of interest.

References

1. Origin of artillery attacks on Ukrainian military positions in Eastern Ukraine between 14 July 2014 and 8 August 2014. https://github.com/Rudiment-Info/arcade/tree/master/sampledata/ukraine. Accessed 03 Mar 2020

2. Sample data of artillery bombardment. https://www.bellingcat.com/news/uk-and-europe/2015/02/17/origin-of-artillery-attacks/. Accessed 03 Apr 2020

3. Abubakar, A., Ugail, H., Bukar, A.M.: Noninvasive assessment and classification of human skin burns using images of Caucasian and African patients. J. Electron. Imaging **29**(4), 041002 (2019)

4. Araújo, T., et al.: Classification of breast cancer histology images using convolutional neural networks. PloS One **12**(6), 1–14 (2017)

5. Brenner, S., Zambanini, S., Sablatnig, R.: Detection of bomb craters in WWII aerial images. In: Welk, M., Urschler, M., Roth, P.M. (eds.) Proceedings of the OAGM Workshop, pp. 94–97 (2018)

6. Chiang, T.C., Huang, Y.S., Chen, R.T., Huang, C.S., Chang, R.F.: Tumor detection in automated breast ultrasound using 3-D CNN and prioritized candidate aggregation. IEEE Trans. Med. Imaging **38**(1), 240–249 (2018)

7. Cohen, J.P., Lo, H.Z., Lu, T., Ding, W.: Crater detection via convolutional neural networks. arXiv preprint arXiv:1601.00978 (2016)

8. Ding, W., et al.: Subkilometer crater discovery with boosting and transfer learning. ACM Trans. Intell. Syst. Technol. (TIST) **2**(4), 1–22 (2011)

9. Emami, E., Ahmad, T., Bebis, G., Nefian, A., Fong, T.: On crater classification using deep convolutional neural networks. In: Lunar and Planetary Science Conference, vol. 49 (2018)

10. Emami, E., Bebis, G., Nefian, A., Fong, T.: On crater verification using mislocalized crater regions. In: 2017 IEEE Winter Conference on Applications of Computer Vision (WACV), pp. 1098–1104. IEEE (2017)

11. Fayek, H.M., Lech, M., Cavedon, L.: Evaluating deep learning architectures for speech emotion recognition. Neural Netw. **92**, 60–68 (2017)

12. Habibzadeh, M., Jannesari, M., Rezaei, Z., Baharvand, H., Totonchi, M.: Automatic white blood cell classification using pre-trained deep learning models: ResNet and inception. In: Tenth International Conference on Machine Vision (ICMV 2017), vol. 10696, p. 1069612. International Society for Optics and Photonics (2018)

13. He, K., Zhang, X., Ren, S., Sun, J.: Deep residual learning for image recognition. In: Proceedings of the IEEE Conference on Computer Vision and Pattern Recognition, pp. 770–778 (2016)

14. Hornik, K., Stinchcombe, M., White, H., et al.: Multilayer feedforward networks are universal approximators. Neural Netw. **2**(5), 359–366 (1989)

15. Jakkula, V.: Tutorial on support vector machine (SVM). School of EECS, Washington State University 37 (2006)

16. Krizhevsky, A., Sutskever, I., Hinton, G.E.: ImageNet classification with deep convolutional neural networks. In: Advances in Neural Information Processing Systems, pp. 1097–1105 (2012)

17. Le Cun, Y., et al.: Handwritten zip code recognition with multilayer networks. In: [1990] Proceedings. 10th International Conference on Pattern Recognition, vol. 2, pp. 35–40. IEEE (1990)

18. LeCun, Y., Bengio, Y., Hinton, G.: Deep learning. Nature **521**(7553), 436–444 (2015)

19. Manual, F.: Manuals Combined: Tactics, Techniques, and Procedures for Field Artillery Meteorology And Field Artillery Target Acquisition. Jeffrey Frank Jones (2002). https://books.google.co.uk/books?id=OXJQDwAAQBAJ

20. Pan, S., Yang, Q.: A survey on transfer learning. IEEE Trans. Knowl. Data Eng. **22**, 1345–1359 (2010)

21. Russakovsky, O., et al.: ImageNet large scale visual recognition challenge. Int. J. Comput. Vis. **115**(3), 211–252 (2015). https://doi.org/10.1007/s11263-015-0816-y

22. Simonyan, K., Zisserman, A.: Very deep convolutional networks for large-scale image recognition. arXiv preprint arXiv:1409.1556 (2014)
23. Szegedy, C., et al.: Going deeper with convolutions. In: Proceedings of the IEEE Conference on Computer Vision and Pattern Recognition, pp. 1–9 (2015)
24. Wetzler, P.G., Honda, R., Enke, B., Merline, W.J., Chapman, C.R., Burl, M.C.: Learning to detect small impact craters. In: 2005 Seventh IEEE Workshops on Applications of Computer Vision (WACV/MOTION 2005)-Volume 1, vol. 1, pp. 178–184. IEEE (2005)
25. Wichrowska, O., et al.: Learned optimizers that scale and generalize. In: Proceedings of the 34th International Conference on Machine Learning-Volume 70, pp. 3751–3760. JMLR. org (2017)
26. Wilcoxon, F., Katti, S., Wilcox, R.A.: Critical values and probability levels for the Wilcoxon rank sum test and the Wilcoxon signed rank test. Sel. Tables Math. Stat. **1**, 171–259 (1970)
27. Woolson, R.: Wilcoxon signed-rank test. In: Wiley Encyclopedia of Clinical Trials, pp. 1–3 (2007)
28. Yu, K.H., et al.: Classifying non-small cell lung cancer histopathology types and transcriptomic subtypes using convolutional neural networks, p. 530360. bioRxiv (2019)

Efficient Text Processing via Context Triggered Piecewise Hashing Algorithm for Spam Detection

Alexey Marchenko[✉], Alexey Utki-Otki, and Dmitry Golubev

Kaspersky, 39A/2 Leningradskoe Shosse, Moscow 125212, Russian Federation
{Alexey.Marchenko,Alexey.Utki-Otki,
Dmitry.S.Golubev}@kaspersky.com

Abstract. In our research, we have examined millions of spam messages and have developed a technology called Spam Term Generator. This technology uses mix of CTPH (Context Triggered Piecewise Hashing), DBSCAN (Density-Based Spatial Clustering of Applications with Noise) and LCS algorithm (Longest Common Substring) to automatically determine almost similar spam messages and extract repetitive text pieces in large collections of spam texts.

During implementation of the technology, we have encountered multiple performance issues, which we have efficiently solved by using different tricks to escape "brute force" analysis of the original texts. These performance solutions are generic and applicable to any other tasks, which include CTPH and clustering.

In this paper, we will show the inside of our technology, talk about performance issues and solutions and finally demonstrate the results of running the technology over real-life spam collections.

Keywords: Spam detection · Hashing · Clustering · Algorithms and programming techniques · CTPH · DBSCAN

1 Introduction

Spam email detection is a high trending problem for every governmental and commercial organization. Spam messages may affect business continuity by overflowing employee mailboxes and lead to financial losses through malware attachments and phishing links, which they often contain [1].

There are multiple ways of detecting spam emails: email message headers, sender info and email message text. The first two options are a good way to go [10] for cases when spammers use their own resources for delivery. However, it is not efficient enough for spam messages sent through legitimate looking channels, such as bot nets and feedback forms. In this case, it may be impossible to recognize spam message depending on message headers or sender info only [2]. The thing that clearly defines such messages as spam is the content they carry – the text of these messages. That is why recognizing spam emails depending on its text is a better option.

© Springer Nature Switzerland AG 2020
G. Nicosia et al. (Eds.): LOD 2020, LNCS 12566, pp. 448–456, 2020.
https://doi.org/10.1007/978-3-030-64580-9_37

However, spammers do know about this option and often obfuscate or modify spam texts in many ways to prevent antispam solutions from detecting their messages. For a human eye, it is easy to see that messages, which belong to the same spam campaign, are all similar except for small changes (word replacement, obfuscation, etc.). The reason is that these messages still contain text pieces, which stay precisely the same in different texts. The question we asked ourselves – is it possible to detect spam messages with similar texts and extract repetitive text pieces to use them for easy detection of other similar messages in the future?

Formally, original task sounds as follows: we have a collection of already known spam messages (the *support* collection) obtained by other methods of spam detection or collected from spam traps. In addition, we have a flow of yet unknown spam/non-spam messages (the *target* collection). What we have to do is to use the knowledge about texts in the *support* collection to detect all spam texts in the *target* collection. After that, we have to extract distinctive text pieces (*spam terms*) from every detected message to use it for future detection of other similar messages.

In this paper, we would like to introduce a technology called Spam Term Generator, which solves the designated task by processing texts in three consecutive steps.

On the first step, we convert the original texts to CTPH (Context Triggered Piecewise Hashing) [3] hashes and perform preliminary grouping according to the distance between hashes. The idea of CTPH is to calculate traditional hashes for small pieces of the original text and then concatenate them into one string. This allows comparison between different CTPH hashes depending on how many hashed pieces they have in common. The result of the comparison shows the level of similarity of the original texts.

On the second step, we use DBSCAN (Density-Based Spatial Clustering of Applications with Noise) algorithm [4] for each group of hashes, determined on the first step, to find clusters with high density and filter noise out.

On the third step, we fall back to the original texts related to hashes, to extract LCS (Longest Common Substring) for each cluster and validate that it fits for future spam detection without risk of false detection.

Combination of the first two steps allow us to escape time-consuming analysis of the original texts as well as escape "brute force" clustering of the whole original collection and run DBSCAN only for some subsets of the collection.

We have evaluated experiments for Spam Term Generator on the collections of real-life spam messages. The results show that this approach is suitable for spam detection as well as for gathering *spam terms*.

2   Related Work

First spam messages appeared almost at the same time the e-mail proved itself as a way for personal and business communication [5]. That is why the problem of spam detection had a long history of evolution.

During the last decades, almost every possibility to detect spam messages was studied in numerous scientific articles. No surprise, CTPH was also considered as a way of determining spam messages similarity for various tasks.

For example, in the work [6], authors use CTPH to split their collection into spam campaigns so they may analyze bot nets behind them. The results show that this approach is suitable for the task.

In addition, one of spam filtering systems introduced a feature [7] based on shingles [11] called "fuzzy check module". The idea is that user somehow obtains examples of spam messages (via spam complains or spam traps) and put them into a database. For every incoming message, the system calculates shingles for its text and accesses the database to find similar spam messages.

In our technology, we use a different approach to obtain precise detection by applying clustering after straightforward grouping of spam texts. In addition, we introduce *spam terms* as a way of simplifying detection for similar spam messages in the future.

3   Method

In sections below, we explain the architecture of Spam Term Generator. Firstly, we introduce the high-level architecture. After that, we dive into the details of text hashing, clustering and common substrings search. In the last section, we overview performance aspects of the above steps.

3.1   High Level Architecture

Figure 1. represents actual architecture of the technology. The processing of texts is a pipeline with the following stages:

1. Preprocessing of the original texts
2. Conversion of the original texts to CTPH hashes. The relation between the original texts and their hashes is stored for later use in "Step 6".
3. Dividing hashes into groups by their similarity to the *target* text hashes and filtering out *source* text hashes that are too distant from any *target* text hash.
4. Running DBSCAN in each group to filter out noise and find high-density clusters, which contain *target* text hashes.
5. Filtering out clusters that are too small. The rest are marked as *spam clusters* and *target* text hashes in these clusters are marked as a detected spam.
6. Finding the original texts in the storage (see "Step 2") for every *spam cluster* and running LCS algorithm over these texts to form list of possible *spam terms*.
7. Validating of possible *spam terms* to filter out strings that are not specific to spam texts but rather specific to e-mail messages in common (automatic signatures, antivirus reports for attachments, etc.)

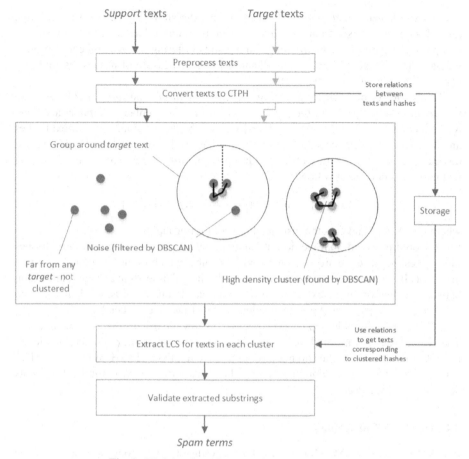

Fig. 1. High level architecture of spam term generator

In sections below, we will explain the details of the above steps.

3.2 Text Preprocessing

Preprocessing includes deletion of all punctuation, extra newlines, spaces and other separators in the original texts; replacement of some Unicode letters to their closest ASCII analogs and lowercasing of all letters. After all actions above, we remove any duplicated texts in the collections.

3.3 CTPH and Preliminary Grouping

We use *ssdeep* library [3] to generate CTPH hashes for the original texts. *Ssdeep* library implements an algorithm, which uses rolling hash based on Alder32 algorithm to determine pieces of original data for traditional hashing and FNV hash to calculate actual checksum for every piece. *Ssdeep* also provides a function for calculating similarity of two hashes named *match score*, which represents the similarity between the original texts. It does not satisfy triangle inequality rule (e.g. not a distance function), so it is inappropriate to use it as a metric for grouping and clustering. To overcome the problem we have used weighted Levenshtein distance [9] as a measure of hashes dissimilarity instead (e.g. as a distance function).

On the step of preliminary grouping, we mix both *target* and *support* collections to split them into groups for further analysis. Let N_S be the number of *support* texts and N_T be the number of *target* texts, let $S = \{S_1..S_{N_S}\}$ be the set of all *support* text hashes and $T = \{T_1..T_{N_T}\}$ be the set of all *target* text hashes. Finally let $Dist(H_1, H_2)$ be the distance function between two hashes. Group G_i of *support* text hashes around *target* text hash T_i is defined as follows:

$$G_i = \{T_i\} \cup \{S_j | Dist(S_j, T_i) < D, j \in [1..N_S]\}, \tag{1}$$

where $i \in [1..N_T]$ and D is the distance threshold for grouping.

It should be noted, that we do not calculate *Dist* between the *support* text hashes themselves, because we are not interested in groups, which do not contain *target* text hashes at all. The same is true for the *target* text hashes. This leads to a linear relationship between number of *target* texts and *Dist* calculations. In other words, we do not have to perform $(N_S + N_T)(N_S + N_T - 1)/2$ calculations of *Dist*, but rather $N_S * N_T$.

What is more, on this step we do not need to know the exact value of *Dist* function. Therefore, instead of using global threshold D, we can stick to the approach mentioned in [12]: if two *ssdeep* hashes do not have at least ROLLING_HASH_WINDOW_SIZE consecutive characters in common, then related texts do not have similar fragments determinable by *ssdeep*.

3.4 DBSCAN Clustering

In this step, we use DBSCAN [4] to cluster hashes inside each group G_i obtained on the previous step. The purpose of clustering is to find the area with high density around T_i.

Note that groups obtained on the previous step may intersect with each other (e.g. contain the same *support* text hashes). For clustering, we will have to calculate *Dist* function for all pairs of text hashes in each group. To escape duplicated calculations, we store every computed *Dist* in the matrix object, so we can reuse it in the case of groups intersection. Let N_S^i be the number of the *support* text hashes in group G_i. The total number of *Dist* calculation for clustering has the upper bound:

$$\sum_{i=1}^{N_T} \left[N_S^i (N_S^i + 1)/2 \right] \tag{2}$$

which is, due to the reuse of *Dist* calculations in cases of group intersection, has the upper bound:

$$N_S(N_S - 1)/2 + N_T * N_S \tag{3}$$

Although, due to the fact, that according to our experiments $\sum_{i=1}^{N_T} N_S^i \ll N_s$ the actual number of calculations is much less than that. Still, we have a linear relationship between the number of *target* texts and the number of *Dist* calculations.

We have chosen DBSCAN algorithm for clustering as the one that fits the following requirements:

1. The algorithm must not rely on any objects in space different from those provided in the input set (concepts such as an "average" are not applicable for hashes).
2. The algorithm must be able to work efficiently for arbitrary cluster shapes.
3. The algorithm should be robust to noise.

DBSCAN has two parameters: *Eps* – the minimal distance between two points to consider them as neighbors and *MinSamples* – the number of samples in a neighborhood for a point to be considered as a core point.

After this step, we have obtained a set of clusters with text hashes that form an area with high density around each *target* text hash. Note that in each group there is only one cluster containing *target* text hash, which is valuable for detection. Other clusters, which consist of *support* text hashes only, are not used. This means that it is possible to use simplified algorithm instead of DBSCAN, which finds only one cluster – the one containing *target* text hash. We have not made use of this optimization and the performance impact is unknown, although it should be researched in the future.

For each cluster, we count a number of *support* text hashes in it. We have considered the cluster to be a *spam cluster* if it has at least $C_S = 2$ *support* text hashes in it. The value of C_S was obtained empirically during the experiments but could be set to a higher value for higher precision.

3.5 Longest Common Substring Extraction and Validation

For each *spam cluster*, we extract LCS from the texts related to the hashes in this cluster. The extraction is performed by means of suffix trees [8].

Validation of the extracted strings is crucial, because in some cases these strings may not represent spam campaigns, but rather represent expressions common for email conversations in general. Usage of such strings for detection is unpromising due to a high probability of false detection.

Our validation phase consists of two types of checks. The first type is heuristics over the string length: the minimum number of words and characters in string. These checks are very fast and should be performed at the first place. The second type is specificity test. Let $C = \{C_1..C_N\}$ be the set of all *spam clusters*, let $L = \{L_1..L_N\}$ be the set of LCS extracted from the corresponded clusters. Finally, let $Sub(C_j, L_i)$ be the indicator that L_i is a substring for any text in the cluster C_j. Test passes if:

$$count(\{C_j : C_j \in C, Sub(C_j, L_i) = 1, j \in [1..N], j \neq i\}) < F = 1, \qquad (4)$$

where F is a threshold with the value obtained empirically.

3.6 Performance

Let us quickly run through performance aspects of the previous sections:

- On the step of preliminary grouping, we have performed $N_S * N_T$ calculations of *Dist* function.
- On the step of clustering we have performed $N_S(N_S - 1)/2 + N_T * N_S$ calculations of *Dist* in the worst scenario (each group has all *support* text hashes in it).
- On the step of LCS extraction, we have performed N_T string extractions in the case that every *target* text hash was included in *spam cluster*.
- During the validation phase of probable *spam terms*, we have performed $N_T + N_S$ substring tests for validation.

The above points lead to the total linear relationship between the number of *target* texts and the number of *Dist* estimations/calculations, LCS algorithm runs and substring checks. Note that "brute force" clustering of *target* and *support* collections would have given us $(N_S + N_T)(N_S + N_T - 1)/2$ calculations of *Dist* function and would have led to a square relationship between the number of *target* texts and the number of *Dist* calculations.

4 Experiments

For an evaluation of our technology, we collected 120000 spam messages with text length greater than 50 characters gathered from spam traps. We also collected 45000 legitimate mailing messages with text lengths greater than 50 from public mailing lists for the same period of time. Our method is not applicable to messages with body text less than 50 characters, which are 47% of the total number of spam messages in our spam traps.

We randomly split spam messages into the *support* collection (100000 messages) and the *target* collection (20000 messages). We also added the legitimate messages to *target* collection. The experiment goals were to detect the spam texts in the *target* collection without false detection of the legitimate messages texts in it. For any *target* message we scored up detection in case we were able to obtain *spam cluster*, which contained this *target* message. For any *target* message, we scored up *spam term* generation if we were able to create valid *spam term*, based on this *target* message.

The following table shows detection precision and recall obtained during test runs (DBSCAN *MinSamples* = 3):

In the following chart, we show the distribution of cluster sizes for spam messages in *target* collection (DBSCAN *Eps* = 60). The first bar represents non-*spam clusters*,

DBSCAN *Eps*	Precision	Recall	FP count
10	1	0.2033	0
20	1	0.3054	0
30	1	0.4981	0
40	0.9998	0.6320	2
50	0.9996	0.74025	5
60	0.9992	0.78075	11

Fig. 2. Precision/Recall

e.g. those with less than $C_S + 1 = 3$ elements in them. The number of such clusters represents the number of *target* spam messages, which were not detected:

Finally, we would like to show the percentage of cases, in which a valid *spam term* was obtained after detection:

5   Results

This report describes Spam Term Generator – a method for spam detection based on CTPH, clustering and longest common substring extraction.

The results of the experiments reveal that this technology suits well for spam detection because of a great detection rate and a low false detection rate. In addition, we have showed that usually it is possible to obtain a *spam term* for easy detection of similar messages in the future (Figs. 2, 3 and 4).

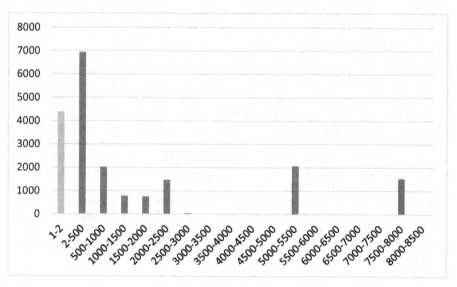

Fig. 3.   Cluster sizes

DBSCAN *Eps*	**%, valid spam terms**
10	65,87
20	72,70
30	56,66
40	47,38
50	45,54
60	45,21

Fig. 4.   Percentage of successful spam term generation for the detected texts

Further investigation is required to determine how the size of support collection influences the detection rate as well as possibility of an online adjunction of the *support* collection by detected texts from the *target* collection.

References

1. Damage caused by spam, Kaspersky IT Encyclopedia, https://encyclopedia.kaspersky.com/knowledge/damage-caused-by-spam/. Accessed 10 Jun 2020
2. Spam delivered through your company website, Kaspersky business blog, https://www.kaspersky.com/blog/contact-form-spam/27880/. Accessed 10 Jun 2020
3. Kornblum, J.: Identifying almost identical files using context triggered piecewise hashing, In: Proceedings of DFRWS 2006, pp. 91–97, Lafayette, Indiana (2006)
4. Ester, M., Kriegel, H.-P., Sander, J., and Xu, X.: A density-based algorithm for discovering clusters in large spatial databases with noise. In: Simoudis, E., Han, J., Fayyad, U. (eds.) Proceedings of the 2nd International Conference on Knowledge Discovery and Data Mining, pp. 226–231, AAAI Press, Portland, Oregon (1996)
5. Reaction to the DEC Spam of 1978, https://www.templetons.com/brad/spamreact.html. Accessed 10 Jun 2020
6. Chen, J., Fontugne, R., Kato, A., Fukuda, K.: Clustering spam campaigns with fuzzy hashing, In: Proceedings of AINTEC Asian Internet Eng. Conf., pp. 66, Chiang Mai, Thailand (2014)
7. Usage of fuzzy hashes, https://rspamd.com/doc/fuzzy_storage.html. Accessed 10 Jun 2020
8. Gusfield, D.: Algorithms on Strings, Trees, and Sequences: Computer Science and Computational Biology. Cambridge University Press, New York (1997)
9. Algorithms and Theory of Computation Handbook, CRC Press LLC, 1999, "Levenshtein distance", Dictionary of Algorithms and Data Structures [online], Paul E. Black, ed. 15 May 2019, https://www.nist.gov/dads/HTML/Levenshtein.html. Accessed 10 Jun 2020
10. Benkovich, N., Dedenok, R., Golubev, D.: Deepquarantine for suspicious mail, arXiv 305 preprint arXiv:2001.04168 (2020)
11. Broder, A.Z.: On the resemblance and containment of documents, In: Proceedings of Compression and Complexity of SEQUENCES 1997, pp. 21–29, IEEE Computer Society, Los Alamitos, California (1998)
12. Wallace, B.: Optimizing ssDeep for use at scale, Grooten, M. (ed.), Virus Bulletin, Cylance, USA (2015)

Machine Learning for Big Data Analysis in Drug Design

Galina Samigulina[1] and Zarina Samigulina[2(✉)]

[1] Institute of Information and Computing Technologies, Almaty, Kazakhstan
kz.galinasamigulina@gmail.com
[2] Kazakh-British Technical University, Almaty, Kazakhstan
samigulinaresearch@gmail.com

Abstract. Researches are devoted to the urgent problem of creating a highly effective technology for predicting the properties of chemical compounds, to create new drugs based on the latest achievements of artificial intelligence and computer technology. Computer-based drugs prediction with desired properties allows reducing costs during the selection of promising chemical compounds that can act as candidates for further research. There was developed a multi-agent system for predicting the "structure-property" dependence of drug compounds based on modified algorithms of artificial immune systems. The article presents the architecture of a multi-agent system, as well as an example of an analysis of a descriptors database of drug compounds of sulfanilamides based on intelligent and statistical algorithms. A comparative analysis of the models effectiveness prediction based on various modified AIS algorithms has been carried out.

Keywords: Big data · Machine learning and statistical methods · Drug design · Descriptors · Modified immune network algorithm

1 Introduction

In the pharmaceutical industry, interest in artificial intelligence (AI) has grown over the past five years. According to the BiopharmaTrend resource, modern IT companies such as Google, Apple, Amazon, Facebook, Microsoft were able to clearly demonstrate the capabilities of modern machine learning methods for solving various applied problems. In this connection, the most famous pharmacological companies (GSK, Sanofi, Abbvie, Genetech, etc.) began to take an active part in the use of modern intelligent algorithms for new drugs creation. The development of AI and the need for large computing capacity led to the development of modern supercomputers, such as NVIDIA DGX-1 AI for data processing in medical institutions. In turn, IBM is partnering with pharmaceutical giant Pfizer to use the Watson supercomputer in order to search for anti-cancer drugs. At the beginning of 2019, several large pharmacological companies signed cooperation agreements with research organizations in the field of AI, for example, the American Charles River Labs and the Canadian Atomwise companies in order to develop pharmaceutical compounds based on their structure. Therefore, the research in the application of modern

© Springer Nature Switzerland AG 2020
G. Nicosia et al. (Eds.): LOD 2020, LNCS 12566, pp. 457–469, 2020.
https://doi.org/10.1007/978-3-030-64580-9_38

AI approaches for the development of new drugs is an actual task and can reduce costs by hundreds of times at choosing promising candidates for chemical compounds with desired pharmacological properties.

Bio-inspired AI methods, such as: neural networks (NN), genetic algorithms (GA), artificial immune systems (AIS) and swarm intelligence (SI) are became widespread in solving the problem of predicting the "structure-property" dependence of drug compounds. In work [1] there was proposed a Quantitative Structure – Activity Relationship (QSAR) modeling approach based on a combination of the Simple Competitive Learning (SCL) neural network and the Radial Basis Function (RBF) neural network. Researches [2] are devoted to the consideration of the advantages and disadvantages of neural networks in the field of chemoinformatics, as well as the prospects for the use of neural networks with deep training for the synthesis of new drugs. An article [3] discusses an interpreted universal deep neural network for predicting chemical properties. Since chemical databases store information in the Simplified Molecular Input Line Entry System (SMILES) text format, which is a universal standard in cheminformatics, then structural information, that can be used to predict complex chemicals, is encrypted in each SMILES line. The authors developed the SMILES2vec model designed to study automatically the functions of SMILES to predict toxicity, activity, solvation energy, solubility, etc. based on neural networks. The work [4] presents the use of GA for 4D-QSAR analysis and pharmacophore modeling. The article [5] is dedicated to the optimal management of the stochastic HIV model using drug therapy. The model demonstrates the effect of antiretroviral drugs at different stages of infection. Optimal model control is achieved using the genetic algorithm (GA). In [6], QSAR models are considered for predicting the physicochemical properties of certain barbiturate derivatives using molecular descriptors and a multiple linear regression genetic algorithm. The researches [7] are focused on the use of the PSO-SVR (Particle Swarm Optimization - Support Vector Regression) algorithm for molecular docking in predicting the antimalarial activity of enantiomeric cycloguanil analogues. The work [8] presents a method for classifying biomedical information based on the algorithm of particle swarm with inertia weight and mutation (PSO with Inertia Weight and Mutation). The article [9] is devoted to a comparative analysis of multi-linear regression algorithms, particle swarm algorithm, artificial neural network and genetic programming in the production of mini-tablets. Research [10] presents the use of swarm intelligence, the CEPSO algorithm (Chaos-embedded particle swarm optimization) for docking and virtual screening of protein ligands.

Along with other intelligent algorithms, artificial immune systems are of particular interest. AIS algorithms are widely used in the field of medicine, bioinformatics and pharmacology. The work [11] presents the use of the Negative Selection Classification Algorithm of encephalogram signals responsible for the movement of human limbs. The researches [12] are devoted to studying the effectiveness of the process of artificial vaccination of training systems, where AIS memory cells and their antibodies are introduced during the training process. The article [13] presents the use of AIS for medical diagnostic systems for the disease recognition. The researches [14] are focused on the use of AIRS for the early detection of breast cancer, as the results of examinations such as mammography, ultrasonography, and magnetic resonance imaging do not always accurately diagnose the disease. In [15], the AIRS algorithm is used to diagnose tuberculosis disease based on real data about the epicrisis of patients. The work [16] is devoted to a comparative analysis of the AIS and other heuristic algorithms for predicting the structure of proteins. The research uses the principle of clonal selection for the prediction of protein structure. The article [17] considers a new algorithm based on the support vector method and artificial immune systems for the diagnosis of tuberculosis disease. Modeling was carried out on the basis of WEKA software. The work [18] is devoted to predicting the sequence of genes based on AIRS. There is proposed a method for gene selection based on local selection of features using AIRS in order to find the optimal biological sequences. Nowadays, a promising area of research is the development of effective modified algorithms. For example, in the work [19], a modified particle swarm optimization algorithm is considered to evaluate the parameters of biological systems.

Therefore, the analysis of the literature shows the relevance of research on the creation of new modified artificial intelligence algorithms for solving the problems of bioinformatics and pharmacology.

The following structure of the article is proposed. The second section describes the formulation of the research problem. The third section is devoted to the development of a multi-agent system for predicting the structure-property dependence of drug compounds. The fourth section contains the results of modeling and experiments. In conclusion, the main advantages of the developed system are described.

2 Statement of the Problem

The purpose of these studies is the computer molecular design of new drugs with desired properties to reduce time and financial costs.

The research problem statement is formulated as follows: it is necessary to develop a multi-agent system for analyzing databases and predicting the "structure-property" dependence of new drug compounds based on modified algorithms of artificial immune systems using machine learning methods and statistical data analysis.

3 Approaches and Methods

3.1 Multi-agent Systems

The implementation of the proposed QSAR predicting technology is carried out in the class of multi-agent systems [20]. Nowadays, multi-agent systems (MAS) are widely used in the creation of intelligent systems for various purposes. A multi-agent system is a set of autonomous agents that operate in a software environment and can interact with each other. Such advantages of MAS as decentralization, rational use of computing resources, flexibility and multifunctionality, self-organization and scalability determine the success of this approach. The technology of multi-agent systems has found wide use in various applications. For example, the work [21] presented a multi-agent artificial immune system for a computer security system, which, as a result of experiments, showed the promise of using this approach. The research [22] is focused on the use of multi-agent technology in order to solve the problems of personalized medicine. The article [23] considers a model of a multi-agent system for diagnosis of personality types. A particular interest has the development of MAS using the ontological models. In a series of works [24, 25], there is considered the possibility of using ontologies at the creation of a software for multi-agent systems based on a model-oriented approach. There is given the rationale for the use of ontologies, since MAS consists of many components and many complex aspects of their interaction must be taken into account.

3.2 Artificial Immune Systems Approach

At the end of the eighties of the last century, there are appeared the first scientific works [26], which laid the foundation for the research in the field of immune systems and machine learning. The bio-inspired AIS approach was formed in the nineties of the last century [27] on the basis of the principles of the human immune system functioning in order to solve various applied problems [28]. AIS solve a fairly wide range of optimization problems, data analysis, pattern recognition and prediction [29], classification, clustering, machine learning, adaptive control, etc. Artificial immune systems are adaptive, distributed and self-organizing systems that are able to learn, remember information and can make decisions in an unfamiliar situation. There has been developed AIS that implements the mechanisms of the human immune system based on clonal selection [30], negative selection [31], and on the immune network model [32]. The main aspects of the promising direction of immune computing are presented in the encyclopedia [33]. There is given the definition of AIS given by de Castro and Timmis in 2002: AIS are adaptive systems based on theoretical immunology and observed immune functions, on principles and models that are used to solve various problems. Very promising is the AIS approach for QSAR modeling. There has been published a series of works [34] on the use of AIS for drug design based on an artificial immune recognition system (AIRS) algorithm. The researches have shown that solving the classification problem using AIS has great prospects in the development of drugs, as well as for virtual screening.

Therefore, nowadays, there have been received many modifications of the well-known AIS algorithms for various specific applications [35], but there is no comprehensive research in order to create a universal AIS software for QSAR, which is publicly available. The development of the general theory of artificial immune systems based on an integrated approach and the systematization of numerous currently existing modified AIS algorithms is relevant.

4   Multi-agent Technology for Computer Molecular Design

A huge amount of experimental data has been accumulated in the field of organic chemistry, molecular biology and medicine, which cannot be processed without modern information technologies through the use of high-performance computing systems and biological process modeling. The intensive development of artificial intelligence approaches contributes to the formation of new trends in QSAR research. The main stages in the computer molecular design of drugs with desired properties (including those with a given pharmacological activity) are: a description of the chemical compound structure; a development of a database with an optimal set of descriptors; a selection of an algorithm for predicting pharmacological properties/activity; a selection of candidates for chemical compounds with desired properties for further research. During the multi-agent technology development for the synthesis of new drugs, the following features of the research in the field of QSAR prediction should be taken into account: the specificity of QSAR research; a huge amount of chemical information, which leads to an increase in calculations at the hardware-software implementation of simulation technology; poor data; class imbalance during the creation of prediction models based on bio-inspired and statistical approaches, which affects the quality of the forecast. A key component of the proposed intellectual multi-agent technology is the use of modified algorithms of artificial immune systems in order to solve the problem of QSAR prediction [35]. There is given an algorithm of the multi-agent prediction technology operation.

Algorithm 1.

1. Connection of a descriptors database of structural chemical information characterizing the considered drug compounds.
2. Choosing a method for selecting informative descriptors and constructing an optimal data set: Weight by PCA, Weight by Gini Index, Weight by SVM, Weight by tree importance and, etc.
3. Creation of an optimal immune model based on a selected set of descriptors [35].
4. Choosing an artificial immune system algorithm: based on clonal selection, CLONALG, CSA; artificial immune recognition system algorithm (AIRS1), immune network modeling algorithm [35].
5. AIS training on standards compiled by experts from drug descriptors with well-known properties.
6. Pattern recognition based on the selected AIS algorithm.
7. Pattern recognition based on other machine learning algorithms to conduct a comparative analysis of the effectiveness of AIS models (Generalized Linear Model, Fast Large Margin, Deep Learning, Gradient Boosted trees).
8. Prediction of the structure-property dependence of chemical compounds.
9. Evaluation of the model prediction effectiveness based on metrics: Accuracy, Classification Error, Precision, Recall, f-measure, AUC, Total Time.
10. Comparison of the effectiveness of prediction models.
11. Selection of the best algorithms and prediction based on them. Selection of candidates for new chemical compounds with desired properties for further research.

Algorithm 1 uses the developed modified immune network modeling algorithm [35] based on homologous proteins, which allows to solve the problem of pattern recognition at the boundary of selected classes for chemical compounds with almost the same structure, but with different pharmacological properties. The use of this modified AIS algorithm to search for drug compounds candidate is being applied for the first time.

Figure 1 shows the main components of intellectual multi-agent technology for QSAR prediction.

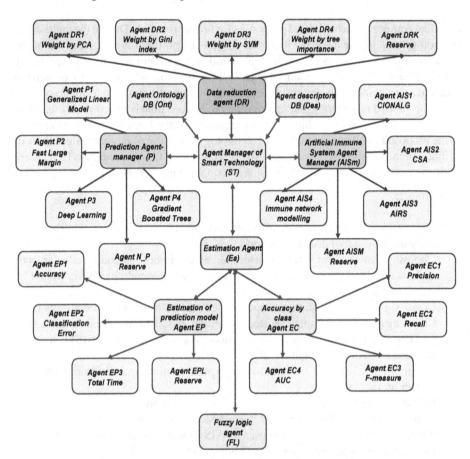

Fig. 1. Multi-agent technology for computer molecular design of drugs

The specification of the functions of the agents shown on Fig. 1 is described in Table 1. According to the modeling results, the modified algorithm is selected that has the best predictive results for the studied database of drug descriptors.

A multi-agent "structure-property" prediction system for drug compounds is implemented using the JADE software product. Developed agents are stored in containers distributed according to their functional purpose. The main container contains system agents: Agent Management System and Directory Facilitator. Other containers consist of agents for: data preprocessing, prediction, evaluating the effectiveness of algorithms, etc.

Table 1. The fragment of agents specification of multi-agent Smart-technology

Name of the agent	Symbol	Functions
Agent manager	SS	This agent monitors the operation of the entire multi-agent system and promotes a phased algorithmic solution of tasks by additional agents (data reduction agents, prediction, assessment, ontology database agents and descriptor databases)
Agent ontology	Ont	The agent contains ontological domain models developed by specialists on QSAR modeling, as well as ontologies of algorithms used for the research
Agent descriptors DB	Des	The agent consists of descriptor databases of the chemical compounds that contain the following information: molecular descriptor values, topological indices, etc.
Data reduction Agent	DRA	The agent performs the reduction of uninformative descriptors
Artificial immune system agent manager	AISm	The agent manager of artificial immune system control algorithms
Agent CLONALG	AIS1	The agent implements a prediction model of artificial immune systems based on clonal selection CLONALG
Agent CSA	AIS2	The agent implements a prediction model based on artificial immune systems CSA
Agent AIRS	AIS3	The agent contains AIRS artificial immune system algorithm
Agent Immune Network Modeling	AIS4	The agent contains an immune network modeling algorithm based on the principles of molecular recognition
Estimation Agent	Ea	The agent acting as a manager of algorithm efficiency assessment
Accuracy by class	EC	The agent for prediction models effectiveness assessment for each class
…	…	…
Estimation of prediction model agent	EP	The agent of prediction model assessment

5   Experimental Researches

The effectiveness of the prediction models largely depends on the nature of the initial data. A feature of QSAR databases is the large dimensionality and heterogeneity of data. Therefore, the experimental researches presented in this article are aimed at considering database for predicting the "structure-property/activity" dependence of chemical compounds by various statistical and bio-inspired artificial intelligence methods (firstly, without uninformative descriptors reduction, and then after uninformative descriptors reduction) with the aim of substantiating the main provisions of the proposed Smart-technology based on a multi-agent approach (Sects. 3 and 4).

5.1 Data Sets Description

Let consider the author sulfanilamide descriptor database compiled on the basis of the Mol-Instincts resource. Sulfanilamides are a group of antimicrobials of various durations of action. In order to predict the QSAR dependence, experts identified 3 classes: class 1 - short duration of action, less than 10 h; class 2 - average duration of action, 10–24 h; class 3 - long duration of action, 24–48 h. Each substance is described by 2005 descriptors. Table 2 presents a fragment of a sulfonamide database containing the following information: nAT - amount of atoms; MW - Molecular weight; AMW - average molecular weight; Sv - sum of atomic van der Waals volumes (scaled on Carbon atom), etc.

Table 2. A fragment of a Sulfanilamide database

SMILES	nAT	MW	AMW	Sv	...	Class
COc2cc(NS(=O)(=O)c1ccc(N)cc1)nc(OC)n2	27.0	250.275	9.270	17.88	...	1
Cc2cc(C)nc(NS(=O)(=O)c1ccc(N)cc1)n2	33.0	278.328	8.440	21.08	...	1
Cc2noc(NS(=O)(=O)c1ccc(N)cc1)c2C	31.0	267.301	8.620	19.59	...	1
Cc2nnc(NS(=O)(=O)c1ccc(N)cc1)s2	27.0	270.324	10.010	17.97	...	1
Cc2cc(NS(=O)(=O)c1ccc(N)cc1)no2	46.0	311.44	6.77	26.67	...	1
Cc2cc(NS(=O)(=O)c1ccc(N)cc1)nc(C)n2	33.0	278.32	8.44	21.08	...	1
Nc1ccc(cc1)S(N)(=O)=O	46.0	311.44	6.77	26.67	...	1
...	...	...	...	...	...	...
CC(=O)NS(=O)(=O)c1ccc(N)cc1	24.0	214.238	8.930	15.00	...	2

5.2 Simulation Results

Next, let consider the effectiveness of using modified artificial intelligence algorithms [35, 36] to process the database of an intellectual medical expert system based on the considered characteristics. Let consider modeling of the presented databases using the following statistical algorithms and artificial intelligence methods according to the developed multi-agent technology (Fig. 1, Table 3): Generalized Linear Model, Fast Large Margin, Deep Learning, Gradient Boosted Trees.

Table 3. A comparative analysis of the simulation results by various methods

Performance	Generalized Linear Model		Fast Large Margin		Deep Learning		Gradient Boosted Trees	
	B	A	B	A	B	A	B	A
Accuracy	71.8%	72.7%	66.4%	68.4%	70.5%	75.2%	84.7%	89.7%
Classification error	28.2%	27.3%	33.6%	31.6%	29.5%	24.8%	15.3%	10.3%
Precision	66.9%	72.7%	–	54.3%	63.6%	67.2%	79.9%	94.0%
Recall	33.3%	30.6%	–	36.7%	30.4%	52.0%	74.3%	74.3%
F-measure	43.9%	42.8%	–	43.2%	40.8%	58.6%	76.5%	82.9%
AUC	0.776	0.779	0.521	0.720	0.742	0.774	0.917	0.902
Total time	3 min 53 s	1 min 37 s	3 min 12 s	1 min 18 s	6 min 17 s	3 min 45 s	6 min 23 s	3 min 34 s
Training time	31 s	25 s	17 s	23 s	8 s	4 s	8 s	4 s
Scoring time	23 ms	18 ms	17 ms	5 ms	22 ms	12 ms	24 ms	9 ms

Since the structure and dimension of the database affects the quality of the prediction, the QSAR prediction efficiency will be assessed before the reduction of uninformative descriptors based on the data preprocessing algorithms (Before, B) and after (After, A). The following indicators were used as criteria for models assessment: accuracy; classification error; sensitivity; specificity; area under the error curve (AUC); time characteristics (Total time, Training time, Scoring time). On the Fig. 2 there is considered ROC comparative analysis of the algorithms effectiveness before and after data reduction. The closer the line trajectory to the upper left corner (tends to 1), the more effective the prediction model for solving a specific problem.

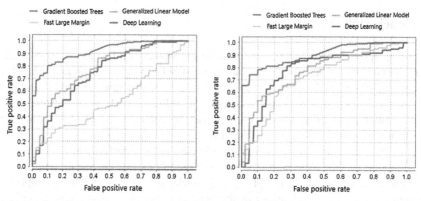

Fig. 2. ROC analysis of the prediction models effectiveness before the selection of informative descriptors and after it

On the basis of obtained data, it can be seen that the predictive ability of the considered models varies depending on the nature of data, and also does not exceed 89,7%. The Table 4 shows the modeling results based on artificial immune systems. The predictive ability of modified AIS algorithms is higher than considered in Table 3.

Table 4. Prediction based on artificial immune systems

Performance	CLONALG		CSCA		AIRS1		AIS	
	B	A	B	A	B	A	B	A
Correctly Classified Instances	76.96%	87.27%	80%	86.66%	75.76%	83.63%	86.66%	92.72%
Incorrectly Classified Instances	23.03%	12.72%	20%	13.33%	24.24%	16.36%	13.33%	7.27%
Precision	0.767	0.872	0.886	0.896	0.793	0.856	0.867	0.929
Recall	0.77	0.873	0.8	0.867	0.758	0.836	0.867	0.927
F-measure	0.768	0.871	0.802	0.865	0.765	0.842	0.867	0.927
ROC area	0.823	0.902	0.864	0.875	0.825	0.887	0.909	0.949
Mean absolute error	0.153	0.084	0.133	0.088	0.161	0.109	0.088	0.048
Root mean squared error	0.391	0.291	0.365	0.298	0.402	0.330	0.298	0.220
Runtime	16 s	0.28 s	57 s	0.03 s	168 ms	0.25 s	2 s	0.27 s

On the basis of the analysis of experiments, modified AIS algorithms give better results for the studied database of sulfanilamide descriptors in comparison with other artificial intelligence algorithms and can be used to develop new drugs.

6 Conclusion

The advantages of the developed multi-agent system are: an integrated approach and interdisciplinarity of research, affecting many fields of science: organic chemistry, molecular biology, medicine, chemoinformatics, computer engineering, IT – technologies, artificial intelligence approaches, and many others; a high quality of the training sample, which most fully describes the molecular structure of a chemical compound with a minimum set of descriptors; the development of technological chains based on modified AIS algorithms adapted to a specific data set; the performance, which allows the prediction of the properties/activity of a large number of chemical compounds in the required time; the possibility of using for various classes of chemical compounds; an assessment of the effectiveness of the used modified AIS algorithms; a sufficient accuracy in predicting the chemical compounds properties.

This work was financially supported by the SC of the MES of the RK under grant №AP05130018 on the theme: "Development and analysis of databases for the information system for predicting the "structure-property" dependence of drug compounds based on artificial intelligence algorithms" (2018–2020).

References

1. Sheikhpour, R., Sarram, A., Rezaeian, M., Sheikhpour, E.: QSAR modelling using combined simple competitive learning networks and RBF neural networks. Int. SAR and QSAR Environ. Res. **29**, 257–276 (2018)
2. Ghasemi, F., Mehridehnavi, A., Perez-Garrido, A., Perez-Sanchez, H.: Neural network and deep-learning algorithms used in QSAR studies: merits and drawbacks. Int. Drag disc. Today. **23**(10), 1784–1790 (2018)
3. Goh, G., Hodas, N., Siegel, C., Vishnu, A.: An interpretable general-purpose deep neural network for predicting chemical properties. In: Proceedings of ACM SIGKDD Conference, pp. 1–8 (2018)
4. Tüzün, B., Yavuz, S., Sarıpınar, E.: 4D-QSAR analysis and pharmacophore modeling: propoxy methylphenyl oxasiazole derivatives by electron conformatitional-genetic algorithm method. Int. J. Phys. Theor. Chem. **14**(2), 139–164 (2018)
5. Saeedizadeh, F., Moghaddam, R. Optimal control of HIV stochastic model through genetic algorithm. In: Proceedings of 7th International Conference on Computer and Knowledge Engineering, pp. 401–405 (2017)
6. Shafiei, F., Esmaeili, E.: QSAR models to predict physico-chemical Properties of some barbiturate derivatives using molecular descriptors and genetic algorithm- multiple linear regressions. Int. Iranian Chem. Commun. **7**(2), 170–179 (2019)
7. Inthajak, K., Toochinda, P., Lawtrakul, L.: Application of molecular docking and PSO–SVR intelligent approaches in antimalarial activity prediction of enantiomeric cycloguanil analogues. Int. J. **29**(12), 957–974 (2018)
8. Li, M., Zhang, M., Chen, H., Lu, S.: A method of biomedical information classification based on particle swarm optimization with inertia weight and mutation. Int. Open Life Sci. **13**(1), 355–373 (2018)
9. Barmpalexis, P., Karagianni, A., Karasavvaides, G., Kachrimanis, K.: Comparison of multi-linear regression, particle swarm optimization artificial neural networks and genetic programming in the development of mini-tablets. Int. J. Pharmaceutics **551**(1), 166–176 (2018)
10. Tai, H.K., Jusoh, S.A., Siu, S.W.I.: Chaos-embedded particle swarm optimization approach for protein-ligand docking and virtual screening. J. Cheminform. **10**(1), 1–13 (2018). https://doi.org/10.1186/s13321-018-0320-9
11. Rashid, N., Igbal, J., Mahmood, F., Abid, A., Khan, U., Tiwana, M.: Artificial immune system-negative selection classification algorithm (NSCA) for four class electroencephalogram (EEG) signals. Int. Front. Hum. Neurosci. **12**(439), 1–12 (2018)
12. Ahmad, W., Narayanan, A.: Principles and methods of artificial immune system vaccination of learning systems. In: Liò, P., Nicosia, G., Stibor, T. (eds.) ICARIS 2011. LNCS, vol. 6825, pp. 268–281. Springer, Heidelberg (2011). https://doi.org/10.1007/978-3-642-22371-6_24
13. Babu, M.S., Katta, S.: Artificial immune recognition system in medical diagnosis. In: Proceedings of the 6th IEEE International Conference on Software Engineering and Service, pp. 1082–1087 (2015)
14. Katsis, C.D., Gkogkou, I., Papadopoulos, C.A., Goletsis, Y., Boufounou, P.V., Stylios, G.: Using Artificial Immune Recognition Systems In Order To Detect Early Breast Cancer. Int. Intell. Syst. Appl. **2**(34), 40 (2013)

15. Shamshirband, S., Hessam, S., Javindnia, H., Amiribesheli, M., Vandat, S., Petkovic, D., Gani, A., Laiha, M.K.: Tuberculosis disease diagnosis using artificial immune recognition system. Int. J. Med. Sci. **11**(5), 508–514 (2014)
16. Mansour, R.F., Al-Ghamdi, F.: Comparison between artificial immune system and other heuristic algorithms for protein structure prediction. Int. Am. J. Bioinform. Res. **2**(4), 61–67 (2012)
17. Saybani, M., Shamshirband, S., Hormozi, S., Wah, T., Aghabozorgi, S., Pourhoseingholi, M., Olariu, T.: Diagnosing tuberculosis with a novel support vector machine-based artificial immune recognition·system. Int. Iran Red Crescent Med. J. **17**(4), 1–8 (2015)
18. Batur, C., Diri, B.: Identifying predictive genes for sequence classification using artificial immune recognition system. Int. J. Sci. Technol. **8**(4), 58–66 (2018)
19. Mosayebi, R., Bahrami, F.: A modified particle swarm optimization algorithm for parameter estimation of a biological system. Theor. Biol. Med. Modell. **15**(17), 1–10 (2018)
20. Samigulina, G.A., Massimkanova, Zh.A.: Multiagent system of recognize on the basis of modified algorithms of swarm intelligence and immune network modeling. In: Proceedings of the 12th International Conference Agents and Multi-agent Systems: Technologies and Applications, pp. 199–208 (2018)
21. Asa, A., Seo, H., Hassanien, A.E. Multi-agent artificial immune system for network intrusion detection and classification. In: International Joint Conference. Advances in Intelligent Systems and Computing, vol. 299, pp. 145–154 (2014)
22. Ivanović, M., Semnic, M. The role of agent technologies in personalized medicine. In: Proceedings of 5th International Conference on Systems and Informatics, pp. 1–10 (2018)
23. Ramírez, M.R., Ramírez Moreno, H.B., Rojas, E.M., Hurtado, C., Núñez, S.O. Multi-Agent System Model for Diagnosis of Personality Types. In: Proceedings of the 12th International Conference Agents and Multi-agent Systems: Technologies and Applications. Smart Innovation, Systems and Technologies book series (SIST), vol. 96, 209–214 (2018)
24. Freitas, A., Bordini, R.H., Vieira, R.: Model-driven engineering of multi-agent systems based on ontologies Int. Appl. Ontol. **12**, 157–188 (2017)
25. Freitas, A., Bordini, R.H., Vieira, R.: Designing multi-agent systems from ontology models. Lecture Notes in Computer Science, **11375**, 76–95 (2018)
26. Farmer, J.D., Packard, N., Perelson, A.: The immune system, adaptation and machine learning. Phys. D Nonlinear Phenom. **2**, 187–204 (1986)
27. Kephart, J.A.: A biologically inspired immune system for computers. Artificial Life IV. In: Proceedings of the Fourth International Workshop on Synthesis and Simulation of Living Systems, pp. 130–139 (1994)
28. Dasgupta, D.: Artificial immune systems and their applications, p. 306 (1999)
29. Cutello, V., Nicosia, G., Pavone, M., Timmis, J.: An immune algorithm for protein structure prediction on lattice models. IEEE Trans. Evol. Comput. **1**, 101–117 (2007)
30. Zhu, H., Wu, J., Gu, J.: Studies on immune clonal selection algorithm and application of bioinformatics. Int. J. Intell. Eng. Syst. **8**(1), 10–16 (2015)
31. Gao, X.Z., Ovaska, S.J., Wang, X., Chow, M.Y.: Clonal optimization-based negative selection algorithm with applications in motor fault detection. Int. Neural Comput. Appl. **18**(7), 719–729 (2009)
32. Tarakanov, A.O., Borisova, A.V.: Formal immune networks: self-organization and real-world applications. Int. Adv. Appl. Self-organizing Syst. 321–341 (2013)
33. Timmis, J.: Artificial immune systems. In: Sammut, C., Webb, G.I. (eds.) Encyclopedia of Machine Learning and Data Mining. Springer, Boston (2017)
34. Ivanciuc, O.: Artificial immune system classification of drug-induced torsade de pointes with AIRS. Int. J. Molecular Des. **5**, 488–502 (2006)

35. Samigulina, G.A., Samigulina, Z.I.: Modified immune network algorithm based on the Random Forest approach for the complex objects control. Artif. Intell. Rev. **52**(4), 2457–2473 (2018). https://doi.org/10.1007/s10462-018-9621-7
36. Samigulina, G.A., Sami Samigulina, G.A., Samigulina, Z.I.: Ontological model of multi-agent Smart-system for predicting drug properties based on modified algorithms of artificial immune systems. Theor. Biol. Med. Modell. **17**(12), 1–22 (2020)

Pareto-Weighted-Sum-Tuning: Learning-to-Rank for Pareto Optimization Problems

Harry Wang[1]([⊠])(iD) and Brian T. Denton[2]([⊠])(iD)

[1] Computer Science and Engineering, University of Michigan,
Ann Arbor, MI 48109, USA
`harrydw@umich.edu`
[2] Industrial and Operations Engineering, University of Michigan,
Ann Arbor, MI 48109, USA
`btdenton@umich.edu`

Abstract. The weighted-sum method is a commonly used technique in Multi-objective optimization to represent different criteria considered in a decision-making and optimization problem. Weights are assigned to different criteria depending on the degree of importance. However, even if decision-makers have an intuitive sense of how important each criteria is, explicitly quantifying and hand-tuning these weights can be difficult. To address this problem, we propose the Pareto-Weighted-Sum-Tuning algorithm as an automated and systematic way of trading-off between different criteria in the weight-tuning process. Pareto-Weighted-Sum-Tuning is a configurable online-learning algorithm that uses sequential discrete choices by a decision-maker on sequential decisions, eliminating the need to score items or weights. We prove that utilizing our online-learning approach is computationally less expensive than batch-learning, where all the data is available in advance. Our experiments show that Pareto-Weighted-Sum-Tuning is able to achieve low relative error with different configurations.

Keywords: Machine learning · Multi-objective optimization · Information retrieval · Online learning

1 Introduction

Multi-objective optimization (also known as multicriteria optimization or Pareto optimization) involves the optimization of multiple objective criteria simultaneously. Often times, levels of importance of different criteria are difficult to quantify explicitly or directly and can be dependent on latent or subjective preferences [7,10,11]. It may be possible to assign levels of importance in simple problems and contexts with a low number of criteria with decent accuracy. However, this becomes harder as the number of criteria increases or as the nature of the criteria become complex and difficult to compare.

© Springer Nature Switzerland AG 2020
G. Nicosia et al. (Eds.): LOD 2020, LNCS 12566, pp. 470–480, 2020.
https://doi.org/10.1007/978-3-030-64580-9_39

In many fields, such as structural engineering, transportation, medicine, business, or finance, there are independent and conflicting criteria to consider [2]. These could include risk, reward, cost, side effects, and others. When making decisions that pertain to such fields, it is necessary to quantify the relative degree of importance for each of these criteria. This allows decision-makers to properly weigh each criteria in a Multi-objective (Pareto) optimization problem to the right degree.

We consider m criteria with objective function coefficients $[c_1, c_2, ..., c_m]$. Thus, there are m different criteria to consider for our decision-making problem. We also define $[\alpha_1, \alpha_2, ..., \alpha_m]$ where α_i is the importance level given to criteria c_i. This is necessary to define the weighted-sum objective function of a Pareto optimization problem: $\sum_{i=1}^{m} \alpha_i c_i$, $\alpha_i \in (0, 1)$. The weighted sum method is often described as being one of the simplest ways of solving a Pareto optimization problem. However, it requires a decision-maker to know the criteria weights $[\alpha_1, \alpha_2, ..., \alpha_m]$ a priori [21].

Hand-tuning criteria weights is known to be a challenging problem in Pareto optimization. It often requires background knowledge on the context of a specific problem, as well as information on the preferences of a specific user. Studies in inverse optimization have found that decision-makers may be unaware of their exact weighting schemes even when they select one decision over another [1,6, 8,12,13]. In other words, decision-makers may not know exactly how much they prioritize one criteria over another. While methods have been developed to learn weights, they often require decision-makers to explicitly score items rather than the relative weights themselves. However, techniques from information retrieval have utilized learn-to-rank methods rather than scores to discover preferences of a decision-maker [5,14,15]. We leverage this fact to eliminate the need for decision-makers to score items directly. This allows us to automate the process of weight-tuning.

In this study, we propose the Pareto-Weighted-Sum-Tuning (PWST) algorithm, and discuss how it addresses the problem of quantifying the importance of different criteria to elicit objective function coefficients from decision-makers. We will also illustrate how it can be applied to a well known problem involving stock investment.

The remainder of this paper is organized as follows. In Sect. 2, we discuss Related Work. In Sect. 3, we discuss the Method Developed by us. In Sect. 4, we discuss a Model Application that PWST has been applied to. In Sect. 5, we discuss the Experimental Results of applying PWST to our Model Application. Finally, in Sect. 6, we describe the most important conclusions that can be drawn from this study.

2 Related Work

Several methods exist for assisting decision-makers in selecting criteria weights. One of the most prominent ones is the Analytical Hierarchy Process (AHP). This consists of decision-makers performing pairwise comparisons between nodes in a decision tree. Decision-makers must roughly quantify how much more important one node is over another using a rating system [9].

Another method is CODA, which was proposed in the context of structural engineering. In this scheme, an executive will score a structural-design decision based on multiple criteria. This is an iterative process that generates decisions based on feedback from the executive [2]. Several other methods exist that can be broadly categorized as rating or categorization methods. These methods rely on users to directly assign weights and group criteria based on general levels of importance respectively. In all of these methods, the decision-maker is required to make intermediate quantifications on weight criteria [16].

Evidence shows that decision-maker's objective and subjective preferences for criteria weights can be inferred from choices and comparisons. The Multi-Nomial Logit (MNL) model has been deployed successfully in marketing and transportation, natural-language processing, and computer vision. This shows how data-driven models can be effective in learning decision-maker's preferences. Recent advances have framed these models as learning parameters to best explain the data generated from decision-makers [18]. Inverse optimization has also explored ways to explain decision-makers' underlying behaviors using observed data [1,6,8,12,13].

Similarly, studies in information retrieval have utilized learn-to-rank methods to predict the degree of relevance of documents with respect to specific queries made by users [5,14,15]. Each document is given a score based on its predicted relevance to a query. Parallels between information retrieval and Pareto optimization can be drawn. Both aim to optimize the the fit between coefficients learned and feedback from users. We frame queries as decisions epochs and document relevance rankings as criteria tuple rankings. We also frame the relevance score of a document as the total objective value of a decision.

To the best of the authors' knowledge, this is the first attempt to learn objective functions in Pareto optimization problems using an online learn-to-rank approach while eliminating explicit quantifications from a decision maker.

3   Method Developed

The PWST algorithm we propose analyzes discrete choices and comparisons made sequentially by decision-makers. This eliminates the need for the decision-maker to explicitly quantify criteria-weights.

We utilize a Ranking-Support-Vector-Machine [14] that learns how to rank data-points by fitting a function f to user inputs based on minimizing a loss function. The Ranking-SVM will output $[\alpha_1, \alpha_2, ..., \alpha_m]$, the vector of importance levels for the m criteria. The Ranking-SVM takes in parameters $[t_1, t_2, ..., t_x]$, and $[r_1, r_2, ..., r_x]$ where t_i is a decision represented by its resulting tuple of criteria values and r_i is the rank given to t_i. A higher r_i value indicates a more favorable ranking for t_i, For example, a decision-maker is given $[(1, 2), (0, 3)]$, where $t_1 = (1, 2)$ and $t_2 = (0, 3)$. The tuple t_1 has a c_1 value of 1 and a c_2 value of 2. The tuple t_2 has a c_1 value of 0 and a c_2 value of 3. If the decision maker favors t_2 over t_1, they will respond with $[1, 2]$. In the case where a decision-maker is given $[t_1, t_2, t_3]$, and the decision-maker believes t_2 is the most favorable tuple and t_3 is the least favorable tuple, the decision maker will respond with $[2, 3, 1]$. Note: the higher the r_i value, the more favorable t_i is.

We define r_{true} as the *target ranking*. The target ranking is the decision-maker's true ranking of a certain set of data-points. It is used to calculate the loss function in PWST. Optimizing the loss function allows PWST to generate the ideal decision boundary based on the data provided to it.

The pseudo code in Algorithm 1 summarizes the main steps of the iterative PWST algorithm.

Algorithm 1. Pareto-Weighted-Sum-Tuning

Require: n data-points, x data-points at each iteration where $x \leq n$

  α-vectors $= \emptyset$

  $i = 0$

  $[t_1, t_2, ..., t_x] \leftarrow x$ data-points sampled from the data-set and displayed to decision-maker in random order

  while $i \leq \lfloor \frac{n}{x} \rfloor$ **do**

    (I) $[r_1, r_2, ..., r_x] \leftarrow$ ranking given by decision-maker for the x sampled data-points; this action is equivalent to "correcting" the order predicted by the algorithm

    (II) $[\alpha_1, \alpha_2, ..., \alpha_m] \leftarrow$ Ranking-SVM($[t_1, t_2, ..., t_x], [r_1, r_2, ..., r_x]$)

    (III) $[t_1, t_2, ..., t_x] \leftarrow x$ data-points sampled from the data-set (Note: in this step, Pareto-Weighted-Sum-Tuning uses the α-values learned to predict r_{true} and displays $[t_1, t_2, ..., t_x]$ according to its predicted order based on $\sum_{i=1}^{m} \alpha_i c_i$, where a higher sum corresponds to a better rank)

  end while

  $\alpha = [\bar{\alpha}_1, \bar{\alpha}_2, ..., \bar{\alpha}_m]$ where $\bar{\alpha}_i$ is the mean value of α_i of vectors in α-vectors

  return α

3.1 Theoretical Analysis

From step (III) of the algorithm, we define the set of ordered pairs that are generated by PWST as r_{pred}. A will be used to represent the number of ordered pairs that exist in both r_{true} and r_{pred}. B will be used to represent the number of ordered pairs that exist in r_{pred}, but not r_{true}. This allows us to calculate Kendall's τ, which is commonly used to measure discrepancies between two sets of rankings [5,14,15]. It is defined as follows:

$$\tau(r_{true}, r_{pred}) = \frac{A - B}{A + B} \tag{1}$$

The Ranking-SVM embedded in the PWST algorithm aims to learn a ranking function f that minimizes the loss function $-\tau(r_{true}, r_{pred})$ for the x data-points sampled at each iteration. The function f will contain our desired coefficients $[\alpha_1, \alpha_2, ..., \alpha_m]$ [5,14,15].

The fact that PWST is an online-learning algorithm contributes to its efficiency over a batch-learning approach.

Proposition 1. *Splitting the data-set into $\lfloor \frac{n}{x} \rfloor$ mini-batches (where $x < n$) reduces the number of pairwise comparisons needed from the decision-maker*

Proof. Assuming decision-makers decompose the ranking decisions into pairwise comparisons (a commonly used heuristic for ranking), if all n data-points were given to the user at once, the user would have to perform $\frac{n!}{(n-2)!2}$ pairwise comparisons.

If x data-points were given to the user $\lfloor \frac{n}{x} \rfloor$ times, the user would have to perform $\lfloor \frac{n}{x} \rfloor(\frac{x!}{(x-2)!2})$ pairwise comparisons. The propositions follows because

$$\lfloor \frac{n}{x} \rfloor (\frac{x!}{(x-2)!2}) < (\frac{n}{x})(\frac{x!}{(x-2)!2}) < \frac{n!}{(n-2)!2} \tag{2}$$

Where the latter inequality is true because

$$\frac{x!}{x(x-2)!} < \frac{(n-1)!}{(n-2)!} \tag{3}$$

$$\frac{(x-1)!}{(x-2)!} < \frac{(n-1)!}{(n-2)!} \tag{4}$$

The last inequality follows from the fact that $x < n$. ∎

Thus, decreasing the value of x would decrease the computational cost of performing the PWST algorithm. However, this may also affect the relative error of the PWST algorithm in learning the appropriate $[\alpha_1, \alpha_2, ..., \alpha_m]$ values. The value x is a hyper-parameter that must be set for this algorithm. We evaluate alternative choices of x for the specific examples we consider in Sect. 5.

4   Model Application

The PWST algorithm provides a generalizable framework to address Multi-objective (Pareto) optimization problem where the user cannot exactly quantify the degree of importance for different criteria. This is particularly useful for systematically analyzing multiple decision-makers who have different preferences and thus potentially different objective functions.

As an example, we will consider a problem that involves deciding how many shares of each stock to purchase in the context of two criteria based on optimistic and pessimistic future outcomes. Consider a scenario that a financial advisor faces with multiple stock-trading clients. Different clients may have different degrees of risk-aversion based on various factors, some of which could be subjective and latent.

The problem has the following characteristics that define the optimization model for which the financial advisor wishes to learn their clients' weights:

- Stocks behave in a stochastic manner with price uncertainty that can be modeled with a binomial pricing tree [17,19].

$n = 0$ $n = 1$ $n = 2$

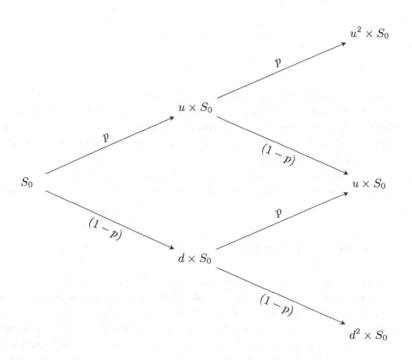

Fig. 1. Binomial Pricing Tree that shows possible stock price transitions and the associated probabilities of a stock price increase (p) and decrease ($1 - p$)

- there are s different stocks indexed by $i = 1, ..., s$, and stock i has price s_i
- each stock i has a different pessimistic expected value (p_i) and optimistic expected value (o_i)
- a budget limit b limits the total stock that can be purchased
- decision variable x_i is 1 if stock i is purchased and 0 otherwise
- the financial clients' weight on the pessimistic stock return is denoted by α and the optimistic stock return is denoted by $1 - \alpha$
- we have access to a data-set of pessimistic and optimistic expected values of various stocks at various time-points

The optimization problem described above can be expressed as the following *Knapsack Problem* :

$$\max \sum_{i=1}^{s} x_i f_i(\alpha) \tag{5}$$

$$s.t. \sum_{i=1}^{s} s_i x_i \leq b \tag{6}$$

$$x_i \in \{0,1\} \tag{7}$$

Where $f_i(\alpha)$ is the expected utility of purchasing stock i and $f_i(\alpha) = \alpha \times o_i + (1 - \alpha) \times p_i$. This is the well known *Discrete Knapsack Problem*. However, α is an unknown value that must be determined to solve the problem. Given access to a data-set of various stocks at various time points, we can use the PWST algorithm to solve this problem.

Once α is found, integer programming methods, such as branch-and-bound or dynamic programming can be used to solve the remainder of the problem [1,3,4].

5   Experimental Results

The following experiments were performed with data-points generated from the binomial pricing tree in Fig. 1. The binomial pricing tree had a height (n) of 1, an initial stock price (S_0) of 30, a factor change (u) of 1.3, and a risk free interest rate per annum (r) of 0.25. The final prices generated from these binomial pricing tree parameters were added to the data-set. This process was repeated for 300 iterations. At each iteration, noise was added to produce variations in final stock prices [20].

A virtual decision-maker was created with a true (latent) α-value and a tolerance level. The tolerance level introduces a certain degree of variation in the α-value that the decision-maker uses to rank data-points at each iteration. We created a virtual decision-maker with a true α-value of 0.3 and a tolerance level of 0.05 (low tolerance). We also created a second virtual decision-maker with the same true α-value but a tolerance of 0.15 (high tolerance). The user's α-value varies from iteration to iteration with uniform distribution based on the tolerance level and the user's true (base) α-value. In the experiments we ran with the low tolerance $\alpha = 0.3$ virtual decision-maker, α could take a value uniformly in the range [0.25, 0.35] on each iteration due to the 0.05 tolerance level. In the experiments we ran with the high tolerance $\alpha = 0.3$ virtual decision-maker, α could take a value uniformly in the range [0.15, 0.45] on each iteration due to the 0.15 tolerance level. This was meant to simulate the variation and subjectivity in tendencies that most decision-makers have in real-life. While decision-makers might vary their α-value from one iteration to the next, the base (true) α-value remains the same in our experiments. For each iteration, we sampled data-points at random. No data-point was repeated in the random sampling process. For random sampling, multiple trials were run for each configuration to obtain a confidence interval and mean.

5.1 Low Tolerance Sample Decision-Maker

We first tested the PWST algorithm on a sample decision-maker with a low tolerance level (0.05). Relative error is defined as $\frac{|\alpha_{true} - \alpha_{pred}|}{\alpha_{true}} \times 100$.

Table 1. Sample Decision-Maker with Tolerance of 0.05 where Data-points are Sampled at Random where x is the number of data-points sampled at each iteration and RE_i describes the relative error achieved by PWST at the ith iteration (e.g. RE_1 is the relative error at the 1st iteration)

x	RE_1	RE_5	RE_{10}	RE_{15}
3	4.534 ± 0.394	2.083 ± 1.001	1.198 ± 0.998	1.284 ± 1.071
5	3.051 ± 0.984	3.119 ± 0.389	2.595 ± 0.513	2.119 ± 0.754
7	1.806 ± 0.653	0.3092 ± 0.014	0.5517 ± 0.197	0.8105 ± 0.701
9	0.8024 ± 0.291	0.3468 ± 0.022	0.3501 ± 0.1171	0.7652 ± 0.159
11	1.634 ± 0.376	3.38 ± 0.384	2.985 ± 0.793	2.181 ± 0.851
13	1.777 ± 0.029	1.2 ± 0.005	1.737 ± 0.286	1.843 ± 0.428
15	0.2684 ± 0.0111	0.1808 ± 0.049	0.2019 ± 0.179	0.06107 ± 0.000001
17	2.307 ± 0.346	1.27 ± 0.0291	1.089 ± 0.0544	1.002 ± 0.00733

A mean RE_{15} value of 1.258 was achieved with the lower tolerance decision-maker across the different x values. Sampling 3, 7, 9, 13, 15, or 17 data-points at each iteration was able to achieve a relative error of below 2% after 15 iterations.

5.2 High Tolerance Sample Decision-Maker

We then tested the PWST algorithm on a sample decision-maker with a high tolerance level (0.15). Below are the results with this sample decision-maker.

A mean RE_{15} value of 2.764 was achieved with the higher tolerance decision-maker across the different x values. Sampling 9 data-points at each iteration was able to achieve a relative error of below 2% after 15 iterations.

5.3 Analysis of Experiments

The PWST algorithm achieved lower relative error on the true α-value for the sample decision-maker with lower tolerance. This was likely due to the fact that a decision-maker with a high tolerance level changes their desired α-value more

Table 2. Sample Decision-Maker with Tolerance of 0.15 where Data-points are Sampled at Random where x is the number of data-points sampled at each iteration and RE_i describes the relative error achieved by PWST at the ith iteration (e.g. RE_1 is the relative error at the 1st iteration)

x	RE_1	RE_5	RE_{10}	RE_{15}
3	0.4316 ± 0.0913	1.816 ± 0.00701	2.046 ± 1.365	2.086 ± 1.365
5	1.9 ± 0.299	2.81 ± 2.198	2.872 ± 0.279	2.953 ± 1.599
7	6.594 ± 3.531	6.185 ± 4.949	5.699 ± 1.735	5.279 ± 2.826
9	2.89 ± 0.778	0.1696 ± 0.0335	0.8039 ± 0.016	1.067 ± 0.752
11	5.971 ± 2.044	5.168 ± 2.501	2.419 ± 0.03288	2.073 ± 0.523
13	4.638 ± 0.3217	3.178 ± 2.181	3.001 ± 1.339	2.726 ± 0.374
15	2.613 ± 1.306	0.01872 ± 0.001	1.88 ± 1.447	2.356 ± 0.869
17	4.585 ± 0.1492	3.798 ± 0.388	3.891 ± 1.836	3.57 ± 1.246

from one iteration to the next, introducing more noise. In the higher tolerance case, there was higher fluctuation in the α-value from one iteration to the next. This makes it harder for the algorithm to learn the base α-value. The mean RE_{15} from Table 2 was 2.8, which was higher than the mean RE_{15} from Table 1, which was 1.3.

Relative error in estimating the decision-maker's α value decreases with more iterations in many cases. Of the 16 experiments configurations we conducted, 12 yielded a lower RE_{15} than RE_1. For the low tolerance sample decision-maker, sampling 3, 7, 9, 13, 15, or 17 data-points at each iteration was able to achieve a relative error of below 2% after 15 iterations. For the high tolerance sample decision-maker, sampling 9 data-points at each iteration was able to achieve a relative error of below 2% after 15 iterations. This is likely due to the fact that the algorithm is analyzing a greater amount of data as the number of iterations and points sampled at each iteration increases. We did not find a strictly increasing or decreasing trend in RE_{15} as the number of data-points sampled at each iteration increased.

However, trade-offs could occur with increasing the number of iterations. Increasing the amount of data-points analyzed in total introduces more potential for the algorithm to over-fit. Noisy data from the decision-maker (due to tolerance) could cause more variance in the algorithm we have proposed. Furthermore, with more data-points sampled at each iteration, the decision-maker must make more pairwise comparisons, increasing the computational complexity of the tasks required of them.

6   Conclusions

We proposed the PWST algorithm for learning criteria weights in Pareto optimization problems. PWST automates the process of selecting criteria and eliminates the need for hand-tuning weights. It utilizes online learning-to-rank to analyze sequential feedback from decision-makers in a way that does not require explicit scoring from them. PWST can be configured by altering the number of data-points presented to the user at each iteration and altering how these data-points are sampled. We proved that utilizing an online-learning approach is computationally less expensive than a batch approach, where all the data is available in advance. Our experiment found that the PWST algorithm can quickly estimate the decision-maker's latent α values with relatively low error given a proper configuration. The algorithm's accuracy improves with more iterations. Further work is needed to find how to configure the number of data-points at each iteration, as well as how to sample these data-points more effectively. There is also a need to explore the effectiveness of this approach on other types of optimization problems to see how well the PWST algorithm generalizes to other contexts. The initial work presented in this article helps lay the foundation for these future studies.

References

1. Ahmadian, S., Bhaskar, U., Sanità, L., Swamy, C.: Algorithms for inverse optimization problems. In: 26th Annual European Symposium on Algorithms, Dagstuhl Publishing, Germany (2018)
2. Beck, J., Chan, E., Irfanoglu, A., Papadimitriou, C.: Multi-criteria optimal structural design under uncertainty. Earthquake Eng. Struc. Dyn. **28**(7), 741–761 (1999)
3. Bertsimas, D., Demir, R.: An approximate dynamic programming approach to multidimensional knapsack problems. Manage. Sci. **48**(4), 550–565 (2002)
4. Bettinelli, A., Cacchiani, V., Malaguti, E.: A branch-and-bound algorithm for the knapsack problem with conflict graph. INFORMS J. Comput. **29**(3), 457–473 (2017)
5. Bruch, S.: An Alternative Cross Entropy Loss For Learning-To-Rank. arXiv.org (2020). https://arxiv.org/abs/1911.09798. Accessed 27 May 2020
6. Bärmann, A., Martin, A., Pokutta, S., Schneider, O.: An Online-Learning Approach To Inverse Optimization (2020). arXiv.org. https://arxiv.org/abs/1810.12997. Accessed 27 May 2020
7. Chang, K.: E-Design. Elsevier, Amsterdam (2016)
8. Dong, C., Chen, Y., Zeng, B.: Generalized inverse optimization through online learning. In: 32nd Conference on Neural Information Processing Systems (NeurIPS 2018). Montréal, Canada (2018)
9. Dyer, J.: Remarks on the analytic hierarchy process. Manage. Sci. **36**(3), 249–258 (1990)
10. Engau, A., Sigler, D.: Pareto solutions in multicriteria optimization under uncertainty. Euro. J. Oper. Res. **281**(2), 357–368 (2020)
11. Gensch, D., Recker, W.: The multinomial, multiattribute logit choice model. J. Market. Res. **16**(1), 124 (1979)
12. Ghobadi, K., Lee, T., Mahmoudzadeh, H., Terekhov, D.: Robust inverse optimization. Oper. Res. Lett. **46**(3), 339–344 (2018)
13. Heuberger, C.: Inverse combinatorial optimization: a survey on problems, methods, and results. J. Combinatorial Optimiz. **8**(3), 329–361 (2004)
14. Joachims, T.: Optimizing search engines using clickthrough data. In: KDD 2002: Proceedings of the Eighth ACM SIGKDD International Conference on Knowledge Discovery and Data Mining, pp. 133–142 (2002)

15. Liang, X., Zhu, L., Huang, D.: Multi-task ranking SVM for image cosegmentation. Neurocomputing **247**, 126–136 (2017)
16. Marler, R., Arora, J.: The weighted sum method for multi-objective optimization: new insights. Struc. Multidisciplinary Optimizat. **41**(6), 853–862 (2009)
17. Muzzioli, S., Torricelli, C.: The pricing of options on an interval binomial tree. an application to the DAX-index option market. Euro. J. Oper. Res. **163**(1), 192–200 (2005)
18. Negahban, S., Oh, S., Thekumparampil, K., Xu, J.: Learning from comparisons and choices. J. Mach. Learn. Res. **19**(1–95) (2018)
19. Shvimer, Y., Herbon, A.: Comparative empirical study of binomial call-option pricing methods using S and P 500 index data. North Am. J. Econ. Finance **51**, 101071 (2020)
20. Wang, H.: Pareto-Weighted-Sum-Tuning (2020). https://github.com/harryw1248/Pareto-Weighted-Sum-Tuning
21. Yang, X.: Nature-Inspired Optimization Algorithms. [Place of publication not identified]: Elsevier (2016)

Fast Hyperparameter Tuning for Support Vector Machines with Stochastic Gradient Descent

Marcin Orchel[1]([⊠])[iD] and Johan A. K. Suykens[2][iD]

[1] Department of Computer Science, AGH University of Science and Technology,
Kraków, Poland
morchel@agh.edu.pl
[2] ESAT-STADIUS, KU Leuven, 3001 Leuven (Heverlee), Belgium
johan.suykens@esat.kuleuven.be

Abstract. We propose a fast training procedure for the support vector machines (SVM) algorithm which returns a decision boundary with the same coefficients for any data set, that differs only in the number of support vectors and kernel function values. The modification is based on the recently proposed SVM without a regularization term based on stochastic gradient descent (SGD) with extreme early stopping in the first epoch. We realize two goals during the first epoch: we decrease the objective function value, and we tune the margin hyperparameter M. Experiments show that a training procedure with validation can be speed up substantially without affecting sparsity and generalization performance.

Keywords: Support vector machines · Stochastic gradient descent

We solve a classification problem by using SVM [14]. The SVM have been shown effective in many applications including computer vision, natural language, bioinformatics, and finance [12]. There are three main performance measures for SVM : the generalization performance, sparsity of a decision boundary and computational performance of learning. SVM are in the group of the most accurate classifiers and are generally the most efficient classifiers in terms of overall running time [16]. They may be preferable due to its simplicity compared to deep learning approach for image data, especially when training data are sparse. One of the problem in the domain of SVM is to efficiently tune two hyperparameters: the cost C which is a trade-off between the margin and the error term; and σ which is a parameter of a Gaussian kernel, also called the radial basis function (RBF) kernel [14]. The grid search is the most used in practice due to its simplicity and feasibility for SVM , where only two hyperparameters are tuned. The generalization performance of sophisticated meta-heuristic methods for hyperparameter optimization for SVM , like genetic algorithms, particle swarm optimization, estimation of distribution algorithms is similar to simpler

© Springer Nature Switzerland AG 2020
G. Nicosia et al. (Eds.): LOD 2020, LNCS 12566, pp. 481–493, 2020.
https://doi.org/10.1007/978-3-030-64580-9_40

random search and grid search [9]. The random search can have some advantages over grid search when more hyperparameters are considered like for neural networks [1]. The random search still requires considerable fraction of the grid size. The problem with a grid search method is high computational cost due to exhaustive search of a discretized hyperparameter space.

In this article, we tackle the problem of improving performance of hyperparameter search for the cost C in terms of computational time while preserving sparsity and generalization. In [4], authors use a general approach of checking fewer candidates. They first use a technique for finding optimal σ value, then they use a grid search exclusively for C with an elbow method. The potential limitation of this method is that it still requires a grid search for C, and there is an additional parameter, tolerance for an elbow point. In practice, the number of checked values has been reduced to 5 from 15. In [3], authors use an analytical formula for C in terms of a jackknife estimate of the perturbation in the eigenvalues of the kernel matrix. However, in [9] authors find that tuning hyperparameters generally results in substantial improvements over default parameter values. Usually, a cross validation is used for tuning hyperparameters which additionally increases computational time.

Recently, an algorithm for solving SVM using SGD has been proposed [10] with interesting properties. We call it Stochastic Gradient Descent for Support Vector Classification (SGD-SVC) for simplicity. Originally, it was called OLLAWV. It always stops in the first epoch, which we call *extreme early stopping* and has a related property of not using a regularization term. The SGD-SVC is based on iterative learning. Online learning has a long tradition in machine learning starting from a perceptron [12]. Online learning methods can be directly used for batch learning. However, the SGD-SVC is not a true online learning algorithm, because it uses knowledge from all examples in each iteration. The SGD-SVC due to its iterative nature is similar to many online methods having roots in a perceptron, like the Alma Forecaster [2] that maximizes margin. Many perceptron-like methods have been kernelized, some of them also related to SVM like kernel-adatron [14]. In this article, we reformulate slightly the SGDSVC by replacing a hyperparameter C with a margin hyperparameter M. This parameter is mentioned as a desired margin in [14], def. 4.16. The margin plays a central role in SVM and in a statistical learning theory, especially in generalization bounds for a soft margin SVM. The reformulation leads to simpler formulation of a decision boundary with the same coefficients for any data set that differs only in kernel function values and the number of support vectors which is related to the margin M. Such simple reformulation of weights is close in spirit to the empirical Bayes classifier, where all weights are the same. It has been inspired by fast heuristics used by animals and humans in decision-making [6]. The idea of replacing the C hyperparameter has been mentioned in [13] and proposed as ν support vector classification (ν-SVC). The problem is that it leads to a different optimization problem and is computationally less tractable. The ν-SVC has been also formulated as ν being a direct replacement of $C = 1/(n\nu)$ in [14], where n is the number of examples, with the same optimization problem as support vector classification (SVC). The margin classifier has been mentioned in [15], however, originally it has been artificially converted to the classifier with the regularization term. The statistical bounds for the margin classifier has been given in [5], but without proposing a solver based on these bounds. There is also a technique of solution/regularization path with a procedure of computing a solution for some values of C using a piecewise linearity property. However, the approach is complicated and requires solving a system of equations and several checks of $O(n)$ [7]. In the proposed method, we use one solution for a particular M for generating all solutions for remaining values of M.

The outline of the article is as follows. First, we define a problem, then the methods and update rules. After that, we show experiments on real world data sets.

1   Problem

We consider a classification problem for a given sample data $\boldsymbol{x}_i$ mapped respectively to $y_i \in \{-1, 1\}$ for $i = 1, \ldots, n$ with the following decision boundary

$$f(\boldsymbol{x}) \equiv \boldsymbol{w} \cdot \varphi(\boldsymbol{x}) = 0, \tag{1}$$

where $\boldsymbol{w} \in \mathbb{R}^m$ with the feature map $\varphi(\cdot) \in \mathbb{R}^m$, $f(\cdot)$ is a decision function. We classify data according to the sign of the left side $f(\boldsymbol{x})$. This is the standard decision boundary formulation used in SVM with a feature map and without a free term b. The primal optimization problem for (C-SVC) is

Optimization problem (OP) 1.

$$\min_{\boldsymbol{w}} \frac{1}{2} \|\boldsymbol{w}\|^2 + C \sum_{i=1}^{n} \max\{0, 1 - y_i(\boldsymbol{w} \cdot \varphi(\boldsymbol{x}_i))\}, \tag{2}$$

where $C > 0$, $\varphi(\boldsymbol{x}_j) \in \mathbb{R}^m$.

The first term in (2) is known as a *regularization term* (regularizer), the second term is an *error term*. The $\boldsymbol{w}$ can be written in the form

$$\boldsymbol{w} \equiv \sum_{j=1}^{n} \beta_j \varphi(\boldsymbol{x}_j), \tag{3}$$

where $\boldsymbol{\beta} \in \mathbb{R}^n$. We usually substitute (3) to a decision boundary and we get

$$\sum_{j=1}^{n} \beta_j \varphi(\boldsymbol{x}_j) \cdot \varphi(\boldsymbol{x}) = 0. \tag{4}$$

The optimization problem OP 1 is reformulated to find β_j parameters.

The SGD procedure for finding a solution of SVM proposed in [10], called here SGD-SVC is to update parameters β_k iteratively using the following update rule for the first epoch

$$\beta_k \leftarrow -\eta_k \begin{cases} -Cy_{w(k)}, & \text{if } 1 - y_{w(k)} \sum_{j=1}^{k-1} \beta_j \varphi(\boldsymbol{x}_{w(j)}) \cdot \varphi(\boldsymbol{x}_{w(k)}) \geq 0 \\ 0, & \text{otherwise} \end{cases}, \tag{5}$$

where η_k is a learning rate set to $\eta_k = 1/\sqrt{k}$ for $k = 1, \ldots, n$, all β_k are initialized with 0 before the first epoch. We set $w(1) = 1$. We always stop in the first epoch, either when the condition in (5) is violated, or when we updated all parameters β_k. The $w(k)$ is used for selection of an index using the worst violator technique. It means that we look for the example among all remaining examples, with the worst value of the condition in (5). We check the condition only for the examples not being used in the iteration process before. The worst violators are searched among all remaining examples, so when one wants to use this method for online learning, it is still required to train the model in a batch for optimal performance. We use a version of SVM without a free term b for simplicity, which does not impact any performance measures. We update each parameter maximally one time. Finally, only parameters β_k for the fulfilled condition during the iteration process have nonzero values. The remaining parameters β_k have zero values. In that way, we achieve sparsity of a solution. The number of iterations n_c for β_k parameters with the fulfilled condition is also the number of support vectors. The derivation of an update rule has been already given in [10]. We call the algorithm that stops always in the first epoch as *extreme early stopping*.

The idea that we want to explore is to get rid of the C hyperparameter from the update rule and from the updated term for β_k (5).

2   Solution – Main Contribution

The decision boundary (4) for SGD-SVC can be written as

$$\sum_{k=1}^{n_c} C y_{w(k)} \eta_k \varphi\left(\boldsymbol{x}_{w(k)}\right) \cdot \varphi\left(\boldsymbol{x}\right) = 0, \tag{6}$$

where $n_c \leq n$ is the number of support vectors. In the same way, we can write the margin boundaries

$$\sum_{k=1}^{n_c} C y_{w(k)} \eta_k \varphi\left(\boldsymbol{x}_{w(k)}\right) \cdot \varphi\left(\boldsymbol{x}\right) = \pm 1. \tag{7}$$

When we divide by C, we get

$$\sum_{k=1}^{n_c} y_{w(k)} \eta_k \varphi\left(\boldsymbol{x}_{w(k)}\right) \cdot \varphi\left(\boldsymbol{x}\right) = \pm 1/C. \tag{8}$$

The left side is independent of C, the right side is a new margin value. The new decision boundary can be written as

$$\sum_{k=1}^{n_c} y_{w(k)} \eta_k \varphi\left(\boldsymbol{x}_{w(k)}\right) \cdot \varphi\left(\boldsymbol{x}\right) = 0. \tag{9}$$

We propose a classifier based on a margin solving the following optimization problem

OP 2.

$$\min_{\boldsymbol{w}} \frac{1}{2} \|\boldsymbol{w}\|^2 + \sum_{i=1}^{n} \max\left\{0, M - y_i\left(\boldsymbol{w} \cdot \varphi\left(\boldsymbol{x}_i\right)\right)\right\}, \tag{10}$$

where $M > 0$ is a desired *margin* – a hyperparameter that replaces the C hyperparameter. We call it M Support Vector Classification (M-SVC). The classifier with explicitly given margin has been investigated in [14]. In our approach, we tune a margin, unlike for standard SVM when the margin is optimized, see [14] page 220. We have the following proposition.

Proposition 1. *The OP 2 is equivalent to OP 1.*

Proof. We can write (10) as

$$\min_{\boldsymbol{w}} \frac{1}{2} \|\boldsymbol{w}\|^2 + M \sum_{i=1}^{n} \max\left\{0, 1 - y_i\left(\frac{\boldsymbol{w}}{M} \cdot \varphi\left(\boldsymbol{x}_i\right)\right)\right\}. \tag{11}$$

When we substitute $\boldsymbol{w}' \to \boldsymbol{w}/M$, we get

$$\min_{\boldsymbol{w}'} \frac{1}{2} \|\boldsymbol{w}'M\|^2 + M \sum_{i=1}^{n} \max\left\{0, 1 - y_i\left(\boldsymbol{w}' \cdot \varphi\left(\boldsymbol{x}_i\right)\right)\right\}, \tag{12}$$

So we get

$$\min_{\boldsymbol{w}'} \frac{1}{2} \|\boldsymbol{w}'\|^2 + \frac{1}{M} \sum_{i=1}^{n} \max\left\{0, 1 - y_i\left(\boldsymbol{w}' \cdot \varphi\left(\boldsymbol{x}_i\right)\right)\right\}. \tag{13}$$

The M is related to C by

$$M = 1/C. \tag{14}$$

It is a similar term as for ν-SVC classifier given in [14], where $C = 1/(n\nu)$ and $\nu \in (0, 1]$. Because the optimization problems are equivalent, generally all properties of SVM in the form OP 2 applies also for M-SVC. In [14], page 211, authors stated an SVM version, where the margin M is automatically optimized as an additional variable. However, they still have the constant C. From the statistical learning theory point of view, the original bounds [14], page 211 applies for a priori chosen M.

We can derive the update rules for M-SVC similar as for SGD-SVC. The new update rules called (SGD-M-SVC) are

$$\beta_k \leftarrow -\eta_k \begin{cases} -y_{w(k)}, & \text{if } M - y_{w(k)} \sum_{j=1}^{k-1} \beta_j \varphi\left(\boldsymbol{x}_{w(j)}\right) \cdot \varphi\left(\boldsymbol{x}_{w(k)}\right) \geq 0 \\ 0, & \text{otherwise} \end{cases} \tag{15}$$

In the proposed update rules, there is no hyperparameter in the updated value, only in the condition, in opposite to (5). It means that for different values of a margin M, we get solutions that differ only in the number of terms. The corresponding values of parameters β_k are the same for each M value, so the ordering of corresponding parameters is the same. It means that we do not need to tune values of parameters β_k, only the stopping criterion and thus the number of terms in a solution. When we have a set of M values, and we have a model for the $M_{\max}$, we can generate solutions for all remaining M values just by removing the last terms in the solution for $M_{\max}$. We have a correspondence between M value and the number of support vectors n_c stated as follows.

Proposition 2. *After running the SGD-M-SVC for any two values M_1 and M_2, such as $M_1 > M_2$, the number of support vectors n_c is bigger or equals for M_1.*

Proof. The n_c is the number of support vectors and also the number of terms. The stopping criterion is the opposite for the update condition (15) for the k-th iteration. Due to the form $M < \cdot$, it is fulfilled earlier for M_2. There is a special case when stopping criterion would not be triggered for both values, then we get the same model with n terms. Another special case is when only one condition is triggered, then we get model for M_2 and for M_1 with all n terms.

3   Theoretical Analysis

The interesting property of the new update rules is that we realize two goals with update rules: we decrease the objective function value (10) and simultaneously, we generate solutions for a set of given different values of a hyperparameter M, and all is done in the first epoch. We can say, that we solve a discrete non-convex optimization problem OP 2 where we can treat M as a discrete variable to optimize. The main question that we want to address is *how is it possible, that we can effectively generate solutions for different values of M in the first epoch.* First, note due to convergence analysis of a stochastic method, we expect that we improve the objective function value of (10) during the iteration process. We provide an argument that we are able to generate solutions for different values of M. The SVM can be reformulated as solving a multiobjective optimization problem [11] with two goals, a regularization term, and the error term (2). The SVM is a weighted (linear) scalarization with the C being a scalarization parameter. For the corresponding multiobjective optimization problem for OP 2, we have the M scalarization parameter instead. Due to convexity of the two goals, the set all solutions of SVM for different values of C is a Pareto frontier for the multiobjective optimization problem. We show that during the iteration process, we generate *approximated Pareto optimal solutions.* The error term for the t-th iteration of SGD-M-SVC for the example to be added $\boldsymbol{x}_{w\,(t+1)}$ can be written as

$$\sum_{\substack{i=1 \\ i \neq t+1}}^{n} \max\left\{0, M - y_{w(i)} f_t\left(\boldsymbol{x}_{w\,(i)}\right)\right\} + \max\left\{0, M - y_{w(t+1)} f_t\left(\boldsymbol{x}_{w\,(t+1)}\right)\right\}, \quad (16)$$

where $f_t(\cdot)$ is a decision function of SGD-M-SVC after t-th iteration. After adding $t+1$-th parameter, we get an error term

$$\sum_{\substack{i=1 \\ i \neq t+1}}^{n} \max \left\{ 0, M - y_{w(i)} f_t \left(\boldsymbol{x}_{\boldsymbol{w}(i)} \right) - y_{w(i)} y_{w(t+1)} \frac{1}{\sqrt{t+1}} \varphi \left(\boldsymbol{x}_{\boldsymbol{w}(t+1)} \right) \cdot \varphi \left(\boldsymbol{x}_{\boldsymbol{w}(i)} \right) \right\}$$

$$+ \max \left\{ 0, M - y_{w(t+1)} f_t \left(\boldsymbol{x}_{\boldsymbol{w}(t+1)} \right) - \frac{1}{\sqrt{t+1}} \right\} \tag{17}$$

assuming that we replace a scalar product with an RBF kernel function. The update for the regularization term from (10) is

$$\| \boldsymbol{w}_{t+1} \|^2 = \sum_{i=1}^{t+1} \sum_{j=1}^{t+1} y_{w(i)} y_{w(j)} \frac{1}{\sqrt{i}} \frac{1}{\sqrt{j}} \varphi \left(\boldsymbol{x}_{\boldsymbol{w}(i)} \right) \cdot \varphi \left(\boldsymbol{x}_{\boldsymbol{w}(j)} \right). \tag{18}$$

So we get

$$\| \boldsymbol{w}_{t+1} \|^2 = \| \boldsymbol{w}_t \|^2 + 2 y_{w(t+1)} \frac{1}{\sqrt{t+1}} f_t \left(\boldsymbol{x}_{\boldsymbol{w}(t+1)} \right) + \frac{1}{\sqrt{t+1}} \frac{1}{\sqrt{t+1}}. \tag{19}$$

The goal of analysis is to show that during the iteration process, we expect decreasing value of an error term and increasing value of a regularization term. It is the constraint for generating Pareto optimal solutions. Due to Proposition 2, we are increasing value of M, which corresponds to decreased value of C due to (14). For SVM, oppositely, we are increasing value of a regularization term, when C is increased. We call this property a *reversed scalarization for extreme early stopping*. First, we consider the error term. We compare the error term after adding an example (17) to the error term before adding the example (16). The second term in (17) stays the same or it has smaller value due to the update condition for the $t + 1$-th iteration

$$M - y_{w(t+1)} f_t \left(\boldsymbol{x}_{\boldsymbol{w}(t+1)} \right) \geq 0 \tag{20}$$

and due to the positive $1/\sqrt{t+1}$. Moreover, the worst violator selection technique maximizes the left side of (20) among all remaining examples, so it increases the chance of getting smaller value. Now regarding the first term in (16). After update (17), we decrease a value of this term for examples already processed with same class so for which $y_{w(i)} = y_{w(t+1)}$ for $i \leq t$. However, we increase particular terms for remaining examples with the opposite class. The worst violators will likely be surrounded by examples for an opposite class. So we expect bigger similarities to the examples with the opposite class, thus we expect $\varphi \left(\boldsymbol{x}_{\boldsymbol{w}(t+1)} \right) \cdot \varphi \left(\boldsymbol{x}_{\boldsymbol{w}(i)} \right)$ to be bigger.

Regarding showing increasing values of (19) during the iteration process. The third term in (19) is positive. The second term in (19) can be positive or negative. It is closely related to the update condition (20). During the iteration process, we expect the update condition to be improved, because, we have an improved model. During the iteration process, the update condition starts to improving and there is a point for which

$$y_{w(t+1)} f_t \left(x_{w(t+1)}\right) > -1/\sqrt{t+1}. \tag{21}$$

Then the update for (19) becomes positive. We call this point a *Pareto starter*. So we first optimize the objective function value by minimizing the regularization term and minimizing the error term, then after Pareto starter we generate approximated Pareto optimal solutions, while still improving the objective function value by minimizing only the error term.

3.1   Bounds for M

We bound M by finding bounds for the decision function $f(\cdot)$. Given σ, we can compute the lower and upper bound for $f(\cdot)$ for the RBF kernel for a given number of examples as follows

$$l = (-1)\exp\left(0/\left(-2\sigma^2\right)\right)\sum_{i=1}^{n}\frac{1}{\sqrt{i}}, u = \exp\left(0/\left(-2\sigma^2\right)\right)\sum_{i=1}^{n}\frac{1}{\sqrt{i}} = \sum_{i=1}^{n}\frac{1}{\sqrt{i}}. \tag{22}$$

It holds that $l \leq f(\cdot) \leq u$. In the lower bound, we assume all examples with a class -1. The upper bound is a harmonic number $H_{n,0.5}$. The bounds capture cases when margin functions have all examples on the same side. For random classes for examples (with the Rademacher distribution), the expected value of l (with replaced -1 with classes for particular examples) is 0. We also consider the case with one support vector with class 1 for capturing the error term close to 0. We have the error term $1 - \exp\left(1/\left(-2\sigma^2\right)\right)$. Given σ_{max} arbitrarily, we can compute σ_{min} assuming one support vector according to the numerical precision. For simplicity, we can use one value, σ_{max} for computing l_2. Overall, we can compute bounds for $f(\cdot)$ as the lower bound based on l_2 and the upper bound based on u. For example, for $n = 100000$ and $\sigma_{max} = 2^9$, we get after rounding powers to integers $\sigma_{min} = 2^{-4}$, $M_{min} = 2^{-19}$, $M_{max} = 2^{10}$.

4   Method

The SGD-M-SVC returns the equivalent solutions as SGD-SVC. However, it is faster for validating different values of M. First, we run a prototype solver SGD-M-SVC with $M = M_{max} = 2^{10}$ with provided a list of sorted M_i values as a parameter and with particular σ value. During the iteration process in the prototype solver, we store margin values defined as $m_k \equiv y_{w(k)} f_{k-1}\left(x_{w(k)}\right)$.

Algorithm 1. SGD-M-SVC for M_i

Input: M_i value to check, σ, m_k values, validation errors v, a map (M_i, k) of margin
   values to indices
Output: v
   1: index $= (M_i, k)$.get(M_i) //get the index k from a map (M_i, k) for M_i
   2: v $= v_k(k)$ // get a validation error for the found k index from a list of v_k values

---

Because sometimes it may happen that $m_k < m_{k-1}$, then we copy a value of
m_{k-1} to m_k, so we have always a sorted sequence of m_k values. The size of m_k
is $n - 1$ at most. We also store validation errors v_k that are updated in each
iteration. The size of this list is the size of a validation set. We also update the
map of M_i values to k indices during the iteration process. Then, we use a solver
returning a validation error for particular M_i as specified in Algorithm 1. Given
validation error, we can compare solutions for different hyperparameter values.

4.1   Computational Performance

The computational complexity of SGD-SVC based on update rules (5) is $O(n_c n)$,
when n_c is the number of iterations. It is also the number of support vectors. So
sparsity influences directly computational performance of training. The requirement for computing the update rule for each parameter is a linear time. The
update rules (5) are computed in each iteration in a constant time. However, a
linear time is needed for updating values of a decision function for all remaining
examples. The procedure of finding two hyperparameter values σ and C using
the cross validation, a grid search method and SGD-SVC has the complexity
$O(n_c(n - n/v)v|C||\sigma| + n_c n|C||\sigma|)$ for v-fold cross validation, where $|C|$ is the
number of C values to check, $|\sigma|$ is the number of σ values to check, n_c is the
average number of support vectors. For each fold, we train a separate model. The
first term is related to training a model. The second term is related to computing
a validation error. The complexity of SGD-M-SVC is

$$O(n_{c,p}(n - n/v)v|\sigma| + n_{c,p}n|\sigma|), \tag{23}$$

where $n_{c,p}$ is the number of support vectors for a prototype solver. We removed
a multiplier $|C|$ from the first term, that is related to the training complexity,
and from the second term, that is related to the computation of a validation
error.

5   Experiments

The M-SVC returns equivalent solutions to C-SVC. However, it is faster for
validating different values of M. We validate equally distributed powers of 2

as M values from 2^{-19} to 2^{10} for integer powers, based on the analysis in the Sect. 3.1. We use our own implementation of both SGD-SVC and SGD-MSVC. We compare performance of both methods for real world data sets for binary classification. More details about data sets are on the LibSVM site ([8]). We selected all data sets from this site for binary classification. For all data sets, we scaled every feature linearly to $[0, 1]$. We use the RBF kernel in a form $K(x, z) = \exp(-\|x - z\|^2 / (2\sigma^2))$. The number of hyperparameters to tune is 2, σ and M for SGD-M-SVC, and σ and C for SGD-SVC. For all hyperparameters, we use a grid search method for finding the best values. The σ values are integer powers of 2 from 2^{-4} to 2^9. We use the procedure similar to repeated double cross validation for performance comparison. For the outer loop, we run a modified k-fold cross validation for $k = 15$, with the optimal training set size set to 80% of all examples with maximal training set size equal to 1000 examples. We limit a test data set to 1000 examples. We limit all read data to 35000. When it is not possible to create the next fold, we shuffle data and start from the beginning. We use the 5-fold cross validation for the inner loop for finding optimal values of the hyperparameters. After that, we run the method on training data, and we report results on a test set.

The observations based on experimental results are as follows. The proposed method SGD-M-SVC is about 7.6 times faster SGD-SVC (see Table 1) for binary classification, with the same generalization performance and the number of support vectors. We have 30 values of M to tune. Some authors tune value of C with fewer values. Then the effect of this speed improvement may be smaller. We generally expect the accuracy performance to degrade slowly for smaller number of values of M. We also implemented the method for multiclass classification with similar results, however we do not report it here due to space constraints. We validated also theoretical results. We check Pareto frontier every 10 iterations. The results is that the approximated Pareto frontier is generated from almost the beginning of a data set after processing 0.05% examples on average (column pS in Table 1). Approximated Pareto frontier is generated perfectly for some data sets (1.0 in a column pU), on average in 75% updates. While we check Pareto updates every 10 iterations, it may be worth to check them only for selected solutions for given M, which are distributed differently. From the practical point of view, we recommend to use SGD-M-SVC instead of SGD-SVC due to speed performance benefits.

Table 1. Experiment 1. The numbers in descriptions of the columns mean the methods: 1 - SGD-SVC, 2- SGD-M-SVC. Column descriptions: dn – data set, $size$ – the number of all examples, dim – dimension of the data set, err – misclassification error, sv – the number of support vectors, t – average training time per outer fold in seconds, the best time is in bold (last row is a sum), pU – Pareto optimal solutions ratio (last row is an average), pS – Pareto starter ratio (last row is an average).

dn	size	dim	err1	err2	sv1	sv2	t1	t2	pU	pS
aa	34858	123	0.149	0.149	333	333	98	**9**	0.59	0.02
australian	690	14	0.146	0.146	219	219	35	**3**	0.55	0.08
avazu-app	35000	25619	0.133	0.133	673	673	90	**10**	0.57	0.01
avazu-site	35000	27344	0.211	0.211	444	444	100	**10**	0.44	0.02
cod-rna	35000	8	0.063	0.063	356	356	104	**8**	0.72	0.04
colon-cancer	62	2000	0.19	0.19	27	27	**0**	**0**	1.0	0.21
covtype	35000	54	0.292	0.292	672	672	121	**9**	0.77	0.01
criteo.kaggle2014	35000	662923	0.222	0.222	581	581	108	**11**	0.65	0.01
diabetes	768	8	0.236	0.236	334	334	40	**3**	0.42	0.06
duke	44	7129	0.222	0.222	21	21	**0**	**0**	0.87	0.31
epsilon_normalized	35000	2000	0.269	0.269	716	716	133	**11**	0.65	0.02
fourclass	862	2	0.001	0.001	556	556	46	**4**	1.0	0.01
german.numer	1000	24	0.257	0.257	411	411	67	**6**	0.55	0.03
gisette_scale	7000	4971	0.039	0.039	414	414	130	**17**	1.0	0.01
heart	270	13	0.17	0.17	100	100	6	**0**	0.76	0.12
HIGGS	35000	28	0.448	0.448	890	890	129	**10**	0.54	0.01
ijcnn1	35000	22	0.09	0.09	235	235	64	**9**	0.72	0.05
ionosphere_scale	350	33	0.081	0.082	92	89	8	**0**	0.8	0.13
kdd12	35000	54686452	0.04	0.04	819	819	74	**10**	0.39	0.01
kdda	35000	20216664	0.145	0.145	639	639	100	**12**	0.77	0.01
kddb	35000	29890095	0.144	0.144	849	849	97	**12**	0.97	0.0
kddb-raw-libsvm	35000	1163024	0.144	0.144	789	789	89	**10**	0.6	0.01
leu	72	7129	0.062	0.062	22	22	**0**	**0**	1.0	0.24
liver-disorders	341	5	0.394	0.394	202	202	10	**0**	0.34	0.07
madelon	2600	500	0.332	0.332	963	963	132	**10**	1.0	0.0
mushrooms	8124	112	0.001	0.001	842	842	103	**8**	1.0	0.01
news20.binary	19273	1354343	0.151	0.151	760	760	196	**52**	0.83	0.01
phishing	5772	68	0.059	0.059	346	346	111	**9**	0.96	0.03
rcv1.binary	35000	46672	0.064	0.064	608	608	129	**13**	0.87	0.01
real-sim	35000	20958	0.103	0.103	435	435	112	**12**	0.68	0.02
skin_nonskin	35000	3	0.011	0.011	132	132	89	**8**	0.99	0.05
sonar_scale	208	60	0.122	0.122	97	97	3	**0**	0.98	0.03
splice	2989	60	0.12	0.12	674	674	120	**9**	0.99	0.0
SUSY	35000	18	0.281	0.281	630	630	123	**9**	0.42	0.03
svmguide1	6910	4	0.04	0.04	173	173	97	**8**	0.87	0.08
svmguide3	1243	21	0.186	0.186	383	383	90	**9**	0.51	0.04
url_combined	35000	3230439	0.044	0.044	302	302	112	**10**	0.8	0.04
wa	34686	300	0.02	0.02	348	348	72	**9**	0.88	0.05
websam_trigram	35000	680715	0.044	0.044	201	201	502	**155**	0.75	0.05
websam_unigram	35000	138	0.07	0.07	260	260	101	**8**	0.71	0.03
All							3767	**495**	0.75	0.05

6 Conclusion

We proposed a novel method for SVC based on tuning margin M instead of C, with an algorithm SGD-M-SVC which improves substantially tuning time for the margin M hyperparameter compared to tuning the cost C in SGD-SVC. We provided theoretical analysis of an approximated Pareto frontier for this solver, which confirms the ability to generate solutions for different values of M during the first epoch.

Acknowledgments. The theoretical analysis of the method is supported by the National Science Centre in Poland, project id 289884, UMO-2015/17/D/ST6/04010, titled "Development of Models and Methods for Incorporating Knowledge to Support Vector Machines" and the data driven method is supported by the European Research Council under the European Union's Seventh Framework Programme. Johan Suykens acknowledges support by ERC Advanced Grant E-DUALITY (787960), KU Leuven C1, FWO G0A4917N. This paper reflects only the authors' views, the Union is not liable for any use that may be made of the contained information.

References

1. Bergstra, J., Bengio, Y.: Random search for hyper-parameter optimization. J. Mach. Learn. Res. **13**, 281–305 (2012)
2. Cesa-Bianchi, N., Lugosi, G.: Prediction, learning, and games. Cambridge University Press (2006). https://doi.org/10.1017/CBO9780511546921
3. Chang, C., Chou, S.: Tuning of the hyperparameters for l2-loss svms with the RBF kernel by the maximum-margin principle and the jackknife technique. Pattern Recognition **48**(12), 3983–3992 (2015). https://doi.org/10.1016/j.patcog.2015.06.017
4. Chen, G., Florero-Salinas, W., Li, D.: Simple, fast and accurate hyper-parameter tuning in gaussian-kernel SVM. In: 2017 International Joint Conference on Neural Networks, IJCNN 2017, Anchorage, AK, USA, May 14–19, 2017, pp. 348–355 (2017). https://doi.org/10.1109/IJCNN.2017.7965875
5. Cristianini, N., Shawe-Taylor, J.: An introduction to support vector machines : and other kernel-based learning methods. Cambridge University Press, 1 edn. (March 2000)
6. Gigerenzer, G., Todd, P., Group, A.R.: Simple Heuristics that Make Us Smart. Oxford University Press, Evolution and cognition (1999)
7. Hastie, T., Rosset, S., Tibshirani, R., Zhu, J.: The entire regularization path for the support vector machine. J. Mach. Learn. Res. **5**, 1391–1415 (2004)
8. Libsvm data sets (2011). www.csie.ntu.edu.tw/~cjlin/libsvmtools/datasets/
9. Mantovani, R.G., Rossi, A.L.D., Vanschoren, J., Bischl, B., de Carvalho, A.C.P.L.F.: Effectiveness of random search in SVM hyper-parameter tuning. In: 2015 International Joint Conference on Neural Networks, IJCNN 2015, Killarney, Ireland, July 12–17, 2015. pp. 1–8 (2015). https://doi.org/10.1109/IJCNN.2015.7280664
10. Melki, G., Kecman, V., Ventura, S., Cano, A.: OLLAWV: online learning algorithm using worst-violators. Appl. Soft Comput. **66**, 384–393 (2018)

11. Orchel, M.: Knowledge-uncertainty axiomatized framework with support vector machines for sparse hyperparameter optimization. In: 2018 International Joint Conference on Neural Networks, IJCNN 2018, Rio de Janeiro, Brazil, July 8–13, 2018, pp. 1–8 (2018). https://doi.org/10.1109/IJCNN.2018.8489144
12. Sammut, C., Webb, G.I. (eds.) Encyclopedia of Machine Learning and Data Mining. Springer (2017). https://doi.org/10.1007/978-1-4899-7687-1
13. Schölkopf, B., Smola, A.J.: Learning with Kernels: Support Vector Machines, Regularization, Optimization, and Beyond. MIT Press, Cambridge (2001)
14. Shawe-Taylor, J., Cristianini, N.: Kernel methods for pattern analysis. Cambridge University Press, Cambridge (2004). https://doi.org/10.1017/CBO9780511809682
15. Vapnik, V.N.: Statistical Learning Theory. Wiley-Interscience, September 1998
16. Zhang, C., Liu, C., Zhang, X., Almpanidis, G.: An up-to-date comparison of state-of-the-art classification algorithms. Expert Syst. Appl. **82**, 128–150 (2017). https://doi.org/10.1016/j.eswa.2017.04.003

PlattForm: Parallel Spoken Corpus of Middle West German Dialects with Web-Based Interface

Aynalem Tesfaye Misganaw[✉] and Sabine Roller

Universitäte Siegen, Hölderlinstraße 3, 57068 Siegen, Germany
{aynalem.misganaw,sabine.roller}@uni-siegen.de

Abstract. In this research, we contribute to the preservation of dialects by producing a searchable parallel audio corpus. We constructed a parallel spoken corpus for dialects in middle west Germany with a web-based search interface in order to look up the database of the spoken documents. The audio documents are initially collected as part of the DMW (*Dialektatlas Mittleres Westdeutschland*) project for showing which dialectal or dialect-related varieties currently exist in Central West Germany and describes them in their important characteristics. Selected people in particular places in the region are interviewed to answer a list of questions prepared by linguists. The collected data are systematically processed, analyzed and documented on a phonetic-phonological, morphological, syntactic and lexical level. Thus, this research utilizes these processed audio data and the corresponding standard German text to create a parallel spoken corpus of dialects in middle west Germany and standard German. Finally, we created a web-based search interface that accepts standard German text and returns its translation into the nine dialects.

As a result, we created a parallel spoken corpus for nine dialects, each having a parallel audio data of size 550 for each dialect, where size is measured by the count of words or phrases. The outcome of this work could be used in computer-supported language learning, language variety identification as well as speech recognition.

Keywords: Audio parallel corpus · Endangered languages · Dialect · Less-resourced languages

1 Introduction

Nearly 97% of the world population speak about 4% of world languages. In other words, 96% of the world languages are spoken by about 3% of the world population (Bernard, 1996). This shows that the heterogeneity of the world languages relies on a very small number of people (UNESCO, 2003). Dialects are among less-resourced languages in the context of applying written or spoken document-intensive NLP (Natural Language Processing) technologies. Written documents in dialects are very scarce. This hinders them from making the most out of language technologies which might have helped them out from getting endangered. Although dialects are not rich in written resources, recent developments have been seen on producing written online documents for dialects e.g. Plattdeutsch. Plattdeutsch is one of the dialects of Standard German spoken mainly in

© Springer Nature Switzerland AG 2020
G. Nicosia et al. (Eds.): LOD 2020, LNCS 12566, pp. 494–503, 2020.
https://doi.org/10.1007/978-3-030-64580-9_41

northwest Germany. The availability of online written documents in dialects, in this case, Plattdeutsch, has led to exploiting the web for construction of parallel corpus for Standard German and Plattdeutsch (Misganaw and Sabine, 2018).

Dialects were assumed old-fashioned and attributed to low social status. However, through generations, the trend of learning and speaking dialects has changed. There are now groups, like pop-group called "Tüdelband" from Hamburg (Koneva and Gural, 2015) advocating the use of dialects through their musical works. This motivates the young to learn the language. Furthermore, as dialects are just variants of a language, one can learn relatively fast if one knows the language which the dialect is a variant of.

In this study, we tried to exploit the existing technologies and resources to contribute to the preservation of dialects by producing parallel audio corpus in the context of DMW (*Dialektatlas Mittleres Westdeutschland*)[1] project that aims at collecting and documenting the various dialects in middle west German. The result of the study could be used in various use cases such as computer-supported language learning, language variety identification as well as speech recognition.

The rest of the paper is organized as follows. Related works are briefly reviewed in Sect. 2. In Sect. 3 the data collection phase of this work is described in detail. Section 4 discusses the process of data extraction taking relevance into account is made. The process of converting the standard German text into audio by applying the text-to-speech technology is described in Sect. 5. After preparing all the relevant data, follows the process of searching which is explained in Sect. 6. Based on the search word/phrase given in Sect. 6, retrieval of the corresponding audios is explained in 7. Results and discussions are explained in Sects. 8.1 and 8.2 respectively. Finally, we provide recommendations for future works in Sect. 9.

2 Related Works

Several researches have been conducted in the area of preserving endangered languages. According to UNESCO Atlas of World Languages, about 14 dialects in Germany are classified under vulnerable, severely endangered, definitely endangered or extinct (Moseley 2010). According to UNESCO language classification, a language (dialect) is endangered if most children speak the language but restricted to some aspect of their lives e.g. home.

Dialects are among the list of less-resourced languages. This could be attributed to the fact that dialects are not taught in schools which in turn results in being denied of an appropriate level of attention (Lothian, et al., 2019). Preserving a dialect starts with making use of native words when speaking it. This, in fact, applies to any language. Koneva and Gural (2015) indicated that the purity of a language can be kept by:

1. *consciously using the existing words in a language*
2. *creating new words on the basis of already existing ones"* (Koneva and Gural, 2015, p 251)

---

[1] https://www.dmw-projekt.de/.

The most commonly used language preserving techniques as listed in (Chang-Castillo and Associates, 2019) are; creating recorded and printed resources, preparing language courses, using digital and social media outlets, and insisting on speaking their own language. Merely collecting resources does not help preserve endangered languages by itself. Only when the resources are made available for those interested could help to preserve the language since it will either be used to conduct further researches focusing on the same objective, or people could use the resource to learn the language. Such an effort is shown by Castelruiz et al. (2004) which focuses on developing a web-based interface with search forms for a multimedia archive of audio and video recordings of the Basque language.

Another similar work by Barbiers et al. (2007) describes the methods employed to collect audio data of the Dutch dialect in Dutch Syntactic Atlas of the Dutch Dialects (SAND) project. In SAND project, the researchers were able to produce audio data for 267 dialects through oral and telephone interviews and postal surveys. In addition, the SAND project has also a search engine for creation of online maps.

Another notable study, GreeD (Karasimos et al., 2008), is a database of dialectal linguistic and meta-linguistic corpora of the Greek dialects which aims at digitizing, cataloging and encoding Greek dialects.

3 Data Collection

DMW is a project aimed at collecting and documenting the various dialects in middle west Germany. The main data collection method involved is interview. Interviews are made with respondents for whom questionnaire has been sent to identify whether they are potential candidates for the interview. Potential candidates are selected based on the fact that they speak the dialect in their respective region.

Before conducting the interviews, 1183 geographic locations are identified in the central west Germany. In addition, informants[2] are classified into two clusters based on their age groups. Group 1 is composed of informants aged older than 59 and in group 2 are those aged between 28 and 45. This grouping is made for the purpose of studying whether (or how) the usage of the dialect evolved over different generations. In each group and for each selected geographical area, whenever possible, two informants (one male and one female) are selected and interviewed. At the end of 2019, from the selected 1183 places, 240 informants from 120 locations are already interviewed. During the time of writing this paper, no informant from group 2 is interviewed. Accordingly, all the data represented in this work is from group 1.

The audios are collected by asking informants a series of predefined questions prepared by linguists. The questions are designed to extract dialectically relevant data from the answers that the informants would give. The questions are stored in relational database (MySQL) and the corresponding audio files collected through interviews for each informant are stored in file systems.

Interviews are made by experts through asking respondents a series of questions which aim at collecting dialectically relevant words and phrases. Some of these questions

---

[2] Informant is a person who is interviewed.

are designed in such a way that respondents look at a picture or watch a snippet of video before their answer is recorded. This is to enable them understand and describe the activities and situations. One example of this type is *"Welches Tier sehen Sie auf dem Bild?* (what do you see on the picture?)"

This research utilizes these interview data collected for the DMW project for the purpose of building the parallel corpus for dialects found in middle west Germany.

4 Extraction of Relevant Data

The audio snippets corresponding to the questions are further processed in order to crop out only the relevant part of the audio. For example, the respondent might be asked to speak out a whole sentence *"Das Wetter ist schön geworden"* (The weather has become nice). However, the dialectically relevant phrase in the sentence is *"schön geworden"* (become nice). Therefore, further processing is necessary to crop only *"schön geworden"* out of the audio containing the whole sentence. This process creates an alignment between the standard German text stored in a database (DB) and the audio files.

Searching and retrieving the audio files is performed via the corresponding textual information as shown in Fig. 1. Figure 1 depicts where the information is stored, i.e., in relational DB and file system, and the interaction between them for searching and displaying the result back. The DB contains meta-data about the audio files, i.e., the natural language representation of the words and/or phrases uttered by informants. In addition, it contains information regarding the location the informant represents and the

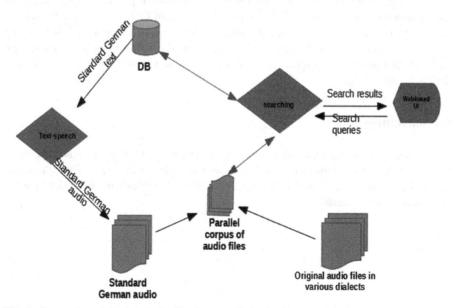

Fig. 1. Interaction of the data in the Database, audio in the file system and the web-based search interface.

dialect the informant speaks. In file system, the preprocessed audios which match with the text stored in the DB are stored.

A PHP script and relational database querying are used to extract these data from their corresponding storage and to keep their link, i.e., which speaker corresponds to which audio, and which place does the informant represent (this in turn provides us with which dialect the informant speaks). The questions and the informant are identified by unique ID and these IDs are used to uniquely identify the audio files corresponding to each speaker.

5 Text-to-Speech Conversion for Audio Alignment

Interviews are made to collect audio for various dialects using standard German texts, i.e., the interview questions are prepared in standard German. The corresponding question used in the interview for the phrase shown in the search box in Fig. 2 is: *"Bitte übersetzen Sie in Ihren Dialekt/Ihr Platt: Das Wetter ist schön geworden."* (Please translate in your dialect: The weather has become nice)

The dialectically relevant phrase of this question in this text, i.e., *"schön geworden"* (become nice) which is in standard German, is stored in the database. By applying text-to-speech conversion using Google API[3], the equivalent standard German audio is created. This enables us to come up with a parallel audio corpus of dialects and standard German. Figure 3 shows this list of audios for selected dialectically relevant words/phrases with their corresponding dialect names.

For the text-to-speech conversion, the gTTS google API is used together with a python script. Individual standard German text are feed to the python script which gives back the audio versions of the text. Since each Standard German text and the audios collected from informants are linked with IDs, the standard German audios from the text-to-speech conversion are automatically linked to the audios in dialects. Thus, each standard German text is aligned with the corresponding standard German audio and audios in different dialects.

6 Searching

For searching a word/phrase, we created an interface in which as one starts typing, a minimized list of available words will be listed in alphabetical order. This avoids the unnecessary *'the word is not found'* message if the word/phrase the user is looking for is not found in our database.

---

[3] https://cloud.googl.com/text-to-speech/.

As one types in the search box, a shorter list in the auto-complete field on the interface shown in Fig. 2 will be displayed for selection. This helps the user to see whether a word in search already exists or not, rather than waiting for system response after hitting search.

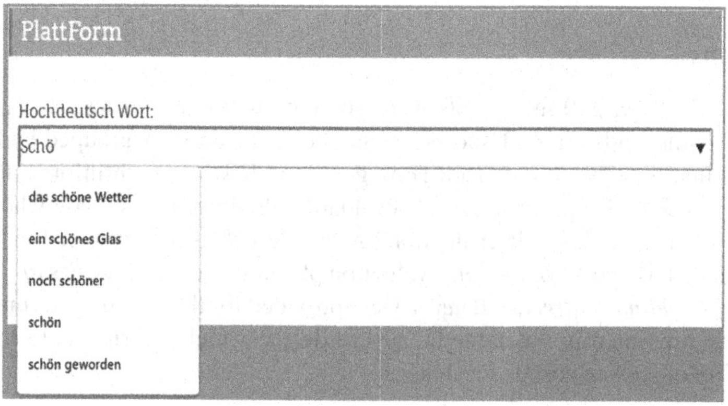

Fig. 2. Searching for a word with auto-complete.

In Fig. 2, it is shown that as a user starts typing '*schö*', the system lists a minimized list of words/phrases which contains '*schö*'.

7 Getting Audios for Dialects

After the user has picked a word/phrase, the corresponding audios in the available dialects are listed. Figure 3 shows a sample list of audio files for the phrase '*schön geworden*'

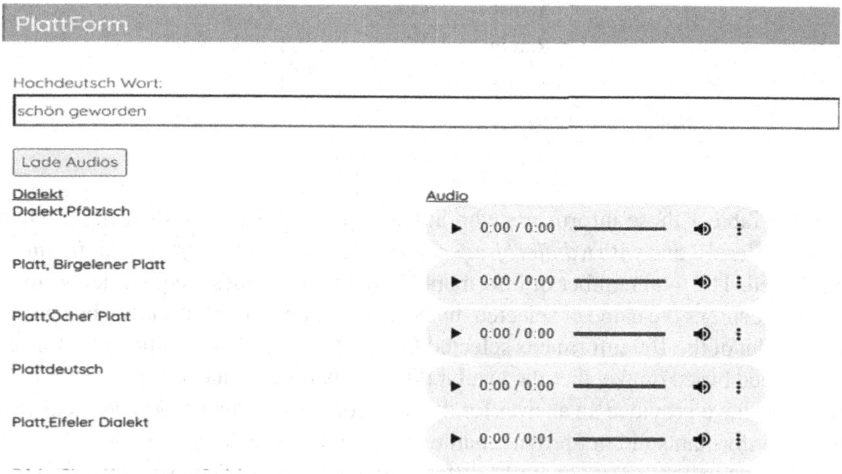

Fig. 3. Sample list of dialect names with corresponding audio players.

in different dialects. One can use the audio player to listen to the audio of the selected phrase. The column '*Dialekt*' shows the name of the dialect the informants speak with his/her family members and/or in the residence place they live in.

8 Results and Discussions

8.1 Results

At the end of 2019, 240 informants were already interviewed. These 240 informants are used for this study. Table 1 shows the number of informants grouped by how good and how often they speak a dialect. How good and how often an informant speaks a particular dialect is taken from the self-evaluation the informants made when they fill out the questionnaire. The self-evaluation for "how good" is in a scale of 1 to 6, where 1 is *excellent* and 6 is *bad*. For the self-evaluation of "how often", the values *always, often, frequently, sometimes, rarely*, and *never* were provided for the informant as options. The information presented in this table is used to decide which informants to select for a good quality of spoken corpus for dialects.

Table 1. Number of informants grouped into how often and how good they speak a dialect.

Dialect rating	Dialect frequency	No. Informants
1	Always, Frequently, Often	104
1	Sometimes	6
2	Always, Frequently, Often	65
2	Sometimes, Rarely	29
3	Often, Frequently	9
3	Sometimes, Rarely	19
4	Sometimes, Rarely	4
5	Rarely	1
6	Never	1
Not given	Not given	2

From Table 1 those informants who at least speak a very good dialect, i.e., dialect rating of at least 2 and with a *dialect frequency* values of '*always*','*often*' and '*frequently*' are selected. The total number of informants found to fulfill these requirements are 169.

Table 2 shows the name of selected dialects and number of informants who speak the dialects. Out of the 169 informants selected from Table 1, further selection of informants is performed based on whether the availability of a particular dialect name. That is, 167 informants have provided an answer for the question "which dialect do you speak?" and only two informants did not provide name of the dialect she/he speaks.

As shown in Table 2, there are 13 different dialect names. However, the words *Platt, Mundart* and *Dialekt* all mean dialect while *Umgangssprache* and *Hochdeutsch* mean

Table 2. Number of informants per dialect.

Dialect	No. Informant
Birgelener	1
Boker	1
Dialekt	4
Eifeler	4
Hochdeutsch	8
Mundart	5
Muster	1
Not given	2
Öcher	1
Pfälzisch	2
Platt	135
Plattdeutsch	1
Umgangssprache	2
Wildener	1
Wingeshauser	1

colloquial language and standard German, respectively. Since the focus of this study is to collect spoken corpus for dialects, *Platt, Mundart, Dialekt, Umgangssprache* and *Hochdeutsch* are not taken into consideration as they do not specify the dialect name. Accordingly, the nine selected dialects are: *Birgelener, Boke, Eifele, Muster, Öcher, Pfälzisch, Plattdeutsch, Wildene* and *Wingeshauser.*

8.2 Discussions

The name of the dialect is collected from the questionnaire that the informant filled out. Sometimes, even if the informants are from the same geographic location, they did not use the same name to refer to their respective dialects. This makes the process of categorizing the audios extracted into their respective dialect not to be the same, i.e., even if they are the same, they fall into different dialects. For example, the words *Dialekt,* and *Mundart* are used by the informants to indicate that they speak the dialect of the region they live in. But these words all mean the same – dialect and do not specify a dialect name. As a result, most of the audios collected are not used for the corpus construction.

9   Conclusion and Future Works

In this study, we tried to exploit the existing technologies and resources to contribute to the preservation of dialects by producing parallel audio corpus. A parallel spoken corpus for nine dialects, each having a parallel audio data of size 550 for each dialect, where size is measured by the count of words or phrases was created. The outcome of this work could be used in various use cases such as computer-supported language learning, language variety identification, dialect identification as well as speech recognition. Furthermore, these data can be utilized in training a neural network for translation of German dialects into standard German (or vice versa).

As a future work, it can be extended by including speech to text technology to come up with a parallel corpus of written word/phrases. Moreover, during the linguistic analysis process, linguists transcribe the audios in IPA (International Phonetic Alphabet) and POP (Popular) notations. These notations are the phonetic transcriptions of the audio items for dialects. These notations can be used to construct text-based parallel corpus of dialects – standard German. This could be achieved by transforming the notations into standard language text.

One of the problems we faced during the audio extraction phase is that some informants have described their dialect as *Mundart* or *Umgangssprache* which mean dialect and colloquial speech, respectively. These descriptions of the informants' dialect do not give us precise information about which dialect the informant speaks. This needs further work to automatically identify the dialect from the information the informants provided in the questionnaire.

Since there are several projects focused on the preservation and linguistic study of the various dialects in Germany, this research could be extended to include other dialects that are not part of the DMW project.

In this research, a prototype for a web-based interface for searching through the parallel corpus is developed. However, in this era of hand-held devices, since many people use smart phones, and tablets this can be scaled up by developing a mobile App.

Acknowledgement. We would like to acknowledge the informants who have participated in the recordings and linguistic experts in the DMW project who have worked on audio processing.
linguistic experts in the DMW project who have worked on audio processing.

References

Aynalem, T.M., Sabine, R.: Towards Parallel corpora construction: standard German –Plattdeutsch. In: Parallel Corpora: Creation and Applications International Symposium PaCor, Madrid (2018)

Barbiers, S., Leonie, C., Kunst, J.P.: The Syntactic Atlas of the Dutch Dialects (SAND): a corpus of elicited speech and text as an online dynamic atlas. In: Creating and Digitizing Language Corpora, pp. 54–90, Palgrave Macmillan, London (2007)

Bernard, H.R.: Preserving language diversity. Human Organization **51**(1), 82–89 (1996). In: Indigenous Literacies in the Americas: Language Planning from the Bottom up. ed. by Hornberger, N.H., pp. 139-156. Mouton de Gruyter, Berlin (1992)

Castelruiz, A., Sánchez, J., Zalbide, X., Navas, E., Graminde. I.: Description and design of a web accessible multimedia archive. In: Proceedings of the 12th IEEE Mediterranean Electrotechnical Conference (IEEE Cat. No. 04 CH37521), Vol. 2, pp. 681–684. IEEE (2004)

Chang-Castillo and Associates. https://ccalanguagesolutions.com/language-preservation-how-countries-preserve-their-languages/. Accessed 1 Oct 2019

Koneva, E.V., Gural, S.K.: The role of dialects in German society. In: The XXVI Annual International Academic Conference, Language and Culture, pp. 248–252 (2015)

Karasimos, A., Melissaropoulou, D., Papazachariou, D., Assimakopoulos, D.: GreeD: cataloguing and encoding modern greek dialectal oral corpora. In: Proceedings of CatCod, Orleans, France (2008)

Lothian, D., Akcayir, G., Carrie, D.E.: Accommodating Indigenous People When Using Technology to Learn Their Ancestral Language. At Edmonton, Alberta (2019)

Moseley, C., (ed.): Atlas of the World's Languages in Danger, 3rd edn. UNESCO Publishing, Paris (2010). http://www.unesco.org/culture/en/endangeredlanguages/atlas. Accessed 01 Oct 2016

UNESCO: Language Vitality and Endangerment Ad Hoc Expert Group on Endangered Languages, International Expert Meeting on UNESCO Programme Safeguarding of Endangered Languages (2003)

Quantifying Local Energy Demand Through Pollution Analysis

Cole Smith[1]([📧])[iD], Andrii Dobroshynskyi[1][iD], and Suzanne McIntosh[1,2][iD]

[1] Courant Institute of Mathematical Sciences, New York University,
New York, NY 10012, USA
{css,andrii.d}@nyu.edu, mcintosh@cs.nyu.edu
[2] Center for Data Science, New York University, New York, NY 10012, USA
https://cs.nyu.edu/home/index.html
https://cds.nyu.edu/

Abstract. In this paper, we explore the process of quantifying energy demand given measurable emissions of airborne pollutants throughout the United States, from 2014–2019. Airborne pollutants often take the form of criteria gasses, including Ozone, NO2, Carbon Monoxide, Sulfur Dioxide, and toxic compounds. Using daily emission measurements throughout the United States from the U.S. Environmental Protection Agency (EPA), and spot prices for oil and gas products from the U.S. Energy Information Administration (EIA), we define a mapping between these two universes. As a basis of analysis for the ecological consequences of oil and gas pricing on a daily scale, we define a demand score, based upon the Time-Lagged Cross Correlation (TLCC) between pollution levels, and spot prices. We have developed a scalable, pure-Apache Spark implementation of our TLCC algorithm. We then provide a reproducible justification for the predominantly negative correlation of pollution measures and oil prices on a local scale.

Keywords: Air quality · Oil and gas · Apache spark · Time series

1 Introduction

Spot prices of petroleum, natural gas, and energy are dependent on a variety of influencing factors, and major factors are likely geopolitical in nature. However, there does exist a regularity in these markets such that an increase in demand in the United States (which consumes about 20% of the world's oil and gas [1]) often prompts a spike in prices by the Organization of the Petroleum Exporting Countries (OPEC).

It is also known that Ozone and NO2 gasses are produced by the combustion of fossil fuels, those being oil, natural gas, and coal. By exploring daily pollution data, we may derive "hot spots" of high energy demand, relative to locations, and demand spikes, relative to time.

We approach the problem of quantifying demand by constructing multiple time-series of pollution data and spot prices, and then analyzing the moving

© Springer Nature Switzerland AG 2020
G. Nicosia et al. (Eds.): LOD 2020, LNCS 12566, pp. 504–515, 2020.
https://doi.org/10.1007/978-3-030-64580-9_42

trends between both to measure these "hot spots." We structure our problem architecturally as a pipeline of operations, which can provide automated analysis, thus addressing a common problem in this field of reproducibility.

The main contributions of our work are as follows:

- A novel application for quickly creating on-demand subsets and transformations on all of U.S. EIA's 350+ gigabyte AirData dataset
- A parallelized pure-Spark implementation of the Time-Lagged Cross Correlation algorithm for identifying relationships on arbitrarily large datasets, with significant speedup over sequential implementations
- A region-sensitive scoring measure that relates pollution output to energy demand

The remaining parts of the paper are organized as follows: Sect. 2 covers select related work, Sect. 3 describes our data, and Sect. 4 the analytic algorithms used. Sections 5 and 6 provide our application design for performing analysis and the specific area-based scoring analytic, respectively. Finally, Sect. 7 shows visualizations of the scoring metrics.

2   Related Work

2.1   Modeling Oil Spot Price Distributions

Multiple works have attempted to accurately model the price of commodities such as petroleum using forecasting methods. [2] presents a framework for time series prediction when dealing with irregular spikes in the data. The paper specifically tests on crude oil price prediction and focuses on utilizing neural networks for forecasting on small spikes in the time series. [3] explains a similar approach with wavelets, but with a focus on identifying market inefficiencies. [4] and [5] show promising results on modelling electricity usage, with [4] also using wavelet transforms for feature selection. Across the works, emphasis is placed on accounting for non-stationarity for more accurate modelling of the resource's price movement, as well as on averaging multiple models to better generalize to irregular movements, i.e. spikes that were the focus of paper [2].

Overall, the emphasis of relevant work on oil spot price forecasting has been on efficient utilization of the oil price signal itself for accurate prediction. Furthermore, the datasets used for forecasting are only individual price signals. Thus, much of the emphasis is placed on forecasting strategy rather than the data.

2.2   Learning from Air Pollution Data

In [6], the authors are able to accurately predict the emissions and air pollution of oil and gas production in Utah from 2015 to 2019. They do so by measuring Volatile Organic Compounds (VOCs), greenhouse gasses, and methane, against drilling activity, and validating with Monte-Carlo methods. Their results come within 5% of the actual predicted activity for drilling and hydraulic fracturing. Their approach exhibits the feasibility of our approach, although we will consider data from throughout the United States, rather than constraining ourselves to narrow air quality information centered only around gas production.

[7] shows a direct correlation between a reduction in CO_2 emission levels and energy demand by studying country-level emissions for various levels of confinement index (CI). CI captures the extent to which different COVID-19 confinement policies affected CO_2 emissions. By encouraging, or in some cases requiring, working and attending classes from home, the confinement policies caused reduced consumption. According to [8], human activity is responsible for almost all of the increase in greenhouse gases over the past 150 years. [9] states that, in 2018, the primary sources of emissions were from burning fossil fuels for electricity (26.9"%"), transportation (28.2%), and industry (22%).

[10] shows that oil prices are predictable using economic signals alone. This is a natural approach, and one that the authors had success with. They express that the price of oil could be a successful indicator of Gross National Product (GNP). The authors then predict daily oil prices using Gene Expression Programming (GEP) and Neural Networks. The goal of our approach is to relate this information at a local level. The cited work attempts to forecast oil price or pollution signals using the signal itself, while our work is concerned with discovering a relationship between these two signals geographically.

Energy demand in the United States varies from one area to the next, and quantifying it continues to be an ongoing research problem that is not easily solved. In our approach, we will use different data than was used in the prior art cited. We have verified that the modeling approach described in prior art can be applied successfully. Further, we found that the GEP model outperforms existing Neural Network solutions to the same problem. We discuss translating this technique for distributed, arbitrarily large datasets in the Future Work section.

[11] outlines the predicted energy demand, and the geographical indicators that occur with a (predicted) increase in oil drilling and fracking. Moreover, it introduces the models that are used to predict the probability of drilling at certain sites, but it does not describe the details of those models. Our heatmap visualizations are consistent with [11], namely, we find that these production areas in the western United States are significant hot spots, according to our local scoring measure.

[12] introduces a use of the Time-Lagged Cross Correlation algorithm for meteorology. Similarly, [13] uses a height-modified version of the TLCC algorithm for detecting PM2.5 fluctuations in neighboring cities in China, where PM2.5 describes fine inhalable particles having diameters of 2.5 micrometers or less. Both works provide sufficient numerical justification for their results. However, they do not discuss in detail the implementation of their respective algorithms, nor the scalability of their respective algorithms to process arbitrarily large data sets common in meteorological studies.

3   Datasets

We employ two datasets for analyzing energy demand: (1) a fine-grained measure of air data that allows us to quantify pollution, and (2) spot price data for oil that we use to measure demand.

3.1   AirData

Key Fields:

- Longitude, Latitude
- Criteria Gas Levels (Ozone, NO2 [Nitrogen Dioxide], CO [Carbon Monoxide], SO2 [Sulfur Dioxide])
- Toxic Compound Levels (NOy, NO, NOx, NMOC)

Metadata:

- File size: ~350GB
- Time span: 1985–2019 (Our analysis is from 2014–2019)
- Source: https://aqs.epa.gov/aqsweb/airdata/download_files.html#Raw

The `AirData` dataset gives us detailed features for air quality based on location. Key features (columns) we used include longitude, latitude, Ozone levels, and NO2 levels. Gaseous and toxic compound levels are measured in parts per million (ppm). We have a scalable Extract Transform Load (ETL) pipeline in place that transforms multiple measures from this dataset into normalized quantities independent of units.

For our analysis, the most important group from the `AirData` datasets is the criteria gasses group. We begin by assessing the trends in air pollution by considering all spatial data points together, grouped by day of year, over nearly five years (1-Jan-2014 to 29-May-2019) to visualize seasonality of pollution. At the time of analysis, this was the latest possible data provided by the EPA. In particular, we notice in Fig. 1 that criteria gasses are seasonally correlated, with Ozone exhibiting negative correlation while the rest exhibit positive correlation.

Lastly, we assess the L2 Norm of all normalized criteria gas levels, grouped by day of year and averaged over five years (1-Jan-2014 to 29-May-2019). We see that the signal itself appears to maximally correlate with the discrete Ozone signal. Thus, Fig. 2 illustrates that the majority of total pollution magnitude for a given day of the year can be explained by the Ozone level on that day.

3.2   Petroleum Data

Key Fields:

- Spot Price
- Date (`YYYYmmDD` format)

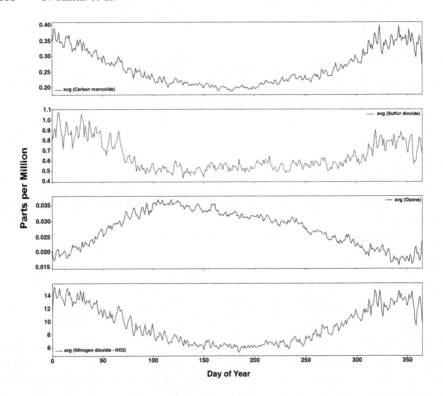

Fig. 1. Daily average pollution levels from 1-Jan through 31-Dec. The daily averages include data from 2014 through 2019.

Metadata:

– File size: ~50MB
– Time span: 2-Jan-1986 to 15-Jun-2020
– Source: www.eia.gov/opendata/qb.php?category=241335&sdid=PET. RWTC.D

In order to quantify demand for the oil resource, we use a dataset of daily spot prices for crude oil, or in other words, the market price that somebody is willing to pay for the resource. The data, which we organize as a `PetroleumData` dataset, is available in daily granularity, and is measured in a dollar amount per barrel of the resource. The ETL pipeline in place for `PetroleumData` transforms the price column into a normalized quantity and computes extra columns based on the date column of the spot price.

In conjunction with `AirData`, we are able to align the normalized price quantities to normalized gaseous air compound levels based on time. We construct the `criteria_gas_levels` and `petroleum_price_levels` time series that are the subject of analysis in Sect. 6.

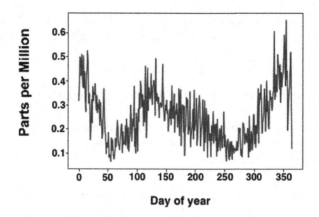

Fig. 2. L2 Norm of pollution vectors (Normalized PPM) for average of each day of year, 2014 through 2019

4   Description of Analytic, Algorithms

We store and process both data sets in the Apache Hadoop Distributed File System (HDFS), on the Apache Hadoop/Spark cluster, Dumbo, administered by New York University (NYU) High Performance Computing (HPC). The Dumbo cluster is an on-premises cluster comprised of 44 worker nodes running Linux CentOS 6.9. Each worker has two 8-core Intel Haswell CPUs, 128GB of memory, and sixteen 2TB hard drives for HDFS.

Our main analytic application is built using the Simple Build Tool (SBT), version 1.2.8 with Spark version 2.2, and Scala version 2.11.8.

4.1   Time-Lagged Cross Correlation

We provide a ground-up implementation in pure Spark and Scala of a Time-Lagged Cross Correlation (TLCC) framework using Spark's MLLib package. The TLCC class is the root of our analysis of the relationship between pollution and energy demand.

To motivate, we first define two collections of one-dimensional signals of length n, the number of periods in the signal. Let those be the sets:

$$x \in X : x = (x_1, ..., x_n) \in \mathbb{R}^n \tag{1}$$

$$y \in Y : y = (y_1, ..., y_n) \in \mathbb{R}^n \tag{2}$$

It is not necessary for the cardinality of X to equal that of Y, however, the vector length n must equal. We now define a set of integers L of lag-periods, which express how many periods to shift each $x \in X$. Let the function $f(x, l)$ shift x by l-periods forward if l is positive, or backwards if l is negative. The shift is not padded such that the length of $x_l = f(x, l)$ is $n - l$.

The TLCC statistic is easily parallelizable. We combine X, Y and L in an all-play-all manner using Cartesian Product ($\times$) to emit all the possible triplets:

$$X \times Y \times L = \{(x, y, l) \mid x \in X, \ y \in Y, \ l \in L\} \tag{3}$$

These triplets are sent in a distributed manner as keys to the reduction function $R(x, y, l)$ which computes Pearson's Correlation between x and y by truncating y at the start or end, by the signed l, such that non-existent (padded) values of x are ignored in the correlation. (Our TLCC implementation nevertheless also supports padding.)

$$T(y, l) = \begin{cases} (y_i)_{l \leq i < n} & l > 0 \\ (y_i)_{0 \leq i < |l|} & l < 0 \\ y & y = 0 \end{cases} \tag{4}$$

$$R(x, y, l) = \frac{cov(f(x, l), T(y, l))}{\sigma_{f(x,l)} \cdot \sigma_{T(y,l)}} \tag{5}$$

where $cov(x, y)$ is the covariance of x and y, and σ_x is the standard deviation of x. Since all the keys of $X \times Y \times L$ are independent of each other, all $R(x, y, l) \in X \times Y \times L$ can be executed in parallel using MapReduce-like operations in Spark.

It is worth noting that a variety of analytics can be defined for $R(x, y, l)$. These can be linear or non-linear so long as the analytic is not dependent on the result of another pair. We chose Pearson's Correlation Coefficient for our study since it represents a common baseline for a linear relationship between two signals.

4.2   TLCC Performance

To test the performance of our parallelized TLCC algorithm. We compute the auto-correlation of a subset of criteria from the EPA AirData dataset. We found that our Spark implementation is highly scalable, given the independence of each correlation computation from another. The timing analysis provided in Table 1 shows that our TLCC algorithm enjoys a 30% speedup compared to sequential performance for 64 correlations between vectors of 1947 elements in length. In particular, our data preparation pipeline achieves significant speedup. This pipeline includes normalizing the columns, sorting, and extracting the columns as vectors. For smaller vector sizes, we achieve very little speedup because the overhead of the Spark framework cannot be readily amortized. We also see a speedup for a small number of total correlations because Spark is able to automatically allocate its executors based on the problem size.

4.3   TLCC Results

In order to find the lag interval with the highest correlation between pollution and price signals, we deliver the columns from AirData as the X set in TLCC, the signals which will be shifted. These are then compared against the EIA's

Table 1. Timing analysis of data preparation and TLCC running time (seconds).

Executors	Correlations	Elements	Total time	TLCC time	Tot. speedup	TLCC speedup
1	8	121	52.41	0.04		
1	8	1947	17.68	0.04	1.0	1.0
1	64	121	102.07	0.10		
1	64	1947	24.90	0.11		
30	8	121	51.48	0.10	1.02	0.41
30	8	1947	2.48	0.07	**7.13**	0.59
30	64	121	99.70	0.11	1.02	0.85
30	64	1947	1.57	0.09	**15.83**	**1.30**

Table 2. Best correlation to petroleum spot price for 2019

Criteria gas	Corr.	Toxic compound	Corr.
CO	−0.81	**NOy**	−0.32
SO2 (60 Day Lag)	−0.43	**NOx** (30 Day Lag)	−0.47
Ozone (10 Day Lag)	0.85	**NO**	−0.31
NO2 (10 Day Lag)	−0.65	**NMOC** (30 Day lag)	−0.53

benchmark for market oil prices, in TLCC, the Y set. We found that the criteria gasses (Ozone, NO2, Sulfur Dioxide, and Carbon Monoxide) were generally more correlated with oil spot-prices at 0, 10, and 30-day lags than toxic compounds (NOy, NO, NOx, NMOC) were at the same lag values. We considered lags of $\pm$ 60, 30, 10, and 0 days. The criteria gasses are shown on the left side of Table 2, toxic compounds are shown on the right.

CO has the highest correlation to price, although the values of CO had the lowest variance of all the other signals. Ozone had the highest correlation with price signals. The correlations were also computed on the time span of 2014 through 2019 to receive a baseline of correlation in the long-run. As expected, these are quite low, and the average correlation across all time lags was 0.14. This is likely because we introduce more confounding factors that cannot be easily controlled in the long-run. We therefore believe the correlation results in the short-run for 2019 alone to be statistically significant given this mean.

These negative-correlation results are consistent with our expectations– As prices of petroleum increase, we expect quantity demanded to subsequently decrease, and therefore exhibit a negative relationship with the resulting pollution. [14] shows that oil prices often exhibit a similar trend due to production saturation by OPEC. This speculation however cannot be proven with our results alone. While our work provides a quantitative justification for the hypothesis of [14], future work is needed to assess the *causal* nature of our features.

5   Application Design

Our code base is publicly available in two main Scala packages. Namely, we implement the `bdad.etl` and `bdad.model` packages, which provide functionality for loading cleaned data sets from the EPA and EIA, and performing analysis on those datasets, respectively.

Lastly, we compute a heatmap displayed by our User Interface (UI). It reflects the assumed energy demand in local areas throughout the United States.

5.1   Extract, Transform, Load Design

As one of the principal novelties of our contribution, the `AirDataset` class provides an intuitive interface to the otherwise unwieldy 350GB+ EPA AirData data set. In particular, we provide functionality for selecting the required date range and criteria to create specific *Scenarios* for analysis. The recorded measurements are spatial to a given latitude and longitude, the precision of which can be set by the class to average the pollution for either larger or smaller areas.

For the purposes of this paper, we consider the criteria gasses and toxic compounds, although the data contains hundreds of other features. This class will first select only the required files from HDFS, pivot the selected criteria into feature columns, and normalize those selected features (by mean and standard deviation) into a single `Vector` column for use in modeling.

Similar to the functionality of the `AirDataset` class, the `PetroleumDataset` class provides an interface for interacting with the data on crude oil spot prices from the EIA data repository. This interface helps with processing the Price column and generating new columns from the timestamp, such as `year` and `dayofyear` for the purposes of joining when computing correlation analytics.

Finally, the `Scenarios` class provides specific *implementations* of the above ETL classes.

6   Area Scoring Analytic

Given the results of our TLCC performance analysis described in Sect. 4.3, we now deduce a scoring measure for any given region with the current petroleum spot price(s) and past air quality data. We first define a measure of overall pollution as the L2 Norm of each (normalized) criteria gas moving average for a given latitude/longitude pair, on a given day. The motivation is that we interpret the *magnitude* of the pollution or price vector as this L2 norm:

$$\|X\| = \sqrt{\sum_{i=0}^{n} X_i^2} \tag{6}$$

Before computing the L2 norm, the air quality moving averages are computed by taking the Simple Moving Average (SMA) over n periods, where n is the lag value l for a feature x that yielded the highest correlation with another measure y in the TLCC analysis. We assume this final value to be the *magnitude of increasing pollution*.

This is then divided by the petroleum spot price for the current day to get the dollar-impact of pollution. If there are multiple petroleum prices considered, they are again the L2 Norm of the normalized price signals. We consider this the *magnitude of current demand*.

Local Pollution Scoring Measure

- Let $X :=$ Criteria Gas signals by Time and Location
- Let $X_{\text{SMA}_n} :=$ SMA of X over n periods
- Let $Y :=$ Petroleum Price signals by Time (for each Location)

For each row i in X and Y,

$$score := \frac{\|X_{\text{SMA}_n i}\|}{\|Y_i\|} \tag{7}$$

We consider this final score to model the magnitude of an increase in environmental damage per unit of current demand.

Using this score for a region, we can deduce the trend of the overall scoring signal for a given area of interest by computing the historical scores for that area. Moreover, one can compare regions by energy demand using this measure by comparing their scores. Notice that this is the same as comparing the regions by the pollution values themselves, since the numerator of the score can be the same for all regions. This is discussed visually in the following section.

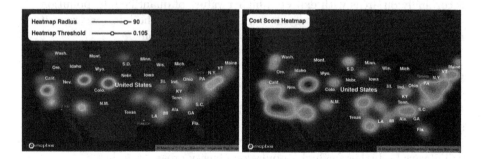

Fig. 3. HeatMap visualization of Geo-Spatial Pollution Data (L2 Norm)

7   Analysis and Visualization

Our application includes a front-end web application built to visualize the data output from our **Heatmap** model analytic. The visualization renders spatial detail of AirData information per longitude and latitude. It also renders data from our CostRepresentation model that generates local pollution scores.

The web application includes tools for controlling the rendering parameters for the two heatmaps, and the geo-map layers are fully interactive maps where one can zoom in/out for finer-level inspection of the heatmap patterns.

Figure 3 is an example of a scoring heatmap. In Fig. 3, we notice that the scoring heatmap appears similar to our pollution-only heatmap, but places more focus on major metropolitan areas. This is likely the case for a few reasons. The pollution heatmap is averaged over all of 2019, while our score in this instance only considers our best found lag time of 10 days for simplicity. Moreover, division by price will allow areas with less overall pollution to be represented by the heatmap. This allows one to consider more subtle movements in pollution scoring over 10-day periods.

8   Conclusion

In this work, we have constructed a series of processes for quantifying localized energy demand from measurable air quality data across the United States. Moreover, we address a common problem of reproducibility through careful orchestration of these processes.

We have shown a dominant inverse correlation between market oil spot prices and gaseous compound emission metrics. We have implemented a parallelized TLCC model as the foundation of our analysis, which offers significant scalability in processing large datasets. We provide accompanying libraries for traversing the large AirData dataset. By identifying the lag interval of highest correlation between oil price and pollution of 10 days, our cost-score analytic justifies existing beliefs of oil pricing and pollution relationships, placing more weight accordingly on areas with expected higher values of quantified air pollution.

For reproducibility, we provide our source code at the following repository: https://github.com/css459/energy-demand-and-pollution

9   Future Work

Future work will need to be conducted on the feasibility of our scoring measure for analysis on a per-capita basis. We deliver gross scores for a given latitude/longitude pair, but as we see in our cost-score heatmap, this gross value will always be higher for high population-density areas.

We also consider looking at the cost-score distributions over time as future work. We think it may be useful to study the patterns of a cost-score given to a latitude/longitude pair changing over time per specific time lags and to measure consistency of pollution quantified by our metric across regions. Moreover, nonlinear regression methods can be implemented using our TLCC algorithm by replacing the reduction function. This algorithm can be adjusted to shift signals for use in predictive GEP or Neural Networks.

Acknowledgment. We thank NYU HPC and the Cloudera Academic Partnership for providing the Big Data tools that we used in developing our analytic application.

References

1. How much oil is consumed in the United States? Frequently Asked Questions (FAQs) - U.S. Energy Information Administration (EIA) (2020). https://www.eia.gov/tools/faqs/faq.php?id=33&t=6

2. Yu, L., Lai, K.K., Wang, S., He, K.: Oil Price forecasting with an EMD-based multiscale neural network learning paradigm. In: Shi, Y., van Albada, G.D., Dongarra, J., Sloot, P.M.A. (eds.) ICCS 2007. LNCS, vol. 4489, pp. 925–932. Springer, Heidelberg (2007). https://doi.org/10.1007/978-3-540-72588-6_148

3. Youse, S., Weinreich, I., Reinarz, D.: Wavelet-based prediction of oil prices. Chaos, Solitons & Fractals **25**(2), 265–275 (2005)

4. Yang, Z., Ce, L., Lian, L.: Electricity price forecasting by a hybrid model, combining wavelet transform, ARMA and kernel-based extreme learning machine methods. Appl. Energy **190**, 291–305 (2017)

5. Weron, R., Misiorek, A.: Forecasting spot electricity prices with time series models. In: The European Electricity Market EEM 2005 (2005)

6. Wilkey, J., et al.: Predicting emissions from oil and gas operations in the Uinta Basin, Utah. J. Air & Waste Mgmt. Assoc. **66**(5), 528–545 (2016). https://doi.org/10.1080/10962247.2016.1153529

7. Le Quéré, C., et al.: Temporary reduction in daily global CO2 emissions during the COVID-19 forced confinement. In: Nat. Clim. Chang. (2020). https://www.nature.com/articles/s41558-020-0797-x#Tab2

8. Solomon, S., et al.: Summary for Policymakers. In: Climate Change 2007: The Physical Science Basis. Contribution of Working Group I to the Fourth Assessment Report of the Intergovernmental Panel on Climate Change. In: AR4 Climate Change 2007: The Physical Science Basis (2007)

9. Sources of Greenhouse Gas Emissions (2020). https://www.epa.gov/ghgemissions/sources-greenhouse-gas-emissions#t1fn1

10. Mostafa, M., El-Masry, A.: Oil price forecasting using gene expression programming and artificial neural networks. Econ. Model. **54**, 40–53 (2016). http://www.sciencedirect.com/science/article/pii/S0264999315004101

11. Wolaver, B.D., Pierre, J.P., Ikonnikova, S.A., Andrews, J.R., McDaid, G., Ryberg, W.A., Hibbitts, T.J., Duran, C.M., Labay, B.J., LaDuc, T.J.: An improved approach for forecasting ecological impacts from future drilling in unconventional shale oil and gas plays. Environ. Manag. **62**(2), 323–333 (2018). https://doi.org/10.1007/s00267-018-1042-5

12. Shen, C.: Analysis of detrended time-lagged cross-correlation between two nonstationary time series. Phys. Lett. A (Mar. 2015). https://doi.org/10.1016/j.physleta.2014.12.036

13. Wang, F., Wang, L., Chen, Y.: Detecting PM2.5's correlations between neighboring cities using a time-lagged cross-correlation coefficient. Nat. News (2017). https://www.nature.com/articles/s41598-017-10419-6

14. Cavallo, A.: Elephant in the room: how OPEC sets oil prices and limits carbon emissions. Bull. Atom. Sci. **69**(4), 18–29 (2013). https://doi.org/10.1177/0096340213493583

On Graph Learning with Neural Networks

Zahra Jandaghi[✉] and Liming Cai

Department of Computer Science, University of Georgia, Athens, GA 30602, USA
{zahra,liming}@uga.edu

Abstract. Graphs are ideal for modeling natural systems where relations may be intrinsic among data objects. With massive data available, learning graph models from data has become potentially feasible as well as necessary. Yet from the traditional machine learning perspective, learning structural topology of an unknown graphical model remains challenging. In particular, it is computationally intractable to learn graph topologies beyond a tree structure. Nevertheless, deep learning with neural networks, showing great potentials in visual imagery and other application domains, offers an alternative venue for effective machine learning on graphs. In this review, we discuss graph (structure) learning with deep neural networks. In particular, we examine graph neural networks (GNNs) from the task-based and the architecture-based perspectives, respectively.

Keywords: Graph learning · Graph neural networks · Neural networks · Machine learning

1 Introduction

Graphs are often used to characterize complex systems quantified from many real-world applications, where data objects and their relationships can conveniently be modeled with vertices and edges, respectively. Developing effective graph models from observed data is a typical machine learning process, in which either a set of probabilistic parameters in a graph model or structure of the graph or both are learned [3, 26]. In terms of statistical and traditional optimization methods, tasks of graph learning have different levels of challenges. In particular, parameter learning can be solved with the maximum likelihood method on complete data and with the expectation-maximization on incomplete data. However, in general, machine learning pertaining to unknown structure could be much more challenging [23].

Machine learning tasks on an unknown graph model, e.g., classification of a set of observed objects, may unavoidably reveal the structure topology (i.e., the dependency over a set of random variables). This is a notoriously difficult learning task [14, 23]. Typical efforts have been made on restricted categories of topology; maximum likelihood [20] or information divergence method [25]

© Springer Nature Switzerland AG 2020
G. Nicosia et al. (Eds.): LOD 2020, LNCS 12566, pp. 516–528, 2020.
https://doi.org/10.1007/978-3-030-64580-9_43

can be applied through graph optimization. Among few successes is the earlier work [7] in tree structure learning from complete, observed data. However, due to the computational intractability in optimizing arbitrary spanning subgraph, structure learning beyond trees cannot guarantee both optimality and efficiency. Existing traditional methods for structure learning have been heuristic and approximate [9,23,43,51]. They are often built upon distribution assumptions or hand-engineered structures, where learning effectiveness becomes an issue due to potential human biases imposed.

Artificial neural networks [18] has proved an effective, alternative machine learning paradigm. Research in deep learning, where neural networks are equipped with more than one hidden layer, has made striking progresses in machine learning problems in visual imagery, natural language processing, and other applications. The great success underscores two essential components of deep learning. Other than the availability of big data, multiple hidden layers of neural networks permit approximation of sophisticated objective functions that may be non-linear or even non-analytical, which may nevertheless suit well the potentially complex problem to be solved. In contrast to the traditional machine learning methods, deep learning does not presume objective functions to be optimized.

Since graphs are almost always adopted to model complex systems, deep learning offers a viable solution to effective machine learning on graphs. In this paper, we give a review on graph learning with neural networks. In particular, we discuss some recent works in graph neural network (GNN) and related techniques. GNNs are categorized according to learning tasks as well as network architectures.

2   Preliminaries

We briefly give some basic concepts in graphical models, graph learning, and neural networks. We refer the reader to references cited in this section for further information.

2.1   Graph Learning

A graph $G = (V, E)$ consists of a set V of objects (called *vertices*) and a binary set $E \subseteq V \times V$ (called *edges*) specifying dependencies or relationships among the objects. Numerical values can often be associated with vertices and edges to quantify the category, potential, relevance, or, typically probability for an object or a relationship of objects to exist. The latter setting results in *graphical models* (Fig. 1 *left*), an important category of probabilistic models for analysis of real-world phenomena and prediction of uncertain events in both homogeneous and heterogeneous systems [23]. Graph learning can be categorized according to two important components in graphical models: *probabilistic parameters θ* and graph *structure topology*. The level of challenges in learning also depends on the nature of the observed data to use. A set of observed data is *complete* if it contains not only raw observations but also structural information about how the observed data might have come to exist, while a set of *incomplete* data often come with missing information, typically about the associated structure.

From complete data D, learning probabilistic parameters θ^* of a graph model with known topology can be accomplished by maximizing the *likelihood* $\mathcal{L}(\theta : D) = Prob(D|\theta)$. The optimization problem can be computed in closed form, e.g., by equating partial derivatives of likelihood function to zero: $\frac{\partial \theta_i}{\partial \mathcal{L}} = 0$, for all $\theta_i \in \theta$. On the other hand, from incomplete data with missing structural information, learning parameters may have to resort to approximated or heuristic methods, e.g., by maximizing the expected likelihood of the overall data structures possibly associated with every observation. However, heuristic methods often do not guarantee the optimality of learned parameters.

Learning tasks that concern revealing (a part of) structure topology of an unknown graphical model pose a greater challenge. Theoretically, it can be approached with minimization of the *KL-divergence* $D_{KL}(P \parallel Q)$, i.e, information loss from the unknown model P to an approximate model Q, calculated as relative entropy of the two [25]. Interestingly, this is equivalent to maximizing the likelihood of topology $\mathcal{L}(Q : D)$ given data D, which in turn is the same as the maximum spanning subgraph problem in graph theory. In particular, the seminal work in [7] shows optimal learning of tree topology is solved by finding the maximum spanning tree in linear time. However, maximizing a spanning subgraph is NP-hard, even for tree-like structures with circles present [4,6]. The computational intractability deems that structure learning cannot guarantee optimality and efficiency at the same time. Practical solutions usually impose known probabilistic distribution or limit model topology details of applications. Potential human biases in hand-engineered treatments may hurt modeling effectiveness.

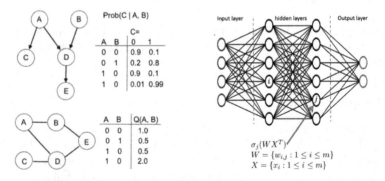

Fig. 1. *Left*: Examples of graphical models: a Bayesian net (top) where edges give conditional probabilities; and a Markov random field (bottom) where edges are associated with potentials Q; *Right*: Schematic illustration of neural networks with multi-hidden layers, where node function σ_j is activated by its input values from nodes in the preceding layer.

2.2 Neural Network

(Artificial) *neural network* [18] is a non-traditional computing paradigm to mimic human brain which can be trained and improved on data. A neural network consists of neuron nodes arranged in layers. Other the input and output layers, there are hidden layers in between. Every node j receives numerical inputs from nodes in the preceding layer and *activates* the associated numerical function σ_j with output to be received by the succeeding layer. Several types of activation functions, e.g., sigmoid $\sigma(x) = \frac{1}{x+e^{-x}}$ and Rectified Linear Unit (ReLU) $\sigma(x) = \max\{0, x\}$, have been proposed and available to neural network design [37]. The connection between two nodes i and j (in two consecutive layers, respectively) is associated with a learnable *weight* $w_{i,j}$ so the typical function of node j is $\sigma(WX^T)$, where $X = \{x_i\}$ is the input values from nodes i in the preceding layers and $W = \{w_{i,j}\}$ is the weights on their connections to the current node j (Fig. 1 *right*). The quality of a network is measured by the *loss function* - the information difference between the real value and the one computed by the network. Before the network is used to compute a task, its weights are adjusted through *training* on observed data through optimization of the loss function.

Neural networks may only approximate desired tasks. However, they can approximate any continuous function, though networks with one hidden layer can only compute linear functions. This superpower of neural networks is made possible by equipping networks with multiple hidden layers; the notion is thus coined *deep learning* [16]. Learning with deep networks has proved viable in various tasks that may require intelligent computation. It especially enjoys successes in application areas where a massive amount of data is available for the training purpose, including visual imagery and natural language processing. Technically, these applications have also furthered deep learning research with several advanced network architectures, typically the following important two.

Convolutional neural network (CNN) [24] has been widely used in computer vision and pattern recognition. Images have a hierarchical pattern in data, where more complex patterns are assembled from (or decomposed into) smaller and simpler patterns. CNNs take this advantage to simulate the hierarchical assembly of images through learning with multiple convolution layers. In particular, a convolution layer reduces a local pattern into small dimensions by filtering the pattern with a kernel matrix, which enhances the desired feature(s) in the pattern. It needs to be pointed out that each convolution may have some feature maps with equal or larger dimensions than that of the input. *Pooling* is the technique to reduce dimensionality to the desirable.

Recurrent neural network (RNN) [34] is an architecture well suited to tasks involving temporal sequences, a data format shared by many applications such as speech recognition and biological sequence analysis. Unlike feedforward connections in most architectures, RNN allows *feedback* connections so nodes can be activated recurrently. Activation paths correspond to the input temporal sequence, so RNNs are able to handle inputs of variable lengths. To account for correlations between different temporal events, an RNN can be equipped with additional stored states, which can be replaced by another network or graph that incorporates time delays or has feedback loops.

3 Graph Neural Network

The seemingly endless potential of deep learning is appealing to graph learning, and the notion of *graph neural network* (GNN) [42] has recently been introduced. Generally speaking, GNNs are conventional neural networks equipped with graph representation techniques, so graph learning may be effectively carried out in downstream steps. In this section, we highlight a few critical techniques in GNNs, which are different from other neural networks.

3.1 Representation of Graph Learning

Various representation techniques have been developed based on the following the *encoder-decoder* perspective [17]. On a given graph $G = (V, E)$, every vertex $v \in V$ is encoded into a vector $z_v \in \mathbb{R}^d$, for some $d \ll |V|$, and the information about vertices into a matrix in space $\mathbb{R}^{d \times |V|}$. For a pair of vector (z_u, z_v), it is expected that a decoding scheme converts (z_u, z_v) approximately to similarity degree $sim_G(u, v)$, i.e., preserves the relevance between original vertices u and v.

The above more straightforward *shallow embedding* methods, which has a few drawbacks, can be replaced by *neighborhood aggregation* methods, where encoding a vertex v takes consideration of all (neighborhood) information of v in the graph [17]. Beginning from a shallow embedding $z_v \in \mathbb{R}^d$, layers of neural networks are used to encode vertex v with inputs from the encodings of all v's neighbors in the preceding layer. Neighborhood aggregation requires that an ideal decoder convert encoding z_i into a neighborhood vector approximately the same as v's original neighborhood. Representation of graphs can be learned with the loss function defined as the distance between the decoded neighborhood and original neighborhood overall vertices [17].

3.2 Graph Convolution Layer

Since graphs do not have clear grid structures, convolution operations in CNNs are redefined for GNNs. There are two graph convolution operations: *spectral convolution* and *spatial convolution* [46]. Figure 2 illustrates the difference between image convolution and graph convolution. Spectral convolutions are based on the graph convolution theory that presumes the graph is non-directed. A commonly used spectral convolution is performed through eigen decomposition $L = U \Lambda U^T$ on the normalized Laplacian $L = I - D^{-\frac{1}{2}} A D^{-\frac{1}{2}}$. The latter is transformed from graph adjacency matrix A, where D is the diagonal matrix of the graph node degrees. Then for a chosen filter F, a diagonal matrix, the convolution operation on any vertex information $\mathbf{x}$ of the graph, $\mathbf{x} \in \mathbb{R}^n$, is defined as $F \circ \mathbf{x} = U(F\Lambda)U^T \mathbf{x}$ [54]. On the other hand, like the convolutional layer on images, spatial convolutions directly apply to graph adjacency matrices

and compute convolutions based on graph nodes' local neighborhood information. Section 5 discusses spatial convolutions as well as other proposed spectral
convolution operations.

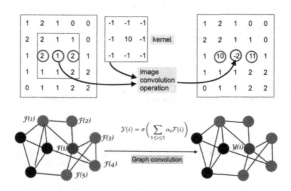

Fig. 2. Illustration of convolution operations. *Top*: Example of image convolution where
the kernel is used to enhance grayscale of pixels, where the circled ones have convolutions computed; *Bottom*: Example of graph convolution where the operation is based
on weight sum of the node's neighborhood. Unlike in images, a graph node's neighbors
are not always "geometrically nearby", and the number of neighbors may vary.

3.3 GNN Categorizations

Many methods and techniques for machine learning on graphs have been proposed. In the next two sections, we will highlight GNNs based on the learning
tasks they have proposed to solve. We will also categorize them based on the
proposed architectures.

4   Task-Based Categorizations

In this section, we review and categorize GNNs based on the graph learning
tasks they were proposed to accomplish. Machine learning on graphs can be to
extract or predict information about various data objects in a graph, including
graph nodes, edges, subgraphs, and the whole structure [46].

4.1 Node-Level

For neural networks, graph learning tasks at node-level are essential to extract
high-level information of each node of interest. Through graph convolution operations, node information can be propagated and classified according to the need.
Compared to other levels, node-level learning can be more straightforward in
architecture building and effective in training. In particular, tasks could be

accomplished in an end-to-end structure. It has been shown that convolutional graph neural networks can accurately classify nodes in a graph after the networks are trained with graphs where only a portion of nodes have labels, a scenario of semi-supervised learning (Fig. 3). The above situation of learning is where the graph topology is given but observed data containing missing information. Traditional learning methods may resort to maximizing expected likelihood over all possible labels on nodes (see Sect. 2), a method often only converges to a local optima.

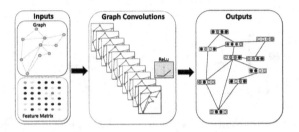

Fig. 3. Illustration of a node-level task accomplished with GNNs, where hidden layers are graph convolution layers. The input is the graph topology and a feature matrix with information about partially labeled nodes. The output is the original graph but with all nodes labeled.

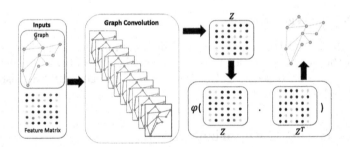

Fig. 4. Illustration of graph-level learning tasks realized by convolutional graph neural network. The network takes graph (A) as input and convolutionalizes them getting the latent representation (Z); then it applies deconvolution to reconstruct graph ($\hat{A}$) [46].

4.2   Edge-Level

At this level, neural networks learn hidden representations for every pair of nodes and predict their connections. This category of learning tasks has been treated from different perspectives. A typical approach is based on the encoder-decoder view introduced in Sect. 3.1 that formulates the problem as unsupervised learning and uses an *autoencoder* (see Sect. 5 for some details) to first encode the graph into latent representation, then decode information about edges and even reconstruct the graph [22,39]. Depending on used encoding methods, shallow embedding or neighborhood aggregation, decoding schemes are different to recover relevance information of neighbors.

4.3 Graph-Level

The output of this category, tasks can be about learning the global structure of the query graph (like tasks at edge-level). More often, they are about graph classification where neural networks obtain a compact (see Fig. 4). A specific task can be to find a label for the entire graph [41,48,52]. It could be accomplished with convolutional graph neural networks, where a combination of graph convolutional layers, graph pooling layers, and readout layers is present. In particular, graph convolutional layers extract a high-level representation of each node and graph pooling down-sample the output. Readout layers can be added to transform node representation into a graph representation. For example, this design has been used to categorize brain analysis and chemical compounds [40].

5 Architecture-Based Categorization

This section will examine GNNs further based on their network architectures. Since there are many different types of architectures [28,46,53], we will ignore some detailed differences and group them into the following five categories.

5.1 Recurrent Graph Neural Network (RecGNNs)

Recurrent graph neural networks are among the first attempts in this area [13,42]. They focus on learning node representation and assume each node continuously has information exchange with its neighbor connections to reach a stable balance [8,32].

Spatial-Temporal Graph Neural Network (STGNN): The goal of this GNN is to learn hidden patterns in spatial-temporal graphs whose node/edge attributes may vary with time. It has began to play critical roles in traffic forecasting [31], driver maneuver anticipation [19], and human action recognition [47]. They may need to be implemented using RNN with spatial convolution.

5.2 Convolutional Graph Neural Network (ConvGNN)

Convolutional graph neural networks follow the idea of convolution operation and aggregate each node features with it's neighbors/connections features.

Spectral-Based ConvGNNs: As mentioned in Sect. 3.1, all spectral-based graph convolutions follow the operation $U(F\Lambda)T^T\mathbf{x}$, for graph information $x \in \mathbb{R}^n$, with the flexibility to choose the diagonal matrix filter F. However, there are drawbacks with spectral-based graph convolutions, such as graph specificity of learned filters as well as high computation cost. Developments in ConvGNN have tried to address these issues. ChebNet [11] and GCN [21] used approximation techniques in order to reduce computational complexity. *Adaptive Graph Convolutional Network* (AGCN) [30] attempted to improve the performance by introducing a residual graph adjacency matrix, which can be learned through a distance function of two nodes' features. Two graph convolution in parallel, sharing parameters between layers was proposed as *Dual Graph Convolution Networks* (DGCN) [55] to integrate local and global topological information.

Spatial-Based ConvGNNs: This architecture follows a very similar approach to RecGNNs by passing messages between nodes. Neural network for Graphs (NN4G) [35] is considered the first study of graph neural network with a spatial-based graph convolution. Spatial convolutions are more efficient than spectral convolution since they involve less computations and may directly employ adjacency matrix of the graph.

Diffusion Convolutional Neural Network (DCNN): This architecture was introduced based on the assumption that traveling information from one node to another carries a transition probability that eventually reaches an equilibrium [1]. Similar assumptions were used in other architectures. For example, graph convolutions are treated as message exchanging processes between nodes along edges by the Message Passing Neural Network (MPNN) [15]. Mixture Model Network (MoNet) [36] uses relative neighbors' positions to give weights to each neighbor of a node. PATCHY-SAN [38] sorts node's neighbors by their label.

5.3   Graph Autoencoder (GAEs)

Graph autoencoders are in unsupervised class of GNNs, aiming to learn latent representation of graphs. GAEs could be used to learn graph structure in network embedding DNGR [5], VGAE [22], and DNRE [44] or graph generation GraphRNN [50] and RGVAE [33] which have high value in drug discovery.

5.4   Graph Reinforcement Learning

Efforts in graph learning have adopted *reinforcement learning* (RL) which is useful in dealing with non-differentiable equations. GCPN [49] uses RL for generating molecular graphs where an RL agent is responsible for adding nodes and edges, and its actions receive rewards as a link prediction task. MoIGAN [10] followed the same idea except for the graph generation that happened once. This study was a suitable approach for small molecules. GTPN [12] uses RL for chemical reaction prediction; it employs GCN to learn each node representation and RNN to remember previous predictions.

5.5   Attention-Based Models

Such a model considers a special priority for relevant information to integrate into the representation. Models in [28] consider a function to map each node/edge

(depends on the task) to a relevance score, which gives the degree to which it is important. There are three main types of attention, with basic ideas as follows:
General Attention: It assigns a relevance score a node/edge by considering its features compared to the target node/edge [45].

Similarity-Based Attention: It pays attention to the object whose feature has the most similarity with the target feature. This approach was first studied in the Natural Language Processing (NLP) [2] following the idea of more relevant data provides more information for the task.
Attention-Guided Walk: It uses an RNN to encode the visited node and build a subgraph embedding. Attention function uses the embedding to prioritize the steps [27].

6   Conclusions

We have reviewed neural networks for graph learning. We categorized significant approaches and techniques in graph learning with neural networks based on the problem, network structure, and chosen strategy to tackle the problem. While it is optimistic that neural networks will continue to offer benefits to graph learning problems, there are some limitations in the current graph neural networks that need to be pointed out.

Graph Types: In most of the available GNNs, it is assumed the graph is homogeneous, and it would be challenging to directly apply existing works on heterogeneous graphs where more than one type of nodes or edges may be present in the graph. Besides, many applications can be modeled with graphs that change over time or based on another parameter, e.g., social media networks. These are some of the types of graph that need to be considered in GNN development.

Network Depth: While it is a common belief that the more layers in neural networks, the more accurate result we get, many experiments show this is not true especially for GNNs; performance usually drops significantly after adding more layers. In theory, a graph's representation converges to a single point with an unlimited number of convolution layers [29]. It is crucial to retain this factor in the design of GNNs.

Interpretability: Many GNN applications are in essential research areas yet remote to computational sciences; the importance of accurately interpreting computer models cannot be overemphasized. For example, GNNs have benefited in solving several problems in pharmaceutical research, where a very detailed and accurate understanding of both data and GNN process is required. Addressing this issue needs to be a part of GNN studies.

References

1. Atwood, J., Towsley, D.: Diffusion-convolutional neural networks. In: Advances in Neural Information Processing Systems, pp. 1993–2001 (2016)

2. Bahdanau, D., Cho, K., Bengio, Y.: Neural machine translation by jointly learning to align and translate. arXiv preprint arXiv:1409.0473 (2014)
3. Bishop, C.M.: Pattern Recognition and Machine Learning. Springer, New York (2006)
4. Cai, L., Maffray, F.: On the spanning k-tree problem. Discret. Appl. Math. **44**(1–3), 139–156 (1993)
5. Cao, S., Lu, W., Xu, Q.: Deep neural networks for learning graph representations. In: 13th AAAI Conference on Artificial Intelligence (2016)
6. Chickering, D.M., Geiger, D., Heckerman, D., et al.: Learning Bayesian networks is NP-hard. Technical report. Citeseer (1994)
7. Chow, C., Liu, C.: Approximating discrete probability distributions with dependence trees. IEEE Trans. Inf. Theor. **14**(3), 462–467 (1968)
8. Dai, H., Kozareva, Z., Dai, B., Smola, A., Song, L.: Learning steady-states of iterative algorithms over graphs. In: International Conference on Machine Learning, pp. 1106–1114 (2018)
9. Daly, R., Shen, Q., Aitken, S.: Learning Bayesian networks: approaches and issues. Knowl. Eng. Rev. **26**(2), 99 (2011)
10. De Cao, N., Kipf, T.: MolGAN: an implicit generative model for small molecular graphs. arXiv preprint arXiv:1805.11973 (2018)
11. Defferrard, M., Bresson, X., Vandergheynst, P.: Convolutional neural networks on graphs with fast localized spectral filtering. In: Advances in Neural Information Processing Systems, pp. 3844–3852 (2016)
12. Do, K., Tran, T., Venkatesh, S.: Graph transformation policy network for chemical reaction prediction. In: Proceedings of the 25th ACM SIGKDD International Conference on Knowledge Discovery & Data Mining, pp. 750–760 (2019)
13. Gallicchio, C., Micheli, A.: Graph echo state networks. In: The 2010 International Joint Conference on Neural Networks (IJCNN), pp. 1–8. IEEE (2010)
14. Ghahramani, Z.: Probabilistic machine learning and artificial intelligence. Nature **521**(7553), 452–459 (2015)
15. Gilmer, J., Schoenholz, S.S., Riley, P.F., Vinyals, O., Dahl, G.E.: Neural message passing for quantum chemistry. In: Proceedings of the 34th International Conference on Machine Learning, vol. 70, pp. 1263–1272. JMLR.org (2017)
16. Goodfellow, I., Bengio, Y., Courville, A.: Deep Learning. MIT Press, Cambridge (2016)
17. Hamilton, W.L., Ying, R., Leskovec, J.: Representation learning on graphs: methods and applications. arXiv preprint arXiv:1709.05584 (2017)
18. Hassoun, M.H., et al.: Fundamentals of Artificial Neural Networks. MIT Press, Cambridge (1995)
19. Jain, A., Zamir, A.R., Savarese, S., Saxena, A.: Structural-RNN: deep learning on spatio-temporal graphs. In: Proceedings of the IEEE Conference on Computer Vision and Pattern Recognition, pp. 5308–5317 (2016)
20. Karger, D.R., Srebro, N.: Learning Markov networks: maximum bounded treewidth graphs. In: SODA, pp. 392–401 (2001)
21. Kipf, T.N., Welling, M.: Semi-supervised classification with graph convolutional networks. arXiv preprint arXiv:1609.02907 (2016)
22. Kipf, T.N., Welling, M.: Variational graph auto-encoders. arXiv preprint arXiv:1611.07308 (2016)
23. Koller, D., Friedman, N.: Probabilistic Graphical Models: Principles and Techniques. MIT Press, Cambridge (2009)

24. Krizhevsky, A., Sutskever, I., Hinton, G.E.: ImageNet classification with deep convolutional neural networks. In: Advances in Neural Information Processing Systems (NIPS) (2012)
25. Kullback, S., Leibler, R.A.: On information and sufficiency. Ann. Math. Stat. **22**(1), 79–86 (1951)
26. Lauritzen, S.L.: Graphical Models, vol. 17. Clarendon Press, Oxford (1996)
27. Lee, J.B., Rossi, R., Kong, X.: Graph classification using structural attention. In: Proceedings of the 24th ACM SIGKDD International Conference on Knowledge Discovery & Data Mining, pp. 1666–1674 (2018)
28. Lee, J.B., Rossi, R.A., Kim, S., Ahmed, N.K., Koh, E.: Attention models in graphs: a survey. ACM Trans. Knowl. Discov. Data (TKDD) **13**(6), 1–25 (2019)
29. Li, Q., Han, Z., Wu, X.M.: Deeper insights into graph convolutional networks for semi-supervised learning. In: 32nd AAAI Conference on Artificial Intelligence (2018)
30. Li, R., Wang, S., Zhu, F., Huang, J.: Adaptive graph convolutional neural networks. In: 32nd AAAI Conference on Artificial Intelligence (2018)
31. Li, Y., Yu, R., Shahabi, C., Liu, Y.: Diffusion convolutional recurrent neural network: data-driven traffic forecasting. arXiv preprint arXiv:1707.01926 (2017)
32. Li, Y., Tarlow, D., Brockschmidt, M., Zemel, R.: Gated graph sequence neural networks. arXiv preprint arXiv:1511.05493 (2015)
33. Ma, T., Chen, J., Xiao, C.: Constrained generation of semantically valid graphs via regularizing variational autoencoders. In: Advances in Neural Information Processing Systems, pp. 7113–7124 (2018)
34. Mandic, D., Chambers, J.: Recurrent Neural Networks for Prediction: Learning Algorithms, Architectures and Stability. Wiley, Hoboken (2001)
35. Micheli, A.: Neural network for graphs: a contextual constructive approach. IEEE Trans. Neural Netw. **20**(3), 498–511 (2009)
36. Monti, F., Boscaini, D., Masci, J., Rodola, E., Svoboda, J., Bronstein, M.M.: Geometric deep learning on graphs and manifolds using mixture model CNNs. In: Proceedings of the IEEE Conference on Computer Vision and Pattern Recognition, pp. 5115–5124 (2017)
37. Nair, V., Hinton, G.E.: Rectified linear units improve restricted Boltzmann machines. In: International Conference on Machine Learning (ICML) (2010)
38. Niepert, M., Ahmed, M., Kutzkov, K.: Learning convolutional neural networks for graphs. In: International Conference on Machine Learning, pp. 2014–2023 (2016)
39. Pan, S., Hu, R., Long, G., Jiang, J., Yao, L., Zhang, C.: Adversarially regularized graph autoencoder for graph embedding. arXiv preprint arXiv:1802.04407 (2018)
40. Pan, S., Wu, J., Zhu, X., Long, G., Zhang, C.: Task sensitive feature exploration and learning for multitask graph classification. IEEE Tran. Cybern. **47**(3), 744–758 (2016)
41. Pan, S., Wu, J., Zhu, X., Zhang, C., Philip, S.Y.: Joint structure feature exploration and regularization for multi-task graph classification. IEEE Trans. Knowl. Data Eng. **28**(3), 715–728 (2015)
42. Scarselli, F., Gori, M., Tsoi, A.C., Hagenbuchner, M., Monfardini, G.: The graph neural network model. IEEE Trans. Neural Netw. **20**(1), 61–80 (2008)
43. Teyssier, M., Koller, D.: Ordering-based search: a simple and effective algorithm for learning Bayesian networks. arXiv preprint arXiv:1207.1429 (2012)
44. Tu, K., Cui, P., Wang, X., Yu, P.S., Zhu, W.: Deep recursive network embedding with regular equivalence. In: Proceedings of the 24th ACM SIGKDD International Conference on Knowledge Discovery & Data Mining, pp. 2357–2366 (2018)

45. Veličković, P., Cucurull, G., Casanova, A., Romero, A., Lio, P., Bengio, Y.: Graph attention networks. arXiv preprint arXiv:1710.10903 (2017)
46. Wu, Z., Pan, S., Chen, F., Long, G., Zhang, C., Philip, S.Y.: A comprehensive survey on graph neural networks. IEEE Trans. Neural Netw. Learn. Syst., 1–21 (2020)
47. Yan, S., Xiong, Y., Lin, D.: Spatial temporal graph convolutional networks for skeleton-based action recognition. In: 32nd AAAI Conference on Artificial Intelligence (2018)
48. Ying, Z., You, J., Morris, C., Ren, X., Hamilton, W., Leskovec, J.: Hierarchical graph representation learning with differentiable pooling. In: Advances in Neural Information Processing Systems, pp. 4800–4810 (2018)
49. You, J., Liu, B., Ying, Z., Pande, V., Leskovec, J.: Graph convolutional policy network for goal-directed molecular graph generation. In: Advances in Neural Information Processing Systems, pp. 6410–6421 (2018)
50. You, J., Ying, R., Ren, X., Hamilton, W.L., Leskovec, J.: GraphRNN: generating realistic graphs with deep auto-regressive models. arXiv preprint arXiv:1802.08773 (2018)
51. Yuan, C., Malone, B.: Learning optimal Bayesian networks: a shortest path perspective. J. Artif. Intell. Res. **48**, 23–65 (2013)
52. Zhang, M., Cui, Z., Neumann, M., Chen, Y.: An end-to-end deep learning architecture for graph classification. In: 32nd AAAI Conference on Artificial Intelligence (2018)
53. Zhang, Z., Cui, P., Zhu, W.: Deep learning on graphs: a survey. IEEET Trans. Knowl. Data Eng. (2020)
54. Zhou, J., et al.: Graph neural networks: a review of methods and applications. arXiv preprint arXiv:1812.08434 (2018)
55. Zhuang, C., Ma, Q.: Dual graph convolutional networks for graph-based semi-supervised classification. In: Proceedings of the 2018 World Wide Web Conference, pp. 499–508 (2018)

On Bayesian Search for the Feasible Space Under Computationally Expensive Constraints

Alma Rahat[1](✉)[ID] and Michael Wood[2][ID]

[1] Department of Computer Science, Swansea University, Swansea, UK
a.a.m.rahat@swansea.ac.uk
[2] ACT Acoustics, Exeter, UK
m.wood@actacoustic.co.uk

Abstract. We are often interested in identifying the feasible subset of a decision space under multiple constraints to permit effective design exploration. If determining feasibility required computationally expensive simulations, the cost of exploration would be prohibitive. Bayesian search is data-efficient for such problems: starting from a small dataset, the central concept is to use Bayesian models of constraints with an acquisition function to locate promising solutions that may improve predictions of feasibility when the dataset is augmented. At the end of this sequential active learning approach with a limited number of expensive evaluations, the models can accurately predict the feasibility of any solution obviating the need for full simulations. In this paper, we propose a novel acquisition function that combines the probability that a solution lies at the boundary between feasible and infeasible spaces (representing exploitation) and the entropy in predictions (representing exploration). Experiments confirmed the efficacy of the proposed function.

Keywords: Active learning · Feasible region · Feasible design exploration · Gaussian processes · Constrained problems

1 Introduction

In engineering applications, we are often interested in determining the feasible design space for a given problem. This requires estimating a set of decision variables that does not violate given conditions. This is a challenging task, particularly if the constraints cannot be expressed analytically. In these cases, computationally expensive simulations or physical experiments are required to explore the design space. For instance, in some nuclear power applications, keeping the neutron production ratio below a critical level is essential for safe operation [1]. This presents a significant design challenge without analytical constraints. It

Electronic supplementary material The online version of this chapter (https://doi.org/10.1007/978-3-030-64580-9_44) contains supplementary material, which is available to authorized users.

© Springer Nature Switzerland AG 2020
G. Nicosia et al. (Eds.): LOD 2020, LNCS 12566, pp. 529–540, 2020.
https://doi.org/10.1007/978-3-030-64580-9_44

is not practical to test each set of plant parameters by simulation, since each evaluation of the simulator takes between 5 and 30 min. Hence, a classifier that can accurately predict feasibility can allow operators to explore the design space rapidly before setting up the plant obviating the need for simulations. The challenge is then how to train this classifier with few expensive evaluations.

In this context, surrogate-assisted Bayesian search method has been shown to be a data-efficient approach [2]. This method starts with a small training set of independent parameters. These parameters are expensively evaluated with a set of constraint functions. The resulting dataset is used to train a Bayesian regression model (in this case, a Gaussian process, $\mathcal{GP}$) for each constraint [3]. Together, these models estimate the probability that a given solution is feasible. In this way, the combination of models act as a *binary classifier*. The idea is then to locate a candidate sample for evaluating expensively such that adding this sample to the training dataset would achieve the greatest improvement in the feasible space estimation. This candidate is located by maximising an *acquisition function* (often referred to as an infill criterion or a utility function). We keep adding new samples until the budget on additional expensive evaluations is exhausted.

We understand that using this method of Bayesian search for feasible region identification and design exploration is new with Knudde *et al.* publishing the first acquisition function recently [2]. This function considers the loss of entropy of the posterior predictive distribution from adding a new sample in the training dataset. With the aim of providing alternative acquisition functions, the novel contributions of this paper are:

- A new acquisition function $\alpha_{PBE}(\cdot)$ based on the probability of a solution residing at the boundary (representing exploitation) and the entropy of predictive distribution (representing exploration).
- Adaptation of a range of alternative acquisition functions (that are originally used in reliability engineering for rapidly estimating the volume of the *infeasible* space) for the purpose of data-efficient construction of a feasibility classifier.
- A full investigation of these acquisition functions in a set of constrained problems.

In Sect. 2, we discuss necessary concepts focusing on using $\mathcal{GP}$s to model constraints functions, and the standard Bayesian search framework. Then we propose a range of acquisition functions suitable for Bayesian search of the feasible space in Sect. 3. We present our results in Sect. 4. Finally, we finish with general conclusions in Sect. 5.

2   Background

Consider, a design vector $\mathbf{x}$ in a design space $\mathcal{X} \in \mathbb{R}^n$. Without loss of generality, a constrained problem with L constraints can be defined as:

$$G(\mathbf{x}) = (g_1(\mathbf{x}), \dots, g_L(\mathbf{x}))^\top \leq \mathbf{t} = (t_1, \dots, t_L)^\top, \tag{1}$$

where, $g_l : \mathbb{R}^n \to \mathbb{R}$ is the lth constraint function with a threshold for feasibility t_l. To deal with equality constraints, we can add a small fixed constant ϵ. This converts the equation to an inequality constraint [4].

The lth constraint function $g_l(\mathbf{x})$ generates a feasible space $\mathcal{F}_l \subseteq \mathcal{X}$. The infeasible set of solutions for this constraint is therefore $\mathcal{I}_l = \mathcal{X} \setminus \mathcal{F}_l$. The total infeasible set of solutions becomes $\mathcal{I} = \bigcup_{l=1}^{L} \mathcal{I}_l$. If all constraints are considered, the feasible space is at the intersection of all feasible sets: $\mathcal{F} = \bigcap_{l=1}^{L} \mathcal{F}_l$.

If constraint functions are cheap to evaluate, we can determine feasibility by *brute force* using Monte Carlo methods [5]. However, where each constraint function evaluation $g_l(\mathbf{x})$ requires a computationally expensive simulation, this approach would be prohibitively slow.

2.1 Modelling Constraints with Gaussian Processes

Gaussian processes ($\mathcal{GP}$) are commonly used to construct surrogate models for constraints $g_l(\mathbf{x})$. $\mathcal{GP}$s produce a Normal predictive distribution for any arbitrary solution, producing a mean and standard deviation.

In essence, a $\mathcal{GP}$ is a field of joint Gaussian distributions [3]. Consider a $\mathcal{GP}$ trained for lth constraint function $g_l(\mathbf{x})$ on dataset $\mathcal{D}_l = \{(\mathbf{x}_m, g_l(\mathbf{x}_m))\}_{m=1}^{M}$ evaluated at M locations. The predictive probability for g_l at $\mathbf{x}$ is a Gaussian distribution with mean $\mu_l(\mathbf{x})$ and variance $\sigma_l^2(\mathbf{x})$ is:

$$p(g_l \mid \mathbf{x}, \mathcal{D}_l) = \mathcal{N}(\mu_l(\mathbf{x}), \sigma_l^2(\mathbf{x}) \mid \mathbf{x}, \mathcal{D}_l). \tag{2}$$

The efficacy of $\mathcal{GP}$s is conferred by a flexible kernel function. We use a Matern 5/2 kernel, as recommended for modelling realistic functions [6]. We refer the reader to [3] for full documentation on how the $\mathcal{GP}$ is trained and interrogated.

We train a model for each constraint independently. Thus, the combined posterior predictive distribution across all component models is a multi-variate Gaussian:

$$p(G \mid \mathbf{x}, \mathcal{D}) = \mathcal{N}(\boldsymbol{\mu}(\mathbf{x}), \Sigma(\mathbf{x})) = \prod_{l=1}^{L} p(g_l \mid \mathbf{x}, \mathcal{D}_l), \tag{3}$$

where, the training dataset is $\mathcal{D} = \{(\mathbf{x}_m, g_1(\mathbf{x}_m)), \ldots, g_L(\mathbf{x}_m))\}_{m=1}^{M}$, the mean prediction vector is $\boldsymbol{\mu}(\mathbf{x}) = (\mu_1(\mathbf{x}), \ldots, \mu_L(\mathbf{x}))^\top$, and the predictive covariance matrix is $\Sigma(\mathbf{x}) = diag(\sigma_1^2(\mathbf{x}), \ldots, \sigma_L^2(\mathbf{x}))$. There are no cross-covariances as each model is independent.

2.2 Classifying the Feasible Space

Since the predictive distribution is Gaussian, we can compute the probability of violation of each constraint. For the lth constraint, the *probability of feasibility* is [7-9]:

$$p(\mathbf{x} \in \mathcal{F}_l) = p(\, p(g_l \mid \mathbf{x}, \mathcal{D}_l) \leq t_l \,) = \Phi(\tau_l), \tag{4}$$

where $\tau_l = \frac{t_l - \mu_l(\mathbf{x})}{\sigma_l(\mathbf{x})}$ and $\Phi(\cdot)$ is the standard Gaussian cumulative distribution function. The overall probability of feasibility is therefore:

$$p(\mathbf{x} \in \mathcal{F}) = \prod_{l=1}^{L} p(\mathbf{x} \in \mathcal{F}_l) = \prod_{l=1}^{L} \Phi(\tau_l). \tag{5}$$

Due to symmetry, the probability of infeasibility is $p(\mathbf{x} \in \mathcal{I}) = 1 - p(\mathbf{x} \in \mathcal{F})$. Using these probabilistic estimations, a decision vector $\mathbf{x}$ is feasible *iff* $p(\mathbf{x} \in \mathcal{F}) > p(\mathbf{x} \in \mathcal{I})$. Figure 1 illustrates the predicted feasible spaces for two constraints modelled with two $\mathcal{GP}$s.

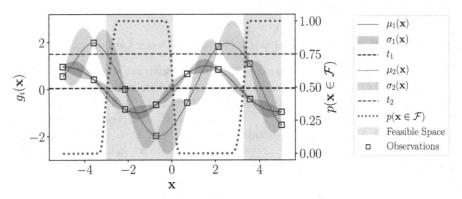

Fig. 1. Illustration of probability of feasibility (dotted red line) with two surrogate $\mathcal{GP}$s trained on observations at regular intervals (depicted by black squares). The grey shaded areas represent the true feasible space due to the threshold vector $\mathbf{t} = (t_1, t_2)^{\top} = (0.05, 1.5)^{\top}$ on constraint functions $g_1(\mathbf{x}) = \sin(\mathbf{x})$ and $g_2(\mathbf{x}) = 2\sin(\mathbf{x}-1)$. The shaded areas around the mean predictions $\mu_l(\mathbf{x})$ (blue for $\mu_1(\mathbf{x})$ and green for $\mu_2(\mathbf{x})$) show the uncertainty estimations $2\sigma_l(\mathbf{x})$ from respective $\mathcal{GP}$s. With only a few carefully selected observations, the models have well approximated the feasible space. In this paper, instead of sampling at regular intervals, our aim is to sequentially select the training data to construct the best possible classifier of feasibility with the smallest budget of expensive evaluations. (Color figure online)

2.3   Bayesian Search Framework

Bayesian search is a surrogate-assisted active learning framework. This method takes inspiration from Efficient Global Optimisation (EGO), first proposed by Kushner [10] and later improved by Jones *et al.* [11]. The framework can be used to minimise the mean squared error in the sequential design of experiments, and is particularly useful where there are few observations [12]. It has also been used to compute the volume of infeasible space [1,9,13], and to *locate* the feasible space under multiple constraints [2].

Bayesian search is a global search strategy. It sequentially samples the design space to determine the boundary of the feasible space. The algorithm has two stages: initial sampling, and sequential improvement.

During the first stage, we sample the parameters using a space filling design, typically with Latin Hypercube sampling (LHS) [14]. These parameters are then evaluated by the true function. The LHS parameter samples and their true-function output create the initial training set. Each design set is used to create a set of models, one for each constraint, $\hat{G} = \{\hat{g}_1, \ldots, \hat{g}_L\}$.

For the sequential improvement phase, we can use $\hat{G}$ to locate promising samples. For an arbitrary design vector $\mathbf{x}$, the function $\hat{G}$ provides a multi-dimensional posterior distribution $p(G \mid \mathbf{x}, \mathcal{D})$ with a mean prediction (a vector) and uncertainty (a covariance matrix). The predictive distribution permits a closed form calculation of probability. We use this predictive distribution to estimate the likelihood that a constraint function value will exceed a threshold. Since our goal is to minimise the uncertainty around the threshold that bounds the infeasible space, we can design our *acquisition function* $\alpha(\mathbf{x}, \hat{G}, \mathbf{t})$ accordingly. The aim is to strike a balance between exploitation (through the probability of a solution residing at the boundary) and global exploration (through the prediction uncertainty). In this way, the acquisition function will drive the search towards the areas we are interested in. The proposed acquisition functions are presented in Sect. 3.

The most promising solution is defined where $\mathbf{x}^* = \text{argmax}_{\mathbf{x}} \, \alpha(\mathbf{x}, \hat{G}, \mathbf{t})$. We then determine $\mathbf{x}^*$ expensively, and use the results to augment the data and retrain $\hat{G}$. We repeat this process until we exhaust the simulation budget. When training is complete, we use $\hat{G}$ to estimate the feasible space. For an arbitrary $\mathbf{x}$ a probability of feasibility is returned using (5). Algorithm 1 summarises the method.

Algorithm 1. Bayesian search framework.

Inputs

    M : Number of initial samples
    T : Budget on expensive function evaluations
    $\mathbf{t}$: Threshold vector

Steps

1: $X \leftarrow \text{LatinHypercubeSampling}(\mathcal{X})$ ▷ Generate initial samples
2: $\Gamma \leftarrow \{G(\mathbf{x} \in X)\}$ ▷ Expensively evaluate all initial samples
3: **for** $i = M \rightarrow T$ **do**
4:     $\hat{G} \leftarrow \text{Train}\mathcal{GPs}(X, \Gamma)$ ▷ Train a mono- or multi-surrogate model of constraints
5:     $\mathbf{x}^* \leftarrow \text{argmax}_{\mathbf{x}} \, \alpha(\mathbf{x}, \hat{G}, \mathbf{t})$ ▷ Optimise acquisition function
6:     $X \leftarrow X \cup \{\mathbf{x}^*\}$ ▷ Augment data set with $\mathbf{x}^*$
7:     $\Gamma \leftarrow \Gamma \cup \{G(\mathbf{x}^*)\}$ ▷ Expensively evaluate $\mathbf{x}^*$
8: **end for**
9: **return** $\hat{G}$ ▷ Return trained models for feasibility classification using (5)

---

3 Acquisition Functions

For Bayesian search of the feasible space, the acquisition function's aim is to locate the boundary between feasible and infeasible spaces: $\mathcal{F}_l$ and $\mathcal{I}_l$. A solution based on $\hat{g}_l$ is often identified with the probability of being at the boundary: $p(\mathbf{x} \in \mathcal{F}_l)p(\mathbf{x} \in \mathcal{I}_l)$. If we add the sample at this to the training set, the estimation of feasibility with $\hat{g}_l$ is maximally improved. In this way, we achieve maximal *exploitation* of the latest knowledge of the model.

When data is limited, the uncertainty in predictions may be high, especially in areas of the design space that have a sparse number of samples. We should, therefore, promote the *exploration* of these areas.

However, if we only prioritise sparsely populated areas, we may miss areas near the threshold of interest. We therefore need to consider areas where *both* the uncertainty and the boundary probability are high. We test both a range of existing and a novel acquisition function for balancing the above requirements.

The first acquisition function for feasible region discovery was proposed by Knudde *et al.* [2]. It is designed to maximise the loss in entropy of the posterior distribution around the boundary for adding a solution to the training dataset. For *l*th constraint, it can be expressed as:

$$\alpha_{K[l]}(\mathbf{x}, \hat{g}_l, t_l) = \frac{1}{2} \ln(2\pi e \sigma_l^2(\mathbf{x})) - \ln(\Phi(\tau_l)(1 - \Phi(\tau_l))). \tag{6}$$

A detailed simplification of this equation is provided in the Supplementary materials.

Here, the first term is the predictive entropy $\mathbb{H}(p(g_l \mid \mathbf{x}, \mathcal{D}))$ (representing areas of high uncertainty where exploration should be maximised). The second term is the natural logarithm of the probability of being at the boundary (representing exploitation of the knowledge of the boundary); c.f. Eq. (5).

This function was originally designed for a single constraint. For multiple constraints, they proposed to combine these component utilities across all constraints as a sum:

$$\alpha_K(\mathbf{x}, \hat{G}, \mathbf{t}) = \sum_{l=1}^{L} \alpha_{K[l]}(\mathbf{x}, \hat{g}, t_l). \tag{7}$$

The summation formulation of the acquisition function permits a situation where one of the components $\alpha_{K[l]}$ can dominate the overall utility. For instance, if an arbitrary solution $\mathbf{x}$ is expected to gain significantly more information than other constraints, it may still maximise the acquisition utility and thus get evaluated. This biases the search towards maximal individual gain without allowing existing information to influence over the value of the acquisition function. As a result, the overall progress of this acquisition function may be slow.

We also **adapt** a range of alternative acquisition functions from the field of reliability engineering. These acquisition functions were originally developed for a single constraint, and were first used in an active learning framework by

Ranjan *et al.* [15] and Bichon *et al.* [16], and later popularised by Picheny *et al.* [1,17] for computing the volume of the infeasible space.

The most popular acquisition functions for **single** constraint are:

$$\alpha_{T[l]}(\mathbf{x}, \hat{g}_l, t_l) = \sigma(\mathbf{x})\phi(z), \tag{8}$$

$$\alpha_{B[l]}(\mathbf{x}, \hat{g}_l, t_l) = \sigma(\mathbf{x})[z^+\Phi(z^+) + z^-\Phi(z^-) + \phi(z^+) + \phi(z^-) - 2z\Phi(z) - 2\phi(z)], \tag{9}$$

$$\alpha_{R[l]}(\mathbf{x}, \hat{g}_l, t_l) = \sigma^2(\mathbf{x})[z^2(\Phi(z^-) - \Phi(z^+)) + z^+\phi(z^-) - z^-\phi(z^+)]. \tag{10}$$

Here, $z = \frac{\mu_l(\mathbf{x}) - t_l}{\sigma(\mathbf{x})}$, $z^+ = z + 1$, $z^- = z - 1$, and $\phi(\cdot)$ is the standard Gaussian probability density function. $\alpha_{T[l]}(\cdot)$ is the targeted mean squared error and was defined by Picheny *et al.* [9]. $\alpha_{B[l]}(\cdot)$ and $\alpha_{R[l]}(\cdot)$ are functions that compute a form of average positive difference between uncertainty and the predictive distance from the threshold, defined by Bichon *et al.* [16] and Ranjan *et al.* [15]. Further details of these can be found in [9,13].

A similar acquisition function proposed by Echard *et al.* that can also be used [18]. This is written as [18,19]:

$$\alpha_{E[l]}(\mathbf{x}, \hat{g}_l, t_l) = -\frac{|\mu_l(\mathbf{x}) - t_l|}{\sigma_l(\mathbf{x})}. \tag{11}$$

This is the negative of the probability of wrongly predicting feasibility. Maximising this function finds solutions that reduce the misclassification error.

To determine areas of system failure under multiple constraints, a composite-criterion approach is commonly taken. This approach calculates the acquisition function for each model, selecting a single model based on the best individual mean prediction [18–20]. We reformulated this approach into a generalised version appropriate for Bayesian search (without requiring a large number of Monte Carlo samples):

$$\alpha_Y(\mathbf{x}, \hat{G}, \mathbf{t}) = \alpha_{Y[k]}(\mathbf{x}, \hat{g}_k, t_k) \mid k = \operatorname*{argmax}_{l=1}^{L}(\mu_l(\mathbf{x}) - t_l), \tag{12}$$

where, $\alpha_{Y[k]}(\mathbf{x}, \hat{g}_k, t_k)$ is the acquisition function for kth constraint $g_k(\mathbf{x})$, with $Y \in \{T, B, R, E\}$.

Using the acquisition function (12) improves the learning of the individual boundaries between feasible and infeasible spaces for each constraint $g_k(\mathbf{x})$. However, this approach does not directly account for the true boundary under multiple constraints. For multiple constraints, *any* violation is treated as infeasible, and since Eq. (12) may sample infeasible space, it will likely introduce unnecessary redundancy. A further weakness is that the model selection term $k = \operatorname{argmax}_{l=1}^{L}(\mu_l(\mathbf{x}) - t_l)$ does not consider prediction uncertainty. The result can therefore be misleading. The scale of the function value in each constraint can also cause problems, since the magnitude differences in $\mu_l(\mathbf{x}) - t_l$ may be inverse to relative importance. Our new acquisition function aims to solve these shortcomings.

3.1 Probability of Being at the Boundary and Entropy (PBE)

We have discussed how single-constraint acquisition functions can be combined to create an acquisition function for multiple constraints. However, since our aim is to find solutions with a high probability of being at the boundary of the feasible space (*exploitation*), whilst minimising the overall uncertainty in the models (*exploration*), we combine these two objectives as a product.

The probability that a solution is at the boundary β between the feasible and infeasible spaces, given a multi-surrogate model $\hat{G}$, is:

$$p(\mathbf{x} \in \beta) = p(\mathbf{x} \in \mathcal{F}) \, p(\mathbf{x} \in \mathcal{I}) = \prod_{l=1}^{L} \Phi(\tau_l) - \prod_{l=1}^{L} \Phi^2(\tau_l). \tag{13}$$

If we maximise the probability over the design space, we will locate solutions at the boundary, thereby exploiting the current knowledge.

To evaluate the overall uncertainty for a multi-surrogate model $\hat{G}$, we compute the differential entropy of a multi-variate Gaussian distribution:

$$\mathbb{H}(\mathbf{x} \mid \hat{G}) = \frac{L}{2} \ln(2\pi e) + \frac{1}{2} \ln(\mid \Sigma \mid) \propto \ln \left(\prod_{l=1}^{L} \sigma_l^2(\mathbf{x}) \right). \tag{14}$$

The extremes of the above equation identify the solutions with most overall uncertainty across the models. These extremes identify the most informative samples.

To maximise both quantities, we combine these two measures together as a product. This is a somewhat greedy approach that ensures that a solution that improves all components is selected. Our multi-surrogate acquisition function – the Probability of Boundary and Entropy (PBE) – is defined as:

$$\alpha_{PBE}(\mathbf{x}, \hat{G}, \mathbf{t}) = p(\mathbf{x} \in \beta) \, \mathbb{H}(\mathbf{x} \mid \hat{G}). \tag{15}$$

This function addresses the true boundary β, which is at the intersection of all component constraints' feasible spaces $\mathcal{F} = \bigcap_{l=1}^{L} \mathcal{F}_l$, directly. It is particularly useful, since no explicit model selection is required. Further, since the probability and entropy are being computed via an intra-constraint model (rather than between constraints), we expect it to perform better for unscaled function responses.

4   Experiments

To test the performance of our approach, we used the test suite for constrained single-objective optimisation problems from CEC2006 [4].

We selected a range of problems with only non-linear inequality constraints (for simplicity) and varying proportional volume of the feasible space between 0.5% and 45% (Table 1). Note that we merely use these problems as example constrained problems for *design exploration*, and we do not seek to locate the global optimum.

We ran each method starting from an initial sample size of n where n is the dimension of the decision space to avoid modelling an under-determined system. We set our budget on expensive evaluations to $11n$ to allow each method to gather $10n$ samples after initial sampling. This budget is less than the number of evaluations used by Knudde *et al.* [2] and Yang *et al.* (in reliability engineering) [19] for reporting their results.

Table 1. A range of test problems with a feasible space volume $\rho = \frac{|\mathcal{F}|}{|\mathcal{F}|+|\mathcal{I}|} \times 100\% \geq$ 0.5% from the test suite defined in [4] and implemented in PyGMO [21]. Here, n is the dimension of the decision space, and L is the number of constraints.

ID	n	$\rho(\%)$	L
G4	5	26.9953	6
G8	2	0.8727	2
G9	7	0.5218	4
G19	15	33.4856	5
G24	2	44.2294	2

We ran each method on each problem 21 times[1]. The initial evaluations are matched between acquisition functions, i.e. for each pair of problem and simulation run, the same initial design was used. The exception to this is the LHS with $11n$ samples.

Since the acquisition function landscape is (typically) multi-modal, we used Bi-POP-CMA-ES to search the space, as it is known to solve multi-modal problems effectively [22]. We set the maximum number of evaluations of the acquisition function to $5000n$.

We use *informedness* as a performance indicator for the classifier. The informedness estimates the probability that a prediction is informed, compared to a chance guess. We chose informedness (instead of F1 measure used in [2]) as it performs well for imbalanced class sizes, which are common when comparing the sizes of feasible and infeasible spaces for real-world constrained problems [23,24]. To ensure that we get an accurate estimation, we used 10000 uniformly random samples from the decision space for validation. We keep these validation sets constant across different methods for a specific simulation number and a specific problem.

We used the one-sided Wilcoxon Signed Rank test with Bonferroni correction to test statistical equivalence to the best median performance, due to matched samples. We identify the best method at the level of $p \leq 0.05$ [25]. We used Mann-Whitney-U test to compare the LHS to the other methods (Table 2). We provide box plots for performance comparison in the supplementary sections.

---

[1] Python code for Bayesian search will be available at: bitbucket.org/arahat/lod-2020.

Table 2. Performance of different acquisition functions in terms of median informedness (%) and the median absolute deviation from the median (MAD). The red cells show the best median performance, while the blue cells depict the equivalent methods to the best.

		LHS	α_K	α_T	α_B	α_R	α_E	α_{PBE}
G4	Median	99.83%	99.66%	99.95%	99.94%	99.93%	99.95%	99.99%
	MAD	7.9×10^{-4}	1.3×10^{-3}	2.7×10^{-4}	2.5×10^{-4}	4.3×10^{-4}	2.1×10^{-4}	3.6×10^{-4}
G8	Median	97.85%	93.51%	99.99%	98.85%	98.85%	98.86%	100%
	MAD	1.4×10^{-2}	8.9×10^{-2}	6.0×10^{-3}	1.0×10^{-2}	1.0×10^{-2}	9.0×10^{-3}	5.9×10^{-3}
G9	Median	20.26%	35%	81.62%	80.55%	76.19%	97.95%	81.24%
	MAD	2.0×10^{-1}	1.5×10^{-1}	1.7×10^{-1}	2.3×10^{-1}	2.5×10^{-1}	1.6×10^{-2}	5.6×10^{-2}
G19	Median	99.89%	99.68%	99.92%	99.91%	99.92%	99.94%	99.91%
	MAD	4.0×10^{-4}	7.8×10^{-4}	2.7×10^{-4}	2.4×10^{-4}	2.1×10^{-4}	2.5×10^{-4}	2.6×10^{-4}
G24	Median	99.59%	59.75%	99.66%	99.66%	99.63%	99.63%	99.71%
	MAD	1.4×10^{-3}	7.4×10^{-2}	1.1×10^{-3}	6.1×10^{-4}	8.5×10^{-4}	1.8×10^{-2}	3.9×10^{-4}

The results show that the acquisition functions proposed in this paper outperform naive LHS and the acquisition function α_K proposed by Knudde *et al.* [2]. G9 has the worst median performance of 20.26% for LHS, where the volume of the feasible space is extremely small (about 0.5218%). Here too α_K performs worse than our acquisition function α_{PBE}. The acquisition function α_E from Echard *et al.* outperforms all other methods with a median informedness of 97.95%. In **three out of the five problems**, α_{PBE} achieves the best median performance, while α_E performs best in the rest of the problems. The best median for any problem is at least 97.95% with small MAD, demonstrating the efficacy of the proposed and adapted methods over the current state-of-the-art.

5   Conclusions

This paper has examined the problem of feasible space identification for computationally expensive problems. We have demonstrated an active learning approach using Bayesian models (Bayesian search) and developed a range of acquisition functions for this purpose. Our experiments show that our proposed acquisition function for Bayesian search outperforms naive LHS, and the current state-of-the-art α_K. We propose that future work focusses on batch Bayesian search when it is possible to evaluate multiple solutions in parallel.

Acknowledgements. We acknowledge the support of the Supercomputing Wales project, which is part-funded by the European Regional Development Fund (ERDF) via Welsh Government.

References

1. Chevalier, C., Bect, J., Ginsbourger, D., Vazquez, E., Picheny, V., Richet, Y.: Fast parallel kriging-based stepwise uncertainty reduction with application to the identification of an excursion set. Technometrics **56**(4), 455–465 (2014)

2. Knudde, N., Couckuyt, I., Shintani, K., Dhaene, T.: Active learning for feasible region discovery. In: 2019 18th IEEE International Conference on Machine Learning and Applications (ICMLA), pp. 567–572. IEEE (2019)
3. Rasmussen, C.E., Williams, C.K.I.: Gaussian Processes for Machine Learning. The MIT Press, Cambridge (2006)
4. Liang, J.J., et al.: Problem definitions and evaluation criteria for the CEC 2006 special session on constrained real-parameter optimization. J. Appl. Mech. **41**(8), 8–31 (2006)
5. Mori, Y., Ellingwood, B.R.: Time-dependent system reliability analysis by adaptive importance sampling. Struct. Saf. **12**(1), 59–73 (1993)
6. Snoek, J., Larochelle, H., Adams, R.P.: Practical Bayesian optimization of machine learning algorithms. In: Advances in Neural Information Processing Systems, pp. 2951–2959 (2012)
7. Hughes, E.J.: Evolutionary multi-objective ranking with uncertainty and noise. In: Zitzler, E., Thiele, L., Deb, K., Coello Coello, C.A., Corne, D. (eds.) EMO 2001. LNCS, vol. 1993, pp. 329–343. Springer, Heidelberg (2001). https://doi.org/10.1007/3-540-44719-9_23
8. Fieldsend, J.E., Everson, R.M.: Multi-objective optimisation in the presence of uncertainty. In: The 2005 IEEE Congress on Evolutionary Computation, vol. 1, pp. 243–250. IEEE (2005)
9. Chevalier, C., Picheny, V., Ginsbourger, D.: KrigInv: an efficient and user-friendly implementation of batch-sequential inversion strategies based on kriging. Comput. Stat. Data Anal. **71**, 1021–1034 (2014)
10. Kushner, H.J.: A new method of locating the maximum point of an arbitrary multipeak curve in the presence of noise. J. Basic Eng. **86**(1), 97–106 (1964)
11. Jones, D.R., Schonlau, M., Welch, W.J.: Efficient global optimization of expensive black-box functions. J. Glob. Optim. **13**(4), 455–492 (1998)
12. Sacks, J., Welch, W.J., Mitchell, T.J., Wynn, H.P.: Design and analysis of computer experiments. Stat. Sci. **4**, 409–423 (1989)
13. Bect, J., Ginsbourger, D., Li, L., Picheny, V., Vazquez, E.: Sequential design of computer experiments for the estimation of a probability of failure. Stat. Comput. **22**(3), 773–793 (2012)
14. McKay, M.D., Beckman, R.J., Conover, W.J.: A comparison of three methods for selecting values of input variables in the analysis of output from a computer code. Technometrics **42**(1), 55–61 (2000)
15. Ranjan, P., Bingham, D., Michailidis, G.: Sequential experiment design for contour estimation from complex computer codes. Technometrics **50**(4), 527–541 (2008)
16. Bichon, B.J., Eldred, M.S., Swiler, L.P., Mahadevan, S., McFarland, J.M.: Efficient global reliability analysis for nonlinear implicit performance functions. AIAA J. **46**(10), 2459–2468 (2008)
17. Picheny, V., Ginsbourger, D., Roustant, O., Haftka, R.T., Kim, N.-H.: Adaptive designs of experiments for accurate approximation of a target region. J. Mech. Des. **132**(7), 071008 (2010)
18. Echard, B., Gayton, N., Lemaire, M.: AK-MCS: an active learning reliability method combining Kriging and Monte Carlo Simulation. Struct. Saf. **33**(2), 145–154 (2011)
19. Yang, X., Mi, C., Deng, D., Liu, Y.: A system reliability analysis method combining active learning Kriging model with adaptive size of candidate points. Struct. Multidisc. Optim **60**(1), 137–150 (2019). https://doi.org/10.1007/s00158-019-02205-x
20. Fauriat, W., Gayton, N.: AK-SYS: an adaptation of the AK-MCS method for system reliability. Reliab. Eng. Syst. Saf. **123**, 137–144 (2014)

21. Biscani, F., et al.: esa/pagmo2: pagmo 2.15.0, April 2020
22. Hansen, N., Auger, A., Ros, R., Finck, S., Pošík, P.: Comparing results of 31 algorithms from the black-box optimization benchmarking BBOB-2009. In: Proceedings of the 12th Annual Conference Companion on Genetic and Evolutionary Computation, pp. 1689–1696. ACM (2010)
23. David Martin Powers: Evaluation: from precision, recall and f-measure to roc, informedness, markedness and correlation (2011)
24. Tharwat, A.: Classification assessment methods. Appl. Comput. Inform. (2018)
25. Fonseca, C.M., Knowles, J.D., Thiele, L., Zitzler, E.: A tutorial on the performance assessment of stochastic multiobjective optimizers. In: 3rd International Conference on Evolutionary Multi-criterion Optimization, EMO 2005, vol. 216, p. 240 (2005)

A Transfer Machine Learning Matching Algorithm for Source and Target (TL-MAST)

Florin Tim Peters$^{(\boxtimes)}$ and Robin Hirt

prenode GmbH, Karlsruhe, Germany
`florin.peters@me.com, robin@prenode.de`

Abstract. Sequentially transferring machine learning (ML) models in a row over several data sets can result in an improvement of performance. Considering an increasing number of data sets, the possible number "transfer paths" of a transfer grows exponentially. Thus, in this paper, we present TL-MAST, a matching algorithm for identifying the optimal transfer of ML models to reduce the computational effort across a number of data sets—determining their transferability. This is achieved by suggesting suitable source data sets for pairing with a target data set at hand. The approach is based on a layer-wise, metric-supported comparison of individually trained base neural networks through meta machine learning as a proxy for their performance in a crosswise transfer scenario. We evaluate TL-MAST on two real-world data sets: a unique sales data set composed of two restaurant chains and a publicly available stock data set. We are able to identify the best performing transfer paths and therefore, drastically decrease computational time to find the optimal transfer.

Keywords: Transfer machine learning · Transferability · Meta machine learning

1 Introduction

Transfer machine learning enables us to derive knowledge in the form of machine learning models from different data sets. Hereby, a model is trained on a source set and transferred and adapted on a target data set [23] leading to better performing models and faster convergence. Additionally, transferring machine learning models can be a means of information transport across potentially sensitive data sets [10,11]. By doing so, one could enable machine learning across multiple sensitive data sets without the need for data centralization - preserving data confidentiality [7].

To train models on multiple, distributed data sets, one could also sequentially train and transfer models across data sets in a peer-to-peer manner to achieve a knowledge transfer across n data sets. Besides the abundance of a centralized

© Springer Nature Switzerland AG 2020
G. Nicosia et al. (Eds.): LOD 2020, LNCS 12566, pp. 541–558, 2020.
https://doi.org/10.1007/978-3-030-64580-9_45

model aggregation, sequentially transferring models can have several benefits, such as flexibility in a growing system of data sets, or usage of existing models when data can not be achieved without temporal synchronization.

Although a peer-to-peer approach yields advantages, it requires a direction of transfer in a system, where the number of data sets can grow. Considering, that models can be transferred crosswise across and over different data sets, the amount of possible transfers grows with a quadratic complexity [31]. Additionally, not all transfers in such a case yield in performance improvement, and only some "paths" of transfer might lead to the desired effect. Thus, we are interested in examining the influence of a model transfer prior to performing it.

In literature, three main research fields in transfer learning are introduced by asking what, how, and when to transfer. "What to transfer?" deals with the topic of discovering which part of knowledge can be transferred from a source to a target domain, taking into account that they have underlying structures in common but differ on many levels of detail. The "how to transfer?" question deals with the issue which algorithm or approach is needed to achieve the desired transfer of knowledge. "When to transfer?" sets the view on situations where transferring is useful and situations where it is not. It is highlighted that the "when" perspective tries to find suitable matching source and target domains that create a transfer benefit while avoiding transfers that worsen the performance. A lot of research focuses on the "how" and "what" aspects which make research in the "when" aspect desired [23].

Due to source and target data being more or less related to each other, a transfer can be also either beneficial or not [6]. One can argue that transferring anything is better than starting with a plain case [36] but with a growing number of potential base nets, exploring every possible transfer is computationally intense and therefore resource and time consuming [31]. Building a crosswise transfer matrix, in which every source is matched to every target, can lead to a good result, as everything is tested, but may at the same time not be the most efficient approach [17].

Measuring the transferability, as a factor of how each source data performs as a basis for each target data is a strict indicator for matching or not combining them [1]. A structured approach that searches source domains to find transferable knowledge for target domains is therefore desired [1]. The use of such an approach can solve the problem of finding the most helpful source-domain from a collection of many data sets [37]. Therefore, we suggest a meta-learning approach for predicting the quality of a transfer, the transferability. This prediction shall serve as a basis for matching transfer sources and target data efficiently.

This work is structured as follows. At first, we elaborate on foundations and related work and then give an overview of the methodological approach of this study. Then, we outline the proposed transfer learning matching algorithm. Consecutively, we report the results of the evaluation along with two cases and, lastly, we conclude by stating limitations and future research perspectives.

2   Foundations and Related Work

In literature, it is discovered that first-layer features can be generalized and therefore transferred to another task. The ability to transfer the network, the transferability, depends on how specialized the higher layers are to their original task. It is found out that transferability shrinks if the source and target tasks diverge. Additionally, they noticed that there are advantages in transferring weights from a less related task than starting with a bare random one. The similar tasks were created by splitting one data set, grouped by classification labels, into two for the learning processes to compare [36].

Research shows that dependent, in their definition correlated tasks result in them being more transferable from one to another. Finding the right transfer combination is important and can achieve advantages especially in a small-data scenario. Data from a target task is therefore reinforced by chosen source data [34].

A cross-validation framework that helps to choose the right model and data sets to fit during the transfer learning process can be deployed. If the source and target models have different underlying data distributions, good matching is not evident. One approach is to overcome the difference between marginal distributions by adjusting it with a factor. This is done through a density ratio weighting as a measurement of a models ability to approximate the true target domain distribution. It is proofed that the resulting weighted loss of the source is similar to the loss of the target-domain and thus helps to select a good source model for a specific target scenario [37].

A correlation of statistical measures between source and target data sets to their ability to transfer from one to another is discovered in research. These correlation measures are an indicator that successful transfer work and a high transferability is present. It is discovered that a correlation between data divergence, T-SNE divergence, PCA divergence, and MDS divergence can be measured for all data sets in a crosswise comparison. Additionally, the source and target network are compared by using the Singular Vector Canonical Correlation Analysis (SVCCA) [24] to assess neural network similarity [11].

An algorithm called TLAC exists which aims at matching transfer source and target domains through the highest similarity. As they work with a time series task they decided not to transfer models but to transfer parts of those similar time series. In detail, to assess which time-series work together, the Pearson correlation coefficient between the time series is used. At a process view, the correlation between each source and target time-series is computed and then the highest matching source time-series components are added to the target domain and after that, the actual forecast is executed [33].

There also exists an eigenvector perspective on transfer learning. For each transfer learning task, one can compute an eigenvector representation of the data at hand which is then compared to the target data set. An eigenvector is a scalar related representation of a structure [20]. They discover that similar eigenvector representations to the target data are important. Optimal transfer combinations are discovered through a graph-based approach [6].

Research highlights the importance of finding learning from a source task that can be reapplied to a target task. They introduce an H-score, an estimator used to predict the performance of a transfer from one task to another. This score has a connection to the error function of the transfer task. The H-score depends on covariances between source and target data. With its help, best transferability scores can be estimated and therefore selected [1].

Other previous work focuses on employing a mapping function between source and target domain for transfer learning. Hereby, the basis for determining the fitness of a transfer is raw data. The function itself takes the maximum mean discrepancy between source and target data into account and recommends promising source data to a target domain [31].

In contrast, in this work, we do not compare raw data itself, but instead treat trained neural networks as a proxy to data similarity. Hereby, we avoid the necessity of centralizing data. Furthermore, we are employing the second layer of machine learning to predict optimal source-to-target matching.

To summarize, in present work, relationships between source and target data in the context of transfer learning is examined to find implications on transferability. There exist approaches that use certain metrics to predict transferability but they are mostly focused on raw data and consider only a few relationship indicators rather than dealing with various measures to ensure feature completeness.

Thus, our contribution is fourfold:

1. We use neural networks and derived metrics as a proxy to compare data sets.
2. We design and evaluate an eclectic meta-machine learning approach to associate optimal sources to targets.
3. We consider and test multiple features and association measures as part of a matching algorithm between source and target data sets.
4. We propose an approach for matching source and target data sets that is suitable for many tasks and scenarios where data is distributed and machine learning models are sequentially transferred.

3   Methodological Approach

At first, we derive the reference points for our algorithm from the literature. We outline the envisioned algorithm for predicting the transfer success between a data source and a data target to optimize a crosswise transfer between multiple data sources. To evaluate our results, we instantiate the envisioned crosswise transfer on two distinct use cases. The first one is the forecasting of daily net sales of different restaurant branches. The second use case is a prediction of the stock value of different companies, where we transfer different models across those.

We perform an experiment in three steps for these two different cases. First, we train base models for each data source that only utilize isolated data. The retrieved performance values serve as a benchmark for the transferred models.

In a second step, we transfer all models crosswise and adapt the models accordingly on each data set. After each transfer, we measure the performance and calculate the percentage improve in comparison to a model trained in an isolated manner. We call that value the transferability. In the third step, we use the retrieved models and performance values to apply the TL-MAST algorithm and evaluate the predicted optimal transfer in both cases against the retrieved actual transferability scores.

4   An Algorithm for Optimizing the Source-Target Matching in a Sequential Transfer

4.1   Transferability

Transferability is a measurement to assess the success of transfer learning between a chosen source and target domain in comparison to solely training on a target domains data [1]. We define transferability for a neural network η_{p_s} trained on a distribution p_s which is transferred to another network η_{p_s,p_t} trained on a distribution p_t. The second network, the target network η_{p_t}, is trained without transfer for comparison on a distribution p_t. The transferability ΔM is measured as a difference in the mean absolute percentage error (MAPE) results. A negative transferability implies lower error rates and is therefore desired. Each neural network results in a performance score M depending on the derivation of their true and predicted labels. We define:

$$\Delta M(\eta_{p_t}, \eta_{p_s,p_t}) = M(\eta_{p_s,p_t}(X^t), V^t) - M(\eta_{p_t}(X^t), V^t)|$$
$$\Delta M(\eta_{p_t}, \eta_{p_s,p_t}) \in \mathbb{R}^N$$

4.2   Sequential Transfer Approach

To compare a sequential transfer across several data sets, we need to set every data set $d \in \mathbb{D}$ as a transfer target and transfer source and combine them crosswise. The transfer is conducted through the initial models trained from each data set. For n data sets we consequently have $n^2 - n$ transfers to conduct.

$$S_i \to T_j, \forall d_{i,j} \in \mathbb{D}$$

As a transfer source and base comparison we chose to not transfer any raw data as a principle of this work. Initially, we use individually trained base models that have not been transferred before. Therefore we need n additional trainings for each data set to create the initial neural networks as a transfer source and for base-performance comparison. For $1, \ldots, s$ sources data sets and $1, \ldots, t$ target data sets we present their 1st degree crosswise transfer matrix measuring the mean average percentage errors, as a balanced performance metric of each possible combination:

$$\begin{pmatrix} M(\eta_{p_1,p_1}(X^1),V^1) & M(\eta_{p_1,p_2}(X^2),V^2) & \cdots & M(\eta_{p_1,p_t}(X^t),V^t) \\ M(\eta_{p_2,p_1}(X^1),V^1) & M(\eta_{p_2,p_2}(X^2),V^2) & \cdots & M(\eta_{p_s,p_t}(X^t),V^t) \\ \cdots & & \cdots & & \ddots & & \cdots \\ M(\eta_{p_s,p_1}(X^1),V^1) & M(\eta_{p_s,p_2}(X^2),V^2) & \cdots & M(\eta_{p_s,p_t}(X^t),V^t) \end{pmatrix}$$

4.3   TL-MAST Algorithm

In the following, we describe the TL-MAST algorithm (Transfer Learning - Matching Algorithm for Source and Target). To transfer knowledge from one entity to another some kind of data exchange is needed. As described above this data exchange takes place on pre-trained neural network models. These neural network are trained on all available data sets and are named according to the perspective either η_{p_s} for resulting source networks trained on p_s and η_{p_t} for target networks trained on p_t. To clarify, at this point we have not yet executed the sequential transfer approach but created trained neural networks for each data set individually.

We now want to make e networks comparable. Therefore, we apply flattening to the pre-trained models through which the neural network layers' structure gets reduced into a one dimensional vector [12]. For a generalized weight representation of a layer the transformation is described in the following with k being the number of layers, i the vertical structural dimension within the layer, j the horizontal structural dimension within the layer, $w_{i,j,k}$ the weight at the specified position. For the set containing all layers $\mathbb{L}$, the flattening transformation can be described as follows:

$$L_k = \begin{pmatrix} w_{1,1,k} & w_{1,2,k} & \cdots & w_{1,j,k} \\ w_{2,1,k} & w_{2,2,k} & \cdots & w_{2,j,k} \\ \vdots & \vdots & \ddots & \vdots \\ w_{i,1,k} & w_{i,2,k} & \cdots & w_{i,j,k} \end{pmatrix} \longrightarrow \begin{pmatrix} w_{1,1,k} \\ w_{1,2,k} \\ \vdots \\ w_{1,j,k} \\ w_{2,1,k} \\ w_{2,2,k} \\ \vdots \\ w_{2,j,k} \\ w_{i,1,k} \\ w_{i,2,k} \\ \vdots \\ w_{i,j,k} \end{pmatrix}, \forall L_k \in \mathbb{L}$$

These flattened weight vectors are used to compute comparison matrices listed in Table 1. Thus, the vectors have to be created for each source and target network, η_{p_s} and η_{p_t}. With the goal of comparing the vectors, we need to find some level of abstraction for them. This is because comparing raw weights is very computationally expensive and would be counterproductive to the purpose of saving resources. Therefore, the following metrics were selected by studying literature on how to compare data distributions. A relationship between the probability distribution of two data sets can be a key indicator for transferability [1]. Hereby, the Kulback-Leibler divergence is used twice for each direction *source* → *target* and *target* → *source*.

Table 1. Comparison metrics used for TL-MAST

Feature ID	Test statistic feature
0	Pearson R [27]
1	Anderson-Darling [25]
2	Kolmogorov-Smirnov [16]
3	Mann-Whitney [19]
4	Levene [21, 26]
5	Wilcoxon [9, 32]
6	Kruskal-Walis [4, 29]
7	Brunner-Munzel [5]
8	Mood's [18]
9	Bartlett's [2]
10	Epps-Singleton [8]
11	Kulback-Leibler $(S||T)$ [3, 15]
12	Kulback-Leibler $(T||S)$ [3, 15]

The feature set $\mathbb{F}$ is created by calculating the comparison metrics for each pairing of η_{p_s} and η_{p_t}. The data d, used to train the networks is be drawn from the set of available data sets $\mathbb{D}$. This layer-wise similarity feature matrix is computed for each layer $L_k \in \mathbb{L}$. The matrix looks as follows for $0, \ldots, p$ computed test statistic features and $0, \ldots, q$ layers:

$$\begin{pmatrix} sim_{1,1} & sim_{1,2} & \cdots & sim_{1,p} \\ sim_{2,1} & sim_{2,2} & \cdots & sim_{2,p} \\ \cdots & \cdots & \ddots & \cdots \\ sim_{q,1} & sim_{q,2} & \cdots & sim_{q,p} \end{pmatrix}$$

Following up, the presented similarity matrix needs to be computed for every data set combination to create a joint sample-based feature matrix with the dimension

$$dim_{features} = (|\mathbb{D}|^2 - |\mathbb{D}|, |\mathbb{L}|, |\mathbb{F}|),$$

which is then used to predict the transferability $\Delta \widetilde{M}(\eta_{p_t}, \eta_{p_s, p_t})$ for each data set combination in a 1st degree transfer scenario. The resulting transferability vector $\mathbb{T}$ in a case with $|\mathbb{D}|$ data sets has the length of $|\mathbb{D}|^2 - |\mathbb{D}|$. Because neural networks make good features extractors [14, 22, 28], we do not have to exclude or pre-filter features from use in the algorithm.

The prediction itself is performed using a convolutional neural network with two convolutional layers as illustrated in Fig. 1. The first convolutional layer has 16 filters and a kernel size of 6, the second layer comes in smaller with 8 filters and a kernel size of 4. As a speciality we use no pooling in this network and connect the fully-connected layer with 16 neurons through a flattening step to the second convolutional network. We want the learning process of the network to be especially sensitive to the boundary values. This aims at giving the later used decision algorithm a solid basis of avoiding very bad pairings and preferring promising ones. Based on this we use not a mean average percentage error or a similar mean-focused loss function but a mean squared error (without rooting) to better account for very high or very low values. The discrepancy d between true transferabilities $\Delta M(\eta_{p_t}, \eta_{p_s,p_t})$ and predicted transferability $\Delta\widetilde{M}(\eta_{p_t}, \eta_{p_s,p_t})$ is to be minimized.

$$min(d) = min(\Delta(\Delta\widetilde{M}(\eta_{p_t}, \eta_{p_s,p_t}), \Delta M(\eta_{p_t}, \eta_{p_s,p_t})))$$

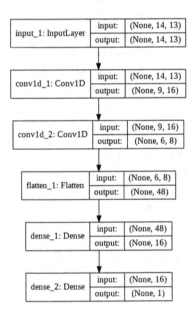

Fig. 1. The convolutional neural network used to predict the transferabilities

For every data set we can choose a matching source partner through the best predicted transferability. The more precise we predict the transferability the better the matching works. As an algorithmic approach we can choose the x

highest transferabilities and their inducing source data set for each data set at hand. For a set of possible partners $\mathbb{D}_{try}$ with $|\mathbb{D}_{try}| = x$ we define

$$\mathbb{D}_{try,i} \cup \operatorname{argmax}(\mathbb{T}), \forall i \in |\mathbb{D}|,$$

which gets repeated after removing the selected index of the prior step x-times.

In practice these x chosen data sets can be trained and evaluated in a transfer learning scenario to see the goodness of the matching. High x values result in a higher testing effort but also in a better chance to match good performing source and target combinations. A lower x value means that we have to train and evaluate less neural networks under the risk of lowering our confidentiality to have the best matching pair.

To measure success of the algorithm we recommended to compute the delta between the true best model performance and the predicted best model performance for each data set in $\mathbb{D}$:

$$\Delta(M_{max}(\eta_{p_s,p_t}(X^t), V^t), \widetilde{M}_{max}(\eta_{p_s,p_t}(X^t), V^t))$$

We provide an overview of the suggested approach in Fig. 2.

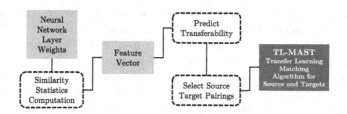

Fig. 2. The TL-MAST approach

5   Evaluation and Results

5.1   Case 1: Predicting Sales for Restaurants

Our data sets for the first part of the evaluation are a list of transactions from 13 restaurants of a similar chain. We focused on the date and the net revenue features (respectively per product). The date serves as our index parameter. We aggregate the net revenue values per transaction to both a daily and a weekly sum. Then we use the daily net values of the current month to predict the sum of the next two week's daily revenues (weeksum): $\text{revenue}_{net,month_{T-1}} \rightarrow \text{weeksum}_{fortnight_T}$.

In restaurant operation forecasting it is an important task to understand demand fluctuation. As a day's revenue correlates to required human and material resources, the knowledge of future demands is critical. As a result of good revenue prediction human and material resources can be utilized more efficiently, which in consequence saves money [13]. In Fig. 3 we show two exemplary data sets plots with their aggregated weekly revenue sums on the ordinate and the corresponding year-week combination on the abscissa.

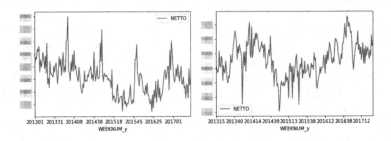

Fig. 3. Plots of two example data sets (censored due to confidentiality reasons)

As mentioned, we have 13 data sets with the first seven coming from company A and the second six ones coming from company B. We choose an almost similar period of test weeks for both companies, with a difference resulting as one had an interruption in opening times. The number of available training weeks can differ highly due to opening times of the specific restaurant. The time horizon of the data sets is heterogeneous as well ranging from 27 to 76 tracked months and sales volumes ranging from 1 to 30 million throughout the tracked periods. The data is sliced into 30 d portions to predict the future 14 d. We again predict the weekly revenue sums of the following week based on the daily revenues of the past week. In Table 2 we present the result of sequential transfers in a crosswise scenario in which every data set is paired with each other.

Table 2. crosswise transfer results (MAPE), $M(\eta_{p_s,p_t}(X^t), V^t)$, for case 1 (rows: transfer targets t, columns: transfer sources s)

Target	1→i	2→i	3→i	4→i	5→i	6→i	7→i	8→i	9→i	10→i	11→i	12→i	13→i
1	0	13.65	15.05	10.81	17.22	10.64	11.71	38.40	49.07	72.04	12.23	51.84	9.98
2	12.21	0	10.76	59.48	9.19	13.64	9.67	182.58	76.07	80.99	13.91	104.47	97.59
3	8.89	8.59	0	31.29	9.15	8.63	7.91	62.92	25.65	53.24	14.27	44.35	130.60
4	11.53	10.62	10.07	0	10.89	11.59	12.04	34.70	9.70	10.04	9.096	17.04	8.29
5	12.40	9.94	10.59	17.91	0	9.69	10.83	66.87	51.89	99.09	98.20	18.84	9.70
6	10.05	10.85	10.23	10.62	9.90	0	9.96	29.71	34.43	72.98	10.70	31.36	16.29
7	11.08	17.61	11.32	34.34	14.39	10.97	0	74.67	76.12	44.20	14.12	34.01	100
8	17.00	23.45	16.04	12.22	16.47	23.15	16.98	0	13.20	13.80	13.47	9.96	13.25
9	6.36	6.01	7.39	7.12	7.66	6.11	5.48	17.53	0	8.91	9.83	9.65	16.81
10	8.51	9.03	9.46	8.94	6.98	12.39	10.25	6.72	7.88	0	8.05	8.03	6.74
11	9.13	9.18	8.84	9.51	8.53	11.52	9.68	7.88	9.86	7.78	0	9.28	7.99
12	24.14	24.27	23.52	30.91	25.47	32.24	24.06	21.17	18.83	12.27	21.15	0	12.74
13	12.91	12.54	12.55	13.03	13.18	12.92	12.41	13.10	13.45	13.69	13.13	13.25	0

The neural network model used for this evaluation case is the attention-based Transformer network [30]. As this model was developed for language tasks we replaced the typical word embedding by adding one convolutional layer in the beginning.

Even though achieving decent MAPE performances in the base case, we discover improvements through transfer learning for every data set as depicted in Table 3.

Table 3. Comparing the models best $MAPE_{base}$, $M_{max}(\eta_{p_t}(X^t), V^t)$, with its best transfer scores, $M_{max}(\eta_{p_s,p_t}(X^t), V^t)$, for case 1; negative values in transferability $\Delta M(\eta_{p_t}, \eta_{p_s,p_t})$ represent lower error rates and thus the best result achieved by transferring

Data set	$M_{max}(\eta_{p_t}(X^t), V^t)$	$\Delta M(\eta_{p_t}, \eta_{p_s,p_t})$
1	17.25	−7.26
2	9.99	−0.79
3	11.04	−3.13
4	8.92	−0.63
5	14.94	−5.25
6	11.65	−1.74
7	13.87	−2.89
8	14.47	−4.50
9	11.17	−5.69
10	8.05	−1.33
11	8.68	−0.90
12	26.50	−14.23
13	13.00	−0.59
ϕ_{mean}		−3.77
$count(< 0)$		13
$\%count(< 0)$		100%

5.2   Case 2: Predicting Stock Values

In our second evaluation case we choose stock data as its publicly available. We randomly selected 7 ticker market data time series from the German stock index TEC-DAX[1]. The time frames start in January 2014 and range to the end of December 2018. There we predict the open price for day t, p_t by using the past 14 d (p_{t-1}-p_{t-14}) as the training data for each slice. Stock data is very

---

[1] Used stocks: Aixtron SE (FRA:AIXA), Cancom SE (FRA:COK), CompuGroup Medical SE (FRA:COP), Dialog Semiconductors plc. (FRA:DLG), Evotec SE (FRA:EVT), Qiagen N.V. (FRA:QIA), Software AG (FRA:SOW).

difficult to predict [35], thus we do not aim for finding the best possible model for this but for creating a scenario to evaluate our algorithm. For the prediction we used a convolutional neural network with two convolutional layers and two fully-connected ones. In Table 4 we present the result of sequential transfers in a crosswise scenario where every data set is paired with each other.

Table 4. crosswise transfer results (MAPE), $M(\eta_{p_s,p_t}(X^t), V^t)$, for case 2 (rows: transfer targets t, columns: transfer sources s)

Target	1→i	2→i	3→i	4→i	5→i	6→i	7→i
1	0	7.45	9.64	8.14	8.28	6.96	9.73
2	10.70	0	11.20	13.68	6.43	7.17	9.43
3	11.56	10.25	0	11.04	7.63	4.90	8.27
4	8.14	7.16	13.19	0	7.39	5.62	10.41
5	9.86	6.50	5.36	7.36	0	4.62	11.78
6	12.94	8.85	9.81	14.06	9.07	0	8.10
7	14.04	11.81	7.68	5.15	4.88	7.91	0

We can improve our semi-performing base cases decently by utilizing transfer learning as presented in Table 5.

Table 5. Comparing the models best MAPE$_{base}$, $M_{max}(\eta_{p_t}(X^t), V^t)$, with its best transfer scores, $M_{max}(\eta_{p_s,p_t}(X^t), V^t)$, for case 1; negative values in transferability $\Delta M(\eta_{p_t}, \eta_{p_s,p_t})$ represent lower error rates and thus the best result achieved by transferring

Data set	$M_{max}(\eta_{p_t}(X^t), V^t)$	$\Delta M(\eta_{p_t}, \eta_{p_s,p_t})$
1	8.92	−1.96
2	13.93	−7.50
3	6.63	−1.73
4	7.93	−2.32
5	27.02	−22.40
6	9.59	−4.47
7	23.45	−9.42
ϕ_{mean}		−7.11
$count(< 0)$		7
$\%count(< 0)$		100%

5.3   TL-MAST Results

For the prediction scenario of the transferability $\Delta\widetilde{M}(\eta_{p_t}, \eta_{p_s,p_t})$ with the convolutional neural network depicted above we used the data sets $1,\ldots,11$ as training data and the data sets $12,13$ as test data for case 1 as well as data sets $1,\ldots 6$ as training data and data set 7 as test for case 2. Every similarity feature is used in the input vector, not only the ones with significant correlation due to the good feature extraction capability of neural networks. We achieved the following mean average percentage errors for the train and test data while predicting transferabilities given layer-weight similarity measures as features:

$$\text{Case 1} \quad \begin{array}{l} \text{Train MAPE 2.14} \\ \text{Test MAPE 4.30} \end{array}$$

$$\text{Case 2} \quad \begin{array}{l} \text{Train MAPE 19.73} \\ \text{Test MAPE 34.71} \end{array}$$

The significantly better prediction for transferability in the first case might be and indicator that the rather large Transformer network used in case 1 has more comparison layers (14) to feed into the transferability prediction than the CNN (4 comparison layers) used in case 2. This has to be evaluated in future studies as discussed later.

To evaluate our source-target matching algorithm we have to first determine an x_{select} as the number of data sets to be evaluated in the original transfer learning scenario. We suggest an $x_{select} = 3$ for our case as it represents less than but close to one fourth of all available data. This selection of x_{select} can be higher or lower depending on the desired outcome considering the trade-off described prior where lower x_{select} means potentially more time savings in the true execution and higher x means more significance on not missing a good performing model. Below in Table 6 we show the results of the comparison between the actual best model, defined through the highest transferability, and the best model among the $x_{select} = 3$ selected ones for case 1 and $x_{select} = 2$ for case 2. The lower x_{select} accounts for case 2 having less data sets. The selection algorithm is again faced with the last two data sets being non-trained test sets for case 1 and the last one for case 2.

To evaluate time savings we utilized a Intel Xeon (1 core, 2.3 GHz) CPU with a Nvidia Tesla K80 (2496 CUDA cores, 12 GB RAM) GPU within shared hardware resources. We ran the crosswise transfer task between all data sets using mainly the graphic processing unit, the similarity measures using the CPU and the transferability prediction again on the GPU as its a neural network driven task. The time span after which the tasks terminated successfully are shown in Table 7.

The presented scenarios achieve time savings of 95% compared to testing all models to find suitable partners. These immense time savings are achieved even though the CPU is very under-powered compared to the GPU at hand. With more balanced hardware the result could have been more drastic. The actual transfer computation of the recommended models are not included in this calculation. We could do this exemplary for case 1 by taking 3 models for the

Table 6. Comparing TL-MAST predicted selection results, $\widetilde{M}_{max}(\eta_{p_s,p_t}(X^t), V^t)$, with the true best models, $M_{max}(\eta_{p_s,p_t}(X^t), V^t)$, for $x_{select} = 3$ values recommendation. The values that are presented are the performance MAPE scores.

Data set	$M_{max}(\eta_{p_s,p_t}(X^t), V^t)$	$\widetilde{M}_{max}(\eta_{p_s,p_t}(X^t), V^t)$	$\Delta(M_{max}, \widetilde{M}_{max})$
	Case 1		
1	7.26	7.26	0
2	0.79	0.79	0
3	3.13	2.40	0.72
4	0.63	0.63	0
5	5.25	4.99	0.25
6	1.74	1.74	0
7	2.89	2.79	0.10
8	4.50	4.50	0
9	5.69	4.05	1.64
10	1.33	1.31	0.02
11	0.90	0.90	0
12	14.2	13.76	0.46
13	0.59	-0.02	0.62
	Case 2		
1	6.96	6.96	0
2	6.43	6.43	0
3	4.90	4.90	0
4	5.62	5.62	0
5	5.36	4.62	0.74
6	8.10	8.10	0
7	7.68	4.88	2.80

Table 7. Computation time results and time savings for case study 1 and 2

	Case 1	Case 2
Crosswise transfer	7 h 35 min	1 h 28 min
Similarity measure matrix	22 min	3 min
Transferability prediction	1 min	1 min
Resulting time savings	95%	95%

$x = 3$ case into account, for each data set to match and transfer. Which results in 39 transfers for all 13 models. Assuming that each model takes the same time to train, this proofing case takes an additional 1 h and 54 min to compute which still would result in time savings of 70% through the same calculation as above.

6   Conclusion

In this work, we address the issue of reducing the transfer complexity by proposing and evaluating the TL-MAST algorithm (Transfer Learning - Matching Algorithm for Source and Target). The algorithm identifies well-performing source data to a specific target data transfer task by predicting the transferability based on a meta-model. The suggested algorithm is not limited to the exemplary presented time-series forecasting tasks or one distinct network architecture—it can be applied to many scenarios thanks to an adjustable design. Using this form of algorithmic approach for transfer learning results in a significant improvement of computation time.

We contribute to the field of machine learning by showing the use of a meta-machine-learning approach to optimize another machine learning problem at hand. Furthermore, we show that the sequential transfer of models across multiple data sets can be optimized and directed by an additional layer of machine learning. This leads to a significant reduction of required computational time. In the field of transfer learning, we provide a structured approach to help solve the problem of finding a suitable source data set for a transfer learning target domain. In the field of distributed private transfer learning, we contribute a source-target matching which we designed in a way so that no raw data needs to be exchanged. Additionally, we provide an algorithm that computes a source data set recommendation on a server for many clients in a system without compromising privacy.

The provided similarity metrics used as the algorithm's input might neither be collectively exhaustive nor mutually exclusive. Continuing with this thought into other relationship structures, it is possible to test the transferability prediction using a step-by-step reduced feature matrix with an even broader input to find out which features are really relevant for the prediction. This can help with computation time (probably very little) but it can do more for scientific interest. It is necessary to further evaluate the suggested approach with a broader range of data from different domains using different machine learning models. This study is needed to discover the capability and boundaries of TL-MAST. Additionally, it can be tried and maybe found a correlation between transferability and the length of source and target training data. Especially the amount of feature-layer input combinations fed into the transferability predicting network has to be evaluated in a larger study. Heterogeneous data sets can also be desired to find out whether transfer among different domains is worth it and moreover, if it can be predicted. Aside from this work we also tested the use of raw neural network weights (without proxy comparison metrics) as an input for the meta-learning model—with little success. This could have been due to too small amounts of data and would be interesting to re-conduct with a larger data basis, although we expect it to be resource intense. As another interesting point for future research we want to highlight that there can be alternative definitions of transferability for which the suggested approach can be executed. We currently compute the transferability as a difference between the transfer and the base performance but for example a quotient-based definition could be interesting to work with as well. Furthermore it can be examined which layers carry relevant information for transferability, an approach for this may be one similar to the above mentioned singular vector canonical correlation analysis. Future research has also the potential to examine the applicability of the suggested approach to further transfers beyond the 1st degree presented in the suggested approach. If we take the transfer learning approach even further it can also be tried to transfer the suggested CNN meta-model to other scenarios. Naturally, the type and architecture of the meta-model can also be iterated.

References

1. Bao, Y., Li, Y., Huang, S., Zhang, L., Zamir, A., Guibas, L.: An information - theoretic metric of transferability for task transfer learning. In: ICLR 2019 (2018)
2. Bartlett, M.S.: Properties of sufficiency and statistical tests. Proc. R. Soc. Lond. Ser. A Math. Phys. Sci. **160**(901), 268–282 (1937). https://doi.org/10.1098/rspa.1937.0109
3. Biroli, G., Bouchaud, J.P., Potters, M.: The Student ensemble of correlation matrices: eigenvalue spectrum and Kullback-Leibler entropy. Quantitative Finance Papers (2007)
4. Breslow, N.: A generalized Kruskal-Wallis test for comparing k samples subject to unequal patterns of censorship. Biometrika **57**(3), 579–594 (1970). https://doi.org/10.1093/biomet/57.3.579
5. Brunner, E., Munzel, U., Puri, M.L.: Rank-score tests in factorial designs with repeated measures. J. Multivar. Anal. **70**(2), 286–317 (1999). https://doi.org/10.1006/JMVA.1999.1821
6. Dai, W., Jin, O., Xue, G.R., Yang, Q., Yu, Y.: EigenTransfer: a unified framework for transfer learning. In: Proceedings of the 26th International Conference on Machine Learning (2009)
7. Dehghani, M., Azarbonyad, H., Kamps, J., de Rijke, M.: Share your model instead of your data: privacy preserving mimic learning for ranking. In: SIGIR 2017 Workshop on Neural Information Retrieval, Neu-IR'17 (July 2017)
8. Epps, T., Singleton, K.J.: An omnibus test for the two-sample problem using the empirical characteristic function. J. Stat. Comput. Simul. **26**(3–4), 177–203 (1986). https://doi.org/10.1080/00949658608810963
9. Gehan, E.A.: A generalized Wilcoxon test for comparing arbitrarily singly-censored samples. Biometrika **52**(1–2), 203–224 (1965). https://doi.org/10.1093/biomet/52.1-2.203
10. Hirt, R., Kühl, N.: Cognition in the era of smart service systems: inter-organizational analytics through meta and transfer learning. In: AIS (2018)
11. Hirt, R., Srivastava, A., Berg, C., Kühl, N.: Transfer Machine Learning in Business Networks: Measuring the Impact of Data and Neural Net Similarity on Transferability. arXiv:2003.13070 (2020)
12. Howard, A.G., et al.: MobileNets: Efficient Convolutional Neural Networks for Mobile Vision Applications (April 2017)
13. Hu, C., Chen, M., McCain, S.L.C.: Forecasting in short-term planning and management for a casino buffet restaurant. J. Travel Tourism Mark. **16**(2–3), 79–98 (2004). https://doi.org/10.1300/J073v16n02_07

14. Kim, Y.: Convolutional neural networks for sentence classification. In: Proceedings of the 2014 Conference on Empirical Methods in Natural Language Processing (EMNLP), pp. 1746–1751 (August 2014)
15. Kullback, S., Leibler, R.A.: On information and sufficiency. Ann. Math. Stat. **22**(1), 79–86 (1951). https://doi.org/10.1214/aoms/1177729694
16. Massey, F.J.: The Kolmogorov-Smirnov test for goodness of fit. J. Am. Stat. Assoc. **46**(253), 68–78 (1951). https://doi.org/10.1080/01621459.1951.10500769
17. Molchanov, P., Tyree, S., Karras, T., Aila, T., Kautz, J.: Pruning convolutional neural networks for resource efficient inference (November 2016)
18. Mood, A.M.: On the asymptotic efficiency of certain nonparametric two-sample tests. Ann. Math. Stat. **25**(3), 514–522 (1954). https://doi.org/10.1214/aoms/1177728719
19. Nachar, N.: The Mann-Whitney U: a test for assessing whether two independent samples come from the same distribution. Tutor. Quant. Meth. Psychol. **4**(1), 13–20 (2008)
20. Nelson, R.: Simplified calculation of eigenvector derivatives. AIAA J. **14**(9), 1201–1205 (1976). https://doi.org/10.2514/3.7211
21. Nordstokke, D.W., Zumbo, B.D.: A New Nonparametric Levene Test for Equal Variances. Technical report (2010)
22. Oquab, M., Bottou, L., Laptev, I., Sivic, J.: Learning and transferring mid-level image representations using convolutional neural networks. In: The IEEE Conference on Computer Vision and Pattern Recognition (CVPR), pp. 1717–1724 (2014)
23. Pan, S.J., Yang, Q.: A survey on transfer learning. IEEE Trans. Knowl. Data Eng. **22**, 1345–1359 (2009). https://doi.org/10.1109/TKDE.2009.191
24. Raghu, M., Gilmer, J., Yosinski, J., Sohl-Dickstein, J.: SVCCA: Singular Vector Canonical Correlation Analysis for Deep Learning Dynamics and Interpretability. Technical report (2017)
25. Scholz, F.W., Stephens, M.A.: K-sample Anderson-Darling tests. J. Am. Stat. Assoc. **82**(399), 918–924 (1987). https://doi.org/10.1080/01621459.1987.10478517
26. Schultz, B.B.: Levene's test for relative variation. Syst. Biol. **34**(4), 449–456 (1985). https://doi.org/10.1093/sysbio/34.4.449
27. Sedgwick, P.: Pearson's correlation coefficient. BMJ **345**, e4483 (2012). https://doi.org/10.1136/bmj.e4483
28. Sharif, A., Azizpour, R.H., Sullivan, J., Carlsson, S.: CNN Features off-the-Shelf: An Astounding Baseline for Recognition. Technical report (2014)
29. Vargha, A., Delaney, H.D.: The Kruskal-Wallis test and stochastic homogeneity. J. Educ. Behav. Stat. **23**(2), 170–192 (1998)
30. Vaswani, A., et al.: Attention is all you need [transformer]. In: The 31st Conference on Neural Information Processing Systems, NIPS 2017 (2017)
31. Wei, Y., Zhang, Y., Yang, Q.: Learning to Transfer. Technical report (2017)
32. Wilcoxon, F.: Individual comparisons by ranking methods. Biom. Bull. **1**(6), 80 (1945). https://doi.org/10.2307/3001968
33. Xiao, J., He, C., Wang, S.: Crude oil price forecasting: a transfer learning based analog complexing model. In: 2012 5th International Conference on Business Intelligence and Financial Engineering, pp. 29–33. IEEE (2012). https://doi.org/10.1109/BIFE.2012.14
34. Xue, Y., Liao, X., Carin, L., Krishnapuram, B.: Multi-Task Learning for Classification with Dirichlet Process Priors. Technical report (2007)
35. Yoon, Y., Swales, G.: Predicting stock price performance: a neural network approach. In: Proceedings of the 24th Annual Hawaii International Conference on System Sciences, vol. 4, pp. 156–162. IEEE (1991)

36. Yosinski, J., Clune, J., Bengio, Y., Lipson, H.: How transferable are features in deep neural networks? Technical report (2014)
37. Zhong, E., Fan, W., Yang, Q., Verscheure, O., Ren, J.: Cross Validation Framework to Choose amongst Models and Datasets for Transfer Learning. Technical report (2010)

Machine Learning and Statistical Models for the Prevalence of Multiple Sclerosis

Nicholas Mandarano[1], Rommel G. Regis[2]([⊠])([iD]), and Elizabeth Bloom[3]

[1] Villanova University, Villanova, PA 19085, USA
nmandara@villanova.edu
[2] Saint Joseph's University, Philadelphia, PA 19131, USA
rregis@sju.edu
[3] Peace Corps, Washington, DC 20526, USA
erb6994@gmail.com

Abstract. Multiple sclerosis is an immune-mediated disease affecting approximately 2.5 million people worldwide. Its cause is unknown and there is currently no cure. MS tends to be more prevalent in countries that are farther from the equator. Moreover, smoking and obesity are believed to increase the risk of developing the disease. This article builds machine learning and statistical models for the MS prevalence in a country in terms of its distance from the equator and the smoking and adult obesity prevalence in that country. To build the models, the center of population of a country is approximated by finding a point on the surface of the Earth that minimizes a weighted sum of squared distances from the major cities of the country. This study compares the predictive performance of several machine learning models, including first and second order multiple regression, random forest, neural network and support vector regression.

Keywords: Multiple sclerosis · Multiple regression · Random forest · Neural network · Support vector regression · Geographic population center · Constrained optimization

1 Introduction and Motivation

Multiple sclerosis (MS) is an immune-mediated disease that affects approximately 2.5 million people worldwide and it is more prevalent among women than men. The cause is unknown and there is currently no cure. However, it has been observed that MS tends to be more prevalent in countries that are farther from the equator [12] and that smoking increases a person's risk of developing MS [4]. Moreover, obesity has been linked to an increased risk of MS [8]. We confirm these observations using data available online and build machine learning and statistical models for predicting the MS prevalence in a country based on the distance of a country's approximate center of population from the equator, the smoking prevalence, and the adult obesity prevalence in a country.

© Springer Nature Switzerland AG 2020
G. Nicosia et al. (Eds.): LOD 2020, LNCS 12566, pp. 559–571, 2020.
https://doi.org/10.1007/978-3-030-64580-9_46

MS is a disease in which the body's immune system attacks the central nervous system. The immune system believes the body to be sick, when in fact it is not. Immune cells will then eat away at the myelin sheath, which coats and protects axons. When the myelin is damaged or destroyed, nerve impulses traveling along axons to and from the brain and spinal cord become distorted and interrupted. Research has shown that people born in an area with a high risk of MS who then migrate to an area with a lower risk before the age of 15 assume the risk of the new area [10]. This finding provides evidence for some environmental factor in the onset of MS. The exact cause of MS is currently unknown, but thought to be triggered in a genetically susceptible individual by a combination of one or more environmental factors.

Geographic location seems to be an important factor that influences the onset and expression of MS. There is the well-known Vitamin D hypothesis, which states that people with low vitamin D levels are more susceptible to MS [9]. This is consistent with the observation that the prevalence of MS is higher in countries farther from the equator where there is less sunlight since the sun is the number one source of daily vitamin D for most people. However, more recent studies suggest that sun exposure and not vitamin D level is associated with the risk of MS [7,13]. In any case, people farther from the equator tend to have a higher risk of developing or being diagnosed with MS [12].

Our paper has two main goals. First, we build several machine learning models for predicting the MS prevalence in a country based on its distance from the equator and the smoking prevalence and adult obesity prevalence in that country, and then we compare their predictive performance. Second, we confirm and quantify the relationship between a country's MS prevalence and its distance to the equator and build linear regression models for predicting MS prevalence. In our analysis, each country will be represented by its geographic population center, which is obtained by solving an equality-constrained optimization problem. The methods compared are first and second order multiple regression, random forest, support vector regression with linear and radial basis function (RBF) kernels, neural network and k-nearest neighbors with some of these methods tuned by cross-validation. The results show that random forest and neural network perform better than the other methods in predicting MS prevalence on hold out test sets. To the best of our knowledge, machine learning regression models for MS prevalence in a country have never been explored in the literature. Hence, our proposed models contribute to the understanding of MS prevalence by country. In addition, our models can be used to estimate the MS prevalence of any connected subregion of any country in terms of the location of its population center and the prevalence of smoking and obesity in that region. Described next is the aforementioned method to find the location in a country that represents its geographic center of population.

2   Finding the Center of Population a Country

Our goal is to build machine learning and statistical models that relate MS prevalence with distance to the equator, smoking prevalence, and adult obesity prevalence. However, since the data on MS prevalence is only available by country, we need to find a location in each country that represents its center of population instead of its geographic center. The procedure we describe can be used to represent a country in any machine learning regression model where the location of a country on the surface of the Earth is a potential predictor variable.

There are different ways of calculating the center of population of a country. We now describe one such procedure. For each country, first determine the latitudes and longitudes of its cities and convert them to Cartesian coordinates (see next section). Now suppose a country has n cities with populations $p_1, p_2, \ldots, p_n$ and that $p^{(1)} = (x_1, y_1, z_1), p^{(2)} = (x_2, y_2, z_2), \ldots, p^{(n)} = (x_n, y_n, z_n)$ are the Cartesian coordinates of these cities. If the cities lie on a flat surface, then the population center can be obtained simply as a weighted mean of the points $p^{(1)}, p^{(2)}, \ldots, p^{(n)} \in \mathbb{R}^3$ where the weight associated with a city is determined by the fraction of the total population that live in that city. That is, the weight for the ith city is $w_i = p_i / \sum_{k=1}^{n} p_k$. However, the cities lie on the surface of the Earth, which we assume to be a perfect sphere. Hence, the weighted mean of their locations in Cartesian coordinates will be under the surface of the Earth, and so, this will not work as the population center of a country.

Let $\bar{p} = (\overline{p_x}, \overline{p_y}, \overline{p_z}) := \sum_{i=1}^{n} w_i p^{(i)}$ be the weighted mean of the locations $p^{(1)}, \ldots, p^{(n)} \in \mathbb{R}^3$ (in Cartesian coordinates) using the weights $w_1, \ldots, w_n$, respectively. It is well-known that the weighted mean $\bar{p}$ is the point $p = (x, y, z) \in \mathbb{R}^3$ that minimizes

$$\sum_{i=1}^{n} w_i \|p - p^{(i)}\|^2 = \sum_{i=1}^{n} w_i[(x - x_i)^2 + (y - y_i)^2 + (z - z_i)^2], \qquad (1)$$

which is the weighted sum of squared distances of p from the points $p^{(1)}, \ldots, p^{(n)}$. Since the population center lies on the surface of the Earth, we impose the constraint that $x^2 + y^2 + z^2 = R^2$, where R is the radius of the Earth, which is approximately 6371 km. Hence, we minimize (1) subject to $x^2 + y^2 + z^2 = R^2$.

A standard approach for such a problem is to use the Method of Lagrange Multipliers. That is, we begin with the Lagrangian:

$$L(x, y, z, \lambda) = \sum_{i=1}^{n} w_i[(x - x_i)^2 + (y - y_i)^2 + (z - z_i)^2] + \lambda(x^2 + y^2 + z^2 - R^2)$$

Calculating the partial derivatives of $L(x, y, z, \lambda)$ and setting them to zero yields:

$$\frac{\partial L}{\partial x} = \sum_{i=1}^{n} 2w_i(x - x_i) + 2\lambda x = 0 \implies x - \overline{p_x} + \lambda x = 0$$

$$\frac{\partial L}{\partial y} = \sum_{i=1}^{n} 2w_i(y - y_i) + 2\lambda y = 0 \implies y - \overline{p_y} + \lambda y = 0$$

$$\frac{\partial L}{\partial z} = \sum_{i=1}^{n} 2w_i(z - z_i) + 2\lambda z = 0 \implies z - \overline{p_z} + \lambda z = 0$$

$$\frac{\partial L}{\partial \lambda} = x^2 + y^2 + z^2 - R^2 = 0$$

Solving for x, y and z from the first three equations and substituting into the last equation yields $\|\overline{p}\|^2/(\lambda+1)^2 = R^2$. Assuming that $\overline{p} \neq 0$ (which is satisfied by the cities in any of the countries considered in the data analysis below), one can show that the global minimum occurs at

$$p = (x, y, z) = R\overline{p}/\|\overline{p}\|, \qquad (2)$$

which is essentially the weighted mean of the locations $p^{(1)}, \ldots, p^{(n)}$ projected onto the surface of the Earth. The formula in (2) was also obtained in [1] using an argument from physics involving the force of gravity instead of a mathematical optimization approach. Now, given the Cartesian coordinates of the population center, we convert back to latitude and longitude (see next section), and then use the latitude to calculate its distance to the equator.

3   Converting Geographic to Cartesian Coordinates

To convert the geographic coordinates (latitude/longitude) of the cities of a country into Cartesian coordinates, we set up a 3-dimensional Cartesian coordinate system where the center of the Earth (represented as a sphere in Fig. 1) is at the origin, the positive x axis passes through the intersection of the equator and the prime meridian (0° longitude), the positive y axis passes through the equator at 90° E longitude, and the z axis passes through the north pole. Now consider a point (x, y, z) with spherical coordinates (R, Θ, Φ) on the surface of the sphere (which is the Earth) in Fig. 1. Note that the angle Φ indicates the longitude while 90° $- \Theta$ provides the latitude.

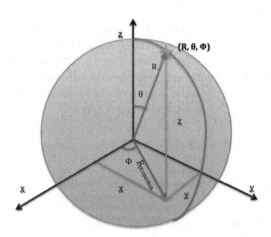

Fig. 1. Converting from spherical to Cartesian coordinates.

Converting from latitude (LAT) and longitude (LON) to Cartesian coordinates are given by (e.g., see [3])

$$x = R\cos(LON)\cos(LAT), \quad y = R\sin(LON)\cos(LAT), \quad z = R\sin(LAT) \quad (3)$$

Once we have the Cartesian coordinates (x, y, z) of the approximate population center of a country, we determine the distance of that country from the equator along the surface of the Earth. To do this, we calculate the latitude (LAT) of this population center using the third equation in (3), giving us

$$LAT = \sin^{-1}(z/R). \tag{4}$$

When $z > 0$, $LAT > 0$ and this corresponds to a northern latitude. When $z < 0$, Eq. (4) results in $LAT < 0$, which corresponds to a southern latitude. Moreover, when $y > 0$, we obtain an eastern longitude from

$$LON = \cos^{-1}\left(x/(R\cos(LAT))\right). \tag{5}$$

Also, when $y < 0$, we obtain a western longitude from Eq. (5). Finally, the distance of this population center to the equator along the surface of the Earth is then given by $|R \cdot LAT|$.

4   Predictive Models for Multiple Sclerosis Prevalence

To build machine learning regression models for MS prevalence, approximate geographic population centers are calculated for 71 countries using the method described in Sect. 2 and their distances to the equator are obtained. For each country, the population center is calculated using the locations and populations of the major cities in the country obtained from the 2015 data at http://simplemaps.com/resources/world-cities-data. Ideally, all inhabited regions of a country should be used in the calculation but gathering the data is difficult, so we only used the cities listed in the aforementioned website and these are the cities with the larger populations. Moreover, the countries included were required to have a Human Development Index (HDI) of at least 0.7 and these are countries with high or very high human development from the 2016 Human Development Reports (http://hdr.undp.org/en/composite/HDI). In underdeveloped countries, many people do not have access to adequate health care and there are usually not enough medical doctors who can accurately diagnose MS patients. For this reason, we restricted the data to only include more developed countries. Data on MS prevalence per 100,000 people for each country can be found in the Atlas of MS (https://www.msif.org/about-us/advocacy/atlas/), which provides data for 2013.

Smoking has been shown to be another risk factor for MS [4]. Data on the prevalence of tobacco smoking by country in 2012 can be found at http://www.who.int/gho/tobacco/use/en/. However, this website provides info on smoking prevalence separately for females and males, so we treat these as two separate predictor variables. Unfortunately, data on MS prevalence by gender does not seem to be available online, so the only response variable in this study is overall MS prevalence. Moreover, since obesity is another risk factor for MS, data on adult obesity prevalence by country was obtained from https://www.cia.gov/library/publications/the-world-factbook/rankorder/2228rank.html.

The data on MS prevalence by country, distance to the equator of the geographic population center of each country we included, prevalence of tobacco smoking among females and males ($\geq$15 years old), and adult obesity prevalence by country are summarized in Table 1. This table also includes the approximate geographic population center of each country obtained using the formula in Sect. 2 and the number of cities included in the calculation. All calculations and analyses were done using the Matlab and R software. More than 70 countries and 4600 cities are included.

To give an idea of the accuracy of our estimate of the geographic population centers, we compare our approximate population center for the US with the location obtained by the US Census Bureau. Table 1 shows that our approximate geographic population center of the US for 2015 and using about 50% of the top US cities is at 38.5 N and 93.4 W. The US Census Bureau reports that from the 2010 US Census this point is at 37.5 N and 92.2 W (https://www.census.gov/geographies/reference-files/time-series/geo/centers-population.html) and these coordinates are reasonably close. For other countries, most of which are smaller than the US, our approximations are expected to be more accurate.

One limitation of our models for MS prevalence is that only one location is used to represent a country whether it is large or small, and whether it is geographically connected or composed of disconnected regions. This is because the data on MS prevalence is only available by country. A large country such as the United States can have different MS prevalences in various regions, so a more accurate model can be developed by replacing these countries with its corresponding regions. Another limitation of the models is that the data used for MS prevalence was obtained in 2013 while the available data on cities populations used to approximate the geographic population center of a country was obtained in 2015. However, it is reasonable to assume that the geographic population center of a country only changed slightly between 2013 and 2015, and so, our analysis is still reasonably accurate.

We evaluate the performance of several machine learning methods for predicting MS prevalence. We begin with the 60 countries in Table 1 that have data for all predictor variables. The machine learning methods compared are first-order multiple regression, second-order (quadratic) multiple regression, k-NN regression, neural network, support vector regression (SVR) with a linear or radial basis function (RBF) kernel, bagged decision tree, and random forest. The main justification for choosing these methods is that they can work well even on small data sets. The experiments were performed in R using the well-known caret, neuralnet, e1071 and randomForest libraries.

The data set is randomly divided into about 80% training set and 20% hold out test set and the RMSE of the various methods are reported on the test set. For some of these methods, hyperparameters are tuned using 10-fold cross-validation on the training set. The values of the variables in the training set were normalized to be between 0 and 1 before training and tuning any of the methods. The random data split into training and test sets is repeated five times and the RMSE values on the test sets are given in Table 2.

Table 1. Data on MS prevalence for 2013 and distance to equator for 71 countries.

Country	MS Prev per 100K	Population center latitude	Population center longitude	Num of cities incl.	Dist. to equator (km)	Smoking prev. (Female)	Smoking prev. (Male)	Obesity prev.
Albania	22.0	41.1999 N	19.8546 E	26	4581.2153	8.20	52.10	21.70
Algeria	20.0	35.2415 N	3.4340 E	51	3918.6771	–	–	27.40
Argentina	18.0	33.3573 S	60.8847 W	158	3709.1669	20.80	31.80	28.30
Australia	95.6	33.3695 S	145.3812 E	224	3710.5159	14.70	18.40	29.00
Austria	140.0	48.0166 N	15.5455 E	10	5339.2074	35.90	37.40	20.10
Bahrain	35.0	26.2361 N	50.5831 E	1	2917.3253	6.90	40.10	29.80
Belgium	100.0	50.8583 N	4.4618 E	10	5655.1798	20.90	28.00	22.10
Bosnia and Herzegovina	60.0	44.1228 N	18.0328 E	6	4906.2276	31.20	49.00	17.90
Brazil	15.0	18.0197 S	45.8153 W	392	2003.7039	12.40	21.00	22.10
Bulgaria	39.1	42.8196 N	24.8807 E	18	4761.3256	30.40	45.20	25.00
Canada	291.0	48.1185 N	86.9176 W	255	5350.5346	14.20	19.50	29.40
China	1.5	33.4708 N	114.7541 E	400	3721.7796	2.00	49.00	6.20
Colombia	4.9	6.0015 N	74.8496 W	72	667.3398	6.90	17.20	22.30
Costa Rica	5.4	9.9841 N	84.1258 W	13	1110.1812	9.20	19.80	25.70
Croatia	59.0	45.1652 N	16.0857 E	10	5022.1367	32.30	39.40	24.40
Cuba	14.0	22.0251 N	79.7440 W	26	2449.0844	19.80	53.20	24.60
Cyprus	175.0	34.9745 N	33.2629 E	6	3888.9921	–	–	21.80
Czech Republic	160.0	49.8145 N	15.8318 E	12	5539.1145	28.90	38.00	26.00
Denmark	227.0	55.8244 N	11.5760 E	12	6207.3880	19.00	21.00	19.70
Ecuador	3.2	1.4357 S	79.3298 W	37	159.6454	3.60	15.10	19.90
Estonia	82.0	59.1694 N	25.4951 E	7	6579.3381	25.40	43.60	21.20
Finland	105.0	61.4206 N	24.9013 E	21	6829.6591	19.60	25.00	22.20
France	94.7	46.3525 N	2.2034 E	71	5154.1639	25.80	31.10	21.60
Germany	149.0	51.0375 N	9.6790 E	57	5675.1120	28.70	33.60	22.30
Greece	70.0	38.5508 N	23.5290 E	30	4286.6581	34.20	54.30	24.90
Hungary	176.0	47.3439 N	19.3124 E	20	5264.4010	26.50	34.50	26.40
Iceland	140.0	64.3301 N	21.4875 W	9	7153.1813	16.90	19.30	21.90
Iran	45.0	34.5908 N	51.9376 E	63	3846.3159	1.10	22.90	25.80
Ireland	140.0	53.1188 N	6.9273 W	15	5906.5361	23.50	24.20	25.30
Israel	62.5	32.1050 N	34.9360 E	6	3569.9126	20.10	41.70	26.10
Italy	113.0	42.7094 N	12.1020 E	56	4749.0664	19.80	29.10	19.90
Japan	8.0	35.7696 N	137.5099 E	69	3977.4025	11.30	36.30	4.30
Jordan	39.0	31.9598 N	35.9268 E	9	3553.7708	10.00	63.60	35.50
Kuwait	83.0	29.3480 N	47.9438 E	4	3263.3513	–	–	37.90
Latvia	89.9	56.7961 N	24.1019 E	6	6315.4413	24.40	50.50	23.60
Lebanon	45.0	33.9184 N	35.5611 E	6	3771.5539	29.40	43.00	32.00
Libya	5.9	32.0976 N	16.2701 E	35	3569.0856	–	–	32.50
Liechtenstein	100.0	47.1337 N	9.5167 E	1	5241.0310	–	–	–
Lithuania	78.0	55.1339 N	24.0519 E	5	6130.6097	21.90	40.50	26.30
Malaysia	2.0	3.9123 N	104.0480 E	32	435.0254	1.60	44.90	15.60
Malta	75.0	35.8997 N	14.5147 E	1	3991.8681	21.50	31.50	28.90
Mexico	15.0	22.0271 N	100.8245 W	189	2449.3050	7.60	23.30	28.90
Mongolia	10.0	48.1346 N	105.2430 E	24	5352.3273	5.80	49.30	20.60
Netherlands	88.0	52.1286 N	4.9413 E	14	5796.4398	24.70	27.50	20.40
New Zealand	73.0	39.1158 S	174.5093 E	33	4349.4749	–	–	30.80
Norway	160.0	60.9413 N	9.7771 E	34	6776.3648	24.90	25.50	23.10
Oman	22.0	22.7639 N	57.6269 E	10	2531.2252	1.00	18.90	27.00

(*continued*)

Table 1. (*continued*)

Country	MS Prev per 100K	Population center latitude	Population center longitude	Num of cities incl.	Dist. to equator (km)	Smoking prev. (Female)	Smoking prev. (Male)	Obesity prev.
Panama	5.2	8.8916 N	79.9693 W	14	988.7030	2.90	12.50	22.70
Peru	5.0	11.3222 S	76.4166 W	90	1258.9659	6.50	–	19.70
Poland	64.0	51.8228 N	19.2429 E	24	5762.4279	26.10	34.90	23.10
Portugal	56.2	39.5005 N	9.3737 W	24	4392.2607	14.00	32.40	20.80
Qatar	64.6	25.2866 N	51.5330 E	1	2811.7367	–	–	35.10
Romania	30.0	45.4898 N	25.4220 E	41	5058.2297	23.40	39.50	22.50
Russia	50.0	57.3366 N	53.7497 E	579	6375.5387	22.50	60.30	23.10
Saudi Arabia	30.0	24.0899 N	43.7906 E	31	2678.6706	3.00	26.30	35.40
Serbia	65.0	44.6478 N	20.5564 E	7	4964.6106	39.80	45.50	21.50
Singapore	3.9	1.2930 N	103.8558 E	1	143.7788	5.20	27.70	6.10
Slovenia	120.0	46.1839 N	14.8114 E	2	5135.4191	18.90	23.60	20.20
South Korea	3.5	36.6322 N	127.5379 E	28	4073.3188	4.40	51.70	4.70
Spain	100.0	39.9327 N	3.0047 W	49	4440.3147	28.00	33.70	23.80
Sri Lanka	4.9	7.5462 N	80.3465 E	14	839.0939	0.50	28.80	5.20
Sweden	189.0	58.9660 N	15.7853 E	34	6556.7254	23.00	22.30	20.60
Switzerland	110.0	46.9704 N	7.5781 E	28	5222.8727	20.90	28.30	19.50
Thailand	0.8	13.7918 N	100.7244 E	79	1533.5792	2.40	42.30	10.00
Tunisia	20.1	35.8846 N	10.1353 E	26	3990.1807	–	–	26.90
Turkey	55.0	39.5709 N	32.4274 E	84	4400.0858	13.50	42.70	32.10
UAE	55.0	25.0589 N	55.2565 E	8	2786.4210	–	–	31.70
UK	164.0	52.6524 N	1.3952 W	57	5854.6794	20.10	21.80	27.80
Uruguay	26.0	33.7291 S	56.2183 W	32	3750.5076	22.00	30.00	27.90
USA	135.0	38.5163 N	93.4226 W	769	4282.8195	16.30	21.00	36.20
Venezuela	6.9	10.0028 N	67.7788 W	50	1112.2606	–	–	25.60

Below are brief overviews of the various methods used. Here, y is the MS prevalence per 100K of a country, x_1 is the distance to the equator of the population center of a country (in km), x_2 and x_3 are the smoking prevalence among females and males (≥ 15 years old), respectively, and x_4 is the adult obesity prevalence in a country. The following methods are used to model y as a function of the input vector $x = [x_1, \ldots, x_p]^T$, where p is the number of predictors.

First-Order Multiple Linear Regression: $f(x) = \beta_0 + \sum_{i=1}^{p} \beta_i x_i$

Quadratic Regression: $f(x) = \beta_0 + \sum_{i=1}^{p} \beta_i x_i + \sum_{i<j} \beta_{i,j} x_i x_j + \sum_{i=1}^{p} \beta_{i,i} x_i^2$

k-NN Regression: Given a new instance $x = (x_1, \ldots, x_p)$, the predicted value is the average of the y values among the k nearest neighbors of $x = (x_1, \ldots, x_p)$ in the training set.

Neural Network for Regression: This is a method that is inspired by a network of biological neurons. Here, we use a multilayer network consisting of an input layer, one hidden layer, and an output layer. Since this is meant for regression, the input layer has $p = 4$ nodes and the output layer has one node. The number of nodes in the hidden layer is a hyperparameter to be determined by cross-validation. The hidden layer uses a sigmoid activation function while the output layer that uses a linear activation function. We use the neuralnet

Table 2. RMSE of machine learning methods on the hold out test sets.

Method	Test set 1	Test set 2	Test set 3	Test set 4	Test set 5
First-order linear regression	49.57	68.83	42.88	37.01	37.71
Quadratic regression	47.24	76.22	40.65	37.27	41.64
k-NN regression (tuned)	48.34	66.13	41.57	38.90	37.95
Neural network (tuned)	44.24	65.67	35.92	34.79	39.86
SVR - linear kernel (tuned)	55.73	70.67	42.05	42.65	43.47
SVR - RBF kernel (tuned)	46.85	60.67	36.55	38.67	40.45
Random forest (mtry = 2)	49.81	59.06	35.51	34.15	37.00
Bagged decision tree	50.49	57.21	36.14	37.55	41.39

library in R, which implements resilient backpropagation (RPROP) [11] with weight backtracking by default to train the neural network.

Random Forest [6] and Bagged Regression Tree [2]: This method constructs B regression trees from B bootstrapped training samples and averages the predictions. Moreover, for each regression tree, each time a split is considered, a random sample of m predictors is chosen from the set of all possible predictors and the split can only use one of these m predictors. A bagged regression tree is a special case of random forest where $m = p$. The hyperparameter m is called mtry in the randomForest R library.

Support Vector Regression (SVR) [14,15]: Let $x = [x_1, \ldots, x_p]^T$ be the vector of predictor variables. Suppose we wish to fit an ϵ-SVR model using n data points $(x^{(1)}, y^{(1)}), \ldots, (x^{(n)}, y^{(n)})$. We wish to find a function $f(x)$ that has at most ϵ deviation from the $y^{(i)}$'s and that is as flat as possible. First, consider the linear case: $f(x) = \beta^T x + \beta_0$, where $\beta = [\beta_1, \ldots, \beta_n]^T$ is the vector of coefficients. One way to ensure flatness is to minimize the 2-norm of β, giving rise to the convex optimization problem:

$$\min_\beta \tfrac{1}{2}\beta^T \beta$$

subject to
$$y^{(i)} - \beta^T x^{(i)} - \beta_0 \leq \epsilon, \quad i = 1, \ldots, n$$
$$\beta^T x^{(i)} + \beta_0 - y^{(i)} \leq \epsilon, \quad i = 1, \ldots, n$$

The above formulation assumes that there exists a linear function $f(x) = \beta^T x + \beta_0$ that approximates all data points $(x^{(i)}, y^{(i)})$ with precision ϵ. However, this is sometimes not the case, so we allow for some errors. That is, we introduce slack variables ξ_i and ξ_i^* to deal with the possibly infeasible constraints. This results in the following optimization formulation [15]:

$$\min_\beta \tfrac{1}{2}\beta^T\beta + C\sum_{i=1}^{n}(\xi_i + \xi_i^*)$$

subject to
$$y^{(i)} - \beta^T x^{(i)} - \beta_0 \le \epsilon + \xi_i, \quad i = 1,\ldots,n$$
$$\beta^T x^{(i)} + \beta_0 - y^{(i)} \le \epsilon + \xi_i^*, \quad i = 1,\ldots,n$$
$$\xi_i, \xi_i^* \ge 0, \quad i = 1,\ldots,n$$

Here, the constant C quantifies the trade-off between the flatness of $f(x)$ and the penalty on the observations that lie outside the ϵ-margin. The above problem is typically solved using its dual formulation. See [14] for more details.

The value of k for k-NN regression was obtained using LOOCV on each training set. The number of hidden nodes in the neural network was tuned on each training set using 10-fold cross-validation. SVR with the linear and RBF kernels were also tuned using 10-fold cross-validation on the training set. For the random forest, we used 100 trees and the mtry parameter was set to 2. Finally, the bagged decision tree also used 100 trees.

Table 2 shows the RMSE of the various machine learning methods on various test sets. The best and second best methods for each test set are highlighted with solid and dashed boxes, respectively. The results show that random forest with mtry = 2 generally provided the smallest RMSEs on the test sets followed by the neural network. In particular, note that first-order and second-order (quadratic) multiple regression models and also SVR and bagged decision tree did not perform as well as the others. Also, the worst RMSEs were generally obtained by SVR that uses a linear kernel. These results indicate that the relationship between MS prevalence and the predictor variables is nonlinear and that a quadratic model is not enough to capture this nonlinearity. In this case, random forest and the neural network provided better predictive performance.

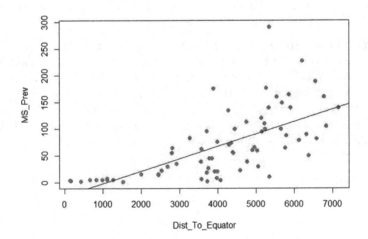

Fig. 2. MS prevalence (per 100K) in a country vs. distance to equator (km)

5   Linear Regression for Multiple Sclerosis Prevalence

In the previous section, machine learning models such as random forest and neural network yielded better predictive performance compared to the other models. However, these models have issues with interpretability. To get a sense of the relationship between MS prevalence and the different predictor variables, we also look at first-order linear regression models involving MS prevalence.

Figure 2 shows a scatterplot and the least-squares regression line for the relationship between MS prevalence and the distance of the population center of a country from the equator. Here, the data points represent the 71 countries in Table 1. Moreover, we performed linear regression in R using all available data points. The scatterplot shows a positive linear relationship with a statistically significant correlation coefficient of 0.66, which is moderately strong. This numerical result is consistent with the observation that countries farther from the equator have higher levels of MS prevalence (e.g., [12]). Moreover, the scatterplot shows an outlier corresponding to Canada, which is known to have the highest MS prevalence in the world. Studies of MS in Canada have indicated strong evidence for other environmental factors affecting MS prevalence [5]. If we set aside this outlier, the correlation coefficient increases to 0.70, indicating a stronger linear relationship between MS prevalence and distance from the equator.

The linear regression model for predicting MS prevalence (per 100,000) of a country (y) given the distance of its population center from the equator in km (x_1) is given by $\hat{y} = -25.63 + 0.023x_1$. The slope of this linear model suggests that MS prevalence (per 100,000) increases by 0.023 on average for every kilometer traveled farther from the equator. The intercept is meaningless since MS prevalence cannot be negative. The R^2 value for the linear model suggests that 44% of the total variation in MS prevalence can be explained by distance of the population center to the equator. The p-value for testing the significance of the slope is 2.84×10^{-10}, so the data provide strong evidence that distance from the equator is statistically useful in predicting MS prevalence. If we set aside the outlier (Canada), the regression model remains statistically significant, R^2 substantially increases to 0.486, and the above slope of the linear model changes only slightly to 0.022. Moreover, we performed standard diagnostic procedures to check the assumptions of a linear model. The residual plot showed a generally random pattern, which is adequate. The histogram of residuals and the normal probability plot show a distribution that is approximately Normal, especially when the outlier (Canada) is set aside. Hence, there does not seem to be any serious violations of the standard regression assumptions.

Further analysis showed a statistically significant moderate positive correlation of 0.49 between MS prevalence and smoking prevalence among females. There are also statistically significant correlations between MS prevalence and smoking prevalence among males (-0.29) and between MS prevalence and adult obesity prevalence (0.27). However, these are both weak and the correlation between MS prevalence and smoking prevalence among males is negative, which is not expected and might indicate the presence of a confounding variable. Using the above notation, a first order model for MS prevalence (per 100,000) of a country (y) given the other predictor variables is $\hat{y} = 21.36 + 0.022x_1 + 0.98x_2 - 2.18x_3 + 0.64x_4$. This model has a p-value of 9.85×10^{-12} for the global F-test, indicating a statistically significant model, and has an adjusted R^2 of 61.6%.

6   Summary and Conclusions

This article built novel machine learning and statistical models for a country's MS prevalence using four predictor variables: the distance of a country from the equator along the surface of the Earth, the prevalence of smoking in that country among females and males ≥ 15 years of age, and the adult obesity prevalence. Each country is represented by its approximate geographic population center obtained by solving a constrained optimization problem using the method of Lagrange multipliers. Our approach can also be used to predict the MS prevalence in a subregion of a country given the same predictor variables.

In this study, we used data from more than 70 countries and over 4600 cities. Comparison of several machine learning methods on test sets indicated that random forest and neural network yielded better RMSE values than the other methods, including first and second-order multiple regression models, SVR and bagged decision tree. The results suggest a nonlinear relationship between MS prevalence and the predictor variables. Moreover, we performed linear regression involving MS prevalence of a country and its distance to the equator. Our analysis yielded a sample correlation of 0.66, thereby confirming the observation that MS prevalence increases with distance from the equator. The resulting linear regression model is statistically significant and did not show any serious departures from the standard regression assumptions, has an R^2 value of 0.44 and suggests that MS prevalence (per 100,000) increases by 0.023 on average for every kilometer traveled farther from the equator. Further analysis yielded a statistically significant first order multiple regression model for MS prevalence in terms of the predictor variables with an adjusted R^2 of 0.616.

References

1. Aboufadel, E., Austin, D.: A new method for computing the mean center of population of the United States. Prof. Geogr. **58**(1), 65–69 (2006)
2. Breiman, L.: Bagging predictors. Mach. Learn. **24**(2), 123–140 (1996)
3. Brundritt, R.: Conversion between spherical and cartesian coordinates systems (2020). https://rbrundritt.wordpress.com/2008/10/14/conversion-between-spherical-and-cartesian-coordinates-systems/. Accessed 3 Jun 2020
4. Degelman, M.L., Herman, K.M.: Smoking and multiple sclerosis: a systematic review and metaanalysis using the Bradford Hill criteria for causation. Multiple Scler. Relat. Disord. **17**, 207–216 (2017)
5. Ebers, G.C.: Environmental factors and multiple sclerosis. Lancet Neurol. **7**(3), 268–277 (2008)

6. Ho, T.K.: The random subspace method for constructing decision forests. IEEE Trans. Pattern Anal. Mach. Intell. **20**(8), 832–844 (1998)

7. Langer-Gould, A., et al.: MS sunshine study: sun exposure but not vitamin D is associated with multiple sclerosis risk in Blacks and Hispanics. Nutrients **10**(3), 268 (2018)

8. Mokry, L.E., Ross, S., Timpson, N.J., Sawcer, S., Smith, G.D., Richards, J.B.: Obesity and multiple sclerosis: a Mendelian randomization study. PLOS Med. **13**(6), e1002053 (2016)

9. Munger, K.L., et al.: Vitamin D intake and incidence of multiple sclerosis. Neurology **62**(1), 60–65 (2004)

10. National Multiple Sclerosis Society: What causes MS? (2020). http://www.nationalmssociety.org/What-is-MS/What-Causes-MS. Accessed 3 Jun 2020

11. Riedmiller, M., Braun, H.: A direct adaptive method for faster backpropagation learning: the RPROP algorithm. In: Proceedings of the IEEE International Conference on Neural Networks (ICNN), pp. 586–591 (1993)

12. Simpson, S., Blizzard, L., Otahal, P., Van der Mei, I., Taylor, B.: Latitude is significantly associated with the prevalence of multiple sclerosis: a meta-analysis. J. Neurol. Neurosurg. Psychiatry **82**(10), 1132–1141 (2011)

13. Simpson, S., et al.: Sun exposure across the life course significantly modulates early multiple sclerosis clinical course. Front. Neurol. **9**, 16 (2018)

14. Smola, A.J., Schölkopf, B.: A tutorial on support vector regression. Stat. Comput. **14**, 199–222 (2004)

15. Vapnik, V.: The Nature of Statistical Learning Theory. Springer, New York (2000). https://doi.org/10.1007/978-1-4757-3264-1

Robust and Sparse Support Vector Machines via Mixed Integer Programming

Mahdi Jammal[1,2]($\boxtimes$), Stephane Canu[1]($\boxtimes$), and Maher Abdallah[2]($\boxtimes$)

[1] INSA de Rouen, Saint-Étienne-du-Rouvray, France
mahdi.jammal@insa-rouen.fr, stephane.canu@insa-rouen.fr
[2] Lebanese University, Beirut, Lebanon
maher.abdallah@ul.edu.lb

Abstract. In machine learning problems in general, and in classification in particular, overfitting and inaccuracies can be obtained because of the presence of spurious features and outliers. Unfortunately, this is a frequent situation when dealing with real data. To handle outliers proneness and achieve variable selection, we propose a robust method performing the outright rejection of discordant observations together with the selection of relevant variables. A natural way to define the corresponding optimization problem is to use the ℓ_0 norm and recast it as a mixed integer optimization problem (MIO) having a unique global solution, benefiting from algorithmic advances in integer optimization combined with hardware improvements. We also present an empirical comparison between the ℓ_0 norm approach, the 0–1 loss and the hinge loss classification problems. Results on both synthetic and real data sets showed that, the proposed approach provides high quality solutions.

Keywords: Robust classification · Sparse classification · SVM · Mixed integer programming

1 Introduction

In support vector machine (SVM) classification, the natural way to quantify the performance of a classifier is via the 0–1 loss. This loss is non-convex and considered to be $\mathcal{NP}$-Hard. To this end, the hinge loss, which is convex, was introduced for the first time with [1]. Since then, it has become one of the most popular classifiers. An important reason behind the popularity of SVM is its significant empirical success in various applications such as data mining, engineering and bio-informatics [2]. In Fig. 1, the difference between the hinge-loss and the 0–1 loss is shown.

Considering training examples $x_i \in \mathbb{R}^p$ with their respective labels $y_i \in \{-1, 1\}$, $i = 1, \ldots, n$. The main goal of SVM is to find a hyperplane (classifier) by introducing hard margins for separable data and soft margins for linearly non-separable data, the purpose of which is to separate data as far as possible from the hyperplane. A decision hyperplane can be defined by an intercept term b and

© Springer Nature Switzerland AG 2020
G. Nicosia et al. (Eds.): LOD 2020, LNCS 12566, pp. 572–585, 2020.
https://doi.org/10.1007/978-3-030-64580-9_47

a decision hyperplane normal vector w which is perpendicular to the hyperplane. This vector is commonly referred to, in the machine learning, literature as the weight vector. To choose among all the hyperplanes that are perpendicular to the normal vector, we specify the intercept term b. Because the hyperplane is perpendicular to the normal vector, all points x on the hyperplane satisfy $w^T x + b = 0$. Let the margin be defined as the distance from the hyperplane to the closest point across both classes. It can be shown that the width of the margin is equal to $\frac{2}{||w||_2}$, thus maximizing this width is equivalent to minimizing the norm $||w||_2^2$ (or $\frac{1}{2}||w||_2^2$). To obtain the optimal hyperplane, one should solve the following optimization problem:

$$\min_{w,\xi} \frac{1}{2}||w||_2^2 + C \sum_{i=1}^{n} \xi_i$$
$$\text{s.t. } y_i(w^T x + b) \geq 1 - \xi_i \ i = 1\ldots n \tag{1}$$
$$\xi_i \geq 0 \qquad\qquad i = 1\ldots n$$

where ξ is a slack variable and C is a parameter controlling the trade-off between a large margin and a less constrained violation. The dual problem can be formulated through the use of Lagrange multipliers:

$$\max_{\alpha} C \sum_{i=1}^{n} \alpha_i - \frac{1}{2} \sum_{i=1}^{n} \sum_{i=1}^{n} \alpha_i \alpha_j y_i y_j x_i^T x_j$$
$$\text{s.t } 0 \leq \alpha_i \leq C \qquad i = 1,\ldots,n$$
$$\sum_{i=1}^{n} \alpha_i y_i = 0$$

Both the primal and dual are convex quadratic optimization problems. Because the dual problem has fewer decision variables, and the majority of these variables tend to be equal to zero, it is typically the problem solved in practice [3].

While algorithmic advances in integer optimization combined with hardware improvements have resulted in an astonishing 200 billion factor speedup in solving Mixed Integer Optimization (MIO) problems [4], this rapid development of MIO enabled [5] to reformulate the 0–1 loss classification problem as a mixed integer optimization problem and use it to solve small-scale classification problems.

In addition to all benefits listed above, SVM suffers from the existence of outliers and the existence of irrelevant features (especially for high dimensional data sets). Indeed, in the past three decades, the dimensionality of the data involved in machine learning and data mining tasks has increased explosively. Data with extremely high dimensionality has presented serious challenges to existing learning methods [3,6]. With the presence of a large number of features, a learning model tends to overfit, resulting in their performance degenerates. Feature selection for SVM has been widely studied. For example, [7] introduced an algorithm based upon finding the features which minimize bounds on the leave-one-out error. The search can be efficiently performed via gradient descent. [8] proposed an approach that takes existing theoretical bounds on the generalization error for SVMs instead of performing cross-validation. This is computationally faster than k-fold cross-validation. Additionally, in general, the error bounds have a higher bias than cross-validation in practical situations they often have a lower variance and can thus reduce the overfitting of the wrapper algorithm. A convex energy-based framework to jointly perform feature selection and SVM parameter learning for linear and non-linear kernels was proposed by [9]. They also showed the equivalence between their approach and the ℓ_1 SVM. In a recent work, [10] developed an efficient method for sparse support vector machines with ℓ_0 norm approximation. The proposed method approximates the ℓ_0 minimization through solving a series of ℓ_2 optimization problems, which can be formulated with dual variables.

Furthermore, in practical applications, training samples are often contaminated by noise and some even have wrong labels [11]. These are usually known as outliers. In order to mitigate the effects of outliers, different approaches have been proposed to improve the robustness of SVM. [12] suggested to use the distance between each training sample and its class center to calculate an adaptive margin so as to reduce the influence of outliers. Weighted SVM (WSVM) or fuzzy SVM was also proposed to deal with outliers [13–15]. In WSVM, different weights are assigned to different training samples which can show their importance in the training data set. Several weight functions have been proposed [13–15]. [16] presented a novel combinatorial technique, which was called random gradient descent (RGD) tree, to identify and remove outliers in SVM and developed a new algorithm called RGD-SVM. [17] proposed the re-scaled hinge loss which is a monotonic, bounded and non-convex loss. Introducing a Ramp Loss function into one-class SVM optimization to reduce outliers influence was suggested by [18]. Then the outliers are identified and removed from the training set. The final classification surface is obtained on the remaining training samples. [19] introduced a new robust loss function (called L_q loss) based on the concept of quantile and correntropy, which can be seen as an improved version of quantile loss function. To deal with label outliers, [20] introduced a variable $\Delta y_i \in \{0, 1\}$ where 1 indicates that the label was incorrect and has in fact been flipped, and 0 indicates that the label was correct. They also introduced a variable Δx_i to deal with uncertainty of features. They proposed the use of mixed integer optimization problems to solve the obtained problem. However, the algorithm is not sparse.

To obtain a sparse and robust least squares support vector machines (SR-LSSVM), [21] proposed the SR-LSSVM algorithm to obtain a sparse solution of the robust least squares SVM (R-LSSVM) [22,23] by applying a low-rank approximation of the kernel matrix.

Contributions:

In this paper, we address the problem of both feature selection and outlier detection using the ℓ_0 norm. We summarize our contributions in this paper below:

- We present an approach jointly performing feature selection and outlier detection for SVM classification;

- We propose to recast the presented problem as a mixed integer optimization problem which allows the use of efficient solvers (Gurobi) to solve it. Note that the sub-optimality (near-optimality) of the obtained solution is guaranteed even if we terminate the algorithm early;
- We present computational results on both real and synthetic datasets and compare the proposed approach with the classical 0–1 loss and hinge loss classification problems. The results show that the proposed approach provides high quality solutions.

The remainder of the paper is organized as follows. In Sect. 2, we present our approach for variable selection and outliers detection using the ℓ_0 norm together with its formulation as a mixed integer optimization problem allowing to obtain the global solution. Section 3 reports empirical evidence on synthetic data sets, while empirical results on real data sets were presented in Sect. 4. Finally, the paper is concluded in Sect. 5.

2   Linear Binary Classification

We have n training points, where each input x_i has p attributes and is in one of two classes $y_i \in \{-1, 1\}$. Under linear assumption, the classification function can be expressed as $f(x, w) = w^T x + b$. The goal is to predict the target class $\hat{y} \in \{-1, 1\}$ which is defined by:

$$\hat{y}_i = \begin{cases} 1 & f(x_i, w) \geq 0 \\ -1 & f(x_i, w) < 0 \end{cases} \tag{2}$$

The natural way to quantify the performance of a classifier is using the 0–1 loss function: for a given instance x and a true binary label $y \in \{-1, 1\}$, we incur a loss of 1 if $sign(yf) < 0$, and 0 otherwise, that is:

$$\mathbb{1}[y \neq sign(f(x, w))] = \begin{cases} 1 & \text{if } y \neq sign(f(x, w)) \\ 0 & otherwise \end{cases} \tag{3}$$

The 0–1 loss classification problem can be written as

$$\min \sum_{i=1}^{n} \mathbb{1}[y_i \neq sign(f(x_i, w))] \tag{4}$$

Problem (4) is non-convex, to this end it has been replaced by a convex surrogate such as the hinge loss. However, advances in integer optimization resulted an impressive speedup in solving mixed integer optimization problems (MIO). To this end, [5] proposed to recast the problem of 0–1 loss classification (4) as a mixed integer optimization problem, that is:

$$\min \sum_{i=1}^{n} l_i$$
$$\text{s.t. } y_i(w^T x_i + b) \geq 1 - M l_i \tag{5}$$
$$l \in \{0, 1\}^n$$

where M is a sufficiently large constant. Since this formulation suffers from infinite number of optimal solutions and it lacks from the generalization ability, [5] proposed a maximum margin 0–1 loss classifier defined as follows:

$$\min \sum_{i=1}^{n} l_i + Cw^T w$$
$$\text{s.t. } y_i(w^T x_i + b) \geq 1 - Ml_i$$
$$l \in \{0,1\}^n$$

(6)

where C is a positive parameter, and showed the efficiency of this approach for small-scale classification problems.

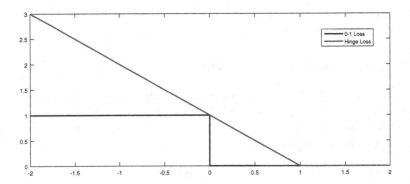

Fig. 1. Illustration of the hinge loss which is a convex surrogate to the 0–1 loss. The 0–1 loss is shown in blue and the hinge loss is shown in red. (Color figure online)

2.1   Introducing Binary Variables

Variable selection involves the ℓ_0 norm function to count the number of useful variables. This counting function can be represented by introducing p binary variables $z_j \in \{0,1\}$ such that

$$\|w\|_0 = \sum_{j=1}^{p} z_j \quad \text{and} \quad z_j = 0 \Leftrightarrow w_j = 0.$$

Different approaches can be used to force $z_j = 0 \Leftrightarrow w_j = 0$ into an optimization problem, such as:

1. Replace w_j by $z_j w_j$ for $j = 1, \ldots, p$,

2. Set $|w_j|(1 - z_j) = 0$ for $j = 1, \ldots, p$ or $\sum_{j=1}^{p} |w_j|(1 - z_j) = 0$,

3. Use a big-M constraint, $|w_j| \leq M_v z_j$ for $j = 1, \ldots, p$ and for some fixed constant M_v large enough (such as $M_v \geq \max_j |w_j^\star|$, $w_j^\star$ being the solution of the optimization problem),

4. Treat $z_j = 0 \Leftrightarrow w_j = 0$ as logical implications (also called indicator constraints or special ordered set SOS-1). Note that this kind of logical implication can be efficiently handled in a branch-and-bound procedure for MIO problems.

We now discuss and give a short overview of the advantages and drawbacks of each approach. The two first approaches involve nonlinear interaction terms between binary and continuous variables. Their interest lies in the possibility of obtaining interesting continuous relaxations. The main advantage of the big M method (approach 3) is that it brings only linear inequality constraints but the value of the M term needs to be chosen carefully since it shows a great deal of practical influence on the solver performance. Logical implications (approach 4) have the advantage of avoiding these types of problems, as they do not rely on a separate constant value. However, they tend to have weaker relaxations, a condition which may lead to longer solve times in a model. In this paper we will use the third approach for our implementation.

2.2 Our Approach

To deal with the problem of outlier detection, we propose to add a variable τ so that Problem (1) becomes:

$$
\begin{aligned}
\min_{w, \xi, \tau} \quad & \tfrac{1}{2}\|w\|_2^2 + C \sum_{i=1}^{n} |\xi_i - \tau_i| \\
\text{s.t.} \quad & y_i(w^T x + b) \geq 1 - \xi_i \quad i = 1 \ldots n \\
& \|w\|_0 \leq k_v \\
& \|\tau\|_0 \leq k_o \\
& \xi_i \geq 0 \quad\quad\quad\quad i = 1 \ldots n
\end{aligned}
\tag{7}
$$

where the ℓ_0 norm of a vector w counts the number of nonzeros in w. In this formulation, k_v represents the number of features to be selected while k_o represents the number of outliers to be detected. We note that in Problem (7), $\tau(i) \neq 0$ means that the observation "i" is an outlier. In Fig. 2 we can see the effect of an outlier on the hinge-loss classifier. Furthermore, it can be also seen that the MIO approach can still recover the true classifier even in the presence of the outlier.

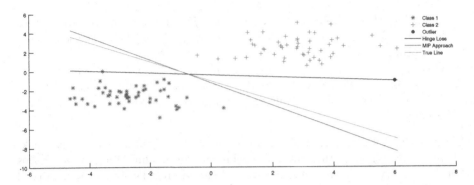

Fig. 2. Example of synthetically generated data in two dimensions to show the effect of an outlier on the Hinge-loss classification. The true generating hyperplane in green, the Hinge-loss hyperplane in blue and the MIO approach hyperplane in red. (Color figure online)

2.3 A MIO Formulation

To solve (7) exactly, we recast it as a mixed integer optimization problem. Two binary variables z and t are introduced to control the sparsity levels for w and τ respectively. The MIO formulation of (7) is as follows:

$$
\min_{w,\xi,\tau,t,z,b} \tfrac{1}{2}\|w\|_2^2 + C\sum_{i=1}^{n}|\xi_i - \tau_i|
$$

$$
\text{s.t.} \quad \sum_{j=1}^{p} z_j \leq k_v
$$

$$
|w_j| \leq z_j M_v \qquad j = 1\ldots p \qquad (8)
$$

$$
\sum_{i=1}^{n} t_i \leq k_o
$$

$$
|\tau_i| \leq t_i M_o \qquad i = 1\ldots n
$$

$$
y_i(w'x_i + b) \geq 1 - \xi_i \qquad i = 1\ldots n
$$

$$
\xi_i \geq 0 \qquad i = 1\ldots n
$$

where $w \in \mathbb{R}^p$, $\tau, \xi \in \mathbb{R}^n$, $t \in \{0,1\}^n$, $z \in \{0,1\}^p$ and $b \in \mathbb{R}$.

When $k_v = 0$ and $k_o = 0$, no feature selection nor outlier detection are performed, the resulting problem is the classical hinge loss classification problem. In the above formula, M_v and M_o are two big values.

2.4 Solving the Problem Using Gurobi

To overcome the absolute value in the objective function, we introduce two new variables α^+ and α^-, such that $\xi_i - \tau_i = \alpha_i^+ - \alpha_i^-$, and $|\xi_i - \tau_i| = \alpha_i^+ + \alpha_i^-$, where $\alpha_i^+, \alpha_i^- \geq 0$ for $i = 1\ldots n$. Then the new obtained problem is as follows:

$$\min_{w,\xi,\tau,t,z,b} \frac{1}{2}\|w\|_2^2 + C\sum_{i=1}^{n}(\alpha_i^+ + \alpha_i^-)$$

$$\text{s.t.} \quad \sum_{j=1}^{p} z_j \leq k_v$$

$$|\beta_j| \leq z_j M_v \qquad\qquad j = 1 \ldots p$$

$$\sum_{i=1}^{n} t_i \leq k_o \qquad\qquad\qquad\qquad (9)$$

$$|\tau_i| \leq t_i M_o \qquad\qquad i = 1 \ldots n$$

$$y_i(w'x_i + b) \geq 1 - \xi_i \qquad i = 1 \ldots n$$

$$\xi_i - \tau_i = \alpha_i^+ - \alpha_i^- \qquad i = 1 \ldots n$$

$$\xi_i \geq 0 \qquad\qquad\qquad i = 1 \ldots n$$

$$\alpha_i^+ \geq 0 \qquad\qquad\qquad i = 1 \ldots n$$

$$\alpha_i^- \geq 0 \qquad\qquad\qquad i = 1 \ldots n$$

2.5 Computational Cost

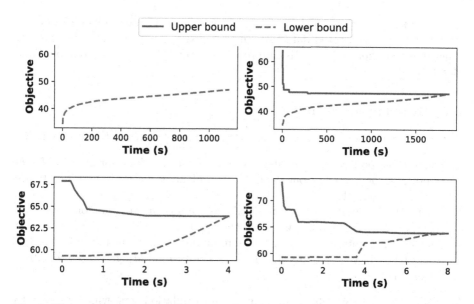

Fig. 3. The evolution of the MIO for the breast cancer prognostic data set with $n = 194$ and $p = 33$. The top panel shows the evolution of upper and lower bounds with time when $k_o = 5\%$, while the bottom panel shows the evolution of upper and lower bounds with time when $k_o = 2.5\%$. The left panel shows the evolution of upper and lower bounds with time when $k_v = p$, while the right panel shows the evolution of upper and lower bounds with time when $k_v = 0.8p$. For all panels, $C = 1$.

In Fig. 3, the left panel shows the evolution of upper and lower bounds with time when $k_v = p$, while the right panel shows this evolution when $k_v = 0.8p$.

By comparing the left and the right panels, we can see that the computational time increased from 1200 s to 1800 s (top panel) and from 4 s to 8 s (bottom panel). This means that the value of k_v has an influence on the computational cost.

Similarly, the top panel shows the evolution of upper and lower bounds with time when $k_o = 5\%$, while the bottom panel shows this evolution when $k_o = 2.5\%$. A simple comparison between the top and the bottom panels sheds the light on how much increasing the value of k_o (percentage of outliers to detect) will increase the time needed to certify optimality. Indeed, decreasing k_o from 5% to 2.5% resulted a significant decrease of the computational cost, that is from 1200 s to only 4 s, and from 1800 s to only 8 s.

We note that optimal solutions are found in a few seconds in the top panel examples, but it takes 20–30 min to certify optimality via the lower bounds. We also note that the computational time depends on the value of C and the big-M values.

3   Experiments on Synthetic Data Sets

To report the robustness of the proposed approach, we evaluated its performance on synthetically generated data sets. In these experiments, we run the classical hinge-loss classifier and the MIO approach to recover the separating hyperplane classifier.

3.1   Experimental Setup

The experiment uses data in $\mathbb{R}^2$. The data are generated synthetically as follows:

1. Twenty-five points are generated as multivariate random normal, $N(3.5e, I)$ where e is the vector of ones and I is the identity matrix. These points are given the label $+1$.
2. Twenty-five points are generated as multivariate random normal, $N(-3.5e, I)$. These points are given the label -1.
3. Ten outlier points are introduced as multivariate random normal $N(0, 3I)$, where 0 is the vector of zeros. The labels are randomly generated as either -1 or $+1$.

We split the data 75%/25% into training and validation sets, which we used to tune the parameters for both methods. To create the test set, we generated 1000 points in the same way as items 1 and 2 above.

An example of a data set generated according to this procedure is shown in Fig. 4. By the symmetry of this data generation process, we can see that the true hyperplane separating the two clusters of points is given by $e^T x = 0$. The goal of the experiment is to show how closely the two methods can recover the truth in the data. We are interested in the following two measures:

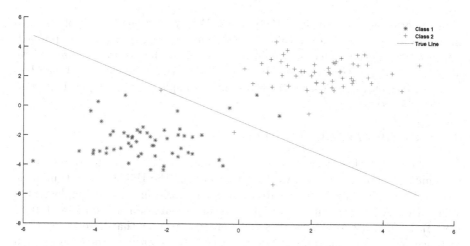

Fig. 4. Example of synthetically generated data in two dimensions alongside the true generating hyperplane

– Accuracy: We measure and evaluate the out-of sample accuracy of the trained classifiers on the test set, defined by:

$$Accuracy = \frac{TP + TN}{TP + FP + TN + FN}$$

where TP and TN represent the quantity of correct positive and correct negative samples, respectively; FN and FP respectively represent the number of misclassification negative and positive samples. The higher the values of the Accuracy, the better the model is.

– Similarity: To evaluate the ability of each method to recover the truth in the data, we measure the cosine of the angle between the separating hyperplane generated by the methods and the true hyperplane.

We recall that the cosine of the angle α between two vectors u and v is given by:

$$cos(\alpha) = \frac{u.v}{||u|| \times ||v||}$$

3.2 Results

Table 1. Performance results for synthetic data experiments

	Accuracy	Similarity
Hinge loss	96.93	0.9428
MIO approach	**97.85**	**0.9813**

This experiment was repeated 1000 times. We present the means of the two measures for each method in Table 1. The results show that the MIO approach improved the performance of classification. In fact, the accuracy increased by about 1% and that it recovered the truth better than the classical hinge loss classifier (cosine value closer to 1 means smaller angle between hyperplanes and thus better recovery).

4   Experiments on Real Data Sets

To evaluate the effectiveness of the proposed method, we carry out numerical simulations on twelve real-world data sets from the University of California Irvine (UCI) Machine Learning Repository. All experiments are implemented using MATLAB-Gurobi interface. The experiment environment is: PC with Intel Core i7 4700MQ (2.40 GHz) with 8 GB memory. We note that for each problem instance, we used a time limit of 15 min for Gurobi to optimize the classification problem.

We recall that to obtain: the hinge-loss classification problem solution we solved Problem (1), the 0–1 loss problem solution we solved Problem (6). The solution of the MIO approach was found by solving Problem (7).

4.1   Experimental Setup

To evaluate the performance of the proposed approach, we considered two scenarios:

1. In the first scenario, 10% of the training and validation sets labels were randomly flipped. The aim is to study the robustness of the mixed integer programming approach.
2. In the second scenario, we wanted to mimic real-world setting, hence data sets were not modified.

For both scenarios, each data set was normalized using the min-max scaling and was split randomly into three parts: the training set (60%), the validation set (20%), and the testing set (20%). The training set was used to train each classifier for a variety of combinations of input parameters. For each combination of parameters, the accuracy on the validation set was calculated, and this was used to select the best combination of parameters for each classifier. Finally, the classifier was trained by using these best parameters on the combined training and validation sets, before reporting the out-of-sample accuracy on the testing set. All methods were trained, validated, and tested on the same random splits, and computational experiments were repeated five times for each data set with different splits. For each data set and classification method, we report the average out-of-sample accuracy across all five splits. C was chosen from the set $[10^{-4}, 10^{-3}, \ldots, 10^4]$, k_v was set to $k_v = p$ for the first scenario, and chosen from the set $[p, 0.8p, 0.6p]$ for the second scenario that is no feature selection was performed, 80% and 60% of features are selected respectively. k_o was chosen from the set $[0.025n, 0.05n, 0.1n]$ that is 2.5%, 5% and 10% of outliers to be detected respectively.

Table 2. Out of sample accuracy averaged across five seeds for each classification method on all data sets. (first scenario)

	n	p	Hinge loss	0–1 loss	MIO approach
Arrythmia	68	280	52.31	**64.62**	**64.62**
Breast cancer coimbra	116	9	65.22	60.87	**72.17**
Breast cancer prognostic	194	33	63.16	78.42	**84.74**
Connections bench sonar	208	60	65.17	72.20	**75.61**
Fertility	100	9	64.00	78.00	**86.00**
Ionosphere	351	33	63.71	84.86	**85.43**
Monks-1	124	6	65.83	**72.50**	71.67
Monks-2	169	6	**65.45**	63.03	60.51
Monks-3	122	6	82.50	77.21	**83.33**
Pima	768	8	56.60	68.37	**76.73**
Spect heart	80	22	65.00	72.50	**75.92**
Spectf heart	80	44	78.75	79.25	**81.25**

Table 3. Out of sample accuracy averaged across five seeds for each classification method on all data sets. (second scenario)

	n	p	Hinge loss	0–1 loss	MIO approach
Arrythmia	68	280	70.76	69.23	**81.53**
Breast cancer coimbra	116	9	**73.04**	70.43	70.43
Breast cancer prognostic	194	33	76.84	78.94	**81.05**
Connections bench sonar	208	60	72.19	**76.58**	**76.58**
Fertility	100	9	86.00	86.00	**88.00**
Ionosphere	351	33	84.28	82.57	**85.14**
Monks-1	124	6	62.50	**67.51**	64.98
Monks-2	169	6	**61.21**	59.79	**61.21**
Monks-3	122	6	79.16	82.50	**82.78**
Pima	768	8	78.30	78.21	**78.82**
Spect heart	80	22	63.75	67.50	**70.83**
Spectf heart	80	44	70.00	71.25	**77.50**

4.2   Results and Discussion

Tables 2 and 3 present the means of the accuracy for each method. We note that n stands for training points and p for attributes. The robustness of the proposed approach is shown in Table 2. In fact, it had a superior performance on 9 data sets, and a tie for one data set, when 10% of labels were flipped. An important remark is that no variable selection was performed during this scenario so the comparison between the MIO approach and the hinge-loss classification is based only on the robustness of the MIO approach. This side by side comparison sheds the light on the significant improvement obtained with the MIO approach.

The second scenario is closer to the real world setting. The data sets are taken without any change or modification. From Table 3, it is clear that the prediction accuracy of our approach is higher than those of the compared algorithms for almost all datasets. We can remark a significant accuracy improvement for some datasets. For example, we obtained about 11% improvement for Arrythmia dataset. In general, it can be seen that the proposed approach provides high quality solutions. We also note that the pairwise comparison of the 0–1 classification against the hinge loss classification shows that none of the two losses dominates the other. Indeed each loss showed better results on six data sets, while a tie was obtained for one data set. An important caveat to emphasize upfront is that the ℓ_0 robust regression algorithm was given 15 min time limit per problem instance per subset size. This practical restriction may have caused this algorithm to under perform in some cases.

5   Conclusion

In this paper, we propose a method for support vector machine which solves the underlying optimization problem that handles both feature selection and outlier detection. We formulate the problem as a mixed integer optimization problem and use an efficient commercial solver (Gurobi) to solve it. Furthermore, we present an empirical comparison between this method, the classical hinge-loss and the 0–1 loss classification methods. The experimental results have verified the superior performance of the proposed method. In terms of computational efficiency, the MIO solution can already be adopted for relatively small data sets. For the high dimensional case, a screening procedure would be suggested to reduce the computational cost.

References

1. Cortes, C., Vapnik, V.: Support-vector networks. Mach. Learn. **20**(3), 273–297 (1995)
2. Lee, Y.: Support vector machines for classification: a statistical portrait. In: Bang, H., Zhou, X., van Epps, H., Mazumdar, M. (eds.) Statistical Methods in Molecular Biology. Methods in Molecular Biology (Methods and Protocols), vol. 620, pp. 347–368. Humana Press, Totowa (2010). https://doi.org/10.1007/978-1-60761-580-4_11
3. Hastie, T., Tibshirani, R., Friedman, J.: The Elements of Statistical Learning. SSS. Springer, New York (2009). https://doi.org/10.1007/978-0-387-84858-7
4. Bertsimas, D., King, A., Mazumder, R., et al.: Best subset selection via a modern optimization lens. Ann. Stat. **44**(2), 813–852 (2016)
5. Tang, Y., Li, X., Xu, Y., Liu, S., Ouyang, S.: A mixed integer programming approach to maximum margin 0–1 loss classification. In: 2014 International Radar Conference, pp. 1–6. IEEE (2014)

6. Liu, H., Motoda, H.: Computational Methods of Feature Selection. CRC Press, Chapman (2007)

7. Weston, J., Mukherjee, S., Chapelle, O., Pontil, M., Poggio, T., Vapnik, V.: Feature selection for SVMs. In: Advances in Neural Information Processing Systems, pp. 668–674 (2001)

8. Frohlich, H., Chapelle, O., Scholkopf, B.: Feature selection for support vector machines by means of genetic algorithm. In: Proceedings of the 15th IEEE International Conference on Tools with Artificial Intelligence, pp. 142–148. IEEE (2003)

9. Minh Hoai Nguyen and Fernando De la Torre: Optimal feature selection for support vector machines. Pattern Recogn. $43(3)$, 584–591 (2010)

10. Liu, Z., Elashoff, D., Piantadosi, S.: Sparse support vector machines with L0 approximation for ultra-high dimensional omics data. Artif. Intell. Med. 96, 134–141 (2019)

11. Frénay, B., Verleysen, M.: Classification in the presence of label noise: a survey. IEEE Trans. Neural Netw. Learn. Syst. $25(5)$, 845–869 (2013)

12. Song, Q., Wenjie, H., Xie, W.: Robust support vector machine with bullet hole image classification. IEEE Trans. Syst. Man Cybern. Part C Appl. Rev. $32(4)$, 440–448 (2002)

13. Yichao, W., Liu, Y.: Adaptively weighted large margin classifiers. J. Comput. Graph. Stat. $22(2)$, 416–432 (2013)

14. Lin, C.-F., Wang, S.-D.: Fuzzy support vector machines. IEEE Trans. Neural Networks $13(2)$, 464–471 (2002)

15. Batuwita, R., Palade, V.: FSVM-CIL: fuzzy support vector machines for class imbalance learning. IEEE Trans. Fuzzy Syst. $18(3)$, 558–571 (2010)

16. Ding, H., Xu, J.: Random gradient descent tree: a combinatorial approach for SVM with outliers. In: Proceedings of the Twenty-Ninth AAAI Conference on Artificial Intelligence (2015)

17. Guibiao, X., Cao, Z., Bao-Gang, H., Principe, J.C.: Robust support vector machines based on the rescaled hinge loss function. Pattern Recogn. 63, 139–148 (2017)

18. Xiao, Y., Wang, H., Wenli, X.: Ramp loss based robust one-class SVM. Pattern Recogn. Lett. 85, 15–20 (2017)

19. Yang, L., Dong, H.: Robust support vector machine with generalized quantile loss for classification and regression. Appl. Soft Comput. 81, 105483 (2019)

20. Bertsimas, D., Dunn, J., Pawlowski, C., Zhuo, Y.D.: Robust classification. INFORMS J. Optim. $1(1)$, 2–34 (2018)

21. Chen, L., Zhou, S.: Sparse algorithm for robust LSSVM in primal space. Neurocomputing 275, 2880–2891 (2018)

22. Wang, K., Zhong, P.: Robust non-convex least squares loss function for regression with outliers. Knowl. Based Syst. 71, 290–302 (2014)

23. Yang, X., Tan, L., He, L.: A robust least squares support vector machine for regression and classification with noise. Neurocomputing 140, 41–52 (2014)

Univariate Time Series Anomaly Labelling Algorithm

Gideon Mbiydzenyuy[(✉)] [iD]

Univeristy of Boras, 501 90 Boras, Sweden
gideon.mbiydzenyuy@hb.se
https://www.hb.se/en/

Abstract. Unsupervised anomaly detection in an n-length univariate time series often comes with high risk. Anomaly contextual dependencies limit the application of binary classification methods. Analyzing the statistical features of data may help enrich the context of anomaly detection. This article proposes a quadratic time algorithm for analyzing possible anomalies in the context of unsupervised learning. Detection of possible anomalies uses Median Absolute Deviation on the residual of a univariate time series. Computation of residuals uses robust STL (Seasonal and Trend decomposition using Loess). Experiments on three datasets (Yahoo, NUMENTA NAB and district-heating substation power profiles) show the ability of the algorithm to enrich anomalies by associating labels such as Certainty, Uncertainty, and Probable, with the probable class indicating a need to further process the anomalies.

Keywords: Univariate time series · Anomaly · Labelling · District heating · Robust

1 Introduction

In the energy sector, it is increasingly possible to collect large volumes of data from infrastructure facilities such as apartment buildings and industrial buildings. Data can be about services such as electricity, water, heating, or internet [14]. Historical data provide an opportunity for analysing the performance of the underlying physical system, e.g., patterns generated from data (unsupervised) can help determine if the underlying physical system exhibits normal or abnormal performance [3,5,7,13]. In most available datasets, normal and abnormal patterns are unknown or unlabelled [19]. This limits the use of models that typically rely on labelled data, e.g., classification models. Large volumes of unlabelled data, e.g., in District Heating Substations, are driving interest in models that can process such data, e.g., in detection of anomalies. The past decade has witnessed several advances in the development and application of models for

The work presented in this article is financed by the Swedish Knowledge Foundation (KKS http://www.kks.se/om-oss/in-english/) under grant no Dnr. 20170182 within the project Data Analytics for Fault Detection in District Heating (DAD).

© Springer Nature Switzerland AG 2020
G. Nicosia et al. (Eds.): LOD 2020, LNCS 12566, pp. 586–599, 2020.
https://doi.org/10.1007/978-3-030-64580-9_48

anomaly detection [3–5]. However, out of the need to validate the performance of anomaly detection models, most of the work has focused on labelled datasets. If an anomaly detection model suggests evidence of abnormality in the labelled data, the problem becomes a classification problem, i.e., using labels in a dataset, a decision maker can reach conclusions about the anomaly. On the other hand, for an unlabelled dataset, it remains crucial for a decision maker to understand whether: 1) the abnormality is a strange but acceptable system behaviour, 2) the abnormality suggests a degradation in system performance or, 3) the abnormality suggests a fault in the system behaviour. Therefore, for anomaly detection models relying on unlabelled datasets, further analysis of suspected anomalies is vital for decision-making.

1.1 The Risk of Rejecting or Accepting Data as Anomalies

Since it is difficult to detect anomalies in an unlabelled dataset with confidence, several methods have emerged that focus on enriching the context of anomalies, e.g., labelling anomalies with ranks, [10], distance measures [13,19], and probabilities [6,11]. For unlabelled datasets, and due to the contextual nature of anomalies, if model parameters are set to lower thresholds levels, this may result in an increase risk of false negatives. On the other hand, higher threshold levels will increase the risk of false positives. To avoid dealing with this binary choice, one strategy may be to design anomaly detection models such that they admit false positives and further analyse suspected anomalies using suitable scoring functions. Thus, in order to understand the risks to the physical system, it is indispensable to find ways through which suspected anomaly data can be analysed.

1.2 Anomaly Labelling Problem Specification

The problem involves detection of potential anomalies. Detected potential anomalies need to be analysed to enrich unlabelled dataset, e.g., given two potential anomalies, it should be possible to separate them based on the extent of their deviations. Given an n-length univariate time series $X = \{x_{it}, i \in 1..n, t \in \tau\}$, measured over a period of time τ, determine $A \subset X$ that consists of all suspected anomalies. The result is a new dataset X_λ (same as X) in which all suspected anomalies (found in A) have labels. The labels are based on the extent of the deviation of each suspected anomaly. The set A can be computed using different anomaly detection methods, e.g., robust statistics with Median Absolute Deviation (MAD):

$$MAD = k * |med(x_{it} - med(X))| \tag{1}$$

where $k = 1.483$ is the scale constant if X has a Gaussian distribution. Determine the following:

$$f : A \rightarrow X_\lambda \tag{2}$$

s.t. $\forall (x_{it}, x_{jt}) \in A$ & $x_{it} > x_{jt} \implies f(x_{it}) > f(x_{jt})$, $\forall (f(x_{it}), f(x_{jt})) \in X_\lambda$.

The anomaly labelling function f transform suspected anomalies from a two dimensional binary space to a multidimensional space (with a continuous anomaly grading scale). $f(x_{it}) > f(x_{jt})$ does not achieve comparison in the strictest mathematical sense because the relative anomaly assessment can rely on several different criteria or behaviours, e.g., an anomaly with a small magnitude and high economic impact. There are several statistical methods for characterising data with respect to central location and distributions such as Chebychev's inequality [8] that estimate the probability of a data point to exist beyond a certain number of standard deviations from the centre. In particular, for univariate time series, candidate scoring and ranking functions have relied on majority voting [10], variants of Euclidean distances [19], and other forms of probability estimation [11]. The rest of the article uses "detected anomalies" in the sense of suspected or potential anomalies. The detection uses a statistical technique in which the measure of the central tendency of the data uses Eq. 1. The design of the algorithm builds on the principles of robust statistics [17] using robust STL (Seasonal and Trend decomposition using Loess) [18]. The technique has been modified to include the use of a Monte-Carlo style simulation to estimate a suitable scale constant for the MAD. In the analysis part (Eq. 2), several anomaly labelling functions are proposed to comprehend the temporal (time-based) but also spatial characteristics of the anomalies.

2   Univariate Time Series Anomaly Labelling (UTAL) Algorithm

Given an n-length univariate time series $X = \{x_{it}, i \in 1..n, t \in \tau\}$, measured over a period of time τ, the UTAL algorithm (see Algorithm 1) proceeds in two phases: 1) An anomaly detection phase helps to identify suspected anomaly data (see Algorithm 1) and, 2) An anomaly analysis phase processes suspected anomalies and assign corresponding labels (see Algorithm 2).

Since time series data consist of time and space dimensions, the second phase uses the time and space dimensions when processing anomalies (spatiotemporal) [16]. The goal is to characterize anomalies along different dimensions such as distances to threshold or percentiles for the space dimension, or deviation from measurements obtained in a specified hour (of the day) for the time dimension. Along each dimension the relative comparison of any two suspected anomalies is estimated. The estimate is used to label the anomaly. Anomaly labels are information for further analysis of the data, e.g., system performance, or how to impute for missing data.

2.1   Detection of Suspected Anomalies

Anomalies are data points that deviate from a threshold value calculated based on MAD of the data with the use of a standard Hampel filter [15]. The simplicity, scalability and transparency of robust statistical methods (e.g., MAD) is helpful in the analysis of anomalies. This is in contrast to advanced black-box methods for which the lack of transparency about the mechanism for detecting anomalies, might limit analysis of anomalies. Moreover MAD is resilient to corrupted data compared to most methods because of it high breaking point (high percentage of anomalies are needed to degrade the effectiveness of the method). To apply MAD, the following conditions are necessary [14,17]:

- data seen as normal have a pattern (belonging to clusters) which separates it from any unexpected data (often in smaller proportions).
- the distribution of the data is close enough to a Gaussian distribution.
- moreover, the detection of anomalies in a rolling time window, remains valid for the complete time series.

To address the first concern, one can repeatedly generate line plots for different data samples. Such plots uncover any patterns in the data, e.g., cycles, seasonality, etc. Seasonal patterns in a time series can sometimes overshadow other important features in the time series. This can make it difficult to detect local anomalies. In the presence of trends and seasonality, for example in the case of power signals from a district heating substation, it becomes important to extract the trend and season so as to focus on the residual [9,18].

With the help of Monte-Carlo simulations, distribution parameters of the data were studied. Supposed the data fit a Gaussian distribution, calculate the dispersion of the data from the sample standard deviation. Conversely, supposed the data does not fit the Gaussian distribution, we then calculate the dispersion of the data using the MAD with a scale constant of 1. A standard deviation measures dispersion from the mean and MAD measures dispersion from the median. By repeatedly sampling the data and measuring the standard deviation and MAD, the ratio of standard deviation to MAD converges eventually in the intuition of a Monte-Carlo simulation, i.e., a Simulated-MAD (SMAD). To detect anomalies, SMAD is then used to replace the standard scale constant of **1.483** which is usually applicable with a Gaussian distribution [17]. A robust scale estimate such as MAD will preserve the characteristics of the data for varying lengths of a rolling window.

The algorithm is initialised with a chosen number of deviations (nd) from the center of the data. Any data point beyond nd deviations is a suspected anomaly. To estimate the value of nd, multiple trials of the algorithm are performed. To simulate SMAD, a sample size (ss) of the univariate time series data is set for each simulation run, i.e., the data is randomly sampled N times. Using a chosen window size (ws), estimated MAD values (SMAD) are logged for all N simulations and the average value is chosen as the MAD value for the ws-sized time series. The SMAD value is used in a standard Hampel filter to detect potential anomalies [15] for the ws-sized time series. This generates the anomaly set A as input for the analysis phase.

2.2   Spatiotemporal Analysis, Scoring and Labelling of Anomalies

As stated earlier, suspected anomaly data need further analysis in the context of unsupervised learning. Such an analysis can rely on a relative comparison of suspected anomalies. For example, if a suspected anomaly has a high influence on the variance of the data relative to the rest of the suspected anomalies (Grubbs measure), it can be evidence that it has a higher chance of being an anomaly. If in addition, the same point is in the extreme end of measurements taken at a specific time in a period, the chance that it is an anomaly further increases. Different criteria result in different scores. In order to compare the results for a given set of anomalies, the scores are min-max normalized. Further insight that leads to labelling anomalies in a way that facilitates decision making under unsupervised learning conditions builds on the following simple logic:

 - identify suspected anomalies that consistently score high for most of the time and space criteria.
 - identify suspected anomalies that consistently score low for most of the time and space criteria.

In the first case, there is an increased assertion of the anomalous nature of such a data point. Thus, the data point will receive a relatively higher confidence label as an anomaly (C for certainly anomaly). In the second case, the opposite is true resulting in the lower confidence label (U for unlikely anomaly). Any other case in between these two is a probable anomaly. The in between case needs to be further processed. A P label (probable anomaly) is given to such a data point. The phrase "consistently score high or low" most be quantified e.g., in at least five criteria both including both space and time function measures.

To address Eq. 2, eight scoring functions are proposed. Each of these functions is mathematically described under Algorithm 2.

Grubbs [1] Change in Sample Variance-*gDist*: Grubb suggested that a suspected anomalous point is relatively more anomalous if its presence leads to a larger change in the variance of the dataset. For each suspected anomaly, estimate the ratio of the variance of a subset $\tilde{X} \subset X$ without the anomaly to the variance X with the anomaly.

Nearest Non-anomaly Distance-*nDist* [14]: If a suspected anomaly is relatively further away from the rest of data set in a defined neighbourhood, then such an anomaly is relatively more anomalous in relation to the rest of the suspected anomalies. For each suspected anomaly, return the maximum of the half-window length for the left and right neighbours in the given time series. Calculate the absolute difference of the obtained maximum with the given suspected anomaly and repeat this procedure for all suspected anomalies.

Percentile Distance to Data Set-*pDist*: If a suspected anomaly is extremely higher than or lower than most of the data (in percentile) for a specific time stamp (hour of the day), then it is relatively more anomalous than those close to the middle of the data. This is similar to the interquartile range technique (IQR) which is a measure of variability, based on quartiles [17] except that it is limited to data for a specific hour in a time series. For each suspected anomaly, estimate the corresponding percentile score in an array of data values with the same time stamp, e.g., all values measured between 7AM to 8AM for in the whole time series.

Algorithm 1 UTAL-Univariate Time Series Anomaly Labelling

1: **procedure** $X=(x_{it}, i \in 1..n, t \in \tau)$ ▶ n-length univariate time series, measured over τ

2:    $nd \leftarrow 3$ ▶ Number of deviations to determine a threshold

3:    $0 < ss < 1$ ▶ Sample size to use in each iteration

4:    $ws \leftarrow 7days$ ▶ Window size for moving window week

5:    $N \leftarrow 1000$ ▶ Number simulation runs

6:    $SM \leftarrow \varnothing$ ▶ Keep track of simulated MAD in each run

7:    $A \leftarrow \varnothing$ ▶ Suspected anomaly set

8:    $X_R = X - X_T - X_S$ ▶ Remove trend X_T and season X_S in the data based on [19]

9:    γ ▶ User defined threshold for reporting anomaly

     Estimate the MAD constant -Monte Carlo

10:    **for** $l \in 1 \cdots N$ **do**

11:      $\check{X}_l \subset X_R : \frac{|X|}{|\check{X}_l|} = ss$ ▶ random sample of the given time series

12:      $SD_l = \sqrt{\frac{1}{|\check{X}_l|-1} \sum_{l=1}^{l=N} (x_{it} - \bar{x})^2}, x_{it} \in \check{X}_l, \bar{x} = mean(\check{X}_l)$ ▶ sample standard deviation

13:      $MAD_l = med(x_{it} - med(\check{X}_l))$ ▶ MAD with constant 1

14:      $SM_l = \frac{SD_l}{MAD_l}$

15:      $SM \bigcup SM_l$

16:    $SM = mean(SM)$ ▶ Mean value of simulated sample MAD

     Detect Suspected Anomalies

17:    **for** $i = 1, ..., n - ws$ **do**

18:      **for** $j = ws, ..., n$ **do**

19:        $\check{X} = \{x_{it}, ..., x_{jt}\}, \check{X} \subset X_R$ ▶ Pick a moving window subset

20:        $medValue = med(\check{X})$

21:        $\beta = nd * MAD(\check{X}, scale = SM)$ ▶ Detection threshold uses SM instead of **1.4826**

22:        **if** $|x_{it} - medValue| > \beta$ **then**

23:        $A \leftarrow x_{its}$

   return $X_\lambda = SASL(X, A, ws, \gamma, \beta)$ ▶ SASL, Algorithm 2, is invoked

---

Chebychev's Weighted Distance from the Threshold-cDist [8]: If a suspected anomaly is relatively further away from the threshold boundary in a defined neighborhood, then such a data point is relatively more anomalous in relation to the rest of the suspected anomalies. Chebychev's inequality helps to determine the weight of the distances. Chebychev's inequality suggests that no more than $1/k^2$ fraction of data in a distribution can be more than k standard deviations from the data centre, e.g., mean and median. This measure converges faster than the percentile distance above. Moreover, it is a consistent test compared to measuring the distance to the closest (non-anomaly) data point.

Deviation from Same Time Stamp Measurement-tDist: If a suspected anomaly is relatively far away from a median value of data for a specific time (e.g., all data from 7AM to 8AM) in a defined neighbourhood, then such a datum is relatively more anomalous in relation to the rest of the suspected anomalies. Unlike the percentile distance, this measure makes a direct comparison with the

50th percentile of the data. For each suspected anomaly, this function returns the absolute difference with the median of a subset of the time series taken over a specific hour (hour of the day).

Algorithm 2. Spatiotemporal Analysis, Scoring and Labelling anomalies (SASL)

1: **procedure** $(X, A, ws, \gamma, \beta)$ ▸ Univariate time series, anomalies, window size and threshold
2:     $ln \leftarrow \frac{ws}{2}$ ▸ Search length = half window size
3:     $F = \{f : f : A \rightarrow X_\lambda\}$ ▸ Set F consist of all scoring functions defined below
4:     $\eta = [1, |F|]$ ▸ Number of score functions
5:     $med(S)$ ▸ Median score across all functions in F
6:     $N_{max} = N_{min} = 0.3 * |A|$ ▸ Fraction of maximum (resp minimum) scores for function f
7:     **for** $x_{it}^i \in A$ **do**
8:         **if** $\tilde{x}_{its} \geq \eta$ & $med(x_{its}) > med(S)$ **then**
9:             $x_{it} = C$ ▸ "C" anomaly with relative certainty
10:        **else**
11:            **if** $\hat{x}_{its} \leq \eta$ & $med(x_{its}) < med(S)$ **then**
12:                $x_{it} = U$ ▸ "U" anomaly with relative uncertainty
13:            **else**
14:                $x_{it} = P$ ▸ "P" probable, neither certain nor uncertain anomaly

$$\tilde{x}_{its} = \sum_{s \in F} f_{max}(x_{its}), \quad f_{max}(x_{its}) = \begin{cases} 1, x_{its} \in N_{max} \\ 0 \quad otherwise \end{cases}$$ ▸ Consistent high scores

$$\hat{x}_{its} = \sum_{s \in F} f_{min}(x_{its}), \quad f_{min}(x_{its}) = \begin{cases} 1, x_{its} \in N_{min} \\ 0 \quad otherwise \end{cases}$$ ▸ Consistent low scores

15:    **if** $\frac{length(D \backslash x_{its}=C)}{length(D)} > \gamma$ **then**   Day D as anomaly
16:
       if $\frac{length(W \backslash x_{its}=C)}{length(W)} > \gamma$ **then**   Week w as anomaly

Computation of score functions:
$gDist : \{1 - \frac{VAR(X)}{VAR(X \backslash x_{it})}\}$ ▸ Grubbs change in sample variance
$nDist : \{arg \max |x_{jt} - x_{it}|, x_{it} \in A, x_{jt} \in X\}$ ▸ Nearest none-anomaly distance

$$pDist : \begin{cases} 50 - \tilde{x}, \tilde{x} = \frac{x_{it}^r}{|X|} < 50 \\ \tilde{x} - 50, \tilde{x} = \frac{x_{it}^r}{|X|} > 50 \end{cases}$$ ▸ Percentile for x_{it}^r in the rth order of X

$cDist : |x_{it} - \beta| * (1 - 1/k^2)$ ▸ Chebychev's weight $(k = |x_{it} - th|/std(X))$
$tDist : |x_{it} - med(A)|_t$ ▸ Deviation from same time stamp measurement
$aDist : \sum_{l=1}^{l=|A|} g(x_{lt}), \quad t = 0 - 23 \quad g(x_{lt}) = \begin{cases} 1, i = j \\ 0 \quad otherwise \end{cases}$ Hourly anomalies
$eDist : \sum_{l=1}^{l=|X|} q(x_{lt}), \quad t = 0 - 23 \quad q(x_{lt}) = \begin{cases} 1, |x_{lt} - x_{lt+1}| \\ 0 \quad otherwise \end{cases}$ Neighboring anomalies
$vDist : x_{its} \in \sum_{l=1}^{l=|X|} R(x_{lt}), \quad t = 0, 1, 2 ... 23 \, R(x_{lt}) = \begin{cases} 1, |x_lts = x_jt| \\ 0 \quad otherwise \end{cases}$ Value count
       return X_λ ▸ Time series with labelled anomalies

Count of Detected Anomalies for a Given Time Stamp in a Cycle-aDist:

A suspected anomaly is relatively more anomalous, if it is detected with a time stamp (hour of the day) at which there is higher pattern or frequency of suspected anomalies. For each suspected anomaly in a given time window, return the count of suspected anomalies for that specific time stamp (hour of the day).

Count of Data Values Equal to a Given Suspected Anomaly Value-vDist: A suspected anomaly is relatively more anomalous, if there are fewer or no normal data points with the same value in a defined subset of the time series. For each suspected anomalous datum in a given time window, return the frequency count of the anomalous value.

Count of Neighboring Suspected Anomalies-eDist: A suspected anomaly in a defined neighbourhood is relatively more anomalous if it has fewer sequentially related neighbours that are themselves suspected anomalies. For each suspected anomaly, return the count of sequentially related neighbours within a given neighbourhood.

2.3   Experiments

An implementation of the algorithm uses a Python3/Spyder4 environment on an HP Elite-Book 830, with modest RAM 16G (with i7 CPU, 1.8 GHz). All experiments begin with initialisation of the number of standard deviations (nd), rolling window size (ws), number of simulations N, and a Boolean variable that determines whether to decompose the time series data or not.

Time Series Data Set: Time series dataset consists of measurements of district heat consumption in Kilowatts-Hour (sampling rate 1 h) from apartment buildings in Western Sweden. Real data from nine substations collected in 2011 are used. In addition, selected time series from Yahoo's S5 labeled anomaly detection dataset, version 1.0 [20] and the NUMENTA NAB dataset [2] are also used to evaluate the proposed algorithm.

Experiment with a Small Modified District Heating Dataset: To understand if the proposed scores are logical with respect to the magnitude of the deviation for each suspected anomaly, the time series is injected with incremental noise before passing it to the algorithm. An analysis of the effect on the quality of the anomaly detection and labelling with 1), a simulated scale constant on the residual of the time series, 2) A repeat of 1) without decomposition of the time series. In both cases, the number of deviations is set to 3 (three-sigma rule). A one week rolling window and $nd = 3$ number of deviations were employed.

Experiment with Open Datasets from Yahoo and NAB: When an anomaly detection model is employed in the context of supervised learning, then it is possible to know those anomalies suggested by the detection model that are in fact not anomalies in the real dataset (False Positives (FP)) but also those that are indeed anomalies in the real dataset (True Positives (TP)). With FP and TP, the performance of a model can be estimated by calculating Precision (PN), Recall (RC) and F-measure (FM) as defined in Eq. 3

$$PN = \frac{TP}{TP + FP}, \quad RC = \frac{FP}{TP + FP}, \quad FM = 2 * \frac{PN * RC}{PN + RN} \tag{3}$$

To obtain the precision-recall curve, labelled datasets from Yahoo [20] and the NUMENTA NAB dataset [2] were selected. UTAL Algorithm 1 was configured with $nd = 3$, and the time series not decomposed both cases. The rolling window for the NAB data was one day while for the Yahoo data it was one week.

Experiment with District Heating Substation Energy Profiles: Data from nine substations gathered during 2011 were decomposed following a rubust STL algorithm proposed by Wen et al. [18] to remove seasonality and trends where one week was taken as a cycle. The UTAL Algorithm 1 was configured with $nd = 3$ while the SASL was configured with $\gamma = 0.2\%$ threshold, with a rolling window of one week.

3   Result

3.1   Experiment with a Small Modified District Heating Dataset

Processing of the corrupted time series suggests that 1) local anomalies are easily detected when seasonality is removed (see right hand side in Fig. 1), 2) proposed functions are consistent with increasing deviations for suspected anomalies (see Fig. 2), 3) some of the functions are only useful in specific anomaly scenarios, for example, counting the number of times an anomaly value exists in the dataset could be useful if the analysis relates to a large dataset (see Fig. 2).

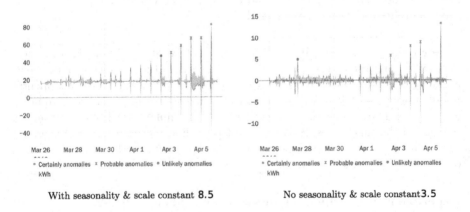

With seasonality & scale constant **8.5**   No seasonality & scale constant **3.5**

Fig. 1. Anomaly detection and labeling for a small univariate time series dataset from a DHS

Removing seasonal and trend components, then estimating a scale constant, results in changes to the scores for suspected anomalies and hence the labels. Scoring based on closest non-anomaly data (nDist), percentile distance (pDist), comparison to data in time (tDist), change of sample variance (gDist), and Chebychev's weighted distance from the thresholdare (cDist) are all stable across both scenarios and monotonically increasing with the size of the anomaly, although at different growth rates. This is noteworthy, since they evaluate different aspects of a given anomaly.

Changes in the number of anomalies for a given time stamp (aDist) in a dataset are proportional to the suspected anomalies in the residual time series. This is the same for the number of neighbouring anomalies (eDist).

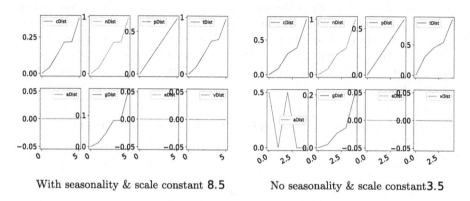

With seasonality & scale constant 8.5   No seasonality & scale constant 3.5

Fig. 2. Anomaly sores for a small univariate time series from a DHS (these anomalies were artificially induced)

3.2   Experiment with Open Source Datasets from Yahoo and NAB

The MAD constants values were 1.77 and 2.16 for NAB and Yahoo datasets respectively. Figure 3 suggests that the anomaly detection with UTAL is capable of differentiating potentially less anomaly data (U_label) from those that are more likely to be anomalies (C_label). Although the labels in both datasets were

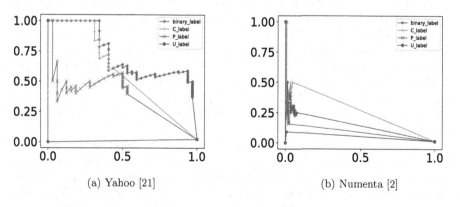

(a) Yahoo [21]   (b) Numenta [2]

Fig. 3. Precision-recall curves for Yahoo and NUMENTA-taxi datasets

constructed with a binary perspective, precision scores for the C_label anomalies correlate with anomalies identified using the binary approach. The NAB taxi dataset pose a challenge due to the non-precise nature of the anomalies, e.g., the complete day with 24 measurements were used as ground-truth instead of specific anomaly data points.

3.3   Experiment with District Heating Substation Energy Profiles

As a result of the time series decomposition cost, the experiments took about 7hours to process an 80000 length time series, 332 of these or 0.004% were anomalies of type C. A summary of the substations, sorted according to the total number of abnormal weeks are shown in Table 1. All the substations had some form of anomaly. The most abnormal substation was shown to record 18/52 abnormal weeks.

Table 1. Detection of anomalies (type C only) using UTAL Algorithm 1,

Substation	Number of weeks	Week number	Number of days
1	18	[1, 5, 12, 13, 16, 24, 25, 32, 33, 35, 36, 40, 41, 42, 43, 46, 48, 51]	33
2	11	[1, 5, 6, 12, 14, 18, 35, 36, 40, 45, 46]	22
3	10	[1, 5, 8, 10, 11, 20, 41, 46, 48, 51]	39
4	7	[23, 27, 29, 32, 30, 33, 34]	20
5	5	[1, 5, 8, 13, 35]	24
6	5	[1, 5, 6, 8, 11]	25
7	3	[11, 35, 40]	62
8	3	[1, 5, 16]	20
9	2	[1, 5]	19

4   Related Work and Discussion

Anomaly detection methods can be based on distance (furthest points are anomalies), density (relatively less dense areas are anomalies) or rankings (based on neighbourhood) [14]. Methods using distance, density, or ranks can either be supervised, semi-supervised, or unsupervised. The current article combines distance-based techniques with ranking [10] in an unsupervised learning perspective, extending similar ideas to the algorithm proposed by Knorr & Ng [12]. The approach in this article is that no single method is capable of objectively and optimally ranking anomalies, therefore multiple criteria can expose different features of an anomaly.

   In a study of the automatic anomaly detection in the cloud via statistical learning, seasonal decomposition is employed to filter the trend and seasonal components of the time series, followed by the use of robust statistical metrics -median and median absolute deviation (MAD) [9]. Hochenbaum, Vallis, and Kejariwal, [9] uses robust STL to minimize the effect of extreme values when separating the trend and seasonal components from the time series and assumes that the underlying dataset follows a Gaussian distribution.

Iwata et al. [11] proposed a supervised method for anomaly detection when precise knowledge about the anomaly is unknown. Wu, He, and Lin [19] argues that the increase of unlabelled data with complex seasonality underscores the need to focus on unsupervised anomaly detection, hence they proposed a method for online anomaly detection based on predictions. They [19] use local sequences in a time series. In the current article, a combination of different functions for assessing anomalies is proposed as shown in the SASL Algorithm 2.

A robust algorithm for seasonal-trend decomposition succeeds in handling long seasonality fluctuations with abrupt changes while minimizing the impact of the anomalies on the decomposition [18]. Due to it ability to manage seasonal fluctuations, the current article builds on the proposed robust STL algorithm [18] when decomposing energy-heat load time series in order to identify potential outliers. Although the algorithm successfully decomposed the time series data, convergence was slow. A review on outlier/anomaly detection in time series data shows that several anomaly detection methods for univariate time series fail to take into consideration the temporal elements of the time series making it difficult to detect outliers that violate the temporal correlation of the series [3]. In analysing, scoring and ranking anomalies, the current article takes into consideration the spatiotemporal characteristics of suspected anomalies.

5   Conclusions

The objective addressed by this article is to determine if the statistical characteristics of data may help to enrich the context of anomaly detection in the data. Statistical properties of data are considered in terms of 1) change in sample variance, 2) distance to nearest non-anomaly, 3) percentiles, 4) Chebyche's distance, 5) deviations from same time stamp measurements, 6) count of anomalies for the same time stamp, 7) count of values equal to a given suspected anomaly value, and 8) count of neighboring suspected anomalies.

In order to determine if the above statistical properties can enrich anomaly detection context, the study implemented a Univariate Time Series Anomaly Labelling (UTAL) algorithm.

Spatiotemporal analysis and scoring of potential anomalies enables a relative measure that differentiate anomalies with "certainty" (C) from "unlikely" anomalies (U). The rest of the anomalies are "probable" (P) since they cannot be labelled as C or U. When suspected anomalies consistently score high relative to the rest of the anomalies across time and space criteria, there is a higher confidence that they are indeed anomalies. If they consistently show low scores across time and space, there is lesser confidence that they are anomalies.

From employing the UTAL algorithm in labelled open datasets, it can be observed that the algorithm does differentiate different classes of anomalies in such a way that the result are consistent with binary classification models (as shown with the Yahoo dataset). The potential of the algorithm in a real-world district heating dataset suggests that district heat management operators can improve their ability to locate abnormal days and weeks in various substations following a sensitivity threshold level of their choice. How well different types of faults correlate with identified anomalies is a subject for a future study.

References

1. Adikaram, K.L.B., Hussein, M.A., Effenberger, M., Becker, T.: Data transformation technique to improve the outlier detection power of Grubbs' test for data expected to follow linear relation. J. Appl. Math. **2015**, 1–9 (2015)
2. Ahmad, S., Lavin, A., Purdy, S., Agha, Z.: Unsupervised real-time anomaly detection for streaming data. Neurocomputing **262**, 134–147 (2017)
3. Blázquez-García, A., Conde, A., Mori, U., Lozano, J.A.: A review on outlier/anomaly detection in time series data. arXiv preprint arXiv:2002.04236 (2020)
4. Braei, M., Wagner, S.: Anomaly detection in univariate time-series: a survey on the state-of-the-art. arXiv preprint arXiv:2004.00433 (2020)
5. Chandola, V., Banerjee, A., Kumar, V.: Anomaly detection: a survey. ACM Comput. Surv. (CSUR) **41**(3), 1–58 (2009)
6. Committee, A.M., et al.: Robust statistics-how not to reject outliers. Part 1. Basic concepts. Analyst **114**(12), 1693–1697 (1989)
7. Freeman, C., Merriman, J., Beaver, I., Mueen, A.: Experimental comparison of online anomaly detection algorithms. In: The Thirty-Second International Flairs Conference (2019)
8. Grechuk, B., Molyboha, A., Zabarankin, M.: Chebyshev inequalities with law-invariant deviation measures. Prob. Eng. Inf. Sci. **24**(1), 145–170 (2010)
9. Hochenbaum, J., Vallis, O.S., Kejariwal, A.: Automatic anomaly detection in the cloud via statistical learning. arXiv preprint arXiv:1704.07706 (2017)
10. Huang, H.: Rank based anomaly detection algorithms (2013)
11. Iwata, T., Toyoda, M., Tora, S., Ueda, N.: Anomaly detection with inexact labels. Mach. Learn. **109**(8), 1617–1633 (2020). https://doi.org/10.1007/s10994-020-05880-w
12. Knox, E.M., Ng, R.T.: Algorithms for mining distance based outliers in large datasets. In: Proceedings of the International Conference on Very Large Data Bases, pp. 392–403. Citeseer (1998)
13. Mehrotra, K.G., Mohan, C.K., Huang, H.: Anomaly Detection Principles and Algorithms. Springer, Cham (2017). https://doi.org/10.1007/978-3-319-67526-8
14. Moustakidis, S., et al.: Innovative technologies for district heating and cooling: indeal project. In: Multidisciplinary Digital Publishing Institute Proceedings, vol. 5, p. 1 (2019)
15. Pearson, R.K., Neuvo, Y., Astola, J., Gabbouj, M.: Generalized Hampel filters. EURASIP J. Adv. Signal Process. **2016**(1), 1–18 (2016)
16. Rao, T.S., Antunes, A.M.C.: Spatio-temporal modelling of temperature time series: a comparative study. In: Time Series Analysis and Applications to Geophysical Systems, pp. 123–150. Springer, Cham (2004). https://doi.org/10.1007/978-3-319-67526-8

17. Rousseeuw, P.J., Hubert, M.: Robust statistics for outlier detection. Wiley Interdisc. Rev. Data Min. Knowl. Disc. **1**(1), 73–79 (2011)
18. Wen, Q., Gao, J., Song, X., Sun, L., Xu, H., Zhu, S.: RobustSTL: a robust seasonal-trend decomposition algorithm for long time series. In: Proceedings of the AAAI Conference on Artificial Intelligence, vol. 33, pp. 5409–5416 (2019)
19. Wu, W., He, L., Lin, W.: Local trend inconsistency: a prediction-driven approach to unsupervised anomaly detection in multi-seasonal time series. arXiv preprint arXiv:1908.01146 (2019)
20. Yahoo: S5 - a labeled anomaly detection dataset, version 1.0(16m). https:// webscope.sandbox.yahoo.com/catalog.php?datatype=s%5c&did=70

Challenges in Real-Life Face Recognition with Heavy Makeup and Occlusions Using Deep Learning Algorithms

Natalya Selitskaya[✉], Stanislaw Sielicki, and Nikolaos Christou

School of Computer Science, University of Bedfordshire, Luton LU1 3JU, UK
nselitsk@gmail.com
https://www.beds.ac.uk/computing

Abstract. We study a Face Recognition problem caused by unforeseen in the training set variations in face images, related to artistic makeup and other occlusions. Existing Artificial Neural Networks (ANNs) have achieved a high recognition accuracy; however, in the presence of significant variations, they perform poorly. We introduce a new data set of face images with variable makeup, hairstyles and occlusions, named BookClub artistic makeup face data, and then examine the performance of the ANNs under different conditions. In our experiments, the recognition accuracy has decreased when the test images include an unseen type of the makeup and occlusions, happened in a real-world scenario. We show that the fusion off the training set with several heavy makeup and other occlusion images can improve the performance.

Keywords: Makeup · Occlusion · Disguise · Face recognition · Benchmark data set

1 Introduction

Artificial Neural Networks (ANNs) and Deep Learning (DL)are capable of learning patterns of interest from data in the presence of noise and natural variations. In recent years the ANNs applied to face recognition problems has achieved an above-human accuracy. However, various edge case conditions thrown into the process drastically reduce the recognition performance. In this paper, we study disruptive and spoofing effecting the face recognition on a novel data set of face images with visual obstacles, such as makeup, face masks, headdress, and eyeglasses. In this study, we introduce BookClub artistic makeup data set built upon the existing publicly available work of collecting and investigating data sets featuring the makeup and other occlusions face disguise techniques.

Possibility of the targeted face spoofing using makeup was studied in. An unspecified commercial face-recognising system and a well-established Convolution Neural Network (CNN) VGG-Face implementation are used on two data sets: original Makeup Induced Face Spoofing (MIFS) data set of the non-makeup

© Springer Nature Switzerland AG 2020
G. Nicosia et al. (Eds.): LOD 2020, LNCS 12566, pp. 600–611, 2020.
https://doi.org/10.1007/978-3-030-64580-9_49

and makeup images, and Labeled Faces in the Wild (LFW) [11] spoof target images data set. The researchers conclude that both face recognition algorithms are vulnerable to targeted spoofing attacks [4].

Every-day makeup impact of various intensity on the engineered features Local Binary Patterns (LBP) has been studied on a proprietary data set. The study confirms that even simple makeups affect the facial recognition accuracy for training sets consisting only from the non-makeup images. Infusing makeup images into a training set may have a dual effect, either increasing or decreasing accuracy, depending on the nature of the training data set enrichment. Medium intensity facial cosmetic images infused symmetrically into the training set is observed to increase accuracy [8]. Occlusion influence research [19] also uses an engineered features approach to detect occlusions (Gabor wavelet features, Principal Component Analysis (PCA) dimensionality reduction, and Support Vector Machine (SVM) classification) and LBR for face detection. Facial areas where occlusions were detected are excluded from the face recognition processing. The AR data set was used for the algorithm evaluation [18].

CV Dazzle project concentrates on individually tailored suggestions of the artistic makeup and hairstyle patterns that look like "inconspicuous" fashion details, but still disrupt face recognition algorithms. It is based on the identification of the individual prominent face landmarks that contribute mostly to the algorithms' decisions and then selecting appropriate makeup and hairstyle templates disrupting these features. Four engineered features algorithms (Eigen-Faces, FisherFace, Elastic Bunch Graphic Matching (EBGM), LBP, Gabor) and three commercial face recognition systems (Google+, PicTriev, Baidu) were tested on the data set of celebrity images the authors have collected. Makeup and hairstyle patterns were digitally applied to the images [9].

MIFS data set contains 107 images of the makeup transformations taken from the YouTube makeup tutorials [4]. Majority of the original subjects in the data set are young Caucasian females. The YouTube Makeup Database (YMU), created by the same team, as well as the following two data sets, contains 600 images of 151 Caucasian females taken the before and after makeup application. Additional Makeup in the Wild Database (MIW) contains 154 images corresponding to 125 subjects. Half of the images are with makeup, and a half - without makeup [5]. All three data sets are scraped from the internet. The Virtual Makeup (VMU) database is created by synthetically adding makeup to 51 female Caucasian subjects in the Face Recognition Grand Challenge (FRGC) database [7].

The CyberExtruder data set contains 10,205 images of 1000 people scraped from the internet. It contains large pose, lighting, expression, race and age variation, and a multitude of occlusions such as hats, glasses, and makeup in general [1]. Unfortunately, access is discontinued. Disguised Faces in the Wild (DFW) data set consists of 1000 subjects and 11157 images collected from the internet. Each subject may contain one or two non-makeup images, one or more disguised and one impersonator image. Disguised features may be hairstyles, beard, moustache, glasses, makeup, caps, hats, turbans, veils, masquerades and ball masks.

Images vary in their pose, lighting, expression, background, ethnicity, age, gender, clothing, and camera quality [16].

AR Face Database is a relatively old created in 1998, conditionally public, and, from our experience, not maintained data set. However, it still could have been relevant, at least as an example, to contemporary research of the face recognition with occlusions due to its quality and systematic approach, which many data sets mentioned above lack, in collecting multiple images per session representing a set of lighting conditions, emotions, orientations and occlusions such as sunglasses and scarves. It contains more than 4000 face images of 126 subjects (70 men and 56 women) [18].

A newer AR-like data set partial replacement, Database of Partially Occluded 3D Faces (UMB-DB), is composed of 1473 2D colour images and 3D depth images of 143 subjects which include males and females of an age ranging between 19 and 50, almost all of them are Caucasians. Each photo session has a minimum of nine images with three types of emotions [6]. Another benchmark data sets where makeup and other occlusions can be found, usually include them as additional features, concentrating on other parameters, as in the Indian Movie Face Database (IMFDB) [23].

Motivation for the novel BookClub data set [20] creation was an attempt to fill in the apparent gap in the makeup and occlusions data sets of high quality and resolution images featuring sophisticated original artistic makeup and non-trivial occlusions, made in the controlled environment with a wide variety of the represented emotions, orientations, lighting conditions per subject, for subjects representing age ranges, genders, and races. As the majority of the cited work concentrated on the effects of makeup and occlusions on face recognition using engineered features algorithms here, the behaviour of the modern CNN architectures, represented by AlexNet implementation, was investigated.

The paper is organised as follows. Section 2 briefly introduces ANN algorithms of the model used to evaluate the data set. Section 3 presents detailed description of the BookClub data set. Section 4 lists how the data set was structured in the experiments and how the models were applied; Sect. 5 presents the obtained results, and Sect. 6 draws practical conclusions from the results and states directions of the research of not yet answered questions.

2   Artificial Neural Networks

An ANN model can be viewed as a multivariate piece-wise non-linear regression over high-dimensional data, targeting to map an input-output relationship. Such a model is fitted to the given data to minimise a given error function. Feature selection and multiple models have been capable of efficiently increasing the performance, as discussed in [13–15].

Each layer of neurons can be perceived as a transformation (regression) from a space of inputs $\mathcal{X}$ with one dimensionality m into another space of outputs $\mathcal{Y}$ with another dimensionality n:

$$f : \mathcal{X} \subset \mathbb{R}^m \mapsto \mathcal{Y} \subset \mathbb{R}^n \tag{1}$$

If transformations from one space into another performed by the neuron layers are linear, they could be represented in the matrix form:

$$\mathbf{y} = f(\mathbf{x}) = \mathbf{W}\mathbf{x}, \ \forall \mathbf{x} \in \mathcal{X} \subset \mathbb{R}^m, \ \forall \mathbf{y} \in \mathcal{Y} \subset \mathbb{R}^n, \tag{2}$$

where $\mathbf{W} \in \mathcal{W} \subset \mathbb{R}^n \times \mathbb{R}^m$ is the adjustable coefficient matrix.

However, such a model is limited by ability to learn only linear relations between input signals. To create truly multiple hidden layers capable of learning non-linear relationships, one needs to add "activation functions" - elements of non-linearity between the layers of neurons.

One of the commonly used family of the activation functions are sigmoids such as the logistic function:

$$z = g(y) = \frac{1}{1 + e^{-y}} = \frac{e^y}{e^y + 1}, \ y = \mathbf{w}^T \mathbf{x}. \tag{3}$$

The output of this function can be interpreted as the conditional probability of observations over classes, which is related to a Gaussian probability function with an equal covariance [12], which is a very convenient output for the final ANN layer.

Rectified Linear Unit (ReLU) is another popular family of the activation functions:

$$z = g(y) = y^+ = max(0, y). \tag{4}$$

They address the problem of vanishing gradients of sigmoid activation functions in ANN [3].

The way to fit an ANN model to a real-world problem is adjusting the ANN parameters W_{ij} weights at each layer k. To find out how close ANN transformations fall into the expected neighbourhood of the training data, a metric or distance function is needed, which is usually called in ML as a cost or objective function. The most popular family of the learning algorithms are Gradient Descent (GD):

$$W_{t+1} = W_t - \eta \nabla_{W_t}^l. \tag{5}$$

where t is the sequence member or iteration number, $0 < \eta < 1$ is learning rate, and $\nabla_{W_t}^l$ is the gradient of the cost function $l : \mathcal{Z} \subset \mathbb{R}^k \mapsto \mathcal{L} \subset \mathbb{R}$ in respect to the weight matrix W.

Similarly, back-propagation algorithms define a cost function $l(\mathbf{z}, \hat{\mathbf{z}})$, where $\hat{\mathbf{z}}$ is a training observation vector, activation function $\mathbf{z} = g(\mathbf{y})$, and neuron layer summation function $\mathbf{y} = \mathbf{W}\mathbf{x}$, partial derivatives of the cost function in respect to the activation function results $\frac{\partial l}{\partial z_j}$ are readily available, where j is the index of a neuron in a layer. Using the chain rule for partial derivatives, it is easy to find out the cost function derivative in respect to the summation function for the given j-th neuron:

$$\nabla_{\mathbf{x}}^l = J\left(\frac{\partial \mathbf{y}}{\partial \mathbf{x}}\right)^T \nabla_{\mathbf{y}}^l, \tag{6}$$

where $J(\frac{\partial \mathbf{y}}{\partial \mathbf{x}})^T$ is a transposed Jacobian matrix of the partial derivatives of the vector of the neuron summation function results y_j in respect to the vector of inputs $\mathbf{x}$, and $\nabla_{\mathbf{y}}^l = (\frac{\partial l}{\partial y_1}, \ldots \frac{\partial l}{\partial y_j}, \ldots \frac{\partial l}{\partial y_k})^T$ is a gradient of the cost function l in respect to the vector $\mathbf{y}$.

Similarly, one can express the needed for the learning algorithm cost function derivative in respect to the matrix or tensor of learning parameters flattened to vector $\mathbf{W}$:

$$\nabla_{\mathbf{W}}^l = J(\frac{\partial \mathbf{y}}{\partial \mathbf{W}})^T \nabla_{\mathbf{y}}^l. \tag{7}$$

A natural and straightforward cost function based on the Euclidean distance - Sum of Squared Errors (SSE) is convenient to use with linear transformations. However, if the logistic sigmoid activation function is used, SSE causes problems. Similarly does the 'softmax' generalisation of the logistic activation function applied to the multi-class problem:

$$z = g(y_j) = \frac{e^{y_j}}{\sum_j e^{y_j}}, \mathbf{y} = \mathbf{W}\mathbf{x}, \tag{8}$$

Partial derivatives of SSE cost function in respect to y_j, when logistic sigmoid activation function applied to it, results in polynomials of the third degree which have three roots. Such a gradient $\nabla_{\mathbf{y}}^l$ has multiple local minimums which is inconvenient even for GD algorithms.

To make partial derivatives having one root, a convenient 'cross-entropy' function that, being positive and becoming zero when $z_j = \hat{z}_j$, is suitable for the cost function role for logistic-type activation functions:

$$l(z) = -(\hat{z}_j \ln z_j + (1 - \hat{z}_j) \ln (1 - z_j)). \tag{9}$$

One of the popular ANN architectures for image and signal recognition is a Convolutional Neural Network (CNN) [10,17]. A CNN uses local receptive fields – neighbouring pixel patches that are connected to few neurons in the next layer. Such an architecture hints the CNN to extract locally concentrated features and could be implemented using Hadamard product $\mathbf{y} = (\mathsf{M} \odot \mathsf{W})\mathsf{K}\mathbf{x}$ of the shared weight matrix W and its sliding receptive field binary mask $\mathsf{M} \in \mathcal{M} \subset \mathbb{B}^n \times \mathbb{B}^k$, corresponding kernel mask $\mathsf{K} \in \mathcal{K} \subset \mathbb{R}^k \times \mathbb{R}^m$, where k - length of the combined by receptive fields and flattened input vector. The shared rows weight matrix W can be viewed as a shift and distortion invariant feature extractor synthesised by a CNN, and $\mathbf{y}$ - as a generated feature map. Multiple parallel feature masks and maps ensure learning multiple features.

3   Data Set

The novel BookClub artistic makeup data set contains images of 21 subjects. Each subject's data may contain a photo-session series of photos with no-makeup, various makeup, and images with other obstacles for facial recognition, such as wigs, glasses, jewellery, face masks, or various types of headdress. Overall, the data set features 37 photo-sessions without makeup or occlusions, 40 makeup sessions, and 17 sessions with occlusions. Each photo-session contains circa 168 RAW images of up to 4288×2848 dimensions (available by request) of six basic emotional expressions (sadness, happiness, surprise, fear, anger, disgust), a neutral expression, and the closed eyes photo-shoots taken with seven head rotations at three exposure times on the off-white background.

Subj.1.Sess.MK3, Subj.5.Sess.NM1; Subj.4.Sess.GL1, Subj.5.Sess.NM1

Subj.12.Sess.MK2, Subj.20.Sess.NM1; Subj.14.Sess.HD1, Subj.6.Sess.NM1

Subj.20.Sess.MK1, Subj.17.Sess.NM1; Subj.21.Sess.MK1, Subj.9.Sess.NM1

Fig. 1. Sample images from the BookClub data set sessions grouped by pairs from left to right. The first (left) image is from the test session, the second (right) - from the session misidentified by AlexNet retrained on only non-makeup images. For example, Session MK3 of Subject 1 in the left upper corner was misidentified as Subject 5.

Default publicly available downloadable format is a JPEG of the 1072×712 resolution. The subjects' age varies from their twenties to sixties. Race of the subjects is predominately Caucasian and some Asian. Gender is approximately evenly split between sessions. The photos were taken over the course of two months. A few sessions were done later, and some subjects posed at multiple sessions over several week intervals in various clothing with changed hairstyles [20].

Table 1. Subject and Session name, test accuracy, best (wrong) Subject guess and its score for the retrained AlexNet model on the non-makeup BookClub images. Worse case examples with low accuracy and high wrong guess score.

Session label	Accuracy	Guess	Guess score
S1MK3	0.0057	S5	0.8921
S4GL1	0.0069	S5	0.7986
S7FM2	0.0000	S1	0.4970
S10MK4	0.0000	S1	0.5868
S12MK1	0.0000	S1	0.5357
S12MK2	0.0000	S20	0.4880
S14HD1	0.0000	S6	0.5706
S20MK1	0.0000	S17	0.3855
S21MK1	0.0000	S9	0.5061

4   Experiments

For computational experiments in the face recognition accuracy, the well known AlexNet CNN [2] was retrained on images of the BookClub Data set. "Sgdm" learning algorithm with 0.001 learning coefficient, mini-batch size 64, and 30 epochs parameters are used for training.

A commonly used technique for estimating the accuracy of the ML models in laboratory conditions was initially used. The whole BookClub data set was divided into training and test sets with a 4/6 ratio, which translated into approximately 6500 training images. All subjects and sessions were represented in the training set with the randomly selected images from these sessions. The high-level recognition accuracy of a session was calculated as a ratio between the number of correctly identified images and the whole number of images in the session (or, analogously, "guess score" for misidentified sessions). As expected, for such a well-known ML model as AlexNet and large and variable data set, the estimated ML model accuracy was effectively 100% for all sessions.

Table 2. Subject and Session name, test accuracy, best (wrong) Subject guess and its score, test accuracy after infusion and (correct) guess after that, for the retrained AlexNet model on the non-makeup and infused in-subject heavy make up BookClub images. Sessions with verdict changed from wrong to correct.

Session label	Accuracy 1	Guess 1	Guess score 1	Accuracy 2	Guess 2
S1MK8	0.3072	S5	0.4277	1.0000	S1
S7MK1	0.2857	S8	0.6071	0.7024	S7
S7FM1	0.0494	S3	0.5741	1.0000	S7
S10MK2	0.0120	S1	0.4750	1.0000	S10
S10MK3	0.0473	S8	0.5266	1.0000	S10

Table 3. Subject and Session name, test accuracy, (correct) Subject guess, test accuracy after infusion and best (wrong) guess and its guess score, for the retrained AlexNet model on the non-makeup and infused in-subject heavy makeup BookClub images. Sessions with verdict changed from correct to wrong.

Session label	Accuracy 1	Guess 1	Accuracy 2	Guess 2	Guess score 2
S1HD2	1.0000	S1	0.0061	S5	0.7195
S1MK6	1.0000	S1	0.0419	S12	0.8144

To emulate the real-life conditions, when makeup or other occlusions of the subjects may not be predicted and included into training sets beforehand, non-makeup and non-occlusion sessions were selected into the training set, which amounted to roughly 6200 images. Then, the facial recognition accuracy estimation was run with the same as above mentioned training parameters of AlexNet for each makeup and occlusion photo-session. While the majority of the sessions, were correctly identified, scoring high 93–100% accuracy, some sessions scored significantly lower accuracy in the 50–90% range, and some sessions were misidentified. Some sessions with especially high contrast, dark pigment makeup, or created by professional artists scored close to 0% accuracy. Even worse, some of the low accuracy makeup scored high guess scores for the wrong subject (Table 1, Fig. 1), it appears making accuracy <90% effectively unreliable by upper bound.

For a more granular accuracy estimate, descriptive statistics were collected on the distributions of the final 'softmax' layer activations that for practical purposes could be treated as probabilities of the image belonging to the particular class.

To explore hypotheses that including a limited number of sessions with makeup into the test set would improve facial recognition accuracy for the other unexpected makeup for the same or even different subjects, the following makeup and occlusion subsets were independently included into the training set:

1. Light makeups, easily recognised by the non-makeup trained AlexNet: S1MK4, S3MK3, S5MK4, S7MK3, S7FM3, S10MK1, S11MK1, S12MK1
2. Heavy makeups, initially unrecognised by the non-makeup trained AlexNet, to test in-subject effects: S1MK3, S3MK4, S5MK2, S7MK2, S7FM2, S10MK4, S12MK1
3. Heavy makeups, initially unrecognised by the non-makeup trained AlexNet, to test cross-subject effects: S11MK2, S12MK2, S20MK1, S21MK1.

5 Results

Introduction of the light, easily recognisable makeup into the training set gave mixed results with some previously unrecognised sessions became recognised,

Subj.7.Sess.FM1, Subj.3.Sess.NM1, Subj.7, Sess.NM1.Subj.7.Sess.FM2

Subj.10.Sess.MK2, Sub1.5.Sess.NM1, Subj.10.Sess.NM1, Sub1.10.Sess.MK4

Fig. 2. Sample images from the BookClub data set sessions (left) that were initially misidentified (second left), and became identified correctly (second right) by AlexNet retrained on non-makeup and infused in-subject heavy makeup (right) Book-Club images.

while other previously correctly recognised sessions became misidentified. However, wrong or correct estimated accuracy/score for both outcomes stayed relatively low, at the level of unreliable recognition for this model and data set.

Adding some heavy, previously unrecognised makeup to the training set has helped not only to recognise other difficult sessions of the same subject (Table 2, Fig. 2) but in some cases lead to cross-subject recognition improvement Table 4. Similarly to the light makeup introduction into the training set, there were also cases when previously successfully recognised sessions became misidentified (Tables 3, and 5, Fig. 3). However, unlike in the light, already recognised makeup addition, the heavy unrecognised makeup infusions in the majority of cases led to the accuracy improvement to the high confidence numbers (close to 100%). At the same time, the changes that broke initially correct recognition scored an accuracy level that was unreliable for this model/data interval (<90%).

Table 4. Subject and Session name, test accuracy, best (wrong) Subject guess and its score, test accuracy after infusion and (correct) guess after that, for the retrained AlexNet model on the non-makeup and infused cross-subject heavy make up BookClub images. Sessions with verdict changed from wrong to correct.

Session label	Accuracy 1	Guess 1	Guess score 1	Accuracy 2	Guess 2
S1MK3	0.0057	S5	0.8921	0.4091	S1
S1MK8	0.3072	S5	0.4277	1.0000	S1

Statistics of the distribution of the probability of all BookClub images recognised correctly for the non-makeup trained AlexNet (number of correctly recognised images 6379, mean 0.91, standard deviations (SD) 0.17, range 0.19–1.0)

Table 5. Subject and Session name, test accuracy, (correct) Subject guess, test accuracy after infusion and best (wrong) guess and its guess score, for the retrained AlexNet model on the non-makeup and infused cross-subject heavy makeup BookClub images. Sessions with verdict changed from correct to wrong.

Session	Accuracy 1	Guess 1	Accuracy 2	Guess 2	Guess score 2
S1MK2	0.3691	S1	0.4226	S10	0.5298
S5MK3	0.9697	S5	0.3152	S20	0.5030
S7FM3	0.6310	S7	0.3750	S14	0.6191

has compacted and moved to the "right": (number of correctly recognised images 6726, mean 0.95, standard deviations (SD) 0.12, range 0.25–1.0) for the in-subject infusion trained model. However, the same compacting and rightward moving tendencies were also true for the probability distributions of the misidentified images. From the number of the wrongly classified images 3445, the mean is 0.64, SD 0.21, range 0.16–1.0, to 3059, 0.71, 0.22, 0.2–1.0, respectively for the in-subject infusion trained model.

A/B test thresholds between the wrongly and correctly identified distributions for 75th, 90th, and 95th percentiles resulted in 0.8256, 0.9344, 0.9668 probability values for non-makeup trained distributions, and 0.9174, 0.9850, 0.9952 - for the in-subject makeup infusion trained. Corresponding 75%, 90%, and 95% confidence level accuracy was calculated as a ratio of images with probabilities greater than the A/B thresholds to the overall number of images. For the non-makeup trained AlexNet model it was 0.5300, 0.4827, 0.4508, while for the in-subject makeup infusion trained, it was 0.5814, 0.4903, 0.4248, respectively. Which means that the decision either to use makeup infusions for training sets depends on the task's tolerance to false-positive errors. For example, if only 5% and less of the false positives are acceptable, then the heavy makeup infusion worsens trusted accuracy while improving it for lower confidence levels.

6   Discussion and Future Work

Experiments on the novel BookClub data set of artistic makeup and occlusions have demonstrated that even contemporary state-of-the-art face recognition algorithms such as AlexNet (even retrained on a large amount of data) when meeting with real-life obstacles, perform poorly at the desired high confidence levels. Acceptability of their use for applications that have a high intolerance for false-positive errors, such as those that can have legal implications, has to be questioned when test images are significantly distorted or occluded.

Investigation of the easily applicable technique of enriching the training data set with the limited "difficult" cases has shown that the feasibility depends on the target application's tolerance to false positive or false negative errors. If a low confidence level in the decision is acceptable, such a technique noticeably

Subj.1.Sess.MK2, Subj.1.Sess.NM1, Subj.10.Sess.NM1, Subj.12.Sess.MK2

Subj.7.Sess.FM3, Subj.7.Sess.NM1, Subj.14.Sess.NM1, Subj.20.Sess.MK1

Fig. 3. Sample images from the BookClub data set sessions (left) that were originally recognised correctly (second left), then became misidentified (second right) by AlexNet retrained on non-makeup and infused cross-subject heavy makeup (right) BookClub images.

improved the model's performance for the populations represented by the Book-Club data set.

However, a relatively small data set and sparsity of the makeup types suggests future work in an extension of the data set to cover more racial and age variability. The use of informative statistic hypothesis within a Bayesian framework, such as [21, 22] could also extend the future work.

References

1. Face Matching Data Set | Biometric Data | CyberExtruder (Dec 2019). https://cyberextruder.com/face-matching-data-set-download. Accessed 8 Dec 2019
2. Transfer Learning Using AlexNet Example (Dec 2019). https://www.mathworks.com/help/deeplearning/examples/transfer-learning-using-alexnet.html. Accessed 8 Dec 2019
3. Agarap, A.F.: Deep Learning Using Rectified Linear Units (RELU) (2018)
4. Chen, C., Dantcheva, A., Swearingen, T., Ross, A.: Spoofing faces using makeup: an investigative study. In: 2017 IEEE International Conference on Identity, Security and Behavior Analysis (ISBA), pp. 1–8 (Feb 2017)
5. Chen, C., Dantcheva, A., Ross, A.: Automatic facial makeup detection with application in face recognition. In: 2013 International Conference on Biometrics (ICB), pp. 1–8 (2013)
6. Colombo, A., Cusano, C., Schettini, R.: UMB-DB: a database of partially occluded 3d faces. In: 2011 IEEE International Conference on Computer Vision Workshops (ICCV Workshops), pp. 2113–2119 (Nov 2011)
7. Dantcheva, A., Chen, C., Ross, A.: Can facial cosmetics affect the matching accuracy of face recognition systems? In: 2012 IEEE Fifth International Conference on Biometrics: Theory, Applications and Systems (BTAS), pp. 391–398 (Sep 2012)
8. Eckert, M., Kose, N., Dugelay, J.: Facial cosmetics database and impact analysis on automatic face recognition. In: 2013 IEEE 15th International Workshop on Multimedia Signal Processing (MMSP), pp. 434–439 (Sep 2013)

9. Feng, R., Prabhakaran, B.: Facilitating fashion camouflage art. In: Proceedings of the 21st ACM International Conference on Multimedia MM 2013, pp. 793–802. ACM, New York, NY, USA (2013)

10. Goodfellow, I., Bengio, Y., Courville, A.: Deep Learning. MIT Press, Cambridge (2016)

11. Huang, G.B., Mattar, M., Berg, T., Learned-Miller, E.: Labeled faces in the wild: a database for studying face recognition in unconstrained environments. In: Workshop on Faces in 'Real-Life' Images: Detection, Alignment, and Recognition. Erik Learned-Miller and Andras Ferencz and Frédéric Jurie, Marseille, France (Oct 2008)

12. Izenman, A.J.: Modern Multivariate Statistical Techniques: Regression, Classification, and Manifold Learning. Springer Publishing Company Incorporated, New York (2008)

13. Jakaite, L., Schetinin, V., Maple, C.: Bayesian assessment of newborn brain maturity from two-channel sleep electroencephalograms. Comput. Math. Methods Med. pp. 1–7 (2012)

14. Jakaite, L., Schetinin, V., Maple, C., Schult, J.: Bayesian decision trees for EEG assessment of newborn brain maturity. In: The 10th Annual Workshop on Computational Intelligence UKCI 2010 (2010)

15. Jakaite, L., Schetinin, V., Schult, J.: Feature extraction from electroencephalograms for Bayesian assessment of newborn brain maturity. In: 24th International Symposium on Computer-Based Medical Systems (CBMS), pp. 1–6. Bristol (June 2011)

16. Kushwaha, V., Singh, M., Singh, R., Vatsa, M., Ratha, N., Chellappa, R.: Disguised faces in the wild. In: 2018 IEEE/CVF Conference on Computer Vision and Pattern Recognition Workshops (CVPRW), pp. 1–18 (June 2018)

17. Lecun, Y., Bengio, Y.: Convolutional Networks For Images, Speech. And Time-series. MIT Press, Cambridge (1995)

18. Martinez, A., Benavente, R.: The ar face database. Technical report. 24, Computer Vision Center, Bellatera (Jun 1998)

19. Min, R., Hadid, A., Dugelay, J.L.: Improving the recognition of faces occluded by facial accessories. In: FG 2011, 9th IEEE Conference on Automatic Face and Gesture Recognition, March 21–25, 2011, Santa Barbara, CA, USA (03 2011)

20. N. Selitskaya, S. Sielitsky, M.K.: Bookclub data set (Jun 2019). https://drive.google.com/file/d/18nH8m_ST-kqcAWZp_vjdo-nS3cESCBOO/view. Accessed 15 June 2019

21. Schetinin, V., Jakaite, L., Krzanowski, W.: Bayesian averaging over decision tree models: an application for estimating uncertainty in trauma severity scoring. Int. J. Med. Inform **112**, 6–14 (2018)

22. Schetinin, V., Jakaite, L., Krzanowski, W.: Bayesian averaging over decision tree models for trauma severity scoring. Artif. Intell. Med. **84**, 139–145 (2018). (2.0), (2.0)

23. Setty, S., Husain, M., Beham, P., Gudavalli, J., Kandasamy, M., Vaddi, R., Hemadri, V., Karure, J.C., Raju, R., Rajan, B., Kumar, V., Jawahar, C.V.: Indian movie face database: A benchmark for face recognition under wide variations. In: 2013 Fourth National Conference on Computer Vision, Pattern Recognition, Image Processing and Graphics (NCVPRIPG). pp. 1–5 (Dec 2013)

Who Accepts Information Measures?

Gail Gilboa-Freedman$^{(\boxtimes)}$ ⓘ, Yair Amichai Hamburger ⓘ, and Dotan Castro ⓘ

The Interdisciplinary Center Herzliya, Herzliya, Israel
gail.gilboa@idc.ac.il

Abstract. Are people intuitions regarding the conceptual notion of information in agreement with the properties of measures that are common in the information theory literature? We capture the abstract notion of "informative mechanism" by a simple model of broadcasting a 'signal' that is associated with a single binary 'secret'. The mechanism is more informative when there is a larger distance between the two distributions of the signal conditioned by each value of the secrets. We chose the f-divergence as a common metric for this distance. We consider an axiomatic model that characterize f-divergence orders, and conduct an empirical study to test whether each of the axioms in this model are in agreement with human preferences. As the main result of this study, we identify three monotone increasing relations between someone's scores of three psychological tests (internal, powerful others, and chance locus of control) and the number of axioms that are in agreement with his preferences.

Keywords: Information gain · Kullback-Leibler divergence · Decision making · Axiomatic approach · Internal powerful others · Locus of control

1 Introduction

Inspired by the abundance of empirical choice studies that investigate the acceptability of the axioms that characterize rational choices over lotteries, our study follows the same approach for investigating the acceptability of axioms that characterize information-wise choices. To motivate this study, consider a marketer who needs to buy data for revealing some personal information on potential customers. The information is not implicitly included in the data set, but instead is associated with it. Choosing the right company to buy data from is a prime consideration. We are interested in studying 'how he would have ranked' in comparison with 'how he should rank' two data sets in this context.

We describe an "information gain mechanism", as a process of observing signals that are associated with the value of a secret. For the binary case, when there are two optional secret values, we formalize a mechanism by a 2-rows stochastic matrix, where each row is a distribution over some abstract space of signals, conditional on the value of the secret. The further away distributions are, the more valuable is to observe the signal when guessing the type. Thus, the question of

© Springer Nature Switzerland AG 2020
G. Nicosia et al. (Eds.): LOD 2020, LNCS 12566, pp. 612–616, 2020.
https://doi.org/10.1007/978-3-030-64580-9_50

comparing information-gain-mechanisms fits itself to a preference order over the set of 2-rows stochastic matrices. Taking an information-theory approach induces a preference order that is represented by a function that quantifies the distance between the two distributions in the mechanism rows. A standard measure for this distance is the f-divergence, which is the expected value of the function f, taken at the likelihood ratio of the signals.[1]

Our study is related to the question of how to measure information [8] which is an open question on both axiomatic and empirical ground [5]. In a companion paper, Gilboa-Freedman and Smorodinsky [10] take an axiomatic approach and show that a collection of four axioms uniquely captures the behavioral implications of prioritizing "information gain mechanisms" by f-divergence. The current study explores whether human preferences satisfy these properties. An overview of different axiomatizations is introduced by Csiszar [7], and a recent study of information cost functions can be found in Caplin et al. [4]. We conduct a series of four experiments, each designed to examine the tendency of human choices to satisfy one of the four axioms that characterize f-divergence orders. Moreover, we study the impact of *internal powerful others and chance locus of control* in those people who behave in disagreement with these axioms. Studying the impact of these descriptors is a recognized device.[2]

We identify three monotone increasing relations between someone's scores of three psychological test (internal, powerful others, and chance of control) and the number of axioms that are in agreement with his choices. Statistics of the tendency to satisfy different combinations of axioms are also provided. Moreover, our methodology of building an axiomatic model that characterizes the preference order that is represented by a measure, and then performing an experiment to test whether human accept these axioms, may set an example of bridging the gap between analytical and behavioral approaches.

2   Behavioral Experiment

The f-divergence preference order is characterized by a set of four axioms: "Independence", "Post Processing", "Monotonicit Under Certainty", and "continuity". The curious reader is referred to Gilboa-Freedman and Smorodinsky [10] for farther particulars of the axiomatic model. We conducted an experimental study to test if preferences of a typical human subject satisfy each of the axioms. We invited psychology BA students from a private university in Israel in exchange for credit. We received replies from 99 participants (mean age = 22.53, sd =1.39)

---

[1] The f-divergence, introduced independently by Csiszar [6] and Ali and Silvey [1], is a family of functions widely used and studied in the context of Information Theory as a natural metric between two distributions (the reader is referred to [12] for a survey of this). Familiar example of f-divergence is the KL-divergence [11].

[2] It is assessed through the use of relevant and important personality theory – Rotter's of control [13]. The impact of personality is familiar in the digital behavior, demonstrated by Amichai-Hamburger in [2] , and by Amichai-Hamburger & Hayat in [3].

from whom we had 95 valid and full responds. Participants were presented with a questionnaire via web based application (Qualtrics) where they could read short scenarios in neutral context (flowers). The experiment includes 4 scenarios, each represents an axiom in our model. Scenarios are independent and presented in a random order. For each scenario, subjects are told that their purpose is to reveal a species of a flower by making efficient choice over two information sets that are useful for this task. A mechanism that publishes the selected piece of information is more attractive when it is more informative in this context.

3   Results

3.1   Monotone Influence of Psychological Scores

Our research data includes the results of our experiment [9]. For each participant, it include the following processed data: a vector of four binary values, representing whether the participant behavior is in agreement with each of the 4 axioms; and three scores of the participant in the psychological tests. We partition the participant into groups by the number of axioms they accept, and for each group we calculated the average scores of the psychological tests. The following figure (Fig. 1) demonstrates the relation between the scores and the number of accepted axioms, showing a clear monotone increasing relation.

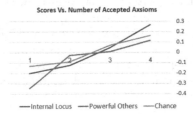

3.2   Acceptability of Each Axiom, Separately

For each scenario in the experiment, we calculated the percentage of participants whose preference is in agreement with the corresponding axiom. Then, we used *Chi-squared test* to compare the observed data with the expected result according to a null-hypothesis that subjects have an arbitrary preferences over the possible sets of information. For the first axiom ("Independence") 88%, for the Second axiom ("Post Processing") 76%, and for the third axiom ("Monotonicit Under Certainty") 66% of the choices are in line with the axiom. The corresponding χ-squared values are 58, 24.3, and 10.54. Thus, for each of these axioms the observable preference is in agreement with the corresponding axiom. This agreement is highly significant (all p-values are all lower than 0.01), so we reject the null hypothesis that subjects have an arbitrary preferences in favor of the alternative hypothesis that their choices are in agreement with the corresponding axioms in our model. For axiom1, Continuity, we found that only 40% of the choices are in line with the axiom. This property is satisfied when comparing optional information-gain mechanisms by the computation of their f-divergence, but our experiment demonstrates that people's behavior is not in agreement with this. It means that people are not sensitive enough to interpret information gain regarding that axiom. Similar result was found in other studies, showing that people do not act according to Continuity axiom when making decisions. The result leads to recommendation on warning human decision maker against their wrong privacy-related decisions (or at least paying their attention toward their wrong intuition when making decisions that involve minor changes of the probabilities of the outcome).

4   Conclusions

As the main result of this study, we identify three monotone increasing relations between someone's scores of the psychological test (internal, powerful others, and chance locus of control) and the number of axioms (out of four axioms that characterize f-divergence) that are in agreement with his choices. This is intuitive because people with strong internal locus tend to reinforce their expectancies by internal rather than external factors, which may lead to their difficulties in ranking external stimuli. We also identify that disagreement is demonstrated mainly because people violate "Continuity" axiom, being insensitive enough to interpret the influence of minor changes of outcome probabilities. Similar result was found in other studies, showing that people do not act according to continuity axiom when making decisions.

Our study suggests that wrong decisions in the context of identifying informative data is more prevalent in people with strong internal locus of control. This observation is potential to be useful for pricing information and data sets.

References

1. Ali, M.S., Silvey, D.: A general class of coefficients of divergence of one distribution from another. Statist. Soc. **28**, 131–140 (1966)
2. Amichai-Hamburger, Y.: Internet and personality. Comput. Human Behav. **18**, 1–10 (2002)
3. Amichai-Hamburger, Y., Hayat, Z.: Personality and the Internet, in The social net: understanding our online behavior. OUP, Oxford, Chennai (2013)
4. Caplin, A., D.M., Leahy, J.: Rationally inattentive behavior: Characterizing and generalizing shannon entropy. Nat. Bur. Econ. Res. (2017)
5. Crupi, V., Nelson, J.D., Meder, B., Cevolani, G., Tentori, K.: Generalized information theory meets human cognition. Cogn. Sci. **42**(5), 1410–1456 (2018)
6. Csiszar, I.: Information measures: a critical survey. In: 7th Prague Conference on Information Theory, Academia Prague, pp. 73–86 (1974)
7. Csiszár, I.: Axiomatic characterizations of information measures. Entropy **10**(3), 261–273 (2008)
8. Frankel, A., Kamenica, E.: Quantifying information and uncertainty. Am. Econ. Rev. **109**(10), 3650–80 (2019)
9. Gilboa-Freedman, G., A.H.Y.C.D.: Who accepts information measures (2019)

10. Gilboa Freedman, G., Smorodinsky, R.: On the properties that characterize privacy. J. Math. Soc. Sci. **103**, 59–68 (2020)
11. Kullback, S., Leibler, R.: On information and sufficiency. Math. Statist. pp. 79–86 (1951)
12. Renyi, A.L.F.R.P.E.D.: On measures of entropy and information. In: Fourth Berkeley Symposium on Mathematical Statistics And Probability (1961)
13. Rotter, J.: Internal versus external control of reinforcement: a case history of a variable. Am. Psychol. **45**, 489–493 (1990)

An Error-Based Addressing Architecture for Dynamic Model Learning

Nicolas Bach$^{(\boxtimes)}$, Andrew Melnik, Federico Rosetto, and Helge Ritter

CITEC, Bielefeld University, Bielefeld, Germany
nbach@techfak.uni-bielefeld.de, andrew.melnik.papers@gmail.com

Abstract. We present a distributed supervised learning architecture, which can generate trajectory data conditioned by control commands and learned from demonstrations. The architecture consists of an ensemble of neural networks (NNs) which learns the dynamic model and a separate addressing NN that decides from which NN to draw a prediction. We introduce an error-based method for automatic assignment of data subsets to the ensemble NNs for training using the loss profile of the ensemble. Our code is publicly available (Code: https://github.com/NicoBach/distributed-dynamics-model).

1 Introduction

In this work, we present a data-driven learning architecture that can learn a dynamic model of an agent in an environment while staying responsive to control commands. The proposed model consists of two parts, an ensemble of neural networks and an addressing neural network. The ensemble consists of neural networks (NNs) that learn situational models [1], i.e. each such model is tuned for only a subset of the training data. This is akin to a divide-and-conquer-strategy and is especially useful for problems with repeating cycles of state trajectories, where the ensemble NNs can specialize on different phase ranges, for example, a cyclic motion. Our approach is motivated by the phase-functioned neural networks (PFNN) approach [2] for developing a real-time controller for a virtual agent. This approach can generate accurate kinematic trajectories for locomotion learned from real-world motion capture data. The PFNN architecture generates weights of a regression network using a numerical phase value as input. This phase value is a time-resolving variable which labels each time step and helps to handle ambiguous data in the locomotion cycle. The required phase labeling of the training data is done in a semi-automatic and therefore time-consuming preprocessing step [2]. In contrast to this, our method implements an automated, error-based clustering mechanism of training data points, using the loss profiles of the ensemble NN across the training data points and, therefore, does neither require a handcrafted labeling preprocessing step nor a phase value.

N. Bach and A. Melnik—Equal contribution.

© Springer Nature Switzerland AG 2020
G. Nicosia et al. (Eds.): LOD 2020, LNCS 12566, pp. 617–630, 2020.
https://doi.org/10.1007/978-3-030-64580-9_51

Deep ensembles have been empirically shown to be a promising approach for improving accuracy, and out-of-distribution robustness of deep learning models [3]. One possible explanation is that deep ensembles tend to explore diverse modes in function space.

The problem of deriving a controller from motion capture data, similarly to PFNN, is also addressed by other approaches. Quaternet [4] uses a quaternion-based recurrent model to generate sequences of human poses. Others use deep reinforced learning (DRL) methods to produce motion control over a 3D virtual skeleton in kinematic or physical environments, [5,6]. [7] propose an architecture based on hierarchical reinforcement learning to develop a model capable of locomotion. Modularization of end-to-end learning is another opportunity for improvement [8]. The MOSAIC architecture [9] is a modular architecture for motor learning and control based on multiple pairs of forward (predictor) and inverse (controller) models. The architecture simultaneously learns the multiple inverse models necessary for control as well as how to select the set of inverse models appropriate for a given environment. Similarly, [10] implement a biologically inspired, modular architecture for controlling a six-legged robot.

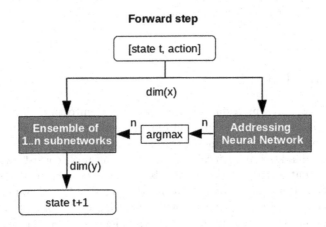

Fig. 1. The architecture consists of an ensemble of n fully connected networks and an addressing network, which chooses the network from the ensemble to predict the output y given the input x.

2   Methods

The structure of the architecture (Fig. 1) can be divided into two parts - the addressing NN and an ensemble of n networks. The task of the addressing NN is to choose a network from the ensemble. The addressing network is a fully connected network with the same input x as each subnetwork in the ensemble. The output dimension of the addressing network is equal to the number of ensemble subnetworks n. The *argmax* of the addressing NN output identifies the subnetwork from the ensemble to predict y from the *state* x. The input dimensions of all ensemble subnetworks are equal to $dim(x)$, and their task is to predict the next *state* y. With this input-dependent addressing, the architecture performs a regression step to predict the next state given the current state and action. The number of ensemble subnetworks is a hyperparameter and is chosen depending on the task. Thus, a forward step of the architecture consists of two steps, first, prediction by the addressing NN, second, prediction by the chosen subnetwork from the ensemble.

2.1 Clustering of Training Data

In this section, we describe the automatic clustering algorithm (Algorithm 1; Fig. 3) for the training data, i.e. how the training set can be partitioned into n approximately equally-sized (in terms of the number of data points) bins for the n ensemble networks. Each subnetwork will be trained only on data points from its bin. An equal number of samples in bins ensures good competitiveness of subnetwork specialization. Not providing equally sized bins for the ensemble networks will preclude specialization if at one point an ensemble network becomes strictly better than all the others and then wins all training data for itself.

Algorithm 1. Clustering of training data using the loss profile space of the ensemble

---

Input: Training data points $\mathbf{x} \in X$, target $\mathbf{y} \in Y$, ensemble of n models

1: **for** $\mathbf{x}$ in X **do**
2:     Perform forward-step with every ensemble model on $\mathbf{x}$
3:     Compute losses between outputs of $\mathbf{x}$ for every model and target $\mathbf{y}$
4:     Form loss profile L of the data point
5: **end for**
6: Compute norms $||\mathbf{l}||^2$ for $\mathbf{l} \in L$
7: Sort loss profiles in L regarding norms from highest to lowest ($Ls = \text{sort}(L)$).
8: **while** $\text{len}(Ls) < 0$ **do**
9:     $\mathbf{x} = Ls.\text{pop}()$
10:     $processing = True$
11:     **while** $processing$ **do**
12:         **if** $\text{len}(bins[argmin(\mathbf{x})]) < \text{len}(X) / \text{len}(bins)$ **then**
13:             $bins[argmin(\mathbf{x})].\text{append}(x)$
14:             $processing = False$
15:         **else**
16:             $\mathbf{x}[argmin(\mathbf{x})] = inf$
17:         **end if**
18:     **end while**
19: **end while**

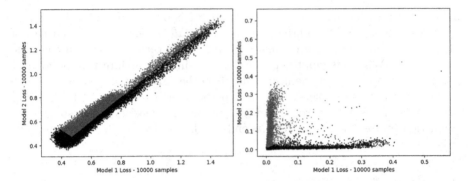

Fig. 2. Loss profile for 2-NNs ensemble in the Acrobot environment. On the left are loss profiles of data points of untrained subnetworks in the ensemble. We apply clustering of these data points into two bins (green and black). On the right, in the alignment of the data points along the axes (loss profile) is reflecting the successful specialization of the subnetworks after training of the ensemble of the two subnetworks (green and black). (Color figure online)

In the first step of the clustering algorithm, each of n ensemble networks produces predictions for all data points in the training data. Next, for each data point, we collect the prediction losses for each subnetwork into a n-dimensional vector that we refer to as the *loss profile* of the data point. Third, we sort the data points in descending order of the euclidean norms of their loss profiles. Data points sorted towards the end of this ordering have small loss norms, indicating that they pose no particular challenge for any of the n subnetworks and, thus, have little information value for the specialization of the subnetworks. Conversely, data points near the beginning of the sequence are challenging and thus they should guide the optimization of clustering more strongly (Fig. 2). Therefore, we assign the data points into n bins by the following sequential process: we take a data point with the highest norm value from the sorted list of data points and delete this data point from the list. The *argmin* of the loss profile of this data point determines its destination bin if the bin still is filled below its maximal capacity (*bin size = number of all data points/n bins*). Otherwise, if the bin is already full, the destination bin is chosen by the index of the next lowest loss entry in the data point's loss profile.

2.2   Training

We train the addressing NN and the ensemble separately in two sequential phases (Fig. 3). In the first phase, the ensemble of $1..n$ subnetworks is trained for a fixed number of epochs. At the beginning of each epoch, we first apply the clustering algorithm described in Algorithm 1 to fill n data bins. Each data bin corresponds to one of the subnetworks and is used to train the corresponding subnetwork. Thus, for each subnetwork, we sample data points from its corresponding bin in some random order, until all training samples of the bin have been used once.

Then, we continue to the next subnetwork and its corresponding bin. When all subnetworks have been traversed, the current epoch is finished and the next epoch is started.

When a fixed number of epochs is completed, we start the second phase of training (Fig. 3), where we train the addressing NN for several epochs. We first compute an updated set of loss profiles that reflects the training progress of the ensemble networks during phase one. Then, for each data point (x, y) the *argmin* of its loss profile determines the target output of the addressing network for input x (i.e. the addressing network's target output is the k-unit vector, where k identifies the smallest component of the n-dimensional loss profile of (x, y)) With this choice of target output, the addressing network is adapted towards predicting the ensemble subnetwork with the lowest loss for an input $x \in X$. To this end, we train the addressing NN for several epochs, each consisting of sampling batches of a certain size from the training set and optimizing the network using these batches.

Finally, when both ensemble and addressing networks have been trained, the forward step consists of two simple parts: a prediction of the addressing NN which subnetwork to use given input x, and computing the prediction of this subnetwork for obtaining the output y.

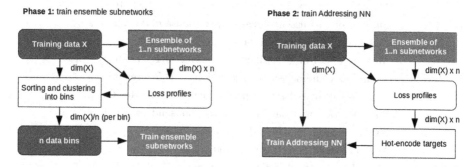

Fig. 3. Data flow graphs. Phase 1: Sorting and clustering of training data into equal-sized bins (one bin for each subnetwork in the ensemble) and ensemble training. Phase 2: Training of the Addressing NN using targets created by the loss profiles of the data points.

3   Results

3.1   Mountaincar Environment

For the first experiment, we train a model to simulate the dynamics of the mountaincar environment [11] shown in Fig. 4a. Its observation space and action space consists of two dimensions (position and velocity) and three dimensions (actions:

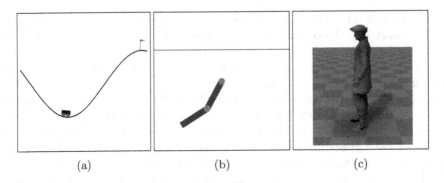

Fig. 4. Environments of the experiments: Mountaincar (a), acrobot (b) and the virtual environment used in the phase-functioned neural network paper [2] (c).

Table 1. Addressing and losses in the mountaincar environment.

Mean loss of selected and not selected subnetworks	
Mean subnetwork 1 loss selected	0.2 * 1e−6
Mean subnetwork 1 loss not selected	262.6 * 1e−6
Mean subnetwork 2 loss selected	1.2 * 1e−6
Mean subnetwork 2 loss not selected	1861.3 * 1e−6
Subnetwork 1 missclassification per thousand data points	
Number of missclassified datapoints	27
Median loss of missclassified predictions	0.4 * 1e−6
Mean loss of missclassified predictions	3.5 * 1e−6
Max. loss of missclassified predictions	65.2 * 1e−6
Subnetwork 2 missclassification per thousand data points	
Number of missclassified datapoints	1
Median loss of missclassified predictions	7.8 * 1e−6
Mean loss of missclassified predictions	7.8 * 1e−6
Max. loss of missclassified predictions	7.8 * 1e−6

push left, push right, no push), respectively. We encode the actions via one-out-of-k-hot-encoding for feeding it into the network, which extends the input dimension of our networks to five dimensions (state plus action). Output dimensionality amounts to two for each of the ensemble models as we predict the next state. Training data is derived from Deep Q-network training, where we recorded the data points, which are generated by the policy stepping in the environment during training and amounts to $T = 1,000,000$ data points. Generating the next state is realized by one of the n subnetworks, mapping $f(s_t, a) = s_{t+1}$, while the addressing NN $g(s_t, a) = argmax(\{f_1, ..., f_n, \})$ predicts which model to use. We used an ensemble with of two subnetworks. Each subnetwork has two hidden layers with 128 neurons. The addressing NN is configured with three layers (128, 128 and 64 neurons). We use the mean-squared-error as loss-function in this experiment. We used ADAM [12] optimizer in this experiment with standard-setting of learning rate $\mu = 1e^{-4}$ and $\beta = [0.9, 0.99]$.

The output of the environment and the states produced by the architecture are nearly identical. The mean squared error for an episode of 1000 steps between environment steps and predicted states amounts to $2.225e^{-5}$ for position prediction and $4.859e^{-8}$ for velocity prediction. The performance in terms of loss of both subnetworks can be seen in Table 1. The addressing network is able to correctly distribute the samples the subnetworks are specialised on. In autoregressive runtime, the architecture could generate the same dynamics as the original environment. In contrast, a model with only one fully connected NN with 3 hidden layers with 512 neurons each was unable to learn the dynamics of the agent (Fig. 5).

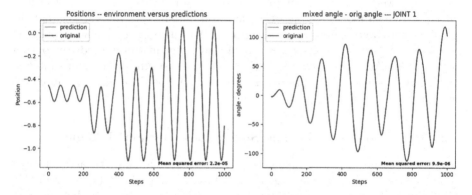

Fig. 5. Left: Comparison between ground truth and one-step-predictions in terms of position in the mountaincar environment. Right: Comparison between ground truth and one-step-predictions in terms of angle for the first joint in the acrobot environment.

3.2   Acrobot Environment

The acrobot environment [13], shown in Fig. 4b, consists of two rotational joints with angles α_1, α_2 linked in a row. The six dimensional observation space consists of the tuple $\{\cos\alpha_1, \sin\alpha_1, \cos\alpha_2, \sin\alpha_2, \omega_1, \omega_2\}$, where ω_1, ω_2 are angle velocities for joint one and two, respectively. Its action space consists again of three actions. These three actions correspond to fixed torque values $(-1, 0, 1)$. We used one-out-of-k-hot-encoding of the actions to feed it to the network. We again use the learning process of a Deep-Q network to record 150,000 data points during training. In addition to that, the training data consists of 50,000 data points, in which the Acrobot was recorded while applying zero torque to highlight behavior in the low-velocity range. We use an ensemble of eight NNs. We organize the input into 6 sequential observations, thus requiring 54 input neurons, while the output dimension for each subnetwork amounts to four $(\cos \alpha_1, \sin \alpha_1, \cos \alpha_2, \sin \alpha_2)$. The addressing NN has the same input as the ensemble models, while its output dimension is equal to the number of ensemble networks $(n = 8)$. We build subnetworks with three hidden layers consisting of 256 neurons each. Two layers with 512 neurons are provided for the addressing NN. Furthermore, we equip the addressing NN with 8 heads of two layers with 256 neurons each. The output dimension of each of the eight heads is one and a softmax function is applied on the outputs of all eight heads. For optimization we chose to use batches of 32 data points to perform updates in both ensemble and addressing NN training. As activation functions for all neurons in all networks we use hyperbolic tangent. The loss is the mean absolute error. The optimzer is ADAM with standard setting of learning rate $\mu = 1e^{-4}$ and $\beta = [0.9, 0.99]$.

The architecture was able to predict the next state in terms of cosine and sine with high precision. The prediction of sine and cosine values plus the computation of the weighted angle using those exhibits an error of $1e^{-6}$. The velocities which were also fed to the network were computed using the current and the previous state and exhibit a higher error in the range of $1e^{-2}$. The errors were computed by comparing the output of the environment to the predicted output of the state. At autoregressive runtime, the architecture with an ensemble of eight networks was able to produce similar behavior as the original environment exhibits.

3.3 Prediction of Cosine and Sine Values - Weighted Angle

Predicting cosine and sine values is necessary for modeling the acrobot environment as described in Sect. 3.2. Using a function approximator for predicting cosine and sine values and constructing an angle from these predictions isn't straightforward due to imprecisions. Especially the edges of the intervals, that the inverse trigonometric functions arccosine and arcsine map cosine and sine values to, can cause problems. Further, both intervals for cosine translated to radians $[0, .., \pi]$ and for sine translated to radians $[-\frac{\pi}{2}, ..., \frac{\pi}{2}]$ are displaced by $\frac{\pi}{2}$ and together contain the information, in which quadrant of the unit circle the angle is located. Therefore, we use a weighted sum of both angles, which complementary uses the predictions to cover the edges of each other's intervals:

$$w = \frac{c}{c + s} \tag{1}$$

$$\alpha = (1 - w) * \arccos(c) + w * \arcsin(s) \tag{2}$$

c represents the network's prediction of the cosine of an angle of a joint, while s is the network's prediction of the sine of the angle of the same joint. Variable w represents the weight, which is calculated by formula 1. When cosine angle is around 0 or π (refers to cosine of 1 and -1 respectively), then sine prediction is used and vice versa. In Fig. 6, the reconstruction of the angle using sine and cosine values can be observed. We use this weighted angle at runtime in the forward step after training is done.

3.4 Phase-Functioned Neural Network Database

We trained our architecture on the training set provided in the phase-functioned neural network implementation [2]. For a detailed description of the training set, refer to [2]. The database consists of recorded motion capture data and provides about four million data samples. The input dimension of a sample is 342 dimensions and consists largely of positional information, as well as velocities in three different directions (x, y, z coordinates) for 31 joints which make up the agent. The output dimension per sample has 311 dimensions and also predicts position and velocities, plus rotational values of each joint.

The phase function of PFNN generates weights of a regression network from a number of trained NNs, which then predicts the next state. In the code implementation, the PFNN consists of four pre-trained networks and uses a cubic catmull-rom spline as the phase function. For the comparison between our architecture and PFNN, we choose the same number of parameters in the network ensemble as the implementation of PFNN provides (four ensemble networks, with two hidden layers and 512 neurons per layer). In addition to that, the addressing network consists of two hidden layers with 512 neurons each. Our approach trained for forty epochs, each with about 80,000 training steps. The batch size was chosen to be 32.

The comparison was performed on a test set of the PFNN database. The size of the test set amounts to 40,000 data samples. The mean absolute error over the training set for PFNN amounts to 2.21. Our architecture is able to fit

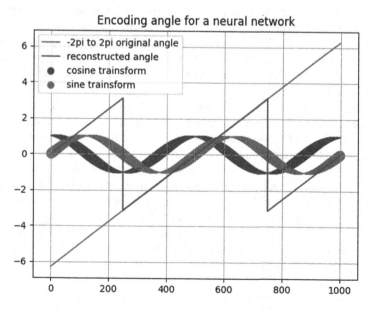

Fig. 6. Construction of angle by mixture of cosine and sine prediction. For illustrating, which portion of the angle relies on cosine prediction and which on the sine prediction, we highlighted the corresponding areas of sine and cosine by the curve width.

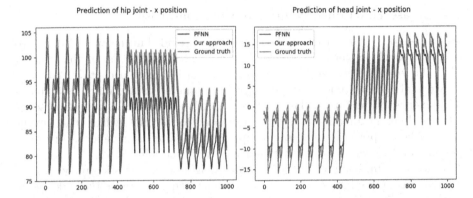

Fig. 7. Predictions of hip (left) and head joint (right) of trained model on a test set of the PFNN database.

the same data, producing a much lower mean absolute error of 0.69. A series of one-step-predictions over 1000 steps of the hip joint in terms of forward-direction and for prediction of the head joint can be seen in Fig. 7.

4   Discussion

We presented an error-based addressing architecture for dynamic model learning, which is able to generate trajectory data conditioned by control commands. The key part is an addressing network that automates the selection of specialized models implemented by an ensemble of subnetworks. Specialization of the subnetworks during training is achieved through a self-organized partitioning of the training data set that is based on the loss profiles of the data points with the subnetworks as predictors.

We successfully simulated the dynamics of agents conditioned by control commands in three simulated environments (Fig. 4). The architecture successfully generated a well-working simulation of environmental dynamics. In direct comparison to the environment, it can approximate these dynamics with an error of about $1e^{-6}$. In autoregressive runtime, the proposed architecture generates agent's behavior in a similar fashion, when comparing it to the original environments. In the latter we could determine, that the addressing NN's choice which model to use is in line with the lowest loss for 77.5% of all training data points at the end of a training process. The addressing NN predicted the best subnetwork from the ensemble with a high accuracy. The loss values for the data points, which were chosen for a specific model, were low compared to those not chosen (see Appendix A).

Furthermore, we showed that the proposed architecture performs well on the training set of the PFNN database [2]. In comparison to the PFNN model provided in the original implementation, the mean absolute error of the present model was lower than for the PFNN model. However, the PFNN in the code implementation uses regularization (dropout layers after each hidden layer with 0.7 dropout rate) and we still need to measure the performance of the proposed architecture in the autoregressive mode in the virtual environment provided by the PFNN implementation.

As a future work we can investigate several methods to perform a forward-step prediction. In the default case, we compute the output $g(X) = (b_1, ..., b_n)$, where b_k is an estimation value representing model $k \in f_1, ..., f_n$ and apply $argmax(b_1, ..., b_n)$ to determine which model to choose. We can also use $(b_1, ..., b_n)$ as a probability of which model to choose to further the fluctuation between subnetworks. Another variant is using output $(b_1, ..., b_n)$ as the weight for outputs computed by the subnetworks to create a mixture of states. At last, to create a pseudo-model by creating a weighted mixture of parameters of each subnetwork $f_1, ..., f_k$.

A   Appendix - Performance of the Architecture in the Acrobot Environment

The prediction generated by the addressing NN and the losses computed between the output of the ensemble models and the target outputs correlated as designed, resulting in a 77.5% overlap between addressing NN prediction and minimal loss. Figure 8 shows this correlation. The distribution of about 200,000 data points is approximately uniform, excluding *model* 3 and *model* 7. In theory, each model should perform on 25,000 data points, which is mostly the case. The addressing network performs slightly worse in terms of uniformity, having also *model* 1 as an additional outlier. The reasons for that are hypothesized in the discussion.

In Fig. 11, we can observe the prediction of the addressing NN when distributing data points over models. On the x-axis the predicted values are divided into

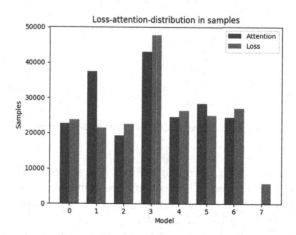

Fig. 8. data points per model in terms of distribution by minimal loss and by prediction.

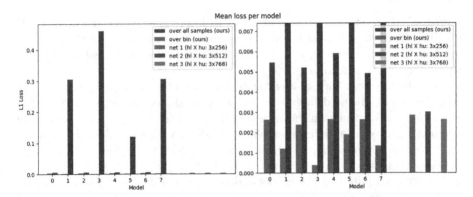

Fig. 9. Mean losses of all data points per ensemble model (right column) and of the distributed data points per ensemble model (left column). Additionally, mean losses for three fully connected neural networks. (right: focus on the distributed losses)

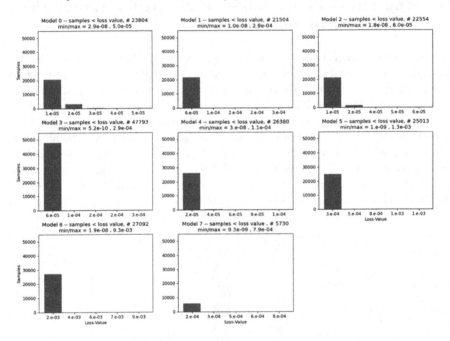

Fig. 10. Losses of the distributed data points.

intervals of $[0, ..., 0.1]$, $[0.1, ..., 0.2]$ and so forth. The shown x-value represents the upper bound of the interval. The accumulated number of points in the corresponding interval is shown on the y-axis. Most of the data points get distributed between a value of 0.9 to 1.0 in all models.

Additionally, Fig. 10 shows the losses of the distributed data points. After the data points get distributed in clusters by the addressing NN, the chosen models perform a forward step on these data points and the L1-loss is computed. They again get subdivided into ten equally sized intervals, ranging from lowest to highest loss per model. As we can see, most data points fall into the first interval with the lowest upper bound.

Next, we calculate the mean loss over the training data, as well as the mean loss over the clusters of training data provided by the sorting process for each ensemble network. For comparing the performance of the architecture to traditionally used feed-forward NNs, we also trained three fully connected models with three hidden layers of different sizes per model on the same training data. The training duration and optimizer settings were the same as for the architecture, but the activation function was changed to ReLu. Then the mean loss was computed for each of the three models over the whole training data. In Fig. 9, this can be observed. The performance of each ensemble model over the whole training data is worse than the fully connected models, although they exhibit a lower mean loss on each of the sorted training clusters. Furthermore, a low loss on the specific training cluster corresponds to a high loss on all training data.

In Fig. 9 we show, that the architecture is able to outperform basic NNs, which are trained on all training data points. Specialized models had a bigger mean loss over the complete training set. However, the mean loss over the clusters of training data provided by the sorting process for each subnetwork could be shown to be lower (see Fig. 9). The training process of the NN architecture could, therefore, be interpreted as more sample efficient as the commonly used

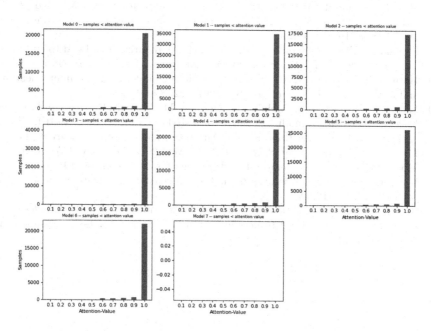

Fig. 11. Prediction values of the addressing neural network for data points for distributing them over the 8 models.

training procedure because the subnetworks were trained on a lower number of data points due to the sorting mechanism. The here developed technique could also dynamically subdivide training data into coherent clusters, such that the subnetworks learned specific parts of the dynamics, which an environment can provide.

References

1. Van Dijk, T.A., Kintsch, W., et al.: Strategies of Discourse Comprehension. Academic Press, New York (1983)
2. Holden, D., Komura, T., Saito, J.: Phase-functioned neural networks for character control. ACM Trans. Graph. (TOG) **36**(4), 42 (2017)
3. Fort, S., Hu, H., Lakshminarayanan, B.: Deep ensembles: a loss landscape perspective (2019). arXiv preprint arXiv:1912.02757
4. Pavllo, D., Grangier, D., Auli, M.: Quaternet: a quaternion-based recurrent model for human motion (2018). arXiv preprint arXiv:1805.06485
5. Peng, X.B., Abbeel, P., Levine, S., van de Panne. , M.: Deepmimic: example-guided deep reinforcement learning of physics-based character skills. ACM Trans. Graph. (TOG), **37**(4), 1–14 (2018)
6. Kidziński, Ł., et al.: Learning to run challenge solutions: adapting reinforcement learning methods for neuromusculoskeletal environments. In: The NIPS'17 Competition: Building Intelligent Systems, pp. 121–153. Springer (2018)
7. Schilling, M., Melnik, A.: An approach to hierarchical deep reinforcement learning for a decentralized walking control architecture. In: Samsonovich, A.V. (ed.) BICA 2018. AISC, vol. 848, pp. 272–282. Springer, Cham (2019). https://doi.org/10.1007/978-3-319-99316-4_36
8. Melnik, A., Fleer, S., Schilling, M., Ritter, H.: Modularization of end-to-end learning: case study in arcade games (2019). arXiv preprint arXiv:1901.09895
9. Haruno, M., Wolpert, D.M., Kawato, M.: Mosaic model for sensorimotor learning and control. Neural Comput. **13**(10), 2201–2220 (2001)
10. Konen, K., Korthals, T., Melnik, A., Schilling, M.: Biologically-inspired deep reinforcement learning of modular control for a six-legged robot. In: 2019 IEEE International Conference on Robotics and Automation Workshop on Learning Legged Locomotion Workshop, (ICRA) 2019, 20–25 May 2019, Montreal, CA (2019)
11. Moore, A.W.: Efficient memory-based learning for robot control (1990)
12. Kingma, D.P., Ba, J.: Adam: a method for stochastic optimization (2014). arXiv preprint arXiv:1412.6980
13. Sutton, R.S.: Generalization in reinforcement learning: successful examples using sparse coarse coding. In: Advances in Neural Information Processing Systems, pp. 1038–1044 (1996)

Learn to Move Through a Combination of Policy Gradient Algorithms: DDPG, D4PG, and TD3

Nicolas Bach[(⊠)], Andrew Melnik[(⊠)], Malte Schilling, Timo Korthals, and Helge Ritter

CITEC, Bielefeld University, Bielefeld, Germany
nbach@techfak.uni-bielefeld.de, andrew.melnik.papers@gmail.com

Abstract. Deep Reinforcement Learning has recently seen progress for continuous control tasks, driven by yearly challenges such as the *NeurIPS Competition Track*. This work combines complementary characteristics of two current state of the art methods, Twin-Delayed Deep Deterministic Policy Gradient and Distributed Distributional Deep Deterministic Policy Gradient, and applied this in the state-of-the-art *Learn to move—Walk Around* locomotion control challenge which was part of the *NeurIPS 2019 Competition Track*. The combined approach showed improved results and achieved the 4th place in this competition. The article presents this combination and evaluates the performance.

1 Introduction

The *NeurIPS 2019: Learn to Move—Walk Around*[1] challenge [1,2] poses a continuous control task for a physiologically plausible 3D walking agent in the physics-based OpenSim environment [3] that is to be controlled by activation of muscle fibers attached to the agent (see Fig. 1). The agent is supposed to follow a prescribed 2D velocity vector. The task became incrementally harder compared to the previous *NeurIPS 2018: AI for Prosthetics* challenge, in which the provided 1D velocity vector had always the same direction and only the absolute value was changing.

We solved the task by combining Twin-delayed Deep Deterministic Policy Gradient (TD3) [4] and Distributed Distributional Deep Deterministic Policy Gradient (D4PG) [5] algorithms. Both algorithms are extensions of the Deep Deterministic Policy Gradient (DDPG) [6] and implement several improvements (see Table 1). This solution showed to score an improvement compared to the two algorithms individually and scored fourth place out of 310 teams in this competition. In this paper, we evaluate the feasibility and performance of combining these improvements and compare it to the performance of the two original algorithms in the *NeurIPS 2019: Learn to Move—Walk Around* challenge. The

---

[1] https://www.aicrowd.com/challenges/neurips-2019-learn-to-move-walk-around.

N. Bach and A. Melnik—Equal contribution.

© Springer Nature Switzerland AG 2020
G. Nicosia et al. (Eds.): LOD 2020, LNCS 12566, pp. 631–644, 2020.
https://doi.org/10.1007/978-3-030-64580-9_52

combined algorithm is tested against its components, TD3 and D4PG, in two experiments. Other top ranked solutions for this and previous years challenge variants are described in [2,5,7,8]. Deep Reinforcement Learning methods has been successfully applied in an increasing number of areas, ranging from computer games towards robotic control [9–15].

2   Methods

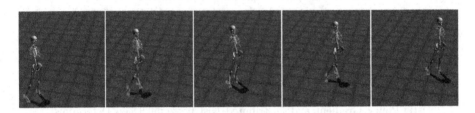

Fig. 1. The task of the competition: Developing a controller capable of locomotion for the skeleton, which can only be controlled via activation of its muscles on its legs. The figure shows a movement of the agent in a sequence of five time steps. Active muscles are shown in red, inactive muscles are shown in blue. (Color figure online)

2.1   Combination of Algorithms

Our algorithm is based on DDPG and combines all improvements (see Table 1 for an overview) introduced by TD3 and D4PG. The implementation of TD3 and D4PG improvements is mostly straightforward (compartmentalization of both algorithms can be seen in algorithm 1; red highlights D4PG and green TD3). The improvements themselves do not intertwine with each other, except for clipped double Q-Learning of TD3 and the distributional value function of D4PG (in algorithm 1 computing the Q-value with twin critics and choosing their minimum is part of the distributional update steps). The combined algorithm should provide a more stable learning signal, while offering the same scalability as found in D4PG. Overall, we assume that it should lead to an improvement, first, compared to TD3 as the sample efficiency is better. Second, in comparison with D4PG, as improvements to stabilize the learning process are deployed.

2.2   Comparison of Deterministic Policy Gradient Algorithms

In the following, we describe experiments that demonstrate the viability of the combination of deterministic policy gradient algorithms for the *NeurIPS 2019: Learn to Move—Walk Around* challenge. We compare the combined approach to its' combinational parts, TD3 and D4PG. Further, we elaborate on the details regarding the challenge, especially the reward function, which had to be optimized, as this appeared detrimental to a reinforcement learning problem.

Algorithm 1 Combination of TD3 (green) and D4PG (red)

Input: batch size M, trajectory length N, number of actors K, replay size R, exploration constant ϵ, initial learning rates α_0 and β_0

1: Initialize critic networks Q_{θ_1}, Q_{θ_2} and actor network π_ϕ replicating network weights to each of the K actors with random network weights θ_1, θ_2, ϕ
2: Initialize target networks $\theta'_1 \leftarrow \theta_1$, $\theta'_2 \leftarrow \theta_2$, $\phi' \leftarrow \phi$
3: **for** $t = 1, ..., T$ **do**
4: Sample mini-batch of M transitions $(\mathbf{x}_{i:i+N}, \mathbf{a}_{i:i+N-1}, r_{i:i+N-1})$
   of length N for replay buffer with priority p_i
5: Generate target action $\tilde{a} \leftarrow \pi'_\phi(s_{i:i+N-1}) + \epsilon, \quad \epsilon \sim clip(\mathcal{N}(0, \tilde{\delta}), -c, c)$
6: Construct target distributions

$$Y_i = (\sum_{n=0}^{N-1} \gamma^n r_{i+n}) + \gamma^N min_{j=1,2} Z^j_{w'}(\mathbf{x}_{i+N}, \pi_{\theta'}(\mathbf{x}_{i+N}))$$

   for critics j = 1,2
7: Compute the actor and critics updates

$$\delta_{\theta_i} = \frac{1}{M} \sum_i \nabla_w (Rp_i)^{-1} d(Y_i, argmin_{Z_j} Z^j_w(\mathbf{x}_i, \mathbf{a}_i))\Big|_{j=1,2}$$

$$\delta_\phi = \frac{1}{M} \sum_i \nabla_\theta \pi_\theta(\mathbf{x}_i) \mathbb{E}[\nabla_{\mathbf{a}} argmin_{Z_j} Z^j_w(\mathbf{x}_i, \mathbf{a})]\Big|_{\mathbf{a}=\pi_\theta(\mathbf{x}_i), j=1,2}$$

8: Update critics $\theta_i \leftarrow \theta_i + \beta_t(\delta_{\theta_i})$
9: **if** t mod d **then**
10:   Update $\phi \leftarrow \phi + \alpha_t \delta_\phi$ and replicate network weights to the actors
11:   Update target networks θ'_i and ϕ'
12: **end if**
13: **end for**
14: **return:** policy parameters ϕ

Actor

1: **repeat**
2: Sample action $\mathbf{a} = \pi_\theta(\mathbf{x}) + \epsilon \mathcal{N}(0, 1)$
3: Execute action $\mathbf{a}$, observe reward r and state $\mathbf{x}'$
4: Store $(\mathbf{x}, \mathbf{a}, r, \mathbf{x}')$
5: **until** learner finishes

The same configuration for all three algorithms was used to evaluate their characteristics. Actor and critic neural networks (TD3 and our algorithm operated with a pair of critics) were given three hidden layers with sizes $(512, 512, 256)$ in all three cases. We used Gaussian noise for exploration. Further, for each algorithm we deployed a trainer thread to perform update steps and 22 sampler threads to produce samples in parallel, which were necessary for the updates, and store them in a shared replay buffer. Although originally TD3 has no distributed training framework, we extended the algorithm for better evaluation of the other improvements. A sampler is a copy of the policy network, which acts on the environment, whereas the trainer contains the algorithm, which optimizes the policy and value function, and copies the weights of the policy functions to the sampler threads every 500 update steps.

Table 1. List of components of the deterministic policy gradients. A detailed description of each component can be found in the appendix. The components of our combined approach are shown in column four.

Component	Approach			
	DDPG	TD3	D4PG	**Ours**
Deterministic policy gradient	✓	✓	✓	✓
Target policy and value networks	✓	✓	✓	✓
Explorative noise	✓	✓	✓	✓
Experience replay buffer	✓	✓		
Clipped Double Q-Learning		✓		✓
Delayed update of policy networks		✓		✓
Target policy smoothing		✓		✓
Multiple sampler			✓	✓
Distributional critic			✓	✓
N-step returns			✓	✓
Prioritized experience replay buffer			✓	✓

The implementation of prioritized experience replay [16] that is employed by D4PG and our approach uses the absolute TD-errors as sample-weighing-strategy and produces batches dependant on those weights, favoring more important samples. TD3 was implemented with a regular replay buffer, which is sampled uniformly. For realizing the distributional critic used in D4PG and our algorithm, we used a quantile distribution [17] consisting of 101 atoms. N-step returns were set to five. Delayed updates of policy networks were executed in a ratio of two critic updates for every actor update. For target policy smoothing we used Gaussian noise.

2.3   OpenSim Environment

The *NeurIPS 2019: Learn to Move—Walk Around* challenge poses the task to control a physiologically plausible 3D walking agent in the physics-based OpenSim environment [3] only by activation of muscle fibers attached to the agent. The activation range of the muscles spans the continuous space between 0 and 1. The agent has 22 muscles distributed over its lower body, so the action space amounted to twenty-two dimension. The agent is supposed to follow a provided 2D velocity vector field. This vector field V is a $2 \times 11 \times 11$ tensor of 2D velocities in forward and leftward direction of the agent. It spans a 11×11 grid within 5 m around the agent with the agent at its center. The distance between each discrete point in the grid amounts to 0.5 m (as can be seen in Fig. 2). The vector field is one part of the observation space the agent could access. The second part of the provided observation space is a dictionary of 97 observations for pelvis state, ground reaction forces, joint angles, and velocities, as well as muscle states,

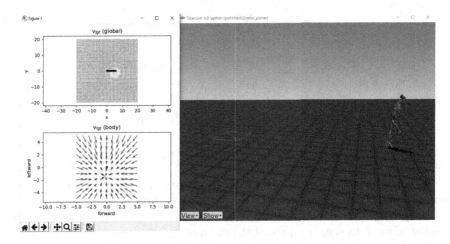

Fig. 2. The competition's environment. Based on OpenSim it provides a 3D environment, in which the agent should be controlled, and a velocity field to determine the trajectory the agent should follow. (source: [18])

such as their length. Therefore, the accessible observation space amounts to 341 dimensions. Our solution took only into account the actual target velocity in the agents position, as well as the difference between target velocity and real velocity, resulting in an observation space of 103 dimensions for our agent.

The environment provided two different reward functions for round one and two of the *NeurIPS 2019: Learn to Move—Walk Around* on which the agent was optimized. We used the reward function of the second round to conduct the experiments described in Sect. 3, which was provided by the competition's environment[2]. It was not shaped in any form. The environment returned reward in each step (dense reward). The total reward $J(\pi)$ is described as a sum of reward for staying alive and reward for performing footsteps, where the latter was defined as bridging a minimum distance between contact with the ground, while traveling in the right direction and using minimal effort in terms of muscle activation. The maximum number of steps in the first round was set to 1000 and in round two to 2500 steps per episode.

$$J(\pi) = R_{alive} + R_{step} \tag{1}$$

$$= \sum_i r_{alive} + \sum_{step_i} (w_{step}r_{step} - w_{vel}c_{vel} - w_{effort}c_{effort}) \tag{2}$$

In Eq. (2) w_{step}, w_{vel} and w_{effort} refers to a constant weight of the stepping reward as well as to the weights for effort and velocity costs. The costs and rewards are defined as

$$r_{alive} = 0.1 \tag{3}$$

---

[2] https://github.com/stanfordnmbl/osim-rl.

$$r_{step} = \sum_{i \ in \ step_i} \Delta t_i = \Delta t_{step_i} \qquad (4)$$

$$c_{vel} = \left\| \sum_{i \ in \ step_i} (v_{pelvis} - v_{vectorfield}) \Delta t_{step_i} \right\| \qquad (5)$$

$$c_{effort} = \left\| \sum_{i \ in \ step_i} \sum_{m}^{muscles} A_m^2 \Delta t_{step_i} \right\| \qquad (6)$$

with $\Delta t_i = 0.01\,s$ is the simulation time step. v_{pelvis} and $v_{vectorfield}$ are the velocity of the pelvis and the target velocity and A_m are muscle activations.

A bonus of 500 was given by successfully standing near the target for a time period (two to four seconds). This bonus could be achieved twice, as after achieving the first reward a new velocity field spawned, which the agent also needed to solve to successfully end the episode.

3   Experiments

In the first experiment, we ran a test agent with no exploration noise every 500 updates of policy weights and collected reward values for that episode. For each algorithm, we repeated the training process with a different seed three times. The average result of three tests for each algorithm is plotted in Fig. 3. One repetition of the training process amounted to 1,000,000 update steps of the trainer thread. As described above, during training the episode length was constrained to 1000 steps per episode. This limits the reward, which could be achieved, by the number of steps.

For the second experiment, we tested the performance of the trained networks from the first experiment. We ran each of the 9 agents (3 algorithms × 3 seeds) on the same 50 episodes after the absolved training process (1,000,000 updates steps). The maximum length of an episode was set to 2500 steps. As mentioned above, this was established in the second round of the learning to run challenge, where the agent's task was to solve two velocity fields in one episode. Thereby, the reward was limited. The networks were compared in terms of reward earned and steps taken.

3.1   Results of the Experiments

The results for experiment 1, which are depicted in Fig. 3, show that a combination of algorithm converges faster to a well-performing policy, enabling the agent to achieve the second round's reward of 500. Due to restricted episode length of 1000 during training the agent was not able to solve whole episodes of the environment (default settings for difficulty 2 of the environment are 2500 steps per episode). Further, the reward is maxed out by around 350 for an episode. This relates to the spikes in Fig. 3. The agent was not able to achieve the second bonus, as the episode length was constrained to 1000 steps per episode during training. The spikes amount to a reward of around 350 as they are averaged over

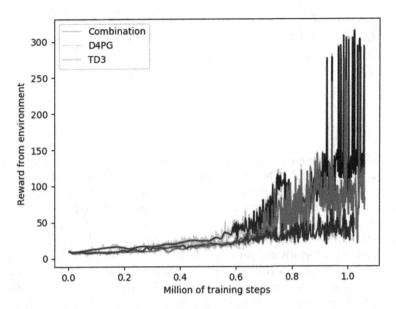

Fig. 3. Results of experiment 1. In later stages of training the agent was able to achieve a bonus reward of 500 by standing at its target for multiple seconds, resulting in spikes in the later stages of training. However, the rewards of the episodes only amount to about 350 in average because of their occurrence rate and difference in time during training runs.

episodes, where the agent did not achieve bonus due to different occurrence rates in all seeds. Moreover, we can also observe a smoother trend of the curve until convergence (around 600,000 steps) for the combination of algorithms in comparison to D4PG. In later stages of the training process we could also observe, that our approach has less low-reward outliers as D4PG. TD3 was not able to produce a policy, which was able to score more than a reward of around 50 and therefore wasn't able to produce any high-reward outliers by scoring the bonus reward. D4PG was able to score the round two bonus, but at a later stage of the training than our proposed algorithm.

In experiment 2 (see Table 2), we observed similar results. The trained policy of our combined approach was able to outperform both TD3 and D4PG. After the finished training, the TD3 algorithm scored worse compared to D4PG and our approach in terms of average steps and reward. Our proposed combined approach was able to produce a policy, which scores about 30% higher on average than D4PG in terms of average reward and average steps taken in episodes. We were also able to decrease the standard deviation in reward by about 20% and in steps by about 15%, which implies less proneness to the failure mode.

Table 2. Results of second experiment.

Approach	Reward - mean	Reward - std	Steps - mean	Steps - std
Ours	354.9	195.1	2070.5	703.8
D4PG	265.1	242.3	1448.0	847.3
TD3	25.6	1.9	259.4	6.0

4   Discussion

We found that combining the algorithms improved the results. TD3 couldn't solve the task at hand. The policy produced was not able to exhibit walking behavior and finished each episode abruptly with falling down in runtime with a frozen model. During training, TD3 was also not able to improve its behavior, such that the standing bonus could never be achieved. In general, TD3 fell off behind D4PG and our approach. This could be due to the fact, that certain improvements like prioritized sampling or n-step returns are helpful features for solving the challenge posed in this particular environment. D4PG was able to exhibit better performance compared to TD3. On runtime, a frozen D4PG policy was able to move around and in some episodes earn the bonus by standing in the middle of the target for an amount of time.

The combined approach was able to perform even better than D4PG. It maximized reward faster than D4PG and showed to be more stable, as the training curve of our algorithm has less low-reward outliers than D4PG. It also scored higher in the second experiment than TD3 and D4PG (see Table 2), while having a smaller standard deviation of reward and steps than D4PG. However, it was not able to fully solve an environment of the second round. This might be due to the fact that we chose to reduce the episode length to 1000 steps during training.

All in all, in the comparison to its components, we could not find any unfavorable repercussions for the integrated approach in the two experiments by combining the here mentioned improvements of D4PG and TD3.

A   Appendix

A.1   Deterministic Policy Gradient Algorithms

Environments with continuous action spaces come a bit closer to the reality. Although, they are often more difficult to solve, it is necessary to be able to solve problems in this group, when we try to make progress towards algorithms we can deploy in the real world to e.g. build controllers for robots. In the *NeurIPS 2019: Learn to Move—Walk Around* challenge using a continuous action space as design choice is justified by imitating a humanoid 3D model more realistically. This humanoid is controlled by activation of the muscles on his legs. The action space consists of activation intervals for different muscles, from which a value can be sampled from. The correct sampling of a value to produce a behaviour pattern is the goal of the environment provided by the challenge. One of the current state-of-the-art algorithms to solve such types of environments is the model-free off-policy algorithm Deep Deterministic Policy Gradient (DDPG,[6]) as well as its improved versions, twin-delayed DDPG (TD3, [4]) and distributed, distributional DDPG (D4PG, [5]). In the following we describe these algorithms in more detail.

Deep Deterministic Policy Gradient (DDPG). In this work we use DDPG as baseline algorithm for solving locomotion in reinforcement learning problems. DDPG is an off-policy, model-free algorithm and it is able to solve problems in environments with continuous action spaces. It can be seen as a variant of Deep Q-networks, as it combines Deterministic Policy Gradients (DPG, [19]) with Q-Learning and other extensions, namely experience replay and target value and policy networks. DDPG is furthermore an actor-critic-algorithm and consists of 4 different neural networks in total: The actor, the critic and both target actor and target critic. The actor is also called the policy function $\pi(s) = a$, which computes an action for a given state. The critic, also refered to as Q-value function, computes the Q-value, a numerical value that represents the discounted future reward for a state-action-pair. The critic is also the main objective to be optimized, such that we find the maximal, real Q-value for given state-action-pairs. We can derive the optimal Q-Value function $Q^*(s, a)$ by minimizing the loss between the output of the function approximator and the bellman-equation:

$$Q^*(s_t, a_t) = \mathbb{E}[r(s_t, a_t) + \gamma max_{a_{t+1}} Q'^*(s_{t+1}, \pi'(s_t))] \tag{7}$$

This computes the Q-value for a given time step t. The discount rate γ diminishes additional reward of steps into the future. This has the effect, that immediate reward is given a preference over future reward. The value and policy function on the right-hand site of the equation are the target value and policy function. Their functions are discussed in Sect. A.1.

Given the target function and the neural networks as function approximators (given by actor and critic networks) we now can derive the loss function L:

$$L(\theta^Q) = \mathbb{E}[(Q(s_t, a_t | \theta^Q) - Y_t)^2] \tag{8}$$

where Y_t is the target in a supervised learning sense and is computed by using the Bellman-equation as intermediate optimum:

$$Y_t = r(s_t, a_t) + \gamma Q(s_{t+1}, \mu(s_{t+1} | \theta^Q) \tag{9}$$

θ^Q are the function parameters for policy μ and value function Q. In the next sections we discuss how to update the parameters of a policy by the optimization step of the value function and the components experience replay buffer, exploration noise and target policy and value networks.

Deterministic Policy Gradients. In an environment which provides a continuous action space we can derive a deterministic policy by using Deterministic policy gradient (DPG). Rather than returning a probability distribution over actions $\mathcal{A}$ given a state, a deterministic policy $\mu(s) = a$ returns a single action in a deterministic way. The main objective $J(\theta)$ in an off-policy actor-critic algorithm, which mainly optimizes the value function is defined as:

$$J(\theta) = \int_{\mathcal{S}} \rho^{\mu}(s)Q(s, \mu_{\theta}(s))ds \tag{10}$$

where θ are the parameters and $\mathcal{S}$ is the state space.

$$\rho^{\mu}(s') = \int_{\mathcal{S}} \sum_{k=1}^{\inf} \gamma^{k-1} \rho_0(s)\rho^{\mu}(s \to s', k)ds \tag{11}$$

is defined as the discounted sum of state visitation probability density at state s'. $\rho^{\mu}(s \to s', k)$ gives us the probability density from state s to state s' after moving k steps by using policy μ. $\rho_0(s)$ is the initial distribution over states.

We can now compute the gradient of $J(\theta)$ using the Deterministic policy gradient theorem.

$$\nabla_{\theta}J(\theta) = \int_{\mathcal{S}} \rho^{\mu}(s)\nabla_a Q^{\mu}(s, a)\nabla_{\theta}\mu_{\theta}(s)|_{a=\mu_{\theta}(s)}ds \tag{12}$$

$$= \mathbb{E}_{s \approx \rho^{\mu}}[\nabla_{\theta}\mu_{\theta}(s)\nabla_a Q^{\mu}(s, a)|_{a=\mu_{\theta}(s)}] \tag{13}$$

First, the chainrule yields the gradient of Q $\nabla_a Q^{\mu}(s, a)$ with respect to a. Second, we derive the gradient of the deterministic policy $\nabla_{\theta}\mu_{\theta}(s)$ with respect to *theta*, which optimizes our policy. As an example to show how to compute updates, consider DPG in combination with on-policy actor-critic policy $SARSA$. First, we compute the TD-error in $SARSA$:

$$\delta_t = R_t + \gamma Q_w(s_{t+1}, a_{t+1}) - Q_w(s_t, a_t) \tag{14}$$

The parameter update of the value function is defined as:

$$w_{t+1} = w_t + \alpha_w \delta_t \nabla_w Q_w(s_t, a_t) \tag{15}$$

Then, we can use the Deterministic policy gradient theorem to compute policy parameter updates of θ using Eq. 12:

$$\theta_{t+1} = \theta_t + \alpha_{\theta}\nabla_a Q^{\mu}(s, a)\nabla_{\theta}\mu_{\theta}(s)|_{a=\mu_{\theta}(s)} \tag{16}$$

One problem of using DPG is exploration because of the deterministic nature of the policy we optimize. On way to prevent this is to add noise to the parameter space or action space, which in this case would result in an off-policy nondeterministic policy.

Exploration Noise. As mentioned in Sect. A.1, DPG updates could inhibit exploration depending on the environment. To ensure exploration in the continuous action space, DDPG uses an exploration policy μ', in which noise is added to the actions of the policy network μ.

$$\mu'(s_t) = \mu(s_t) + \mathcal{N} \tag{17}$$

$\mathcal{N}$ denotes noise sampled from a noise generating process, such as Gaussian noise. The authors of the DDPG paper suggest using the Ornstein-Uhlenbeck process [20] for exploring physical environments, as it allows temporally correlated exploration.

Target Value and Policy Networks. DDPG utilizes frozen copies of value and policy function to compute the target Y_t (Eq. 9). More specifically, they are used to compute the right-hand site of the bellman-equation, as it was found that the learn process gets less stable, when not using copies due to the change of weights during optimization. Thus, the learning process consists of the following steps: first, a batch of training data is sampled from the experience buffer. Second, the loss L (Eq. 8) is computed using the target value and policy networks to generate Y_t. After update steps of value and policy networks, the target networks get softupdated by:

$$\theta^{Q'} \leftarrow \tau\theta^Q + (1 - \tau)\theta^{Q'} \tag{18}$$

$$\theta^{\mu'} \leftarrow \tau\theta^\mu + (1 - \tau)\theta^{\mu'} \tag{19}$$

θ^Q and θ^μ are the parameters of the value network and the policy network, $\theta^{Q'}$ and $\theta^{\mu'}$ are the parameters of the target value and the policy network. The constant $\tau \ll 1$ is a hyperparameter, that realizes the soft update by scaling down the update step, so that the parameters of the target networks change slower than those of the actor and critic networks.

Experience Replay Buffer. An experience sample typically consists of the tuple $s = \{s_n, a, r, d, s_{n+1}\}$, where s_n is the current and s_{n+1} the next state, a is the action, r is the reward and d is a boolean indicating, whether an episode is over or not. DDPG makes use of an experience replay buffer, in which samples generated by the interaction of policy and environment are stored and from which batches are sampled to perform updates using the value function, the bellman-equation and DPG.

A.2   Twin-delayed Deep Deterministic Policy Gradient (TD3)

One common problem of DDPG is the overestimation of the Q-value, which in turn results in policy-breaking. Twin-delayed Deep Deterministic Policy Gradient is able to diminish this effect by extending DDPG algorithm with three additional improvements. The first improvement is introducing a second value function network (as in twin-critics) to learn two q-functions. Second, it updates the policy network less frequently than the value networks. The third extension consists of target policy smoothing, i.e. adding a small amount of noise to the output of the target policy network. All these mentioned extensions provide more stability for approximating the optimal policy.

Clipped Double Q-Learning. Addressing overestimation of the Q-value, i.e. a state-action-pair is incorrectly valued too high, the first improvement of TD3 over DDPG is implemented by using two critics or value function networks instead of one (which also means two target critics). The two value functions are optimized with one target Q-function, which uses the minimum of the Q-values estimated by both target functions:

$$Y_t = r(s_t, a_t) + \gamma \min_{i=1,2} Q_{\theta_i}(s_{t+1}, \mu_{\theta_i'}(s_{t+1})) \tag{20}$$

By always choosing the minimum Q-value, it is more difficult for the value functions to develop an overestimation of Q-value for certain inputs.

Delayed Policy Network Updates. Less frequent updates of the policy network ensures, that the value function has a harder time converging on the failure mode, where it overestimates actions incorrectly. In a scenario, where the value function would start overestimating the outputs of a poor policy, additional updates of the value network while keeping the same policy could lead to overcoming the incorrect estimation of the poor performing policy.

Target Policy Smoothing. The third improvement of TD3 is also an improvement of the target Y_t. The action produced by the target policy network, which is utilized in the target Q-function, gets modified by adding a small amount of noise, which is also clipped into an interval. This has the effect of covering a clipped area around the action in the action space, instead of predicting a deterministic action. In case, that the value-function produces a Q-value incorrectly to large for a certain action, adding a clipped amount of noise to the action acts as a regularizer, as the high-valued action gets smoothed by the noise.

A.3   Distributed Distributional Deep Deterministic Policy Gradient (D4PG)

D4PG, similar to TD3, is an extended version of DDPG. It implements 4 additional improvements, which overall address stability and scalability of DDPG. The first improvement, a distributional value function, provides a more stable estimation of the Q-value. Second, the process of gathering experiences is distributed over a number of in parallel acting policy networks, which store their experiences in a shared experience replay buffer. The third improvement, prioritized experience replay, weighs the produced experiences, so that important experiences are more often sampled than others. The last improvement is n-step returns. When computing the TD-error n-step-returns allows a more confident estimation of a state-action-pair by producing a reward over n steps into the future.

Multiple Sampler. To address the sample-inefficiency problem of model-free reinforcement learning, multiple copys of the policy network run in parallel to produce samples and store them in a shared experience buffer. The copys are updated at the same time and the number of sampler can be chosen as required.

Distributional Value Function. D4PG uses a distributional version of critic updates. This means, that expected Q-value is modeled as a random variable, thus the value function maps the input, a state-action-pair, to a distribution Z_w, which is distributed over w. Given $Q_w(s, a) = \mathbb{E}Z_w(x, a)$, the loss for the distributional function is given by minimizing the distance between two distributions $L(w) = \mathbb{E}[d(\mathcal{T}_{\mu_\theta}, Z_{w'}(s, a), Z_w(s, a)]$, where $\mathcal{T}_{\mu_\theta}$ is the Bellman operator. As [21] show, this improvement results in a more stable learning signal.

N-Step Returns. When constructing the target and doing the forward step of the value network for computing the loss, this improvement incorporates computing the sum of rewards of n-steps instead of having a one-step reward. The target incorporating n-step returns is computed by:

$$Y_t = \sum_{n=0}^{N-1} \gamma^N r_{t+n} + \gamma^N Q_{\theta'}(s_{t+N}, \mu_{\theta'}(s_{t+N})) \tag{21}$$

This estimates future reward more accurately.

Prioritized Experience Replay. Instead of sampling uniformly from the replay buffer, the samples stored in the prioritized experience replay buffer are weighted with an importance weight and are sampled with a non-uniform probability p_i. The weight, which adjust the probability can, e.g. be realized by the TD-error. This would have the effect, that samples with high TD-error get sampled more often than others.

References

1. AIcrowd.com: Neurips 2019: Learn to move - walk around (2019)
2. Kidziński, Ł., et al.: Learning to run challenge solutions: adapting reinforcement learning methods for neuromusculoskeletal environments. In: Escalera, S., Weimer, M. (eds.) The NIPS '17 Competition: Building Intelligent Systems. TSSCML, pp. 121–153. Springer, Cham (2018). https://doi.org/10.1007/978-3-319-94042-7_7. CoRR, abs/1804.00361

3. Seth, A., Sherman, M., Reinbolt, J.A., Delp, S.L.: OpenSim: a musculoskeletal modeling and simulation framework for in silico investigations and exchange. Procedia Iutam **2**, 212–232 (2011)
4. Fujimoto, S., van Hoof, H., Meger, D.: Addressing function approximation error in actor-critic methods. arXiv preprint arXiv:1802.09477 (2018)
5. Barth-Maron, G., et al.: Distributed distributional deterministic policy gradients. arXiv preprint arXiv:1804.08617 (2018)
6. Lillicrap, T.P., et al.: Continuous control with deep reinforcement learning. arXiv preprint arXiv:1509.02971 (2015)
7. Kolesnikov, S., Khrulkov, V.: Sample efficient ensemble learning with catalyst. RL. arXiv preprint arXiv:[WIP] (2019)
8. Kolesnikov, S., Hrinchuk, O.: Catalyst. RL: a distributed framework for reproducible RL research. arXiv preprint arXiv:1903.00027 (2019)
9. Schilling, M., Konen, K., Ohl, F.W., Korthals, T.: Decentralized deep reinforcement learning for a distributed and adaptive locomotion controller of a hexapod robot (2020)
10. Korthals, T., Melnik, A., Leitner, J., Hesse, M.: Multisensory assisted in-hand manipulation of objects with a dexterous hand. juxi.net (2019)
11. Konen, K., Korthals, T., Melnik, A., Schilling, M.: Biologically-inspired deep reinforcement learning of modular control for a six-legged robot. In: 2019 IEEE International Conference on Robotics and Automation Workshop on Learning Legged Locomotion Workshop, (ICRA) 2019, Montreal, CA, 20–25 May 2019 (2019)
12. Melnik, A., Fleer, S., Schilling, M., Ritter, H.: Modularization of end-to-end learning: case study in arcade games. arXiv preprint arXiv:1901.09895 (2019)
13. Schilling, M., Melnik, A.: An approach to hierarchical deep reinforcement learning for a decentralized walking control architecture. In: Samsonovich, A.V. (ed.) BICA 2018. AISC, vol. 848, pp. 272–282. Springer, Cham (2019). https://doi.org/10.1007/978-3-319-99316-4_36
14. Korthals, T., Hesse, M., Leitner, J., Melnik, A., Rückert, U.: Jointly trained variational autoencoder for multi-modal sensor fusion. In: 2019 22th International Conference on Information Fusion (FUSION), pp. 1–8. IEEE (2019)
15. Melnik, A., Bramlage, L., Voss, H., Rossetto, F., Ritter, H.: Combining causal modelling and deep reinforcement learning for autonomous agents in minecraft (2019)
16. Horgan, D., et al.: Distributed prioritized experience replay. arXiv preprint arXiv:1803.00933 (2018)
17. Dabney, W., Rowland, M., Bellemare, M.G., Munos, R.: Distributional reinforcement learning with quantile regression. In: Thirty-Second AAAI Conference on Artificial Intelligence (2018)
18. Neurips 2019 learn to move-environment. http://osim-rl.stanford.edu/docs/nips2019/ (2019)
19. Silver, D., Lever, G., Heess, N., Degris, T., Wierstra, D., Riedmiller, M.: Deterministic policy gradient algorithms (2014)
20. Uhlenbeck, G.E., Ornstein, L.S.: On the theory of the brownian motion. Phys. Rev. **36**(5), 823 (1930)
21. Bellemare, M.G., Dabney, W., Munos, R.: A distributional perspective on reinforcement learning. In: Proceedings of the 34th International Conference on Machine Learning, vol. 70, pp. 449–458. JMLR.org (2017)

ℓ_1 Regularized Robust and Sparse Linear Modeling Using Discrete Optimization

Mahdi Jammal[1,2]($\boxtimes$), Stephane Canu[1], and Maher Abdallah[2]

[1] INSA de Rouen, Rouen, France
`mahdi.jammal@insa-rouen.fr`
[2] Lebanese University, Beirut, Lebanon

Abstract. In regression, feature selection is an effective strategy to handle contaminated data and to deal with high dimensionality while providing better prediction. In addition to the presence of spurious variables, estimators suffer form corrupted, incorrectly measured or misrecorded observations known as outliers. The natural way to select relevant variables and to detect outliers is done by using the ℓ_0 norm for both aspects and recast the obtained optimization problem as a mixed integer optimization (MIO) problem. The ℓ_0 norm estimators perform well when the signal to noise ratio (SNR) is high. However, its performance decreases when the SNR is low due to the overfitting behavior of the ℓ_0 norm when the noise is relatively high. To fix this problem, we propose to regularize the ℓ_0 norm problem for variable selection and outlier detection by adding an ℓ_1 penalty term. We also propose an efficient and scalable non-convex proximal alternate algorithm producing high quality solution in a short time and used as a warm start for the MIO solver. An empirical comparison between the ℓ_0 norm approach and its ℓ_1 regularized extension is presented as well. Results provided that the MIO regularized approach and its discrete first order warm start provide high quality solutions and performs better then the ℓ_0 approach especially for low SNR values.

Keywords: Discrete first order algorithms · Regularization · Mixed integer programming

1 Introduction

We consider the linear regression model:

$$y = X\beta + \epsilon.$$

Where $y \in \mathbb{R}^n$ is the response vector, $X \in \mathbb{R}^{n \times p}$ is the model matrix, $\beta \in \mathbb{R}^p$ is the vector of regression coefficients and $\epsilon \in \mathbb{R}^n$ is the error vector. We assume that the columns of X have been standardized to have zero means and unit ℓ_2-norm.

In high-dimensional regimes i.e $p \gg n$, it is desired to estimate β by a sparse vector, that is a vector with few nonzero elements. To this end, feature selection

© Springer Nature Switzerland AG 2020
G. Nicosia et al. (Eds.): LOD 2020, LNCS 12566, pp. 645–661, 2020.
https://doi.org/10.1007/978-3-030-64580-9_53

has been of great importance in the last few decades [1]. A natural way to compute sparse regression coefficients is to solve the, well known, best subset selection problem:

$$\min_{\beta \in \mathbb{R}^p} \tfrac{1}{2}\|X\beta - y\|_2^2$$
$$\text{s.t. } \|\beta\|_0 \le k_v \tag{1}$$

Where the ℓ_0 norm of a vector β counts the number of nonzeros in β. This classical problem dates back to at least [2,3]. It has been considered as intractable since it is an NP-hard problem [1,4]. To overcome the computational difficulty of the best subset selection, [5] proposed an ℓ_1 relaxation of the cardinality constraint, widely known as the "lasso":

$$\min_{\beta \in \mathbb{R}^p} \tfrac{1}{2}\|X\beta - y\|_2^2$$
$$\text{s.t. } \|\beta\|_1 \le k \tag{2}$$

The popularity of Lasso is due to its computational feasibility with the guarantee of getting a sparse model with good predictive performance. There have been an impressive amount of works studying statistical properties of Lasso and propos- ing algorithms to solve it [6–9] and the books or surveys [10–12]. Throughout the years, researchers thought that best subset selection should be used whenever it is possible. Unfortunately, best subset selection is NP-hard and popular imple- mentations such as the **R** package leaps do not scale to problem sizes larger than $p = 30$. To this end, Problem (1) was considered to be an intractable problem. In a recent work, [13] showed that it is possible to find near optimal solution for high dimensional regimes in minutes (even though it takes hours to prove opti- mality) by formulating it as a mixed integer optimization problem (MIO). They also claimed that the best subset selection performs better than its competitors (lasso for example). However, this claim was refuted by [14], and best subset selection is no more the "holy grail" estimator for sparse modeling in regression: it overperforms other estimators for high signal to noise ratio (SNR) values, while lasso ensures better predictive performance for low SNR values. In fact, best subset selection suffers from overfitting. To overcome the overfitting of the best subset selection, [15] suggested to add an ℓ_q penalty term to the objective function of Problem (1), where $q \in \{1, 2\}$, so that the obtained problem is the following:

$$\min_{\beta \in \mathbb{R}^p} \tfrac{1}{2}\|X\beta - y\|_2^2 + \lambda\|\beta\|_q^q$$
$$\text{s.t. } \|\beta\|_0 \le k_v \tag{3}$$

The proposed method *mitigates, to a large extent, the poor predictive perfor- mance of best-subsets in the low SNR regimes.*

The quality of estimators is known to be very sensitive to the presence of corrupted observations (outliers). Dealing with the presence of outliers can be divided into two categories: (a) the so-called "robust statistics", that is robust- to-outlier loss functions and (b) outlier detection per se which exclude outliers from the training set. In category (a), [16] is a relevant reference (see for instance chapters 3, 6 and 7). However, in category (b), [17–20] are worth mentioning. For both categories (a) and (b) [21,22] offer comprehensive references.

To solve the robust sparse regression problem, [23] proposed the least absolute deviation (LAD) loss to deal with outliers, together with the ℓ_1 Lasso penalty, that is for $\lambda \geq 0$:

$$\min_{\beta} \|X\beta - y\|_1 + \lambda\|\beta\|_1.$$

Later, [24,25] showed that this problem was a particular case of a more general class of problems. To model outliers a sparse variable τ was introduced in [26] for the first time. The idea was to minimize $\|\beta\|_1 + \|\tau\|_1$ s.t $y = X\beta + \tau$. This formulation is the convex relaxation of minimizing $\|\beta\|_0 + \|\tau\|_0$ s.t $y = X\beta + \tau$ as claimed by [27]. The same idea was developped in many works [19,28] for example. Later and for the first time, the use of the ℓ_0 norm for both variables β and τ was proposed by [17]. They proposed the brute force algorithm defined, for some $q \in \{1, 2\}$, by:

$$\min_{\beta \in \mathbb{R}^p, \tau \in \mathbb{R}^n} \frac{1}{q}\|X\beta + \tau - y\|_q^q$$
$$\text{s.t.} \quad \|\beta\|_0 \leq k_v \tag{4}$$
$$\|\tau\|_0 \leq k_o,$$

This formulation allows the selection of relevant variables and the avoidance of outliers, it can be solved by recasting it as a mixed integer optimization problem. Furthermore, as it is consists of double ℓ_0 norms, this formulation performs well for high SNR values, and its performance degrades as the SNR value decreases. To this end, we propose to regularize Problem (4) by adding an ℓ_1 penalty term as done in [15] to overcome the limitation of the ℓ_0 norme when the SNR is low. The proposed problem is defined by:

$$\min_{\beta \in \mathbb{R}^p, \tau \in \mathbb{R}^n} \frac{1}{2}\|y - X\beta - \tau\|_2^2 + \lambda\|\beta\|_1$$
$$\text{s.t} \quad \|\beta\|_0 \leq k_v \tag{5}$$
$$\|\tau\|_0 \leq k_o$$

Contribution:

We summarize the contribution of this paper below:

1. We propose to recast Problem (5) as a mixed integer optimization problem and use an efficient solver to solve it. Note that the suboptimality of the obtained solution is guaranteed even if we terminate the algorithm early,
2. We introduce an algorithm based on a discrete extension of first order continuous optimization methods. This framework is scalable and provides a local solution often close to the global one.
3. We propose to accelerate the MIO by using the discrete first order algorithm as a warm start.
4. We present computational results on both real and synthetic data sets. The results show that the proposed method performs well for low and high SNR values.

The remainder of the paper is organized as follows. In Sect. 2, we present the mixed integer optimization formulations for Problems (4) and (5). Section 3 introduces discrete first order algorithms solving Problems (4) and (5). This is followed by Sects. 4 and 5 reporting empirical evidence on both synthetic and real data sets respectively. Finally, the paper is concluded in Sect. 6.

2   Mixed Integer Optimization Formulation

Binary variables will be introduced to reformulate Problems (4) and (5) as mixed integer binary optimization problems. These binary variables represent whether or not variables and observations are useful.

2.1   Introducing Binary Variables

Variable selection involves the ℓ_0 norm function to count the number of useful variables. This counting function can be represented by introducing p binary variables $z_j \in \{0, 1\}$ such that

$$\|\beta\|_0 = \sum_{j=1}^{p} z_j \quad \text{and} \quad z_j = 0 \Leftrightarrow \beta_j = 0.$$

To force $z_j = 0 \Leftrightarrow \beta_j = 0$ Replace β_j by $z_j\beta_j$ for $j = 1, \ldots, p$. Outlier detection also involves the ℓ_0 norm function to count the number of outliers. This counting function can be represented by introducing n binary variables $t_i \in \{0, 1\}$ such as

$$\|\tau\|_0 = \sum_{i=1}^{n} t_i \quad \text{and} \quad t_i = 0 \Leftrightarrow \tau_i = 0, \ (x_i, y_i) \text{ is not an outlier.}$$

2.2   MIO Formulation of Problem (4)

Introducing binary variables for both variables and outliers with two big M constraints, given appropriate parameters k_v, k_o, M_v and M_o, Problem (4) becomes for some $q \in \{1, 2\}$:

$$
\begin{aligned}
\min_{\beta \in \mathbb{R}^p, \tau \in \mathbb{R}^n, z \in \{0,1\}^p, t \in \{0,1\}^n} \quad & \tfrac{1}{2}\|X\beta + \tau - y\|_2^2 \\
\text{s.t.} \quad & \sum_{j=1}^{p} z_j \leq k_v \quad \text{and} \quad |\beta_j| \leq z_j M_v, \quad j = 1, \ldots, p \\
& \sum_{i=1}^{n} t_i \leq k_o \quad \text{and} \quad |\tau_i| \leq t_i M_o \quad i = 1, \ldots, n.
\end{aligned}
$$
$$(6)$$

This problem is a mixed binary quadratic program.

2.3 MIO Formulation of Problem (5)

To deal with the ℓ_1-norm term in Problem (5), we introduce two variables $\eta^+, \eta^- \in \mathbb{R}^p$ such that $|\beta_j = \eta_j^+ + \eta_j^-|$ and $\beta_j = \eta_j^+ - \eta_j^-$.

$$
\begin{aligned}
\min_{\beta \in \mathbb{R}^p, \tau \in \mathbb{R}^n, \eta^+ \in \mathbb{R}^p, \eta^- \in \mathbb{R}^p, z \in \{0,1\}^P, t \in \{0,1\}^n} \quad & \tfrac{1}{2}\|X\beta + \tau - y\|_2^2 + \lambda \sum_{j=1}^{p}(\eta^+ + \eta^-) \\
\text{s.t.} \quad & \sum_{j=1}^{p} z_j \le k_v \quad \text{and} \quad |\beta_j| \le z_j M_v, \qquad j = 1, \ldots, p \\
& \sum_{i=1}^{n} t_i \le k_o \quad \text{and} \quad |\tau_i| \le t_i M_o \qquad i = 1, \ldots, n \\
& \eta_j^+ - \eta_j^- = \beta_j \qquad\qquad\qquad\qquad\quad j = 1, \ldots, p
\end{aligned}
\tag{7}
$$

3 Discrete First Order Algorithms

In this section we propose discrete first order methods to obtain near optimal solutions for Problems (4) and (5). These first order methods are used as warm starts to the MIO formulations leading to a significant decrease in computational time [13]. Furthermore, as it will be shown in numerical experiments, the first order algorithms provide high quality solutions in a short time (compared to MIO problems). The proposed algorithms borrow ideas from alternating minimization and projected gradient descent methods.

In [29], a proximal alternating linearized minimization (PALM) algorithm for non-convex and non-smooth problems was proposed to solve, under some assumptions, problems of the form:

$$
\min_{x,y} \Psi(x, y) := f(x) + g(y) + H(x, y)
$$

To deal with cardinality constraints, it is appropriate to introduce the problem of finding the projection of a vector $u \in \mathbb{R}^p$ onto the set of $k \le p$ sparse vectors

$$
\begin{aligned}
\min_{v \in \mathbb{R}^p} \quad & \tfrac{1}{2}\|v - u\|^2 \\
\text{s.t.} \quad & \|v\|_0 \le k.
\end{aligned}
\tag{8}
$$

This problem is easy and its solution $v^\star$ is given by sorting on the absolute value of vector $|u|$, that is by a sequence of indices (j) such that $|u_{(1)}| \ge |u_{(2)}| \ge \ldots |u_{(j)}| \ge \cdots \ge |u_{(p)}|$. Using these indices, the projection $v^\star = P_k(u)$ of u is the vector u itself with its smallest coefficients set to 0 that is

$$
v^\star = P_k(u) = \begin{cases} u_j & \text{if } j \in \{(1), \ldots, (k)\} \\ 0 & \text{else.} \end{cases}
$$

We propose to use this projection mechanism, on both β and τ, to get a local solution to the initial Problems (4) and (5) at a low computational cost.

3.1   Discrete First Order Algorithm for Problem (4)

We use $f(\beta, \tau)$ to denote the objective function in (4):

$$\min_{\beta, \tau} \quad F(\beta, \tau) := f(\beta, \tau) \text{ s.t } ||\beta||_0 \leq k_v \ ||\tau||_0 \leq k_o \qquad (9)$$

The partial derivatives of f are given by:

$$\nabla_\beta f(\beta, \tau) = X^t (X\beta + \tau - y) = G_\beta(\beta)$$

and

$$\nabla_\tau f(\beta, \tau) = (X\beta + \tau - y) = G_\tau(\tau)$$

By simple calculation, it can be shown that G_β and G_τ are Lipschitz with modules $L_1 = \sigma_{\max}^2(X)$ and $L_2 = 1$ respectively, where $\sigma_{\max}(X)$ is the maximum singular value of X.

We want to obtain local solutions for Problems (4) and (5) at a low computational cost. A possible way to achieve this goal consists of using the so-called block Gauss-Seidel iteration scheme on variables β and τ, also known as alternating minimization. To this end, a sequence $\{(\beta^\ell, \tau^\ell)\}_\ell \in \mathbb{N}$ is generated starting from some (β^0, τ^0) using the following scheme:

$$\begin{cases} beta^{\ell+1} = \arg\min_{\beta \in \mathbb{R}^p} (\beta - \beta^\ell)^t X^t (X\beta^\ell + \tau^\ell - y) \\ \qquad \text{s.t. } ||\beta||_0 \leq k_v \\ \qquad ||\beta - \beta^\ell||^2 \leq d_v \end{cases} \begin{cases} \tau^{\ell+1} = \arg\min_{\tau \in \mathbb{R}^n} (\tau - \tau^\ell)^t (X\beta^{\ell+1} + \tau^\ell - y) \\ \qquad \text{s.t. } ||\tau||_0 \leq k_o \\ \qquad ||\tau - \tau^\ell||^2 \leq d_o. \end{cases}$$

Where d_v and d_o are two given positive parameters that can be changed each step. The idea of the proximal method is, at each iteration, to minimize a regularized first-order approximation of the cost that can be interpreted as a local trust region mechanism [for details see for instance 30]. This surrogate loss is also a local upper bound of the targeted loss since, for well chosen ρ_v and ρ_o, the Lagrange multipliers associated with the trust region constraints are

$$\begin{cases} \frac{1}{2}||X\beta + \tau - y||^2 \leq \frac{1}{2}||X\beta^\ell + \tau^\ell - y||^2 + (\beta - \beta^\ell)^t X^t (X\beta^\ell + \tau^\ell - y) + \frac{1}{2\rho_v}||\beta - \beta^\ell||^2 \\ \frac{1}{2}||X\beta^{\ell+1} + \tau - y||^2 \leq \frac{1}{2}||X\beta^{\ell+1} + \tau^\ell - y||^2 + (\tau - \tau^\ell)^t (X\beta^{\ell+1} + \tau^\ell - y) + \frac{1}{2\rho_o}||\tau - \tau^\ell||^2. \end{cases}$$

For each iteration, this method introduced by [29] and called the proximal alternating linearized minimization (PALM) algorithm, consists of minimizing the upper bounds as follows:

$$\begin{cases} \beta^{\ell+1} = \arg\min_{\beta \in \mathbb{R}^p, ||\beta||_0 \leq k_v} (\beta - \beta^\ell)^t X^t (X\beta^\ell + \tau^\ell - y) + \frac{1}{2\rho_v}||\beta - \beta^\ell||^2 \\ \tau^{\ell+1} = \arg\min_{\tau \in \mathbb{R}^n, ||\tau||_0 \leq k_o} (\tau - \tau^\ell)^t (X\beta^{\ell+1} + \tau^\ell - y) + \frac{1}{2\rho_o}||\tau - \tau^\ell||^2. \end{cases}$$

That is, after some algebra,

$$\begin{cases} \beta^{\ell+1} = \arg\min_{\beta \in \mathbb{R}^p, ||\beta||_0 \leq k_v} \frac{1}{2}||\beta - \beta^\ell + \rho_v X^t (X\beta^\ell + \tau^\ell - y)||^2 \\ \tau^{\ell+1} = \arg\min_{\tau \in \mathbb{R}^n, ||\tau||_0 \leq k_o} \frac{1}{2}||\tau - \tau^\ell + \rho_o(X\beta^{\ell+1} + \tau^\ell - y)||^2. \end{cases}$$

These two minimization problems are of the same kind as Problem (8) and thus the sequence can be generated by using two ℓ_0 projected gradient, that is:

$$\begin{cases} \beta^{\ell+1} = P_{k_v}\left(\beta^\ell - \rho_v X^t(X\beta^\ell + \tau^\ell - y)\right) \\ \tau^{\ell+1} = P_{k_o}\left(\tau^\ell - \rho_o(X\beta^{\ell+1} + \tau^\ell - y)\right). \end{cases}$$

Algorithm 1 presents the pseudo code of the PALM algorithm.

Algorithm 1. Proximal alternating linearized minimization (PALM) [29]

Data: X, y initialization $\beta, \tau = 0$
Result: β, τ
set $\rho_v \leq \frac{1}{\sigma_M^2}$ and $\rho_o \leq 1$
while *it has not converged* $(||\beta_{n+1} - \beta_n||_2 > 10^{-6})$ **do**

$\quad\quad d \leftarrow \beta - \rho_v X^\top (X\beta + \tau - y)$ $\quad\quad\quad\quad\quad\quad\quad\quad\quad$ variable selection
$\quad\quad \beta \leftarrow P_{k_v}(d)$
$\quad\quad \delta \leftarrow \tau - \rho_o(X\beta + \tau - y)$ $\quad\quad\quad\quad\quad\quad\quad$ eliminating outliers
$\quad\quad \tau \leftarrow P_{k_o}(\delta)$

This algorithm converges towards a local minima of Problem (4) since it fulfills the assumption needed for Theorem 3.1 in [29]. Indeed, the partial gradients $G_\beta(\beta) = X^\top(X\beta + \tau - y)$ and $G_\tau(\tau) = (X\beta + \tau - y)$ are globally Lipschitz with module respectively $\frac{1}{\sigma_M^2}$ and 1, σ_M being the largest singular value of X. Also, to prove the convergence, the stepsizes have to be chosen such that $\rho_v \leq \frac{1}{\sigma_M^2}$ and $\rho_o \leq 1$.

3.2 Discrete First Order Algorithm for Problem (5)

The main difference between Problems (4) and (5) is the presence of an ℓ_1 penalty in the objective function of (5). Let

$$F(\beta) = \tfrac{1}{2}||X\beta + \tau - y||_2^2 + \lambda||\beta||_1 \text{ s.t } ||\beta||_0 \leq k_v$$

$$= f(\beta) + \lambda||\beta||_1 \text{ s.t } ||\beta||_0 \leq k_v$$

The gradient of $f(\beta)$ is Lipschitz and continuous with parameter L_v, in fact:

$$||\nabla f(\beta) - \nabla f(\tilde\beta)||_2 \leq L_v||\beta - \tilde\beta||_2 \quad \forall \beta, \tilde\beta \in \mathbb{R}^p$$

whith $L_v = \sigma_{max}^2(X)$, where $\sigma_{max}(X)$ is the maximum singular value of X. Consequently, for $L \geq L_v$, we have the following upper bound to $f(\beta)$:

$$f(\beta) \leq f(\tilde\beta) + \langle \nabla f(\tilde\beta), \beta - \tilde\beta \rangle + \frac{L}{2}||\beta - \tilde\beta||_2^2 := Q_L(\beta, \tilde\beta) \quad \forall \beta, \tilde\beta \in \mathbb{R}^p$$

Our goal is to minimize this upper bound, that is:

$$\min_{||\beta||_0 \leq k_v} Q_L(\beta, \tilde\beta) + \lambda||\beta||_1 \Leftrightarrow \min_{||\beta||_0 \leq k_v} \frac{L}{2}||\beta - (\tilde\beta - \frac{1}{L}\nabla f(\tilde\beta))||_2^2 + \lambda||\beta||_1$$

Dealing with the ℓ_1 penalty term necessitates the use of the soft-thresholding operator, while dealing with the cardinality constraint requires using the hard-thresholding operator (the projection operator).

To solve our problem, we introduce a thresholding operator [15] combining both ideas:

$$S(u; k_v; \lambda) = \underset{||\beta||_0 \leq k_v}{\arg\min} \frac{1}{2}||\beta - u||_2^2 + \lambda||\beta||_1$$

This threshold operator has the form:

$$\beta_i = \begin{cases} sgn(u_i)max\{0, |u_i - \lambda|\} & i \in \{(1), (2), \ldots, (k_v)\} \\ 0 & \text{otherwise} \end{cases}$$

where $(1), \ldots, (p)$ is a permutation of the indices $1, \ldots, p$.

Summing up all together, we obtain that the β update is given by: By introducing the soft-thresholding operator, it can be shown that the $\beta-$ update is given by:

$$\beta \leftarrow \beta - \rho_v X^t(X\beta + \tau - y)$$
$$\beta \leftarrow sgn(\beta) \max(0, |\beta| - \rho_v\lambda)$$
$$\beta \leftarrow P_{k_v}(\beta)$$

Putting all that together leads to Algorithm 2.

Algorithm 2. Proximal alternating linearized minimization (PALM) for Problem (5)

Data: X, y initialization $\beta, \tau = 0$
Result: β, τ
set $\rho_v \leq \frac{1}{\sigma_M^2}$ and $\rho_o \leq 1$
while *it has not converged* ($||\beta_{n+1} - \beta_n||_2 > 10^{-6}$) **do**
$\quad\mid\quad d \leftarrow \beta - \rho_v X^\top(X\beta + \tau - y)$
$\quad\mid\quad d \leftarrow sign(d)(max(0, |d| - \rho_v\lambda))$
$\quad\mid\quad \beta \leftarrow P_{k_v}(d)$
$\quad\mid\quad \delta \leftarrow \tau - \rho_o(X\beta + \tau - y)$
$\quad\mid\quad \tau \leftarrow P_{k_o}(\delta)$

We note that this algorithm converges towards a local minima of Problem (5) since it fulfills the assumption needed for Theorem 3.1 in [29].

In the rest of the paper, Algorithm 1 will be denoted by PALM_0, while Algorithm 2 will be denoted by $\text{PALM}_{0,1}$.

4   Experiments on Synthetic Data Sets

The aim of the experiments is to shed the light on the superior performance of Problem (5) over Problem (4), espacially for low SNR values. We note that, Problem (5) will be denoted by $L_{0,1}$, Problem (4) will be denoted by L_0, Algorithm 2 will be denoted by $PALM_{0,1}$ and Algorithm 1 will be denoted by $PALM_0$.

4.1   Setup

Given n, p (problem dimensions), s (sparsity level), beta-type (pattern of sparsity), ρ (predictor autocorrelation level), and ν (SNR level) as in [14], the setup can be described as follows:

- We define coefficients $\beta_0 \in \mathbb{R}^p$ with a sparsity level $s = 5$ and with beta-type, as described below.
- We draw the rows of the matrix $X \in \mathbb{R}^{n \times p}$ from $N_p(0, \Sigma)$, where $\Sigma \in \mathbb{R}^{p \times p}$ has entry (i, j) equal to $\rho^{|i-j|}$. We considered two values $\rho = 0.3$ and $\rho = 0.6$
- The columns of X are standardized to have mean zero and unit ℓ_2 norm.
- We then draw the vector $y \in \mathbb{R}^n$ from $N_n(X\beta_0, \sigma^2 I)$, with σ^2 defined to meet the desired SNR level, i.e., $\sigma^2 = \beta_0^T \Sigma \beta_0 / \nu$. We considered three values of SNR $0.5, 1$ and 3.
- The data set (X, y) is used as a training set. A validation set (X_t, y_t) is created in the same way as (X, y) with $n_t = 1000$.
- 5% of outliers were added to the training data set by following a normal $N(50, \sigma)$ instead of $N(0, \sigma)$.
- In order to determine the values of k_v, M_v, k_o, M_o and λ, we run the PALM algorithms for and pick the solution minimizing the error on the testing data, let β_{PALM} denote this solution.
- $M_v = (1 + \alpha)\|\beta_{palm}\|_\infty$, $M_o = (1 + \alpha)\|\tau_{palm}\|_\infty$ with $\alpha = 0.1$, k_v and k_o are set as the number of nonzero elements in the solutions β_{PALM} and τ_{PALM} respectively.
- We considered three configurations:the medium setting with $n = 50, p = 100$ and the high-dimensional setting $n = 100, p = 1000$.

Coefficients: We considered two settings for the coefficients $\beta_0 \in \mathbb{R}^p$:

- beta-type 1: $\beta_0 = (1, \quad 0, \quad 1, \quad 0, \quad 1, \quad 0, \quad 1, \quad 0, \quad 1, \quad 0, \quad \underbrace{0, \ldots, 0}_{p-10 \text{ times}})$;
- beta-type 2: β_0 has its first 5 components equal to 1, and the rest equal to 0;

4.2   Selecting Tuning Parameters

In order to compare Problem (4) (L_0), Problem (5) $(L_{0,1})$, Algorithm 1 $(PALM_0)$ and Algorithm 2 $(PALM_{0,1})$, we considered two different experiments:

Experiment 1: To shed the light on the influence of the ℓ_1 penalty, we run both PALM Algorithms 1 and 2 on the data set (X, y) for $k_v \in \{2, 5, 10, 15\}$ and for $k_o = 5\%$ that is the true percentage of outliers added to the data. In order to determine λ, we run PALM Algorithm 2 for 100 values of λ starting with $\lambda_1 = \|X'y\|_\infty$. $\{\lambda_i\}_{i=1}^{100}$ is a geometrically spaced sequence with $\lambda_{100} = 10^{-4}\lambda_1$.

We pick the solutions $\hat{\beta}_{PALM_{0,1}}(\lambda, k_v)$ and $\hat{\beta}_{PALM_0}(k_v)$ with smallest error $\|X_t\beta - y_t\|_2^2$. These two solutions are used as warm starts for MIO formulations.

We report the prediction error defined by:

$$PE = \|X\beta - X\beta_0\|_2.$$

The results have been averaged over ten different replications of data.

4.3   Experiment 2 on Synthetic Data

In this experiment, we want to mimic real word setting, k_v and k_o are not fixed anymore. To this end we take a $3D$ grid of tuning parameters. $\{\lambda_i\}_1^N$ is a geometrically spaced sequence of 100 values, $\lambda_1 = ||X^t y||_\infty$ and $\lambda_N = 10^{-4}\lambda_1$, $k_v \in \{1,\dots,15\}$ and $k_o \in \{0, 0.025n, 0.05n\}$. We train Algorithm 2 on (X, y) and find $\hat{\beta}(\lambda, k_v, k_o)$ with minimal error on (X_t, y_t). This solution $\hat{\beta}(\lambda, k_v, k_o)$ is used as a warm start for the MIO formulation (5). The same is done for Problem (4) but without λ.

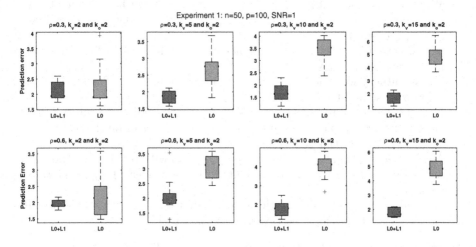

Fig. 1. Experiment 1 showing the effect of the ℓ_1 penalty

4.4   Computational Time

The $PALM_{0,1}$ estimator is tuned over 100 values of λ, 15 values of k_v and 3 values of k_o. We denote by $\beta_{PALM_{0,1}}(\lambda, k_v, k_o)$ the 3D family of solutions obtained. For $n = 50$, $p = 100$ computing the family takes about 1 min. For $n = 100$, $p = 1000$ about 10 min were needed to find this family. The threshold of the discrete first order algorithms is 10^{-6}. Once the family is obtained, the best solution $(\hat{\lambda}, \hat{k}_v, \hat{k}_o)$ on a separate validation set. This solution is used as a warm start for the MIO solver with a time limit of $1000\,$s which seemed to be enough to obtain a global solution for the majority of problems solved. However even if the optimal solution is not certified, a near optimal solution with a high quality is guaranteed. We report the two solutions obtained by the DFO algorithm and the MIO formulation and we denote them by $PALM_{0,1}$ and $L_{0,1}$ respectively. The same is done to obtain the solutions $PALM_0$ and L_0 representing the solutions obtained by Algorithm 1 and Problem (4).

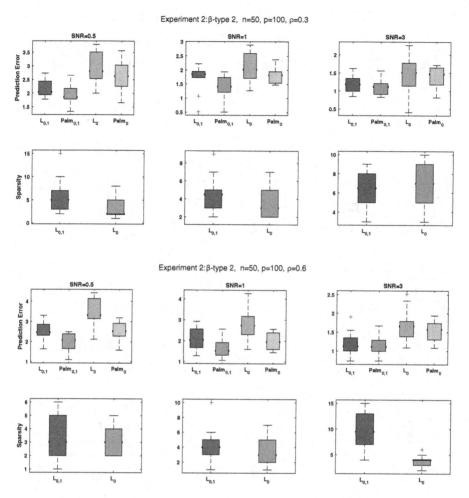

Fig. 2. Experiment 2 β-type 2 simulations with $n = 50$ and $p = 100$. The top two rows display the results for $\rho = 0.3$ while the bottom two rows display the results for $\rho = 0.6$. Prediction error refers to the best predictive models obtained after tuning on separate validation set. Sparsity refers to the corresponding number of nonzero coefficients. Three SNR values were considered.

4.5   Results

We summarize the experimental results below:

– In Fig. 1, it is shown that as k_v gets greater than the true sparsity level, the prediction error of L_0 decreases. However, the performance of $L_{0,1}$ stay stable, which demonstrate the significant importance of using the ℓ_1 penalty term since in practice, if the PALM algorithms are used to tune the parameters by a k-fold cross validation, we remarked that they overestimate the sparsity level especially for low SNR values. Thus using this penalty term, will ensure avoiding the decrease of the performance for overestimated sparsity level s.

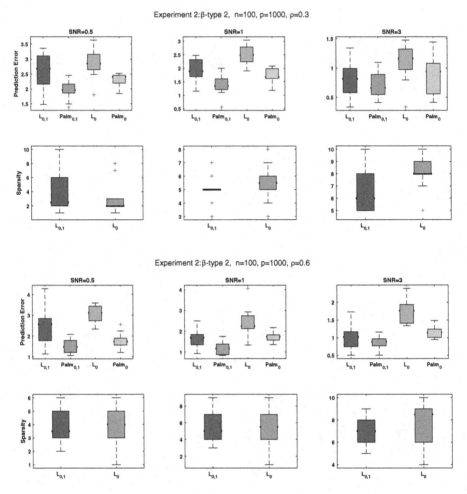

Fig. 3. Experiment 2 β-type 2 simulations with $n = 100$ and $p = 1000$. The top two rows display the results for $\rho = 0.3$ while the bottom two rows display the results for $\rho = 0.6$. Prediction error refers to the best predictive models obtained after tuning on separate validation set. Sparsity refers to the corresponding number of nonzero coefficients. Three SNR values were considered.

- For $n = 50$, $p = 100$ Figs. 4 and 2 show that when SNR $= 0.5$, $PALM_{0,1}$ produces the best solution in terms of prediction error. $L_{0,1}$ is close to $PALM_{0,1}$ and both outperform L_0 and $PALM_0$. As the SNR value increases, the results of all methods get closer. Furthermore, the poor performance of the L_0 problem are noticed for low SNR values because of the overfitting effect. We also note that as ρ increases, the ℓ_1 penalty ensures obtaining better solutions (Figs. 3 and 5).
  In terms of sparsity, the L_0 produced a little bit sparser solutions.

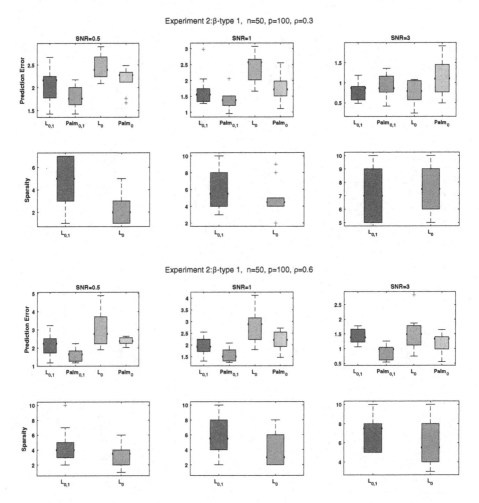

Fig. 4. Experiment 2 β-type 1 simulations with $n = 50$ and $p = 100$. The top two rows display the results for $\rho = 0.3$ while the bottom two rows display the results for $\rho = 0.6$. Prediction error refers to the best predictive models obtained after tuning on separate validation set. Sparsity refers to the corresponding number of nonzero coefficients. Three SNR values were considered.

- For the high dimensional setting $n = 100$ and $p = 1000$, same results are obtained but in this case $PALM$ algorithms have an advantage over their corresponding MIO formulations. This may be explained by the fact the MIO formulations have a time limit of 1000 s.

In general, the ℓ_1 penalty term improved the ℓ_0 formulation for low SNR values, higher correlation ρ values and high dimensional settings. The added shrinkage fixes the overfitting behavior of L_0. Finally, the good performance of of both PALM algorithms is clearly noticed.

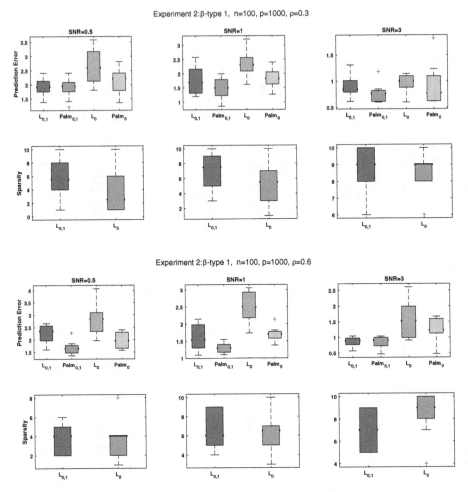

Fig. 5. Experiment 2 β-type 1 simulations with $n = 100$ and $p = 1000$. The top two rows display the results for $\rho = 0.3$ while the bottom two rows display the results for $\rho = 0.6$. Prediction error refers to the best predictive models obtained after tuning on separate validation set. Sparsity refers to the corresponding number of nonzero coefficients. Three SNR values were considered.

5   Experiments on Real Data Sets

We compare the performance of all mentioned methods above on five real data sets from the UCI Machine Learning Repository presented in Table 1.

We split each data set into three parts: the training set (40%), the validation set (40%) and the testing set (20%). The training set was used to train both proximal algorithms for a variety of combinations of input parameters. For each combination of parameters, the mean squared error on the validation set was calculated, and this was used to select the best combination of parameters for

Table 1. Summary of used data sets.

Name of the dataset	Number of instances n	Number of attributes p	Origin
Auto Mpg	398	8	UCI
Concrete Compressive Strength	1030	9	UCI
Concrete Slump Test	103	10	UCI
Diabetes	442	10	stat.ncsu.edu
Forest Fires	517	13	UCI

$PALM_{0,1}$ and $PALM_0$. Finally, all methods were trained by using these best parameters on the combined training and validation sets, before reporting the out-of-sample prediction error, defined below, on the testing set. We note that the columns of X were standardized to have zero means and unit ℓ_2-norms andd that the response vector y was standardized to have zero mean. All methods were trained, validated, and tested on the same random splits, and computational experiments were repeated ten times for each data set with different splits. For each data set and regression method, we report the average of the results and the standard deviation in parentheses for the raw data.

$$\text{Prediction Error}: \ PE = \frac{1}{n_t}||X_{test}\beta - y_{test}||_2$$

where n_t represents the number of observations of X_{test}. Results in Table 2 sheds the light on the good performance of the proposed algorithm. In fact, over five data sets, the ℓ_1 regularized methods performs better than ℓ_0 methods on three data sets. However there is no big difference between results on all methods. This can be explained by the fact that for high SNR values these methods perform almost similarly and the ℓ_1 regularized problem fixes the overfitting problem of the ℓ_0 problems for low SNR values.

Table 2. Prediction error rates (standard deviations) of the $L_0, 1$, $PALM_{0,1}$, L_0 and $PALM_0$ methods on 5 real datasets.

	$L_{0,1}$	$PALM_{0,1}$	L_0	$PALM_0$
Auto MPG	**0.4332 (0.0472)**	0.4379 (0.0483)	0.4436 (0.0476)	0.4365 (0.0477)
Concrete Compressive Strength	0.7366 (0.0342)	0.7465 (0.0347)	**0.7345 (0.0334)**	0.7444 (0.0331)
Concrete Slump Test	0.6015 (0.1136)	**0.5925 (0.1228)**	0.6058 (0.1100)	0.6022 (0.1344)
Diabetes	5.9346 (0.3330)	5.9485 (0.3459)	5.9146 (0.2996)	**5.9133 (0.2962)**
Forest Fires	**5.6359 (3.9369)**	5.6372 (3.9370)	5.6456 (3.9360)	5.6396 (3.9385)

6   Conclusion

In this paper, an ℓ_1 regularized method performing both feature selection and outliers detection in linear regression. The problem is formulated as a mixed integer optimization problem and warm started by a discrete first order algorithm. We present an empirical comparison between the ℓ_0 algorithms and their ℓ_1 cousins. Results on synthetic data sets show that the regularized problems overperformed the ℓ_0 methods for low SNR values and as the SNR increases, these methods become almost similar. Furthermore, results on real data sets show that the proposed method produces solutions with good predictive performance on both real and synthetic data sets. Based on the above we propose to use the proximal Algorithm 2 in practical use if we are not interested in obtaining the global solution, since it performs well for both high and low SNR values in a short time and it can be used for high dimensional problems.

References

1. Miller, A.: Subset Selection in Regression. CRC Press, Boca Raton (2002)
2. Beale, E.M.L., Kendall, M.G., Mann, D.W.: The discarding of variables in multivariate analysis. Biometrika **54**(3–4), 357–366 (1967)
3. Hocking, R.R., Leslie, R.N.: Selection of the best subset in regression analysis. Technometrics **9**(4), 531–540 (1967)
4. Natarajan, B.K.: Sparse approximate solutions to linear systems. SIAM J. Comput. **24**(2), 227–234 (1995)
5. Tibshirani, R.: Regression shrinkage and selection via the lasso. J. Roy. Stat. Soc.: Ser. B (Methodol.) **58**(1), 267–288 (1996)
6. Efron, B., Hastie, T., Johnstone, I., Tibshirani, R., et al.: Least angle regression. Ann. Stat. **32**(2), 407–499 (2004)
7. Friedman, J., Hastie, T., Tibshirani, R.: Regularization paths for generalized linear models via coordinate descent. J. Stat. Softw. **33**(1), 1 (2010)
8. Friedman, J., Hastie, T., Simon, N., Tibshirani, R.: Lasso and elastic-net regularized generalized linear models. R-package version 2.0-5. 2016 (2016)
9. Weijie, S., Bogdan, M., Candes, E., et al.: False discoveries occur early on the lasso path. Ann. Stat. **45**(5), 2133–2150 (2017)
10. Bühlmann, P., Van De Geer, S.: Statistics for High-Dimensional Data: Methods, Theory and Applications. Springer, Heidelberg (2011). https://doi.org/10.1007/978-3-642-20192-9
11. Friedman, J., Hastie, T., Tibshirani, R.: The Elements of Statistical Learning. Springer Series in Statistics, vol. 1. Springer, New York (2001). https://doi.org/10.1007/978-0-387-84858-7
12. Tibshirani, R.: Regression shrinkage and selection via the lasso: a retrospective. J. Roy. Stat. Soc. Ser. B (Stat. Methodol.) **73**(3), 273–282 (2011)
13. Bertsimas, D., King, A., Mazumder, R.: Best subset selection via a modern optimization lens. Ann. Statist. **47**(3), 2324–2354 (2015)
14. Hastie, T., Tibshirani, R., Tibshirani, R.J.: Extended comparisons of best subset selection, forward stepwise selection, and the lasso. arXiv preprint arXiv:1707.08692 (2017)

15. Mazumder, R., Radchenko, P., Dedieu, A.: Subset selection with shrinkage: sparse linear modeling when the SNR is low. arXiv preprint arXiv:1708.03288 (2017)
16. Rousseeuw, P.J., Leroy, A.M.: Robust Regression and Outlier Detection, vol. 589. Wiley, Hoboken (1987)
17. Chen, Y., Caramanis, C., Mannor, S.: Robust sparse regression under adversarial corruption. In: International Conference on Machine Learning, pp. 774–782 (2013)
18. Li, X.: Compressed sensing and matrix completion with constant proportion of corruptions. Constr. Approx. **37**(1), 73–99 (2013)
19. Laska, J.N., Davenport, M.A., Baraniuk, R.G.: Exact signal recovery from sparsely corrupted measurements through the pursuit of justice. In: 2009 Conference Record of the Forty-Third Asilomar Conference on Signals, Systems and Computers, pp. 1556–1560. IEEE (2009)
20. She, Y., Owen, A.B.: Outlier detection using nonconvex penalized regression. J. American Stat. Assoc. **106**(494), 626–639 (2011)
21. Rousseeuw, P.J., Hubert, M.: Anomaly detection by robust statistics. Wiley Interdiscipl. Rev. Data Min. Knowl. Discov. **8**(2), e1236 (2018)
22. Hodge, V., Austin, J.: A survey of outlier detection methodologies. Artif. Intell. Rev. **22**(2), 85–126 (2004)
23. Wang, H., Li, G., Jiang, G.: Robust regression shrinkage and consistent variable selection through the lad-lasso. J. Bus. Econ. Stat. **25**(3), 347–355 (2007)
24. Li, Y., Zhu, J.: L 1-norm quantile regression. J. Comput. Graph. Stat. **17**(1), 163–185 (2008)
25. Wang, L., Yichao, W., Li, R.: Quantile regression for analyzing heterogeneity in ultra-high dimension. J. Am. Stat. Assoc. **107**(497), 214–222 (2012)
26. Wright, J., Yang, A.Y., Ganesh, A., Sastry, S.S., Ma, Y.: Robust face recognition via sparse representation. IEEE Trans. Pattern Anal. Mach. Intell. **31**(2), 210–227 (2009)
27. Wright, J., Ma, Y., Mairal, J., Sapiro, G., Huang, T.S., Yan, S.: Sparse representation for computer vision and pattern recognition. Proc. IEEE **98**(6), 1031–1044 (2010)
28. Nguyen, N.H., Tran, T.D.: Robust lasso with missing and grossly corrupted observations. IEEE Trans. Inf. Theory **59**(4), 2036–2058 (2013)
29. Bolte, J., Sabach, S., Teboulle, M.: Proximal alternating linearized minimization for nonconvex and nonsmooth problems. Math. Program. **146**(1–2), 459–494 (2014)
30. Parikh, N., Boyd, S.P.: Proximal algorithms. Found. Trends Optim. **1**(3), 127–239 (2014)

Author Index

Printed in the United States
By Bookmasters